Office for
National Statistics

Indexes to the UK Standard Industrial Classification of Economic Activities 2007

Editor: Lindsay Prosser
Office for National Statistics

palgrave
macmillan

ISBN 978-0-230-21014-1

A National Statistics publication

National Statistics are produced to high professional standards as set out in the Code of Practice for Official Statistics. They are produced free from political influence.

About us

The Office for National Statistics

The Office for National Statistics (ONS) is the executive office of the UK Statistics Authority, a non-ministerial department which reports directly to Parliament. ONS is the UK government's single largest statistical producer. It compiles information about the UK's society and economy, and provides the evidence-base for policy and decision-making, the allocation of resources, and public accountability. The Director-General of ONS reports directly to the National Statistician who is the Authority's Chief Executive and the Head of the Government Statistical Service.

The Government Statistical Service

The Government Statistical Service (GSS) is a network of professional statisticians and their staff operating both within the Office for National Statistics and across more than 30 other government departments and agencies.

Palgrave Macmillan

This publication first published 2010 by Palgrave Macmillan.

Palgrave Macmillan in the UK is an imprint of Macmillan Publishers Limited, registered in England, company number 785998, of Houndmills, Basingstoke, Hampshire RG21 6XS.

Palgrave Macmillan in the US is a division of St Martin's Press LLC, 175 Fifth Avenue, New York, NY 10010.

Palgrave Macmillan is the global academic imprint of the above companies and has companies and representatives throughout the world.

Palgrave® and Macmillan® are registered trademarks in the United States, the United Kingdom, Europe and other countries.

A catalogue record for this book is available from the British Library.

10 9 8 7 6 5 4 3 2 1
18 17 16 15 14 13 12 11 10 09

Contacts

This publication

For information about the content of this publication, contact the Editor
Tel: 01329 444100
Email: lindsay.

Other customer enquiries

ONS Customer Contact Centre
Tel: 0845 601 3034
International: +44 (0)845 601 3034
Minicom: 01633 812399
Email: info@statistics.gsi.gov.uk
Fax: 01633 652747
Post: Room 1015, Government Buildings,
Cardiff Road, Newport, South Wales NP10 8XG
www.ons.gov.uk

Media enquiries

Tel: 0845 604 1858
Email: press.office@ons.gsi.gov.uk

Publication orders

To obtain the print version of this publication, contact Palgrave Macmillan
Tel: 01256 302611
www.palgrave.com/ons
Price: £60.00

Copyright and reproduction

© Crown copyright 2010
Published with the permission of the Office of Public Sector Information (OPSI)

You may use this publication (excluding logos) free of charge in any format for research, private study or internal circulation within an organisation providing it is used accurately and not in a misleading context. The material must be acknowledged as Crown copyright and you must give the title of the source publication. Where we have identified any third party copyright material you will need to obtain permission from the copyright holders concerned.

For re-use of this material you must apply for a Click-Use Public Sector Information (PSI) Licence from:

Office of Public Sector Information, Crown Copyright Licensing and Public Sector Information, Kew, Richmond, Surrey TW9 4DU
Tel: 020 8876 3444
www.opsi.gov.uk/click-use/index.htm

Printing

This book is printed on paper suitable for recycling and made from fully managed and sustained forest sources. Logging, pulping and manufacturing processes are expected to conform to the environmental regulations of the country of origin.

Printed and bound in Great Britain by Hobbs the Printer Ltd, Totton, Southampton

Typeset by Academic + Technical Typesetting, Bristol

Contents

Preface

This volume contains both an alphabetical and a numerical list of typical activities or, sometimes, products. These lists show the UK SIC (2007) class or subclass to which the activity or product should be classified.

Both lists attempt to provide a link to the last Standard Industrial Classification (UK SIC (2003), which was significantly different in structure, although in some cases it was not possible to do so without further information. The link indicated should, therefore, not be taken to be definitive.

It is important to remember that classification involves identification of the appropriate statistical unit. For example, when a road freight transport section of a manufacturing firm is a separate statistical unit, it will be classified to transport (division 49, group 49.4) but when it is an integral part of the main firm then it will be classified to the principal activity of that firm.

Activities and products may be known under different names and some have been classified in the alphabetical index under several alternative titles. These alternatives also appear in the numerical index. Due to changes and innovation, it has not been possible to list all economic activities within this publication; the indexes remain dynamic and any addendums will be published on the UKSIC webpage www.statistics.co.uk/uksic

Unlike the SIC80 Indexes publication, it has not been necessary to include references to the equivalent European Union classification system, NACE rev 2 today, as the two systems are identical down to and including the four digit class level.

The detail of the classification has substantially increased (from 514 to 615 classes). For service-producing activities, this increase is visible at all levels, including the highest one, while for other activities, such as agriculture, the increase in detail affected mostly the lower level of the classification. One effect of the increase in detail at class level has been a reduced need for detail at the subclass level. The total number of subclasses has decreased from 285 to 191.

Contact

Office for National Statistics
Classifications and Harmonisation Branch
Segensworth Road
Titchfield
Fareham
Hampshire
PO15 5RR

Classifications Helpdesk

Tel: 01329 444970
Email: classifications.helpdesk@ons.gsi.gov.uk

December 2009

Alphabetical Index

Part 1

Alphabetical Index

SIC 2007	SIC 2003	Activity	SIC 2007	SIC 2003	Activity
01160	01110	Abaca and other vegetable textile fibre growing	20140	24140	Acid (organic) (manufacture)
10110	15111	Abattoir (manufacture)	20120	24120	Acid dye (manufacture)
17120	21120	Abrasive base paper (manufacture)	46750	51550	Acids (wholesale)
23910	26810	Abrasive bonded disc, wheel and segment (manufacture)	02300	02010	Acorn gathering
			43290	45320	Acoustical engineering
23910	26810	Abrasive cloth (manufacture)	71129	74204	Acoustical engineering design
23910	26810	Abrasive grain (manufacture)	71200	74300	Acoustics and vibration testing
23910	26810	Abrasive grain of aluminium oxide (manufacture)	20520	24620	Acrylic adhesives (manufacture)
23910	26810	Abrasive grain of artificial corundum (manufacture)	20301	24301	Acrylic paints (manufacture)
23910	26810	Abrasive grain of boron carbide (manufacture)	46120	51120	Acrylic polymers in primary forms (commission agent)
23910	26810	Abrasive grain of silicon carbide (manufacture)	46750	51550	Acrylic polymers in primary forms (wholesale)
08990	14500	Abrasive materials mining and quarrying	20160	24160	Acrylic resins (manufacture)
23910	26810	Abrasive paper (manufacture)	20160	24160	Acrylics (manufacture)
20411	24511	Abrasive soap (manufacture)	20140	24140	Acrylonitrile (manufacture)
23910	26810	Abrasive wheel (bonded) (manufacture)	20160	24160	Acrylonitrile butadiene styrene (abs) polymers (manufacture)
32500	24422	Absorbable haemostatics (manufacture)	32409	36509	Action figures (manufacture)
26511	33201	Absorptiometer (electronic) (manufacture)	20140	24140	Activated and unactivated charcoal (other than wood charcoal) (manufacture)
26513	33202	Absorptiometer (non-electronic) (manufacture)			
23190	26150	Absorption drums made of glass (manufacture)	20590	24660	Activated carbon (manufacture)
46690	51870	AC and DC electrical motors (single or multi-phase) (wholesale)	20140	24140	Activated earths (manufacture)
			97000	95000	Activities of households as employers of domestic babysitters
94120	91120	Academic organisations			
85590	80429	Academic tutoring	97000	95000	Activities of households as employers of domestic butlers
20200	24200	Acaricide (manufacture)			
29320	34300	Accessories and parts for motor vehicles and their engines (manufacture)	97000	95000	Activities of households as employers of domestic caretakers
			97000	95000	Activities of households as employers of domestic chauffeurs
26702	33403	Accessories for photographic equipment (manufacture)			
86101	85112	Accident and emergency service (private sector)	97000	95000	Activities of households as employers of domestic cooks
86101	85111	Accident and emergency service (public sector)	97000	95000	Activities of households as employers of domestic gardeners
65120	66031	Accident insurance			
77390	71320	Accommodation container renting	97000	95000	Activities of households as employers of domestic gatekeepers
55300	55210	Accommodation in protective shelters or plain bivouac facilities for tents and/or sleeping bags			
			97000	95000	Activities of households as employers of domestic governesses
32200	36300	Accordion (manufacture)	97000	95000	Activities of households as employers of domestic laundresses
17230	22220	Account books (manufacture)			
69201	74121	Accountancy services	97000	95000	Activities of households as employers of domestic maids
69201	74121	Accounting activities			
94120	91120	Accounting associations	97000	95000	Activities of households as employers of domestic secretaries
28230	30010	Accounting machine (manufacture)			
77330	71330	Accounting machinery and equipment rental and operating leasing	97000	95000	Activities of households as employers of domestic stable-lads
70221	74142	Accounting systems design	97000	95000	Activities of households as employers of domestic staff
27200	31400	Accumulator (manufacture)			
23190	26150	Accumulator cell cases made of glass (manufacture)	97000	95000	Activities of households as employers of domestic tutors
28990	29560	Accumulator for hydraulic equipment (manufacture)	97000	95000	Activities of households as employers of domestic valets
20140	24140	Acetic acid (manufacture)			
20140	24140	Acetone (manufacture)	97000	95000	Activities of households as employers of domestic waiters
20110	24110	Acetylene (manufacture)			
46690	51870	Acetylene gas generators (wholesale)			
20130	24130	Acid (inorganic) (manufacture)			

SIC 2007	SIC 2003	Activity	SIC 2007	SIC 2003	Activity
33200	28520	Activities of millwrights (manufacture)	93290	92729	Adventure playground
64209	74156	Activities of service trades holding companies not engaged in management	73110	74402	Advertising agencies
			73110	74402	Advertising campaign creation and realisation
93110	92619	Activity centre (sports)	18129	22220	Advertising catalogue printing (manufacture)
90010	92311	Actors	73110	74402	Advertising consultants
66290	67200	Actuarial services	73110	74402	Advertising contractor
27110	31620	Actuator (electro-magnetic positioner) (manufacture)	59111	92111	Advertising film production
28120	29122	Actuator for hydraulic equipment (manufacture)	27400	31500	Advertising light (manufacture)
20140	24140	Acyclic (fatty) alcohols (manufacture)	18140	22230	Advertising mailing literature finishing (manufacture)
46120	51120	Acyclic and cyclic hydrocarbons (commission agent)	22290	25240	Advertising material made of plastic (manufacture)
20140	24140	Acyclic hydrocarbons (saturated and unsaturated) (manufacture)	73110	74409	Advertising material or samples delivery or distribution
28230	30010	Adding machines (manufacture)	58190	22150	Advertising material publishing
46660	51850	Adding machines (wholesale)	58130	22120	Advertising newspaper publishing
28230	30010	Address plate embossing machine (manufacture)	18129	22220	Advertising printed matter printing (manufacture)
28230	30010	Addressing machine (manufacture)	73120	74401	Advertising space or time (sales or leasing thereof)
20520	24620	Adhesive (formulated) (manufacture)	59112	92111	Advertising video production
18140	22230	Adhesive binding of books, brochures, etc. (manufacture)	71111	74201	Advisory and pre-design architectural activities
20520	24620	Adhesive coating (manufacture)	68310	70310	Advisory services in connection with buying, selling and renting of real estate
46180	51180	Adhesive dressings, catgut and similar materials (commission agent)	70229	74149	Advisory, conciliation and arbitration service
46460	51460	Adhesive dressings, catgut and similar materials (wholesale)	70229	74143	Advisory, guidance and operational assistance services concerning business policy and strategy
22290	25240	Adhesive labels of plastic or cellulose (manufacture)	69101	74112	Advocate of the Scottish bar
20520	24620	Adhesive made of urea formaldehyde (manufacture)	25730	28620	Adze (manufacture)
17230	21230	Adhesive paper ready for use (manufacture)	11070	15980	Aerated water (manufacture)
20520	24620	Adhesive paste (manufacture)	28290	29240	Aeration plant for effluent treatment (manufacture)
32500	24422	Adhesive plaster and surgical bandage (manufacture)	26309	32300	Aerial (domestic) (manufacture)
22190	25130	Adhesive repair material made of rubber (manufacture)	26309	32300	Aerial (non-domestic) (manufacture)
			73110	74409	Aerial and outdoor advertising services
22190	25130	Adhesive tape of rubberised textile (manufacture)	49319	60219	Aerial cable-ways operation
25720	28630	Adjustable seat mechanisms (manufacture)	43210	45310	Aerial erection (domestic)
66210	67200	Adjuster (insurance)	43999	45250	Aerial mast (self supporting) erection
84230	75230	Administration and operation of administrative civil and criminal law courts	74202	74813	Aerial photography (other than for cartographic and spatial activity purposes)
84240	75240	Administration and operation of regular and auxiliary police forces supported by public authorities	74202	74206	Aerial photography for cartographic and spatial activity purposes
84240	75240	Administration and operation of special police forces	26309	32300	Aerial reflectors (manufacture)
84220	75220	Administration of defence-related research and development policies and related funds	28220	29220	Aerial ropeway and cableway (manufacture)
			26309	32300	Aerial rotors (manufacture)
84220	75220	Administration supervision and operation of engineering, and other non-combat forces and commands	26309	32300	Aerial signal splitters (manufacture)
			71122	74206	Aerial survey
84220	75220	Administration supervision and operation of intelligence and other non-combat forces and commands	30300	35300	Aero engine manufacture (all types) (manufacture)
			30300	35300	Aero engine parts and sub assemblies (manufacture)
			90010	92349	Aerobatic display
84220	75220	Administration supervision and operation of land, sea, air and space defence forces	52230	63230	Aerodrome
84220	75220	Administration supervision and operation of military defence affairs	18130	22250	Aerographing (manufacture)
			81299	74709	Aeroplane cleaning (non-specialised)
84220	75220	Administration supervision and operation of reserve/ auxiliary forces of the defence establishment	30300	35300	Aeroplanes (manufacture)
			25920	28720	Aerosol cans made of metal (manufacture)
84220	75220	Administrative, operational and supervisory services related to civil defence	82920	74820	Aerosol filling on a fee or contract basis
			30300	35300	Aerospace equipment (manufacture)
84220	75220	Administrative, operational and supervisory services related to military defence	20420	24520	After shave lotion (manufacture)
			74909	74879	Agents and agencies in entertainment
88990	85321	Adoption activities (charitable)	74909	74879	Agents and agencies in motion picture theatrical production
88990	85322	Adoption activities (non-charitable)			
85590	80429	Adult education centre	74909	74879	Agents and agencies in placement of artworks with publishers
85590	80429	Adult education residential college			
46420	51421	Adults' fur and leather clothing exporter (wholesale)	74909	74879	Agents and agencies in placement of books with publishers
46420	51421	Adults' fur and leather clothing importer (wholesale)			

A

SIC 2007	SIC 2003	Activity
74909	74879	Agents and agencies in placement of photographs publishers
74909	74879	Agents and agencies in placement of plays with publishers producers
74909	74879	Agents and agencies in sports attractions
23910	26810	Agglomerated abrasives (manufacture)
16290	20520	Agglomerated cork (manufacture)
19100	23100	Agglomeration of coke (manufacture)
01629	01429	Agistment services on a fee or contract basis
46610	51880	Agricultural and forestry machinery (wholesale)
77310	71310	Agricultural and forestry machinery and equipment operational leasing (without operator)
28302	29320	Agricultural and forestry machinery parts (except tractor parts) (manufacture)
28302	29320	Agricultural broadcaster (manufacture)
23490	26250	Agricultural ceramic ware (manufacture)
85320	80220	Agricultural college
01610	01410	Agricultural contracting
71200	74300	Agricultural grain electrophoresis
25730	28620	Agricultural hand tools, not power-driven (manufacture)
25730	28620	Agricultural knife (manufacture)
43120	45110	Agricultural land drainage
68209	70209	Agricultural land letting
01610	01410	Agricultural land maintenance in good agricultural and environmental condition
23520	26520	Agricultural lime processing (manufacture)
46140	51140	Agricultural machinery (commission agent)
28302	29320	Agricultural machinery (manufacture)
46610	51880	Agricultural machinery and accessories and implements, including tractors exporter (wholesale)
46610	51880	Agricultural machinery and accessories and implements, including tractors importer (wholesale)
77310	71310	Agricultural machinery and equipment rental and leasing (without operator)
01610	01410	Agricultural machinery and equipment rental with operator
28302	29320	Agricultural machinery for soil preparation (manufacture)
77310	71310	Agricultural machinery hire (without operator)
28302	29320	Agricultural motor drawn trailer (manufacture)
28302	29320	Agricultural planting machinery (manufacture)
46110	51110	Agricultural raw materials (commission agent)
72190	73100	Agricultural research (other than biotechnological)
28302	29320	Agricultural self-loading semi-trailers (manufacture)
28302	29320	Agricultural self-loading trailers (manufacture)
28302	29320	Agricultural self-unloading semi-trailers (manufacture)
28302	29320	Agricultural self-unloading trailers (manufacture)
84130	75130	Agricultural services administration and regulation (public sector)
82990	74879	Agricultural showground
13940	17520	Agricultural twine (manufacture)
74909	74879	Agricultural valuer
46750	51550	Agro-chemical products (wholesale)
20200	24200	Agro-chemical products n.e.c. (manufacture)
74909	74149	Agronomy consulting
22290	25240	Air beds made of inflatable plastic (manufacture)
52290	63400	Air cargo agents activities
51210	62209	Air charter service (freight)
51102	62201	Air charter service (passenger)
28250	29230	Air cleansing plant (not for air conditioning equipment) (manufacture)
28132	29122	Air compressor (manufacture)
43220	45330	Air conditioning contracting
28250	29230	Air conditioning equipment for aircraft (manufacture)
28250	29230	Air conditioning machines (manufacture)
46690	51870	Air conditioning machines (wholesale)
28250	29230	Air conditioning package (manufacture)
30300	35300	Air cushion vehicle (manufacture)
28250	29230	Air filter for air conditioning equipment (manufacture)
52290	63400	Air freight agent or broker
25400	29600	Air gun (manufacture)
46499	51479	Air heaters and hot air distributors (non-electric) (wholesale)
71200	74300	Air measuring related to cleanness
26511	33201	Air navigation instruments and systems (electronic) (manufacture)
26513	33202	Air navigation instruments and systems (non-electronic) (manufacture)
52242	63110	Air passenger baggage handling services
77351	71231	Air passenger transport equipment rental and leasing without operator
25400	29600	Air pistol (manufacture)
28120	29122	Air preparation equipment for use in pneumatic systems (manufacture)
28131	29121	Air pump (manufacture)
25400	29600	Air rifle (manufacture)
51102	62201	Air taxi service
52230	63230	Air terminal operated by airline
52230	63230	Air traffic control activities
52230	63230	Air traffic control centre
77352	71239	Air transport equipment for freight rental (without operator)
51210	62109	Air transport of freight over regular routes
51102	62201	Air transport of passengers by aero clubs for instruction or pleasure
52230	63230	Air transport supporting activities
29320	34300	Airbags for motor vehicle (manufacture)
46140	51140	Aircraft (commission agent)
30300	35300	Aircraft (manufacture)
46690	51870	Aircraft (wholesale)
25400	29600	Aircraft bomb (manufacture)
30300	35300	Aircraft brake (not brake lining) (manufacture)
28990	35300	Aircraft carrier catapults and related equipment (manufacture)
71200	74300	Aircraft certification
26513	33202	Aircraft engine instruments (non-electronic) (manufacture)
30300	35300	Aircraft galley (manufacture)
77352	71239	Aircraft hire for freight (without crew)
77351	71231	Aircraft hire for passengers without crew
46690	51870	Aircraft launching gear and deck arresters (wholesale)
28990	35300	Aircraft launching gear and related equipment (manufacture)
22290	25240	Aircraft parts and accessories made of plastic (manufacture)
30300	35300	Aircraft parts and sub assemblies (not electric) (manufacture)
30300	35300	Aircraft propeller (manufacture)

SIC 2007	SIC 2003	Activity	SIC 2007	SIC 2003	Activity
52230	63230	Aircraft refuelling services	13200	17220	Alpaca woollen weaving (manufacture)
30300	36110	Aircraft seat (manufacture)	13200	17230	Alpaca worsted weaving (manufacture)
26511	33201	Airfield electronic controls and approach aids (manufacture)	27110	31100	Alternating current (ac) generators (manufacture)
52230	63230	Airfield ground service activities	27110	31100	Alternating current (ac) motors (manufacture)
30200	35200	Airfield mechanical or electro-mechanical signalling, safety, traffic control equipment(manufacture)	29310	31610	Alternator for vehicle (manufacture)
42110	45230	Airfield runway construction	27110	31100	Alternators (not for vehicles) (manufacture)
30300	35300	Airframe (manufacture)	26513	33202	Altimeter (non-electronic) (manufacture)
30300	35300	Airframe parts and sub assemblies (not electric) (manufacture)	08910	14300	Alum mine
56290	55520	Airline catering	24420	27420	Aluminium alloys production (manufacture)
49390	60219	Airline coach service (scheduled)	26200	30020	Aluminium coating inside pc cases (manufacture)
52230	63230	Airport activities	20130	24130	Aluminium compounds (except bauxite and abrasives) (manufacture)
52230	63230	Airport fire fighting and fire-prevention services	24420	27420	Aluminium foil laminates made from aluminium foil as primary component (manufacture)
42110	45230	Airport runway construction	24420	27420	Aluminium from alumina production (manufacture)
81299	90030	Airport runways clearing of snow and ice	24420	27420	Aluminium hardener (manufacture)
49390	60231	Airport shuttles	07290	13200	Aluminium ore (bauxite) mining or preparation
30300	35300	Airscrew (manufacture)	24420	27420	Aluminium oxide (alumina) production (manufacture)
30300	35300	Airship (manufacture)	20301	24301	Aluminium paint (manufacture)
51101	62101	Air-transport equipment renting with operator, for scheduled passenger transportation	20301	24301	Aluminium paste (manufacture)
52230	63230	Airway terminals operation	24420	27420	Aluminium refining (manufacture)
23700	26700	Alabaster bowl cutting (manufacture)	24420	27420	Aluminium semi-manufactures production (manufacture)
08110	14110	Alabaster mine	24420	27420	Aluminium smelting (manufacture)
26520	33500	Alarm clock (manufacture)	24420	27420	Aluminium wire made by drawing (manufacture)
80200	74602	Alarm monitoring activities	23510	26510	Aluminous cement (manufacture)
18129	22220	Album printing (manufacture)	20130	24130	Alums (manufacture)
46170	51170	Alcoholic beverages (commission agent)	32990	36639	Amber turning (manufacture)
47250	52250	Alcoholic beverages (retail)	29100	34100	Ambulance (manufacture)
46342	51342	Alcoholic beverages (wholesale)	86900	85140	Ambulance service
46342	51342	Alcoholic beverages exporter (wholesale)	45111	50101	Ambulances with a weight not exceeding 3.5 tonnes (new) (retail)
46342	51342	Alcoholic beverages importer (wholesale)	45111	50101	Ambulances with a weight not exceeding 3.5 tonnes (new) (wholesale)
11010	15910	Alcoholic distilled potable beverage (manufacture)	45112	50102	Ambulances with a weight not exceeding 3.5 tonnes (used) (retail)
20140	24140	Aldehyde (manufacture)	45112	50102	Ambulances with a weight not exceeding 3.5 tonnes (used) (wholesale)
46120	51120	Aldehyde function compounds (commission agent)	46180	51180	Amides and their derivatives and salts (commission agent)
46750	51550	Aldehyde function compounds (wholesale)	46120	51120	Amine function compounds (commission agent)
11050	15960	Ale brewing (manufacture)	20140	24140	Amines (manufacture)
10910	15710	Alfalfa (Lucerne) meal and pellets (manufacture)	46120	51120	Amino resins, phenolic resins and polyurethanes in primary forms (commission agent)
01190	01110	Alfalfa growing	46750	51550	Amino resins, phenolic resins and polyurethanes in primary forms (wholesale)
03110	05010	Algae gathering	20160	24160	Aminoplastic resins (manufacture)
20160	24160	Alginates (manufacture)	26511	33201	Ammeters (electronic) (manufacture)
84240	75240	Alien registration administration and operation	26513	33202	Ammeters (non-electronic) (manufacture)
20120	24120	Alizarin dye (manufacture)	20150	24150	Ammonia (manufacture)
20130	24130	Alkali (manufacture)	19100	23100	Ammoniacal liquor from coke ovens (manufacture)
46120	51120	Alkili or alkaline earth metals (commission agent)	20150	24150	Ammonium chloride (manufacture)
20301	24301	Alkyd (manufacture)	46120	51120	Ammonium chloride, nitrites and carbonates (commission agent)
20160	24160	Alkyd resins (manufacture)	20150	24150	Ammonium compounds (excluding ammonium nitrate, sulphate and phosphate) (manufacture)
01130	01120	Alliaceous vegetable growing			
25500	28400	Alloy and steel forging roll (manufacture)	20150	24150	Ammonium nitrate (not for explosives) (manufacture)
24100	27100	Alloy bearing steel (manufacture)	20510	24610	Ammonium nitrate for explosives (manufacture)
24100	27100	Alloy pig iron (manufacture)	20150	24150	Ammonium phosphate (manufacture)
24100	27100	Alloy tool steel (manufacture)	20150	24150	Ammonium sulphate (manufacture)
18129	22220	Almanac printing (manufacture)	19100	23100	Ammonium sulphate from coke ovens (manufacture)
10612	15612	Almond grinding (manufacture)			
01250	01139	Almond growing			
13100	17120	Alpaca and mohair spinning on the woollen system (manufacture)			
13100	17130	Alpaca and mohair spinning on the worsted system (manufacture)			

A

A

SIC 2007	SIC 2003	Activity
25400	29600	Ammunition (manufacture)
47789	52489	Ammunition (retail)
20510	24610	Amorce (manufacture)
29100	34100	Amphibious vehicles (manufacture)
26400	32300	Amplifier for audio separates (manufacture)
26309	32300	Amplifier for broadcasting studio (manufacture)
26400	32300	Amplifiers and sound amplifier sets (manufacture)
26110	32100	Amplifying valve (manufacture)
23190	26150	Ampoules made of glass (hygienic and pharmaceutical) (manufacture)
92000	92710	Amusement arcade
46180	51180	Amusement goods (commission agent)
18129	22220	Amusement guide periodical printing (manufacture)
58142	22130	Amusement guide periodical publishing
77390	71340	Amusement machine hire
32401	36501	Amusement machines (manufacture)
93210	92330	Amusement park activities
93210	92330	Amusement park mechanical rides, water rides, games
93210	92330	Amusement park shows, theme exhibits and picnic grounds
20520	24620	Anaerobic adhesive (manufacture)
32500	33100	Anaesthetic equipment (manufacture)
21200	24421	Anaesthetics (manufacture)
86101	85111	Anaesthetist (public sector)
21200	24421	Analgesics (manufacture)
26200	30020	Analogue computer (manufacture)
70221	74142	Analysis of capital investment proposals consultancy services
71200	74300	Analytical chemist
25990	28750	Anchor (manufacture)
50200	61102	Anchor handling services
10130	15139	Andouillettes (manufacture)
28922	29522	Angle-dozers (manufacture)
24420	27420	Angles made of aluminium (manufacture)
24100	27100	Angles of stainless steel (manufacture)
01490	01250	Angora rabbit breeding
08110	14120	Anhydrite mine or quarry
23520	26530	Anhydrite plaster (manufacture)
46750	51550	Aniline (wholesale)
20120	24120	Aniline dye (manufacture)
75000	85200	Animal ambulance activities
10910	15710	Animal compound feed (manufacture)
49390	60239	Animal drawn vehicle transport
10410	15410	Animal fat and oil production (non-edible) (manufacture)
28930	29530	Animal fat or oils extraction and preparation machinery (manufacture)
46210	51210	Animal feed (wholesale)
10910	15710	Animal feed supplement (manufacture)
10110	15112	Animal grease (manufacture)
01430	01220	Animal hair (not carded or combed) including horsehair production
46210	51210	Animal hair (wholesale)
75000	85200	Animal health care and control activities for farm animals
75000	85200	Animal health care and control activities for pet animals
75000	85200	Animal hospital (RSPCA, PDSA, Blue Cross)
75000	85200	Animal hospital run by veterinary surgeon

SIC 2007	SIC 2003	Activity
75000	85200	Animal hospital supervised or run by registered veterinarian
01700	01500	Animal hunting and trapping
10110	15112	Animal offal (inedible) production (manufacture)
10110	15112	Animal offal processing (manufacture)
10410	15420	Animal oil refining (manufacture)
01629	01429	Animal production support activities other than farm animal boarding and care
01629	01429	Animal propagation, growth and output – activities to promote these, on a fee or contract basis
94990	91330	Animal protection organisations
01490	01250	Animal rearing for medical research
01629	01429	Animal rearing for production of serum
77390	71340	Animal rental (herds, racehorses)
90010	92349	Animal training for circuses, etc.
59111	92111	Animated film production
59112	92111	Animated video production
01280	01139	Anise growing
28210	29210	Annealing lehr (manufacture)
25610	28510	Anodising (manufacture)
14131	18221	Anoraks for men and boys (manufacture)
14132	18222	Anoraks for women and girls (manufacture)
20140	24140	Anthracene (manufacture)
46180	51180	Antibiotics (commission agent)
21100	24410	Antibiotics (manufacture)
46460	51460	Antibiotics (wholesale)
43341	45440	Anti-corrosive coatings application work
20301	24301	Anti-corrosive paint (manufacture)
20590	24660	Anti-freeze mixtures (excluding pure ethyl glycol) (manufacture)
46120	51120	Anti-freezing preparations and prepared de-icing fluids (commission agent)
24450	27450	Antifriction metal (manufacture)
30300	35300	Anti-icing equipment and systems for aircraft (manufacture)
21200	24421	Anti-infectives (manufacture)
20590	24660	Anti-knock compounds (manufacture)
46120	51120	Anti-knock preparations and additives for mineral oils and similar products (commission agent)
24450	27450	Antimony (manufacture)
20420	24520	Anti-perspirant (manufacture)
50200	61102	Anti-pollution vessel services
47791	52501	Antique books (retail)
23110	26110	Antique glass (manufacture)
47791	52501	Antiques (retail)
46470	51479	Antiques (wholesale)
29320	34300	Anti-roll bars for motor vehicles (manufacture)
20590	24660	Anti-rust preparations (manufacture)
45200	50200	Anti-rust treatment of motor vehicles
21200	24421	Antiseptics (manufacture)
21200	24421	Antisera and other blood fractions (manufacture)
46180	51180	Antisera and vaccines (commission agent)
46460	51460	Antisera and vaccines (wholesale)
20200	24200	Anti-sprouting products (manufacture)
46750	51550	Anti-sprouting products (wholesale)
25730	28620	Anvils (manufacture)
68100	70120	Apartment buildings buying and selling
68209	70209	Apartment buildings letting
55209	55232	Apartment letting (self catering) other than holiday centres and holiday villages or youth hostels

SIC 2007	SIC 2003	Activity
11010	15910	Aperitif (spirit based) (manufacture)
28990	33403	Apparatus and equipment for automatically developing photographic film
28940	29540	Apparel and leather production machinery (manufacture)
13200	17210	Apparel cloth woven from yarns spun on the cotton system (manufacture)
14200	18300	Apparel made of fur (manufacture)
22290	25240	Apparel made of plastic (if only sealed together, not sewn) (manufacture)
22190	25130	Apparel made of rubber (if only sealed together, not sewn) (manufacture)
22190	25130	Apparel made of sealed rubber (manufacture)
28940	29540	Apparel production machinery (manufacture)
84230	75230	Appeal Committee of the House of Lords
01240	01139	Apple growing
10890	15899	Apple pomace and pectin (manufacture)
11030	15949	Apple wine making
27900	31300	Appliance cords with insulated wire and connectors (manufacture)
74909	74879	Appraiser and valuer (not insurance or real estate)
85320	80220	Apprentice school
01240	01139	Apricot growing
22190	25130	Aprons (manufacture)
14120	18210	Aprons for domestic use (manufacture)
14120	18210	Aprons for industrial use (manufacture)
03210	05020	Aquaculture in salt water filled tanks or reservoirs
03210	05020	Aquaculture in sea or brackish waters
03220	05020	Aquaculture, freshwater
27510	29710	Aquarium heater (electric) (manufacture)
42910	45240	Aqueduct construction
84230	75230	Arbitration of civil actions
70229	74149	Arbitrators between management and labour
69109	74119	Arbitrators legal activities
27400	31500	Arc lamp (manufacture)
32300	36400	Archery equipment (manufacture)
28131	29121	Archimedean screw pump (manufacture)
94120	91120	Architects associations
71111	74201	Architectural activities and related technical consultancy
71111	74201	Architectural draughtsman
58110	22110	Architectural drawing publishing
71129	74204	Architectural engineering activities
23190	26150	Architectural glass (manufacture)
22230	25239	Architrave made of plastic (manufacture)
91012	92510	Archive activities
27110	31100	Arc-welding transformers (manufacture)
20110	24110	Argon (manufacture)
29320	36110	Arm rest for motor vehicle (manufacture)
31090	36110	Armchair (manufacture)
30400	29600	Armoured amphibious military vehicles (manufacture)
29100	34100	Armoured car (except military fighting vehicles) (manufacture)
80100	74602	Armoured car services
25990	28750	Armoured doors (manufacture)
22190	25130	Armoured hose made of rubber (manufacture)
46660	51850	Armoured or reinforced safes, strong boxes and doors made of base metal (wholesale)
25400	29600	Arms (manufacture)
15120	19200	Army accoutrement made of leather (manufacture)

SIC 2007	SIC 2003	Activity
84220	75220	Army establishment (civilian personnel)
84220	75220	Army establishment (service personnel)
94910	91310	Army scripture readers association
01280	01139	Aromatic crops growing
20530	24630	Aromatic distilled waters (manufacture)
20140	24140	Aromatic hydrocarbons (manufacture)
10620	15620	Arrowroot (manufacture)
24450	27450	Arsenic (manufacture)
47781	52486	Art (retail)
94990	91330	Art clubs
90030	92319	Art expert
91020	92521	Art gallery (not dealer)
85520	80429	Art instruction
15120	19200	Art leather work (manufacture)
25990	28750	Art metal work (manufacture)
91020	92521	Art museums
13923	17403	Art needlework (manufacture)
23410	26210	Art pottery (manufacture)
58190	22150	Art publishing
91011	92510	Art work lending and storage
42210	45250	Artesian well contractor
01130	01120	Artichoke growing
32120	36220	Articles for religious use of base metals clad with precious metals (manufacture)
13950	17530	Articles made from non-wovens (manufacture)
17290	21259	Articles made of paper and paperboard n.e.c. (manufacture)
47710	52421	Articles of fur (retail)
46690	51870	Articulated link chain (except for motor vehicles and bicycles) (wholesale)
28150	29140	Articulated link chain (manufacture)
10890	15899	Artificial concentrates (manufacture)
23990	26829	Artificial corundum (manufacture)
32500	33100	Artificial eye (manufacture)
32990	36639	Artificial flowers and fruit made of paper (manufacture)
32990	36639	Artificial flowers and fruit made of plastic (manufacture)
32990	36639	Artificial flowers and fruit made of textiles (manufacture)
46499	51479	Artificial flowers, foliage and fruit (wholesale)
14200	18300	Artificial fur and articles thereof (manufacture)
10890	15620	Artificial honey (manufacture)
26513	33202	Artificial horizon (non-electronic) (manufacture)
01629	01429	Artificial insemination activities on a fee or contract basis
46460	51460	Artificial joints and parts for the body (wholesale)
86900	85140	Artificial kidney unit
32500	33100	Artificial limb (manufacture)
86900	85140	Artificial limb and appliance centre
20150	24150	Artificial manure (manufacture)
32500	33100	Artificial parts for the heart (manufacture)
32500	33100	Artificial respiration equipment (manufacture)
22230	25239	Artificial stone made of plastic (manufacture)
32500	33100	Artificial teeth (manufacture)
46460	51460	Artificial teeth and dental fittings (wholesale)
20412	24512	Artificial waxes (manufacture)
25400	29600	Artillery (manufacture)
25400	29600	Artillery ammunition (manufacture)
90030	92319	Artist

SIC 2007	SIC 2003	Activity
32910	36620	Artists' brush (manufacture)
13960	17549	Artists' canvases (manufacture)
20301	24301	Artists' colours (manufacture)
96090	93059	Artists' model
13960	17549	Artists' tracing cloth (manufacture)
46180	51180	Artists', students' and sign board painters' colours, modifying tints and similar (commission agent)
85320	80220	Arts and crafts school
90040	92320	Arts facilities operation
41201	45211	Arts, cultural or leisure facilities buildings construction
23990	26821	Asbestos carding (manufacture)
23650	26650	Asbestos cement products (manufacture)
23990	26821	Asbestos felting (manufacture)
08990	14500	Asbestos mining and quarrying
23990	26821	Asbestos mixing (manufacture)
23990	26821	Asbestos moulding (manufacture)
39000	45250	Asbestos removal work
23990	26821	Asbestos spinning (manufacture)
23990	26821	Asbestos weaving (manufacture)
39000	90030	Asbestos, lead paint, and other toxic material abatement
32500	33100	Aseptic hospital furniture (manufacture)
38210	14500	Ashes and residues of incineration of mining and quarrying waste
01130	01120	Asparagus growing
23990	26829	Asphalt (manufacture)
08990	14500	Asphalt (natural) mining and quarrying
28923	29523	Asphalt laying plant (manufacture)
23990	26829	Asphalt or similar materials (e.g. Asphalt-based adhesives), articles thereof (manufacture)
42110	45230	Asphalt paving of roads
28923	29523	Asphalt processing plant (manufacture)
42110	45230	Asphalting contractor (civil engineering)
08990	14500	Asphaltites and asphaltic rock mining and quarrying
28930	29530	Aspirator separators (manufacture)
01430	01220	Ass farming and breeding
71200	74300	Assay office
30200	35200	Assembled rail sections (manufacture)
18140	22230	Assembling of books, brochures, etc. (manufacture)
33200	33301	Assembling of electronic industrial process control equipment (manufacture)
33200	33302	Assembling of non-electronic industrial process control equipment (manufacture)
41201	45211	Assembly and erection of prefabricated non-residential constructions on the site
41201	28110	Assembly and installation of self-manufactured commercial buildings of metal on site
41201	25239	Assembly and installation of self-manufactured commercial buildings of plastic on site
41201	20300	Assembly and installation of self-manufactured commercial buildings of wood on site
41202	28110	Assembly and installation of self-manufactured domestic buildings of metal on site
41202	25239	Assembly and installation of self-manufactured domestic buildings of plastic on site
41202	20300	Assembly and installation of self-manufactured domestic buildings of wood on site
87300	85311	Assisted-living facilities for the elderly or disabled (charitable)
87300	85312	Assisted-living facilities for the elderly or disabled (non-charitable)
94120	91120	Association of corporate and certified accountants
94990	91330	Associations for the pursuit of a cultural or recreational activity or hobby
65110	66011	Assurance company (life)
96090	93059	Astrologer
26701	33402	Astronomical equipment (optical) (manufacture)
72190	73100	Astronomy research and experimental development
86101	85112	Asylums (private sector)
86101	85111	Asylums (public sector)
93199	92629	Athletes
14190	18249	Athletic clothing (manufacture)
93120	92629	Athletic club
32300	36400	Athletic equipment (manufacture)
15200	19300	Athletic footwear (manufacture)
18129	22220	Atlas printing (manufacture)
58110	22110	Atlas publishing
28250	29230	Atmospheric pollution control plant (manufacture)
72190	73100	Atomic energy research and experimental development
15120	19200	Attaché case made of leather or leather substitute (manufacture)
69201	74121	Attestations, valuations and preparation of pro forma statements
28250	29710	Attic ventilation fans (manufacture)
69102	74113	Attorney
29100	34100	ATVs, go-carts and similar including race cars (manufacture)
01130	01120	Aubergine (egg-plant) growing
47990	52630	Auction houses (except antiques and second hand goods) (retail)
47630	52450	Audio and video recordings (retail)
58110	22110	Audio book publishing
47430	52450	Audio equipment (retail)
26400	32300	Audio separate (manufacture)
46431	51431	Audio separates (wholesale)
18201	22310	Audio tape recording, except master copies for records or audio material (manufacture)
47630	52450	Audio tapes and cassettes (retail)
47430	52450	Audio/visual cassettes (retail)
47430	52450	Audio/visual equipment (retail)
77390	71340	Audio/visual equipment for professional use hire
32500	33100	Audiometers (manufacture)
69201	74121	Auditing activities
25730	28620	Auger and auger bit (manufacture)
13921	17401	Austrian blinds (manufacture)
90030	92319	Author
26701	33402	Auto correlator (optical) (manufacture)
29310	31610	Auto electrical equipment (manufacture)
29320	34300	Auto spare parts (manufacture)
32500	33100	Autoclaves (manufacture)
30910	35410	Autocycle (manufacture)
77330	71330	Automatic data processing equipment hire
25400	29600	Automatic gun (manufacture)
32200	36300	Automatic pianos (manufacture)
26511	33201	Automatic pilots (electronic) (manufacture)
26513	33202	Automatic pilots (non-electronic) (manufacture)
28120	29130	Automatic process control valves (manufacture)
46690	51870	Automatic regulating or controlling instruments and apparatus (wholesale)
28940	29540	Automatic stop motions (textile machinery) (manufacture)

SIC 2007	SIC 2003	Activity
26200	30020	Automatic teller machines (ATMs) computer terminals not mechanically operated (manufacture)
46660	51850	Automatic typewriters and word-processing machines (wholesale)
94990	91330	Automobile association
52219	50200	Automobile association road patrols
45200	50200	Automobile association service centres
79110	63301	Automobile association touring department
77110	71100	Automobile rental (self drive)
26513	33202	Automotive emissions testing equipment (non-electronic) (manufacture)
71121	74205	Automotive production design
13960	17542	Automotive trimmings (manufacture)
84250	75250	Auxiliary fire brigade services
46640	51830	Auxiliary machinery for use with machines for working textiles (wholesale)
25300	28300	Auxiliary plant for use with steam generators (manufacture)
30300	35300	Auxiliary power unit for aircraft (manufacture)
66210	67200	Average adjuster
65120	66031	Aviation insurance
19201	23201	Aviation spirit (manufacture)
19201	23201	Aviation turbine fuel (manufacture)
01220	01139	Avocado growing
13200	17250	Awning cloth weaving (manufacture)
13922	17402	Awnings (manufacture)
46410	51410	Awnings and sun blinds (wholesale)
47510	52410	Awnings and sunblinds (retail)
22290	25240	Awnings made of plastic (manufacture)
25730	28620	Axe (manufacture)
28132	29122	Axial compressor (manufacture)
28131	29121	Axial flow pump (manufacture)
29320	34300	Axle for motor vehicle (manufacture)
30910	35410	Axle for motorcycle (manufacture)
30200	35200	Axles and wheels for locomotive and rolling stock (manufacture)
30200	35200	Axles for rail and tramway vehicles (manufacture)
30200	35200	Axles made of steel for railway and tramway vehicles (manufacture)
13931	17511	Axminster carpet (manufacture)
20120	24120	Azoic dye (manufacture)
14190	18249	Babies garments (manufacture)
22290	25240	Baby baths made of plastic (manufacture)
30920	36639	Baby carriage (manufacture)
47789	52489	Baby carriages (retail)
46499	51479	Baby carriages (wholesale)
14190	18249	Baby clothing (manufacture)
46380	51380	Baby food (wholesale)
10860	15880	Baby foods (manufacture)
10860	15880	Baby foods (milk based) (manufacture)
14190	18249	Baby linen (manufacture)
13923	17403	Baby napkins made of towelling (manufacture)
28940	29540	Backing and curling machinery (carpet making) (manufacture)
27900	31620	Backward wave oscillator (manufacture)
10130	15131	Bacon curing (manufacture)
10130	15131	Bacon production (manufacture)
10130	15131	Bacon smoking (manufacture)
23440	26240	Bacteria bed tile (manufacture)
71200	74300	Bacteriologist (non medical)

SIC 2007	SIC 2003	Activity
13960	17542	Badges (textile) (manufacture)
25990	28750	Badges made of metal (manufacture)
01280	01139	Badian growing
93120	92629	Badminton club
32300	36400	Badminton shuttlecock (manufacture)
25990	28750	Bag clasp (manufacture)
25990	28750	Bag frame (manufacture)
32409	36509	Bagatelle board (manufacture)
26600	33100	Baggage scanning equipment (manufacture)
13200	17250	Bagging cloth (manufacture)
32200	36300	Bagpipes and reeds (manufacture)
13922	17402	Bags made of canvas (manufacture)
13922	17402	Bags made of canvas or cotton cloth (manufacture)
32130	36610	Bags made of chain (manufacture)
15120	19200	Bags made of leather or leather substitute (manufacture)
17211	21211	Bags made of paper (manufacture)
22220	25220	Bags made of plastic (not designed for prolonged use) (manufacture)
22220	25220	Bags made of plastic for packaging (manufacture)
22220	25220	Bags made of polyethylene (manufacture)
22220	25220	Bags made of transparent regenerated cellulose film (manufacture)
69109	74119	Bailiffs activities
03210	01250	Bait digging
03210	01250	Bait production
13990	17549	Baize (manufacture)
47240	52240	Baker (retail)
10890	15899	Bakers' yeast from distillery (manufacture)
10710	15810	Bakery (baking main activity) (manufacture)
47240	52240	Bakery (selling main activity) (retail)
28930	29530	Bakery machinery and ovens (manufacture)
28930	29530	Bakery moulders (manufacture)
28930	29530	Bakery ovens (industrial) (manufacture)
46690	51870	Bakery ovens (wholesale)
46360	51360	Bakery products (wholesale)
25990	28750	Baking dish, pan and tin (manufacture)
10890	15899	Baking powder (manufacture)
14190	18241	Balaclava helmet (manufacture)
46690	51870	Balances and scales (wholesale)
26511	33201	Balancing machines (electronic) (manufacture)
26513	33202	Balancing machines (non-electronic) (manufacture)
22190	25130	Balata belting (manufacture)
22190	25130	Balata goods (excluding belting) (manufacture)
28940	29540	Bale breakers (manufacture)
28302	29320	Bale handler for agriculture (manufacture)
28302	29320	Baler for hay and straw (manufacture)
13940	17520	Baler twine (manufacture)
28290	29240	Baling press (not agricultural) (manufacture)
46690	51870	Ball and roller bearings (wholesale)
28150	29140	Ball bearing (manufacture)
08120	14220	Ball clay extraction (mine or opencast working)
22190	25130	Ball core (rubber) (manufacture)
28140	29130	Ball valves (manufacture)
46690	51870	Ballasts for discharge lamps or tubes (wholesale)
02300	02010	Balata gathering
90010	92311	Ballet company
85320	80220	Ballet school
15200	19300	Ballet shoe (manufacture)

B

SIC 2007	SIC 2003	Activity	SIC 2007	SIC 2003	Activity
25400	29600	Ballistic missile, except intercontinental ballistic missile (ICBM) (manufacture)	01110	01110	Barley growing
30300	35300	Balloon (not toy) (manufacture)	11060	15970	Barley malting (manufacture)
46499	51479	Balloons, dirigibles and other non-powered aircraft (wholesale)	10611	15611	Barley meal production (manufacture)
			10611	15611	Barley milling (manufacture)
22190	25130	Balloons, rubber (except pilot and sounding balloons, dirigibles and hot-air balloons) (manufacture)	10611	15611	Barley processing (blocked, flaked, puffed or pearled) (manufacture)
			26511	33201	Barometer (electronic) (manufacture)
23190	26150	Ballotini (manufacture)	26513	33202	Barometer (non-electronic) (manufacture)
32990	36631	Ballpoint pen and refill (manufacture)	25920	28720	Barrels made of aluminium (manufacture)
46499	51479	Ball-point, felt-tipped and other porous-tipped pens and markers (wholesale)	25910	28710	Barrels made of iron or steel (manufacture)
			22220	25220	Barrels made of plastic (manufacture)
32300	36400	Balls for all sports (finished) (manufacture)	16240	20400	Barrels made of wood (manufacture)
02300	02010	Balsam gathering	69101	74112	Barrister
31090	36140	Bamboo furniture (other than seating) (manufacture)	77390	71219	Barrow hiring
16290	20520	Bamboo preparation (manufacture)	24420	27420	Bars made of aluminium (manufacture)
10390	15330	Banana ripening and conditioning (manufacture)	24440	27440	Bars made of brass (manufacture)
01220	01139	Bananas and plantain growing	24440	27440	Bars made of copper (manufacture)
90010	92311	Band (musical)	23190	26150	Bars made of glass (manufacture)
74909	74879	Band agency	24100	27100	Bars of stainless steel (manufacture)
13200	17210	Bandage cloth weaving (manufacture)	24430	27430	Bars, rods, profiles and wire made of lead (manufacture)
13990	17542	Banding (woven) (manufacture)			
22190	25130	Bands made of elastic (manufacture)	24430	27430	Bars, rods, profiles and wire made of tin (manufacture)
25930	28730	Bands made of plaited metal (manufacture)			
22190	25130	Bands made of rubber (manufacture)	24430	27430	Bars, rods, profiles and wire made of zinc (manufacture)
25930	28730	Bands made of uninsulated plaited copper (manufacture)	08910	14300	Barytes mine
			08110	14110	Basalt mine
25930	28730	Bands, slings, etc. Made of uninsulated plaited iron or steel (manufacture)	28290	29240	Base exchange plant for water treatment (manufacture)
25730	28620	Bandsaw blades (manufacture)	25990	28750	Base metal articles (manufacture)
64205	65234	Bank holding companies	17120	21120	Base paper for printing and writing paper (manufacture)
28230	30010	Bank note counting machine (manufacture)			
17120	21120	Bank note paper (manufacture)	23190	26150	Basement lights made of glass (manufacture)
18129	22220	Bank note printing (manufacture)	20120	24120	Basic dye (manufacture)
64110	65110	Bank of England	20150	24150	Basic slag (ground) (manufacture)
66110	67110	Bankers' clearing house	01280	01139	Basil growing
64929	65229	Banking institutions in Channel Islands and Isle of Man activities (not in UK banking sector)	32300	36400	Basins for swimming and paddling pools (manufacture)
28230	30010	Banknote dispensing machine (manufacture)	23650	26650	Basins made of fibre-cement (manufacture)
64191	65121	Banks (deposit taking)	25990	28750	Basins made of metal (manufacture)
64929	65229	Banks not authorised by the FSA as deposit taking	31090	36110	Basket chair (manufacture)
13922	17402	Banner (making up) (manufacture)	31090	36140	Basket furniture (manufacture)
16230	20300	Bannister rails made of wood (manufacture)	16290	20520	Baskets made of materials (other than plastic) (manufacture)
56210	55520	Banquet catering			
94910	91310	Baptist church	22290	25240	Baskets made of plastic (manufacture)
82990	74879	Bar code imprinting services	16290	20520	Basketware (manufacture)
46740	51540	Barbed wire and stranded wire (wholesale)	23690	26660	Bas-relief and haut-relief made of concrete, plaster, cement or artificial stone (manufacture)
25930	28730	Barbed wire made of steel (manufacture)			
96020	93020	Barber	13200	17250	Bast fibres and special yarns weaving (manufacture)
46460	51460	Barbers' and similar chairs (wholesale)	13100	17170	Bast fibres preparation and spinning (manufacture)
26200	30020	Barcode readers and other optical readers (manufacture)	63110	72300	Batch processing
			30920	35430	Bath chair (manufacture)
26110	32100	Bare printed circuit boards (manufacture)	20420	24520	Bath preparations (manufacture)
30110	35110	Barge (manufacture)	20420	24520	Bath salts (manufacture)
50400	61209	Barge lessee or owner (freight)	13923	17403	Bath towel (manufacture)
50300	61201	Barge lessee or owner (passenger) (inland waterway service)	22290	25240	Bathing caps made of plastic (manufacture)
50200	61102	Barge transport (except for inland waterway) freight service	22190	25130	Bathing caps of rubber (manufacture)
			93110	92619	Bathing pool proprietor
16100	20100	Bargeboard (manufacture)	46730	51530	Baths (wholesale)
08910	14300	Barium sulphate (natural) mining	23420	26220	Baths made of ceramic (manufacture)

SIC 2007	SIC 2003	Activity
22230	25239	Baths made of fibre glass (manufacture)
25990	28750	Baths made of metal (manufacture)
22230	25239	Baths made of plastic (manufacture)
32300	36400	Bats (manufacture)
23200	26260	Bats made of ceramic (manufacture)
16210	20200	Battenboard (manufacture)
27200	31400	Batteries for vehicles (manufacture)
27110	31100	Battery charger (manufacture)
27200	31400	Battery for car (manufacture)
27200	31400	Battery for flash lamp (manufacture)
28990	29560	Battery making machine (manufacture)
29100	34100	Battery powered electric commercial vehicle (manufacture)
14120	18210	Battledress for men (manufacture)
14120	18210	Battledress for women (manufacture)
01280	01139	Bay growing
25710	28750	Bayonet (manufacture)
77210	71401	Beach chairs and umbrellas renting and leasing
93290	92729	Beach facilities rental (bathhouses, lockers, chairs, etc.)
15200	19300	Beach footwear (manufacture)
93290	92729	Beach hut proprietor
14190	18249	Beachwear for women and girls (manufacture)
30110	35110	Beacon for shipping (manufacture)
30110	35110	Beacons for ships (manufacture)
16100	20100	Beaded wood (manufacture)
16230	20300	Beading made of wood (manufacture)
16230	20300	Beadings and mouldings made of wood (manufacture)
23190	26150	Beads made of glass (manufacture)
16290	20510	Beads made of wood (manufacture)
17220	21220	Beakers made of paper (manufacture)
28940	29540	Beaming machinery (textile) (manufacture)
16230	20300	Beams (manufacture)
10612	15612	Bean grinding (manufacture)
01110	01120	Bean growing
10612	15612	Bean milling (manufacture)
10612	15612	Bean splitting (manufacture)
96020	93020	Beard trimming
28150	29140	Bearing housings (manufacture)
46690	51870	Bearing housings and plain shaft bearings (wholesale)
28150	29140	Bearing, gearing and driving element parts (manufacture)
28150	29140	Bearings (manufacture)
20420	24520	Beauty and make-up preparations (manufacture)
96020	93020	Beauty parlour
96020	93020	Beauty specialist
96020	93020	Beauty treatment activities
46180	51180	Beauty, make-up and skin-care preparations including sun tan preparations (commission agent)
46160	51160	Bed and table linen (commission agent)
31090	36140	Bed heads made of metal (manufacture)
13923	17403	Bed linen (manufacture)
47510	52410	Bed linen (retail)
31090	36110	Bed settee (manufacture)
47510	52410	Bedding (retail)
32990	36639	Bedfolder (manufacture)
13200	17210	Bedford cord (not worsted) weaving (manufacture)
13200	17230	Bedford cord worsted weaving (manufacture)

SIC 2007	SIC 2003	Activity
31030	36150	Beds (mattress and mattress support) (manufacture)
47599	52440	Beds (retail)
14310	17710	Bedsock (manufacture)
13923	17403	Bedspread (manufacture)
13923	17403	Bedspreads made of lace (manufacture)
31090	36140	Bedsteads made of metal (manufacture)
31090	36140	Bedsteads made of wood (manufacture)
01490	01250	Bee keeping
10130	15139	Beef extract (manufacture)
10130	15139	Beef paste (manufacture)
10130	15139	Beef pickling (manufacture)
16290	20510	Beehives made of wood (manufacture)
28302	29320	Bee-keeping machinery (manufacture)
46610	51880	Bee-keeping machines (wholesale)
11050	15960	Beer (non-alcoholic) (manufacture)
47250	52250	Beer (retail)
11050	15960	Beer brewing (manufacture)
56302	55402	Beer gardens (independent)
56302	55404	Beer gardens (managed)
56302	55403	Beer gardens (tenanted)
56302	55402	Beer halls (independent)
56302	55404	Beer halls (managed)
56302	55403	Beer halls (tenanted)
46110	51110	Beer or distilling dregs for animal feed (commission agent)
46342	51342	Beer, wines and liqueurs (wholesale)
10810	15830	Beet pulp (manufacture)
01190	01110	Beet seed growing
10810	15830	Beet sugar (manufacture)
01610	01410	Beet thinning on a fee or contract basis
01130	01120	Beetroot growing
27900	31620	Bell apparatus (other than telegraphic or telephonic) (manufacture)
24540	27540	Bell founding (manufacture)
25940	28740	Belleville washer (manufacture)
22190	25130	Bellows made of rubber (manufacture)
27900	31620	Bells (other than telephone type) (electric) (manufacture)
25990	28750	Bells for pedal cycles (manufacture)
26301	32201	Bells for telephones (manufacture)
25990	28750	Bells made of base metals (manufacture)
13200	17210	Belting duck weaving (manufacture)
22190	25130	Belting for domestic appliances made of rubber (manufacture)
22210	25210	Belting made of plastic (manufacture)
22190	25130	Belting made of rubber (manufacture)
15120	19200	Belts made of leather or leather substitute (manufacture)
31090	36110	Bench seat (manufacture)
25730	28620	Bench vice (manufacture)
28410	29420	Bending machines (metal forming) (manufacture)
24200	27220	Bends made of steel (manufacture)
88990	85321	Benevolent society (charitable services)
65120	66031	Benevolent society (insurance)
10410	15410	Benniseed crushing (manufacture)
10410	15420	Benniseed oil refining (manufacture)
16100	20100	Bent timber (manufacture)
31090	36140	Bentwood furniture (manufacture)
20140	24140	Benzene (manufacture)

B

B

SIC 2007	SIC 2003	Activity
35210	40210	Benzole (crude) from gas works
14190	18241	Beret (knitted) (manufacture)
14190	18241	Beret (not knitted) (manufacture)
02300	01139	Berries, gathering from the wild
52220	63220	Berthing activities
24450	27450	Beryllium (manufacture)
32910	36620	Besom (manufacture)
91020	92521	Bethnal Green Museum
92000	92710	Betting activities
92000	92710	Betting shop
82920	74820	Beverage bottling on a fee or contract basis
01270	01139	Beverage crop growing other than wine grapes
28930	29530	Beverage processing machinery (manufacture)
46170	51170	Beverages (commission agent)
11040	15950	Beverages (non-distilled, fermented) (manufacture)
46390	51390	Beverages (non-specialised) (wholesale)
11030	15949	Beverages made of fermented fruit n.e.c. (manufacture)
13990	17542	Bias binding (manufacture)
17120	21120	Bible paper (manufacture)
94910	91310	Bible society
28150	29140	Bicycle chain
77210	71401	Bicycle hire
25720	28630	Bicycle locks with or without keys (manufacture)
52219	63210	Bicycle parking operations
28131	29121	Bicycle pumps (manufacture)
46150	51150	Bicycles (commission agent)
30920	35420	Bicycles (non-motorised) (manufacture)
47640	52485	Bicycles (retail)
30920	35420	Bicycles and parts (manufacture)
46499	51479	Bicycles and their parts and accessories (wholesale)
32409	36509	Bicycles for children (manufacture)
23420	26220	Bidets made of ceramic, fireclay, etc. (manufacture)
25940	28740	Bifurcated rivet (manufacture)
10110	15112	Bile processing by knackers (manufacture)
74909	63400	Bill auditing and freight rate information
64991	65233	Bill broker on own account (other than discount house)
66120	67122	Bill broking on behalf of others (other than discount house)
82911	74871	Bill collecting
73110	74402	Bill posting agency
24420	27420	Billets made of aluminium (manufacture)
24440	27440	Billets made of brass (manufacture)
24440	27440	Billets made of copper (manufacture)
24100	27100	Billets made of semi-finished steel (manufacture)
15120	19200	Billfolds made of leather (manufacture)
32401	36501	Billiard ball (manufacture)
32401	36501	Billiard cue (manufacture)
93110	92629	Billiard room or saloon
32401	36501	Billiard table (manufacture)
13990	17549	Billiard table cloth (manufacture)
93120	92629	Billiards and snooker club
22220	25220	Bin liners made of plastic (manufacture)
13940	17520	Binder twine (manufacture)
13990	17542	Binding (woven) (manufacture)
18140	22230	Binding and finishing of books, brochures, magazines, catalogues etc. (manufacture)
13300	17300	Binding and mending of textiles (manufacture)

SIC 2007	SIC 2003	Activity
18140	22230	Binding and related services (manufacture)
28230	30010	Binding machine (manufacture)
92000	92710	Bingo hall
26701	33402	Binoculars (manufacture)
25990	28750	Bins made of metal (manufacture)
22290	25240	Bins made of plastic (manufacture)
26511	33201	Biochemical analysers (electronic) (manufacture)
20200	24200	Biocides (manufacture)
19201	23201	Biofuels from blending of alcohols with petroleum, e.g. Gasohol (manufacture)
72110	73100	Biology research and experimental development
21200	24421	Biotech pharmaceuticals (manufacture)
72110	73100	Biotechnology research and experimental development
32910	36620	Birch broom (manufacture)
10920	15720	Bird food (manufacture)
01490	01250	Bird raising, other than poultry
01700	01500	Bird skin production (from hunting)
01490	01250	Bird skin production from ranching operation
28930	29530	Biscuit making machinery and ovens (manufacture)
23310	26300	Biscuit tile (manufacture)
10720	15820	Biscuits (manufacture)
94990	91330	Bishopsgate Institute
24450	27450	Bismuth (manufacture)
25730	28620	Bit stock drill (manufacture)
19201	23201	Bitumen (manufacture)
23990	26829	Bitumen and flax felts for roofing and damp-proof courses (manufacture)
08990	14500	Bitumen mining and quarrying
28923	29523	Bitumen spreaders (manufacture)
17120	21120	Bituminised building board (manufacture)
46130	51130	Bituminous construction materials (commission agent)
06100	11100	Bituminous or oil shale and sand extraction
20301	24301	Bituminous paint (manufacture)
23990	26829	Bituminous sealants (manufacture)
03210	05020	Bivalves cultured in sea water
11050	15960	Black beer brewing (manufacture)
20510	24610	Black powder (manufacture)
10130	15139	Black pudding (manufacture)
01250	01139	Blackberry (cultivated) growing
28990	29560	Blackboard chalk making machinery (manufacture)
28230	36631	Blackboards (manufacture)
01250	01139	Blackcurrant growing
24100	27100	Blackplate (manufacture)
25620	28520	Blacksmith (not including farriers) (manufacture)
25730	28620	Blacksmiths' tools (manufacture)
08110	14110	Blackstone quarry
32990	36639	Bladder dressing (manufacture)
10110	15112	Bladder processing (manufacture)
46630	51820	Blades for bulldozers and angle-dozers (wholesale)
10890	15899	Blancmange powder (manufacture)
46520	51431	Blank audio and video tapes and diskettes, magnetic and optical disks (CDs, DVDs) (wholesale)
26800	24650	Blank diskettes (manufacture)
26800	24650	Blank optical discs (manufacture)
47630	52450	Blank tapes and discs (retail)
13923	17403	Blanket making up, outside weaving or knitting establishment (manufacture)
27510	29710	Blankets (electric) (manufacture)

B

SIC 2007	SIC 2003	Activity	SIC 2007	SIC 2003	Activity
13923	17403	Blankets including travelling rugs (manufacture)	22210	25210	Blocks made of plastic (manufacture)
13923	17403	Blankets made of cotton and man-made fibres (manufacture)	86900	85140	Blood banks
			96090	93059	Blood pressure machine operation (coin operated)
13923	17403	Blankets made of wool (manufacture)	21100	24410	Blood processing (manufacture)
23190	26150	Blanks for corrective spectacle lens (manufacture)	10130	15139	Blood pudding (manufacture)
24420	27420	Blanks made of aluminium (manufacture)	86900	85140	Blood transfusion service
52102	63121	Blast freezing for air transport activities	21200	24421	Blood-grouping reagents (manufacture)
52103	63121	Blast freezing for land transport activities	24440	27440	Blooms made of copper (manufacture)
52101	63121	Blast freezing for water transport activities	24100	27100	Blooms made of semi-finished steel (manufacture)
28210	29210	Blast furnace (manufacture)	17120	21120	Blotting paper (manufacture)
43120	45110	Blasting and associated rock removal work	14142	18232	Blouses for women and girls (manufacture)
43120	45110	Blasting of construction sites	25730	28620	Blow lamp (manufacture)
20510	24610	Blasting powder	28960	29560	Blow moulding machine for rubber or plastic (manufacture)
14131	18221	Blazers for men and boys (manufacture)			
14132	18222	Blazers for women and girls (manufacture)	23110	26110	Blown glass (manufacture)
13300	17300	Bleach works (manufacture)	28921	29521	Blow-out prevention apparatus (manufacture)
17110	21110	Bleached paper pulp made by chemical dissolving (manufacture)	28490	29430	Blowpipes (hand held, high or low pressure) (manufacture)
17110	21110	Bleached paper pulp made by mechanical processes (manufacture)	28940	29540	Blowroom machinery (textile) (manufacture)
			23320	26400	Blue brick (manufacture)
17110	21110	Bleached paper pulp made by non-dissolving processes (manufacture)	23520	26520	Blue lias lime kiln (manufacture)
			08110	14110	Blue pennant stone quarry
17110	21110	Bleached paper pulp made by semi-chemical processes (manufacture)	01250	01139	Blueberry growing
13300	17300	Bleaching and dyeing of fabrics (manufacture)	82190	74850	Blueprinting
			32409	36509	Board game (manufacture)
13300	17300	Bleaching and dyeing of textile articles, including wearing apparel (manufacture)	28950	29550	Board making machinery (except chipboard) (manufacture)
13300	17300	Bleaching and dyeing of textile fibres (manufacture)	55900	55239	Boarding house
13300	17300	Bleaching and dyeing of yarns (manufacture)	96090	93059	Boarding of pet animals
28940	29540	Bleaching machinery (textile) (manufacture)	55900	55239	Boarding school accommodation
13300	17300	Bleaching of jeans (manufacture)	23990	26821	Boards made of asbestos (manufacture)
13300	17300	Bleaching of wool cop, hank, warp, etc. (manufacture)	23610	26610	Boards made of concrete (manufacture)
			23140	26140	Boards made of glass fibre (manufacture)
27510	29710	Blenders for domestic use (electric) (manufacture)	23620	26620	Boards made of plaster (manufacture)
28930	29530	Blenders for the food industry (manufacture)	77342	71229	Boat hire for freight (without crew)
28930	29530	Blenders for the grain milling industry (manufacture)	77341	71221	Boat hire for passengers (without crew) (not linked with recreational service)
11010	15910	Blending and bottling of whisky (manufacture)	30120	35120	Boat kits for assembly (manufacture)
11010	15910	Blending of distilled spirits (manufacture)	50100	61101	Boat rental for passenger conveyance with crew (except for inland waterway services)
11010	51342	Blending of distilled spirits by wholesalers (manufacture)	50300	61201	Boat rental for passenger conveyance with crew (inland waterway service)
13100	17130	Blending of fibres on the worsted system (manufacture)	50200	61102	Boat rental for transport of freight with crew (except for inland waterway service)
13921	17401	Blinds (soft furnishings) (manufacture)	50400	61209	Boat rental for transport of freight with crew (inland waterway service)
13922	17402	Blinds made of canvas (manufacture)			
17290	21259	Blinds made of paper (manufacture)	30120	35120	Boatbuilding (manufacture)
22230	25239	Blinds made of plastic (manufacture)	47640	52485	Boats (retail)
16230	20300	Blinds made of wood (excluding shop blinds) (manufacture)	28940	29540	Bobbins for textile machinery (manufacture)
			17290	21259	Bobbins made of paper and paperboard (manufacture)
24440	27440	Blister copper (manufacture)			
82920	74820	Blister packaging foil-covered packaging	16290	20510	Bobbins made of wood (not textile accessory) (manufacture)
23320	26400	Block flooring made of clay (manufacture)	17290	21259	Bobbins, spools and cops made of paper and paperboard (manufacture)
13300	17300	Block printing of textiles (manufacture)			
16210	20200	Blockboard (manufacture)	29201	34201	Bodies (coachworks) for motor vehicles (manufacture)
18130	22250	Blocking (printing)			
28990	29560	Blocking machine (manufacture)	29201	34201	Body building for motor vehicles (manufacture)
28110	29110	Blocks for industrial engines (manufacture)	29201	34201	Body for bus (manufacture)
16290	36639	Blocks for the manufacture of smoking pipes (manufacture)	29201	34201	Body for car (manufacture)
23610	26610	Blocks made of breeze (manufacture)	29201	34201	Body for coach (manufacture)
23610	26610	Blocks made of concrete (manufacture)			
23200	26260	Blocks made of graphite (manufacture)			

SIC 2007	SIC 2003	Activity	SIC 2007	SIC 2003	Activity
29201	34201	Body for commercial vehicle (manufacture)	20520	24620	Bone glue (manufacture)
30200	35200	Body for locomotive (manufacture)	10110	15112	Bone meal (manufacture)
96090	93059	Body piercing studios	10110	15112	Bone meal from knackers (manufacture)
30920	35430	Body shell for invalid carriage (manufacture)	10410	15410	Bone oil (manufacture)
29201	34201	Body shell for motor vehicle made of fibre glass (manufacture)	32500	33100	Bone plates and screws (manufacture)
29201	34201	Body shell for motor vehicle made of plastic (manufacture)	32500	24422	Bone reconstruction cements (manufacture)
93130	92629	Body-building clubs and facilities	10110	15112	Bone scraping by knackers (manufacture)
80100	74602	Bodyguard activities	10110	15112	Bone sorting by knackers (manufacture)
30200	35200	Bogie for locomotive (manufacture)	32990	36639	Bone working (manufacture)
30200	35200	Bogies (manufacture)	13200	17210	Book cloth weaving (manufacture)
10130	15131	Boiled ham production (manufacture)	28230	30010	Book keeping machine (manufacture)
10822	15842	Boiled sweet (manufacture)	91011	92510	Book lending and storage
25300	28300	Boiler (nuclear powered) (manufacture)	18129	22220	Book printing (manufacture)
23200	26260	Boiler block (manufacture)	58110	22110	Book publishing
81223	74705	Boiler cleaning and scaling	77210	71401	Book rental
32990	36639	Boiler covering (not asbestos or slag wool) (manufacture)	18140	22230	Bookbinding (manufacture)
25300	28300	Boiler drum (manufacture)	46690	51870	Book-binding and book-sewing machinery (wholesale)
25300	28300	Boiler feed water heater (manufacture)	28990	29560	Bookbinding machine (manufacture)
25210	28220	Boiler for central heating (manufacture)	31010	36120	Bookcase (non-domestic) (manufacture)
27520	29720	Boiler for domestic use (oil) (manufacture)	31090	36140	Bookcase made of wood (manufacture)
27520	29720	Boiler for domestic use (solid fuel) (manufacture)	69202	74122	Book-keeping activities
25300	28300	Boiler for marine applications (manufacture)	92000	92710	Bookmaker
25300	28300	Boiler fuel economiser (manufacture)	47610	52470	Books (retail)
25300	28300	Boiler fuel handling plant (manufacture)	46499	51479	Books (wholesale)
25300	28300	Boiler house plant (manufacture)	77299	71409	Books, journals and magazine renting and leasing
65120	66031	Boiler insurance	15200	19300	Boot (manufacture)
32990	36639	Boiler packing (not asbestos or slag wool) (manufacture)	17120	21120	Boot and shoe board (manufacture)
23990	26821	Boiler packing made of asbestos (manufacture)	74100	74872	Boot and shoe designing
14120	18210	Boiler suit (manufacture)	15200	19300	Boot closing (manufacture)
25300	28300	Boilers and associated equipment and parts (manufacture)	13990	17542	Boot lace (manufacture)
28110	29110	Boiler-turbine sets (manufacture)	16290	20510	Boot or shoe lasts and trees made of wood (manufacture)
10130	15139	Bolognas (manufacture)	15200	19300	Boot stiffener (manufacture)
13921	17401	Bolster (manufacture)	15200	19300	Boot upper (manufacture)
13923	17403	Bolster case (manufacture)	96090	93059	Bootblack
25940	28740	Bolt (manufacture)	22190	19300	Bootee (rubber protective) (manufacture)
25730	28620	Bolt cropper (manufacture)	47721	52431	Boots and shoes (retail)
25940	28740	Bolt end (manufacture)	08910	14300	Borates (natural) mining
13960	17549	Bolting cloth (manufacture)	84240	75240	Border guard administration and operation
25400	29600	Bomb fuse (manufacture)	43130	45120	Borehole drilling
46690	51870	Bombs, missiles and other projectiles (wholesale)	71122	74206	Borehole surveying
17290	21259	Bonbon paper (manufacture)	28921	29521	Borers (mining machinery) (manufacture)
23910	26810	Bonded abrasives (manufacture)	43999	45250	Boring (civil engineering)
13939	17519	Bonded fibre carpets (manufacture)	46630	51820	Boring and sinking machinery (wholesale)
13950	17530	Bonded fibre fabric (manufacture)	28410	29420	Boring machine (metal cutting) (manufacture)
52102	63129	Bonded store, vault or warehouse for air transport activities	24100	27100	Boron steel (manufacture)
52103	63129	Bonded store, vault or warehouse for land transport activities	91040	92530	Botanical gardens
52101	63129	Bonded store, vault or warehouse for water transport activities	21200	24421	Botanical products for pharmaceutical use (manufacture)
13300	17300	Bonding of fabric to fabric (manufacture)	13100	17130	Botany spinning (manufacture)
10110	15112	Bone boiling by knackers (manufacture)	81299	74709	Bottle cleaning
10110	15112	Bone crushing by knackers (manufacture)	28290	29240	Bottle cleaning and or drying machinery (manufacture)
10110	15112	Bone degreasing by knackers (manufacture)	22220	25220	Bottle crates made of plastic (manufacture)
10110	15112	Bone flour from knackers (manufacture)	23130	26130	Bottle stoppers made of glass (manufacture)
			25920	28720	Bottle tops made of metal (manufacture)
			46120	51120	Bottled gas (commission agent)
			23130	26130	Bottles made of glass or crystal (manufacture)

14

SIC 2007	SIC 2003	Activity	SIC 2007	SIC 2003	Activity
22220	25220	Bottles made of plastic (manufacture)	26200	30020	Braille pads and other output devices (manufacture)
28290	29240	Bottling machinery (manufacture)	18129	22220	Braille printing (manufacture)
10850	15330	Bottling of fruit and vegetables (manufacture)	23990	26821	Brake linings made of asbestos (manufacture)
82920	74820	Bottling on a fee or contract basis	29320	34300	Brakes and parts (excluding linings for motor vehicles) (manufacture)
71122	74206	Boundary surveying activities			
10110	15111	Bovine hides and skins production from knackers (manufacture)	30200	35200	Brakes and parts of brakes for locomotive and rolling stock (manufacture)
01420	01210	Bovine semen production	30200	35200	Braking systems for locomotives (manufacture)
28990	29560	Bow thruster (manufacture)	10611	15611	Bran (manufacture)
93110	92629	Bowling alley	28930	29530	Bran cleaners (manufacture)
28990	36501	Bowling alley equipment (manufacture)	28930	29530	Bran cleaners for the grain milling industry (manufacture)
93120	92629	Bowling clubs			
93110	92619	Bowling lanes operation	11010	15910	Brandy (manufacture)
32300	36400	Bowls and bowls equipment (manufacture)	24440	27440	Brass powder (manufacture)
93120	92629	Bowls club	14142	18232	Brassiere (manufacture)
23130	26130	Bowls made of glass (manufacture)	13922	17402	Brattice cloth (manufacture)
22290	25240	Bowls made of plastic (manufacture)	10130	15139	Brawn (manufacture)
32300	36400	Bows (manufacture)	25730	28620	Brazing lamp (manufacture)
29202	34202	Bowsers (tanks on wheels) (manufacture)	28490	29430	Brazing machines (electric) (manufacture)
15110	19100	Box and willow calf leather (manufacture)	28490	29430	Brazing machines (gas) (manufacture)
28210	29210	Box furnace (manufacture)	47240	52240	Bread (retail)
16240	20400	Box pallet (manufacture)	46360	51360	Bread (wholesale)
31030	36150	Box spring mattress (manufacture)	10710	15810	Bread and flour confectionery baking (manufacture)
16100	20100	Boxboard (manufacture)	10710	15810	Bread baking including rolls (manufacture)
32409	36509	Boxed game (manufacture)	10710	15810	Bread rolls (manufacture)
17230	21230	Boxed stationery (manufacture)	16290	20510	Breadboards made of wood (manufacture)
25910	28710	Boxes and other containers made of iron or steel, of capacity not exceeding 300 litres (manufacture)	28930	29530	Breading rolls or mills (manufacture)
			29100	34100	Breakdown lorry (manufacture)
25920	28720	Boxes made of aluminium (manufacture)	10612	15612	Breakfast cereal (cooked) (manufacture)
17219	21213	Boxes made of corrugated cardboard (manufacture)	10612	15612	Breakfast cereal (uncooked) (manufacture)
17219	21213	Boxes made of corrugated paper (manufacture)	32500	33100	Breathing apparatus for diving (manufacture)
25910	28710	Boxes made of iron or steel (manufacture)	46460	51460	Breathing appliances for medical use (wholesale)
25920	28720	Boxes made of metal (collapsible) (manufacture)	14131	18221	Breeches (manufacture)
17219	21214	Boxes made of non-corrugated cardboard (manufacture)	94990	91330	Breed society
			11050	15960	Brewery (beer and other brewing products) (manufacture)
17219	21214	Boxes made of non-corrugated paper (manufacture)			
22220	25220	Boxes made of plastic (manufacture)	28930	29530	Brewing machinery and plant (manufacture)
17219	21214	Boxes made of rigid board (manufacture)	20590	24660	Brewing preparations (excluding yeast) (manufacture)
17219	21213	Boxes made of rigid corrugated board (manufacture)	32990	36639	Briar pipe (manufacture)
16240	20400	Boxes made of wood (manufacture)	43999	45250	Brick furnace construction
16240	20400	Boxes made of wood (wirebound) (manufacture)	43999	45250	Brick kiln construction
93199	92629	Boxing	28990	29560	Brick making machinery (manufacture)
93110	92619	Boxing arena	43999	45250	Bricklaying
93120	92629	Boxing clubs	23200	26260	Bricks and blocks made of refractory ceramic (manufacture)
32300	36400	Boxing glove (manufacture)			
93199	92629	Boxing promoter	23200	26260	Bricks and mouldings made of magnesite (manufacture)
94990	91330	Boy scouts			
94990	91330	Boys brigade	23200	26260	Bricks made for refractory insulating (manufacture)
14190	18249	Braces (not made of leather or leather substitute) (manufacture)	23200	26260	Bricks made of alumina (manufacture)
			23200	26260	Bricks made of bauxite (manufacture)
47710	52424	Braces (retail)	23320	26400	Bricks made of ceramic (manufacture)
46420	51429	Braces (wholesale)	23200	26260	Bricks made of chrome (manufacture)
14190	18249	Braces made of leather (manufacture)	23200	26260	Bricks made of chromite (manufacture)
25990	28750	Brackets made of base metal (manufacture)	23320	26400	Bricks made of clay (manufacture)
13960	17542	Braid made of elastic (manufacture)	23610	26610	Bricks made of concrete (manufacture)
13960	17542	Braid made of elastomeric (manufacture)	23200	26260	Bricks made of dolomite (manufacture)
13960	17542	Braid made of non-elastic (manufacture)	23200	26260	Bricks made of gannister (manufacture)
13960	17542	Braid made of textile material (manufacture)	23190	26150	Bricks made of glass (manufacture)
18140	22230	Braille copying (manufacture)	23200	26260	Bricks made of high alumina (manufacture)
			23200	26260	Bricks made of magnesite (manufacture)

B

SIC 2007	SIC 2003	Activity
23200	26260	Bricks made of magnesite chrome (manufacture)
23200	26260	Bricks made of refractory (manufacture)
23610	26610	Bricks made of sand lime (manufacture)
23200	26260	Bricks made of silica (manufacture)
23200	26260	Bricks made of siliceous (manufacture)
23200	26260	Bricks made of sillimanite (manufacture)
42130	45213	Bridge construction
85510	92629	Bridge instructor
52219	63210	Bridge operation
28921	29521	Bridge plugs (manufacture)
30400	29600	Bridgelayer (tracked military) (manufacture)
26301	32201	Bridges for telecommunications (manufacture)
32500	33100	Bridges made in dental labs (manufacture)
16230	20300	Bridges made of wood (manufacture)
15120	19200	Bridle cutting (manufacture)
14141	18231	Briefs for men and boys (manufacture)
14142	18232	Briefs for women and girls (manufacture)
24310	27310	Bright steel bars (manufacture)
08930	14400	Brine pit
08930	14400	Brine production
19201	10103	Briquette solid fuel production (manufacture)
32910	36620	Bristle dressing for brushes (manufacture)
10110	15112	Bristles from knackers (manufacture)
17120	21120	Bristol board (manufacture)
24430	27430	Britannia metal (manufacture)
24540	27540	Britannia metal founding (manufacture)
32130	36610	Britannia metalware (manufacture)
52230	63230	British Airports Authority
94120	91120	British Association for the Advancement of Science
94990	91330	British Board of Film Classifications
60100	92201	British Broadcasting Corporation (radio broadcasting)
60200	92202	British Broadcasting Corporation (television broadcasting)
94120	91120	British Computer Society
94120	91120	British Dental Association
94910	91310	British Humanist Association
94910	91310	British Jews Society
94990	91330	British Legion (other than social clubs)
91011	92510	British Library
94120	91120	British Medical Association
91020	92521	British Museum
94990	91330	British Safety Council
79909	63309	British Tourist Authority
28410	28620	Broach for metal working machine tools (manufacture)
28410	29420	Broaching machine (metal cutting) (manufacture)
01110	01120	Broad bean growing
28302	29320	Broadcaster for agricultural use (manufacture)
82990	74879	Broadcasting transmission rights agent
86101	85111	Broadmoor Hospital
13200	17210	Brocade weaving (manufacture)
01130	01120	Broccoli growing
18129	22220	Brochure printing (manufacture)
58110	22110	Brochure publishing
20130	24130	Bromine and bromides (manufacture)
24540	27540	Bronze founding (manufacture)
32910	36620	Broom (manufacture)
16290	20510	Broom handles made of wood (manufacture)
46499	51479	Brooms and brushes for domestic use (wholesale)

SIC 2007	SIC 2003	Activity
32910	36620	Brooms and brushes for household use (manufacture)
10890	15891	Broth containing meat or vegetables or both (manufacture)
15120	19200	Brown saddlery (manufacture)
23410	26210	Brown stone pottery (manufacture)
27510	29710	Brush (electric) (manufacture)
32910	36620	Brush (manufacture)
32910	36620	Brush (not electrical) for machines (manufacture)
16290	20510	Brush back made of wood (manufacture)
15120	19200	Brush case (not of leather or plastics) (manufacture)
32910	36620	Brush for cosmetics (manufacture)
16290	20510	Brush head made of wood (manufacture)
16290	20510	Brush top made of wood (manufacture)
16290	20510	Brush wood ware (manufacture)
22190	36620	Brushes of rubber (manufacture)
20420	24520	Brushless shaving cream (manufacture)
13931	17511	Brussels carpet (manufacture)
01130	01120	Brussel sprout growing
28922	29522	Bucket for construction machinery (manufacture)
28220	29220	Bucket wheel reclaimer (manufacture)
13922	17402	Buckets made of canvas (manufacture)
25990	28750	Buckets made of metal (manufacture)
22290	25240	Buckets made of plastic (manufacture)
22190	25130	Buckets made of rubberised fabric (manufacture)
25910	28710	Buckets made of steel (manufacture)
16240	20400	Buckets made of wood (manufacture)
46690	51870	Buckets, grabs, shovels and grips for cranes, excavators and the like (wholesale)
25990	28750	Buckles made of metal (manufacture)
13960	17549	Buckram (manufacture)
13960	17549	Buckram and similar stiffened textile fabrics (manufacture)
14190	18241	Buckram shape (manufacture)
15110	19100	Buckskin (manufacture)
70221	74142	Budgetary control procedures design
15120	19200	Buff and mop made of leather (manufacture)
01410	01210	Buffalo milk, raw
15120	19200	Buffalo pickers made of leather (manufacture)
01420	01210	Buffalo raising and breeding for meat
30200	35200	Buffers and buffer parts (manufacture)
56101	55301	Buffet (licensed)
27400	31620	Bug zappers without light (manufacture)
32130	36610	Buhl cutting (manufacture)
41201	45211	Builder and contractor for commercial buildings
41202	45212	Builder and contractor for domestic buildings
43341	45440	Builder and decorator (own account)
43320	45420	Builder and joiner
25120	28120	Builders' carpentry and joinery made of metal (manufacture)
16230	20300	Builders' carpentry and joinery made of wood (manufacture)
46730	51530	Builders' carpentry and joinery of metal (wholesale)
28220	29220	Builders' hoist (manufacture)
25730	28620	Builders' knife (manufacture)
22230	25239	Builders' ware made of plastic (manufacture)
16230	20300	Builders' woodwork such as window frames etc. (manufacture)
43910	45220	Building and roofing contractor

SIC 2007	SIC 2003	Activity
23650	26650	Building boards made of asbestos (manufacture)
23650	26650	Building boards made of asbestos cement (manufacture)
16210	20200	Building boards made of fibre (manufacture)
17120	21120	Building boards made of paper (manufacture)
16210	20200	Building boards made of wood waste (manufacture)
81210	74701	Building cleaning activities
43390	45450	Building completion work
25990	28750	Building components of zinc e.g. Gutters, roof capping (manufacture)
38110	90020	Building debris removal
43110	45110	Building demolition and wrecking
71111	74201	Building design and drafting
41201	45211	Building maintenance and restoration commercial buildings
41202	45212	Building maintenance and restoration domestic buildings
46130	51130	Building materials (commission agent)
23320	26400	Building materials made of clay (non-refractory) (manufacture)
23650	26650	Building materials made of vegetable substances (manufacture)
23650	26650	Building materials of vegetable materials agglomerated with cement, plaster etc. (manufacture)
47520	52460	Building materials such as bricks, wood, sanitary equipment (retail)
30110	35110	Building of commercial vessels (manufacture)
30110	35110	Building of ferry-boats (manufacture)
30110	35110	Building of fish-processing factory vessels (manufacture)
30120	35120	Building of motor boats (manufacture)
30110	35110	Building of passenger vessels (manufacture)
30120	35120	Building of recreation-type hovercraft (manufacture)
30120	35120	Building of sailboats with or without auxiliary motor (manufacture)
23520	26530	Building plaster (manufacture)
22230	25239	Building products made of plastic (manufacture)
72190	73100	Building research and experimental development
68100	70120	Building sales and purchase
43120	45110	Building site drainage
43120	45110	Building sites clearance
64192	65122	Building societies
64921	65221	Building societies' personal finance subsidiaries activities
71129	74204	Building structure design for ancillary services
43341	45440	Buildings painting
24100	27100	Bulb flats made of steel (manufacture)
27400	31500	Bulb for flash lamp (manufacture)
01130	01120	Bulb growing
46220	51220	Bulbs (wholesale)
23130	26130	Bulbs for vacuum flask inners (manufacture)
23190	26150	Bulbs made of glass (manufacture)
30110	35110	Bulk carrier (cargo ship) (manufacture)
49410	60249	Bulk haulage in tanker trucks
49410	60249	Bulk haulage milk collection at farms
52102	63122	Bulk liquid and gases storage services for air transport activities
52103	63122	Bulk liquid and gases storage services for land transport activities
52101	63122	Bulk liquid and gases storage services for water transport activities
49410	60249	Bulk road haulage
28922	29522	Bulldozer and angle-dozer blades (manufacture)
28922	29522	Bulldozers (manufacture)
46630	51820	Bulldozers and angle-dozers (wholesale)
64991	65233	Bullion broker on own account
66190	67130	Bullion broking on behalf of others
64991	65233	Bullion dealer in investment grades
29320	34300	Bumpers for motor vehicles (manufacture)
16240	20400	Bungs made of wood (manufacture)
31090	36140	Bunk beds made of metal (manufacture)
31090	36140	Bunk beds made of wood (manufacture)
52220	63220	Bunkering services
25290	28210	Bunkers made of heavy steel plate exceeding 300 litres (manufacture)
13922	17402	Bunting (making-up) (manufacture)
13200	17210	Bunting made of cotton (manufacture)
13200	17220	Bunting woollen weaving (manufacture)
16290	20520	Buoyancy apparatus made of cork (except cork life preservers) (manufacture)
30110	35110	Buoys (not plastic) (manufacture)
30110	35110	Buoys made of plastic (manufacture)
66120	67130	Bureaux de change activities
23190	26150	Burettes made of glass (manufacture)
56103	55303	Burger bar (take-away)
56102	55302	Burger bar restaurant (unlicensed)
56103	55304	Burger stand
26301	31620	Burglar alarm and system (manufacture)
77390	71340	Burglar alarm hire
80200	74602	Burglar and fire alarm monitoring including installation and maintenance
46439	51439	Burglar and fire alarms for household use (wholesale)
96030	93030	Burial services
28210	29210	Burners (manufacture)
19201	23201	Burning oil (manufacture)
29100	34100	Bus (manufacture)
52213	63210	Bus and coach passenger facilities at bus and coach stations
27120	31200	Bus bar (switchgear type) (manufacture)
27330	31200	Bus bars, electrical conductors (except switchgear-type) (manufacture)
73110	74402	Bus carding
81299	74709	Bus cleaning (non- specialised)
49319	60219	Bus service
25110	28110	Bus shelters made of metal (manufacture)
52219	63210	Bus station operation
49319	60219	Bus transport (other than inter-city services) (scheduled passenger transport)
49390	60231	Buses operated for transport of employees
28150	29140	Bush for bearing (manufacture)
82990	74879	Business activities n.e.c.
62012	72220	Business and domestic software development
94110	91110	Business and employers membership organisations
74909	74879	Business brokerage activities
74909	74879	Business brokerage and appraisal activities
70229	74143	Business consultancy activities
70229	74143	Business consultant
18129	22220	Business form printing (manufacture)
18140	22230	Business forms finishing (manufacture)
94110	91110	Business organisations

SIC 2007	SIC 2003	Activity
70229	74143	Business systems consultant
74909	74879	Business transfer agent
70229	74143	Business turnaround consultancy services
70221	74142	Business valuation services prior to mergers and/or acquisitions
20140	24140	Butadiene (manufacture)
19201	23201	Butane (manufacture)
46711	51511	Butane (wholesale)
06200	11100	Butane extraction from natural gas
47220	52220	Butchers shop (retail)
24200	27220	Butt welding fittings made of steel (manufacture)
46330	51331	Butter (wholesale)
10512	15512	Butter blending (manufacture)
28930	29530	Butter churns (manufacture)
25990	28750	Butter dishes made of metal (manufacture)
25710	28610	Butter knife (manufacture)
10512	15512	Butter milk (manufacture)
10512	15512	Butter oil (manufacture)
10512	15512	Butter production (manufacture)
28930	29530	Butter workers (manufacture)
10512	15512	Butterfat (manufacture)
28140	29130	Butterfly valves (manufacture)
10822	15842	Butterscotch (manufacture)
22290	36639	Button and button moulds made of plastic (manufacture)
13990	36639	Button carding (manufacture)
13990	36639	Button covering (manufacture)
32990	36639	Buttons (manufacture)
22290	36639	Buttons and button bases (not metal or glass) (manufacture)
32990	36639	Buttons made of glass (manufacture)
25990	36639	Buttons made of metal (manufacture)
46342	51342	Buying of wine in bulk and bottling without transformation (wholesale)
11020	51342	Buying wine in bulk with blending, purification and bottling of wine (manufacture)
49320	60220	Cab hire
01130	01120	Cabbage growing
31090	36140	Cabinet case made of wood (manufacture)
22290	25240	Cabinet components made of plastic (manufacture)
31010	36120	Cabinets (non-domestic) (manufacture)
31020	36130	Cabinets for kitchens (manufacture)
31090	36140	Cabinets for sewing machines (manufacture)
31090	36140	Cabinets for televisions (manufacture)
25990	28750	Cabinets made of metal (not designed for placing on the floor) (manufacture)
27320	31300	Cable accessory (manufacture)
23320	26400	Cable conduit made of clay (manufacture)
25990	28750	Cable drum made of metal (manufacture)
16240	20400	Cable drums made of wood (manufacture)
27320	31300	Cable jointing material (manufacture)
42220	45213	Cable laying
13940	17520	Cable made of textile materials (manufacture)
25930	28730	Cable made of uninsulated aluminium (manufacture)
28990	29560	Cable making machine (manufacture)
61100	64200	Cable service
25930	28730	Cable sheathing made of aluminium (manufacture)
30110	35110	Cable ship (manufacture)
25930	28730	Cable strands made of aluminium (manufacture)
42120	45230	Cable supported transport systems construction
26309	32202	Cable television equipment (manufacture)
50200	61102	Cable-laying vessel services
28220	29220	Cableway (manufacture)
28220	29220	Cableway excavator (manufacture)
29201	34201	Cabs for motor vehicles (manufacture)
10822	15842	Cachous (manufacture)
24450	27450	Cadmium (manufacture)
56101	55301	Cafeteria (licensed)
56102	55302	Cafeteria (unlicensed)
28921	29521	Cage plant for mining (manufacture)
30110	35110	Caisson (manufacture)
17220	21220	Cake board (manufacture)
28930	29530	Cake depositing machines (manufacture)
10611	15611	Cake mixture (manufacture)
10710	15810	Cakes (manufacture)
10720	15820	Cakes (preserved) (manufacture)
47240	52240	Cakes (retail)
23510	26510	Calcareous cement (manufacture)
23520	26520	Calcined dolomite (manufacture)
23520	26530	Calcined sulphate plaster (manufacture)
20130	24130	Calcium and calcium compounds (manufacture)
20130	24130	Calcium carbide (manufacture)
46660	51850	Calculating and accounting machines (wholesale)
26513	33202	Calculating instruments (non-electronic) (manufacture)
28230	30010	Calculating machine (manufacture)
47410	52482	Calculating machines (retail)
28230	30010	Calculator (electronic) (manufacture)
18140	22230	Calendar finishing (manufacture)
18129	22220	Calendar printing (manufacture)
28290	29240	Calender for rubber or plastics working (manufacture)
28290	29240	Calendering machinery for textiles (manufacture)
13300	17300	Calendering of textiles (manufacture)
28290	29240	Calendering or other rolling machine parts (for plastic or rubber) (manufacture)
74100	74872	Calico printers' designing
18140	22230	Calico printers' engraving (manufacture)
13300	17300	Calico printing (manufacture)
13200	17210	Calico weaving (manufacture)
82200	74860	Call centres undertaking market research or public opinion polling
82200	74860	Call centres working on a fee or contract basis
32200	36300	Call horns (manufacture)
32200	36300	Calliopes (manufacture)
26513	33202	Callipers (manufacture)
27520	29720	Calorifier (manufacture)
10130	15139	Calves' foot jelly (manufacture)
94910	91310	Calvinistic Methodist Church
28110	34300	Camshaft for motor vehicle engine (manufacture)
13200	17210	Cambric weaving (manufacture)
22110	25110	Camel back strips for retreading tyres (manufacture)
13100	17120	Camel hair spinning on the woollen system (manufacture)
13100	17130	Camel hair spinning on the worsted system (manufacture)
13200	17220	Camel hair woollen weaving (manufacture)
13200	17230	Camel hair worsted weaving (manufacture)
01440	01250	Camelid raising and breeding
01440	01250	Camels (dromedary) raising and breeding

SIC 2007	SIC 2003	Activity	SIC 2007	SIC 2003	Activity
26309	32202	Camera for television (manufacture)	22220	25220	Caps for bottles made of plastic (manufacture)
77210	71401	Camera hire	29320	34300	Caps for petrol, oil or radiator for motor vehicle (manufacture)
26702	33403	Cameras (manufacture)			
31090	36140	Camp furniture made of wood (manufacture)	14190	18241	Caps made of cloth (manufacture)
77390	71211	Campers (transport) rental (self drive)	28220	29220	Capstan (manufacture)
13922	17402	Camping goods (manufacture)	25920	28720	Capsules made of metal (manufacture)
47640	52485	Camping goods (retail)	28290	29240	Capsuling machinery (manufacture)
55300	55220	Camping sites	45112	50102	Car auctions
45190	50101	Camping vehicles (retail)	45320	50300	Car batteries (retail)
45190	50102	Camping vehicles (used) (retail)	29320	34300	Car body parts (manufacture)
45190	50102	Camping vehicles (used) (wholesale)	26520	33500	Car clock (manufacture)
45190	50101	Camping vehicles (wholesale)	29320	34300	Car components (manufacture)
28150	29140	Camshaft (not for motor vehicle) (manufacture)	49410	60249	Car delivery service (by independent contractors)
50400	61209	Canal carrier (freight)	49410	60249	Car delivery service (by motor manufacturers)
50300	61201	Canal carrier (passenger)	46770	51570	Car dismantlers (wholesale)
30120	35120	Canal cruiser (manufacture)	77110	71100	Car hire (self drive)
52220	63220	Canal maintenance	77110	71100	Car leasing
52220	63220	Canal operation	13922	17402	Car loose covers (manufacture)
72190	73100	Cancer research and experimental development (other than biotechnological)	52219	63210	Car park
			41201	45230	Car park construction
27400	31500	Candelabra made of base metal (manufacture)	20412	24512	Car polish (manufacture)
10822	15842	Candied peel (manufacture)	77110	71100	Car rental (self drive)
32990	36639	Candle (manufacture)	49320	60220	Car rental with driver
46499	51479	Candles and tapers (wholesale)	47421	52488	Car telephones (retail)
27400	31500	Candlestick (manufacture)	45200	50200	Car valeting
31090	36110	Cane chair (manufacture)	45200	50200	Car wash
31090	36140	Cane furniture (manufacture)	93110	92619	Car, dog and horse racetrack operation
16290	20520	Cane preparation (manufacture)	10890	15620	Caramel (not sweets) (manufacture)
16290	20520	Cane splitting and weaving (manufacture)	10822	15842	Caramel sweets (manufacture)
16290	20520	Cane working (manufacture)	29203	34203	Caravan (manufacture)
25990	28750	Canisters made of metal (manufacture)	45190	50102	Caravan (used) (retail)
22220	25220	Canisters made of plastic (manufacture)	45190	50102	Caravan (used) (wholesale)
10890	15891	Canned broth containing meat or vegetables or both (manufacture)	29203	34203	Caravan chassis (manufacture)
			55300	55220	Caravan holiday site operator, owner or proprietor
10890	15891	Canned soup containing meat or vegetables or both (manufacture)	68209	70209	Caravan residential site letting
			55300	55220	Caravan sites
28290	29240	Canning machinery (manufacture)	29203	34203	Caravan trailers (manufacture)
10390	15330	Canning of fruit and vegetables (except fruit juices and potatoes) (manufacture)	52219	63210	Caravan winter storage
			45190	50101	Caravans (retail)
32500	33100	Cannulae (manufacture)	77390	71211	Caravans (touring) rental
30120	35120	Canoe (manufacture)	45190	50101	Caravans (wholesale)
77210	71401	Canoes and sailboats renting and leasing	27400	31500	Carbide lanterns (manufacture)
25920	28720	Cans and boxes made of aluminium (manufacture)	25400	29600	Carbine (manufacture)
25910	28710	Cans and boxes made of iron or steel (manufacture)	20130	24130	Carbon (manufacture)
			13200	17250	Carbon and aramid thread weaving (manufacture)
25920	28720	Cans and boxes made of tin (manufacture)	20130	24130	Carbon black (manufacture)
25920	28720	Cans for food products (manufacture)	27900	31620	Carbon brush (manufacture)
25920	28720	Cans made of aluminium (manufacture)	20110	24110	Carbon dioxide (manufacture)
25910	28710	Cans made of blackplate (manufacture)	20130	24130	Carbon disulphide (manufacture)
25910	28710	Cans made of steel (manufacture)	46690	51870	Carbon electrodes and other articles of graphite or other carbon for electrical purposes (wholesale)
01130	01120	Cantaloupe growing			
13922	17402	Canvas goods (manufacture)	27900	31620	Carbon or graphite electrodes (manufacture)
13960	17549	Canvas prepared for use by painters (manufacture)	17120	21120	Carbon paper in large sheets (manufacture)
13200	17250	Canvas weaving (manufacture)	17120	21120	Carbon paper in rolls (manufacture)
26110	32100	Capacitor for electronic apparatus (manufacture)	17230	21230	Carbon paper ready for use (manufacture)
14190	18241	Capeline felt (manufacture)	17230	21230	Carbon paper stencil ready for use (manufacture)
01130	01120	Capers growing	23990	26829	Carbon products (except carbon paper and electrical carbon) (manufacture)
14200	18300	Capes made of fur (manufacture)			
70221	74142	Capital structure consultancy services			
22220	25220	Caps and closures made of plastic (manufacture)	32990	36631	Carbon ribbon (manufacture)

SIC 2007	SIC 2003	Activity	SIC 2007	SIC 2003	Activity
08910	14300	Carbonate (barytes and witherite) mining	13931	17511	Carpet weaving (manufacture)
11070	15980	Carbonated soft drink (manufacture)	46160	51160	Carpets (commission agent)
46120	51120	Carbonates (commission agent)	13939	17519	Carpets (other than woven or tufted) (manufacture)
46750	51550	Carbonates (wholesale)	47530	52481	Carpets (retail)
17120	21120	Carbonising base paper (manufacture)	46470	51479	Carpets (wholesale)
17230	21230	Carbonless copy paper ready for use (manufacture)	13939	17519	Carpets and rugs, other than woven or tufted (manufacture)
20140	24140	Carboxylic acid (manufacture)	13939	17519	Carpets made of jute (manufacture)
22220	25220	Carboy made of plastic (manufacture)	49410	60249	Carrier (for general hire or reward)
23130	26130	Carboys made of glass (manufacture)	26301	32201	Carrier equipment (manufacture)
28110	34300	Carburettor and parts for motor vehicle (manufacture)	01130	01120	Carrot growing
28110	29110	Carburettors and such for all internal combustion engines (manufacture)	32990	36639	Carry cot (manufacture)
			29100	34100	Cars (manufacture)
28110	29110	Carburettors for industrial engines (manufacture)	13922	17402	Cart cover made of canvas (manufacture)
93120	92629	Card clubs	49410	60249	Cartage contractor
17230	21230	Card cutting for index cards (manufacture)	71122	74206	Cartographic and spatial information activities
18140	22230	Card embossing (manufacture)	28950	29550	Carton making machinery (manufacture)
85510	92629	Card game instruction	28290	29240	Cartoning machinery (manufacture)
28940	29540	Card tape reader for textile machinery (manufacture)	17219	21215	Cartons and similar containers for carrying liquids (unwaxed) (manufacture)
17120	21120	Cardboard (manufacture)	17219	21215	Cartons and similar containers for carrying liquids (waxed) (manufacture)
28950	29550	Cardboard box making machine (manufacture)			
23990	26821	Carded asbestos fibre (manufacture)	17219	21213	Cartons made of corrugated board (manufacture)
13100	17120	Carded sliver preparation for textiles industry (manufacture)	17219	21213	Cartons made of corrugated paper (manufacture)
			17219	21214	Cartons made of non-corrugated board (manufacture)
28940	29540	Carders (manufacture)	17219	21214	Cartons made of non-corrugated paper (manufacture)
14390	17720	Cardigans (knitted) (manufacture)			
28940	29540	Carding machinery for textiles (manufacture)	59111	92111	Cartoon film production
13960	17542	Carding of trimmings (manufacture)	59112	92111	Cartoon video production
32200	36300	Cards for automatic mechanical instruments (manufacture)	90030	92319	Cartoonists
			25400	29600	Cartridge case (manufacture)
52242	63110	Cargo handling for air transport activities	25400	29600	Cartridge primer (manufacture)
52243	63110	Cargo handling for land transport activities	32990	36631	Cartridge refill for fountain pen (manufacture)
52241	63110	Cargo handling for water transport activities	46690	51870	Cartridges and other ammunition (wholesale)
30110	35110	Cargo ship (manufacture)	25400	29600	Cartridges for riveting guns (manufacture)
13940	17520	Cargo sling (manufacture)	27400	31500	Case for flash lamp (manufacture)
52220	63220	Cargo superintendent	25610	28510	Case hardening (manufacture)
52220	63220	Cargo terminal	17120	21120	Case making materials (manufacture)
32990	36639	Carnival article (manufacture)	25730	28620	Case opener (manufacture)
94990	91330	Carnival clubs	28290	29240	Case packing machinery (manufacture)
43320	45420	Carpenter n.e.c.	20520	24620	Casein based adhesive (manufacture)
43999	45220	Carpenter on building site	10519	15519	Casein production (manufacture)
25730	28620	Carpenter's drill (manufacture)	20160	24160	Casein resins (manufacture)
43320	45420	Carpentry (not structural)	13200	17210	Casement cloth weaving (manufacture)
43999	45220	Carpentry (structural)	25120	28120	Casements made of metal (manufacture)
96010	93010	Carpet cleaning	26520	33500	Cases for clocks and watches (manufacture)
43330	45430	Carpet fitter	15120	19200	Cases for cutlery (not wooden) (manufacture)
25720	28630	Carpet fittings made of metal (manufacture)	15120	19200	Cases for jewellery (not wooden) (manufacture)
28940	29540	Carpet making machinery (manufacture)	15120	19200	Cases for musical instruments (not wooden) (manufacture)
13100	17110	Carpet pile yarn spun on the cotton system (manufacture)			
13100	17120	Carpet pile yarn spun on the woollen system (manufacture)	17219	21213	Cases made of corrugated cardboard (manufacture)
13100	17130	Carpet pile yarn spun on the worsted and semi-worsted systems (manufacture)	17219	21213	Cases made of corrugated fibreboard (manufacture)
			17219	21213	Cases made of corrugated paper (manufacture)
28990	29560	Carpet shampoo appliance (not domestic electric) (manufacture)	15120	19200	Cases made of leather for cutlery, instruments, etc. (manufacture)
20411	24511	Carpet soap (manufacture)	17219	21214	Cases made of non-corrugated cardboard (manufacture)
28990	29560	Carpet sweeper (industrial) (manufacture)			
47530	52481	Carpet tiles (retail)	17219	21214	Cases made of non-corrugated paper (manufacture)
22210	25210	Carpet underlay made of plastic (manufacture)	22220	25220	Cases made of plastic (manufacture)
22190	25130	Carpet underlay made of rubber (manufacture)			

SIC 2007	SIC 2003	Activity	SIC 2007	SIC 2003	Activity
16290	20510	Cases made of wood, for musical instruments (manufacture)	24530	27530	Casting of titanium products (manufacture)
46390	51390	Cash and carry predominantly food (wholesale)	24530	27530	Casting of yttrium products (manufacture)
28230	30010	Cash and credit card imprinting and embossing machine (manufacture)	23200	26260	Casting pot (manufacture)
			01110	01110	Castor bean growing
25990	28750	Cash boxes made of metal (manufacture)	10410	15420	Castor oil processing (manufacture)
28230	30010	Cash dispenser (manufacture)	10410	15410	Castor seed crushing (manufacture)
28230	30010	Cash register (manufacture)	10810	15830	Castor sugar (manufacture)
77330	71330	Cash register hire	01490	01250	Cat and dog raising and breeding
46660	51850	Cash registers and similar machines incorporating a calculating device (wholesale)	10920	15720	Cat food (manufacture)
			58190	22150	Catalogue publishing
01250	01139	Cashew nut growing	91011	92510	Cataloguing and preservation of collections
13200	17220	Cashmere woollen weaving (manufacture)	20590	24660	Catalysts (manufacture)
13200	17230	Cashmere worsted weaving (manufacture)	46120	51120	Catalytic preparations (commission agent)
28921	29521	Casing hangars (manufacture)	29320	34300	Catalyzers (manufacture)
24200	27220	Casing made of steel (manufacture)	30120	35120	Catamaran (manufacture)
10110	15112	Casings for sausages (manufacture)	28990	35300	Catapult for launching aircraft (manufacture)
92000	92710	Casino	56210	55520	Catering contractor
28490	29430	Cask assembly machines (manufacture)	28930	29530	Catering equipment (electric) (manufacture)
16240	20400	Cask heads made of wood (manufacture)	77390	71340	Catering equipment hire
16290	20510	Caskets (except burial caskets) and cases made of wood (manufacture)	84130	75130	Catering trade services administration and regulation (public sector)
25920	28720	Casks made of aluminium (manufacture)	32990	36639	Catgut (manufacture)
25910	28710	Casks made of iron or steel (manufacture)	46460	51460	Catgut and similar materials (wholesale)
16240	20400	Casks made of wood (manufacture)	32500	33100	Catheter (manufacture)
01130	01110	Cassava growing	26110	32100	Cathode ray tube (manufacture)
26400	32300	Cassette player (manufacture)	46520	51860	Cathode-ray oscilloscopes and cathode-ray oscillographs (wholesale)
26400	32300	Cassette type recorders (manufacture)	46520	51860	Cathode-ray television picture tubes, television camera tubes, other cathode-ray tubes (wholesale)
14120	18210	Cassock (manufacture)			
23610	26610	Cast concrete products (manufacture)	94910	91310	Catholic Apostolic Church
23110	26110	Cast glass (manufacture)	96090	93059	Cats' home
23610	26610	Cast stone units made of precast concrete (manufacture)	23190	26150	Catseye reflector (manufacture)
			10840	15870	Catsup (manufacture)
78101	92721	Casting activities for motion pictures, television and theatre	20200	24200	Cattle dip (manufacture)
			01420	01210	Cattle farming
78101	74500	Casting agencies and bureaux	15110	19100	Cattle hide leather (manufacture)
28910	29510	Casting machines (manufacture)	01420	01210	Cattle raising and breeding for meat
28910	29510	Casting machines for foundries (manufacture)	01130	01120	Cauliflower growing
24530	27530	Casting of aluminium (manufacture)	20301	24303	Caulking compounds and similar non-refractory filling or surfacing preparations (manufacture)
24530	27530	Casting of aluminium products (manufacture)			
24530	27530	Casting of beryllium products (manufacture)	32500	33100	Cautery and light unit (manufacture)
24510	27510	Casting of ferrous metal (manufacture)	10200	15209	Caviar (manufacture)
24510	27510	Casting of ferrous patterns (manufacture)	10200	15209	Caviar substitute (manufacture)
24510	27510	Casting of grey iron (manufacture)	43290	45320	Cavity wall insulation
24540	27540	Casting of heavy metal and precious metal (manufacture)	47430	52450	CD, DVD recorders (retail)
			77220	71404	CDs and disks rental
24510	27510	Casting of iron (manufacture)	22230	25239	Ceiling coverings made of plastic (manufacture)
24510	27510	Casting of iron products (finished or semi-finished) (manufacture)	27400	31500	Ceiling rose (manufacture)
			22230	25239	Ceiling tiles made of plastic (manufacture)
24530	27530	Casting of light metal products (manufacture)	08910	14300	Celestine pit
24530	27530	Casting of light metals (manufacture)	32200	36300	Cello (manufacture)
24530	27530	Casting of magnesium products (manufacture)	13200	17210	Cellular cloth weaving from yarn spun on the cotton system (manufacture)
24540	27540	Casting of non-ferrous base metal (manufacture)			
24540	27540	Casting of ornamental brass (manufacture)	61200	64200	Cellular network operations
24540	27540	Casting of other non-ferrous metals (manufacture)	26301	32202	Cellular phones (manufacture)
24530	27530	Casting of scandium products (manufacture)	22190	25130	Cellular rubber products (manufacture)
24510	27510	Casting of spheroidal graphite iron (manufacture)	47421	52488	Cellular telephones (retail)
24520	27520	Casting of steel (manufacture)	16210	20200	Cellular wood panel (manufacture)
24520	27520	Casting of steel products (finished or semi-finished) (manufacture)	20160	24160	Cellulose (manufacture)

SIC 2007	SIC 2003	Activity
20160	24160	Cellulose acetate (manufacture)
22290	25240	Cellulose adhesive tape (manufacture)
20520	24620	Cellulose based adhesive (manufacture)
20160	24160	Cellulose ester and ether ester (manufacture)
17120	21120	Cellulose fibre webs (manufacture)
23650	26650	Cellulose fibre-cement articles (manufacture)
20160	24160	Cellulose nitrate (manufacture)
20301	24301	Cellulose paint (manufacture)
20301	24301	Cellulose varnish (manufacture)
17120	21120	Cellulose wadding (manufacture)
17220	21220	Cellulose wadding products (manufacture)
23510	26510	Cement (manufacture)
46730	51530	Cement (wholesale)
23610	26610	Cement articles for use in construction (manufacture)
20301	24301	Cement based paint (manufacture)
28990	29560	Cement block making machine (manufacture)
23200	26260	Cement made of dolomite (manufacture)
23200	26260	Cement made of fireclay (manufacture)
23200	26260	Cement made of high alumina (manufacture)
23200	26260	Cement made of silica and siliceous (manufacture)
28210	29210	Cement processing kiln (manufacture)
23610	26610	Cement products (manufacture)
23690	26660	Cement products n.e.c. (manufacture)
23200	26260	Cement refractory jointing (manufacture)
23610	26610	Cement wood products (manufacture)
30110	35110	Cementing of ships (manufacture)
96030	93030	Cemetery
84110	75110	Central government administration
28990	29560	Central greasing systems (manufacture)
25210	28220	Central heating boiler parts (manufacture)
94120	91120	Central midwives board
26200	30020	Central processing units for computers (manufacture)
86101	85112	Central sterile supply department (private sector)
86101	85111	Central sterile supply department (public sector)
81100	75140	Centralised supply and purchasing services (public sector)
28990	29560	Centrifugal clothes dryer (manufacture)
28132	29122	Centrifugal compressor (manufacture)
28131	29121	Centrifugal pump (manufacture)
28290	29240	Centrifuge (except laboratory type) (manufacture)
32500	29240	Centrifuge (laboratory type) (manufacture)
28290	29240	Centrifuge parts (except laboratory type) (manufacture)
46690	51870	Centrifuges (wholesale)
46130	51130	Ceramic articles used in construction (commission agent)
23440	26240	Ceramic chemical products (manufacture)
20301	24301	Ceramic colours (manufacture)
20301	24301	Ceramic glaze (manufacture)
28990	29560	Ceramic making machine (manufacture)
28990	29560	Ceramic pastes production machinery (shaped) (manufacture)
43330	45430	Ceramic stove fitting
18130	22250	Ceramic transfer litho engraving (manufacture)
23410	26210	Ceramic ware for domestic use (manufacture)
46690	51870	Cereal and dried vegetable milling or working machinery (wholesale)
10612	15612	Cereal breakfast foods (manufacture)

SIC 2007	SIC 2003	Activity
10720	15820	Cereal for sausage filler (manufacture)
01110	01110	Cereal grains growing, except rice
10611	15611	Cereal grains, flour, groats, meal or pellets (manufacture)
77390	71340	Cereal milling and working machinery renting (non agricultural)
37000	90010	Cesspools emptying and cleaning
25930	28740	Chain (manufacture)
25930	28740	Chain (non-precision) (manufacture)
25930	28740	Chain (not articulated transmission) (manufacture)
28150	29140	Chain (precision) (manufacture)
28220	29220	Chain pulley block (manufacture)
28240	29410	Chain saw parts (manufacture)
28240	29410	Chain saws (manufacture)
22290	25240	Chains made of plastic (manufacture)
25730	28620	Chainsaw blades (manufacture)
31090	36110	Chair (non-upholstered) (manufacture)
31090	36110	Chair (upholstered) (manufacture)
31090	36110	Chair frames made of wood (manufacture)
31090	36110	Chair seating made of cane or wicker (manufacture)
31010	36120	Chairs and seats for offices, workrooms, hotels, restaurants and public premises (manufacture)
31090	36110	Chairs made of metal (manufacture)
31090	36110	Chairs made of plastic (manufacture)
31090	36110	Chaise longue (manufacture)
55201	55231	Chalets in holiday centres and holiday villages (provision of short-stay lodging in)
55209	55232	Chalets, not holiday centres, holiday villages or youth hostels (self catering short-stay lodging)
08910	14300	Chalk (ground) production
32990	36631	Chalk for drawing or writing (manufacture)
08110	14120	Chalk pit or quarry
94110	91110	Chamber of agriculture
94110	91110	Chambers of commerce organisations
16100	20100	Chamfered wood (manufacture)
15110	19100	Chamois leather (manufacture)
10831	15861	Chamomile herb infusions (manufacture)
27400	31500	Chandeliers (manufacture)
28131	29121	Channel impeller pump (manufacture)
64929	65229	Channel Islands and Isle of Man banking institutes
20140	24140	Charcoal (other than wood charcoal) (manufacture)
46719	51519	Charcoal (wholesale)
20140	24140	Charcoal burning (manufacture)
02200	02010	Charcoal production in the forest using traditional methods
99000	99000	Charge d'affaires
88990	85321	Charity administration
18129	22220	Chart printing (manufacture)
58110	22110	Chart publishing
47620	52470	Chart seller (retail)
51210	62209	Charter flights (freight)
51102	62201	Charter flights (passenger)
94120	91120	Chartered institute of secretaries
79110	63301	Chartered rail travel
70221	74142	Chartered secretary (firm acting as)
71122	74206	Chartered surveyor
49390	60231	Charters, excursions and other occasional coach services
29320	34300	Chassis and parts for coaches (manufacture)
29100	34100	Chassis fitted with engine (manufacture)

SIC 2007	SIC 2003	Activity	SIC 2007	SIC 2003	Activity
30200	35200	Chassis for locomotive (manufacture)	08990	14500	Chert quarry
29100	34100	Chassis for motor vehicles (manufacture)	01130	01120	Chervil growing
30920	35430	Chassis for powered invalid carriage (manufacture)	32409	36509	Chess (electronic) (manufacture)
29100	34100	Chassis with engine for commercial vehicle (manufacture)	93120	92629	Chess clubs
			85510	92629	Chess instructor
49320	60220	Chauffeur driven service	32409	36509	Chess set (manufacture)
15120	19200	Check strap made of leather (manufacture)	31090	36140	Chest of drawers (manufacture)
64921	65221	Check trader activities	01250	01139	Chestnut growing
28140	29130	Check valves (manufacture)	16240	20400	Chests made of wood (manufacture)
26511	33201	Checking instruments and appliances (electronic) (manufacture)	10822	15842	Chewing gum (manufacture)
			12000	16000	Chewing tobacco (manufacture)
26513	33202	Checking instruments and appliances (non-electronic) (manufacture)	01110	01120	Chick pea growing
10512	15512	Cheese (manufacture)	10120	15120	Chicken cuts (fresh, chilled or frozen) (manufacture)
46330	51331	Cheese (wholesale)	01470	01240	Chicken farm (battery rearing)
28930	29530	Cheese making machines (manufacture)	10910	15710	Chicken food (manufacture)
28930	29530	Cheese moulding machinery (manufacture)	10130	15139	Chicken paste (manufacture)
28930	29530	Cheese press (manufacture)	01470	01240	Chicken raising and breeding
28930	29530	Cheese pressing machinery (manufacture)	10850	15139	Chicken ready to eat meals (manufacture)
14120	18210	Chefs' clothing (manufacture)	01130	01120	Chicory growing
46180	51180	Chemical contraceptive preparations based on hormones or spermicides (commission agent)	10612	15612	Chicory root drying (manufacture)
			13200	17210	Chiffon weaving from yarn spun on the cotton system (manufacture)
21200	24421	Chemical contraceptive products (manufacture)			
20130	24130	Chemical elements (except metals) (manufacture)	88910	85321	Child day-care activities (charitable)
46120	51120	Chemical elements in disk form and compounds doped for use in electronics (commission agent)	88910	85322	Child day-care activities (non-charitable)
			88990	85321	Child guidance centre (charitable)
26110	24660	Chemical elements in disk form for use in electronics (manufacture)	88990	85322	Child guidance centre (non-charitable)
			55209	55239	Children's and other holiday homes
71129	74204	Chemical engineering design projects	30920	35420	Children's bicycles and tricycles (non-motorised) (manufacture)
19201	23201	Chemical feedstock (manufacture)			
46750	51550	Chemical glues (wholesale)	87900	85311	Children's boarding homes and hostels (charitable)
08910	14300	Chemical minerals mining	87900	85312	Children's boarding homes and hostels (non-charitable)
46180	51180	Chemical preparations and sensitized emulsions for photographic use (commission agent)			
			30920	36639	Children's carriage (manufacture)
28410	29420	Chemical process machine tool (metal working) (manufacture)	46420	51422	Children's clothing (exporter) (wholesale)
			46420	51422	Children's clothing (importer) (wholesale)
46750	51550	Chemical products (wholesale)	46420	51422	Children's clothing (wholesale)
46120	51120	Chemical products and residual products of the chemical or allied industries (commission agent)	87900	85311	Children's home (charitable)
			87900	85312	Children's home (non-charitable)
46750	51550	Chemical products exporter (wholesale)	86101	85112	Children's hospital (private sector)
46750	51550	Chemical products importer (wholesale)	86101	85111	Children's hospital (public sector)
28302	29320	Chemical seed dresser for agricultural use (manufacture)	47710	52422	Children's wear (retail)
			91040	92530	Children's zoos
37000	90010	Chemical toilets servicing	35300	40300	Chilled water for cooling purposes production and distribution
17110	21110	Chemical woodpulp (manufacture)			
46120	51120	Chemically modified animal or vegetable fats and mixtures (commission agent)	01280	01139	Chilli growing
			01280	01139	Chillies and peppers capsicum sop. Growing
21100	24410	Chemically pure sugars (manufacture)	81223	74705	Chimney cleaning
46460	51460	Chemically pure sugars, sugar ethers and sugar esters and their salts (wholesale)	43999	45250	Chimney construction
			23320	26400	Chimney liners made of clay (manufacture)
09900	14300	Chemicals and fertiliser minerals mining support services provided on a fee or contract basis	23320	26400	Chimney pots made of ceramic (manufacture)
			23320	26400	Chimney pots made of clay (manufacture)
20590	24660	Chemicals specially prepared for laboratory use (manufacture)	25110	28110	Chimneys made of steel (manufacture)
71200	74300	Chemist	47599	52440	China (retail)
72190	73100	Chemistry research and experimental development	46440	51440	China (wholesale)
13200	17210	Chenille (manufacture)	08120	14220	China clay (ground) production
18129	22220	Cheque book printing (manufacture)	08120	14220	China clay pit
28230	30010	Cheque writing and signing machine (manufacture)	08120	14220	China stone mine
12000	16000	Cheroot (manufacture)	46440	51440	China, glassware, wallpaper and cleaning materials exporter (wholesale)
11010	15910	Cherry brandy (manufacture)			
01240	01139	Cherry growing			

SIC 2007	SIC 2003	Activity	SIC 2007	SIC 2003	Activity
46440	51440	China, glassware, wallpaper and cleaning materials importer (wholesale)	94910	91310	Church Missionary Society
			94910	91310	Church of Christ Scientist
13300	17300	Chintz glazing (manufacture)	94910	91310	Church of England
13200	17210	Chintz weaving from yarn spun on the cotton system (manufacture)	94910	91310	Church of Ireland
			94910	91310	Church of Scotland
16210	20200	Chipboard agglomerated with non-mineral binding substances (manufacture)	85100	80100	Church schools at nursery and primary level
			85310	80210	Church schools at secondary level
28490	29430	Chipboard press (manufacture)	41201	45211	Churches and other ecclesiastical buildings construction
28923	29523	Chippers for road surfacing (manufacture)			
86900	85140	Chiropodist (NHS)	25920	28720	Churns made of aluminium (manufacture)
86900	85140	Chiropodist (private)	25910	28710	Churns made of iron or steel (manufacture)
86900	85140	Chiropractor clinic (own account)	16240	20400	Churns made of wood (manufacture)
20510	24610	Chlorate explosive (manufacture)	10390	15330	Chutney (manufacture)
20301	24301	Chlorinated rubber based paint (manufacture)	11030	15941	Cider (alcoholic) (manufacture)
28290	29240	Chlorination plant for water treatment (manufacture)	11070	15980	Cider (non-alcoholic) (manufacture)
			01240	01139	Cider apple growing
20130	24130	Chlorine and chloride (manufacture)	28930	29530	Cider making machinery (manufacture)
10821	15841	Chocolate (manufacture)	46342	51342	Cider merchant (wholesale)
28930	29530	Chocolate and cocoa making machinery (manufacture)	10890	15899	Cider pectin (manufacture)
			11030	15941	Cider perry (manufacture)
46360	51360	Chocolate and sugar confectionery (wholesale)	12000	16000	Cigar (manufacture)
47240	52240	Chocolate and sweets (retail)	16240	20400	Cigar box made of wood (manufacture)
10821	15841	Chocolate confectionery (manufacture)	22290	36639	Cigar holder (manufacture)
10821	15841	Chocolate couverture (manufacture)	46350	51350	Cigar importer (wholesale)
26110	31100	Choke and coil (electronic) (manufacture)	28930	29530	Cigar making machinery (manufacture)
28120	29130	Choke manifolds (manufacture)	46350	51350	Cigar merchant (wholesale)
56101	55301	Chop house (licensed)	12000	16000	Cigarette (manufacture)
23140	26140	Chopped roving and strand made of glass fibre (manufacture)	25990	28750	Cigarette cases made of metal (manufacture)
			22290	36639	Cigarette holder (manufacture)
25710	28610	Choppers (manufacture)	46350	51350	Cigarette importer (wholesale)
18129	22220	Christmas card printing (manufacture)	32990	36639	Cigarette lighter (manufacture)
17290	36639	Christmas cracker (manufacture)	46499	51479	Cigarette lighters (wholesale)
17290	36639	Christmas decorations made of paper (manufacture)	28930	29530	Cigarette making machinery (manufacture)
01290	02010	Christmas tree growing	46350	51350	Cigarette merchant (wholesale)
27400	31500	Christmas tree lights (manufacture)	17219	21219	Cigarette packets (manufacture)
46470	51439	Christmas tree lights (wholesale)	17120	21120	Cigarette paper (uncut in rolls) (manufacture)
28140	29130	Christmas trees and other assemblies of valves (manufacture)	17290	21259	Cigarette paper in booklets (manufacture)
			17290	21259	Cigarette tube (manufacture)
26511	33201	Chromatographs (electronic) (manufacture)	47260	52260	Cigarettes (retail)
24450	27450	Chrome alloys (manufacture)	12000	16000	Cigarillo (manufacture)
23200	26260	Chrome magnesite shape (manufacture)	47260	52260	Cigars (retail)
07290	13200	Chrome ore mining and preparation	46170	51170	Cigars, cheroots, cigarillos and cigarettes (commission agent)
25610	28510	Chrome plating (manufacture)			
24450	27450	Chrome production and refining (manufacture)	26702	33403	Cine camera (manufacture)
15110	19100	Chrome tanning (manufacture)	59140	92130	Cine club
24450	27450	Chrome wire made by drawing (manufacture)	59140	92130	Cinema
23200	26260	Chromite articles (manufacture)	59140	92130	Cinema club
24450	27450	Chromium (manufacture)	47110	52113	Cinema kiosk (retail)
20130	24130	Chromium compounds (excluding prepared pigments) (manufacture)	26702	33403	Cinematographic equipment (manufacture)
			59120	92119	Cinematographic film colouring, developing, printing or repairing
20120	24120	Chromium pigment (manufacture)			
25610	28510	Chromium plating (manufacture)	20590	24640	Cinematographic sensitized film (manufacture)
46750	51550	Chromium, manganese, lead and copper oxides and hydroxides (wholesale)	01280	01139	Cinnamon growing
			24420	27420	Circles made of aluminium (manufacture)
86101	85112	Chronic sick hospital (private sector)	24440	27440	Circles made of brass (manufacture)
86101	85111	Chronic sick hospital (public sector)	24440	27440	Circles made of copper (manufacture)
26520	33500	Chronometer (manufacture)	27120	31200	Circuit breaker (moulded case) (manufacture)
94910	91310	Church Army	27120	31200	Circuit breaker for power (manufacture)
94910	91310	Church Commission	26110	31200	Circuit protection device (electronic) (manufacture)
94910	91310	Church in Wales			
94990	91330	Church Lads' Brigade			

SIC 2007	SIC 2003	Activity
82190	74850	Circular addressing
24200	27220	Circular hollow sections made of steel (manufacture)
28240	29410	Circular or reciprocating saws (manufacture)
25730	28620	Circular sawblades (manufacture)
25730	28620	Circular saws for all materials (manufacture)
28131	29121	Circulator pump (manufacture)
90010	92349	Circus
77390	71340	Circus tent rental
22230	25239	Cistern floats made of plastic (manufacture)
25290	28210	Cistern made of metal exceeding 300 litres (manufacture)
22230	25239	Cisterns made of plastic (manufacture)
88990	85321	Citizens Advice Bureau
94990	91330	Citizens initiative or protest movements
20140	24140	Citric acid (manufacture)
85320	80220	City and Guilds of London Institute
94110	91110	City guild (goldsmiths' company, stationers' company, etc.)
94910	91310	City mission
56101	55301	Civic restaurant (licensed)
94990	91330	Civic trust
84220	75220	Civil defence administration
42990	45213	Civil engineering construction
42220	45213	Civil engineering constructions for long-distance communication
42220	45213	Civil engineering constructions for power lines
42220	45213	Civil engineering constructions for power plants
42220	45213	Civil engineering constructions for urban communication
42990	45213	Civil engineering contractor
46630	51820	Civil engineering machinery and equipment (wholesale)
77320	71320	Civil engineering machinery and equipment rental (without operator)
72190	73100	Civil engineering research and experimental development
43341	45440	Civil engineering structure painting
85320	80220	Civil service college
23610	26610	Cladding wall panels made of precast concrete (manufacture)
43999	45250	Claddings (external)
43330	45430	Claddings (internal)
96090	93059	Clairvoyant
25730	28620	Clamp (manufacture)
52219	63210	Clamping and towing away of vehicles
28290	29240	Clarification plant for water treatment (manufacture)
25990	28750	Clasps (manufacture)
46130	51130	Clay (commission agent)
46730	51530	Clay (wholesale)
09900	14220	Clay and kaolin mining support services provided on a fee or contract basis
08120	14220	Clay extraction for brick, pipe and tile production
08120	14220	Clay mining
08120	14220	Clay quarrying
96010	93010	Clean towel company
20412	24512	Cleaning and polishing preparations (manufacture)
13950	17530	Cleaning cloth (non-woven) (manufacture)
13923	17403	Cleaning cloth (not of bonded fibre fabric) (manufacture)
47789	52489	Cleaning materials (retail)
46440	51440	Cleaning materials (wholesale)
81222	74704	Cleaning of heat and air ducts
81229	74709	Cleaning of industrial machinery
81299	74709	Cleaning of the inside of road and sea tankers
20412	24512	Cleaning powder (other than detergents and scouring powder) (manufacture)
81210	74701	Cleaning service for factory, office or shop
81222	74704	Cleaning services for computer rooms
81222	74704	Cleaning services for hospitals
81299	74709	Cleaning services n.e.c.
39000	90030	Cleaning up oil spills and other pollutions in ocean and seas including coastal areas
39000	90030	Cleaning up oil spills and other pollutions in surface water
39000	90030	Cleaning up oil spills and other pollutions on land
28290	29240	Cleaning, filling, packing or wrapping machine parts (manufacture)
28930	29530	Cleaning, sorting or grading machines for seeds, grain or dried leguminous vegetables (manufacture)
46640	51830	Cleaning, wringing, ironing, pressing, dyeing, etc. Machines for textile yarn and fibres (wholesale)
17220	21220	Cleansing tissues (manufacture)
10822	15842	Clear gum confectionery (manufacture)
20590	24640	Clearing agents for photographic use (manufacture)
66110	67110	Clearing house (banking)
30120	35120	Cleat for pleasure boat (manufacture)
25710	28610	Cleavers (manufacture)
01230	01139	Clementine growing
14120	18210	Clerical vestment (manufacture)
14131	18221	Climbing clothing for men and boys (weatherproof) (manufacture)
14132	18222	Climbing clothing for women and girls (weatherproof) (manufacture)
32300	36400	Climbing frame (manufacture)
86900	85140	Clinic (health service)
75000	85200	Clinico-pathological and other diagnostic activities pertaining to animals
23510	26510	Clinkers and hydraulic cement (manufacture)
96090	93059	Cloakroom (not railway, etc.)
14131	18221	Cloaks for men and boys (manufacture)
14132	18222	Cloaks for women and girls (manufacture)
26520	33500	Clock (electric) (manufacture)
26520	33500	Clock (manufacture)
23190	26150	Clock and watch glass (manufacture)
26520	33500	Clock case made of wood (manufacture)
47770	52484	Clocks (retail)
46480	51479	Clocks (wholesale)
15200	19300	Clog (manufacture)
26309	32202	Closed circuit television equipment (CCTV) (manufacture)
25500	28400	Closed die forging (manufacture)
28290	29240	Closing machinery (manufacture)
25920	28720	Closures made of metal (manufacture)
22220	25220	Closures made of plastic (manufacture)
46410	51410	Cloth (wholesale)
13300	17300	Cloth beetling (manufacture)
13300	17300	Cloth crease resisting treatment (manufacture)
13300	17300	Cloth degreasing (manufacture)
13300	17300	Cloth dressing (manufacture)
13300	17300	Cloth dyeing (manufacture)
13300	17300	Cloth embossing (manufacture)
13300	17300	Cloth ending and mending (manufacture)

C

SIC 2007	SIC 2003	Activity
13300	17300	Cloth finishing (manufacture)
13300	17300	Cloth fireproofing (manufacture)
13200	17210	Cloth made of cotton and similar man made fibres (manufacture)
25930	28730	Cloth made of wire (manufacture)
13300	17300	Cloth mercerising (manufacture)
46410	51410	Cloth merchant (wholesale)
13300	17300	Cloth piece goods printing (manufacture)
13300	17300	Cloth proofing (manufacture)
13300	17300	Cloth rot proofing (manufacture)
13300	17300	Cloth shrinking (manufacture)
13300	17300	Cloth waterproofing
27510	29710	Clothes airer (electric) (manufacture)
32910	36620	Clothes brush (manufacture)
74100	74872	Clothes designer
16290	20510	Clothes hangers made of wood (manufacture)
77299	71409	Clothes hire
25990	28750	Clothes hook (manufacture)
16290	20510	Clothes horse made of wood (manufacture)
22290	25240	Clothes pegs made of plastic (manufacture)
16290	20510	Clothes pegs made of wood (manufacture)
46439	51439	Clothes washing and drying machines for domestic use (wholesale)
46160	51160	Clothing (commission agent)
46420	51429	Clothing (wholesale)
14190	18249	Clothing accessories (manufacture)
46420	51429	Clothing accessories (wholesale)
47710	52421	Clothing accessories made of fur or leather (retail)
46420	51421	Clothing accessories made of fur or leather (wholesale)
47710	52422	Clothing accessories, children's, other than of fur or leather (retail)
47710	52424	Clothing accessories, men's, other than of fur or leather (retail)
47710	52423	Clothing accessories, women's, other than of fur or leather (retail)
46420	51429	Clothing exporter (wholesale)
47510	52410	Clothing fabrics (retail)
46420	51429	Clothing importer (wholesale)
23990	26821	Clothing made of asbestos (manufacture)
14200	18300	Clothing made of sheepskin (manufacture)
46420	51429	Clothing outfitter (wholesale)
14190	18249	Clothing pad (manufacture)
46410	51410	Clothing textiles (wholesale)
10511	15511	Clotted cream (manufacture)
01280	01139	Clove growing
01190	01110	Clover growing
32300	36400	Clubs (manufacture)
28150	29140	Clutch (not for motor vehicle) (manufacture)
29320	34300	Clutch and parts for motor vehicles (manufacture)
23990	26821	Clutch linings made of asbestos (manufacture)
46690	51870	Clutches and shaft couplings including universal joints (excluding motor vehicles) (wholesale)
29100	34100	Coach (manufacture)
25940	28740	Coach bolts and screws (manufacture)
29100	34100	Coach engine (manufacture)
49390	60231	Coach hire with driver
49390	60231	Coach services (non-scheduled)
13960	17542	Coach trimming (manufacture)
85510	92629	Coaches of sport

SIC 2007	SIC 2003	Activity
29201	34201	Coachwork for motor vehicles (manufacture)
46120	51120	Coal (commission agent)
46719	51519	Coal (wholesale)
47789	52489	Coal and coke (retail)
19100	23100	Coal carbonisation (manufacture)
05101	10101	Coal cleaning, sizing, grading and pulverising (hard, deep mined)
05102	10102	Coal cleaning, sizing, grading and pulverising (hard, opencast)
05102	10102	Coal contractor (opencast)
28921	29521	Coal cutter (manufacture)
46719	51519	Coal depot (wholesale)
46120	51120	Coal factor (commission agent)
46719	51519	Coal merchant (wholesale)
05101	10101	Coal mine (deep or drift)
46630	51820	Coal or rock cutters and tunnelling machinery (wholesale)
28921	29521	Coal plough (manufacture)
05101	10101	Coal preparation (deep mined)
05102	10102	Coal preparation (opencast)
28921	29521	Coal preparation plant (manufacture)
49500	60300	Coal pumping station
05102	10102	Coal recovery from dumps, tips etc.
05102	10102	Coal recovery of from culm banks
05102	10102	Coal site (opencast)
52103	63129	Coal stockyard for land transport activities
52101	63129	Coal stockyard for water transport activities
35210	40210	Coal tar (crude) from gas works
19201	10103	Coal tar (crude) from manufactured fuel plants (manufacture)
20140	24140	Coal tar (refined) (manufacture)
20140	24140	Coal tar distillation (manufacture)
20140	24140	Coal tar naphtha (manufacture)
23990	26829	Coal tar pitch, articles thereof (manufacture)
05101	10101	Coal washing (deep mined)
05102	10102	Coal washing (opencast)
84240	75240	Coast guards administration and operation
42910	45240	Coastal defence construction
50200	61102	Coastal water transport (freight)
50100	61101	Coastal water transport for passengers
16290	20510	Coat and hat racks made of wood (manufacture)
22290	25240	Coat hangers made of plastic (manufacture)
16290	20510	Coat hangers made of wood (manufacture)
23910	26810	Coated abrasives (manufacture)
25930	28730	Coated electrodes for electric arc-welding (manufacture)
25930	28730	Coated or cored wire (manufacture)
08120	14210	Coated roadstone production
08120	14210	Coated tarmacadam production
28990	29560	Coating machine for bookbinding or paper working (manufacture)
28940	29540	Coating machinery for textiles (manufacture)
13300	17300	Coating of purchased garments (manufacture)
13200	17220	Coating woollen weaving (manufacture)
13200	17230	Coating worsted weaving (manufacture)
14131	18221	Coats for men and boys (manufacture)
14132	18222	Coats for women and girls (manufacture)
31010	36120	Coat stand (non-domestic) (manufacture)
46439	51439	Co-axial cable and co-axial conductors for domestic use (wholesale)

SIC 2007	SIC 2003	Activity
46690	51870	Co-axial cable and co-axial conductors for industrial use (wholesale)
24450	27450	Cobalt (manufacture)
07290	13200	Cobalt mining and preparation
16240	20400	Cock made of wood (manufacture)
03110	05010	Cockle gathering
28140	29130	Cocks for industrial use (manufacture)
31090	36140	Cocktail cabinet (manufacture)
10821	15841	Cocoa (manufacture)
46370	51370	Cocoa (wholesale)
10821	15841	Cocoa bean roasting and dressing (manufacture)
01630	01139	Cocoa beans, peeling and preparation
10821	15841	Cocoa butter (manufacture)
10821	15841	Cocoa fat (manufacture)
01270	01139	Cocoa growing
10821	15841	Cocoa oil (manufacture)
10821	15841	Cocoa powder (manufacture)
10821	15841	Cocoa products (manufacture)
10390	15330	Coconut flakes including desiccated but (not sugared) (manufacture)
01260	01139	Coconut growing
10410	15420	Coconut oil refining (manufacture)
93210	92330	Coconut shy
10410	15420	Cod liver oil refining (manufacture)
10832	15862	Coffee (manufacture)
46370	51370	Coffee (wholesale)
10832	15862	Coffee and chicory essence and extract (manufacture)
10832	15862	Coffee bags (manufacture)
56102	55302	Coffee bar, room or saloon (unlicensed)
10832	15862	Coffee blending (manufacture)
10832	15862	Coffee essence and extract (manufacture)
10832	15862	Coffee extracts and concentrates (manufacture)
10832	15862	Coffee grinding and roasting (manufacture)
01270	01139	Coffee growing
46110	51110	Coffee husks and skins (commission agent)
27510	29710	Coffee or tea makers (electric) (manufacture)
27510	29710	Coffee percolator (electric) (manufacture)
10832	15862	Coffee processing (manufacture)
28930	29530	Coffee processing machinery (manufacture)
10832	15862	Coffee products (manufacture)
10832	15862	Coffee roasting (manufacture)
10832	15862	Coffee substitutes (manufacture)
31090	36140	Coffee table (manufacture)
46370	51370	Coffee, tea, cocoa and spices exporter (wholesale)
46370	51370	Coffee, tea, cocoa and spices importer (wholesale)
30110	35110	Coffer-dam construction (manufacture)
32990	20510	Coffin board (manufacture)
13100	17120	Coffin cloth (manufacture)
13960	17542	Coffin frilling (manufacture)
32990	20510	Coffins (manufacture)
96020	93020	Coiffeur
29310	31610	Coil ignition (manufacture)
25930	28740	Coil springs (not for motor vehicle suspension) (manufacture)
24440	27440	Coils made of copper (manufacture)
32401	36501	Coin operated games (manufacture)
28230	30010	Coin sorting, wrapping and counting machines (manufacture)
32110	36210	Coin striking (manufacture)
92000	92710	Coin-operated gambling machine establishments
93290	92729	Coin-operated games arcade (other than gaming machines)
32110	36210	Coins (manufacture)
47789	52485	Coins (retail)
46719	51519	Coke (wholesale)
35210	40210	Coke gas
46719	51519	Coke merchant (wholesale)
46120	51120	Coke or semi-coke of coal (commission agent)
19100	23100	Coke oven gas (manufacture)
19100	23100	Coke oven products (manufacture)
19201	23201	Coke petroleum (manufacture)
19100	23100	Coke production (manufacture)
10410	15420	Cola oil refining (manufacture)
11070	15980	Cola production (manufacture)
25990	28750	Colanders made of metal (manufacture)
22290	25240	Colanders made of plastic (manufacture)
26110	32100	Cold cathode valve or tube (manufacture)
25730	28620	Cold chisel (manufacture)
24340	27340	Cold drawing or stretching of steel wire (manufacture)
24310	27310	Cold drawn steel bars (manufacture)
24310	27310	Cold drawn steel sections (manufacture)
24310	27310	Cold finished steel bars (manufacture)
24330	27330	Cold formed steel angles (manufacture)
24330	27330	Cold formed steel channels (manufacture)
24330	27330	Cold formed steel sections (manufacture)
25500	28400	Cold pressing of base metals (manufacture)
24320	27320	Cold reduced steel slit strip <600 mm (manufacture)
24100	27100	Cold reduced steel slit strip ≥600 mm (manufacture)
24100	27100	Cold reduced wide steel strip ≥600 mm (cold reduced coil) (manufacture)
24320	27320	Cold rolled narrow steel strip <600 mm (manufacture)
24100	27100	Cold rolled narrow steel strip ≥600 mm (manufacture)
24100	27100	Cold rolled steel plate (manufacture)
24100	27100	Cold rolled steel sheet (manufacture)
24100	27100	Cold rolled steel slit strip ≥600 mm (manufacture)
24100	27100	Cold rolled wide steel strip ≥600 mm (cold rolled wide coil) (manufacture)
25500	28400	Cold stamping of base metals (manufacture)
28250	29230	Cold storage equipment (manufacture)
52102	63121	Cold store for air transport activities
52103	63121	Cold store for land transport activities
52101	63121	Cold store for water transport activities
24330	27330	Cold-folded ribbed sheets and sandwich panels (manufacture)
24330	27330	Cold-formed ribbed sheets and sandwich panels (manufacture)
30120	35120	Collapsible boat (not inflatable dinghy) (manufacture)
16240	20400	Collapsible box made of wood (manufacture)
32990	36639	Collar stud (manufacture)
14190	18249	Collars for men and boys (manufacture)
28990	29560	Collating machine for bookbinders (manufacture)
28230	30010	Collating machinery (manufacture)
18140	22230	Collating of books, brochures, etc. (manufacture)
46770	51570	Collecting, sorting, separating, stripping of used goods to obtain reusable parts (wholesale)

C

SIC 2007	SIC 2003	Activity
65120	66031	Collecting society
38110	90020	Collection and removal of non hazardous debris and rubble
38120	23300	Collection and treatment of radioactive nuclear waste
38120	90020	Collection of bio-hazardous waste
86900	85140	Collection of female human urine for hormone extraction
38110	90020	Collection of non hazardous construction and demolition waste
38110	90020	Collection of non hazardous recyclable materials
38110	90020	Collection of non hazardous waste output of textile mills
38120	90020	Collection of nuclear waste
36000	41000	Collection of rain water
38110	90030	Collection of refuse in litter bins in public places
38120	90020	Collection of used oil from shipment or garages
36000	41000	Collection of water from rivers, lakes, wells
46499	51479	Collectors stamps and coins (wholesale)
85320	80220	College of agriculture
85320	80220	College of art
85421	80302	College of higher education (degree level)
85320	80220	College of music
85410	80301	College of nursing
85320	80220	College of technology
46120	51120	Colloidal precious metals (commission agent)
18129	22220	Collotype printing (manufacture)
20420	24520	Colognes (manufacture)
26511	33201	Colorimeters (electronic) (manufacture)
26513	33202	Colorimeters (non-electronic) (manufacture)
32500	33401	Colour filter (unmounted) (manufacture)
20120	24120	Colour lake (manufacture)
26110	32100	Colour television tubes (manufacture)
23110	26110	Coloured glass (manufacture)
46750	51550	Colouring matter (wholesale)
20120	24120	Colours for food and cosmetics (manufacture)
20120	24120	Colours in dry, liquid or paste form (manufacture)
25110	28110	Column (fabricated structural steelwork) (manufacture)
25300	28300	Column (process plant) (manufacture)
01110	01110	Colza growing
10410	15410	Colza oil production (manufacture)
28940	29540	Comb for textile machinery (manufacture)
13100	17120	Combers' shoddy for woollen industry (manufacture)
13940	17520	Combination rope (manufacture)
28302	29320	Combine harvester (manufacture)
81100	70320	Combined facilities support activities
82110	74850	Combined office admin. Services (e.g. Reception, billing, record keeping, personnel, mail services)
82110	74850	Combined secretarial activities
13100	17130	Combing and slubbing of wool on a commission basis (manufacture)
15110	19100	Combing leather (manufacture)
28940	29540	Combing machinery for textiles (manufacture)
27510	29710	Combs (electric) (manufacture)
32990	36639	Combs (other than of hard rubber, plastic or metal) (manufacture)
22190	36639	Combs of hard rubber (manufacture)
25990	36639	Combs of metal (manufacture)
22290	36639	Combs of plastic (manufacture)
46499	51479	Combs, hair-slides, hairpins, curling pins (wholesale)

SIC 2007	SIC 2003	Activity
47781	52486	Commercial art gallery (retail)
73110	74402	Commercial artist
41201	45211	Commercial buildings construction
81210	74701	Commercial cleaner
77390	71340	Commercial machinery rental and operating leasing
17230	22220	Commercial notebooks (manufacture)
78200	74500	Commercial or industrial workers (supply) (temporary employment agency)
18129	22220	Commercial printed matter printing (manufacture)
85320	80220	Commercial school
84130	75130	Commercial services administration and regulation (public sector)
17230	22220	Commercial stationery (manufacture)
17230	22220	Commercial stationery binders (manufacture)
17230	22220	Commercial stationery business forms (manufacture)
17230	22220	Commercial stationery registers (manufacture)
77120	71219	Commercial vehicle (light) hire (without driver)
77120	71219	Commercial vehicle (medium and heavy type) contract hire (without driver)
77120	71219	Commercial vehicle (medium and heavy type) hire (without driver)
49410	60249	Commercial vehicle hire with driver
52219	63210	Commercial vehicle park
29100	34100	Commercial vehicles (manufacture)
28290	29240	Comminution plant for effluent treatment (manufacture)
13300	17300	Commission mending of textiles (manufacture)
52220	63220	Commissioners of northern lighthouses
64991	65233	Commodities dealing for investment purposes
35230	40220	Commodity and transport capacity exchanges for gaseous fuels
66120	67122	Commodity contracts brokerage
66110	67110	Commodity contracts exchanges administration
93290	92729	Common (local authority or municipally owned)
99000	99000	Commonwealth Armed Forces
99000	99000	Commonwealth Government Service
99000	99000	Commonwealth Institute
99000	99000	Commonwealth Secretariat
99000	99000	Commonwealth War Graves Commission
52230	63230	Communication centre (civil air)
26301	32201	Communication devices using infrared signal (e.g. Remote controls) (manufacture)
77390	71340	Communication equipment rental and operating leasing
42220	45213	Communication lines construction
84130	75130	Communication services administration and regulation (public sector)
61900	64200	Communications telemetry
84120	75120	Community amenity services administration (public sector)
88990	85321	Community and neighbourhood activities (charitable)
88990	85322	Community and neighbourhood activities (non-charitable)
94990	91330	Community centre
86230	85130	Community dental service clinics
86900	85140	Community health service
35300	40300	Community heating plant
87900	85311	Community homes for children (charitable)
87900	85312	Community homes for children (non-charitable)
86900	85140	Community medical service clinics
94990	91330	Community organisations

SIC 2007	SIC 2003	Activity
86900	85140	Community psychiatric nurse (NHS)
26400	32300	Compact disc players (manufacture)
47430	52450	Compact disc players (retail)
18201	22310	Compact disc reproduction from master copies (manufacture)
59200	22140	Compact disc sound recording publishing
46431	51431	Compact discs (recorded) (wholesale)
84230	75230	Companies court
66190	67130	Company promoting
66110	67110	Company registration agent
69202	74122	Company secretary
26511	33201	Comparators (electronic) (manufacture)
26513	33202	Comparators (non-electronic) (manufacture)
26513	33202	Compass (drawing) (manufacture)
26513	33202	Compass (magnetic) (manufacture)
69201	74121	Compilation of financial statements
43320	45420	Completion of ceilings
26520	33500	Components for clocks and watches (manufacture)
28990	29560	Composing room equipment (manufacture)
20590	24660	Composite diagnostic or laboratory reagents (manufacture)
71200	74300	Composition and purity testing and analysis
23990	26821	Composition asbestos (manufacture)
16290	20520	Composition cork (manufacture)
18130	22240	Composition for printing (manufacture)
15110	19100	Composition leather (manufacture)
10910	15710	Compound animal feed (manufacture)
46210	51210	Compound feed stuff (wholesale)
20150	24150	Compound fertiliser (manufacture)
20530	24630	Compound flavour (blended flavour concentrates) (manufacture)
46120	51120	Compound plasticisers and stabilisers for rubber or plastics (commission agent)
20590	24660	Compound plasticisers for rubber or plastics (manufacture)
46750	51550	Compounds of rare earth metals (wholesale)
46120	51120	Compounds of rare earth metals, yttrium or scandium (commission agent)
46750	51550	Compounds of yttrium and scandium (wholesale)
46120	51120	Compounds with nitrogen functions (commission agent)
85310	80210	Comprehensive schools
35300	40300	Compressed air production and distribution
20110	24110	Compressed industrial gases (manufacture)
26513	33202	Compressibility testing equipment (non-electronic) (manufacture)
28110	29110	Compression ignition engines for industrial use (manufacture)
28110	29110	Compressor engine (manufacture)
28132	29122	Compressor for air or other gas (manufacture)
28132	29122	Compressor for refrigerators (manufacture)
28132	29122	Compressor parts (manufacture)
46690	51870	Compressors for refrigerating equipment (wholesale)
46690	51870	Compressors for use in civil aircraft (wholesale)
84300	75300	Compulsory social security administration concerning family and child benefits
84300	75300	Compulsory social security administration concerning government employee pension schemes
84300	75300	Compulsory social security administration concerning permanent loss of income due to disablement
84300	75300	Compulsory social security administration concerning retirement pensions
84300	75300	Compulsory social security administration concerning unemployment benefits
84300	75300	Compulsory social security administration of sickness, maternity or temporary disablement benefits
26200	30020	Computer (electronic) (manufacture)
26200	30020	Computer (manufacture)
62020	72100	Computer audit consultancy services
62020	72220	Computer consultancy (software)
96090	93059	Computer dating agency
26800	24650	Computer discs and tapes (unrecorded) (manufacture)
46140	51140	Computer equipment (commission agent)
62030	72300	Computer facilities management activities
47410	52485	Computer games (retail)
62011	72220	Computer games design
58210	72210	Computer games publishing
62020	72100	Computer hardware acceptance testing services
72190	73100	Computer hardware research and experimental development
18203	22330	Computer media reproduction (manufacture)
46620	51810	Computer numerically controlled (CNC) machine tools (wholesale)
28410	29420	Computer numerically controlled (CNC) metal cutting machines (manufacture)
26200	30020	Computer peripheral equipment (manufacture)
47410	52482	Computer peripheral equipment (retail)
17230	21230	Computer print-out paper (manufacture)
26200	32300	Computer projectors (video beamers) (manufacture)
62090	72600	Computer related activities (other)
85320	80220	Computer repair training
62020	72100	Computer site planning services
26200	30020	Computer store (manufacture)
26200	30020	Computer system (manufacture)
26200	30020	Computer terminal unit (manufacture)
18130	22240	Computer to plate CTP processing of plates for relief printing (manufacture)
18130	22240	Computer to plate CTP processing of plates for relief stamping
85590	80429	Computer training
46620	51810	Computer-controlled machine tools (wholesale)
46640	51830	Computer-controlled machinery for sewing and knitting machines (wholesale)
46640	51830	Computer-controlled machinery for the textile industry (wholesale)
47410	52482	Computers and non-customised software (retail)
46510	51840	Computers and peripheral equipment (wholesale)
77330	71330	Computing machinery and equipment rental and operating leasing
10910	15710	Concentrated animal feed and feed supplements (manufacture)
10519	15519	Concentrated dried milk (manufacture)
90040	92320	Concert halls operation
32200	36300	Concertina (manufacture)
90010	92311	Concerts production
23690	26660	Concrete articles n.e.c. (manufacture)
28990	29560	Concrete block making machinery (manufacture)
25610	28510	Concrete coating of metals (manufacture)
23630	26630	Concrete dry mix (manufacture)
49410	60249	Concrete haulage by a unit which is (not the manufacturer)

C

C

SIC 2007	SIC 2003	Activity	SIC 2007	SIC 2003	Activity
28923	29523	Concrete mixer (manufacture)	41201	45211	Construction of airport buildings
28923	29523	Concrete placing machinery (manufacture)	41201	45211	Construction of arts, cultural or leisure facilities buildings
28131	29121	Concrete pumps (manufacture)	41201	45211	Construction of assembly plants
28923	29523	Concrete surfacing machinery (manufacture)	42990	45213	Construction of chemical plants (except buildings)
43999	45250	Concrete work (building)	42210	45213	Construction of civil engineering constructions for long-distance and urban pipelines
29100	34100	Concrete-mixer lorries (manufacture)			
06200	11100	Condensate extraction	41201	45211	Construction of commercial buildings
20160	24160	Condensation, polycondensation and polyaddition products (plastic material) (manufacture)	41202	45212	Construction of domestic buildings
			30110	35110	Construction of drilling platforms, floating or submersible (manufacture)
10511	15511	Condensed milk (manufacture)	41201	45211	Construction of hospitals
25300	28300	Condenser (steam) (manufacture)	41202	45212	Construction of housing association and local authority housing
25300	28300	Condenser (vapour) (manufacture)			
28250	29230	Condenser for air conditioning equipment (air or water cooled) (manufacture)	41201	45211	Construction of indoor sports facilities
			42990	45213	Construction of industrial facilities (except buildings)
28250	29230	Condenser unit for refrigerator (manufacture)	42210	45240	Construction of irrigation systems (canals)
25990	28750	Condiment set made of metal (manufacture)	41202	45212	Construction of multi-family buildings, including high-rise buildings
10840	15870	Condiments (manufacture)			
25930	28730	Conductor cable made of steel reinforced aluminium (manufacture)	30110	35110	Construction of non-recreational inflatable rafts
			41201	45211	Construction of office buildings
23320	26400	Conduits made of ceramic (manufacture)	42110	45230	Construction of other vehicular and pedestrian ways
23320	26400	Conduits made of clay (manufacture)	42990	45230	Construction of outdoor sports facilities (except buildings)
10822	15842	Confectioner's novelty (manufacture)			
47110	52111	Confectioners, tobacconists and newsagents (CTN's) (retail)	43999	45211	Construction of outdoor swimming pools
			41201	45211	Construction of parking garages
46170	51170	Confectionery (commission agent)	41201	45211	Construction of primary, secondary and other schools
10822	15842	Confectionery (medicated) (manufacture)			
46360	51360	Confectionery (wholesale)	42210	45213	Construction of pumping stations
28930	29530	Confectionery machines and processing equipment (manufacture)	42990	45213	Construction of refineries (except buildings)
			41201	45211	Construction of religious buildings
10821	15841	Confectionery made of chocolate (manufacture)	41202	45212	Construction of residential buildings:
10822	15842	Confectionery made of sugar (manufacture)	42210	45213	Construction of sewage disposal plants
94110	91110	Confederation of British industry	42210	45213	Construction of sewer systems
68310	70310	Conference centre letting (not self-owned)	41202	45212	Construction of single-family houses
68202	70201	Conference centre letting (self owned)	42130	45213	Construction of tunnels
82302	74874	Conference organisers	42120	45230	Construction of underground railways
17290	36639	Confetti paper (manufacture)	42220	45213	Construction of utility projects for electricity
32990	36639	Conjuring apparatus (manufacture)	42220	45213	Construction of utility projects for telecommunications
90010	92311	Conjuror			
43210	45310	Connecting of electric appliances and household equipment, including baseboard heating	41201	45211	Construction of warehouses
			41201	45211	Construction of workshops
94990	91330	Conservation and Preservation Society	84130	75130	Construction services administration and regulation (public sector)
94920	91320	Conservative and Unionist Party			
94920	91320	Conservative Association	32409	36509	Construction sets (manufacture)
46499	51477	Construction and constructional toys (wholesale)	25110	28110	Construction site huts made of metal (manufacture)
38110	90020	Construction and demolition waste collection	71111	74201	Construction supervision
28923	29523	Construction equipment (manufacture)	42990	45213	Constructional engineering
41201	45211	Construction factories	32409	36509	Constructional toy (manufacture)
46630	51820	Construction machinery (wholesale)	99000	99000	Consular office
77320	71320	Construction machinery and equipment rental (without operator)	84210	75210	Consular services abroad administration and operation (public sector)
43999	45500	Construction machinery and equipment rental with operator	71129	74204	Consultant civil or structural engineer
			71121	74205	Consultant design engineer
46630	51820	Construction machinery exporter (wholesale)	71129	74204	Consultant engineer (civil or structural)
46630	51820	Construction machinery importer (wholesale)	71121	74205	Consultant engineer (other than civil or structural)
47520	52460	Construction materials (retail)	94990	91330	Consumer associations
46730	51530	Construction materials (wholesale)	64921	65221	Consumer credit granting company (other than banks or building societies)
46770	51570	Construction materials from demolished buildings (wholesale)			
			94990	91330	Consumers' association
46130	51130	Construction materials made of glass (commission agent)	32500	33401	Contact lens (manufacture)
32409	36509	Construction model (manufacture)			

SIC 2007	SIC 2003	Activity	SIC 2007	SIC 2003	Activity
47782	52487	Contact lenses (retail)	86101	85111	Convalescent home (public sector providing medical care)
27900	31300	Contacts and other electrical carbon and graphite products (manufacture)	87900	85311	Convalescent home without medical care (charitable)
15120	19200	Container for typewriter, radio, etc. Made of leather (manufacture)	87100	85140	Convalescent homes
			87900	85312	Convalescent homes without medical care (non-charitable)
28220	29220	Container handling crane (manufacture)	94910	91310	Convent (not school or orphanage)
52241	63110	Container handling services for water transport activities	85310	80210	Convent schools at secondary level
23190	26150	Container made of glass tubing (hygienic and pharmaceutical) (manufacture)	25400	29600	Conventional missiles (manufacture)
22220	25220	Container made of plastic for closed transit (manufacture)	29203	34203	Conversion of complete vehicles to motor caravans (manufacture)
77390	71219	Container rental	30110	35110	Conversion of ships (manufacture)
30110	35110	Container ship (manufacture)	28910	29510	Converters for hot metal handling (manufacture)
17219	21219	Containers and canisters made of cardboard n.e.c. (manufacture)	31090	36110	Convertible furniture (manufacture)
			27110	31100	Converting machinery (electrical) (manufacture)
46690	51870	Containers designed for carriage by one or more means of transport (wholesale)	26200	30020	Converter for computer (manufacture)
29202	34202	Containers for carriage by one or more modes of transport (manufacture)	46690	51870	Converters, ladles, ingot moulds and casting machines (wholesale)
25290	28210	Containers for compressed or liquefied gases made of metal (manufacture)	22290	25240	Conveyer belts made of plastic (manufacture)
			28220	29220	Conveying plant (hydraulic and pneumatic) (manufacture)
29202	34202	Containers for freight (manufacture)	28220	29220	Conveyor and feeder (not for agriculture or mining) (manufacture)
17219	21219	Containers made of corrugated paper or paperboard n.e.c. (manufacture)	15120	19200	Conveyor bands made of leather (manufacture)
25920	28720	Containers made of foil (manufacture)	13960	17542	Conveyor belting (woven) (manufacture)
23130	26130	Containers made of glass or crystal (manufacture)	22210	25210	Conveyor belts made of plastic (manufacture)
25290	28210	Containers made of metal of a capacity exceeding 300 litres (manufacture)	22190	25130	Conveyor belts made of rubber (manufacture)
			28302	29320	Conveyor for agricultural use (manufacture)
17219	21219	Containers made of paper and paperboard n.e.c. (manufacture)	28921	29521	Conveyor for underground mining (manufacture)
17219	21219	Containers made of solid board n.e.c. (manufacture)	28220	29220	Conveyors (manufacture)
23130	26130	Containers made of tubular glass (manufacture)	10130	15139	Cooked and preserved meat (manufacture)
46130	51130	Containers made of wood (commission agent)	47220	52220	Cooked meats (retail)
16240	20400	Containers made of wood (manufacture)	27510	29710	Cooker (electric) (manufacture)
65120	66031	Contingency insurance	27520	29720	Cooker (gas) (manufacture)
85590	80429	Continuation school	27520	29720	Cooker (oil) (manufacture)
87300	85311	Continuing care retirement communities (charitable)	27520	29720	Cooker (solid fuel) (manufacture)
87300	85312	Continuing care retirement communities (non-charitable)	10720	15820	Cookies (manufacture)
			27520	29720	Cooking and heating appliances for domestic use (gas) (manufacture)
24100	27100	Continuous cast products made of steel (manufacture)	27520	29720	Cooking and heating appliances for domestic use (solid fuel) (manufacture)
24420	27420	Continuous cast rod made of aluminium (manufacture)	28930	29530	Cooking appliance for commercial catering (non-electric) (manufacture)
24440	27440	Continuous cast rod made of copper (manufacture)	27520	29720	Cooking appliances for domestic use (non-electric) (manufacture)
24450	27450	Continuous cast rods of other base non-ferrous metals (manufacture)	28930	29530	Cooking equipment for commercial catering (electric) (manufacture)
20600	24700	Continuous filament yarn of man-made fibres (manufacture)	10420	15430	Cooking fat (compound) (manufacture)
28921	29521	Continuous miner (manufacture)	24420	27420	Cooking foil made of aluminium (manufacture)
17230	21230	Continuous stationery (manufacture)	46690	51870	Cooking or heating equipment for non-domestic use (wholesale)
46460	51460	Contraceptive chemical preparations based on hormones or spermicides (wholesale)	25990	28750	Cooking utensils made of metal (manufacture)
77110	71100	Contract car hire (self drive)	22290	25240	Cooking utensils made of plastic (manufacture)
81210	74701	Contract cleaning service	16290	20510	Cooking utensils made of wood (manufacture)
13300	17300	Contract cutting of textiles (not self-owned) (manufacture)	01630	01410	Cooling and bulk packing of crops for primary market
46630	51820	Contractors' plant (wholesale)	28250	29230	Cooling equipment for non-domestic use (manufacture)
30300	35300	Control surfaces for aircraft (manufacture)			
26200	30020	Control units for computers (manufacture)	47300	50500	Cooling products for motor vehicles
26120	32100	Controllers interface cards (manufacture)	28250	29230	Cooling tower for air conditioning (manufacture)
86101	85112	Convalescent home (private sector providing medical care)	28290	29240	Cooling towers and similar for direct cooling by means of re-circulated water (manufacture)

C

C

SIC 2007	SIC 2003	Activity
01629	01429	Coop cleaning
16240	20400	Cooper's products (manufacture)
16240	20400	Cooper's products reconditioning (manufacture)
16240	20400	Cooper's wood (manufacture)
26511	33201	Co-oxymeters (electronic) (manufacture)
17290	21259	Cop paper (manufacture)
17290	21259	Cop tube (manufacture)
20160	24160	Co-polymer plastics (manufacture)
46720	51520	Copper (wholesale)
47789	52489	Copper goods (retail)
07290	13200	Copper mining and preparation
07290	13200	Copper ore and concentrate extraction and preparation
90030	92319	Copper plate engraver (artistic)
18129	22220	Copper plate printing (manufacture)
24440	27440	Copper refining (manufacture)
24440	27440	Copper smelting (manufacture)
02100	02010	Coppice and pulpwood growing
10410	15410	Copra (coconut) crushing (manufacture)
16290	20510	Cops made of wood (manufacture)
28230	30010	Copying machine (xerographic) (manufacture)
59200	74879	Copyright acquisition and registration for musical compositions
94120	91120	Copyright protection society
69109	74111	Copyrights preparation
73110	74402	Copywriter
03110	05010	Coral gathering
23990	26821	Cord made of asbestos (manufacture)
13990	17542	Cord made of elastic (manufacture)
13990	17542	Cord made of elastomeric material (manufacture)
13940	17520	Cordage made of textile material (manufacture)
47789	52489	Cordage, rope, twine and nets (retail)
11070	15980	Cordial (non-alcoholic) (manufacture)
20510	24610	Cordite (manufacture)
26301	32201	Cordless telephones (except cellular) (manufacture)
13200	17210	Corduroy weaving from yarn spun on the cotton system (manufacture)
71122	74206	Core preparation and analysis activities
43130	45120	Core sampling for construction
13100	17110	Core spun yarn spun on the cotton system (manufacture)
01280	01139	Coriander growing
47599	52440	Cork goods (retail)
46499	51479	Cork goods (wholesale)
32990	20520	Cork life preservers (manufacture)
16290	20520	Cork products (except cork life preservers) (manufacture)
02300	02010	Cork, gathering from the wild
46210	51210	Corn (wholesale)
46210	51210	Corn chandler (wholesale)
46110	51110	Corn exchange (commission agent)
46110	51110	Corn factor (commission agent)
46210	51210	Corn merchant (wholesale)
10620	15620	Corn oil (manufacture)
10611	15611	Corn or other cereal grains (manufacture)
10612	15612	Cornflake (manufacture)
10611	15611	Cornflour (manufacture)
84230	75230	Coroners court
66120	67122	Corporate finance companies

SIC 2007	SIC 2003	Activity
56210	55520	Corporate hospitality catering
84230	75230	Correctional services
32500	33401	Corrective glasses (manufacture)
26701	33402	Correlator (optical) (manufacture)
85590	80429	Correspondence college (not leading to degree level qualifications)
85421	80302	Correspondence college specialising in higher education courses (degree level)
71129	74204	Corrosion engineering activities
17219	21213	Corrugated packing case (manufacture)
17211	21211	Corrugated paper (manufacture)
17211	21211	Corrugated paper board (manufacture)
24420	27420	Corrugated plate, sheet or strip made of aluminium (manufacture)
23650	26650	Corrugated sheets made of fibre-cement (manufacture)
24100	27100	Corrugated sheets made of steel (manufacture)
14142	18232	Corselet (manufacture)
14142	18232	Corset (manufacture)
14142	18232	Corset belt (manufacture)
13200	17210	Corset cloth weaving (from yarn spun on the cotton system) (manufacture)
13990	17542	Corset lace (manufacture)
47710	52423	Corsetiere (retail)
20420	24511	Cosmetic soap (manufacture)
46180	51180	Cosmetics (commission agent)
20420	24520	Cosmetics (manufacture)
47750	52330	Cosmetics (retail)
46450	51450	Cosmetics (wholesale)
85320	80220	Cosmetology and barber schools
69201	74121	Cost accountant
70221	74142	Cost accounting programmes design
69201	74121	Cost draughtsman (legal)
74902	74203	Cost draughtsman (quantity surveyor)
74100	74872	Costume designing
32130	36610	Costume jewellery (manufacture)
32130	36610	Costume or imitation jewellery (manufacture)
14132	18222	Costumes for women and girls (manufacture)
13923	17403	Cot blanket (manufacture)
31090	36140	Cot frame (manufacture)
31030	36150	Cot mattress (manufacture)
13923	17403	Cot quilt (manufacture)
28940	29540	Cots (textile machinery accessory) (manufacture)
55201	55231	Cottages and cabins without housekeeping services, provided in holiday centres and holiday villages
55201	55231	Cottages in holiday centres and holiday villages (provision of short-stay lodging in)
55209	55232	Cottages, not in holiday centres, holiday villages, youth hostels (self catering short-stay lodging)
25940	28740	Cotter pin (manufacture)
46210	51210	Cotton (wholesale)
46110	51110	Cotton broker (commission agent)
13100	17110	Cotton carding (manufacture)
13100	17110	Cotton combing (manufacture)
13300	17300	Cotton cord and velveteen finishing (manufacture)
13100	17110	Cotton doubling (manufacture)
13100	17110	Cotton drawing (manufacture)
13300	17300	Cotton fabric bleaching, dyeing or otherwise finishing (manufacture)
01630	01410	Cotton ginning

SIC 2007	SIC 2003	Activity
28940	29540	Cotton gins (manufacture)
01160	01110	Cotton growing
13100	17110	Cotton lap, sliver, rovings and other intermediate bobbin (manufacture)
10410	15410	Cotton linters production (manufacture)
13100	17110	Cotton opening (manufacture)
13923	17403	Cotton patch quilt (manufacture)
46770	51570	Cotton rags (wholesale)
13100	17110	Cotton reeling (manufacture)
10410	15410	Cotton seed crushing including delinting or cleaning (manufacture)
10410	15410	Cotton seed oil production (manufacture)
10410	15420	Cotton seed oil refining (manufacture)
46750	51550	Cotton size (wholesale)
13100	17110	Cotton sorting (manufacture)
13100	17110	Cotton spinning (manufacture)
28940	29540	Cotton spreaders (manufacture)
13100	17160	Cotton thread mill (manufacture)
52102	63129	Cotton warehouse for air transport activities
52103	63129	Cotton warehouse for land transport activities
52101	63129	Cotton warehouse for water transport activities
13100	17110	Cotton warp (manufacture)
46770	51570	Cotton waste (wholesale)
13300	17300	Cotton waste bleaching, dying or otherwise finishing (manufacture)
13100	17110	Cotton waste spinning (manufacture)
13200	17210	Cotton weaving (manufacture)
32500	24422	Cotton wool and tissues (manufacture)
13100	17110	Cotton yarn (manufacture)
13100	17110	Cotton yarn doubling (manufacture)
13300	17300	Cotton yarn gassing (manufacture)
13300	17300	Cotton yarn polishing (manufacture)
13300	17300	Cotton yarn printing (manufacture)
13100	17110	Cotton yarn twisting (manufacture)
13100	17110	Cotton yarn warping (manufacture)
13100	17110	Cotton yarn winding (manufacture)
13923	17403	Cotton, silk, etc. embroidering (except lace and apparel) (manufacture)
17110	21110	Cotton-linters pulp (manufacture)
01110	01110	Cottonseed growing
13200	17210	Cotton-type fabrics (manufacture)
13100	17110	Cotton-type yarn (manufacture)
31090	36110	Couch (manufacture)
85590	80429	Council for Accreditation of Correspondence Colleges
85421	80302	Council for National Academic Awards
13923	17403	Counterpane (manufacture)
31010	36120	Counters for shops (manufacture)
28230	30010	Counting and dating machines (manufacture)
26513	33202	Counting instruments (non-electric) (manufacture)
84230	75230	County court
29320	34300	Coupling for articulated motor vehicle (manufacture)
24200	27220	Couplings and flange adapters made of iron or steel (manufacture)
01130	01120	Courgette growing
79901	63303	Courier (travel)
53201	64120	Courier activities (other than national post activities) licensed
53202	64120	Courier activities (other than national post activities) unlicensed

SIC 2007	SIC 2003	Activity
93199	92629	Coursing
84230	75230	Court of appeal
84230	75230	Court of protection
84230	75230	Court of session (Scotland)
84230	75230	Court of the Lord Lyon
82990	74850	Court reporting services
10730	15850	Couscous (manufacture)
13922	17402	Covers made of waterproofed canvas (manufacture)
22230	25239	Coving made of plastic (manufacture)
01110	01120	Cow pea growing
01410	01210	Cows' milk (raw) production
28940	29540	Crabbing machinery for textiles (manufacture)
25300	28300	Cracker for process plant (manufacture)
94990	91330	Craft and collectors' clubs
47789	52489	Craftwork (retail)
25730	28620	Cramp (manufacture)
28220	29220	Crane (manufacture)
77320	71320	Crane hire (without operator)
25990	28750	Crane hook (manufacture)
77320	71320	Crane lorries rental (without operator)
29100	34100	Crane lorry (manufacture)
30200	35200	Crane vans (manufacture)
30920	35420	Crank wheel for pedal cycle (manufacture)
28150	29140	Cranks (manufacture)
28150	29140	Crankshaft (not for motor vehicle engine) (manufacture)
28110	34300	Crankshaft for motor vehicle (manufacture)
16240	20400	Crates made of wood (manufacture)
28290	29240	Crating and de-crating machinery (manufacture)
14190	18249	Cravats (manufacture)
14200	18300	Cravats made of fur (manufacture)
28922	29522	Crawler loader (manufacture)
28922	29522	Crawler tractor (manufacture)
32990	36631	Crayon (manufacture)
10511	15511	Cream (sterilised) (manufacture)
46330	51331	Cream (wholesale)
10511	15511	Cream from fresh homogenized liquid milk (manufacture)
10511	15511	Cream from fresh pasteurized liquid milk (manufacture)
10511	15511	Cream production (manufacture)
28930	29530	Cream separator for industrial use (manufacture)
11070	15980	Cream soda (manufacture)
10511	15511	Creamery (not farm or retail shop) (manufacture)
28990	29560	Creasing machine for bookbinding (manufacture)
73110	74402	Creating and placing advertising
73110	74402	Creation of stands and other display structures and sites
88910	85321	Crèche (charitable)
88910	85322	Crèche (non-charitable)
88990	85321	Credit and debt counselling services (charitable)
88990	85322	Credit and debt counselling services (non-charitable)
82912	74871	Credit bureau
64921	65221	Credit card issuer (sole activity – requiring full payment at end of credit period)
66190	67130	Credit or finance broking
82912	74871	Credit rating
47990	52630	Credit trader (retail)
64921	65221	Credit unions

C

SIC 2007	SIC 2003	Activity
96030	93030	Cremation board
96030	93030	Cremation services
96030	93030	Crematorium
20140	24140	Creosote (manufacture)
13200	17240	Crepe weaving (manufacture)
17120	21120	Creped paper (manufacture)
01130	01120	Cress growing
20140	24140	Cresylic acid (manufacture)
20160	24160	Cresylic resins (manufacture)
13200	17210	Cretonne weaving (manufacture)
32300	36400	Cricket ball and equipment (manufacture)
93120	92629	Cricket club
84240	75240	Criminal investigation department
17120	21120	Crinkled paper (manufacture)
10720	15820	Crispbread (manufacture)
14390	17720	Crocheted articles (manufacture)
13910	17600	Crocheted fabric (manufacture)
13910	17600	Crocheted fabrics (manufacture)
47599	52440	Crockery (retail)
77299	71409	Crockery hire
01610	01410	Crop and grass drying plant operation by contractor
01630	01410	Crop drying and disinfecting for primary market
01610	01410	Crop establishing for subsequent crop production, on a fee or contract basis
01500	01300	Crop growing in combination with farming of livestock
01610	01410	Crop harvesting and preparation
01630	01410	Crop preparation for primary market (cleaning, trimming, grading, etc.)
01610	01410	Crop production support activities
01610	01410	Crop spraying on a fee or contract basis
01610	01410	Crop treatment on a fee or contract basis
01630	01410	Crop wax covering, polishing and wrapping for primary market
28940	29540	Cropping machinery for textiles (manufacture)
93120	92629	Croquet club
20520	24620	Cross linking adhesive (manufacture)
32300	36400	Crossbows (manufacture)
26511	33201	Cross-talk meters (electronic) (manufacture)
99000	99000	Crown agents for overseas governments and administrations
25920	28720	Crown cork (manufacture)
46690	51870	Crown corks and stoppers (wholesale)
84230	75230	Crown court
84230	75230	Crown prosecution service
23200	26260	Crucibles made of fireclay (manufacture)
23200	26260	Crucibles made of fireclay or graphite (manufacture)
23200	26260	Crucibles made of graphite (manufacture)
23200	26260	Crucibles made of refractory ceramic (manufacture)
19100	23100	Crude benzole from coke ovens (manufacture)
19100	23100	Crude coal tar from coke ovens (manufacture)
19100	23100	Crude coal tar production (manufacture)
06200	11100	Crude gaseous hydrocarbon production (natural gas)
20411	24511	Crude glycerol (manufacture)
20150	24150	Crude natural phosphates (manufacture)
46711	51511	Crude oil (wholesale)
71122	74206	Crude oil exploration
06100	11100	Crude oil extraction
19201	23201	Crude oil refining (manufacture)
06100	11100	Crude oils obtained by decantation processes
06100	11100	Crude oils obtained by dehydration processes
06100	11100	Crude oils obtained by desalting processes
06100	11100	Crude oils obtained by stabilisation processes
46711	51511	Crude petroleum (wholesale)
06100	11100	Crude petroleum extraction
19201	23201	Crude petroleum jelly (at refinery) (manufacture)
06100	11100	Crude petroleum production
06100	11100	Crude petroleum production from bituminous shale and sand
10410	15410	Crude vegetable oil production (manufacture)
10710	15810	Crumpet making (manufacture)
94910	91310	Crusaders' union
20120	24120	Crushed pigment colours (manufacture)
28930	29530	Crusher (food or drink machinery) (manufacture)
08110	14110	Crushing and breaking of stone
28921	29521	Crushing machine for mining (manufacture)
28923	29523	Crushing plant (not for mines) (manufacture)
38320	37200	Crushing, cleaning and sorting of demolition waste to obtain secondary raw materials
38320	37200	Crushing, cleaning and sorting of glass
38320	37200	Crushing, cleaning and sorting of glass to produce secondary raw materials
10200	15209	Crustacean and mollusc canning (manufacture)
10200	15209	Crustacean and mollusc preservation by drying (manufacture)
10200	15209	Crustacean and mollusc products (manufacture)
10200	15209	Crustacean and mollusc salting (manufacture)
10200	15201	Crustacean freezing (manufacture)
10200	15209	Crustacean preservation (other than by freezing) (manufacture)
47230	52230	Crustaceans (retail)
46380	51380	Crustaceans (wholesale)
03210	05020	Crustaceans cultured in sea water
32500	33100	Crutches (manufacture)
23130	26130	Crystal articles (manufacture)
10822	15842	Crystallised fruit (manufacture)
28990	29560	Crystalliser for chemical industry (manufacture)
26600	33100	CT scanners (manufacture)
01130	01120	Cucumber growing
32401	36501	Cue for billiards or snooker (manufacture)
32130	36610	Cuff link (not of, or clad in, precious metal, or of precious/semi precious stones) (manufacture)
14190	18249	Cuffs for men and boys (manufacture)
23130	26130	Culinary glassware (manufacture)
46719	51519	Culm (wholesale)
28302	29320	Cultivator (manufacture)
28302	29320	Cultivator tine (manufacture)
94120	91120	Cultural associations
85520	80429	Cultural education
94120	91120	Cultural organisation (professional)
94990	91330	Cultural organisations (hobby)
84120	75120	Cultural services administration (public sector)
20140	24140	Cumene (manufacture)
31090	36140	Cupboard (manufacture)
31020	36130	Cupboards for kitchens (manufacture)
23410	26210	Cups and saucers made of china or porcelain (manufacture)
17220	21220	Cups made of paper (manufacture)
22290	25240	Cups made of plastic (manufacture)

SIC 2007	SIC 2003	Activity
10512	15512	Curd production (manufacture)
28940	29540	Curing machinery for textiles (manufacture)
47789	52489	Curios (retail)
27510	29710	Curlers (electric) (manufacture)
01250	01139	Currant growing
66120	67130	Currency broking
26511	33201	Current checking instruments (electronic) (manufacture)
10840	15870	Curry powder (manufacture)
15110	18300	Currying of fur skins and hides with the hair on (manufacture)
96010	93010	Curtain cleaning
96010	93010	Curtain cleaning (not lace dressing)
22290	25240	Curtain hooks, rings and runners made of plastic (manufacture)
13990	17542	Curtain loop (manufacture)
47530	52440	Curtain material (retail)
25720	28630	Curtain rail and runners made of metal (manufacture)
22290	25240	Curtain rail, rollers and fittings made of plastic (manufacture)
25120	28120	Curtain walling made of metal (manufacture)
13921	17401	Curtains (made-up) (manufacture)
47530	52440	Curtains (retail)
46160	51160	Curtains, drapes and interior blinds (commission agent)
13921	17401	Cushion covers (manufacture)
22190	25130	Cushioning for upholstery made of rubber (manufacture)
13921	17401	Cushions (manufacture)
10890	15899	Custard powder (manufacture)
66120	67122	Custodians and settlement services
62012	72220	Custom software development
14131	18221	Custom tailored outerwear for men and boys (manufacture)
14131	18221	Custom tailoring for men and boys (manufacture)
14132	18222	Custom tailoring for women and girls (manufacture)
84110	75110	Customs administration
52290	63400	Customs clearance agents activities
99000	99000	Customs co-operation council
01190	01120	Cut flowers and flower bud production
15200	19300	Cut soles for footwear (manufacture)
14190	18243	Cut, make and trim on a fee or contract basis (manufacture)
32990	18243	Cut, make, trim of fire-resistant and protective safety clothing, fee or contract basis (manufacture)
25710	28750	Cutlasses (manufacture)
46150	51150	Cutlery (commission agent)
25710	28610	Cutlery (electro plated nickel silver) (manufacture)
25710	28610	Cutlery (manufacture)
47599	52440	Cutlery (retail)
46499	51440	Cutlery (wholesale)
15120	19200	Cutlery case (not leather or plastics) (manufacture)
16290	20510	Cutlery case made of wood (manufacture)
25710	28610	Cutlery for domestic use (manufacture)
32990	36639	Cutlery handles made of horn, ivory, tortoise shell, etc. (manufacture)
77299	71409	Cutlery hire
22290	25240	Cutlery made of plastic (manufacture)
32120	36220	Cutlery made of precious metal (manufacture)
32500	33100	Cutter for dental use (manufacture)
28490	29430	Cutter for wood (manufacture)
25730	28620	Cutting blades for machines or mechanical appliances (manufacture)
28990	29560	Cutting machine for bookbinding (manufacture)
28410	29420	Cutting machine for metal (manufacture)
28940	29540	Cutting machine for textile fibres (manufacture)
28921	29521	Cutting machinery for coal or rock (manufacture)
28940	29540	Cutting machinery for textile fabrics (manufacture)
25620	28520	Cutting of metals by laser beam (manufacture)
19209	23209	Cutting oil (manufacture)
28490	29430	Cutting torch (manufacture)
18140	22230	Cutting, cover laying, gluing, collating books, brochures, magazines, catalogues etc. (manufacture)
23700	26700	Cutting, shaping and finishing of stone for use as roofing (manufacture)
23700	26700	Cutting, shaping and finishing of stone for use in cemeteries (manufacture)
23700	26700	Cutting, shaping and finishing of stone for use in construction (manufacture)
23700	26700	Cutting, shaping and finishing of stone for use on roads (manufacture)
46750	51550	Cyanides, cyanide oxides and complex cyanides (wholesale)
20520	24620	Cyanoacrylate adhesive (manufacture)
21100	24410	Cyclamates (manufacture)
47640	52485	Cycle accessories (retail)
47640	52485	Cycle agent (retail)
15120	19200	Cycle bags made of leather (manufacture)
93120	92629	Cycle club
30920	35420	Cycles (non-motorised) (manufacture)
46499	51479	Cycles (wholesale)
30920	35420	Cycles and parts (manufacture)
30910	35410	Cycles fitted with an auxiliary engine (manufacture)
20140	24140	Cyclic alcohols (manufacture)
20140	24140	Cyclic hydrocarbons (saturated and unsaturated) (manufacture)
46750	51550	Cyclic hydrocarbons (wholesale)
20140	24140	Cyclohexane (manufacture)
30920	35420	Cyclometer (manufacture)
28930	29530	Cyclone separators (manufacture)
27900	31620	Cyclotron (manufacture)
18130	22240	Cylinder engraving for gravure printing (manufacture)
18130	22240	Cylinder etching for gravure printing (manufacture)
28120	29122	Cylinder for hydraulic equipment (manufacture)
28120	29122	Cylinder for pneumatic control equipment (manufacture)
28110	29110	Cylinder heads for industrial engines (manufacture)
28110	29110	Cylinder inserts for industrial engines (manufacture)
28110	34300	Cylinder liner for motor vehicle (manufacture)
28110	29110	Cylinder liners for industrial engines (manufacture)
17290	21259	Cylinder made of hardened paper (manufacture)
17219	21219	Cylinders made of board (open ended for posting documents) (manufacture)
28150	29140	Cylindrical roller bearing (manufacture)
26511	33201	Cytometers (electronic) (manufacture)
10910	15710	Dairy concentrate (animal feed) (manufacture)
46610	51880	Dairy farm machinery (wholesale)
01410	01210	Dairy farming
47290	52270	Dairy grocer's shop (retail)
28930	29530	Dairy industry machinery (manufacture)

D

SIC 2007	SIC 2003	Activity
46690	51870	Dairy machinery (not farm) (wholesale)
28930	29530	Dairy machinery and plant (not agricultural) (manufacture)
77390	71340	Dairy machinery rental (non agricultural)
28930	29530	Dairy moulding machinery (manufacture)
10512	15512	Dairy preparation of cheese and butter (manufacture)
10511	15511	Dairy preparation of milk and cream (manufacture)
10519	15519	Dairy preparation of milk products n.e.c. (manufacture)
46330	51331	Dairy produce exporter (wholesale)
46330	51331	Dairy produce importer (wholesale)
46330	51331	Dairy produce n.e.c (wholesale)
47290	52270	Dairy products (retail)
47290	52270	Dairyman (retail)
42910	45240	Dam construction
66210	67200	Damage evaluators activities
13200	17210	Damask weaving (not woollen or worsted) (manufacture)
13200	17220	Damask woollen weaving (manufacture)
13200	17230	Damask worsted weaving (manufacture)
43999	45220	Damp proofing of buildings
90010	92311	Dance band
93290	92341	Dance hall
90010	92311	Dance productions
85520	92341	Dancing academy (ballroom)
85520	92341	Dancing master
85520	92341	Dancing school
85520	92341	Dancing schools and dance instructor activities
10832	15862	Dandelion coffee (manufacture)
26702	33403	Dark room equipment (manufacture)
32401	36501	Dart (manufacture)
32401	36501	Dartboard (manufacture)
29310	31610	Dashboard instruments (electric) (manufacture)
62012	72220	Data analysis consultancy services
62090	72220	Data archiving and backup services
60100	92201	Data broadcasting integrated with radio
60200	92202	Data broadcasting integrated with television
63110	72300	Data conversion
18130	22240	Data files preparation for multi-media printing on CD- ROM (manufacture)
18130	22240	Data files preparation for multi-media printing on internet applications (manufacture)
18130	22240	Data files preparation for multi-media printing on paper (manufacture)
61100	64200	Data network management and support services (wired telecommunications)
63110	72300	Data preparation services
63110	72300	Data processing
26200	30020	Data processing equipment (electronic (other than electronic calculators)) (manufacture)
28230	30010	Data processing equipment (non-electronic) (manufacture)
63110	72400	Data storage services
61100	64200	Data transmission (via cables, broadcasting, relay or satellite)
26301	32201	Data transmission link line (manufacture)
63110	72400	Database running activities
62012	72400	Database structure and content design
32990	36631	Date sealing stamps (manufacture)
32990	36631	Date stamp and accessories (manufacture)
01220	01139	Dates growing
96090	93059	Dating services
88910	85321	Day care for disabled children (charitable)
88910	85322	Day care for disabled children (non-charitable)
88100	85321	Day centres for the elderly, the physically or the mentally ill (charitable)
88100	85322	Day centres for the elderly, the physically or the mentally ill (non-charitable)
85590	80429	Day continuation school
88910	85321	Day nursery (charitable)
88910	85322	Day nursery (non-charitable)
28290	29240	De-aeration plant for water treatment (manufacture)
64991	65233	Dealer in stocks and shares on own account
66120	67122	Dealing in finance markets for others (e.g. Stock broking), related activities (not fund management)
66120	67122	Dealing in securities on behalf of others
82911	74871	Debt collector
64992	65222	Debt purchasing
10832	15862	De-caffeinated coffee (manufacture)
18129	22220	Decal printing (manufacture)
28940	29540	Decatising machinery for textiles (manufacture)
28990	35300	Deck arresters for aircraft (manufacture)
31090	36110	Deck chairs made of wood (manufacture)
30110	35110	Deck for oil platform (manufacture)
30110	35110	Decking for ships (manufacture)
28990	29560	Decompression chamber (manufacture)
39000	90030	Decontamination and cleaning up of surface water following accidental pollution
39000	90030	Decontamination of industrial plants or sites
39000	90030	Decontamination of nuclear plants and sites
39000	90030	Decontamination of soil and groundwater
39000	90030	Decontamination of soils and groundwater at the place of pollution using biological methods
39000	90030	Decontamination of soils and groundwater at the place of pollution using chemical methods
39000	90030	Decontamination of soils and groundwater at the place of pollution using mechanical methods
39000	90030	Decontamination of surface water
81222	74704	Decontamination services
23700	26700	Decorated building stone (manufacture)
43341	45440	Decorating of buildings
31010	35500	Decorative restaurant carts, dessert cart, food wagons (manufacture)
23310	26300	Decorative tile made of glazed earthenware (manufacture)
22210	25210	Decorative unsupported polyvinyl chloride film and sheet (manufacture)
20520	24620	Decorators' size (manufacture)
61100	64200	Dedicated business telephone network services (wired telecommunications)
25990	28750	Deed box (manufacture)
69109	74119	Deeds preparation
05101	10101	Deep coal mines
27510	29710	Deep freeze unit for domestic use (manufacture)
84220	75220	Defence activities
29310	31610	Defrosting and demisting equipment for vehicles (manufacture)
13100	17120	Degreasing and carbonising of wool (manufacture)
43999	45250	De-humidification of buildings
10390	15330	Dehydrating fruit for human consumption (manufacture)

SIC 2007	SIC 2003	Activity	SIC 2007	SIC 2003	Activity
10390	15330	Dehydrating of vegetables for human consumption (manufacture)	47190	52120	Department stores (retail)
			20420	24520	Depilatory (manufacture)
30300	35300	De-icing equipment for aircraft (manufacture)	46750	51550	Depleted uranium and thorium and their compounds (wholesale)
20590	24660	De-icing fluid (manufacture)			
26301	31620	Delay lines and networks (manufacture)	66190	67130	Deposit broker
47290	52270	Delicatessen shop (retail)	09100	11200	Derrick erection in situ, repairing and dismantling
22190	25130	Delivery hose made of rubber (manufacture)	28220	29220	Derricks (manufacture)
28131	29121	Delivery pump (manufacture)	46690	51870	Derricks, cranes, mobile lifting frames (wholesale)
30920	35420	Delivery tricycles (non-motorised) (manufacture)	19201	23201	Derv (manufacture)
29310	31610	Demisters (electrical) (manufacture)	28290	29240	Desalination plant for water treatment (manufacture)
43110	45110	Demolition contracting	36000	41000	Desalting of sea or ground water to produce water
43999	45500	Demolition equipment rental with operator	20590	24660	Desiccants (chemical) (manufacture)
43110	45110	Demolition or wrecking of buildings and other structures	23190	26150	Desiccator made of glass (manufacture)
20140	15920	Denatured ethyl alcohol (manufacture)	71129	74204	Design consultant for civil and structural engineering
13200	17210	Denim weaving (manufacture)	71121	74205	Design consultant for industrial process and production
16210	20200	Densified wood (manufacture)			
26511	33201	Density measuring optical equipment (electronic) (manufacture)	71129	74204	Design office for civil and structural engineering
			71121	74205	Design office for industrial process and production
26513	33202	Density measuring optical equipment (non-electronic) (manufacture)	90030	92319	Designing (artistic)
			62011	72400	Designing of structure and content of an interactive leisure and entertainment software database
86230	85130	Dental activities in operating rooms			
32500	33100	Dental brush (manufacture)	62012	72400	Designing of structure and content of business and domestic software databases
32500	24422	Dental cement (manufacture)			
32500	33100	Dental chair (manufacture)	28940	29540	De-sizing machinery for textiles (manufacture)
20420	24520	Dental cleansing preparation (manufacture)	25990	28750	Desk tray made of metal (manufacture)
86230	85130	Dental clinic	31010	36120	Desks (manufacture)
86230	85130	Dental clinic (health service)	26200	30020	Desktop computers (manufacture)
85421	80302	Dental college or school	10519	15519	Desserts with a milk base (manufacture)
32500	33100	Dental drill engines (manufacture)	27900	31620	Detection apparatus (manufacture)
32500	24422	Dental filling (manufacture)	84230	75230	Detention centres
86101	85112	Dental hospital (private sector)	20411	24511	Detergent (soapless, formulated) (manufacture)
86101	85111	Dental hospital (public sector)	20411	24511	Detergent (synthetic) (manufacture)
86900	85140	Dental hygienist	20510	24610	Detonating fuse (manufacture)
32500	33100	Dental instrument (manufacture)	20510	24610	Detonator (manufacture)
32500	29240	Dental laboratory furnaces (manufacture)	41100	70110	Developing building projects for commercial buildings hotels, stores, shopping malls, restaurants
32500	33100	Dental laboratory instruments and equipment (manufacture)			
			43120	45110	Development and preparation of mineral properties and sites
32500	33100	Dental mirror (manufacture)			
86230	85130	Dental practice activities	64303	65235	Development capital company
86230	85130	Dental receptionist	68209	70209	Development for building projects for own operation
84120	75120	Dental service administration (public sector)	41100	70110	Development of building projects for residential buildings
86230	85130	Dental surgeon (not employed full time by a hospital)			
			32990	36639	Devotional article (manufacture)
32500	33100	Dental surgical instruments and equipment (manufacture)	10110	15113	De-woolling (manufacture)
			10620	15620	Dextrin (manufacture)
86900	85140	Dental therapist	20520	24620	Dextrin based adhesive (manufacture)
20590	24660	Dental wax (manufacture)	10620	15620	Dextrose (manufacture)
32500	24422	Dental wax and other dental plaster preparations (manufacture)	10860	15880	Diabetic food (manufacture)
			21200	24421	Diagnostic reagents (manufacture)
46120	51120	Dental wax and other preparations for use in dentistry with a basis of plaster (commission agent)	46180	51180	Diagnostic reagents and other pharmaceutical products (commission agent)
			46460	51460	Diagnostic reagents n.e.c. (wholesale)
20420	24520	Dentifrices (manufacture)	26301	32201	Dial for telephone (manufacture)
86230	85130	Dentist	61900	64200	Dial-up internet access provision
32500	33100	Denture (manufacture)	28290	29240	Dialysis plant for water treatment (manufacture)
20420	24520	Denture fixative preparations (manufacture)	46180	51180	Diamond broker (commission agent)
20420	24520	Deodorant (manufacture)	32120	36220	Diamond cutting (manufacture)
20412	24512	Deodoriser for household use (manufacture)	43999	45250	Diamond drilling of concrete and asphalt
20412	24512	Deodorisers (manufacture)	23910	26810	Diamond impregnated disc and wheel (manufacture)
24420	27420	Deoxidiser made of aluminium (manufacture)	25730	28620	Diamond tipped tool (manufacture)

D

SIC 2007	SIC 2003	Activity
32120	36220	Diamond working (manufacture)
96010	93010	Diaper supply services
28131	29121	Diaphragm pump (manufacture)
28140	29130	Diaphragm valves (manufacture)
18129	22220	Diary printing (manufacture)
32500	33100	Diathermy apparatus (manufacture)
08990	14500	Diatomite bed
28230	32300	Dictating machines (manufacture)
58110	22110	Dictionary publishing
25730	28620	Die (press tool) (manufacture)
28910	29510	Die casting machines for foundries (manufacture)
24530	27530	Die casting of aluminium (manufacture)
24540	27540	Die casting of copper or copper alloy (manufacture)
24540	27540	Die casting of non-ferrous base metal (manufacture)
25730	28620	Die for machine tools (manufacture)
25500	28400	Die forging of ferrous metals (manufacture)
25730	28620	Die pellet (manufacture)
18130	22250	Die sinking of stationery (manufacture)
18140	22230	Die sinking or stamping finishing activities (manufacture)
28410	29420	Die stamping machines (metal forming) (manufacture)
18130	22250	Die stamping of stationery (manufacture)
28210	29210	Dielectric heating equipment (manufacture)
30200	35200	Diesel electric locomotive (manufacture)
28110	29110	Diesel engines for industrial use (manufacture)
46711	51511	Diesel fuel (wholesale)
30200	35200	Diesel locomotive (manufacture)
19201	23201	Diesel oil (manufacture)
10860	15880	Dietary foods for special medical purposes (manufacture)
10860	15880	Dietetic food (excluding milk based) (manufacture)
10860	15880	Dietetic food with a milk base (manufacture)
46380	51380	Dietetic foods (wholesale)
20140	24140	Diethyl phenylamine diamine sulphate (chlorine tablets) (manufacture)
29320	34300	Differential unit for motor vehicle (manufacture)
26511	33201	Diffraction apparatus (electronic) (manufacture)
28302	29320	Digger (elevator and shaker) (manufacture)
26200	30020	Digital computer (manufacture)
18130	22240	Digital imposition (manufacture)
26200	30020	Digital machines (manufacture)
71122	74206	Digital mapping activities
26702	32300	Digital photographic cameras (manufacture)
71122	74206	Dimensional survey activities
30120	35120	Dinghy made of rubber (manufacture)
31090	36110	Dining chair (manufacture)
56102	55302	Dining room (unlicensed)
31090	36140	Dining table (manufacture)
32120	36220	Dinnerware flatware hollowware of base metals clad with precious metals (manufacture)
28921	29521	Dinting machine for mining (manufacture)
26110	32100	Diode (manufacture)
46520	51860	Diodes, transistors, thyristors, diacs and triacs (wholesale)
46120	51120	Diols, polyalcohols, cyclical alcohols and their derivatives (commission agent)
46750	51550	Diols, polyalcohols, cyclical alcohols and their derivatives (wholesale)
99000	99000	Diplomatic missions

SIC 2007	SIC 2003	Activity
28210	29210	Direct arc furnace (manufacture)
27110	31100	Direct current (dc) generator sets (manufacture)
27110	31100	Direct current (dc) motors and generators (manufacture)
20120	24120	Direct dye (manufacture)
82190	74850	Direct mailing
47990	52630	Direct selling of firewood to the customers premises (retail)
47990	52630	Direct selling of fuel to the customers premises (retail)
47990	52630	Direct selling of heating oil to the customers premises (retail)
25400	29600	Directed energy weapons (manufacture)
46690	51870	Direction finding compasses and other navigational instruments and appliances (wholesale)
46520	51860	Direction finding compasses, other navigational instruments and appliances (electronic) (wholesale)
90020	92349	Direction, production and support activities to circus performances
09100	11200	Directional drilling services
24100	27100	Directly reduced iron (manufacture)
58120	22110	Directories and compilations (in print) publishing
90020	92311	Directors (theatre)
18129	22220	Directory printing (manufacture)
30300	35300	Dirigibles (manufacture)
93199	92629	Dirt track racing
86900	85140	Disablement services centres
88990	85321	Disaster relief organisations (charitable)
88990	85322	Disaster relief organisations (non-charitable)
29320	34300	Disc brakes (manufacture)
28490	29430	Disc cutting machines for stone, ceramic, asbestos cement and similar (not portable) (manufacture)
28240	29410	Disc cutting machines for stone, ceramics, asbestos-cement or similar (portable) (manufacture)
28302	29320	Disc harrow (manufacture)
26702	33403	Discharge lamp (electronic) and other flashlight apparatus (manufacture)
27400	31500	Discharge lamp (manufacture)
87900	85311	Discharged prisoners' hostel (charitable)
87900	85312	Discharged prisoners' hostel (non-charitable)
56301	55401	Discotheques (licensed to sell alcohol)
64992	65222	Discount company (e.g. Debt factoring)
64191	65121	Discount houses (monetary intermediation)
32200	36300	Discs for automatic mechanical instruments (manufacture)
24420	27420	Discs made of aluminium (manufacture)
23910	26810	Discs made of bonded abrasives (manufacture)
24440	27440	Discs made of brass (manufacture)
17290	21259	Discs made of cardboard (manufacture)
24440	27440	Discs made of copper (manufacture)
28290	29240	Dish washing machine parts (industrial type) (manufacture)
46690	51870	Dish washing machines for commercial use (wholesale)
46439	51439	Dish washing machines for domestic use (wholesale)
13923	17403	Dish-cloths and similar articles (manufacture)
17220	21220	Dishes made of paper (manufacture)
22290	25240	Dishes made of plastic (manufacture)
16290	20510	Dishes made of wood (manufacture)
27510	29710	Dishwasher for domestic use (manufacture)
28290	29240	Dishwashing machine for commercial catering (manufacture)

SIC 2007	SIC 2003	Activity
20411	24511	Dish-washing preparations (manufacture)
20200	24200	Disinfectant (manufacture)
46120	51120	Disinfectants (commission agent)
20200	24200	Disinfectants for agricultural and other use (manufacture)
81291	74703	Disinfecting of dwellings and other buildings
38310	37100	Dismantling of automobile wrecks for materials recovery
46770	51570	Dismantling of automobiles, computers, and other equipment for re-sale of usable parts (wholesale)
38310	37100	Dismantling of computers for materials recovery
33200	28520	Dismantling of large-scale machinery and equipment (manufacture)
38310	37100	Dismantling of ship wrecks for materials recovery
38310	37100	Dismantling of televisions for materials recovery
38310	37100	Dismantling of wrecks for materials recovery
47730	52310	Dispensing chemist (retail)
47782	52487	Dispensing ophthalmic optician (retail)
47782	52487	Dispensing optician (retail)
47782	52487	Dispensing optometrist (retail)
20120	24120	Disperse dye (manufacture)
20160	24160	Dispersions of synthetic resin (manufacture)
28250	29230	Display cabinet (refrigerated) (manufacture)
31090	36140	Display cabinet for domestic use (manufacture)
31010	36120	Display cases for shops (manufacture)
26110	32100	Display components (plasma, polymer, LCD) (manufacture)
16230	20300	Display stand (manufacture)
17220	21220	Disposable baby napkins made of paper or cellulose wadding (manufacture)
17220	21220	Disposable bed linen made of paper or cellulose wadding (manufacture)
14190	18249	Disposable clothing (manufacture)
38210	90020	Disposal of non-hazardous waste by combustion or incineration or other methods
38220	23300	Disposal of nuclear waste
38220	90020	Disposal of sick or dead animals (toxic)
38220	90020	Disposal of used goods such as refrigerators to eliminate harmful waste
32500	33100	Dissecting instrument (manufacture)
17110	21110	Dissolving chemical wood pulp (manufacture)
20301	24301	Distemper (manufacture)
32910	36620	Distemper brush (manufacture)
46120	51120	Distilled water (commission agent)
20130	24130	Distilled water (manufacture)
46750	51550	Distilled water (wholesale)
20140	15920	Distillery draft production (manufacture)
28290	29240	Distilling machinery for potable spirits (manufacture)
28290	29240	Distilling or rectifying plant (manufacture)
46690	51870	Distilling or rectifying plant (wholesale)
28290	29240	Distilling or rectifying plant for beverage industries (manufacture)
28290	29240	Distilling or rectifying plant for chemical industries (manufacture)
28290	29240	Distilling or rectifying plant for petroleum refineries (manufacture)
35220	40220	Distribution and supply of gaseous fuels of all kinds through a system of mains
36000	41000	Distribution of water through mains, by trucks or other means
84130	75130	Distribution services administration and regulation (public sector)
86210	85120	District community physician
86900	85140	District nurse
31030	36150	Divan bed (mattress and mattress support) (manufacture)
14132	18222	Divided lightweight skirt (dress made) (manufacture)
28490	28620	Dividing heads and other special attachments for machine tools (manufacture)
52220	63220	Diving contracting (non leisure)
28990	29560	Diving equipment (excluding breathing apparatus) (manufacture)
09100	11200	Diving services incidental to oil and gas exploration
22190	25130	Diving suit made of rubber (manufacture)
47520	52460	DIY equipment (retail)
47520	52460	DIY materials (retail)
28940	29540	Dobbies (textile machinery) (manufacture)
52220	63220	Dock authority
28220	29220	Dock leveller (manufacture)
28220	29220	Dockside cranes (manufacture)
86210	85120	Doctor (unspecified)
86210	85120	Doctors receptionist
28230	30010	Document copying equipment (manufacture)
82190	74850	Document copying service
28230	30010	Document handling machine (manufacture)
82190	74850	Document preparation
28230	30010	Document shredder (manufacture)
18129	22220	Documents of title printing (manufacture)
10920	15720	Dog biscuit (manufacture)
93199	92629	Dog breeding (for greyhound racing)
10920	15720	Dog food (manufacture)
15120	19200	Dog lead made of leather (manufacture)
93199	92629	Dog racing
20411	24511	Dog soap (manufacture)
80100	74602	Dog training for security purposes
96090	93059	Dogs' home
17220	21220	Doilies made of paper (manufacture)
22290	25240	Doilies made of plastic (manufacture)
13923	17403	Doilies made of textiles (manufacture)
77299	71401	Do-it-yourself machinery and equipment hire
32409	36509	Dolls (manufacture)
46499	51477	Dolls (wholesale)
32409	36509	Dolls and doll garments, parts and accessories (manufacture)
46499	51477	Dolls' carriages (wholesale)
32409	36509	Dolls' clothes (manufacture)
32409	36509	Dolls' cots (manufacture)
32409	36509	Dolls' houses (manufacture)
32409	36509	Dolls made of rubber (manufacture)
32409	36509	Dolls' prams (manufacture)
23700	26700	Dolomite (ground) (manufacture)
22230	25239	Dome lights made of plastic (manufacture)
78200	74500	Domestic agency (temporary employment agency)
46150	51150	Domestic electrical appliances (commission agent)
47540	52450	Domestic electrical appliances (retail)
47599	52440	Domestic furniture (retail)
25990	28750	Domestic hollow ware made of metal (manufacture)
46499	51479	Domestic ironmongery (wholesale)
46439	51439	Domestic machinery (wholesale)
27520	29720	Domestic non-electric cooking equipment (manufacture)

SIC 2007	SIC 2003	Activity	SIC 2007	SIC 2003	Activity
27520	29720	Domestic non-electric cooking ranges (manufacture)	46410	51410	Draper (wholesale)
27520	29720	Domestic non-electric heating equipment (manufacture)	96010	93010	Drapery cleaning
			93120	92629	Draughts clubs
25990	28750	Domestic utensils made of aluminium (manufacture)	32409	36509	Draughts set (manufacture)
16290	20510	Domestic woodware (manufacture)	71121	74205	Draughtsman for industrial process and production
93120	92629	Domino clubs	28410	29420	Draw bench (metal forming) (manufacture)
30990	35500	Donkey-carts (manufacture)	25730	28620	Draw knife (manufacture)
25720	28630	Door and window catches (manufacture)	31010	36120	Drawers (non-domestic) (manufacture)
25720	28630	Door fittings made of metal (manufacture)	20590	24660	Drawing ink (manufacture)
25120	28120	Door frames made of metal (manufacture)	26513	33202	Drawing instrument (manufacture)
22230	25239	Door frames made of plastic (manufacture)	16290	20510	Drawing instruments case made of wood (not containing instruments) (manufacture)
16230	20300	Door frames made of wood (manufacture)			
22230	25239	Door furniture for buildings (handles, hinges, knobs, etc.) Made of plastic (manufacture)	28410	29420	Drawing machine (metal forming) (manufacture)
			28940	29540	Drawing machinery for textiles (manufacture)
25720	28630	Door hardware for buildings, furniture and vehicles (manufacture)	17120	21120	Drawing paper (manufacture)
			25930	28730	Drawing pin (manufacture)
25120	28120	Doors (other than safe doors) made of metal (manufacture)	24420	27420	Drawn products made of aluminium (manufacture)
			24440	27440	Drawn products made of copper (manufacture)
47520	52460	Doors (retail)	23110	26110	Drawn sheet glass (manufacture)
46730	51530	Doors (wholesale)	30110	35110	Dredger (manufacture)
30300	35300	Doors for aircraft (manufacture)	42910	45240	Dredging contractor
29320	34300	Doors for motor vehicles (manufacture)	42910	45240	Dredging for water projects
22230	25239	Doors made of plastic (manufacture)	42910	45240	Dredging of waterways
16230	20300	Doors made of wood (manufacture)	14132	18222	Dress and jacket knitted ensemble (manufacture)
47990	52630	Door-to-door sales (retail)	14190	18249	Dress belts (not made of leather or leather substitute) (manufacture)
20590	24660	Doped compounds for use in electronics (manufacture)			
			13990	17542	Dress binding (manufacture)
28131	29121	Dosing and proportioning pump (manufacture)	13200	17210	Dress fabric (woven (not wool)) (manufacture)
28290	29240	Dosing plant for water treatment (manufacture)	14190	18249	Dress gloves made of fabric (manufacture)
32200	36300	Double bass (manufacture)	13200	17220	Dress goods woollen weaving (manufacture)
10511	15511	Double cream (manufacture)	13200	17230	Dress goods worsted weaving (manufacture)
47520	52460	Double glazing (retail)	47510	52410	Dress materials (retail)
25120	28120	Double glazing made of metal (manufacture)	14190	18249	Dress shield (manufacture)
22230	25239	Double glazing made of plastic (manufacture)	47710	52423	Dress shop (retail)
23140	26140	Doubled glass fibre (manufacture)	13100	17140	Dressed line made of flax (manufacture)
23140	26140	Doubled yarn made of glass fibre (manufacture)	31020	36130	Dressers for kitchens (manufacture)
28940	29540	Doubling machinery for textiles (manufacture)	14132	18222	Dresses for women and girls (manufacture)
28930	29530	Dough dividers (manufacture)	15110	18300	Dressing and dying of furskins and hides with the hair on (manufacture)
28930	29530	Dough making machinery (manufacture)			
16290	20510	Dowel pin (manufacture)	15120	19200	Dressing case made of leather or leather substitute (manufacture)
13100	17170	Down of vegetable origin (manufacture)			
10120	15120	Down production (manufacture)	13990	17542	Dressing gown cord and girdle (manufacture)
74202	74813	Downhole photography services	14142	18232	Dressing gown for women and girls (manufacture)
09100	11200	Downhole-fishing services	14141	18231	Dressing gowns for men and boys (manufacture)
09100	11200	Downhole-milling services	15110	19100	Dressing of leather (manufacture)
13200	17210	Downproof cloth weaving (manufacture)	13300	17300	Dressing of textiles and textile articles, including wearing apparel (manufacture)
26513	33202	Drafting tables and machines (manufacture)			
46690	51870	Drafting tables and other drawing, marking out or mathematical calculating instruments (wholesale)	31090	36140	Dressing table (manufacture)
			47710	52423	Dressmaker (retail)
93199	92629	Drag hounds	14132	18222	Dressmaking (manufacture)
28220	29220	Drag scraper (manufacture)	10890	15899	Dried egg (manufacture)
28922	29522	Dragline excavator (manufacture)	46380	51380	Dried fish (wholesale)
43120	45110	Drainage of agricultural or forestry land	01190	01120	Dried flower production
09100	11200	Draining and pumping services incidental to oil and gas extraction, on a fee or contract basis	10390	15330	Dried fruit (except field dried) (manufacture)
			46380	51380	Dried fruit (wholesale)
46740	51540	Drainpipes (wholesale)	10390	15330	Dried fruit cleaning (manufacture)
23320	26400	Drainpipes and fittings made of clay (manufacture)	10840	15870	Dried herbs (except field dried) (manufacture)
22210	25210	Drainpipes and fittings made of plastic (manufacture)	01110	01110	Dried leguminous vegetables growing
37000	90010	Drains maintenance	10519	15519	Dried milk (manufacture)
47510	52410	Draper (retail)			

SIC 2007	SIC 2003	Activity	SIC 2007	SIC 2003	Activity
10390	15330	Dried vegetables (except field dried) (manufacture)	28930	29530	Dryers for agriculture (manufacture)
10130	15139	Dried, salted or smoked meat (manufacture)	28990	29560	Dryers for wood, paper pulp, paper or paperboard (manufacture)
28240	29410	Drill (powered portable) (manufacture)	46690	51870	Dryers for wood, paper pulp, paper or paperboard (wholesale)
28921	28620	Drill bits for well drilling (manufacture)			
28302	29320	Drill for agricultural use (manufacture)	28990	29560	Drying machine for the chemical industry (manufacture)
24200	27220	Drill pipe made of steel (manufacture)			
25730	28620	Drill tools (interchangeable) (manufacture)	28940	29540	Drying machinery (commercial) for textiles (manufacture)
13200	17210	Drill weaving from yarn spun on the cotton system (manufacture)	28940	29540	Drying machines for laundries or dry cleaners (manufacture)
09100	11200	Drilling contractor for offshore oil or gas well	46690	51870	Drying machines with a capacity exceeding 10 kgs (wholesale)
28921	29521	Drilling jars for mining (manufacture)			
28410	29420	Drilling machine (metal cutting) (manufacture)	13300	17300	Drying of textiles and textile articles, including wearing apparel (manufacture)
20590	24660	Drilling mud (manufacture)	16100	20100	Drying of timber (manufacture)
28490	29430	Drilling or milling machines for stone, ceramics, asbestos-cement and similar articles (manufacture)	43999	45250	Drying out of buildings (incl. Water damage)
			13923	17403	Duchesse set (manufacture)
09100	11200	Drilling services to oil and gas extraction wells	10120	15120	Duck (fresh, chilled or frozen) slaughter and dressing (manufacture)
30110	35110	Drilling ship (manufacture)			
28240	29410	Drills and hammer drills (manufacture)	01470	01240	Duck farming
28930	29530	Drink processing including combined processing and packaging or bottling machinery (manufacture)	01470	01240	Duck raising and breeding
			13200	17210	Duck weaving (manufacture)
10821	15841	Drinking chocolate (manufacture)	25300	28300	Duct of heavy steel plate (manufacture)
23130	26130	Drinking glass (manufacture)	22230	25239	Ducting made of plastic (manufacture)
11010	15910	Drinks mixed with distilled alcoholic beverages (manufacture)	18130	22250	Dummies for presentation (manufacture)
10110	15112	Dripping (manufacture)	29100	34100	Dump truck (manufacture)
29320	34300	Drive shaft for motor vehicles (manufacture)	28922	34100	Dumpers for off road use (manufacture)
15120	19200	Driving belts made of leather (manufacture)	14120	18210	Dungarees (manufacture)
28150	29140	Driving elements (manufacture)	20590	24660	Duplicating ink (manufacture)
85530	80410	Driving instruction	28230	30010	Duplicating machines (excluding copiers) (manufacture)
85530	80410	Driving school activities			
85320	80220	Driving schools for occupational drivers e.g. Of trucks, buses, coaches	77330	71330	Duplicating machines rental and operating leasing
			17230	21230	Duplicating paper (cut to size) (manufacture)
25500	28400	Drop forging of ferrous metals (manufacture)	82190	74850	Duplicating service
28410	29420	Drop hammers (manufacture)	17230	21230	Duplicator stencils ready for use (manufacture)
25500	28400	Drop stamping of base non-ferrous metals (manufacture)	13923	17403	Dust cloths (manufacture)
			28250	29230	Dust collector for air conditioning equipment (manufacture)
25500	28400	Drop stamping of ferrous metals (manufacture)			
01629	01429	Droving services	13923	17403	Dust sheet (manufacture)
21200	24421	Drug (medicinal) (manufacture)	24430	27430	Dust, powder and flakes made of zinc (manufacture)
01280	01110	Drug and narcotic crops growing	25990	28750	Dustbins made of metal (manufacture)
47730	52310	Drug store (retail)	22290	25240	Dustbins made of plastic (manufacture)
47730	52310	Druggist (retail)	13923	17403	Duster (cleaning cloth (not of bonded fibre fabric)) (manufacture)
46460	51460	Druggists' sundries (wholesale)			
46460	51460	Druggists' sundriesman (wholesale)	38110	90020	Dustman
47730	52310	Drugs (retail)	25990	28750	Dustpan made of metal (manufacture)
46460	51460	Drugs (wholesale)	22290	25240	Dustpans made of plastic (manufacture)
32200	36300	Drum (musical instrument) (manufacture)	84110	75110	Duty and tax collection
22220	25220	Drums (containers) made of plastic (manufacture)	13923	17403	Duvet (manufacture)
16240	20400	Drums and similar packings made of wood (manufacture)	59132	92120	Dvd distribution
			47430	52450	Dvd players (retail)
25920	28720	Drums made of aluminium (manufacture)	26400	32300	Dvd recorders and players (manufacture)
25910	28710	Drums made of iron or steel (manufacture)	77220	71405	Dvd rental
10720	15820	Dry bakery products (manufacture)	46431	51431	Dvds (recorded) (wholesale)
27200	31400	Dry battery (non-rechargeable) (manufacture)	47630	52450	Dvds (retail)
96010	93010	Dry cleaner	68100	70120	Dwellings buying and selling
28940	29540	Dry cleaning machinery (manufacture)	68209	70209	Dwellings letting
46690	51870	Dry cleaning machines and laundry type washing machines (wholesale)	20120	24120	Dye (manufacture)
			28940	29540	Dye cycle controller (textile machinery) (manufacture)
42910	45240	Dry dock construction			
27510	29710	Dryers (electric) (manufacture)	15110	18300	Dyed lamb including beaver lamb (manufacture)

SIC 2007	SIC 2003	Activity	SIC 2007	SIC 2003	Activity
28940	29540	Dyeing machinery for textiles (manufacture)	01250	01139	Edible nuts growing
13100	17120	Dyeing of wool fleece (manufacture)	10110	15112	Edible offal (processed) production (manufacture)
28230	30010	Dyeline copying machine (manufacture)	28930	29530	Edible oil and fat processing machinery (manufacture)
96010	93010	Dyer and cleaner			
46750	51550	Dyes (wholesale)	46330	51333	Edible oils and fats (wholesale)
20120	24120	Dyes and pigments from any source in basic or concentrated forms (manufacture)	46330	51333	Edible oils and fats exporter (wholesale)
			46330	51333	Edible oils and fats importer (wholesale)
20120	24120	Dyes for food, drink and cosmetics (manufacture)	46330	51333	Edible oils and fats of animal or vegetable origin (wholesale)
20120	24120	Dyes modified for dying acrylic fibres (manufacture)	03210	05020	Edible seaweed growing
28940	29540	Dyesprings (textile machinery accessory) (manufacture)	10110	15112	Edible tallow production (manufacture)
			96090	93059	Educational agency
13300	17300	Dyework (manufacture)	94120	91120	Educational association
42910	45240	Dyke construction	85600	74149	Educational consulting
42910	45240	Dykes and static barrages construction	85600	74149	Educational guidance counselling activities
20510	24610	Dynamite (manufacture)	17230	22220	Educational notebooks (manufacture)
27110	31100	Dynamo (not for vehicle) (manufacture)	72200	73200	Educational research and experimental development
29310	31610	Dynamo for vehicle (manufacture)	84120	75120	Educational services administration (public sector)
29310	31610	Dynamo lighting set for cycles (manufacture)	17230	22220	Educational stationery (manufacture)
32990	36400	Ear and noise plugs (e.g. for swimming and noise protection) (manufacture)	17230	22220	Educational stationery binders (manufacture)
			17230	22220	Educational stationery business forms (manufacture)
86101	85112	Ear, nose and throat hospital (private sector)	17230	22220	Educational stationery registers
86101	85111	Ear, nose and throat hospital (public sector)	46900	51900	Educational supplies (except furniture) (wholesale)
86220	85120	Ear, nose and throat specialist (private practice)	85600	74149	Educational support activities
86101	85111	Ear, nose and throat specialist (public sector)	85600	74149	Educational testing activities
26400	32300	Earphone (manufacture)	85600	74149	Educational testing evaluation activities
28921	29521	Earth boring machine (manufacture)	02300	02010	Eel grass gathering
08910	14300	Earth colours and fluorspar mining	56103	55303	Eel pie shop
28922	29522	Earth leveller (manufacture)	47230	52230	Eels (retail)
28922	29522	Earth mover (construction equipment) (manufacture)	46380	51380	Eels (wholesale)
43999	45500	Earth moving equipment rental with operator	28290	29240	Effluent treatment plant (manufacture)
43120	45110	Earth moving excavation	16240	20400	Egg box made of wood (manufacture)
28922	29522	Earth moving machinery (manufacture)	25990	28750	Egg boxes made of metal (manufacture)
72190	73100	Earth sciences research and experimental development (other than biotechnological)	17290	21259	Egg boxes made of paper (manufacture)
			22220	25220	Egg boxes made of plastic (manufacture)
47599	52440	Earthenware (retail)	28302	29320	Egg cleaning, sorting and grading machines (manufacture)
46499	51440	Earthenware (wholesale)			
23410	26210	Earthenware for domestic use (manufacture)	16290	20510	Egg cup made of wood (manufacture)
43120	45110	Earthmoving contractor	22290	25240	Egg cups made of plastic (manufacture)
77320	71320	Earthmoving equipment hire (without operator)	10890	15899	Egg drying (manufacture)
31010	36120	Easel (manufacture)	46330	51332	Egg grading and packing (wholesale)
22190	25130	Ebonite, vulcanite or hard rubber goods (manufacture)	01470	01240	Egg hatchery
			46330	51332	Egg packing station (wholesale)
26511	33201	Echo sounders (manufacture)	10890	15899	Egg pickling (manufacture)
26600	33100	Echocardiographs (manufacture)	01130	01120	Egg plant growing
94990	91330	Ecological movements	01470	01240	Egg production
84210	75210	Economic aid missions accredited to foreign governments (public sector)	01470	01240	Egg production from poultry
			46330	51333	Egg products (wholesale)
84110	75110	Economic and social planning administration (public sector)	10890	15899	Egg products and egg albumin (manufacture)
			10890	15899	Egg substitute (manufacture)
72200	73200	Economic and Social Research Council	17290	21259	Egg trays and other moulded pulp packaging products (manufacture)
25300	28300	Economic boiler (manufacture)			
84130	75130	Economic services administration and regulation (public sector)	10890	15899	Eggs (powdered or reconstituted) (manufacture)
			47290	52270	Eggs (retail)
72200	73200	Economics, research and experimental development	46330	51332	Eggs (wholesale)
25300	28300	Economisers (manufacture)	46330	51332	Eggs exporter (wholesale)
70229	74143	Economist	46330	51332	Eggs importer (wholesale)
10110	15112	Edible fats of animal origin rendering (manufacture)	13923	17403	Eiderdowns (manufacture)
10612	15612	Edible nut flour or meal production (manufacture)	28131	29121	Ejector pump (manufacture)
47210	52210	Edible nuts (retail)			
46310	51310	Edible nuts (wholesale)			

SIC 2007	SIC 2003	Activity
30300	35300	Ejector seat for aircraft (manufacture)
13910	17600	Elastic and elastomeric fabric (manufacture)
13960	17542	Elastic fabric (not more than 30 cm wide) (manufacture)
13910	17600	Elastic or elastomeric knitted or netted fabric more than 30 cm wide (manufacture)
26513	33202	Elasticity testing equipment (non-electronic) (manufacture)
13960	17542	Elastomeric fabric (not more than 30 cm wide) (manufacture)
24200	27220	Elbows made of steel (manufacture)
27200	31400	Electric accumulators including parts thereof (manufacture)
28210	29210	Electric and other industrial and laboratory furnaces (manufacture)
28210	29210	Electric and other industrial and laboratory incinerators (manufacture)
28210	29210	Electric and other industrial and laboratory ovens (manufacture)
27900	31620	Electric bells (manufacture)
46439	51439	Electric blankets (wholesale)
27320	31300	Electric cable (manufacture)
27120	31200	Electric control and distribution boards (manufacture)
28410	29420	Electric discharge metal working tool (manufacture)
27400	31500	Electric fireplace logs (manufacture)
28210	29710	Electric household heating equipment (permanently mounted) (manufacture)
28210	29210	Electric household type furnaces (manufacture)
27400	31500	Electric insect lamps (manufacture)
27400	31500	Electric lanterns (manufacture)
27400	31500	Electric lighting equipment (manufacture)
27110	31100	Electric motors, generators (manufacture)
27120	31200	Electric power switches (manufacture)
30200	35200	Electric rail locomotives (manufacture)
28210	29210	Electric swimming pool heaters (permanently mounted) (manufacture)
27510	29710	Electric tea makers (manufacture)
27510	29710	Electricaire unit (manufacture)
71129	74204	Electrical and electronic engineering design projects
29310	31610	Electrical and electronic equipment for motor vehicles (manufacture)
46690	51870	Electrical apparatus for line telephony or telegraphy (wholesale)
27120	31200	Electrical apparatus for switching or protecting electrical circuits (manufacture)
27510	29710	Electrical appliances for domestic use (manufacture)
47540	52450	Electrical appliances, accessories and fittings (retail)
46690	51870	Electrical appliances, accessories and fittings for industry (wholesale)
27900	31620	Electrical base metal conduit and fittings (manufacture)
27900	32100	Electrical capacitors (manufacture)
26110	32100	Electronic condensers (manufacture)
27900	31620	Electrical carbon (manufacture)
23430	26230	Electrical ceramic fittings (manufacture)
27900	32100	Electrical condensers and similar components (manufacture)
24200	27220	Electrical conduit tube made of steel (manufacture)
27900	31620	Electrical conduit tubing, base metal (manufacture)
43210	45310	Electrical contractor (construction)
77299	71409	Electrical domestic appliance rental and leasing
27900	31620	Electrical door opening and closing devices (manufacture)
29310	31610	Electrical equipment for engines and vehicles (manufacture)
29310	31610	Electrical equipment for vehicles and aircraft (manufacture)
46439	51439	Electrical heating appliances (wholesale)
46439	51439	Electrical household appliances (excluding radios, televisions, etc) (wholesale)
47540	52450	Electrical household appliances (retail)
46439	51439	Electrical installation equipment for domestic use (wholesale)
23430	26230	Electrical insulating components made of ceramic (manufacture)
27900	31620	Electrical insulators (except glass or porcelain) (manufacture)
46690	51870	Electrical insulators and insulating fittings for electrical machines or equipment (wholesale)
23190	26150	Electrical insulators made of glass (manufacture)
46690	51870	Electrical machinery and apparatus and materials for professional use (wholesale)
46690	51870	Electrical motors (wholesale)
27330	31200	Electrical outlets or sockets (manufacture)
17120	21120	Electrical paper (manufacture)
27900	31620	Electrical pedestrian signalling equipment (manufacture)
71121	74205	Electrical power systems instrumentation design activities
27120	31200	Electrical relays (manufacture)
47599	52482	Electrical security alarm systems e.g. Safes and vaults (not installed or maintained) (retail)
24100	27100	Electrical sheet steel (not finally annealed) (manufacture)
27900	31620	Electrical signs (manufacture)
27900	29430	Electrical soldering equipment (manufacture)
24100	27100	Electrical steel (manufacture)
27900	31620	Electrical traffic lights (manufacture)
27900	29430	Electrical welding equipment (manufacture)
43210	45310	Electrical wiring of buildings
29100	34100	Electrically powered commercial vehicles (manufacture)
35130	40130	Electricity distribution
27120	31200	Electricity distribution and control apparatus (manufacture)
35130	40130	Electricity distribution operations by lines, poles, meters, and wiring
35110	40110	Electricity generation
35110	40110	Electricity generation by gas turbine
26511	33201	Electricity meter (electronic) (manufacture)
26513	33202	Electricity meter (non-electronic) (manufacture)
35140	40130	Electricity power agents
35140	40130	Electricity power brokers
35110	40110	Electricity production
35110	40110	Electricity production from diesel and renewables generation facilities
35110	40110	Electricity production from hydroelectric generation facilities
35110	40110	Electricity production from thermal generation facilities
35140	40130	Electricity sales
35140	40130	Electricity sales agents
35140	40130	Electricity sales to the user
35120	40120	Electricity transmission

E

SIC 2007	SIC 2003	Activity
26600	33100	Electro medical equipment (manufacture)
26600	33100	Electro medical pacemaker (manufacture)
26600	33100	Electro medical stimulator (manufacture)
25710	28610	Electro plated nickel silver cutlery (manufacture)
28210	29210	Electro slag furnace (manufacture)
24100	27100	Electro zinc coated sheet steel (manufacture)
26600	33100	Electrochemical apparatus for industrial use (manufacture)
28410	29420	Electrochemical metal working machine tools (manufacture)
20301	24301	Electrocoats paint (manufacture)
27900	31620	Electrodes for welding (manufacture)
26600	33100	Electro-diagnostic apparatus (manufacture)
46460	51460	Electro-diagnostic apparatus for medical use (wholesale)
26600	33100	Electro-encephalographs (manufacture)
96020	93020	Electrolysis specialist
28490	31620	Electrolytic chemical process plant (manufacture)
24100	27100	Electrolytic chromium/chromium oxide coated steel (manufacture)
24440	27440	Electrolytic copper (manufacture)
24100	27100	Electrolytically metal coated sheet steel (manufacture)
28131	29121	Electro-magnetic pumps (manufacture)
28410	29420	Electro-magnetic-pulse (magnetic-forming) metal forming machines (manufacture)
27900	31620	Electromagnets (manufacture)
26600	33100	Electromyographs (manufacture)
26511	33201	Electron microscope (manufacture)
26110	32100	Electron tubes (manufacture)
26110	32100	Electronic active components (manufacture)
26511	33201	Electronic aircraft engine instruments (manufacture)
46520	51860	Electronic and telecommunications equipment and parts (wholesale)
26511	33201	Electronic apparatus for testing physical and mechanical properties of materials (manufacture)
26511	33201	Electronic automotive emissions testing equipment (manufacture)
61200	64200	Electronic bulletin board services (wireless telecommunications)
46520	51860	Electronic components (wholesale)
26110	31200	Electronic connectors (manufacture)
26511	33201	Electronic counter (manufacture)
26110	32100	Electronic crystals and crystal assemblies (manufacture)
71121	74205	Electronic design consultant
26511	33201	Electronic environmental controls and automatic controls for appliances (manufacture)
77299	71409	Electronic equipment for household use renting leasing
27900	31620	Electronic filter (manufacture)
26511	33201	Electronic flame and burner control (manufacture)
26511	33201	Electronic flight recorders (manufacture)
26400	36509	Electronic games (domestic) (manufacture)
26511	33201	Electronic GPS devices (manufacture)
26511	33201	Electronic humidistats (manufacture)
26511	33201	Electronic hydronic limit controls (manufacture)
26511	33201	Electronic instruments and appliances for measuring, testing, and navigation (manufacture)
26110	32100	Electronic integrated circuits (manufacture)
46520	51860	Electronic integrated circuits and micro-assemblies (wholesale)

SIC 2007	SIC 2003	Activity
26511	33201	Electronic laboratory analytical instruments (manufacture)
26511	33201	Electronic laboratory incubators and sundry laboratory apparatus for measuring, testing (manufacture)
46520	51860	Electronic machinery, apparatus and materials for professional use (wholesale)
61100	64200	Electronic mail services (wired telecommunications)
18130	22240	Electronic makeup (manufacture)
61200	64200	Electronic message and information services (wireless telecommunications)
26511	33201	Electronic metal detectors (manufacture)
26110	32100	Electronic micro-assemblies of moulded module, micromodule or similar types (manufacture)
27900	31620	Electronic miscellaneous unspecified equipment (manufacture)
26511	33201	Electronic motion detectors (manufacture)
32200	36300	Electronic musical instrument (manufacture)
26110	32100	Electronic passive components (manufacture)
26511	33201	Electronic physical properties testing and inspection equipment (manufacture)
26511	33201	Electronic pneumatic gauges (manufacture)
26511	33201	Electronic polygraph machines (manufacture)
58110	22110	Electronic publishing of books
26511	33201	Electronic pulse (signal) generators (manufacture)
27900	31620	Electronic scoreboards (manufacture)
26511	33201	Electronic tally counters (manufacture)
26511	33201	Electronic testing equipment (manufacture)
26520	33500	Electronic timer (not clock or watch) (manufacture)
26400	36509	Electronic toys and games with replaceable software (manufacture)
26110	32100	Electronic tube (manufacture)
46520	51860	Electronic tubes (wholesale)
26110	32100	Electronic valve (manufacture)
46520	51860	Electronic valves (wholesale)
71200	74300	Electrophoresis
25610	28510	Electroplating (manufacture)
28490	31620	Electroplating equipment (manufacture)
28230	30010	Electrostatic copying machine (manufacture)
28290	29240	Electrostatic precipitator (manufacture)
26600	33100	Electrotherapeutic equipment (manufacture)
27510	29710	Electro-thermic appliances for domestic use (manufacture)
46439	51439	Electro-thermic hair-dressing or hand drying apparatus (wholesale)
18130	22240	Electrotyping (manufacture)
20110	24110	Elemental gases (manufacture)
42130	45213	Elevated highways construction
49311	60213	Elevated railways (scheduled passenger transport)
28220	29220	Elevator (manufacture)
15120	19200	Elevator bands made of leather (manufacture)
22190	25130	Elevator belting made of rubber (manufacture)
28302	29320	Elevator for agricultural use (manufacture)
28921	29521	Elevators with continuous action for underground use (manufacture)
99000	99000	Embassy
18130	22250	Embossing (manufacture)
32990	36631	Embossing devices (hand operated) for labels (manufacture)
28940	29540	Embossing machinery for textiles (manufacture)
28950	29550	Embossing machines for working paper and board (manufacture)

SIC 2007	SIC 2003	Activity	SIC 2007	SIC 2003	Activity
21200	24421	Embrocation (manufacture)	46690	51870	Engineers' plant and stores (wholesale)
13923	17403	Embroidering on made-up textile goods (manufacture)	25500	28400	Engineers' stampings and pressings of base non-ferrous metals (manufacture)
13100	17160	Embroidery cotton (manufacture)	25500	28400	Engineers' stampings and pressings of ferrous metals (manufacture)
13990	17541	Embroidery lace (manufacture)			
47510	52410	Embroidery making materials (retail)	29100	34100	Engines (internal combustion) for motor vehicles (manufacture)
13300	17300	Embroidery on made up textile goods	30910	35410	Engines (internal combustion) for motorcycles (manufacture)
23910	26810	Emery cloth (manufacture)			
08990	14500	Emery extraction	28110	29110	Engines and parts for marine use (manufacture)
23910	26810	Emery paper (manufacture)	28110	29110	Engines and parts for railways (manufacture)
23910	26810	Emery wheel (manufacture)	28110	29110	Engines for agricultural machinery (manufacture)
96090	93059	Emigration agency (not of foreign government, etc.)	30300	35300	Engines for aircraft (manufacture)
65300	66020	Employee benefit plans	46690	51870	Engines for aircraft (wholesale)
94110	91110	Employers organisations	28110	29110	Engines for combine harvesters (manufacture)
78200	74500	Employment agency (temporary)	28110	29110	Engines for construction equipment (manufacture)
78109	74500	Employment consultants	28110	29110	Engines for forklift trucks (manufacture)
78109	74500	Employment placement agencies	28110	29110	Engines for industrial application (manufacture)
88990	85321	Employment rehabilitation centre (charitable)	28110	29110	Engines for lawn mowers (manufacture)
88990	85322	Employment rehabilitation centre (non-charitable)	28110	29110	Engines for locomotives (manufacture)
01490	01250	Emu raising and breeding	28110	29110	Engines for marine use (manufacture)
28131	29121	Emulsion (gas lift) pumps (manufacture)	30910	34100	Engines for motorcycles (manufacture)
20520	24620	Emulsion adhesive (manufacture)	28110	29110	Engines for railway vehicles (manufacture)
20301	24301	Emulsion paint (manufacture)	90030	92319	Engravers
20160	24160	Emulsions of synthetic resin (manufacture)	32120	36220	Engraving (personalised) on precious metal (manufacture)
20301	24301	Enamel (manufacture)			
23190	26150	Enamel glass in the mass (manufacture)	18130	22240	Engraving for printing (manufacture)
23310	26300	Enamelled tile (glazed) (manufacture)	32120	36220	Engraving of personal non-precious metal products (manufacture)
25610	28510	Enamelling of metals including vitreous enamelling (manufacture)			
			80300	74879	Enquiry agency
38220	23300	Encapsulation, preparation and other treatment of nuclear waste for storage	20130	23300	Enriched thorium (manufacture)
			46120	51120	Enriched uranium and plutonium and their compounds (commission agent)
23310	26300	Encaustic tile (manufacture)			
58110	22110	Encyclopaedia publishing	20130	23300	Enriched uranium production (manufacture)
28410	28620	End mill (manufacture)	46719	51519	Enriched uranium supply to nuclear reactors (wholesale)
13300	17300	Ending and mending of textiles (manufacture)			
86230	85130	Endodontic dentistry	14132	18222	Ensembles (manufacture)
26600	33100	Endoscopes (manufacture)	93290	92349	Entertainment activities n.e.c.
21200	24421	Enema preparations (manufacture)	26301	32201	Entrance telephones (manufacture)
11070	15980	Energy drinks (manufacture)	82190	74850	Envelope addressing service
74901	74206	Energy efficiency consultancy activities	28950	29550	Envelope making machine (manufacture)
84130	75130	Energy services administration and regulation (public sector)	28230	30010	Envelope stuffing machine (manufacture)
			82190	74850	Envelope stuffing, sealing and mailing service including for advertising
28110	34300	Engine block for motor vehicle (finished) (manufacture)			
			17230	21230	Envelopes and letter-cards (manufacture)
46770	51570	Engine cleaning waste (wholesale)	16290	20520	Envelopes for bottles made of straw (manufacture)
29100	34100	Engine for motor vehicle (manufacture)	23190	26150	Envelopes made of glass (manufacture)
23990	26821	Engine packing made of asbestos (manufacture)	23190	26150	Envelopes made of glass for light bulbs and electronic valves (manufacture)
77390	71340	Engine rental and operating leasing			
94120	91120	Engineering associations	71129	74204	Environmental consultants
23320	26400	Engineering brick (manufacture)	26513	33202	Environmental controls and automatic controls for appliances (non-electronic) (manufacture)
71129	74209	Engineering contractor responsible for complete process plant			
			71129	74204	Environmental engineering consultancy activities
71129	74204	Engineering design activities for the construction of civil engineering works	94990	91330	Environmental movements
			72190	73100	Environmental pollution research and experimental development
71121	74205	Engineering design services for industrial process and production			
			74901	74206	Environmental project consultancy activities
58110	22110	Engineering drawing publishing	84120	75120	Environmental services administration (public sector)
72190	73100	Engineering research and experimental development	46750	51550	Enzymes (wholesale)
24100	27100	Engineering steel (manufacture)	46120	51120	Enzymes and other organic compounds (commission agent)
71121	74205	Engineers' draughtsman			

SIC 2007	SIC 2003	Activity
20140	24140	Enzymes and other organic compounds (manufacture)
94910	91310	Episcopal Church in Scotland
26702	33403	Episcope (manufacture)
20520	24620	Epoxide adhesive (manufacture)
20160	24160	Epoxide resins (manufacture)
20140	24140	Epoxides (manufacture)
20301	24301	Epoxy paint (manufacture)
28923	29523	Equipment for concrete crushing and screening roadworks (manufacture)
28290	29240	Equipment for dispersing liquids or powders (manufacture)
26520	33500	Equipment for measuring and recording (manufacture)
28290	29240	Equipment for projecting liquids or powders (manufacture)
28290	29240	Equipment for spraying liquids or powders (manufacture)
26513	33202	Equipment for testing physical and mechanical properties of materials (non-electronic) (manufacture)
71121	74205	Equipment layout and other plant design services
22190	25130	Eraser rubber (manufacture)
42110	45230	Erection of roadway barriers
43910	45220	Erection of roofs
28220	29220	Escalator (manufacture)
13922	17402	Escape chute for aircraft (manufacture)
96090	93059	Escort agency
46750	51550	Essential oils (wholesale)
20530	24630	Essential oils and essence (other than turpentine) (manufacture)
46120	51120	Essential oils and mixtures of odiferous substances (commission agent)
46750	51550	Essential oils merchant (wholesale)
68310	70310	Estate agent
29100	34100	Estate car (manufacture)
68209	70209	Estate company (owning and managing)
20140	24140	Esters (but (not polyesters)) (manufacture)
20140	24140	Esters of methacrylic acid (manufacture)
90030	92319	Etchers
18130	22240	Etching for printing (manufacture)
20140	24140	Ethane diol (excluding anti-freeze mixtures) (manufacture)
06200	11100	Ethane extraction from natural gas
19201	23201	Ethane production by refining (manufacture)
20140	24140	Ethanol (synthetic) (manufacture)
46120	51120	Ethers, organic peroxides, epoxides, acetals and hemiacetals and their compounds (commission agent)
46750	51550	Ethers, organic peroxides, epoxides, acetals and hemiacetals and their derivatives (wholesale)
94990	91330	Ethnic and minority group organisations
20140	15920	Ethyl alcohol (non-potable) obtained by fermentation (manufacture)
20140	24140	Ethylene (manufacture)
20140	24140	Ethylene glycol (excluding anti-freeze mixtures) (manufacture)
20160	24160	Ethylene polymers (manufacture)
99000	99000	European Communities Representatives and Information Office
99000	99000	European Community
99000	99000	European Free Trade Association

SIC 2007	SIC 2003	Activity
94910	91310	Evangelists Society
10511	15511	Evaporated milk (manufacture)
25300	28300	Evaporator (manufacture)
28250	29230	Evaporator for refrigeration machinery (manufacture)
43120	45110	Excavation
28922	29522	Excavator (manufacture)
79110	63301	Excursion agency
50100	61101	Excursion, cruise or sightseeing boat operation (except for inland waterway service)
50300	61201	Excursion, cruise or sightseeing boats operation (inland waterway service)
84110	75110	Executive and legislative administration (public sector)
78109	74500	Executive employment placement and search activities
78200	74500	Executive personnel (supply) (temporary employment agency)
78109	74500	Executive recruitment consultant
23990	26829	Exfoliated vermiculite (manufacture)
29320	34300	Exhaust pipes for motor vehicles (manufacture)
45320	50300	Exhaust sales and fitting centre (retail)
29320	34300	Exhaust systems and components for motor vehicles (manufacture)
28110	29110	Exhaust valves for internal combustion engines (manufacture)
68310	70310	Exhibition centre letting (not self-owned)
68202	70201	Exhibition centre letting (self owned)
82301	74873	Exhibition contracting and organising
16230	20300	Exhibition stand (manufacture)
82301	74873	Exhibition stand design
82301	74873	Exhibition stand hire
23990	26829	Expanded clay (manufacture)
25930	28730	Expanded metal (manufacture)
23990	26829	Expanded vermiculite (manufacture)
26511	33201	Expansion analysers (electronic) (manufacture)
22190	25130	Expansion joints made of rubber (manufacture)
26513	33202	Expansion meter (non-electronic) (manufacture)
25290	28210	Expansion tank made of metal exceeding 300 litres (manufacture)
71122	74206	Exploration for gas or oil
72190	73100	Explorer
20510	24610	Explosive signalling flares (manufacture)
20510	24610	Explosives (manufacture)
46750	51550	Explosives (wholesale)
46190	51190	Export confirming house, general or undefined (commission agent)
73200	74130	Export consultant
65120	66031	Export credit guarantee department
64929	65229	Export finance company (other than in banks' sector)
52290	63400	Export packer
46190	51190	Export purchasing, general or undefined (commission agent)
26511	33201	Exposure meter (electric) (manufacture)
49390	60211	Express coach service on scheduled routes
27900	31300	Extension cords made from purchased insulated wire (manufacture)
27900	31300	Extension cords with insulated wire and connectors (manufacture)
81229	74709	Exterior cleaning of buildings
43341	45440	Exterior painting of buildings
43310	45410	Exterior plaster application in buildings or other constructions incl. related lathing materials

SIC 2007	SIC 2003	Activity
43310	45410	Exterior stucco application in buildings or other constructions incl. related lathing materials
81291	74703	Exterminating of insects, rodents and other pests (except agricultural)
10890	15139	Extracts and juices of meat, fish, crustaceans or molluscs (manufacture)
20530	24630	Extracts of aromatic products (manufacture)
24440	27440	Extruded products made of copper (manufacture)
24420	27420	Extruded sections made of aluminium (manufacture)
24420	27420	Extruded tubes made of aluminium (manufacture)
28960	29560	Extruder for rubber or plastics (manufacture)
28940	29540	Extruding machinery for textiles (manufacture)
46640	51830	Extruding, drawing, texturing or cutting machinery for man-made textile materials (wholesale)
20160	24160	Extrusion compounds (plastics) (manufacture)
24420	27420	Extrusion ingots made of aluminium (manufacture)
24420	27420	Extrusions made of aluminium (manufacture)
86101	85112	Eye hospital (private sector)
86101	85111	Eye hospital (public sector)
86220	85120	Eye specialist (private practice)
86101	85111	Eye specialist (public sector)
25990	28750	Eyelet (manufacture)
94920	91320	Fabian Society
23990	26821	Fabric made of asbestos (manufacture)
28940	29540	Fabric processing machinery (manufacture)
17240	21240	Fabric wallcoverings (manufacture)
25110	28110	Fabricated structural steelwork for buildings (manufacture)
47510	52410	Fabrics (retail)
46410	51410	Fabrics (wholesale)
13960	17549	Fabrics coated with gum or amylaceous substances (manufacture)
13960	17549	Fabrics impregnated, coated, covered or laminated with plastics (manufacture)
20420	24520	Face powder or cream (manufacture)
96020	93020	Facial massage
81100	70320	Facilities management
26301	32201	Facsimile transmission apparatus (manufacture)
20170	24170	Factice (manufacture)
64992	65222	Factoring company (buying book debts)
49390	60219	Factory bus service
81210	74701	Factory cleaning contractor
68209	70209	Factory letting
56290	55510	Factory or office canteens
29100	34100	Factory rebuilding of motor vehicle engines (manufacture)
94120	91120	Faculty of actuaries
82301	74873	Fair organiser
93210	92330	Fairground activities
32200	36300	Fairground organs (manufacture)
28990	36639	Fairground rides (manufacture)
93290	92330	Fairs and shows of a recreational nature
28940	29540	Faller (textile machinery accessory) (manufacture)
32990	36639	False beard (manufacture)
32990	36639	False eyebrow (manufacture)
86210	85120	Family doctor service
86900	85140	Family Planning Association clinics (not providing medical treatment)
88990	85321	Family Planning Associations (not clinics)
86220	85120	Family planning centre providing medical treatment without accommodation

SIC 2007	SIC 2003	Activity
88990	85321	Family Welfare Association
27510	29710	Fan (electric, domestic) (manufacture)
22190	25130	Fan belts for motor vehicles (manufacture)
28250	29230	Fan coil unit (manufacture)
23190	26150	Fancy articles and goods made of glass (manufacture)
77210	71401	Fancy dress hire
47789	52489	Fancy goods (retail)
46499	51478	Fancy goods (wholesale)
14310	17710	Fancy hosiery (manufacture)
15120	19200	Fancy leather goods (manufacture)
17120	21120	Fancy paper (manufacture)
10710	15810	Fancy pastry (manufacture)
28250	29710	Fans (domestic) (manufacture)
28250	29230	Fans (non-domestic) (manufacture)
46439	51439	Fans and ventilating or recycling hoods for domestic use (wholesale)
10730	15850	Farinaceous products (manufacture)
01621	01421	Farm animal boarding and care (except pets)
10910	15710	Farm animal feeds produced from slaughter waste (manufacture)
01621	01421	Farm animal pound
55209	55239	Farmhouse short stay accommodation
28302	29320	Farmyard manure spreader (manufacture)
01629	28520	Farriers, on a fee or contract basis
14132	18222	Fashion (manufacture)
82990	74879	Fashion agent
82990	74879	Fashion artist
74100	74872	Fashion designing
32130	36610	Fashion jewellery (manufacture)
74209	74819	Fashion photography
18129	22220	Fashion printing (manufacture)
56102	55302	Fast food outlet with restaurant (unlicensed)
25940	28740	Fasteners made of metal (manufacture)
56101	55301	Fast-food restaurants, licensed
56102	55302	Fast-food restaurants, unlicensed
10410	15410	Fat of marine animals production (manufacture)
10110	15112	Fat recovery from knackers (manufacture)
20140	24140	Fat splitting and distilling (manufacture)
10420	15430	Fats (edible) (manufacture)
20140	24140	Fatty acid (manufacture)
21100	24410	Fatty amines and quaternary ammonium salts (manufacture)
26301	32201	Fax machines (manufacture)
32990	36639	Feather curling (manufacture)
32910	36620	Feather duster (manufacture)
32990	36639	Feather ornament (manufacture)
10120	15120	Feather production (manufacture)
32990	36639	Feather purifying (manufacture)
32990	36639	Feather sorting (manufacture)
46900	51900	Feathers (wholesale)
94110	91110	Federations of business and employers' membership organisations
10910	15710	Feed supplements for animals (manufacture)
28930	29530	Feeders (manufacture)
08990	14500	Feldspar mining and quarrying
10110	15113	Fellmongery (manufacture)
13990	17549	Felt (manufacture)
46640	51830	Felt finishing machinery (wholesale)

F

SIC 2007	SIC 2003	Activity
14190	18241	Felt hat bleaching and dying (manufacture)
14190	18241	Felt hat body making (manufacture)
14190	18241	Felt hat finishing (manufacture)
23990	26821	Felt made of asbestos (manufacture)
23140	26140	Felt made of glass fibre (manufacture)
28940	29540	Felt or non-woven fabric production or finishing machines (manufacture)
32990	36631	Felt tipped pen (manufacture)
17120	21120	Feltboard including felt paper (manufacture)
22190	25130	Felting made of rubber (manufacture)
28940	29540	Felts or non-wovens production and finishing machines (manufacture)
01610	01410	Fencing by agricultural contractor
43290	45340	Fencing contractor (not agricultural)
22230	25239	Fencing made of plastic (manufacture)
25930	28730	Fencing made of steel wire (manufacture)
16230	20300	Fencing made of wood (assembled) (manufacture)
16290	20520	Fenders made of cork (manufacture)
01130	01120	Fennel growing
02300	02010	Fern collecting, cutting, gathering
26110	31100	Ferrite parts for electronic apparatus (manufacture)
24100	27100	Ferro alloys (high carbon ferro manganese) (manufacture)
24100	27100	Ferro aluminium (manufacture)
24100	27100	Ferro chromium (manufacture)
43999	45250	Ferro concrete bar bending and fixing contractor
24100	27100	Ferro molybdenum (manufacture)
24100	27100	Ferro nickel (manufacture)
24100	27100	Ferro niobium (manufacture)
24100	27100	Ferro phosphorus (manufacture)
24100	27100	Ferro titanium (manufacture)
24100	27100	Ferro tungsten (manufacture)
24100	27100	Ferro vanadium (manufacture)
24100	27100	Ferro zirconium (manufacture)
24100	27100	Ferro-alloys (except high carbon ferro-manganese production) (manufacture)
24100	27100	Ferrosilicon (manufacture)
24100	27100	Ferrosilicon chromium (manufacture)
24100	27100	Ferrosilicon manganese (manufacture)
24100	27100	Ferrosilicon titanium (manufacture)
24100	27100	Ferrosilicon tungsten (manufacture)
46720	51520	Ferrous and non-ferrous metal ores (wholesale)
46720	51520	Ferrous and non-ferrous metals in primary forms (wholesale)
46720	51520	Ferrous and non-ferrous semi-finished metal products n.e.c. (wholesale)
24510	27510	Ferrous metal foundry (manufacture)
24100	27100	Ferrous products production by reduction of iron ore (manufacture)
30110	35110	Ferry (manufacture)
50300	61201	Ferry transport for passengers (inland waterway service)
20150	24150	Fertiliser (manufacture)
28302	29320	Fertiliser distributor or broadcaster (manufacture)
08910	14300	Fertiliser minerals mining
46120	51120	Fertilisers (commission agent)
47760	52489	Fertilisers (retail)
46750	51550	Fertilisers (wholesale)
28302	29320	Fertilizing plough machinery (manufacture)
46499	51477	Festive, carnival or other entertainment articles, conjuring tricks and novelty jokes (wholesale)

SIC 2007	SIC 2003	Activity
13921	17401	Festoon blinds (manufacture)
16210	20200	Fibre board (manufacture)
16210	20200	Fibre building board (manufacture)
23650	26650	Fibre cement (manufacture)
13940	17520	Fibre core for wire rope (manufacture)
01160	01110	Fibre crop growing
32910	36620	Fibre dressing for brushes (manufacture)
26701	33402	Fibre optic apparatus (manufacture)
27310	31300	Fibre optic cable for data transmission or live transmission of images (manufacture)
32990	36631	Fibre tipped pen (manufacture)
20600	24700	Fibrillated yarn (manufacture)
93120	92629	Field and track clubs
93110	92619	Field and track stadium
01610	01410	Field preparation on a fee or contract basis
01220	01139	Fig growing
23110	26110	Figured glass (manufacture)
23190	26150	Figurines made of glass (manufacture)
20600	24700	Filament tow (manufacture)
24340	27340	Filament wire made of steel (manufacture)
28410	29420	Filament wire spiralling machines (manufacture)
25730	28620	File (hand tool) (manufacture)
31010	36120	Filing cabinet (manufacture)
25990	28750	Filing cabinet made of metal (not designed to be placed on the floor) (manufacture)
20301	24303	Filling and sealing compounds for painters (manufacture)
28290	29240	Filling machinery (manufacture)
47300	50500	Filling station (motor fuel and lubricants)
94990	91330	Film and photo clubs
22210	25210	Film and sheet of decorated unsupported polyvinyl chloride (manufacture)
52102	63129	Film bonded warehouse for air transport activities
52103	63129	Film bonded warehouse for land transport activities
52101	63129	Film bonded warehouse for water transport activities
59131	92120	Film broker
74203	74814	Film copying (not motion picture)
59120	92119	Film cutting activities
59131	92120	Film distribution
59131	92120	Film distribution rights acquisition
59120	92119	Film editing
59131	92120	Film hiring agency
91011	92510	Film lending and storage
59131	92120	Film library
22210	25210	Film made of cellophane (manufacture)
22210	25210	Film made of plastic (manufacture)
22210	25210	Film made of polyethylene (manufacture)
22210	25210	Film made of polypropylene (manufacture)
22210	25210	Film made of polythene (manufacture)
22210	25210	Film made of polyvinyl chloride (PVC) (manufacture)
74203	74814	Film processing
59120	92119	Film processing activities
59111	92111	Film producer (own account)
59111	92111	Film production for projection or broadcasting
28940	29540	Film reader for textile machinery (manufacture)
59131	92120	Film rental
59120	92119	Film sound track dubbing and synchronisation
59111	92111	Film studios
59120	92119	Film title printing

F

SIC 2007	SIC 2003	Activity
13922	17402	Filter cloth (made-up) (manufacture)
13960	17549	Filter cloth weaving (manufacture)
28120	29122	Filter for pneumatic control equipment (manufacture)
17290	21259	Filter paper and paperboard (cut to size) (manufacture)
17120	21120	Filter paper stock (manufacture)
28930	29530	Filtering and purifying machinery for the industrial preparation of food or drink (manufacture)
28290	29240	Filtering or purifying machinery parts (manufacture)
46690	51870	Filters for oil, petrol and air for internal combustion engines (not motor vehicle) (wholesale)
22290	25240	Filtration elements made of plastic (manufacture)
46690	51870	Filtration equipment and apparatus (wholesale)
28290	29240	Filtration equipment for hydraulic equipment (manufacture)
28290	29240	Filtration equipment for the chemical industry (manufacture)
64929	65229	Finance corporation for industry
64921	65221	Finance house activities (non-deposit taking)
66190	67130	Financial advisor
82912	74871	Financial and credit reporting
64991	65233	Financial futures, options and other derivatives dealing in on own account
64999	65239	Financial intermediation n.e.c.
64910	65210	Financial leasing
70221	74142	Financial management consultancy services (except corporate tax)
66110	67110	Financial markets administration
84110	75110	Financial services (public sector)
66190	67130	Financial transactions centre
32130	36610	Findings and stampings made of base metal for jewellery (manufacture)
32120	36220	Findings and stampings made of precious metals for jewellery (manufacture)
90030	92319	Fine art expert
85520	80429	Fine arts schools (except academic)
12000	16000	Fine cut tobacco (manufacture)
01490	01220	Fine or coarse animal hair, not including sheep or goats wool
13100	17120	Fingering wool (manufacture)
03220	05020	Fingerling production, freshwater
03210	05020	Fingerling production, marine
80100	74602	Fingerprinting services
20590	24660	Finings (manufacture)
26110	24660	Finished or semi-finished dice, semiconductor (manufacture)
26110	24660	Finished or semi-finished wafers, semiconductor (manufacture)
46120	51120	Finishing agents, dye carriers and similar industrial chemical products (commission agent)
28490	29430	Finishing and polishing machine tools for optical, spectacle or clock or watch faces (manufacture)
28940	29540	Finishing machinery for textiles (manufacture)
31090	36110	Finishing of chairs and seats
13300	17300	Finishing of leather wearing apparel (manufacture)
13300	17300	Finishing of wearing apparel n.e.c. (manufacture)
18140	22230	Finishing services for CD-ROMS (manufacture)
27510	29710	Fire (electric) (manufacture)
27520	29720	Fire (gas) (manufacture)
27520	29720	Fire (oil) (manufacture)
26301	31620	Fire alarm and system (manufacture)
77390	71340	Fire alarm hire
26301	31620	Fire alarm systems, sending signals to a control station (manufacture)
71122	74206	Fire and explosion protection and control consultancy activities
84250	75250	Fire authorities
84250	75250	Fire brigades
29100	34100	Fire engine (manufacture)
28290	29240	Fire extinguisher (hand held) (manufacture)
46120	51120	Fire extinguisher charges and preparations (commission agent)
46690	51870	Fire extinguishers (excluding motor vehicle) (wholesale)
28290	29240	Fire extinguishing apparatus (excluding hand operated chemical extinguishers) (manufacture)
20590	24660	Fire extinguishing chemicals (manufacture)
84250	75250	Fire fighting and fire prevention
65120	66031	Fire insurance
16290	20510	Fire logs and pellets, of pressed wood, or of coffee or soybean grounds and the like (manufacture)
24440	27440	Fire refined copper (manufacture)
32990	18100	Fire resistant and protective safety clothing of leather (manufacture)
31090	36140	Fire screen (manufacture)
84250	75250	Fire service activities
29100	34100	Fire tender (manufacture)
25400	29600	Firearms for hunting, sporting or protective use (manufacture)
46690	51870	Firearms, sporting, hunting or target shooting rifles (wholesale)
23200	26260	Firebrick and shape (manufacture)
08120	14220	Fireclay mine or quarry
32990	18249	Fire-fighting protection suits (manufacture)
32990	36639	Firelighter (manufacture)
81223	74705	Fire-places cleaning
43290	45320	Fireproofing work
32990	18249	Fire-resistant and protective safety clothing (manufacture)
47789	52489	Firewood (retail)
20510	24610	Firework (manufacture)
93290	92349	Firework displays
46120	51120	Fireworks (commission agent)
25300	28300	Firing plant for boilers, etc. (manufacture)
46180	51180	First aid boxes (commission agent)
46460	51460	First aid boxes (wholesale)
85421	80302	First-degree level higher education
84110	75110	Fiscal services (public sector)
47230	52230	Fish (retail)
46380	51380	Fish (wholesale)
56103	55303	Fish and chip shop
56103	55304	Fish and chip stand
10410	15410	Fish and marine mammal oil extraction (manufacture)
10200	15209	Fish and other aquatic animal meals and solubles unfit for human consumption (manufacture)
16240	20400	Fish boxes made of wood (manufacture)
03220	01250	Fish breeding, freshwater
03210	05020	Fish breeding, marine
10200	15209	Fish cakes (manufacture)
10200	15209	Fish canning (manufacture)
10200	15209	Fish curing (other than by distributors) (manufacture)

SIC 2007	SIC 2003	Activity
10850	15209	Fish dish (prepared) production (manufacture)
10850	15209	Fish dishes, including fish and chips (manufacture)
46380	51380	Fish distribution (wholesale)
10200	15209	Fish drying (manufacture)
25710	28610	Fish eater (manufacture)
46170	51170	Fish factor (commission agent)
03220	05020	Fish farming in fresh water, including farming of freshwater ornamental fish
03210	05020	Fish farming, marine
10200	15209	Fish fillet production (manufacture)
10850	15209	Fish fingers (manufacture)
03220	05020	Fish fry production, freshwater
03210	05020	Fish fry production, marine
03220	05020	Fish hatcheries and farms service activities, freshwater
03210	05020	Fish hatcheries and farms service activities, marine
03220	05020	Fish hatcheries, freshwater
03210	05020	Fish hatcheries, marine
32300	36400	Fish hook (manufacture)
10410	15410	Fish liver oil (unrefined) production (manufacture)
10410	15420	Fish liver oil refining (manufacture)
10200	15209	Fish meal (manufacture)
10410	15410	Fish oil (crude) production (manufacture)
10200	15209	Fish paste (manufacture)
24100	27100	Fish plates (hot rolled) (manufacture)
24100	27100	Fish plates and sole plates (non-rolled) (manufacture)
25930	28740	Fish plates for arches made of steel (manufacture)
10200	15209	Fish preservation (other than by freezing) (manufacture)
10200	15201	Fish preservation by freezing (manufacture)
10200	15209	Fish processing (not freezing) (manufacture)
28930	29530	Fish processing machines and equipment (manufacture)
10200	15209	Fish products (manufacture)
10200	15209	Fish salting (manufacture)
47810	52620	Fish stall (retail)
70229	05010	Fish stock management consultancy services
10200	15209	Fish, crustacean and mollusc cooking (manufacture)
10200	15209	Fish, crustacean and mollusc preparation and preservation, by immersing in brine (manufacture)
10200	15209	Fish, crustacean and mollusc smoking (manufacture)
93199	92629	Fishing (recreational)
30110	35110	Fishing boats (manufacture)
03110	05010	Fishing by line, except for recreation or sport
03110	05010	Fishing for shellfish
47640	52485	Fishing gear (retail)
03110	05010	Fishing in ocean, sea, coastal or inland waters
13940	17520	Fishing line (manufacture)
13940	17520	Fishing net (manufacture)
13940	17520	Fishing net mending (manufacture)
13100	17110	Fishing net yarn made of cotton (manufacture)
03120	05010	Fishing on a commercial basis in inland waters
03110	05010	Fishing on a commercial basis in ocean and coastal waters
46499	51479	Fishing rods, line fishing tackle and articles for hunting or fishing (wholesale)
03110	05010	Fishing service activities
84130	75130	Fishing services administration and regulation (public sector)
32300	36400	Fishing tackle (manufacture)
47640	52485	Fishing tackle (retail)
30110	35110	Fishing vessel (manufacture)
47230	52230	Fishmonger (retail)
46380	51380	Fishmonger (wholesale)
32300	36400	Fitness centre equipment and appliances (manufacture)
93130	92629	Fitness centre operation
46740	51540	Fittings and fixtures (wholesale)
31010	36120	Fittings and furnishing for hotels (manufacture)
31010	36120	Fittings and furnishings for banks (manufacture)
31010	36120	Fittings and furnishings for bars (manufacture)
31010	36120	Fittings and furnishings for laboratories (manufacture)
31010	36120	Fittings and furnishings for libraries (manufacture)
31010	36120	Fittings and furnishings for museums (manufacture)
31010	36120	Fittings and furnishings for offices (manufacture)
31010	36120	Fittings and furnishings for public houses (manufacture)
31010	36120	Fittings and furnishings for restaurants (manufacture)
31010	36120	Fittings and furnishings for shops (manufacture)
24510	27210	Fittings for tubes made of cast iron (manufacture)
24510	27210	Fittings for tubes made of cast steel (manufacture)
22190	25130	Fittings made of rubber (manufacture)
24200	27220	Fittings made of steel (manufacture)
26309	32202	Fixed transmitters (manufacture)
20590	24640	Fixer for photographic use (manufacture)
13200	17220	Flag fabric (woollen) (manufacture)
13922	17402	Flag making up (manufacture)
23310	26300	Flags made of clay (manufacture)
23310	26300	Flags made of non-refractory ceramic (manufacture)
46730	51530	Flagstone merchant (wholesale)
08110	14110	Flagstone quarry
23610	26610	Flagstones made of precast concrete (manufacture)
24420	27420	Flake made of aluminium (manufacture)
24440	27440	Flake made of copper (manufacture)
10390	15330	Flaked coconut including desiccated but (not sugared) (manufacture)
10611	15611	Flaked maize (manufacture)
26513	33202	Flame and burner control (non-electronic) (manufacture)
25400	29600	Flame throwers (manufacture)
25930	28740	Flange jointing sets (manufacture)
24200	27220	Flanges made of steel (manufacture)
13200	17220	Flannel (manufacture)
13300	17300	Flannel ending (manufacture)
13300	17300	Flannel filling (manufacture)
13300	17300	Flannel finishing (manufacture)
13300	17300	Flannel preparing (manufacture)
13300	17300	Flannel scouring (manufacture)
13300	17300	Flannel shrinking (manufacture)
13300	17300	Flannelette raising and finishing (manufacture)
13200	17210	Flannelette weaving (manufacture)
28120	29130	Flaps, diaphragms and other parts of hydraulic and pneumatic valves (manufacture)
43999	45250	Flare stack and flareboom erection work
27400	31500	Flash lamp case (manufacture)
27400	31500	Flashcubes (manufacture)
26702	33403	Flashlight apparatus (manufacture)
27400	31500	Flashlights (manufacture)
23110	26110	Flat glass (manufacture)

SIC 2007	SIC 2003	Activity
47520	52460	Flat glass (retail)
46730	51530	Flat glass (wholesale)
68209	70209	Flat letting
68310	70310	Flat letting agency
24320	27320	Flat rolled steel products in coils or straight lengths <600 mm (manufacture)
24100	27100	Flat rolled steel products in coils or straight lengths ≥600 mm (manufacture)
29202	34202	Flat trailer (motor drawn) (manufacture)
25990	28750	Flat ware made of base metal (manufacture)
55201	55231	Flats in holiday centres and holiday villages (provision of short-stay lodging in)
55209	55232	Flats, not in holiday centres, holiday villages or youth hostels (self catering short-stay lodging)
41201	45230	Flatwork for sport and recreational installations (commercial buildings)
28940	29540	Flatwork machine for laundry (manufacture)
46750	51550	Flavourings (wholesale)
46210	51210	Flax (wholesale)
13100	17140	Flax carding (manufacture)
13100	17140	Flax deseeding (manufacture)
13100	17140	Flax dressing (manufacture)
01160	01110	Flax growing
13100	17140	Flax hackling (manufacture)
13100	17140	Flax preparing (manufacture)
13100	17140	Flax roughing (manufacture)
13100	17140	Flax sorting (manufacture)
13100	17140	Flax spinning (manufacture)
13100	17140	Flax tow (manufacture)
13100	17140	Flax type yarns (manufacture)
13200	17250	Flax woven cloth (manufacture)
13300	17300	Flax yarn bleaching, dyeing or otherwise finishing (manufacture)
13100	17140	Flax-type fibre preparation and spinning (manufacture)
30110	35110	Fleet tender (manufacture)
17290	21259	Flexible paper packaging (manufacture)
22210	25210	Flexible plastic foam (manufacture)
28290	29430	Flexible shaft drive tool (manufacture)
13922	17402	Flexible ventilating ducting made of textiles (manufacture)
20302	24302	Flexographic ink (manufacture)
18129	22220	Flexographic printing (manufacture)
28990	29560	Flexographic printing machine (manufacture)
18121	21251	Flexographic printing on labels or tags (manufacture)
26511	31620	Flight recorder (electric) (manufacture)
30300	35300	Flight simulator (electronic) (manufacture)
08120	14210	Flint bed, pit or quarry
23910	26810	Flint cloth (manufacture)
32990	36639	Flint for lighters (manufacture)
08120	14210	Flint grit production
23910	26810	Flint paper (manufacture)
23110	26110	Float glass (manufacture)
52220	63220	Floating bridge company
92000	92710	Floating casinos
30110	35110	Floating crane (manufacture)
30110	35110	Floating docks construction (manufacture)
09100	11200	Floating drilling rig operation for petroleum or natural gas exploration or production
30110	35110	Floating harbour (manufacture)

SIC 2007	SIC 2003	Activity
30110	35110	Floating landing stages construction (manufacture)
30110	35110	Floating tanks construction (manufacture)
20130	24130	Flocculating agents (chemical) (manufacture)
46410	51410	Flock (wholesale)
13200	17210	Flock made of cotton (manufacture)
23140	26140	Flock made of glass fibre (manufacture)
17120	21120	Flong paperboard (manufacture)
42910	45240	Floodgates, movable barrages and hydro-mechanical structures construction
23320	26400	Floor and quarry tiles made of unglazed clay (manufacture)
23610	26610	Floor and wall tiles made of concrete and terrazzo (manufacture)
77390	71340	Floor cleaning equipment for industrial use leasing
20412	24512	Floor cleanser (manufacture)
43330	45430	Floor covering laying
22230	36639	Floor coverings (hard surface) (manufacture)
47530	52481	Floor coverings (retail)
46730	51479	Floor coverings (wholesale)
22230	25231	Floor coverings made of plastic (manufacture)
22230	25231	Floor coverings made of printed vinyl (manufacture)
22190	25130	Floor coverings made of rubber (manufacture)
22230	25231	Floor coverings made of supported vinyl (manufacture)
16290	20520	Floor coverings of natural cork (manufacture)
20412	24512	Floor polish (manufacture)
27510	29710	Floor polisher (electric) (manufacture)
13939	17519	Floor rugs made of jute (manufacture)
43999	45250	Floor screeding
20412	24512	Floor seal (manufacture)
32910	36620	Floor sweepers (hand operated mechanical) (manufacture)
47530	52481	Floor tiles (not ceramic) (retail)
23610	26610	Floor units made of precast concrete (manufacture)
43330	45430	Flooring contractor
22190	25130	Flooring made of rubber (manufacture)
16100	20100	Flooring made of wood (not parquet flooring) (manufacture)
16100	20100	Flooring made of wood (unassembled) (manufacture)
25110	28110	Flooring systems made of metal (manufacture)
26800	24650	Floppy disk (manufacture)
26200	30020	Floppy disk drives (manufacture)
47760	52489	Florist (retail)
30110	35110	Flotation vessel of steel, for oil platform (manufacture)
10611	15611	Flour (manufacture)
46380	51380	Flour (wholesale)
28930	29530	Flour and meal manufacturing machinery (manufacture)
10710	15810	Flour confectionery (manufacture)
47240	52240	Flour confectionery (retail)
46360	51360	Flour confectionery (wholesale)
28930	29530	Flour confectionery machinery (manufacture)
10611	15611	Flour milling (manufacture)
10611	15611	Flour mixes and prepared blended flour and dough for biscuits (manufacture)
10611	15611	Flour mixes and prepared blended flour and dough for bread (manufacture)
10611	15611	Flour mixes and prepared blended flour and dough for cakes (manufacture)
10611	15611	Flour mixes and prepared blended flour and dough for pancakes (manufacture)

F

SIC 2007	SIC 2003	Activity
10611	15611	Flour of cereal grains production (manufacture)
10612	15612	Flour of dried leguminous vegetables production (manufacture)
10612	15612	Flour or meal of roots or tubers (manufacture)
10110	15112	Flours and meals of meat (manufacture)
26513	33202	Flow measuring and control instrument (non-electronic) (manufacture)
26511	33201	Flow meters (electronic) (manufacture)
26513	33202	Flow meters (non-electronic) (manufacture)
46220	51220	Flower and plants exporter (wholesale)
46220	51220	Flower and plants importer (wholesale)
01190	01120	Flower growing
22290	25240	Flower pots and tubs made of plastic (manufacture)
23490	26250	Flower pots made of clay (manufacture)
23690	26660	Flower pots made of concrete, plaster, cement or artificial stone (manufacture)
46220	51220	Flower salesman (wholesale)
01190	01120	Flower seed growing
47760	52489	Flowers (retail)
46220	51220	Flowers (wholesale)
77299	71409	Flowers and plants rental and leasing
28120	29122	Flowline assembly (hydraulic equipment) (manufacture)
32910	36620	Flue brush (manufacture)
23320	26400	Flue tiles made of clay (manufacture)
28120	29122	Fluid power equipment (manufacture)
28120	29122	Fluid power systems (manufacture)
22190	25130	Fluid seals made of rubber (manufacture)
09100	11200	Fluid-displacement services
27110	31100	Fluorescent ballasts (i.e. Transformers) (manufacture)
20120	24120	Fluorescent brightening agent (manufacture)
27400	31500	Fluorescent tube (manufacture)
26511	33201	Fluorimeter (electronic) (manufacture)
26513	33202	Fluorimeter (non-electronic) (manufacture)
20130	24130	Fluorine, hydrofluoric acid and fluorides (manufacture)
46760	51560	Fluorspar (wholesale)
08910	14300	Fluorspar mining
22230	25239	Flushing cisterns made of plastic (manufacture)
17120	21120	Fluting paper (manufacture)
46120	51120	Flux (commission agent)
20590	24660	Flux (manufacture)
32300	36400	Fly dressing (manufacture)
20200	24200	Fly paper (manufacture)
93120	92629	Flying club
85320	63230	Flying school (for airline pilots)
85530	80410	Flying school activities (not type rating)
85320	63230	Flying schools for commercial airline pilots
85530	80410	Flying schools not issuing commercial certificates and permits
85320	80220	Flying training for professional pilots
28150	29140	Flywheel (not for motor vehicle engine) (manufacture)
46690	51870	Flywheels and pulleys including pulley blocks (wholesale)
13300	17300	Foam backed fabric finishing (manufacture)
13300	17300	Foam backing (single textile material) (manufacture)
13300	17300	Foam backing (texture material sandwich) (manufacture)
22190	25130	Foam rubber (manufacture)
23990	26829	Foamed slag (manufacture)
46210	51210	Fodder (wholesale)
01190	01110	Fodder maize and other grass growing
28302	29320	Fodder preparing equipment (manufacture)
01190	01110	Fodder root growing
20510	24610	Fog signal (manufacture)
30200	35200	Fog signalling equipment (manufacture)
25990	28750	Foil bags (manufacture)
25920	28720	Foil containers made of aluminium (manufacture)
24420	27420	Foil laminate made of aluminium (manufacture)
24420	27420	Foil made of aluminium (decorated, embossed or cut to size) (manufacture)
24420	27420	Foil made of aluminium (not put up as a packaging product) (manufacture)
24440	27440	Foil made of brass (manufacture)
24440	27440	Foil made of copper (manufacture)
22210	25210	Foil made of plastic (manufacture)
24420	27420	Foil packaging goods, made of aluminium (manufacture)
24420	27420	Foil stock made of aluminium (manufacture)
31090	36140	Folding bed (manufacture)
30120	35120	Folding boat made of rubber (manufacture)
17120	21120	Folding boxboard (manufacture)
17219	21219	Folding boxes made of board (manufacture)
28950	29550	Folding machinery for paper and board (not for office use) (manufacture)
17219	21219	Folding paperboard containers (manufacture)
30920	36639	Folding perambulator (manufacture)
10860	15880	Follow-up milk (manufacture)
10860	15880	Follow-up milk for infants (manufacture)
10822	15842	Fondant (manufacture)
46170	51170	Food (commission agent)
47110	52113	Food (general) (retail)
46390	51390	Food (non-specialised) (wholesale)
46690	51870	Food and beverage machinery (wholesale)
46690	51870	Food and drink preparation and manufacturing machinery for industrial use (wholesale)
28930	29530	Food and drink press (manufacture)
82920	74820	Food bottling and packaging on a fee or contract basis
10860	15880	Food for particular nutritional uses (manufacture)
27510	29710	Food freezer for domestic use (manufacture)
28250	29230	Food freezer over 12 cubic feet capacity (manufacture)
71200	74300	Food hygiene testing activities
27510	29710	Food mixer (electric) (manufacture)
46380	51380	Food n.e.c. including fish, crustaceans and molluscs exporter (wholesale)
46380	51380	Food n.e.c. including fish, crustaceans and molluscs importer (wholesale)
56103	55304	Food preparation in market stalls
28930	29530	Food preparation machinery for hotels and restaurants (manufacture)
28930	29530	Food processing equipment (industrial) (manufacture)
28930	29530	Food processing machinery incl. combined processing, packaging or bottling machinery (manufacture)
10890	15899	Food products enriched with vitamins or proteins (manufacture)
10890	15899	Food supplements (manufacture)
46690	51870	Food, beverage and tobacco industry machinery (wholesale)

SIC 2007	SIC 2003	Activity
46390	51390	Food, beverages and tobacco (non-specialised) exporter (wholesale)
46390	51390	Food, beverages and tobacco (non-specialised) importer (wholesale)
10860	15880	Foods for persons suffering from carbohydrate metabolism disorders (manufacture)
10860	15880	Foods to meet the expenditure of intense muscular effort, especially for sportsmen (manufacture)
42990	45230	Foot and cycle path construction
86900	85140	Foot clinic (NHS)
86900	85140	Foot clinic (private)
32500	33100	Foot support (manufacture)
93199	92629	Football Association
32300	36400	Football case made of leather (manufacture)
93120	92629	Football clubs
93110	92619	Football ground
92000	92710	Football pools
93110	92619	Football stadium
46160	51160	Footwear (commission agent)
15200	19300	Footwear (manufacture)
47721	52431	Footwear (retail)
46420	51423	Footwear (wholesale)
46420	51423	Footwear exporter (wholesale)
46420	51423	Footwear importer (wholesale)
15110	19100	Footwear leather preparation (manufacture)
23990	26821	Footwear made of asbestos (manufacture)
14190	18249	Footwear made of textile fabric with applied soles (manufacture)
14190	17710	Footwear made of textiles without applied soles (manufacture)
28940	29540	Footwear making or repairing machinery (manufacture)
46640	51830	Footwear manufacturing and repairing machinery (wholesale)
22290	19300	Footwear parts and accessories made of plastic (manufacture)
77299	71409	Footwear rental and leasing
46210	51210	Forage (wholesale)
28302	29320	Forage harvester (manufacture)
01190	01110	Forage kale and similar forage products growing
01190	01110	Forage plants seed production including grasses
01190	01110	Forage production
10130	15139	Forcemeat (manufacture)
99000	99000	Foreign armed forces
84210	75210	Foreign economic aid services administration (public sector)
99000	99000	Foreign embassy
66120	67130	Foreign exchange broker
99000	99000	Foreign government service
02100	02010	Forest enterprises
02200	02010	Forest harvesting residues, gathering of these for energy
28302	29320	Forest machinery (manufacture)
02400	74149	Forest management consulting services
02400	02020	Forest pest control
02100	02010	Forest tree nursery operation
02400	02020	Forestry fire protection
02400	02020	Forestry inventories
43120	45110	Forestry land drainage
77310	71310	Forestry machinery and equipment rental and leasing (without operator)

SIC 2007	SIC 2003	Activity
46610	51880	Forestry machinery, accessories and implements (wholesale)
02400	02020	Forestry service activities
84130	75130	Forestry services administration and regulation (public sector)
02400	02020	Forestry support services
02100	02010	Forests and timber tract planting, replanting, transplanting, thinning and conservation
24100	27100	Forged bars (manufacture)
24100	27100	Forged rail accessories (manufacture)
24100	27100	Forged sections (manufacture)
24100	27100	Forged semi-finished products (manufacture)
25730	28620	Forges (manufacture)
25500	28400	Forging (manufacture)
24420	27420	Forging bars made of aluminium (manufacture)
28210	29210	Forging furnace (manufacture)
28410	29420	Forging machine (metal forming) (metal working) (manufacture)
25710	28610	Fork (cutlery) (manufacture)
28220	29220	Forklift truck (manufacture)
46690	51870	Forklift trucks (wholesale)
22290	25240	Forks made of plastic (manufacture)
20140	24140	Formaldehyde (manufacture)
28410	29420	Forming machine (high energy rate) (manufacture)
28990	29560	Forming machine for glass working (multi-head) (manufacture)
28960	29560	Forming machine for rubber or plastics (manufacture)
58190	22150	Forms publishing
20200	24200	Formulated pesticide (manufacture)
43999	45250	Formwork (civil engineering)
93210	92330	Fortune telling (fairground)
96090	93059	Fortune telling (not fairground)
52290	63400	Forwarding agents
14142	18232	Foundation garment (manufacture)
43999	45250	Foundations construction
24540	27540	Founding of non-ferrous base metal (manufacture)
24420	27420	Foundry alloy made of aluminium (manufacture)
20590	24660	Foundry bonding clays (manufacture)
20590	24660	Foundry core binder (manufacture)
19100	23100	Foundry coke (manufacture)
20590	24660	Foundry facing (manufacture)
24420	27420	Foundry ingot made of aluminium (manufacture)
16290	20510	Foundry moulding pattern made of wood (manufacture)
28990	29560	Foundry moulds (for rubber or plastic) production machines (manufacture)
28990	29560	Foundry moulds forming machinery (manufacture)
24100	27100	Foundry pig iron (manufacture)
20590	24660	Foundry preparation (manufacture)
32990	36631	Fountain pen (manufacture)
32990	36631	Fountain pen nib (manufacture)
46499	51479	Fountain pens, Indian ink drawing pens, stylograph pens and other pens (wholesale)
45111	50101	Four wheel drive vehicles with a weight not exceeding 3.5 tonnes (new) (retail)
45111	50101	Four wheel drive vehicles with a weight not exceeding 3.5 tonnes (new) (wholesale)
45112	50102	Four wheel drive vehicles with a weight not exceeding 3.5 tonnes (used) (retail)
45112	50102	Four wheel drive vehicles with a weight not exceeding 3.5 tonnes (used) (wholesale)

F

SIC 2007	SIC 2003	Activity
28950	29550	Fourdrinier (manufacture)
09100	11200	Fracture/stimulation services
28490	29430	Frame and carcass cramps (manufacture)
30910	35410	Frame for motor tricycle (manufacture)
30910	35410	Frame for motorcycle (manufacture)
30920	35420	Frame for pedal cycle (manufacture)
30920	35420	Frame for pedal tricycle (manufacture)
16290	20510	Frames for artists canvases (manufacture)
31090	36140	Frames for mattress support made of wood (manufacture)
22230	25239	Frames made of plastic (manufacture)
74909	74879	Franchisers
28230	30010	Franking machine (manufacture)
94990	91330	Fraternities
90030	92400	Freelance journalist
30920	35420	Free wheel for pedal cycle (manufacture)
68209	70209	Freeholder of leasehold property
94990	91330	Freemasons
08110	14110	Freestone mine or quarry
28250	29230	Freezing industrial equipment, including assemblies of components (manufacture)
51210	62209	Freight air transport (non-scheduled)
51210	62109	Freight air transport (scheduled)
77352	71239	Freight air transport equipment operational leasing (without operator)
51210	62209	Freight aircraft rental services with crew (non-scheduled)
52290	63400	Freight broker
77120	71219	Freight container hire
52290	63400	Freight contractor
50200	61102	Freight ferry (domestic or coastal)
50400	61209	Freight ferry (river or estuary)
50200	61102	Freight ferry (sea going)
50400	61209	Freight ferry transport (inland waterway service)
52290	63400	Freight forwarding
77390	71219	Freight land transport equipment rental (without driver)
50200	61102	Freight shipping service (except for inland waterway service)
50200	61102	Freight shipping service (sea and coastal)
49410	60249	Freight transport by animal-drawn vehicles
49200	60109	Freight transport by inter-urban railways
49410	60249	Freight transport by man-drawn vehicles
49200	60109	Freight transport on mainline rail networks
49410	60249	Freight transport operation by road
50400	61209	Freight vessel rental with crew (inland waterway service)
77342	71229	Freight water transport equipment leasing (without operator)
08990	14500	French chalk production
20301	24301	French polish (manufacture)
31090	36140	French polishing (manufacture)
27110	31100	Frequency converter (not power) (manufacture)
26511	33201	Frequency meter (electronic) (manufacture)
26513	33202	Frequency meter (non-electronic) (manufacture)
46310	51310	Fresh fruit (wholesale)
10710	15810	Fresh pastry (manufacture)
03120	05010	Freshwater aquatic animal taking
03120	05010	Freshwater crustacean and mollusc taking
03220	05020	Freshwater crustaceans, bivalves, other molluscs and other aquatic animals, culture of
03120	05010	Freshwater fishing
03120	05010	Freshwater materials gathering
23990	26829	Friction material and unmounted articles thereof, with a mineral or cellulose base (manufacture)
23990	26829	Friction material made of non-metallic mineral (manufacture)
56103	55303	Fried fish shop
65110	66011	Friendly society (not collecting society)
13200	17220	Frieze cloth (manufacture)
13960	17542	Frilling (manufacture)
13960	17542	Fringe (textile material) (manufacture)
03220	05020	Frog farming
46630	51820	Front-end shovel loaders (wholesale)
52102	63121	Frozen and refrigerated goods storage services, for air transport activities
52103	63121	Frozen and refrigerated goods storage services, for land transport activities
52101	63121	Frozen and refrigerated goods storage services, for water transport activities
47110	52113	Frozen food store (retail)
10850	15139	Frozen meals based on meat (manufacture)
10320	15320	Fruit and vegetable concentrates (manufacture)
46310	51310	Fruit and vegetable exporter (wholesale)
46310	51310	Fruit and vegetable importer (wholesale)
46341	51341	Fruit and vegetable juices (wholesale)
46341	51341	Fruit and vegetable juices exporter (wholesale)
46341	51341	Fruit and vegetable juices importer (wholesale)
46310	51310	Fruit and vegetable market porterage (wholesale)
46170	51170	Fruit and vegetables (commission agent)
46310	51380	Fruit and vegetables (processed) (wholesale)
46310	51310	Fruit and vegetables (unprocessed) (wholesale)
01130	01120	Fruit bearing vegetables growing
16290	20510	Fruit bowls made of wood (manufacture)
10710	15810	Fruit cake baking (manufacture)
28302	29320	Fruit cleaning, sorting or grading machines (manufacture)
11070	15980	Fruit cordial (manufacture)
11070	15980	Fruit drinks (non-alcoholic) (manufacture)
20200	24200	Fruit dropping compound (manufacture)
10390	15330	Fruit freezing (manufacture)
10390	15330	Fruit jelly (preserve) (manufacture)
10320	15320	Fruit juice (manufacture)
28930	29530	Fruit juice preparation machinery (manufacture)
10710	15810	Fruit loaf baking (manufacture)
10390	15330	Fruit or vegetable food products (manufacture)
01630	01410	Fruit packing, for primary market
10822	15842	Fruit peel preserving in sugar (manufacture)
10390	15330	Fruit pickling (manufacture)
10710	15810	Fruit pie making (manufacture)
10390	15330	Fruit preserving (manufacture)
10822	15842	Fruit preserving in sugar (manufacture)
10390	15330	Fruit processing and preserving (except in sugar) (manufacture)
28930	29530	Fruit processing machines and equipment (manufacture)
10390	15330	Fruit pulp (manufacture)
46310	51310	Fruit salesman (wholesale)
01250	01120	Fruit seed growing
20200	24200	Fruit setting compound (manufacture)
47210	52210	Fruit shop (retail)

SIC 2007	SIC 2003	Activity	SIC 2007	SIC 2003	Activity
11070	15980	Fruit squash (manufacture)	96030	93030	Funeral furnishing
47810	52620	Fruit stall keeper (retail)	23700	26700	Funerary stonework (manufacture)
11070	15980	Fruit syrup (manufacture)	32401	36501	Funfair articles (manufacture)
01610	01410	Fruit tree and vine trimming, on a fee or contract basis	32401	36501	Funfair games (manufacture)
			20200	24200	Fungicide (manufacture)
01630	01410	Fruit waxing	46120	51120	Fungicides, rodenticides and similar products (commission agent)
11030	15949	Fruit wines other than cider and perry (manufacture)			
10390	15330	Fruit, nuts or vegetables preserved by immersing in oil (manufacture)	46750	51550	Fungicides, rodenticides and similar products (wholesale)
			49319	60219	Funicular railway
10390	15330	Fruit, nuts or vegetables preserved by immersing in vinegar (manufacture)	22290	25240	Funnels made of plastic (manufacture)
			01490	01250	Fur animal raising
47210	52210	Fruiterer (retail)	46420	51429	Fur articles (wholesale)
46310	51310	Fruiterer (wholesale)	46160	51160	Fur broker (commission agent)
27510	29710	Frying pans (electric) (manufacture)	46420	51421	Fur clothing for adults (wholesale)
25990	28750	Frying pans (non-electric) (manufacture)	15110	18300	Fur dressing (manufacture)
46120	51120	Fuel (commission agent)	15110	18300	Fur dressing and dyeing (manufacture)
20590	24660	Fuel additive (manufacture)	01490	01250	Fur farming
77390	71340	Fuel bunkers leasing	46420	51421	Fur merchant (wholesale)
25290	28210	Fuel bunkers made of metal exceeding 300 litres (manufacture)	14200	18300	Fur skin assemblies including dropped fur skins, plates, mats and strips (manufacture)
28210	29210	Fuel burner (other than oil or gas) (manufacture)	01490	01250	Fur skin production from ranching operation
20130	23300	Fuel elements for nuclear reactors production (manufacture)	28210	29210	Furnace (electric) (manufacture)
			23200	26260	Furnace block and pot (manufacture)
47300	50500	Fuel for motor vehicles and motorcycles	46690	51870	Furnace burners, mechanical stokers and grates and mechanical ash dischargers (wholesale)
19201	23201	Fuel heavy fuel oil (manufacture)			
28131	29121	Fuel injection equipment for industrial engines (manufacture)	81223	74705	Furnace cleaning
			28210	29210	Furnace for strip processing line (manufacture)
19201	23201	Fuel oil (manufacture)	28210	29210	Furnace, furnace burner and industrial oven parts (manufacture)
46711	51511	Fuel oil bulk distribution (wholesale)			
47789	52489	Fuel oil for household use (retail)	28210	29210	Furnaces and furnace burners (manufacture)
28131	29121	Fuel pump for industrial engine (manufacture)	46690	51870	Furnaces, ovens, incinerators, for industrial or laboratory use (excluding bakery ovens) (wholesale)
28131	29121	Fuel pump for internal combustion piston engine, for aircraft (manufacture)			
			13921	17401	Furnishing articles (made-up) (manufacture)
84130	75130	Fuel services administration and regulation (public sector)	46470	51471	Furnishing contractor (wholesale)
			13200	17210	Furnishing fabric (woven (not wool or worsted)) (manufacture)
29320	34300	Fuel tank for motor vehicle (manufacture)			
30300	35300	Fuel tanks for aircraft (manufacture)	13200	17230	Furnishing fabric worsted weaving (manufacture)
47789	52489	Fuel wood (retail)	13200	17220	Furnishing fabrics woollen weaving (manufacture)
02100	02010	Fuel wood production	13990	17541	Furnishing lace (manufacture)
46719	51519	Fuels (other than petroleum) (wholesale)	46150	51150	Furniture (commission agent)
46719	51519	Fuels exporter (other than petroleum) (wholesale)	46470	51471	Furniture (wholesale)
			31090	36140	Furniture components made of wood (manufacture)
46719	51519	Fuels importer (other than petroleum) (wholesale)	13921	17401	Furniture covers (manufacture)
			74100	74872	Furniture designing
08120	14220	Fuller's earth pit	46470	51471	Furniture exporter (wholesale)
13300	17300	Fulling mill (manufacture)	31090	36140	Furniture finishing (except chairs and seats) (manufacture)
46750	51550	Fulminates, cyanates and thiocyanates (wholesale)			
31090	36140	Fume cupboards (manufacture)	25720	28630	Furniture fittings made of metal (manufacture)
20200	24200	Fumigating block (manufacture)	22290	25240	Furniture fittings made of plastic (manufacture)
81291	74703	Fumigation services	31090	36140	Furniture for bedrooms (other than mattresses and mattress supports) (manufacture)
93210	92330	Fun fair			
56101	55301	Function room (licensed)	31010	36120	Furniture for churches (manufacture)
66300	67121	Fund management activities	31010	36120	Furniture for cinemas (manufacture)
84110	75110	Fundamental research administration (public sector)	31010	36120	Furniture for drawing offices (manufacture)
84300	75300	Funding and administration of government provided retirement pensions	31090	36140	Furniture for gardens (manufacture)
			31020	36130	Furniture for kitchens (manufacture)
84300	75300	Funding and administration of government provided sickness, work-accident and unemployment insurance	31010	36120	Furniture for laboratories (manufacture)
			31010	36120	Furniture for libraries (manufacture)
82990	74879	Fundraising organisation services on a contract or fee basis	31090	36140	Furniture for living rooms (manufacture)
96030	93030	Funeral and related activities			
96030	93030	Funeral direction			

SIC 2007	SIC 2003	Activity	SIC 2007	SIC 2003	Activity
32500	33100	Furniture for medical, surgical, dental or veterinary use (manufacture)	46720	51520	Galvanised sheets (wholesale)
31010	36120	Furniture for museums (manufacture)	25610	28510	Galvanising (manufacture)
31090	36140	Furniture for nurseries (manufacture)	26513	33202	Galvanometer (manufacture)
31010	36120	Furniture for offices (manufacture)	92000	92710	Gambling activities
46650	51850	Furniture for offices (wholesale)	47220	52220	Game (retail)
31010	36120	Furniture for public houses (manufacture)	46320	51320	Game (wholesale)
31010	36120	Furniture for restaurants (manufacture)	10120	15120	Game bird (fresh, chilled or frozen) dressing or preparation (manufacture)
31010	36120	Furniture for schools (manufacture)	01490	01250	Game bird farming
46650	51850	Furniture for schools (wholesale)	01700	01500	Game propagation
31010	36120	Furniture for ships (manufacture)	32409	36509	Games and toys (manufacture)
31010	36120	Furniture for shops (manufacture)	47650	52485	Games and toys (retail)
31010	36120	Furniture for workrooms (manufacture)	46499	51477	Games and toys (wholesale)
46470	51471	Furniture importer (wholesale)	47650	52485	Games apparatus (retail)
31090	36140	Furniture kit (manufacture)	32401	36501	Games for professional and arcade use (manufacture)
31090	36140	Furniture made of bamboo (other than seating) (manufacture)	32401	36501	Gaming (automatic slot) machines (manufacture)
23690	26660	Furniture made of concrete, plaster, cement or artificial stone (manufacture)	92000	92710	Gaming board for Great Britain
23650	26650	Furniture made of fibre-cement (manufacture)	92000	92710	Gaming club
31090	36140	Furniture parts made of wood (manufacture)	77390	71340	Gaming machine hire
20412	24512	Furniture polish (manufacture)	21200	24421	Gammaglobulin (manufacture)
49420	60241	Furniture removal	46730	51530	Ganister (wholesale)
77299	71409	Furniture rental and leasing	08990	14500	Ganister extraction
52102	63129	Furniture repository for air transport activities	52219	63210	Garage (parking)
52103	63129	Furniture repository for land transport activities	68209	70209	Garage letting (lock up)
52101	63129	Furniture repository for water transport activities	16230	20300	Garage made of wood (manufacture)
14200	18300	Furrier (manufacture)	45112	50102	Garage selling used motor vehicles (retail)
47710	52421	Furrier (retail)	46690	51870	Garage tools (wholesale)
46420	51421	Furrier (wholesale)	23650	26650	Garages made of asbestos cement and concrete (manufacture)
46160	51160	Furs (commission agent)	38110	90020	Garbage collection
14200	18300	Furskin articles (manufacture)	13940	17520	Garden and horticultural net (manufacture)
14200	18300	Furskin assemblies (manufacture)	47760	52489	Garden centre (retail)
14200	18300	Furskin pouffes, unstuffed (manufacture)	31090	36110	Garden chairs (manufacture)
01700	01500	Furskin production (from hunting)	25730	28620	Garden fork (manufacture)
46240	51241	Furskins (wholesale)	25120	28120	Garden frames made of metal (manufacture)
46240	51241	Furskins exporter (wholesale)	16230	20300	Garden frames made of wood (manufacture)
46240	51241	Furskins importer (wholesale)	22190	25130	Garden hose made of rubber (manufacture)
02100	02010	Furze collecting, cutting or gathering	31090	36110	Garden seating (manufacture)
27120	31200	Fuse and fusegear (power) (manufacture)	47760	52489	Garden seeds and plants (retail)
20510	24610	Fuse for explosives (manufacture)	25730	28620	Garden shears (manufacture)
25400	29600	Fuse for shells and bombs (manufacture)	77299	71401	Garden tool hire
24440	27440	Fuse wire (manufacture)	25730	28620	Garden trowel (manufacture)
27120	31200	Fusebox for domestic use (manufacture)	94990	91330	Gardening clubs
20590	24660	Fusel oil (manufacture)	47520	52460	Gardening tools (retail)
30300	35300	Fuselage for aircraft (manufacture)	46610	51880	Gardening tools (wholesale)
27120	31200	Fuses for domestic use (manufacture)	81300	01410	Gardens and sport installations, planting and maintenance on a fee or contract basis
46690	51870	Fuses, relays and apparatus for protecting electrical circuits (wholesale)	01130	01120	Garlic growing
28940	29540	Fusing presses (manufacture)	14200	18300	Garments made of rabbit fur (manufacture)
13200	17210	Fustian weaving (manufacture)	23910	26810	Garnet abrasives (manufacture)
66110	67110	Futures commodity contracts exchanges administration	28940	29540	Garnetters (manufacture)
			14190	18249	Garter (manufacture)
13200	17210	Gabardine (cotton) weaving (manufacture)	47789	52489	Gas (bottled) (retail)
15200	19300	Gaiters made of leather (manufacture)	35230	40220	Gas agents (mains gas)
10130	15139	Galantines (manufacture)	47599	52440	Gas appliances (retail)
13960	17542	Galloon ribbon (manufacture)	46719	51519	Gas bottling and distribution (wholesale)
15200	19300	Galoshes made of rubber (manufacture)	35230	40220	Gas brokers (mains gas)
24100	27100	Galvanised sheet steel (manufacture)	28210	29210	Gas burner (manufacture)

SIC 2007	SIC 2003	Activity
26513	33202	Gas chromatograph (manufacture)
28250	29230	Gas cleansing plant (manufacture)
28131	29121	Gas combustion pumps (manufacture)
28132	29122	Gas compressor (manufacture)
28140	29130	Gas cylinder outlet valves (manufacture)
06200	11100	Gas desulphurisation
27400	31500	Gas discharge lamp (manufacture)
06200	11100	Gas extraction (natural gas)
09100	11200	Gas extraction service activities
28290	29240	Gas generators (manufacture)
25400	29600	Gas guns (manufacture)
27400	31500	Gas lanterns (manufacture)
27510	29710	Gas lighter (electric) (manufacture)
52219	11100	Gas liquefaction for land transportation purposes
52220	11100	Gas liquefaction for water transportation purposes
23200	26260	Gas mantle ring and rod (manufacture)
13960	17549	Gas mantles and tubular gas mantle fabric (manufacture)
32990	33100	Gas masks (manufacture)
26511	33201	Gas meter (electronic) (manufacture)
26513	33202	Gas meter (non-electronic) (manufacture)
15120	19200	Gas meter diaphragm made of leather (manufacture)
42210	45213	Gas offshore pipeline laying
19201	23201	Gas oil (manufacture)
46719	51519	Gas oil (wholesale)
28290	29240	Gas or water generator parts (manufacture)
24200	27220	Gas pipes made of steel (manufacture)
35210	40210	Gas production for the purpose of gas supply
23200	26260	Gas retort and kiln lining (manufacture)
49500	60300	Gas transport via pipelines
28110	29110	Gas turbine (excluding turbo-jets and turbo-propellers) parts (manufacture)
28110	29110	Gas turbine (industrial engine) (manufacture)
28110	29110	Gas turbine for marine use (manufacture)
35210	40210	Gas works
82990	74879	Gas, water and electricity meter reading
46719	51519	Gaseous fuels (other than petroleum) (wholesale)
46711	51511	Gaseous petroleum fuels (wholesale)
28290	29240	Gasket (manufacture)
46690	51870	Gaskets (excluding motor vehicle) (wholesale)
23990	26821	Gaskets made of asbestos (manufacture)
46711	51511	Gasoline (wholesale)
27400	31500	Gasoline lanterns (manufacture)
19201	23201	Gasoline motor fuel (manufacture)
13300	17300	Gassing yarn (manufacture)
16230	20300	Gate made of wood (manufacture)
28140	29130	Gate valves (manufacture)
25120	28120	Gates made of metal (manufacture)
26301	32201	Gateways for telecommunications (manufacture)
28990	29560	Gathering machine (paper working) (manufacture)
26511	33201	Gauge (electronic) (manufacture)
26513	33202	Gauge (non-electronic) (manufacture)
23190	26150	Gauge glass (manufacture)
32990	18249	Gauntlet (protective) (manufacture)
32500	24422	Gauze (surgical) (manufacture)
13200	17210	Gauze weaving (manufacture)
28150	29140	Gear box (not for motor vehicle) (manufacture)
29320	34300	Gear box for motor vehicle (manual or automatic) (manufacture)

SIC 2007	SIC 2003	Activity
30910	35410	Gear box for motorcycle (manufacture)
28150	29140	Gear cutting (not for motor vehicle) (manufacture)
30920	35420	Gear for pedal cycle (manufacture)
28410	29420	Gear making or finishing machines (metal cutting) (manufacture)
28131	29121	Gear pumps (manufacture)
28150	29140	Geared motor unit (manufacture)
28150	29140	Gearing (manufacture)
28150	29140	Gears (manufacture)
46690	51870	Gears, gearing, ball screws, gear boxes, other speed changers (excluding motor vehicles) (wholesale)
01470	01240	Geese farming
20590	24620	Gelatine (manufacture)
20590	24620	Gelatine derivatives (manufacture)
20510	24610	Gelignite (manufacture)
08990	14500	Gem stones mining and quarrying
96090	93059	Genealogical organisation services
96090	93059	Genealogist
81210	74701	General cleaning of houses or apartments
94120	91120	General Council of the Bar
46900	51900	General dealer (wholesale)
25620	28520	General engineering (manufacture)
86101	85111	General hospital (public sector)
86101	85111	General hospital psychiatric unit (public sector)
33120	28520	General mechanical maintenance and repair of machinery (manufacture)
86210	85120	General medical consultant (private practice)
86101	85111	General medical consultant (public sector)
94120	91120	General Medical Council
86210	85120	General medical practitioner
86101	85112	General medicine consultant (private sector)
52102	63129	General merchandise warehouses for air transport activities
94120	91120	General Nursing Council
84110	75140	General personnel administration and operational services (public sector)
18129	22220	General printing (manufacture)
84110	75110	General public administration activities
47110	52112	General store with predominant sale of food beverages or tobacco products (licensed) (retail)
47110	52113	General store with predominant sale of food beverages or tobacco products (unlicensed) (retail)
47190	52120	General stores in which the sale of food beverages or tobacco products is not predominant (retail)
28290	29240	General-purpose machinery parts (manufacture)
46690	51870	Generating sets (wholesale)
35110	40110	Generating station
28110	29110	Generator engine (manufacture)
27110	31100	Generator set (manufacture)
29310	31610	Generators (dynamos and alternators) (manufacture)
86220	85120	Genito-urinary specialist (private practice)
86101	85111	Genito-urinary specialist (public sector)
71122	74206	Geodetic surveying activities
58110	22110	Geographical publishing
71122	74206	Geological and prospecting activities
71122	74206	Geological surveying for petroleum or natural gas (not geological consultancy)
43130	45120	Geological test drilling, test boring and core sampling
71122	74206	Geologist (consultant)
71122	74206	Geophysical consultancy activities (engineering related)

SIC 2007	SIC 2003	Activity
26511	33201	Geophysical instruments and appliances (electronic) (manufacture)
26513	33202	Geophysical instruments and appliances (non-electronic) (manufacture)
43130	45120	Geophysical test drilling, test boring and core sampling
71122	74206	Geophysical, geologic and seismic surveying
71129	74204	Geotechnical engineering consultancy activities
86101	85112	Geriatric hospital (private sector)
86101	85111	Geriatric hospital (public sector)
86101	85111	Geriatrician (public sector)
24450	27450	Germanium (manufacture)
27330	31200	GFCI (ground fault circuit interrupters) (manufacture)
01130	01120	Gherkin growing
10390	15330	Gherkin pickling (manufacture)
47789	52489	Gift shop (retail)
18140	22230	Gilding (printing service) (manufacture)
25610	28510	Gilding of metals (manufacture)
15110	19100	Gill leather (manufacture)
32130	36610	Gilt (manufacture)
13960	17542	Gimp (manufacture)
11010	15910	Gin (manufacture)
10410	15420	Gingelly oil refining (manufacture)
10410	15410	Gingelly seed crushing (manufacture)
11070	15980	Ginger beer (manufacture)
01280	01139	Ginger growing
13200	17210	Gingham weaving (manufacture)
94990	91330	Girl Guides Association
94990	91330	Girls' Brigade
94990	91330	Girls' Friendly Society
15110	19100	Glace kid (manufacture)
21100	24410	Gland extracts (manufacture)
21100	24410	Gland processing (manufacture)
46180	51180	Glands and other organs, extracts thereof and other human or animal substances (commission agent)
46460	51460	Glands, other organs and their extracts and other human or animal substances n.e.c. (wholesale)
47520	52460	Glass (retail)
23190	26150	Glass ball (manufacture)
23190	26150	Glass ball, bar, rod and tube for processing (manufacture)
25730	28620	Glass cutter (manufacture)
28490	29430	Glass cutting (shaping) machines for faceting or for cut-glass articles (cold glass) (manufacture)
28490	29430	Glass cutting machines of the wheel or diamond type (cold glass) (manufacture)
28490	29430	Glass engraving machines of the grinding wheel or diamond type (manufacture)
32500	33100	Glass eyes (manufacture)
23140	26140	Glass fibre spinning and doubling (manufacture)
13200	17250	Glass fibre woven fabric (manufacture)
23140	26140	Glass fibres (manufacture)
23190	26150	Glass in the mass (manufacture)
23130	26130	Glass inners for vacuum flasks and other vacuum vessels (manufacture)
23120	26120	Glass mirrors (manufacture)
23120	26120	Glass mirrors for motor vehicles (not further assembled) (manufacture)
23910	26810	Glass paper (manufacture)
20412	24512	Glass polish (manufacture)
28490	29430	Glass polishing machines (manufacture)

SIC 2007	SIC 2003	Activity
20301	24301	Glass powder (manufacture)
77299	71409	Glass rental and leasing
23120	26120	Glass shaping and processing (manufacture)
46770	51570	Glass waste (wholesale)
23140	26140	Glass wool (manufacture)
28990	29560	Glass, glassware and glass fibre or yarn production machinery (manufacture)
25110	28110	Glasshouses with metal frame (manufacture)
17120	21120	Glassine paper (manufacture)
23190	26150	Glassware (hygienic and pharmaceutical) (manufacture)
47599	52440	Glassware (retail)
46440	51440	Glassware (wholesale)
23130	26130	Glassware for domestic use (manufacture)
23190	26150	Glassware for laboratory, hygienic or pharmaceutical use (manufacture)
23190	26150	Glassware for technical use (manufacture)
23190	26150	Glassware used in imitation jewellery (manufacture)
23310	26300	Glazed fireplace brick (manufacture)
10612	15612	Glazed rice (manufacture)
23310	26300	Glazed tile (manufacture)
23310	26300	Glazed tiles for fireplaces (manufacture)
20301	24301	Glazes and engobes and similar preparations (manufacture)
43342	45440	Glazing
25120	28120	Glazing bars (manufacture)
43342	45440	Glazing contractor
30300	35300	Glider (manufacture)
93120	92629	Glider club
46499	51479	Gliders and hang-gliders (wholesale)
28140	29130	Globe valves (manufacture)
32990	22110	Globes (manufacture)
23190	26150	Globes made of glass (manufacture)
96010	93010	Glove cleaning
15110	19100	Glove leather preparation (manufacture)
14190	18249	Gloves (other than knitted) (manufacture)
46420	51429	Gloves (wholesale)
22190	25130	Gloves and gauntlets of unstitched rubber (manufacture)
14190	18249	Gloves for children (manufacture)
14190	18249	Gloves made of cloth (manufacture)
14190	18249	Gloves made of fur (manufacture)
47710	52421	Gloves made of fur or leather (retail)
46420	51421	Gloves made of fur or leather (wholesale)
14190	18249	Gloves made of leather (not sports) (manufacture)
14190	18249	Gloves made of textiles for household use (manufacture)
22290	25240	Gloves made of unstitched plastic (manufacture)
47710	52422	Gloves, children's, other than of fur or leather (retail)
47710	52424	Gloves, men's, other than of fur or leather (retail)
47710	52423	Gloves, women's, other than of fur or leather (retail)
29310	31610	Glow plugs (manufacture)
10620	15620	Glucose (manufacture)
10620	15620	Glucose syrup (manufacture)
20520	24620	Glue (manufacture)
16210	20200	Glue laminated wood (manufacture)
18140	22230	Glueing of books, brochures, etc. (manufacture)
16230	20300	Glue-laminated and metal connected prefabricated wooden roof trusses (manufacture)
46120	51120	Glues (commission agent)

SIC 2007	SIC 2003	Activity
28990	29560	Gluing machinery for bookbinders, etc. (manufacture)
28490	29430	Gluing machines (manufacture)
10620	15620	Gluten (manufacture)
10860	15880	Gluten-free foods (manufacture)
46120	51120	Glycerol (commission agent)
20411	24511	Glycerol (manufacture)
46460	51460	Glycoside, vegetable alkaloids and their salts, ethers, esters and other derivatives (wholesale)
21100	24410	Glycosides and their salts, ethers, esters and other derivatives (manufacture)
46180	51180	Glycosides, vegetable alkaloids, their salts, ethers, esters and other derivatives (commission agent)
01450	01220	Goat farming
01450	01220	Goat milk (raw) production
01450	01220	Goat raising and breeding
32500	33401	Goggles (manufacture)
24410	27410	Gold (manufacture)
46720	51520	Gold and other precious metals (wholesale)
32120	36220	Gold and silver braid (manufacture)
24410	27410	Gold and silver bullion (manufacture)
32120	36220	Gold and silver embroidery (manufacture)
32120	36220	Gold and silver mounting (manufacture)
18140	22230	Gold blocking (manufacture)
32120	36220	Gold laceman (manufacture)
32120	36220	Gold leaf (manufacture)
07290	13200	Gold mining and preparation
24410	27410	Gold rolled onto base metals or silver production (manufacture)
18140	22230	Gold stamping (manufacture)
15110	19100	Goldbeaters' skin or bung (manufacture)
32120	36220	Goldsmiths' articles (manufacture)
32120	36220	Goldsmiths' articles of base metals clad with precious metals (manufacture)
32120	36220	Goldsmiths' articles of precious metals (manufacture)
32300	36400	Golf ball (finished) (manufacture)
22190	25130	Golf ball core (manufacture)
29100	34100	Golf carts (manufacture)
93120	92629	Golf club
32300	36400	Golf club (manufacture)
42990	45230	Golf course construction
93110	92619	Golf courses
93110	92619	Golf links
28220	29220	Goliath type crane (manufacture)
52290	63400	Goods agent (not transport authority)
52290	63400	Goods handling operations
52219	63210	Goods handling station operation
46900	51900	Goods n.e.c. exporter (wholesale)
46900	51900	Goods n.e.c. importer (wholesale)
30200	35200	Goods van for railways (manufacture)
28290	29240	Goods vending machines (manufacture)
30200	35200	Goods wagon for railways (manufacture)
10120	15120	Goose (fresh, chilled or frozen) slaughter and dressing (manufacture)
01470	01240	Goose raising and breeding
01250	01139	Gooseberry growing
25730	28620	Gouge (wood frame) (manufacture)
42990	45213	Government department (building and civil engineering works division)
91012	75140	Government records and archives maintenance and storage (public sector)
72190	73100	Government research establishment (other than biotechnological)
85320	80220	Government training centre
14132	18222	Gowns (manufacture)
14120	18210	Gowns for academic, legal or ecclesiastical use (manufacture)
26513	33202	GPS devices (non-electronic) (manufacture)
28922	29522	Grab (manufacture)
28922	29522	Grader (manufacture)
46630	51820	Graders and levellers (wholesale)
85421	80302	Graduate school for business studies
23190	26150	Graduated glassware (manufacture)
46210	51210	Grain (wholesale)
28302	29320	Grain auger (manufacture)
46110	51110	Grain broker (commission agent)
28930	29530	Grain brushing machines (manufacture)
28302	29320	Grain cleaning, sorting and grading machines (manufacture)
28302	29320	Grain drier (manufacture)
01630	01410	Grain drying
01110	01110	Grain maize growing
10611	15611	Grain milling (manufacture)
28930	29530	Grain milling industry machinery (manufacture)
24100	27100	Grain oriented electrical sheet steel (manufacture)
28930	29530	Grain processing machinery and plant (manufacture)
52102	63123	Grain silos operation for air transport activities
52103	63123	Grain silos operation for land transport activities
52101	63123	Grain silos operation for water transport activities
52102	63123	Grain warehouse for air transport activities
52103	63123	Grain warehouse for land transport activities
52101	63123	Grain warehouse for water transport activities
46210	51210	Grain, seeds and animal feeds exporter (wholesale)
46210	51210	Grain, seeds and animal feeds importer (wholesale)
85310	80210	Grammar schools
26400	32300	Gramophone (manufacture)
26400	32300	Gramophone accessory (manufacture)
26400	32300	Gramophone cabinet (manufacture)
59200	22140	Gramophone record publishing
18201	22310	Gramophone record reproduction from master copies (manufacture)
18201	22310	Gramophone records (except master copies) including blanks for cutting) (manufacture)
47630	52450	Gramophone records (retail)
46431	51431	Gramophone records (wholesale)
46431	51431	Gramophone records, recorded tapes, CDs etc. and equipment for playing them, exporter (wholesale)
46431	51431	Gramophone records, recorded tapes, CDs etc. and equipment for playing them, importer (wholesale)
52102	63123	Granary for air transport activities
52103	63123	Granary for land transport activities
52101	63123	Granary for water transport activities
46730	51530	Granite (wholesale)
08110	14110	Granite quarrying (rough trimming and sawing)
23700	26700	Granite working (manufacture)
94990	91330	Grant giving activities by membership organisations or others
94990	91330	Grant making foundations
28990	29560	Granulator for the chemical industry (manufacture)
01210	01131	Grape production
01230	01139	Grapefruit growing

G

SIC 2007	SIC 2003	Activity
74100	74872	Graphic designer
26200	30020	Graphics tablets and other input devices (manufacture)
23990	26829	Graphite (manufacture)
08990	14500	Graphite (natural) mining
28990	29560	Graphite electrodes production machinery (manufacture)
23990	26829	Graphite products (other than block and crucible) (manufacture)
96090	93059	Graphologist
25990	28750	Grapnel (manufacture)
28210	29210	Grates (mechanical) (manufacture)
27520	29720	Grates for domestic use (non-electric) (manufacture)
32500	33401	Grating (mounted, (not photographic)) (manufacture)
32500	33401	Grating (unmounted, optical) (manufacture)
46730	51530	Gravel (wholesale)
08120	14210	Gravel and sand breaking and crushing
09900	14210	Gravel and sand pit support services provided on a fee or contract basis
09100	11200	Gravel packing services
08120	14210	Gravel pit or quarry
24540	27540	Gravity casting of non-ferrous base metal (manufacture)
20302	24302	Gravure ink (manufacture)
28990	29560	Gravure printing machine (manufacture)
18121	21251	Gravure printing on labels or tags (manufacture)
10840	15870	Gravy (manufacture)
01629	01429	Grazing
19209	23209	Grease formulation outside refineries (manufacture)
17120	21120	Greaseproof paper (manufacture)
19201	23201	Greases (at refinery) (manufacture)
46719	51519	Greases (wholesale)
47210	52210	Greengrocer (retail)
16230	20300	Greenhouse made of wood (manufacture)
18129	22220	Greeting card printing (manufacture)
46499	51479	Greeting cards (wholesale)
58190	22150	Greeting cards publishing
47620	52470	Greetings cards (retail)
25400	29600	Grenade (manufacture)
17120	21120	Grey board (manufacture)
93110	92619	Greyhound racing stadium
93110	92619	Greyhound track
93199	92629	Greyhound training
16290	20510	Grids made of wood (manufacture)
25120	28120	Grilles made of metal (not cast) (manufacture)
27510	29710	Grills (electric) (manufacture)
25930	28730	Grills made of wire (manufacture)
27510	29710	Grinders (electric) (manufacture)
25610	28510	Grinding (metal finishing) (manufacture)
28923	29523	Grinding and other mineral processing machinery (manufacture)
28410	29420	Grinding machines (metal cutting) (manufacture)
28990	29560	Grinding machines for glass (manufacture)
28302	29320	Grinding mill for agricultural use (manufacture)
28930	29530	Grinding mills (manufacture)
23910	26810	Grinding paste (manufacture)
28240	29410	Grinding tools (powered portable) (manufacture)
28490	29430	Grinding wheel cutting and dressing machines (manufacture)

SIC 2007	SIC 2003	Activity
28490	29430	Grinding, smoothing, polishing and graining machines for stone, ceramics, asbestos
23910	26810	Grindstones made of bonded abrasives (manufacture)
10611	15611	Grist milling (manufacture)
28290	29240	Grit extraction plant for effluent treatment (manufacture)
28923	29523	Gritting machine (manufacture)
10611	15611	Groats production (manufacture)
47110	52112	Grocer with alcohol licence (retail)
47110	52113	Grocer without alcohol licence (retail)
47810	52620	Grocery stall (retail)
96090	93059	Grooming of pet animals
16100	20100	Grooved wood (manufacture)
10832	15862	Ground coffee (manufacture)
30300	35300	Ground effect vehicles (manufacture)
30300	35300	Ground equipment for spacecraft (excluding electronic or telemetric equipment) (manufacture)
30300	35300	Ground flying trainers (manufacture)
46690	51870	Ground flying trainers (wholesale)
26309	32202	Ground station for relay satellite communication (manufacture)
42110	45230	Ground work contracting
68209	70209	Ground, landlord
10410	15410	Groundnut crushing (manufacture)
01110	01110	Groundnut growing
10410	15420	Groundnut oil refining (manufacture)
22190	25130	Groundsheet made of rubber (manufacture)
93110	92619	Groundsman
22190	25130	Grout packers (manufacture)
43999	45250	Grouting contractor (building)
43999	45250	Grouting contractor (civil engineering)
02300	02010	Growing materials, gathering from the wild
01250	01139	Growing of berries
01230	01139	Growing of citrus fruits
02100	01120	Growing of forest tree seeds
01260	01110	Growing of oleaginous fruits other than olives
46750	51550	Guano (wholesale)
08910	14300	Guano mining
80100	74602	Guard activities
55209	55239	Guest house (licensed)
55209	55239	Guest house (unlicensed)
96090	93059	Guide (other than tourist)
25400	29600	Guided weapon airborne delivery system, not intercontinental ballistic missile (ICBM) (manufacture)
26511	33201	Guided weapon launching gear and launch control post (electronic) (manufacture)
25400	29600	Guided weapon warheads (manufacture)
25400	29600	Guided weapon, except intercontinental ballistic missile (ICBM) (manufacture)
94110	91110	Guilds and similar organisations
01470	01240	Guinea fowl production
01470	01240	Guinea fowl raising and breeding
32200	36300	Guitar (manufacture)
23610	26610	Gullies made of concrete (manufacture)
20520	24620	Gum (manufacture)
17230	21230	Gummed paper ready for use (manufacture)
46750	51550	Gums (wholesale)
25400	29600	Gun (manufacture)
25400	29600	Gun carriage mounting or platform (manufacture)

SIC 2007	SIC 2003	Activity
15120	19200	Gun cases made of leather (manufacture)
20510	24610	Guncotton (manufacture)
26511	33201	Gunnery control instrument (electronic) (manufacture)
26513	33202	Gunnery control instrument (non-electronic) (manufacture)
26701	33402	Gunnery control instrument (optical) (manufacture)
20510	24610	Gunpowder (manufacture)
16290	20510	Gunstock made of wood (manufacture)
32990	36639	Gut for musical instruments and sports goods (manufacture)
32990	36639	Gut scraping and spinning (manufacture)
22190	25130	Gutta percha goods (manufacture)
22230	25239	Gutter and fittings made of plastic (manufacture)
93110	92629	Gymnasium
46499	51479	Gymnasium and athletic articles and equipment (wholesale)
32300	36400	Gymnasium equipment and appliances (manufacture)
85510	92629	Gymnastics instruction
86220	85120	Gynaecologist (private practice)
86101	85111	Gynaecologist (public sector)
08110	14120	Gypsum mine or quarry
23520	26530	Gypsum plaster (manufacture)
23620	26620	Gypsum plaster products (manufacture)
28990	33202	Gyroscope (manufacture)
13960	17542	Haberdashery (narrow fabrics) (manufacture)
47510	52410	Haberdashery (retail)
46410	51410	Haberdashery (wholesale)
25730	28620	Hacksaw blades (manufacture)
07100	13100	Haematite quarry
86101	85111	Haematologist (public sector)
10130	15139	Haggis (manufacture)
10110	15112	Hair (animal by-product) from knackers (manufacture)
32910	36620	Hair brush (manufacture)
27510	29710	Hair clippers (electric) (manufacture)
25710	28610	Hair clippers (manufacture)
22290	36639	Hair comb made of plastic (manufacture)
22290	36639	Hair curler made of plastic (manufacture)
32910	36620	Hair dressing for brushes (manufacture)
13100	17170	Hair dressing for upholsterers (manufacture)
27510	29710	Hair dryer (electric) (manufacture)
13300	17300	Hair dyeing (textile) (manufacture)
25990	36639	Hair grips and pins made of metal (manufacture)
20420	24520	Hair lacquers (manufacture)
14190	18249	Hair nets (manufacture)
14190	18249	Hair nets made of lace (manufacture)
32990	36639	Hair pad making (manufacture)
22190	36639	Hair pins of hard rubber (manufacture)
32990	36639	Hair preparation for wig making (manufacture)
20420	24520	Hair preparations (manufacture)
22190	36639	Hair rollers and similar of hard rubber (manufacture)
32990	36639	Hair slides (manufacture)
13200	17210	Haircord weaving (manufacture)
46450	51450	Hairdressers' sundriesman (wholesale)
96020	93020	Hairdressing activities
29320	34300	Half shaft (manufacture)
87900	85311	Halfway group homes for persons with social or personal problems (charitable)
87900	85312	Halfway group homes for persons with social or personal problems (non-charitable)
87900	85311	Halfway homes for delinquents and offenders (charitable)
87900	85312	Halfway homes for delinquents and offenders (non-charitable)
31090	36140	Hall stand (manufacture)
55900	55239	Halls of residence
46120	51120	Halogen or sulphur compounds of non-metals (commission agent)
46750	51550	Halogen or sulphur compounds of non-metals (wholesale)
20140	24140	Halogenated derivatives of hydrocarbon (manufacture)
20130	24130	Halogens and halides (inorganic) (manufacture)
10130	15131	Ham boiling (manufacture)
10130	15131	Ham cooking or preparing in bulk (manufacture)
10130	15131	Ham curing (manufacture)
10130	15131	Ham production (manufacture)
10130	15131	Ham smoking (manufacture)
25730	28620	Hammer (manufacture)
28240	29410	Hammer (portable, powered) (manufacture)
25500	28400	Hammer forging of steel (manufacture)
46740	51540	Hammers (wholesale)
13940	17520	Hammocks (manufacture)
93290	92729	Hampton Court Gardens and Park
01490	01250	Hamster raising and breeding
13300	17300	Hand block printing of textiles (manufacture)
77390	71219	Hand cart hire
20420	24520	Hand cream (manufacture)
47510	52410	Hand knitting yarns (retail)
46410	51410	Hand knitting yarns (wholesale)
15120	19200	Hand luggage made of leather (manufacture)
17120	21120	Hand made paper (manufacture)
46410	51410	Hand mending yarns (wholesale)
46690	51870	Hand or foot operated air pumps (wholesale)
28131	29121	Hand or foot operated pumps (manufacture)
32990	36631	Hand printing sets (manufacture)
32990	36639	Hand riddles (manufacture)
32990	36639	Hand sieves (manufacture)
25730	28620	Hand tools (manufacture)
46740	51540	Hand tools (wholesale)
28240	29410	Hand tools with non-electric motor parts (manufacture)
28240	29410	Hand tools with self contained motor or pneumatic drive (manufacture)
13923	17403	Hand towel (manufacture)
28220	35500	Hand truck made of metal (manufacture)
47722	52432	Handbags (retail)
46499	51479	Handbags (wholesale)
15120	19200	Handbags and the like of composition leather (manufacture)
15120	19200	Handbags made of leather or leather substitutes (manufacture)
30990	35500	Handcarts (manufacture)
26200	30020	Hand-held computers PDA (manufacture)
16290	20510	Handicraft articles made of wood (manufacture)
47789	52489	Handicrafts shop (retail)
14190	18249	Handkerchief folding (manufacture)
14190	18249	Handkerchief hemming (manufacture)
17220	21220	Handkerchief made of paper (manufacture)

H

SIC 2007	SIC 2003	Activity
14190	18249	Handkerchief made of textile material (manufacture)
30910	35410	Handlebar for motorcycle (manufacture)
30920	35420	Handlebar for pedal cycle (manufacture)
16290	20510	Handles and bodies for brooms made of wood (manufacture)
16290	20510	Handles and bodies for brushes made of wood (manufacture)
16290	20510	Handles and bodies for tools made of wood (manufacture)
16290	36639	Handles for canes, umbrellas and similar (manufacture)
22290	25240	Handles for furniture made of plastic (manufacture)
16290	20510	Handles made of wood (manufacture)
28220	29220	Handling plant (hydraulic and pneumatic) (manufacture)
28220	29220	Hand-operated lifting capstans (manufacture)
28220	29220	Hand-operated lifting hoists (manufacture)
28220	29220	Hand-operated lifting jacks (manufacture)
28220	29220	Hand-operated lifting pulley tackle (manufacture)
28220	29220	Hand-operated lifting winches (manufacture)
25730	28620	Handsaw (manufacture)
13923	17403	Hand-woven tapestries (manufacture)
30300	35300	Hang glider (manufacture)
43330	45430	Hanging or fitting wooden wall coverings
52220	63220	Harbour authority
42910	45240	Harbour construction
52220	63220	Harbour operation
19201	10103	Hard coal agglomeration (manufacture)
09900	10102	Hard coal mining (opencast) support services, provided on a fee or contract basis
05101	10101	Hard coal mining (underground)
09900	10101	Hard coal mining (underground) support services, provided on a fee or contract basis
05102	10102	Hard coal recovery from tips
19100	23100	Hard coke (manufacture)
19100	23100	Hard coke breeze (manufacture)
26200	30020	Hard disk drives (manufacture)
26800	24650	Hard drive media (manufacture)
32990	25240	Hard hats and other personal safety equipment of plastics (manufacture)
25730	28620	Hard metal tipped tools (manufacture)
20411	24511	Hard soap (manufacture)
16210	20200	Hardboard (manufacture)
19201	10103	Hard-coal briquettes (manufacture)
26513	33202	Hardness testing instrument (non-electronic) (manufacture)
46150	51150	Hardware (commission agent)
47520	52460	Hardware (retail)
62020	72100	Hardware consultancy
62020	72100	Hardware disaster recovery services
46740	51540	Hardware equipment and supplies (wholesale)
62020	72100	Hardware installation services
46740	51540	Hardware, plumbing and heating equipment (wholesale)
46740	51540	Hardware, plumbing and heating equipment and supplies exporter (wholesale)
46740	51540	Hardware, plumbing and heating equipment and supplies importer (wholesale)
16220	20300	Hardwood flooring strip (manufacture)
46730	51530	Hardwoods (wholesale)
32200	36300	Harmoniums (manufacture)
32200	36300	Harmoniums with free metal reeds (manufacture)
15120	19200	Harness (manufacture)
15110	19100	Harness and saddlery leather preparation (manufacture)
15120	19200	Harness front and rosette made of leather (manufacture)
32200	36300	Harpsichord (manufacture)
28302	29320	Harrow (manufacture)
28302	29320	Harvesters (manufacture)
46610	51880	Harvesters (wholesale)
28302	29320	Harvesting or threshing machinery harvesters, threshers, sorters (manufacture)
15110	19100	Hat and cap leather preparation (manufacture)
13960	17542	Hat bands (manufacture)
15120	19200	Hat box made of leather or leather substitute (manufacture)
14190	18241	Hat lining (manufacture)
46420	51429	Hat materials (wholesale)
14190	18241	Hat pad (manufacture)
14190	18241	Hat shape (manufacture)
28990	29560	Hatch cover (mechanically operated) (manufacture)
25730	28620	Hatchet (manufacture)
14190	18241	Hats made of cloth (manufacture)
14190	18241	Hats made of felt (manufacture)
14190	18241	Hats made of fur (manufacture)
14190	18241	Hats made of fur fabric (manufacture)
17290	36639	Hats made of paper (manufacture)
14190	18241	Hats made of silk (manufacture)
14190	18241	Hats made of wool (manufacture)
14200	18300	Hatters' fur (manufacture)
49410	60249	Haulage in tanker trucks by road
49410	60249	Haulage of automobiles by road
49410	60249	Haulage of logs by road
77120	71219	Haulage tractors rental (without driver)
28921	29521	Hauling engine for mining (stationary) (manufacture)
13922	17402	Haversack (manufacture)
46210	51210	Hay (wholesale)
28302	29320	Hay making equipment (manufacture)
77320	71320	Hazard warning lamp hire
38120	90020	Hazardous waste collection (e.g. Batteries, used cooking oils, etc.)
01250	01139	Hazelnut growing
70100	74154	Head office of catering company
70100	74153	Head office of construction company
70100	74155	Head office of motor trades company
70100	74159	Head office of other non-financial company
70100	74158	Head office of production company
70100	74157	Head office of retail company
70100	74156	Head office of service trades company
70100	74152	Head office of transport company
70100	74151	Head office of wholesale company
32300	36400	Headgear for sports (manufacture)
23990	26821	Headgear made of asbestos (manufacture)
14190	18241	Headgear made of furskins (manufacture)
22290	25240	Headgear made of plastic (manufacture)
28921	29521	Heading machine for mining (manufacture)
26400	32300	Headphones (manufacture)
26400	32300	Headset (not telecommunication type) (manufacture)
14190	18249	Headsquare (manufacture)
28940	29540	Heald (textile machinery accessory) (manufacture)

SIC 2007	SIC 2003	Activity
84220	75220	Health activities for military personnel in the field
71122	74206	Health and safety and other hazard protection and control consultancy activities
84120	75120	Health care services administration (public sector)
86900	85140	Health centre
93130	93040	Health clubs
96040	93040	Health farm
47290	52270	Health foods (retail)
65120	66031	Health insurance
86900	85140	Health visitor
26600	33100	Hearing aid (electronic) (manufacture)
47741	52321	Hearing aids (retail)
46460	51460	Hearing aids (wholesale)
30990	35500	Hearses drawn by animals (manufacture)
32910	36620	Hearth brush (manufacture)
23310	26300	Hearth or wall tiles made of non-refractory ceramic (manufacture)
23310	26300	Hearth tile made of clay (unglazed) (manufacture)
43220	45330	Heat and air-conditioning installation
27520	29720	Heat emitter (space heating equipment) (non-electric) (manufacture)
46690	51870	Heat exchange units and machinery for liquefying air and other gases (wholesale)
28250	29230	Heat exchange units for air conditioning (manufacture)
25300	28300	Heat exchanger for process plant (manufacture)
46740	51540	Heat insulated tanks for central heating (wholesale)
23200	26260	Heat insulating ceramic goods made of siliceous fossil meals (manufacture)
23990	26829	Heat insulating materials (other than asbestos) (manufacture)
26511	33201	Heat meters (electronic) (manufacture)
26513	33202	Heat meters (non-electronic) (manufacture)
23130	26130	Heat resisting glassware for cooking purposes (manufacture)
22290	25240	Heat sensitive adhesive tape made of plastic (manufacture)
28940	29540	Heat setting machinery for textiles (manufacture)
28210	29210	Heat treatment furnace (manufacture)
10390	15330	Heat treatment of fruit and vegetables (manufacture)
10511	15511	Heat treatment of milk (manufacture)
20590	24660	Heat treatment salts (manufacture)
27510	29710	Heater for motor vehicle (manufacture)
02100	02010	Heath collecting, cutting or gathering
27520	29720	Heating and cooking appliances for domestic use (oil fired) (manufacture)
43220	45330	Heating and plumbing contracting
27520	29720	Heating appliances for domestic use (non-electric) (manufacture)
43220	45330	Heating engineering (buildings)
46740	51540	Heating equipment and supplies (wholesale)
46719	51519	Heating oil (wholesale)
27510	29710	Heating resistors (electric) (manufacture)
43220	45330	Heating service contracting
71129	74204	Heating systems for buildings design activities
28140	29130	Heating taps (manufacture)
28140	29130	Heating valves (manufacture)
28210	29210	Heating/melting high frequency induction or dielectric equipment (manufacture)
25500	28400	Heavy forging (manufacture)
29100	34100	Heavy goods vehicle (manufacture)
49410	60249	Heavy haulage by road
50200	61102	Heavy lift vessel services
25930	28730	Heavy wire (manufacture)
28230	30010	Hectograph (manufacture)
28240	29410	Hedge trimmers (powered portable) (manufacture)
81300	01410	Hedge trimming on a fee or contract basis (except as an agricultural service activity)
28302	29320	Hedgecutter for agricultural use (manufacture)
22190	19300	Heel and sole made of rubber (manufacture)
95230	52740	Heel replacement services (while you wait)
15200	19300	Heels made of leather (manufacture)
28131	29121	Helical rotor pump (manufacture)
25930	28740	Helical springs (manufacture)
28490	29430	Helical-wire cutting machines for stone, ceramics, asbestos-cement or similar (manufacture)
28131	29121	Helicoidal pumps (manufacture)
30300	35300	Helicopter (manufacture)
51210	62209	Helicopter freight services (non-scheduled)
51102	62201	Helicopter passenger services (non-scheduled)
30300	35300	Helicopter rotors for aircraft (manufacture)
46690	51870	Helicopters (wholesale)
43999	45250	Helideck erection work
46210	51210	Hemp (unprocessed) (wholesale)
13100	17170	Hemp carding (manufacture)
13100	17170	Hemp dressing (manufacture)
13940	17520	Hemp rope, cord or line (manufacture)
13100	17170	Hemp sorting (manufacture)
13100	17170	Hemp spinning (manufacture)
13200	17250	Hemp weaving (manufacture)
18140	22230	Heraldic chasing and seal engraving (manufacture)
18140	22230	Heraldic engraving (manufacture)
90030	92319	Heraldic painting
46170	51170	Herb infusions (commission agent)
47210	52210	Herb seller (food) (retail)
10831	15861	Herb tea (manufacture)
47210	52210	Herbalist (food) (retail)
20200	24200	Herbicide (manufacture)
46750	51550	Herbicides and insecticides (wholesale)
01130	01120	Herbs (culinary) growing
46310	51310	Herbs (wholesale)
77390	71340	Herd leasing
01629	01429	Herd testing services
46330	51333	Herring oil (wholesale)
10410	15420	Herring oil refining (manufacture)
46380	51380	Herrings (wholesale)
13200	17250	Hessian (manufacture)
46410	51410	Hessian (wholesale)
46120	51120	Heterocyclic compounds (commission agent)
20140	24140	Heterocyclic compounds (manufacture)
46750	51550	Heterocyclic compounds (wholesale)
18129	22220	Hexachrome printing (manufacture)
46110	51110	Hide and skin broker (commission agent)
10110	15111	Hide degreasing (manufacture)
10110	15111	Hide pickling (manufacture)
46240	51249	Hides (wholesale)
28940	29540	Hides and skins preparation, tanning, working or repairing machinery (manufacture)
10110	15111	Hides and skins production from abattoirs (manufacture)

H

SIC 2007	SIC 2003	Activity
10110	15111	Hides and skins production from knackers (manufacture)
10110	15111	Hides and skins production from slaughterhouses (manufacture)
46240	51249	Hides, skins and leather exporter (wholesale)
46240	51249	Hides, skins and leather importer (wholesale)
26400	32300	Hi-fi equipment (manufacture)
47430	52450	Hi-fi equipment (retail)
24100	27100	High carbon ferro-manganese (carbon over 2%) (manufacture)
84230	75230	High Court of Justice
84230	75230	High court of Justice in Bankruptcy
84230	75230	High Court of Justiciary (Scotland)
24100	27100	High speed tool steel (manufacture)
20600	24700	High tenacity yarn made of viscose rayon (manufacture)
85410	80301	Higher education (sub degree level)
85422	80303	Higher education at post-graduate level
85421	80302	Higher education at the first degree level
81299	90030	Highway cleaning of snow and ice
42110	45230	Highway construction
25720	28630	Hinge (manufacture)
01430	01220	Hinny farming and breeding
64921	65221	Hire purchase company (other than in banks' sector)
94990	91330	Historical club
91020	92521	Historical museums
96090	93059	Historical research
91030	92522	Historical sites and buildings preservation
93290	92729	Hobby instructor (own account)
93120	92629	Hockey club
32300	36400	Hockey stick (manufacture)
93110	92619	Hockey, cricket, rugby stadiums operation
25730	28620	Hoe (manufacture)
28220	29220	Hoists (electric wire or chain) (manufacture)
28220	29220	Hoists (hand operated pulley and sheave block) (manufacture)
28220	29220	Hoists, hydraulic, mechanical, pneumatic (not builders, hand operated and electrical) (manufacture)
64205	65234	Holding company activities for banks
64201	74159	Holding company in agricultural sector
64209	74154	Holding company in catering sector
64203	74153	Holding company in construction sector
64205	65239	Holding company in financial services sector
64204	74155	Holding company in motor trades sector
64202	74158	Holding company in production sector
64209	74159	Holding company in property sector
64204	74157	Holding company in retail sector
64209	74152	Holding company in transport sector
64204	74151	Holding company in wholesale sector
64209	74159	Holding company n.e.c.
28230	30010	Hole punches (manufacture)
55201	55231	Holiday and other short stay accommodation, provided in holiday centres and holiday villages
55201	55231	Holiday camp
55201	55231	Holiday centres and villages
55209	55232	Holiday home (not charitable)
79909	63309	Holiday information centre
24200	27220	Hollow bars made of steel (manufacture)
24100	27100	Hollow drill bars made of steel (manufacture)
23130	26130	Hollow glass (manufacture)

SIC 2007	SIC 2003	Activity
23320	26400	Hollow partition made of clay (manufacture)
24510	27210	Hollow profiles of cast-iron (manufacture)
24420	27420	Hollow sections made of aluminium (manufacture)
25990	28750	Hollow ware (domestic) made of metal (manufacture)
22290	25240	Hollow ware made of plastic (manufacture)
23200	26260	Hollow ware made of refractory (manufacture)
87300	85311	Home for the blind (charitable)
87300	85312	Home for the blind (non-charitable)
87300	85311	Home for the disabled (charitable)
87300	85312	Home for the disabled (non-charitable)
87300	85311	Home for the elderly (charitable)
87300	85312	Home for the elderly (non-charitable)
88100	85321	Home help service (charitable)
88100	85322	Home help service (non-charitable)
86900	85140	Home nurse (NHS)
86900	85140	Homeopath (not registered medical practitioner)
86220	85120	Homeopath (registered medical practitioner)
21200	24421	Homeopathic preparations (manufacture)
87300	85311	Homes for the elderly with minimal nursing care (charitable)
87300	85312	Homes for the elderly with minimal nursing care (non-charitable)
87100	85140	Homes for the elderly with nursing care
10860	15880	Homogenised food preparations (manufacture)
10390	15330	Homogenised fruit and vegetables (manufacture)
10511	15511	Homogenised milk production (manufacture)
28930	29530	Homogenisers (manufacture)
23910	26810	Hones (bonded) (manufacture)
46380	51380	Honey (wholesale)
01490	01250	Honey and beeswax production
01490	01250	Honey processing and packing
28410	29420	Honing machine (metal cutting) (manufacture)
25990	28750	Hook and eye (manufacture)
30200	35200	Hooks and coupling devices (manufacture)
16240	20400	Hoops made of wood (manufacture)
02100	02010	Hoopwood production
10110	15112	Hooves from knackers production (manufacture)
11070	15980	Hop bitters (manufacture)
01280	01110	Hop cones growing
10890	15899	Hop extracts (manufacture)
46210	51210	Hops (wholesale)
09100	11200	Horizontal drilling services
21200	24421	Hormonal contraceptive medicaments (manufacture)
21100	24410	Hormone (not plant hormone) (manufacture)
46180	51180	Hormones and their derivatives (commission agent)
46460	51460	Hormones and their derivatives (wholesale)
32990	36639	Horn and tortoise shell working (manufacture)
32990	36639	Horn pressing (manufacture)
32200	36300	Horns (musical) (manufacture)
29310	31610	Horns for motor vehicle (electric) (manufacture)
29202	34202	Horse box trailers (manufacture)
94990	91330	Horse breeding society
96090	93059	Horse clipping
15120	19200	Horse collars made of leather (manufacture)
30990	35500	Horse drawn trailer (manufacture)
30990	35500	Horse drawn truck (manufacture)
01430	01220	Horse farming and breeding
92000	92710	Horse race betting levy board

SIC 2007	SIC 2003	Activity	SIC 2007	SIC 2003	Activity
93199	92629	Horse training (racehorse)	24100	27100	Hot rolled steel cut lengths (manufacture)
15120	36639	Horse whips (manufacture)	24100	27100	Hot rolled steel fish plates (manufacture)
02300	02010	Horse-chestnut gathering	24100	27100	Hot rolled steel flat bars (manufacture)
46210	51210	Horsehair (wholesale)	24100	27100	Hot rolled steel heavy sections (manufacture)
13100	17170	Horsehair curling (manufacture)	24100	27100	Hot rolled steel hexagonal bars (manufacture)
13100	17170	Horsehair dressing (manufacture)	24100	27100	Hot rolled steel h-sections (manufacture)
13100	17170	Horsehair hackling (manufacture)	24100	27100	Hot rolled steel i-sections (manufacture)
13100	17170	Horsehair sorting (manufacture)	24100	27100	Hot rolled steel joists (manufacture)
13100	17170	Horsehair teasing (manufacture)	24100	27100	Hot rolled steel light sections (manufacture)
92000	92710	Horserace totalisator board	24100	27100	Hot rolled steel narrow strip (manufacture)
46230	51230	Horses (wholesale)	24100	27100	Hot rolled steel plate (manufacture)
46610	51880	Horticultural machinery (wholesale)	24100	27100	Hot rolled steel quarto plate (manufacture)
77310	71310	Horticultural machinery hire (without operator)	24100	27100	Hot rolled steel railway materials (manufacture)
22210	25210	Hose and pipe fittings made of plastic (manufacture)	24100	27100	Hot rolled steel reversing mill plate (manufacture)
13300	17300	Hose bleaching, dyeing or otherwise finishing (manufacture)	24100	27100	Hot rolled steel round bars (manufacture)
22210	25210	Hose made of plastic (manufacture)	24100	27100	Hot rolled steel sections for mining frames (manufacture)
22190	25130	Hose made of rubber (manufacture)	24100	27100	Hot rolled steel sheet (manufacture)
13960	17542	Hosepiping made of textiles (manufacture)	24100	27100	Hot rolled steel sheet piling (manufacture)
14310	17710	Hosiery (knitted and crocheted) (manufacture)	24100	27100	Hot rolled steel sole plates (manufacture)
46420	51429	Hosiery (wholesale)	24100	27100	Hot rolled steel special bars (manufacture)
14310	17710	Hosiery blank (manufacture)	24100	27100	Hot rolled steel special sections (manufacture)
13300	17300	Hosiery finishing (manufacture)	24100	27100	Hot rolled steel square bars (manufacture)
28940	29540	Hosiery knitting machinery (manufacture)	24100	27100	Hot rolled steel t-sections (manufacture)
13300	17300	Hosiery printing (manufacture)	24100	27100	Hot rolled steel u-sections (manufacture)
13300	17300	Hosiery scouring (manufacture)	24100	27100	Hot rolled steel wide strip (hot rolled wide coil) (manufacture)
13300	17300	Hosiery shrinking (manufacture)	28290	29430	Hot spraying electrical machines (metal or metal carbides) (manufacture)
13300	17300	Hosiery trimming	25500	28400	Hot stamping of ferrous metals (manufacture)
86101	85112	Hospice (private sector)	22190	25130	Hot water bottles made of rubber (manufacture)
86101	85111	Hospice (public sector)	43220	45330	Hot water engineer
86101	85112	Hospital activities (private sector)	46740	51540	Hot water heaters (wholesale)
86101	85111	Hospital activities (public sector)	35300	40300	Hot water production and distribution
32500	33100	Hospital beds with mechanical fittings (manufacture)	55100	55101	Hotel (licensed with restaurant)
65120	66031	Hospital contribution scheme	55100	55102	Hotel (unlicensed with restaurant)
96010	93010	Hospital laundry	55100	55103	Hotel without restaurant
65120	66031	Hospital saving association	09100	11200	Hot-tap operation services
85100	80100	Hospital schools at nursery and primary level	68310	70310	House agent
85310	80210	Hospital schools at secondary level	41202	45212	House building and repairing
87900	85311	Hostel for the homeless (charitable)	47599	52440	House furnisher (retail)
87900	85312	Hostel for the homeless (non-charitable)	65120	66031	House insurance
55900	55239	Hostels (not social work)	68209	70209	House letting (private)
30300	35300	Hot air balloons (manufacture)	68310	70310	House letting agency
25610	28510	Hot dip coating (metal finishing) (manufacture)	15200	19300	House shoe (manufacture)
24100	27100	Hot dip metal coated sheet steel (manufacture)	30120	35120	Houseboat (manufacture)
24100	27100	Hot dip zinc coated sheet steel (manufacture)	14132	18222	Housecoats (manufacture)
56103	55304	Hot dog vendor	17220	21220	Household and personal hygiene paper (manufacture)
24200	27220	Hot finished steel tube (manufacture)	47599	52440	Household articles and equipment n.e.c. (retail)
28990	29560	Hot glass working machinery (manufacture)	25990	28750	Household articles made of metal (manufacture)
20520	24620	Hot melt adhesive (manufacture)	17220	21220	Household cellulose wadding paper products (manufacture)
28910	29510	Hot metals handling machinery and equipment (manufacture)	47599	52440	Household furnishing articles made of textile materials (retail)
27510	29710	Hot plates (electric) (manufacture)	47599	52440	Household furniture (retail)
25500	28400	Hot pressing of ferrous metals (manufacture)	46150	51150	Household goods (commission agent)
24100	27100	Hot rolled steel angles (l-sections) (manufacture)	77299	71409	Household goods hire
24100	27100	Hot rolled steel base plates (manufacture)	46499	51479	Household goods n.e.c. exporter (wholesale)
24100	27100	Hot rolled steel beams (manufacture)	46499	51479	Household goods n.e.c. Importer (wholesale)
24100	27100	Hot rolled steel bearing piling (manufacture)			
24100	27100	Hot rolled steel channels (manufacture)			
24100	27100	Hot rolled steel columns (manufacture)			

H

SIC 2007	SIC 2003	Activity
46410	51410	Household linen (wholesale)
47599	52440	Household non-electrical appliances (retail)
46499	51479	Household non-electrical appliances (wholesale)
47190	52120	Household stores (retail)
13923	17403	Household textile made-up articles (manufacture)
13923	17403	Household textiles (manufacture)
47510	52410	Household textiles (retail)
46410	51410	Household textiles (wholesale)
47599	52440	Household utensils (retail)
25990	28750	Household utensils made of metal (manufacture)
22290	25240	Household utensils made of plastic (manufacture)
16290	20510	Household utensils made of wood (manufacture)
77299	71409	Housewares rental and leasing
41100	70110	Housing association (building houses for later sale)
41202	45212	Housing association (building work)
68201	70209	Housing association (social housing for rental)
84120	75120	Housing services administration (public sector)
30110	35110	Hovercraft (manufacture)
50100	61101	Hovercraft operator between UK and international ports (passenger)
25400	29600	Howitzer (manufacture)
61100	64200	Hull telephone service
96030	93030	Human or animal corpse burial or incineration
21200	24421	Human plasma extract (manufacture)
70229	74149	Human resources management consultancy services
78300	74500	Human resources provision on a long term or permanent basis
37000	90010	Human waste water collection and transport by sewers, collectors, tanks and other means of transport
72200	73200	Humanities research and experimental development
28250	29230	Humidifier for air conditioning equipment (manufacture)
26513	33202	Humidistats (non-electronic) (manufacture)
93199	92629	Hunt kennels
93199	92629	Hunt stables
94990	01500	Hunting and trapping (commercial) promotion activities
93199	92629	Hunting for sport or recreation
01700	01500	Hunting or trapping of animals for food
01700	01500	Hunting or trapping of animals for fur
01700	01500	Hunting or trapping of animals for pets
01700	01500	Hunting or trapping of animals for skin
01700	01500	Hunting or trapping of animals for use in research
01700	01500	Hunting or trapping of animals for use in zoos
32300	36400	Hunting requisites (manufacture)
84130	75130	Hunting services administration and regulation (public sector)
01700	01500	Hunting, trapping and related service activities
16290	20510	Hurdles made of wood (manufacture)
16230	20300	Huts made of wood (manufacture)
26200	30020	Hybrid computer (manufacture)
26110	32100	Hybrid integrated circuits (electronic) (manufacture)
26200	30020	Hybrid machines (manufacture)
93290	92729	Hyde Park
23520	26520	Hydrated lime (manufacture)
28120	29122	Hydraulic and pneumatic components (manufacture)
28220	29220	Hydraulic and pneumatic conveying plant (manufacture)
28120	29122	Hydraulic and pneumatic cylinders (manufacture)
28220	29220	Hydraulic and pneumatic handling plant (manufacture)
28120	29122	Hydraulic and pneumatic hoses and fittings (manufacture)
28120	29122	Hydraulic and pneumatic power engine and motor parts (manufacture)
46690	51870	Hydraulic and pneumatic power engines and motors (wholesale)
28120	29130	Hydraulic and pneumatic valves (manufacture)
20590	24660	Hydraulic brake fluid (less than 70% petroleum oil) (manufacture)
46120	51120	Hydraulic brake fluids (commission agent)
28410	29420	Hydraulic brakes (manufacture)
43999	45240	Hydraulic construction (subsurface work)
71129	74204	Hydraulic engineering design projects
28120	29122	Hydraulic equipment for aircraft (manufacture)
28120	29122	Hydraulic exhauster (manufacture)
22190	25130	Hydraulic hose made of rubber (manufacture)
15110	19100	Hydraulic leather (manufacture)
23520	26520	Hydraulic lime (manufacture)
19209	23209	Hydraulic oil formulation outside refineries (manufacture)
28120	29122	Hydraulic power engines and motors (manufacture)
35300	40300	Hydraulic power production and distribution
72190	73100	Hydraulic research station
28120	29140	Hydraulic transmission equipment (manufacture)
20590	24660	Hydraulic transmission liquids (manufacture)
28110	29110	Hydraulic turbine (manufacture)
28110	29110	Hydraulic turbine and water wheel parts (manufacture)
28110	29110	Hydraulic turbines and parts thereof: (manufacture)
46690	51870	Hydraulic turbines and water wheels (wholesale)
55209	55239	Hydro (accommodation)
35110	40110	Hydro electric power station
28120	29122	Hydro pneumatic device (manufacture)
46120	51120	Hydrocarbon derivatives (commission agent)
20140	24140	Hydrocarbon derivatives (sulphated, nitrated or nitrosated) (manufacture)
06200	11100	Hydrocarbon liquids mining, by liquefaction or pyrolysis
20140	24140	Hydrocarbons (not fuels) (manufacture)
20130	24130	Hydrochloric acid (manufacture)
30110	35110	Hydrofoil (manufacture)
20110	24110	Hydrogen (manufacture)
46120	51120	Hydrogen chloride (commission agent)
46750	51550	Hydrogen chloride (wholesale)
20130	24130	Hydrogen peroxide (manufacture)
46750	51550	Hydrogen peroxide (wholesale)
46120	51120	Hydrogen, argon, nitrogen, oxygen and rare gases (commission agent)
26511	33201	Hydrographic instrument and apparatus (electronic) (manufacture)
26513	33202	Hydrographic instrument and apparatus (non-electronic) (manufacture)
71122	74206	Hydrographic surveying activities
71122	74206	Hydrologic surveying activities
26511	33201	Hydrological instrument (electronic) (manufacture)
26513	33202	Hydrological instrument (non-electronic) (manufacture)
26511	33201	Hydrometers (electronic) (manufacture)
26513	33202	Hydrometers (non-electronic) (manufacture)

H

SIC 2007	SIC 2003	Activity	SIC 2007	SIC 2003	Activity
46690	51870	Hydrometers, non-medical thermometers, pyrometers, barometers, hygrometers and the like (wholesale)	13200	17220	Imitation fur of long pile fabrics made on the woollen system (manufacture)
26513	33202	Hydronic limit controls (non-electronic) (manufacture)	32130	36610	Imitation jewellery (manufacture)
			46480	51474	Imitation jewellery (wholesale)
28120	29122	Hydrostatic transmissions (manufacture)	46480	51474	Imitation jewellery exporter (wholesale)
20130	24130	Hydrosulphite (manufacture)	46480	51474	Imitation jewellery importer (wholesale)
81299	90030	Hygiene contracting	14110	18100	Imitation leather clothes for men and boys (manufacture)
22190	25130	Hygienic articles made of rubber (manufacture)			
23190	26150	Hygienic glassware (other than containers) (manufacture)	14110	18100	Imitation leather clothes for women and girls (manufacture)
17120	21120	Hygienic paper (uncut) (manufacture)	32130	36610	Imitation pearls (manufacture)
32500	33100	Hyperbaric chambers (manufacture)	46480	51474	Imitation pearls (wholesale)
09100	11200	Hyperbaric welding services	27510	29710	Immersion heater (electric) (manufacture)
47110	52112	Hypermarket selling mainly foodstuffs with alcohol licence (retail)	46439	51439	Immersion heaters (wholesale)
			16100	20100	Immersion treatment of wood (manufacture)
47110	52113	Hypermarket selling mainly foodstuffs without alcohol licence (retail)	21200	24421	Immunoglobin (manufacture)
			28131	29121	Impeller pumps (manufacture)
46750	51550	Hypochlorites, chlorates and perchlorates (wholesale)	91020	92521	Imperial War Museum
			20412	24512	Impregnated cleaning and polishing cloth (manufacture)
32500	33100	Hypodermic syringe and equipment (manufacture)			
35300	15980	Ice (for human consumption)	13300	17300	Impregnating purchased garments (manufacture)
35300	40300	Ice (not for human consumption)	90010	92311	Impresario
31090	36140	Ice box made of wood (manufacture)	16210	20200	Improved wood (manufacture)
52220	63220	Ice breaking services	82200	74860	Inbound call centres
31090	36140	Ice chest made of wood (manufacture)	13960	17549	Incandescent mantle (manufacture)
10520	15520	Ice cream (manufacture)	20510	24610	Incendiary composition (manufacture)
46360	51360	Ice cream (wholesale)	38220	90020	Incineration of hazardous waste
28250	29230	Ice cream conservator (manufacture)	28210	29210	Incinerator (manufacture)
56103	55303	Ice cream parlour	81223	74705	Incinerator cleaning
10520	15520	Ice cream powder (manufacture)	47791	52501	Incunabula (retail)
56103	55303	Ice cream retailer (take away)	82990	74879	Independent auctioneers
56103	55304	Ice cream vans	60100	92201	Independent Broadcasting Authority (radio broadcasting)
93120	92629	Ice hockey club			
93110	92611	Ice skating rink	66190	67130	Independent financial advisors (not specialising in insurance or pensions advice)
93110	92619	Ice-hockey arenas operation			
32300	36400	Ice-skates (manufacture)	29320	34300	Independent suspension units for motor vehicles (manufacture)
10810	15830	Icing sugar (manufacture)			
08110	14110	Igneous rock quarry	17230	21230	Index card (manufacture)
29310	31610	Ignition coil (manufacture)	20590	24660	Indian ink (manufacture)
29310	31610	Ignition equipment (other than coils and magnetos) (manufacture)	29310	31610	Indicating measuring instrument for vehicles and aircraft (electric) (manufacture)
			27900	31620	Indicator panel (manufacture)
29310	31610	Ignition magnetos (manufacture)	46750	51550	Indigo (wholesale)
23190	26150	Illuminated glassware (manufacture)	32409	36509	Indoor game (manufacture)
27400	31500	Illuminated signs and nameplates (manufacture)	28210	29210	Induction furnace (manufacture)
46690	51870	Illuminated signs and name-plates (wholesale)	46690	51870	Induction or dielectric heating equipment for industrial or laboratory use (wholesale)
22290	25240	Illuminated street furniture made of plastic (manufacture)			
			28250	29230	Induction unit for air conditioning equipment (manufacture)
27400	31500	Illuminated traffic signs (manufacture)			
90030	92319	Illuminating (illustrating)	26110	31100	Inductor (electronic) (manufacture)
26110	32100	Image converters and intensifiers (manufacture)	20520	24620	Industrial adhesives (manufacture)
26702	33403	Image projectors (manufacture)	25300	28300	Industrial air heater for boilers (manufacture)
18130	22240	Image setting for letterpress processes (manufacture)	22190	25130	Industrial belting made of rubber (manufacture)
			19201	23201	Industrial benzole (manufacture)
18130	22240	Image setting for offset printing processes (manufacture)	32910	36620	Industrial broom and mop (manufacture)
			32910	36620	Industrial brush (manufacture)
61100	64200	Image transmission via cables, broadcasting, relay or satellite	56290	55520	Industrial canteen (run by catering contractor)
			28990	29560	Industrial carpet sweeper (manufacture)
60200	72400	Image with sound internet broadcasting	20590	24660	Industrial catalyst (manufacture)
13200	17210	Imitation fur (woven long pile fabrics) (manufacture)	46120	51120	Industrial chemicals (commission agent)
13910	17600	Imitation fur obtained by knitting (manufacture)	46750	51550	Industrial chemicals (wholesale)

SIC 2007	SIC 2003	Activity
81210	74701	Industrial cleaning
20590	24660	Industrial cleaning preparation (manufacture)
14120	18210	Industrial clothing (manufacture)
96010	93010	Industrial clothing hire from laundries
71121	74205	Industrial consultants
71121	74205	Industrial design consultants
71121	74205	Industrial design service
70229	74149	Industrial development consultancy services
46750	51550	Industrial dyes (wholesale)
28110	29110	Industrial engine parts (manufacture)
46140	51140	Industrial equipment (commission agent)
81299	74709	Industrial equipment cleaning (non-specialised)
68209	70209	Industrial estate letting
46120	51120	Industrial fatty acids (commission agent)
46750	51550	Industrial fatty alcohols (wholesale)
20110	24110	Industrial gases (manufacture)
46750	51550	Industrial gases (wholesale)
23190	26150	Industrial glassware (not container) (manufacture)
65120	66031	Industrial insurance
28140	29130	Industrial intake taps (manufacture)
15120	19200	Industrial leather (manufacture)
14110	18100	Industrial leather welders aprons (manufacture)
46900	51900	Industrial materials (general or undefined) (wholesale)
28990	29560	Industrial mixing equipment for the chemical industry (manufacture)
46120	51120	Industrial monocarboxylic fatty acids (commission agent)
46750	51550	Industrial monocarboxylic fatty acids and acid oils from refining (wholesale)
23440	26240	Industrial non-refractory ceramic products (manufacture)
25110	28110	Industrial ossature in metal (manufacture)
43999	45250	Industrial ovens erection
17120	21120	Industrial paper (manufacture)
14200	18300	Industrial polishing cloths made of fur (manufacture)
26512	33301	Industrial process control equipment (electronic) (manufacture)
26514	33302	Industrial process control equipment (non-electronic) (manufacture)
32990	18241	Industrial protective headgear (manufacture)
32120	36220	Industrial quality stones (manufacture)
28140	29130	Industrial regulating valves (manufacture)
28990	29560	Industrial robots for multiple uses (manufacture)
46750	51550	Industrial salt (wholesale)
20411	24511	Industrial soap (manufacture)
28110	29110	Industrial spark and compression ignition engine (manufacture)
28220	29220	Industrial special purpose trucks (manufacture)
19201	23201	Industrial spirit from petroleum (manufacture)
28140	29130	Industrial taps (manufacture)
26520	33500	Industrial timer (manufacture)
28220	29220	Industrial tractor (manufacture)
26110	31100	Industrial transformer (electronic) (manufacture)
16230	20300	Industrialised building component made of timber (manufacture)
10200	15209	Inedible flours, meal and pellets of fish, crustaceans and molluscs production (manufacture)
20110	24110	Inert gases such as carbon dioxide (manufacture)
10860	15880	Infant food (milk based) (manufacture)
10860	15880	Infant food (other than milk based) (manufacture)

SIC 2007	SIC 2003	Activity
10860	15880	Infant formulae (manufacture)
85200	80100	Infant schools
47710	52422	Infants' clothing (retail)
46420	51422	Infants' clothing (wholesale)
86101	85111	Infectious disease hospital (public sector)
86220	85120	Infectious disease specialist (private practice)
86101	85111	Infectious disease specialist (public sector)
86101	85112	Infirmary (private sector)
86101	85111	Infirmary (public sector)
22290	25240	Inflatable air bed made of plastic (manufacture)
30120	35120	Inflatable boats (manufacture)
46499	51479	Inflatable boats for pleasure or sports (wholesale)
22190	25130	Inflatable cushion made of rubber (manufacture)
30120	35120	Inflatable dinghy made of rubber (manufacture)
30120	35120	Inflatable liferaft made of rubber (manufacture)
22190	25130	Inflatable mattress made of rubber (manufacture)
30120	35120	Inflatable motor boats (manufacture)
22290	25240	Inflatable plastic products (excluding playballs) (manufacture)
30120	35120	Inflatable rafts (manufacture)
46499	51479	Inflatable vessels for pleasure or sports (wholesale)
28131	29121	Inflator for cycle type tyres (manufacture)
84210	75210	Information and cultural services abroad administration and operation (public sector)
82990	74879	Information bureau (not tourist)
79909	63309	Information bureau for tourists
26200	30020	Information processing equipment
63990	74879	Information search services on a fee or contract basis
62020	72220	Information systems strategic review and planning services
62020	72220	Information technology consultancy activities
27400	31500	Infrared lamps (manufacture)
26301	32201	Infrared remote controls (manufacture)
26701	33402	Infrared systems for night vision (manufacture)
28910	29510	Ingot mould and bottom (manufacture)
24440	27440	Ingots made of brass (manufacture)
24440	27440	Ingots made of copper (manufacture)
24100	27100	Ingots of stainless steel (manufacture)
24100	27100	Ingots of steel (manufacture)
64302	65232	In-house trust activities
28960	29560	Injection moulding equipment (for rubber or plastic) (manufacture)
65120	66031	Injury insurance
20590	24660	Ink for impregnating ink pads (manufacture)
32990	36631	Ink pad (manufacture)
16290	20510	Inlaid wood (manufacture)
50400	61209	Inland water transport (freight)
50300	61201	Inland water transport (passenger)
30200	35200	Inland waterways signalling, safety or traffic control equipment (manufacture)
28110	29110	Inlet valves for internal combustion engines (manufacture)
22110	25110	Inner tube for tyre (manufacture)
94120	91120	Inns of Court
55209	55239	Inns with letting rooms (short-stay lodgings)
20130	24130	Inorganic acid (manufacture)
46120	51120	Inorganic acids (commission agent)
20130	24130	Inorganic bases (manufacture)
20130	24130	Inorganic chemical (manufacture)
20130	24130	Inorganic compounds (manufacture)

SIC 2007	SIC 2003	Activity	SIC 2007	SIC 2003	Activity
80300	74879	Inquiry agency	33200	31100	Installation of electric motors, generators and transformers (manufacture)
01490	01250	Insect raising	43220	45310	Installation of electric solar energy collectors
20200	24200	Insecticide (manufacture)	33200	31610	Installation of electrical equipment for engines and vehicles n.e.c. (manufacture)
46750	51550	Insecticides (wholesale)			
46120	51120	Insecticides, herbicides, plant growth regulators and anti-sprouting products (commission agent)	43220	45310	Installation of electrical heating systems (except baseboard heating)
18130	22250	Insetting (manufacture)	43210	45310	Installation of electrical systems in cable television wiring, including fibre optic
15200	19300	Insoles made of leather (manufacture)	43210	45310	Installation of electrical systems in computer network, including fibre optic
70221	74142	Insolvency management			
43320	28120	Installation (erection) work of self-manufactured builders' ware of metal	43210	45310	Installation of electrical wiring and fittings
43320	25239	Installation (erection) work of self-manufactured builders' ware of plastic	33200	31200	Installation of electricity distribution and control apparatus (manufacture)
43320	20300	Installation (erection) work of self-manufactured builders' ware of wood	33200	32100	Installation of electronic valves and tubes and other electronic components (manufacture)
80200	74602	Installation and repair of electronic locking devices with monitoring	43290	45310	Installation of elevators
80200	74602	Installation and repair of electronic safes security vaults with monitoring	33200	29110	Installation of engines and turbines (except aircraft, vehicle and cycle) (manufacture)
80200	74602	Installation and repair of mechanical locking devices with monitoring	33200	29523	Installation of equipment for concrete crushing and screening and roadworks (manufacture)
80200	74602	Installation and repair of mechanical safes and security vaults with monitoring	43290	45310	Installation of escalators
43290	45340	Installation in buildings of fittings and fixtures n.e.c.	33200	28750	Installation of fabricated metal products (manufacture)
33200	31400	Installation of accumulators, primary cells and primary batteries (manufacture)	33200	28520	Installation of factory assembly lines (manufacture)
			43330	45430	Installation of false floors and computer floors
43210	45310	Installation of aerials and residential antennas	33200	28740	Installation of fasteners, screw machine products, chain and springs (manufacture)
33200	29320	Installation of agricultural and forestry machinery (manufacture)	43210	45310	Installation of fire alarms
43220	45330	Installation of air conditioning equipment and ducts	80200	45310	Installation of fire and burglar alarm systems if together with monitoring of the same systems
43220	45330	Installation of air conditioning plant			
43210	45340	Installation of airport runway lighting	43220	45330	Installation of fire sprinkler systems
33200	34100	Installation of assemblies and sub-assemblies and the like, into motor vehicles (manufacture)	43220	45330	Installation of furnaces
			33200	29210	Installation of furnaces and furnace burners (manufacture)
43290	45310	Installation of automated and revolving doors	43320	45420	Installation of furniture
33200	29140	Installation of bearings, gears, gearing and driving elements (manufacture)	43220	45330	Installation of gas fittings
			43220	45330	Installation of gas heating systems
43290	45340	Installation of blinds and awnings	43220	45330	Installation of gas meters
33200	36501	Installation of bowling alley equipment (manufacture)	33200	29240	Installation of general purpose machinery n.e.c. (manufacture)
43910	20300	Installation of builders carpentry and joinery (roofing materials)	43342	45440	Installation of glass
43320	45420	Installation of built-in furniture	43220	45330	Installation of heating and ventilation apparatus
43210	45310	Installation of burglar alarm systems	43210	45340	Installation of illuminated road signs and street furniture
43210	45310	Installation of cables			
47421	52488	Installation of car telephones (retail)	43210	45340	Installation of illumination and signalling systems for roads, railways, airports and harbours
33200	29530	Installation of catering equipment (manufacture)			
43320	45420	Installation of ceilings	33200	33100	Installation of industrial irradiation and electromedical equipment (manufacture)
33200	26400	Installation of ceramic pipes, conduit, guttering and pipe fittings (manufacture)	33200	28110	Installation of industrial machinery and equipment (manufacture)
33200	33500	Installation of clocks (manufacture)	33200	30020	Installation of industrial mainframe and similar computers (manufacture)
43220	45330	Installation of coldrooms			
33200	29122	Installation of compressors (manufacture)	33200	33201	Installation of instruments and apparatus for measuring, checking, testing, navigating (manufacture)
43210	45310	Installation of computer network cabling and other telecommunications system cables			
			33200	31300	Installation of insulated wire and cable (manufacture)
43220	45330	Installation of cooling towers	43210	45310	Installation of intelecommunications wiring
42110	45230	Installation of crash barriers	43320	45420	Installation of joinery
43320	45420	Installation of doors	33200	28220	Installation of large scale central heating boilers e.g. For large residential blocks (manufacture)
33200	35110	Installation of drilling platforms and the like (manufacture)			
43220	45330	Installation of duct work	43220	45330	Installation of lawn sprinkler systems
33200	29522	Installation of earth-moving and excavating equipment (manufacture)	33200	29220	Installation of lifting and handling equipment (except lifts and escalators) (manufacture)

SIC 2007	SIC 2003	Activity	SIC 2007	SIC 2003	Activity
43290	45310	Installation of lifts	43220	45330	Installation of plumbing
43210	45310	Installation of lighting systems	33200	32300	Installation of professional radio, television, sound and video equipment (manufacture)
43290	45310	Installation of lightning conductors	33200	29121	Installation of pumps (manufacture)
33200	28630	Installation of locks and hinges (manufacture)	33200	32202	Installation of radio and television transmitters (manufacture)
33200	29420	Installation of machine tools (metal working) (manufacture)	47430	52450	Installation of radios in motor vehicles (retail)
33200	29430	Installation of machine tools (other than metal working) (manufacture)	43220	45330	Installation of refrigeration
33200	29560	Installation of machinery for bookbinding (manufacture)	43210	45340	Installation of roadway traffic monitoring and guidance equipment
33200	29530	Installation of machinery for food, beverage and tobacco processing (manufacture)	43220	45330	Installation of sanitary equipment
33200	29510	Installation of machinery for metallurgy (manufacture)	43210	45310	Installation of satellite dishes
33200	29521	Installation of machinery for mining (manufacture)	33200	36110	Installation of seats in aircraft, ships, trains and the like
33200	29550	Installation of machinery for paper and paperboard production (manufacture)	43210	45310	Installation of security alarms
33200	29560	Installation of machinery for printing (manufacture)	43290	45320	Installation of sound insulation
33200	29540	Installation of machinery for textile, apparel and leather production (manufacture)	33200	29560	Installation of special purpose machinery n.e.c. (manufacture)
33200	29560	Installation of machinery for working rubber or plastics and making products of these (manufacture)	43220	45330	Installation of sprinkler systems
33200	30020	Installation of mainframe and similar computers and peripheral equipment (manufacture)	33200	28300	Installation of steam generators (not central heating boilers) incl. related pipework (manufacture)
33200	33100	Installation of medical and surgical equipment and apparatus (manufacture)	43220	45330	Installation of steam piping
43320	45420	Installation of metal grilles and gates	33200	28710	Installation of steel drums and similar (manufacture)
43320	45420	Installation of metal partitioning	43210	45340	Installation of street lighting and electrical signals
43320	45420	Installation of metal shutters	43320	45420	Installation of suspended ceilings
33200	28210	Installation of metal tanks (manufacture)	33200	29130	Installation of taps and valves (manufacture)
33200	26810	Installation of millstones, grindstones, polishing stones and the like (manufacturing)	43210	45310	Installation of telecommunications wiring systems
43342	45440	Installation of mirrors	33200	32201	Installation of telecommunications equipment
45200	50200	Installation of motor vehicle parts and accessories (not as part of production process)	43210	45310	Installation of telephone lines
45200	50200	Installation of motor vehicle parts and accessories, not part of the manufacturing process	43290	45320	Installation of thermal insulation
43320	45420	Installation of movable wooden partitions	33200	28620	Installation of tools (manufacture)
33200	29230	Installation of non-domestic cooling and ventilating equipment (manufacture)	33200	26150	Installation of tubes and pipes of glass, including installation of glass pipe (manufacture)
33200	29560	Installation of non-domestic machinery for drying wood, paper pulp etc. (manufacture)	33200	25210	Installation of tubes, pipes and hoses, of plastics (manufacture)
43220	45330	Installation of non-electric solar energy collectors	33200	33401	Installation of unmounted lenses (manufacture)
33200	33202	Installation of non-electronic instruments and appliances for measuring, testing, etc. (manufacture)	43290	45310	Installation of vacuum cleaning systems
42110	45230	Installation of non-illuminated road signs, bollards etc.	43220	45330	Installation of ventilation
33200	30010	Installation of office machinery (manufacture)	43290	45320	Installation of vibration insulation
43210	45310	Installation of office switchboards and telephone lines	33200	29600	Installation of weapons and weapon systems (manufacture)
42210	45213	Installation of offshore pipelines from oil or gas wells	43320	45420	Installation of windows made of any material
43220	45330	Installation of oil heating systems	33200	28730	Installation of wire products (manufacture)
33200	33402	Installation of optical precision instruments (manufacture)	43320	45420	Installation of wooden door-frames
33200	31620	Installation of other electrical apparatus (manufacture)	43320	45420	Installation of wooden fitted kitchens
43290	45340	Installation of outdoor pumping or filtration equipment	43320	45420	Installation of wooden shop fittings
43210	45340	Installation of outdoor transformer and other outdoor electrical distribution apparatus	43320	45420	Installation of wooden staircases
62090	30020	Installation of personal computers and peripheral equipment	43320	45420	Installation of wooden wall coverings
33200	33403	Installation of photographic and cinematographic equipment (manufacture)	10832	15862	Instant coffee (manufacture)
			20590	24640	Instant print film (manufacture)
			46439	51439	Instantaneous or storage water heaters (electric) (wholesale)
			46439	51479	Instantaneous or storage water heaters (non-electric) (wholesale)
			32500	33100	Instep support (manufacture)
			94120	91120	Institute of Actuaries
			94120	91120	Institute of British Water Colour Painters
			94120	91120	Institute of Chartered Accountants in England and Wales
			94120	91120	Institute of Chartered Accountants of Scotland
			94120	91120	Institute of Civil Engineers

I

SIC 2007	SIC 2003	Activity
94120	91120	Institute of Cost and Management Accountants
94120	91120	Institute of Hygiene
94120	91120	Institute of Incorporated Photographers
94120	91120	Institute of Mechanical Engineers
85320	80220	Instruction for chefs, hoteliers and restauranteurs
85510	92629	Instructors of sport
16290	20510	Instrument case made of wood (manufacture)
26520	33500	Instrument panel clock (manufacture)
46520	51860	Instruments and apparatus for measuring electrical quantities (electronic) (wholesale)
32500	33100	Instruments and apparatus used for medical, surgical, dental or veterinary purposes (manufacture)
46460	51460	Instruments and appliances for dental science (wholesale)
46460	51460	Instruments and devices for doctors and hospitals (wholesale)
32990	36639	Instruments for educational or exhibition purposes (manufacture)
46520	51860	Instruments for measuring flow, level, pressure etc. of liquids or gassed (electronic) (wholesale)
46690	51870	Instruments for measuring or checking flow, level, pressure etc. of liquids or gases (wholesale)
26511	33201	Instruments for testing physical and mechanical properties of materials (electronic) (manufacture)
31090	36140	Insulated cabinets made of wood (manufacture)
23430	26230	Insulated ceramic fittings (manufacture)
27320	31300	Insulated electrical cable (manufacture)
23190	26150	Insulated fittings made of glass (manufacture)
27320	31300	Insulated mains cable for power distribution (manufacture)
26110	32100	Insulated monolithic, hybrid and passive circuit (manufacture)
22290	25240	Insulated plastic fittings (manufacture)
46690	51870	Insulated winding wire (wholesale)
27320	31300	Insulated wire (manufacture)
27320	31300	Insulated wire and cable made of aluminium (manufacture)
27320	31300	Insulated wire and cable made of copper (manufacture)
27320	31300	Insulated wire and cable made of steel (manufacture)
22230	25239	Insulating (heat and sound) sheet, tiles, blocks and granules made of plastic (manufacture)
22190	25130	Insulating cloth tape (manufacture)
43290	45320	Insulating contractor (buildings)
29310	31610	Insulating fittings (other than ceramic for vehicles and aircraft) (manufacture)
23140	26140	Insulating material made of glass fibre (manufacture)
22190	25130	Insulating material made of rubber (manufacture)
16290	20520	Insulating materials made of cork (manufacture)
19201	23201	Insulating oil (at refineries) (manufacture)
19209	23209	Insulating oil formulation outside refineries (manufacture)
43290	45320	Insulating work activities
23990	26821	Insulation made of asbestos (manufacture)
23430	26230	Insulators made of ceramic (manufacture)
23190	26150	Insulators made of glass (manufacture)
65120	66031	Insurance (non-life)
66220	67200	Insurance agent (not employed by insurance company)
66220	67200	Insurance agents activities
66220	67200	Insurance broker (not employed by insurance company)

SIC 2007	SIC 2003	Activity
66220	67200	Insurance brokers activities
66220	67200	Insurance consultancy services
66210	67200	Insurance risk evaluators activities
26110	32100	Integrated circuits (analogue or digital) (manufacture)
46520	51860	Integrated circuits (wholesale)
71129	74209	Integrated engineering activities for turnkey projects
71200	74300	Integrated mechanical and electrical system testing and analysis
28120	29122	Intensifier for hydraulic equipment (manufacture)
20590	24640	Intensifier for photographic use (manufacture)
28120	29122	Intensifier for pneumatic control equipment (manufacture)
94910	91310	Inter Varsity Fellowship of Evangelical Unions
66190	67130	Interbank (worldwide financial telecommunications society)
25730	28620	Interchangeable tools for dies (manufacture)
22110	25110	Interchangeable tyre flaps for retreading tyres (manufacture)
22110	25110	Interchangeable tyre treads for retreading tyres (manufacture)
49390	60211	Inter-city coach services on scheduled routes
30300	29600	Intercontinental ballistic missiles (ICBM) (manufacture)
64991	65233	Inter-dealer brokers
26120	32100	Interface cards (manufacture)
74100	74872	Interior decor design
74100	74872	Interior decorator activities
74100	74872	Interior designers
43341	45440	Interior painting of buildings
43310	45410	Interior plaster application in buildings or other constructions incl. related lathing materials
31030	36150	Interior sprung mattress (manufacture)
43310	45410	Interior stucco application in buildings or other constructions incl. related lathing materials
13200	17210	Interlining weaving from yarn spun on the cotton system (manufacture)
22230	25239	Intermediate bulk containers (other than drums made of plastic) (manufacture)
23200	26260	Intermediate goods of mined or quarried non-metallic minerals e.g. Sand, gravel, clay (manufacture)
68310	70310	Intermediation in buying selling and renting of real estate
29100	34100	Internal combustion engine for motor vehicles (manufacture)
28110	29110	Internal combustion engines and parts (manufacture)
30910	35410	Internal combustion engines for motorcycles (manufacture)
29100	34100	Internal combustion engines for tractors (manufacture)
28110	29110	Internal combustion piston engines (manufacture)
28110	29110	Internal combustion piston marine engines (manufacture)
28110	29110	Internal combustion piston railway engines (manufacture)
99000	99000	International Labour Office
99000	99000	International Monetary Fund
99000	99000	International organisation (e.g. United Nations, International Labour Office)
84210	75210	International peace keeping forces contribution including assignment of manpower
80300	74601	Internet abuse monitoring

I

SIC 2007	SIC 2003	Activity
61200	64200	Internet access providers (wireless telecommunications)
47910	52630	Internet auctions (retail)
45112	50102	Internet car auctions
60100	72400	Internet radio broadcasting
47910	52610	Internet retail sales (retail)
74300	74850	Interpreter
10620	15620	Inulin (manufacture)
30920	35430	Invalid carriage (electrically propelled) (manufacture)
30920	35430	Invalid carriage (manually propelled) (manufacture)
30920	35430	Invalid carriage (power operated) (manufacture)
30920	35430	Invalid carriage parts and accessories (manufacture)
47749	52329	Invalid carriages with or without motor (retail)
46460	51460	Invalid carriages with or without motor (wholesale)
10860	15880	Invalid food (milk based) (manufacture)
10860	15880	Invalid food (other than milk based) (manufacture)
10810	15830	Invert sugar (manufacture)
80300	74601	Investigation activities
66190	67130	Investment advisory services
66120	67122	Investment broking
28910	29510	Investment casting equipment (manufacture)
64301	65231	Investment fund activities
64301	65231	Investment trusts activities
95290	52740	Invisible mending
64992	65222	Invoice discounting
28230	30010	Invoicing machine (manufacture)
20130	24130	Iodine and iodides (manufacture)
28290	29240	Ion exchange plant for water treatment (manufacture)
20160	24160	Ion-exchangers based on polymers (manufacture)
24410	27410	Iridium (manufacture)
46120	51120	Iron (commission agent)
27510	29710	Iron (electric) (manufacture)
24100	27100	Iron (manufacture)
46720	51520	Iron (wholesale)
24510	27510	Iron foundry (manufacture)
24100	27100	Iron of exceptional purity production by electrolysis or other chemical processes (manufacture)
07100	13100	Iron ore beneficiation and agglomeration
07100	13100	Iron ore calcining
07100	13100	Iron ore crushing
07100	13100	Iron ore mine or quarry
09900	13100	Iron ore mining support services provided on a fee or contract basis
07100	13100	Iron ore preparation
07100	13100	Iron ore sintering
07100	13100	Iron ore washing
24100	27100	Iron powder production (manufacture)
08910	14300	Iron pyrites extraction (not for iron production)
20130	23300	Iron pyrites roasting (manufacture)
28921	29521	Iron roughnecks (manufacture)
24100	27100	Iron shot (manufacture)
46720	51520	Iron yard (wholesale)
27510	29710	Ironing machine for domestic use (electric) (manufacture)
28940	29540	Ironing machine for non-domestic use (manufacture)
47520	52460	Ironmonger (retail)
46740	51540	Ironmonger (wholesale)
46150	51150	Ironmongery (commission agent)

SIC 2007	SIC 2003	Activity
26600	33100	Irradiation equipment (manufacture)
28930	29530	Irradiators (manufacture)
42210	45240	Irrigation system construction
01610	01410	Irrigation systems operation on a fee or contract basis
20590	24620	Isinglass (manufacture)
26110	31200	Isolating and make or break switches (electronic) (manufacture)
20110	24110	Isolating gases (manufacture)
86101	85111	Isolation hospital (public sector)
46120	51120	Isotopes and compounds thereof (commission agent)
28990	29560	Isotopic separation machinery or apparatus (manufacture)
66190	67130	Issuing house
32990	36639	Ivory working (manufacture)
28220	29220	Jack for motor vehicle (manufacture)
30110	35110	Jacket leg of steel plate for oil platform (manufacture)
71129	74204	Jacket substructure design and other foundation design services
14131	18221	Jackets for men and boys (manufacture)
14132	18222	Jackets for women and girls (manufacture)
28220	29220	Jacks (other than for motor vehicles) (manufacture)
46690	51870	Jacks and hoists of a kind used for raising vehicles (wholesale)
28940	29540	Jacquard machinery for carpet making (manufacture)
28940	29540	Jacquard textile machinery (manufacture)
10390	15330	Jam (manufacture)
22220	25220	Jam pot covers made of plastic (manufacture)
81210	74701	Janitorial services
25610	28510	Japanning (metal finishing) (manufacture)
23490	26250	Jars made of ceramic (manufacture)
23650	26650	Jars made of fibre-cement (manufacture)
23130	26130	Jars made of glass (manufacture)
22220	25220	Jars made of plastic (manufacture)
13200	17210	Jeans cloth weaving (manufacture)
14131	18221	Jeans for men and boys (manufacture)
14132	18222	Jeans for women and girls (manufacture)
56103	55303	Jellied eel shop
10390	15330	Jelly (table) (manufacture)
10390	15330	Jelly powder (manufacture)
22220	25220	Jerry can made of plastic (manufacture)
14390	17720	Jerseys (knitted) (manufacture)
01130	01120	Jerusalem artichoke growing
30300	35300	Jet engine (manufacture)
08910	14300	Jet mine
32120	36220	Jet ornaments and jewellery (manufacture)
28120	29121	Jet pump (manufacture)
23410	26210	Jet ware (pottery) (manufacture)
15120	19200	Jewel case (not wood or metal) (manufacture)
46499	51479	Jewellers' materials (wholesale)
20120	24120	Jewellers' rouge (manufacture)
46180	51120	Jewellery (commission agent)
32130	36610	Jewellery (gilded and silvered) (manufacture)
32120	36220	Jewellery (gold or silver plated) (manufacture)
47770	52484	Jewellery (retail)
46480	51473	Jewellery (wholesale)
32130	36610	Jewellery containing imitation gem stones (manufacture)
74100	74872	Jewellery designing

J

SIC 2007	SIC 2003	Activity
32120	36220	Jewellery engraving (not distributive trades) (manufacture)
46480	51473	Jewellery exporter (wholesale)
46480	51473	Jewellery importer (wholesale)
32130	36610	Jewellery made of ceramic (manufacture)
23190	26150	Jewellery made of glass (manufacture)
32120	36220	Jewellery made of platinum (manufacture)
32120	36220	Jewellery made of precious metal (manufacture)
32120	36220	Jewellery made of semi-precious stones (manufacture)
32120	36220	Jewellery of base metal clad with precious metal (manufacture)
32120	36220	Jewellery polishing (manufacture)
77299	71409	Jewellery rental and leasing
32120	36220	Jewellery with precious stones (manufacture)
88990	85321	Jewish board of family and children's services
94910	91310	Jewish synagogue
32409	36509	Jigsaw puzzle (manufacture)
28290	29430	Jigs (manufacture)
46620	51810	Jigs and gauges (wholesale)
96010	93010	Job dyeing
18129	22220	Job printing (manufacture)
96090	93059	Jobbing waiter
93199	92629	Jockey
93120	92629	Jockey club
94110	91110	Joint organisation of employers and trade unions
28290	29240	Jointing (precision component) (manufacture)
23990	26821	Joints made of asbestos (manufacture)
32990	36639	Jokes and novelties (manufacture)
58142	22130	Journal and periodical publishing
77210	71401	Journal rental
90030	92400	Journalists
94120	91120	Journalists associations
84230	75230	Judge
84230	75230	Judge advocates
93199	92629	Judges of sport
27510	29710	Juice squeezers (electric) (manufacture)
10822	15842	Jujube (manufacture)
77210	71401	Juke boxes leasing
26400	32300	Jukeboxes (manufacture)
14390	17720	Jumpers (knitted) (manufacture)
27330	31200	Junction box (manufacture)
85200	80100	Junior schools
10519	15519	Junket powder (manufacture)
84230	75230	Justice of the Peace
13300	17300	Jute calendering (manufacture)
13200	17250	Jute cloth (manufacture)
13300	17300	Jute fabrics bleaching, dyeing or otherwise finishing (manufacture)
13200	17250	Jute sacking (manufacture)
13100	17170	Jute sorting (manufacture)
13100	17170	Jute spinning (manufacture)
01160	01110	Jute textile bast fibre growing
13100	17170	Jute tow (manufacture)
13200	17250	Jute weaving (manufacture)
13100	17170	Jute winding (manufacture)
13100	17170	Jute yarn (manufacture)
87900	85311	Juvenile correction homes (charitable)
87900	85312	Juvenile correction homes (non-charitable)
47710	52422	Juvenile outfitter (retail)
08120	14220	Kaolin mining
10410	15410	Kapok seed crushing (manufacture)
10410	15420	Kapok seed oil refining (manufacture)
13100	17170	Kapok willowing (manufacture)
26400	32300	Karaoke machines (manufacture)
30120	35120	Kayak building (manufacture)
29100	34100	KD sets for cars at least 50% of value of complete vehicle (manufacture)
29100	34100	KD sets for commercial vehicles at least 50% of value of complete vehicle (manufacture)
29320	34300	KD sets for vehicles if the value is less than half the value of the complete vehicle (manufacture)
23510	26510	Keene's cement (manufacture)
25920	28720	Kegs made of aluminium (manufacture)
25910	28710	Kegs made of iron or steel (manufacture)
22220	25220	Kegs made of plastic (manufacture)
16240	20400	Kegs made of wood (manufacture)
03110	05010	Kelp collecting, cutting and gathering (uncultivated)
01160	01110	Kenaf and other textile bast fibre growing
93199	92629	Kennel master
96090	93059	Kennels (not racing)
93199	92629	Kennels and garages (racing)
93290	92729	Kensington Gardens
23610	26610	Kerbs and edging made of pre-cast concrete (manufacture)
23700	26700	Kerbstone (not concrete) (manufacture)
10410	15410	Kernel crushing (manufacture)
19201	23201	Kerosene (manufacture)
46719	51519	Kerosene (wholesale)
27400	31500	Kerosene lanterns (manufacture)
10840	15870	Ketchup (manufacture)
46120	51120	Ketone and quinone function compounds (commission agent)
46750	51550	Ketone and quinone function compounds (wholesale)
20140	24140	Ketones (manufacture)
27510	29710	Kettle (electric) (manufacture)
25990	28750	Kettles (non-electric) (manufacture)
91040	92530	Kew gardens
25720	28630	Key (manufacture)
25720	28630	Key blank (manufacture)
96090	52740	Key cutting services (while you wait)
15120	19200	Key tags and cases made of leather (manufacture)
32200	36300	Keyboard instruments (manufacture)
32200	36300	Keyboard pipe organs with free metal reeds (manufacture)
32200	36300	Keyboard stringed instruments (manufacture)
26200	30020	Keyboards for computers (manufacture)
23200	26260	Kiln furniture (manufacture)
23200	26260	Kiln lining (manufacture)
85100	80100	Kindergartens
10200	15209	Kipper (manufacture)
17220	21220	Kitchen cloth made of paper (manufacture)
24420	27420	Kitchen foil made of aluminium (manufacture)
25710	28610	Kitchen knife (manufacture)
13923	17403	Kitchen linen (manufacture)
31020	36110	Kitchen seating (fitted) (manufacture)
17220	21220	Kitchen towels made of paper (manufacture)
47599	52440	Kitchen units (retail)

K

SIC 2007	SIC 2003	Activity	SIC 2007	SIC 2003	Activity
23410	26210	Kitchenware made of ceramics (manufacture)	25730	28620	Knives for horticultural use (manufacture)
23130	26130	Kitchenware made of glass (manufacture)	25730	28620	Knives for industrial use (manufacture)
22290	25240	Kitchenware made of plastic (manufacture)	25730	28620	Knives for machines (manufacture)
16290	20510	Kitchenware made of wood (manufacture)	25730	28620	Knives for tradesmen (manufacture)
77299	71409	Kitchenware rental and leasing	22290	25240	Knives made of plastic (manufacture)
30300	35300	Kite (not toy) (manufacture)	22290	25240	Knobs for furniture made of plastic (manufacture)
46160	51160	Kits for embroidery, etc. (commission agent)	13939	17519	Knotted carpets (manufacture)
01250	01139	Kiwi fruit growing	28940	29540	Knotted net, tulle, lace, braid etc. making machines (manufacture)
26110	32100	Klystron (manufacture)	13940	17520	Knotted netting of twine, cordage or rope (manufacture)
14142	18232	Knickers (manufacture)	17120	21120	Kraft wrapping and packaging paper (manufacture)
25710	28610	Knife (cutlery) (manufacture)	28230	30010	Labelling machine for office use (manufacture)
27510	29710	Knife sharpener (electric) (manufacture)	28290	29240	Labelling machinery (not for office use) (manufacture)
28140	29130	Knife valves (manufacture)	82920	74820	Labelling, stamping and imprinting on a fee or contract basis
25710	28610	Knife with folding blade (manufacture)	13960	17542	Labels (manufacture)
96090	93059	Knifegrinder (travelling)	22290	25240	Labels (not self-adhesive) made of plastic (manufacture)
14190	18241	Knitted bonnet (manufacture)	18121	21251	Labels (printed) made of gummed paper (manufacture)
14190	18249	Knitted bootees (manufacture)	18121	21251	Labels (printed) made of paper (manufacture)
14132	18222	Knitted dress and jacket ensemble (manufacture)	17290	21252	Labels (unprinted) made of gummed paper (manufacture)
14132	18222	Knitted dresses for women and girls (manufacture)	17290	21252	Labels (unprinted) made of paper (manufacture)
13910	17600	Knitted fabrics (manufacture)	13960	17542	Labels made of textiles (manufacture)
14190	18249	Knitted gloves (manufacture)	13960	17542	Labels made of woven fabric (manufacture)
13300	17300	Knitted goods finishing (manufacture)	26513	33202	Laboratory analytical instruments (non-electronic) (manufacture)
13300	17300	Knitted goods printing (manufacture)	31010	36120	Laboratory benches, stools, and other laboratory seating (manufacture)
13300	17300	Knitted goods scouring (manufacture)	28210	29210	Laboratory furnace (manufacture)
13300	17300	Knitted goods shrinking (manufacture)	31010	36120	Laboratory furniture, cabinets and tables (manufacture)
13300	17300	Knitted goods trimming (manufacture)	26513	33202	Laboratory incubators and sundry lab apparatus for measuring, testing (non-electronic) (manufacture)
14190	18249	Knitted mittens and mitts (manufacture)	23440	26240	Laboratory non-refractory ceramic products (manufacture)
14141	18231	Knitted nightwear for men and boys (manufacture)	32500	29240	Laboratory sterilisers (manufacture)
14142	18232	Knitted nightwear for women and girls (manufacture)	32500	29240	Laboratory type distilling apparatus (manufacture)
13910	17600	Knitted or netted elastic over 30 cm wide (manufacture)	26511	33201	Laboratory type sensitive balances (electronic) (manufacture)
14131	18221	Knitted outerwear for men and boys (manufacture)	26513	33202	Laboratory type sensitive balances (non-electronic) (manufacture)
14132	18222	Knitted outerwear for women and girls (manufacture)	32500	29240	Laboratory ultrasonic cleaning machinery (manufacture)
14190	18249	Knitted scarf (manufacture)	84130	75130	Labour affairs services administration and regulation (public sector)
14190	18249	Knitted shawl (manufacture)	94200	91200	Labour organisations
14132	18222	Knitted skirts (manufacture)	94920	91320	Labour party
14132	18222	Knitted suits for women and girls (manufacture)	78109	74500	Labour recruitment
14190	18249	Knitted swimwear (manufacture)	02300	02010	Lac gathering
14190	18249	Knitted swimwear for infants (manufacture)	13990	17541	Lace (manufacture)
14190	18249	Knitted ties (manufacture)	13990	17541	Lace and embroidery in the piece, in strips or in motifs (manufacture)
14190	18249	Knitted underclothing for infants (manufacture)	13990	17541	Lace bleaching (not on commission) (manufacture)
14141	18231	Knitted underwear for men and boys (manufacture)	13300	17300	Lace bleaching, dyeing and dressing (on commission) (manufacture)
14142	18232	Knitted underwear for women and girls (manufacture)	96010	93010	Lace cleaning and mending (not net mending)
14141	18231	Knitted vests for men and boys (manufacture)	13990	17541	Lace clipping (manufacture)
14142	18232	Knitted vests for women and girls (manufacture)	13921	17401	Lace curtains (manufacture)
28940	29540	Knitting machine (manufacture)	74100	74872	Lace designing
47540	52450	Knitting machines (retail)			
46640	51830	Knitting machines (wholesale)			
25930	28730	Knitting needles made of metal (manufacture)			
22290	25240	Knitting needles made of plastic (manufacture)			
47510	52410	Knitting yarn (retail)			
13100	17120	Knitting yarn made of wool (manufacture)			
13100	17130	Knitting yarn made of worsted (manufacture)			
13100	17110	Knitting yarns (cotton) (manufacture)			
13100	17110	Knitting yarns (man made fibres) (manufacture)			
14390	17720	Knitwear (manufacture)			

74

SIC 2007	SIC 2003	Activity
13990	17541	Lace drawing (manufacture)
13990	17541	Lace dressing (manufacture)
13990	17541	Lace dyeing (not on commission) (manufacture)
13990	17541	Lace edging (manufacture)
13990	17541	Lace embroidery (manufacture)
13990	17541	Lace ending (manufacture)
13990	17541	Lace finishing (manufacture)
13990	17541	Lace flouncing (manufacture)
13990	17541	Lace mending (manufacture)
13990	17541	Lace net (manufacture)
13990	17541	Lace scalloping (manufacture)
13923	17403	Lace tablecloth (manufacture)
13990	17541	Lace trimming (manufacture)
13990	17542	Laces for boots and shoes (manufacture)
25610	28510	Lacquering (metal finishing) (manufacture)
31090	36140	Lacquering, varnishing and gilding of furniture (manufacture)
47520	52460	Lacquers (retail)
20140	24140	Lactones (coumarin, methylcoumarins and ethylcoumarins) (manufacture)
21100	24410	Lactones (other than coumarin, methylcoumarins and ethylcoumarins) (manufacture)
46460	51460	Lactones (wholesale)
10519	15519	Lactose production (manufacture)
77320	71320	Ladder hire
13960	17542	Ladder tape (textile material) (manufacture)
25990	28750	Ladders made of metal (manufacture)
25990	28750	Ladders made of metal for fire-fighting vehicles (manufacture)
16290	20510	Ladders made of wood (manufacture)
14190	18249	Ladies fan (manufacture)
15120	19200	Ladies handbags made of leather (manufacture)
47710	52423	Ladies' outfitter (retail)
25710	28610	Ladle (manufacture)
28910	29510	Ladles for handling hot metals (manufacture)
11050	15960	Lager brewing (manufacture)
23990	26821	Lagging rope made of asbestos (manufacture)
20120	24120	Lake (pigment) (manufacture)
50300	61201	Lake steamer service
14200	18300	Lambskin clothing (manufacture)
32500	24422	Laminaria (manufacture)
22210	25210	Laminate made wholly of plastics and/or transparent regenerated cellulose film (manufacture)
23120	26120	Laminated glass (manufacture)
22210	25210	Laminated plastic film (manufacture)
22210	25210	Laminated thermosetting plastics sheet (manufacture)
16210	20200	Laminated veneer wood (manufacture)
16210	20200	Laminated wood products (manufacture)
17120	21120	Laminates and foils laminated with paper or paperboard (manufacture)
22210	25210	Laminates made of plastic (manufacture)
24420	27420	Laminates of aluminium foil with other materials (manufacture)
18130	22250	Laminating (manufacture)
28230	30010	Laminating machine for office use (manufacture)
28950	29550	Laminating machinery (paper working) (manufacture)
13300	17300	Laminating of textile material (manufacture)
16210	20200	Laminboard (manufacture)

SIC 2007	SIC 2003	Activity
23190	26150	Lamp chimneys made of glass (manufacture)
27330	31200	Lamp holder (electric) (manufacture)
28990	29560	Lamp making machine (manufacture)
27400	31500	Lamps (manufacture)
27400	31610	Lamps for cycles (manufacture)
23190	26150	Lamps made of glass (manufacture)
27400	31500	Lampshades (not of glass or plastics) (manufacture)
47599	52440	Lampshades (retail)
22290	25240	Lampshades, reflectors, covers and diffusers made of plastic (manufacture)
68310	70310	Land agent
41100	70110	Land and building company
68100	70120	Land buying and selling
28922	29522	Land clearing equipment and machinery (manufacture)
43120	45110	Land drainage contractor
41100	70110	Land investment company
68209	70209	Land letting
43120	45110	Land reclamation work
84130	75130	Land registry
42990	45230	Land subdivision with land improvement (e.g. Adding of roads etc.)
71122	74206	Land surveying activities
71122	74206	Land surveyor (not valuer)
68310	70310	Land valuer or surveyor
43120	45110	Landfill for construction
38210	90020	Landfill for the disposal of refuse and waste
30300	35300	Landing gear for aircraft (manufacture)
32300	36400	Landing nets (manufacture)
52220	63220	Landing stage
30110	35110	Landing stage (floating) (manufacture)
55900	55239	Landlord (boarding house)
68209	70209	Landlord of real estate
71112	74202	Landscape architecture
81300	01410	Landscape contracting
81300	01410	Landscape gardening
81300	01410	Landscape measures for protecting the environment
85590	80429	Language instruction and conversational skills instruction
72200	73200	Languages research and experimental development
10410	15410	Lanolin recovery (manufacture)
28410	29420	Lapping machine (metal cutting) (manufacture)
13300	17300	Lapping of textiles (manufacture)
25730	28620	Lapping tools (manufacture)
26200	30020	Laptop computers (manufacture)
20301	24301	Lacquer (manufacture)
46330	51333	Lard (wholesale)
10110	15112	Lard from knackers (manufacture)
10410	15410	Lard oil (manufacture)
10110	15112	Lard refining (manufacture)
26701	33402	Laser (excluding complete equipment using laser components) (manufacture)
28410	29420	Laser cutting or welding machine tools (metal working) (manufacture)
46510	51840	Laser printers (wholesale)
32500	33100	Laser surgical apparatus (manufacture)
16290	20510	Lasts made of wood (manufacture)
25720	28630	Latch (manufacture)
22190	25130	Latex foam (manufacture)
28410	29420	Lathe (metal cutting) (manufacture)

SIC 2007	SIC 2003	Activity
28410	29420	Lathe chuck (manufacture)
25730	28620	Lathe tool (manufacture)
50200	61102	Launch barge services
30300	35300	Launch vehicle for spacecraft (manufacture)
28990	35300	Launching gear for aircraft (manufacture)
96010	93010	Launderette
96010	93010	Laundry
20120	24120	Laundry blue (manufacture)
32910	36620	Laundry brush (manufacture)
28940	29540	Laundry machinery (manufacture)
96010	93010	Laundry receiving office
10620	15620	Laundry starch (manufacture)
46690	51870	Laundry-type washing and dry-cleaning machines (wholesale)
22230	25239	Lavatory pans made of plastic (manufacture)
13923	17403	Lavatory seat cover (manufacture)
03210	05020	Laver gathering (cultivated)
03210	05020	Laver growing
69109	74119	Law agent
84240	75240	Law and order administration and operation
85421	80302	Law college
58190	22150	Law publishing
72200	73200	Law research and experimental development
94120	91120	Law Society
69109	74119	Law writing
28302	29320	Lawn mower (manufacture)
47520	52460	Lawn mowers (retail)
46610	51880	Lawn mowers however operated (wholesale)
20150	24150	Lawn sand (manufacture)
69102	74113	Lawyer
43330	45430	Laying or fitting carpets and linoleum floor coverings including of rubber or plastic
43330	45430	Laying or fitting other wooden floor coverings
96030	93030	Laying out the dead
43330	45430	Laying tiling or fitting marble, granite or slate floor coverings
43330	45430	Laying tiling or fitting terrazzo, marble, granite or slate wall coverings
43330	45430	Laying, tiling, hanging or fitting ceramic wall or floor tiles
43330	45430	Laying, tiling, hanging or fitting concrete stone wall or floor tiles
43330	45430	Laying, tiling, hanging or fitting cut stone wall or floor tiles
43330	45430	Laying, tiling, hanging or fitting floor and wall covering
18130	22250	Layouts for presentation (manufacture)
24430	27430	Lead (manufacture)
46720	51520	Lead (wholesale)
27200	31400	Lead acid batteries (manufacture)
24100	27100	Lead coated steel sheet (manufacture)
07290	13200	Lead mining and preparation
07290	13200	Lead ore and concentrate extraction and preparation
20301	24301	Lead paint (manufacture)
24430	27430	Lead wire made by drawing (manufacture)
23190	26150	Leaded light (manufacture)
25930	28740	Leaf springs (manufacture)
58110	22110	Leaflet publishing
01130	01120	Leafy or stem vegetables growing
71200	74300	Leak testing and flow monitoring activities

SIC 2007	SIC 2003	Activity
58141	22130	Learned journal publishing
94120	91120	Learned societies
85590	80429	Learning centres offering remedial courses
77400	74879	Leasing of intellectual property and the like (not copyrighted works e.g. Books and software)
77400	74879	Leasing of non-financial intangible assets
46110	51110	Leather (commission agent)
46240	51249	Leather (wholesale)
15120	19200	Leather articles for use in machinery or mechanical appliances (manufacture)
15120	19200	Leather belting for use in machinery (manufacture)
14110	18100	Leather clothes (manufacture)
47710	52421	Leather clothing for adults (retail)
46420	51421	Leather clothing for adults (wholesale)
15110	19100	Leather dressing (manufacture)
15110	19100	Leather dying (manufacture)
15110	19100	Leather enamelling (manufacture)
15200	19300	Leather fillings (manufacture)
14110	18100	Leather garments for men and boys (manufacture)
14110	18100	Leather garments for women and girls (manufacture)
15110	19100	Leather gilding (manufacture)
46160	51160	Leather goods (commission agent)
15120	19200	Leather goods (not industrial) (manufacture)
47722	52432	Leather goods (retail)
46499	51479	Leather goods (wholesale)
14110	18100	Leather industrial work accessories (manufacture)
15110	19100	Leather proofing (manufacture)
15120	19200	Leather shoe-laces (manufacture)
15110	19100	Leather tanning and dressing (manufacture)
15110	19100	Leather trimmings (manufacture)
14110	18100	Leather wearing apparel (manufacture)
28940	29540	Leather working machine (manufacture)
46640	51830	Leather, hides and skins working machinery (wholesale)
13960	17549	Leathercloth (manufacture)
13960	17549	Leathercloth made of polyvinyl chloride (manufacture)
13990	17541	Leavers lace (manufacture)
25930	28740	Leaves for springs (manufacture)
31010	36120	Lectern (manufacture)
90010	92319	Lecturers
01130	01120	Leek growing
25990	28750	Left luggage lockers (manufacture)
69109	74119	Legal activities n.e.c.
94990	91330	Legal Aid Society
94120	91120	Legal associations
69109	74119	Legal documentation and certification activities
69109	74119	Legal examiner activities
72200	73200	Legal sciences research and experimental development
69109	74119	Legal services in connection with the disposal of assets by auction
99000	99000	Legation
15200	19300	Leggings and gaiters made of cloth (manufacture)
15200	19300	Leggings made of leather (manufacture)
01110	01120	Leguminous crops growing
93110	92619	Leisure centres
30120	35120	Leisure craft made of rubber (manufacture)
01230	01139	Lemon growing
11070	15980	Lemonade powder (manufacture)

L

SIC 2007	SIC 2003	Activity
11070	15980	Lemonade production (manufacture)
91011	92510	Lending and storage of books, periodicals
91011	92510	Lending and storage of CDs, DVDs
91011	92510	Lending and storage of maps music
13200	17210	Leno fabric weaving (manufacture)
32500	33401	Lens (mounted, (not photographic)) (manufacture)
23190	26150	Lens (pressed or moulded, unworked, (not coloured glass for traffic signals)) (manufacture)
32500	33401	Lens (unmounted) (manufacture)
23190	26150	Lens made of coloured glass for rail and road signals (not optically worked) (manufacture)
26701	33401	Lenses (except ophthalmic) (manufacture)
01110	01110	Lentil growing
10612	15612	Lentil splitting, grinding or milling (manufacture)
86101	85111	Leprosaria (public sector)
52219	63210	Lessee of tolls
17230	21230	Letter card (manufacture)
17219	21219	Letter file (manufacture)
28230	30010	Letter opening machine (manufacture)
82190	74850	Letter or resume writing
20302	24302	Letterpress ink (manufacture)
18129	22220	Letterpress printing (manufacture)
28990	29560	Letterpress printing machine (manufacture)
58190	22150	Letterpress publishing
01130	01120	Lettuce growing
30200	35200	Level crossing control gear (manufacture)
26511	33201	Level gauges (electronic) (manufacture)
26513	33202	Level gauges (non-electronic) (manufacture)
26511	33201	Level measuring and control instruments (electronic) (manufacture)
26513	33202	Level measuring and control instruments (non-electronic) (manufacture)
28922	29522	Levellers (manufacture)
43120	45110	Levelling and grading of construction sites
28290	33202	Levels (manufacture)
65120	66031	Liability insurance
94920	91320	Liberal democrat party
91011	92510	Libraries
91011	92510	Library access to IT facilities including internet
91011	92510	Library training courses (IT, information literacy, basic skills)
90030	92319	Librettist
56302	55402	Licensed bars (independent)
56302	55404	Licensed bars (managed)
56302	55403	Licensed bars (tenanted)
53201	64120	Licensed carriers
96090	93059	Licensed porter
56302	55402	Licensed victualler (independent)
56302	55404	Licensed victualler (managed)
56302	55403	Licensed victualler (tenanted)
58290	72210	Licensing for the right to reproduce, distribute and use computer software
02300	02010	Lichen gathering
65110	66011	Life assurance
65201	66012	Life re-insurance
72190	73100	Life sciences research and experimental development (other than biotechnological)
32990	25130	Life vests (manufacture)
32990	25130	Lifebelts (manufacture)
32990	20520	Lifebelts made of cork (manufacture)
30110	35110	Lifeboat (manufacture)
32990	20520	Lifebuoy made of cork (manufacture)
30110	35110	Lifebuoy made of rubber (manufacture)
85590	80429	Lifeguard training
32990	25130	Lifejacket (manufacture)
13922	17402	Lifejacket made of canvas (manufacture)
32990	20520	Lifejacket made of cork (manufacture)
30110	35110	Liferaft (not rubber inflatable) (manufacture)
28220	29220	Lift (manufacture)
28220	29220	Lift and escalator maintenance (manufacture)
28220	29220	Lift for motor vehicle (manufacture)
49390	60239	Lift operating company
49390	60239	Lift operator
28220	29220	Lifting and handling equipment (manufacture)
46690	51870	Lifting and handling equipment (wholesale)
28220	29220	Lifting and handling equipment parts (manufacture)
46690	51870	Lifts, skip hoists, escalators and moving walkways (wholesale)
27400	31500	Light bulb (manufacture)
27400	31500	Light bulbs including fluorescent and neon tubes (manufacture)
26110	32100	Light emitting diodes (LED) (manufacture)
46520	51860	Light emitting diodes (wholesale)
22290	25240	Light fittings made of plastic (manufacture)
19201	23201	Light fuel oil (manufacture)
46690	51870	Light metal containers (wholesale)
25920	28720	Light metal packaging (manufacture)
77110	71100	Light motor vehicle (not exceeding 3.5 tonnes) renting or leasing
14132	18222	Light outerwear for women and girls (manufacture)
26200	30020	Light pens (manufacture)
46120	51120	Light, medium and heavy petroleum oils (commission agent)
30110	35110	Lighter (ship) (manufacture)
32990	36639	Lighter fuel in containers not exceeding 300 cc (liquid or liquefied gas) (manufacture)
52220	63220	Lighter lessee or owner
52220	63220	Lighterage activities
52220	63220	Lighthouse activities
52220	63220	Lighthouse Authority
27400	31500	Lighting equipment (manufacture)
47599	52440	Lighting equipment (retail)
46470	51439	Lighting equipment (wholesale)
27400	31610	Lighting equipment for aircraft (manufacture)
27400	31610	Lighting equipment for boats (manufacture)
27400	31610	Lighting equipment for motor vehicles (manufacture)
27400	31500	Lighting fitting (other than glassware) (manufacture)
22290	25240	Lighting fittings parts made of plastic (manufacture)
27400	31500	Lighting fixture of table lamps (manufacture)
27330	31200	Lightning arresters (manufacture)
52220	63220	Lightship
30110	35110	Lightship (manufacture)
14132	18222	Lightweight jackets for women and girls (manufacture)
05200	10200	Lignite (brown coal) mining including mining through liquefaction methods
19201	10200	Lignite fuel briquettes (manufacture)
05200	10200	Lignite mining
09900	10200	Lignite mining support services provided on a fee or contract basis

L

SIC 2007	SIC 2003	Activity
46120	51120	Lignite or peat (commission agent)
19100	23100	Lignite tars production (manufacture)
05200	10200	Lignite washing, dehydrating, pulverising etc. To improve quality or facilitate transport or storage
86900	85140	Limb fitting centre
20150	24150	Lime (ammonium nitrate) (manufacture)
23520	26520	Lime (manufacture)
01230	01139	Lime growing
28210	29210	Lime processing kiln (manufacture)
23700	26700	Limestone (ground) (manufacture)
46730	51530	Limestone (wholesale)
08110	14120	Limestone including dolomite mine or quarry
08110	14120	Limestone quarrying, crushing and breaking for constructional purposes
23700	26700	Limestone working (manufacture)
09900	14120	Limestone, gypsum and chalk quarrying support services provided on a fee or contract basis
25940	28740	Linchpin (manufacture)
17240	21240	Lincrusta (manufacture)
26301	32201	Line apparatus (carrier, duplex and repeater) (manufacture)
24200	27220	Line pipe made of steel (manufacture)
26301	32201	Line telegraphy apparatus (manufacture)
46520	51860	Line telegraphy or telegraphy apparatus (wholesale)
26301	32201	Line telephony apparatus (manufacture)
13100	17140	Line yarn made of flax (manufacture)
13940	17520	Line yarn made of hard fibre (manufacture)
46690	51870	Linear acting (cylinders) hydraulic and pneumatic power engines and motors (wholesale)
32990	19200	Linemen's safety belts and other belts for occupational use (manufacture)
46410	51410	Linen and linen goods (wholesale)
13200	17250	Linen and union cloth (manufacture)
13300	17300	Linen bleaching (manufacture)
13960	17549	Linen buckram weaving (manufacture)
13300	17300	Linen dyeing (manufacture)
96010	93010	Linen hire (associated with laundry service)
13300	17300	Linen printing (manufacture)
13200	17250	Linen weaving (manufacture)
30110	35110	Liner (ship) (manufacture)
28921	29521	Liner hanger equipment (manufacture)
22220	25220	Liner made of non-woven polyethylene (manufacture)
14142	18232	Lingerie (manufacture)
72200	73200	Linguistics research and experimental development
14190	18241	Linings for hats (manufacture)
25930	28740	Link chain (welded) (manufacture)
47530	52481	Lino tiles (retail)
22230	36639	Linoleum (manufacture)
46730	51479	Linoleum (wholesale)
43330	45430	Linoleum laying
10410	15410	Linseed crushing (manufacture)
01110	01110	Linseed growing
10410	15410	Linseed oil production (manufacture)
10410	15420	Linseed oil refining (manufacture)
46180	51180	Lip make-up and eye make-up preparations (commission agent)
20420	24520	Lipstick (manufacture)
09100	11100	Liquefaction and regasification of natural gas for transport

SIC 2007	SIC 2003	Activity
46120	51120	Liquefied gas for motor purposes (commission agent)
20110	24110	Liquefied or compressed industrial gases (manufacture)
20110	24110	Liquefied or compressed industrial or medical refrigerant gases (manufacture)
20110	24110	Liquefied or compressed medical gases (manufacture)
46711	51511	Liquefied petroleum gases (wholesale)
11010	15910	Liqueurs (manufacture)
46342	51342	Liqueurs (wholesale)
46120	51120	Liquid and compressed air (commission agent)
46750	51550	Liquid and compressed air (wholesale)
19201	23201	Liquid butane gas (manufacture)
10832	15862	Liquid coffee (manufacture)
46690	51870	Liquid dielectric transformers (wholesale)
28120	29121	Liquid elevator parts (manufacture)
28131	29121	Liquid elevators (manufacture)
46719	51519	Liquid fuels (other than petroleum) (wholesale)
06200	11100	Liquid hydrocarbon fractions draining and separation
20301	24301	Liquid lustres (manufacture)
20110	24110	Liquid or compressed air (manufacture)
46499	51479	Liquid or liquefied-gas fuels for lighters in containers (300 cc or more) (wholesale)
46711	51511	Liquid petroleum fuels (wholesale)
19201	23201	Liquid propane gas (manufacture)
20411	24511	Liquid soap (manufacture)
24100	27100	Liquid steel (manufacture)
10810	15830	Liquid sugar (manufacture)
26513	33202	Liquid supply meter (non-electronic) (manufacture)
82920	74820	Liquids bottling on a fee or contract basis
49500	60300	Liquids transport via pipelines
10822	15842	Liquorice (manufacture)
28230	30010	Listing machine (manufacture)
74909	74879	Literary agency
94990	91330	Literature and book club
27200	31400	Lithium batteries (manufacture)
20590	24640	Litho plate making (sensitized) (manufacture)
18130	22240	Litho plate making (unsensitized) (manufacture)
23700	26700	Litho stone working (manufacture)
90030	92319	Lithographic artist (own account)
20302	24302	Lithographic ink (manufacture)
18129	22220	Lithographic printing (manufacture)
18121	21251	Lithographic printing on labels or tags (manufacture)
18130	22240	Lithographic stones and wood blocks (manufacture)
18130	22250	Lithography (manufacture)
32500	33100	Lithotriptors (manufacture)
38110	90020	Litter box refuse collection (public)
46230	51230	Live animal exporter (wholesale)
46230	51230	Live animal importer (wholesale)
46110	51110	Live animals (commission agent)
46230	51230	Live animals (wholesale)
46230	51230	Live poultry (wholesale)
28220	29220	Live storage rack (manufacture)
14120	18210	Livery (manufacture)
46230	51230	Livestock (wholesale)
65120	66031	Livestock insurance
71200	74300	Lloyd's Register of Shipping
65120	66031	Lloyd's underwriter (non-life)

L

SIC 2007	SIC 2003	Activity
26120	32100	Loaded electronic boards (manufacture)
26120	32100	Loaded printed circuit boards (manufacture)
28921	29521	Loader for mining (manufacture)
52243	63110	Loading and unloading of freight railway cars
52242	63110	Loading and unloading of goods travelling via air transport
52243	63110	Loading and unloading of goods travelling via rail transportation
52241	63110	Loading and unloading of goods travelling via water transport
52242	63110	Loading and unloading of passengers' luggage travelling via air transport
52241	63110	Loading and unloading of passengers' luggage travelling via water transport
52243	63110	Loading and unloading passengers' luggage travelling via rail transportation
28922	29522	Loading shovel (manufacture)
13940	17520	Loading slings (manufacture)
64921	65221	Loan company (other than in banks' sector)
70210	74141	Lobbying activities
28131	29121	Lobe pump (manufacture)
03220	05020	Lobsterling production, freshwater
03210	05020	Lobsterling production, marine
91020	92521	Local authority art galleries and museums
96040	93040	Local authority baths (hot water and sauna)
52220	63220	Local authority canal services
52219	63210	Local authority car parks
96030	93030	Local authority cemeteries
88990	85322	Local authority citizen's advice bureau
42990	45213	Local authority civil engineering department
87900	85312	Local authority community homes (children)
90040	92320	Local authority concert halls and theatres
96030	93030	Local authority crematoriums
52220	63220	Local authority docks and harbours
37000	90010	Local authority drainage services
42990	45213	Local authority engineer's department
84250	75250	Local authority fire brigade services
93110	92619	Local authority football and other sports grounds
50400	61209	Local authority freight ferry services on rivers, canals and lakes
96030	93030	Local authority funeral services
42110	45230	Local authority highways construction and maintenance
88100	85322	Local authority home help service
87300	85312	Local authority homes for the disabled and the elderly
41202	45212	Local authority house building and maintenance
93110	92619	Local authority leisure centres
52220	63220	Local authority lighthouse service
87900	85312	Local authority lodging houses
52230	63230	Local authority municipal airport
84230	75230	Local authority observation and assessment centres
41202	45212	Local authority or new town direct labour department (domestic dwellings)
93290	92729	Local authority parks and gardens
50300	61201	Local authority passenger ferry services on rivers, canals and lakes
81291	74703	Local authority pest control department
88990	85322	Local authority probation service
93110	92619	Local authority recreational facilities
38210	90020	Local authority refuse disposal
56102	55302	Local authority restaurants, cafes, snack bars, etc. (unlicensed)
42110	45230	Local authority road construction and major repairs
49319	60219	Local authority road passenger transport services
84240	75240	Local authority school crossing patrols
56290	55510	Local authority school meals service
37000	90010	Local authority sewage services
88990	85322	Local authority social services department
93110	92619	Local authority sports facilities (incl. Football and other sports grounds, swimming baths, etc.)
43210	45340	Local authority street lighting
93110	92619	Local authority swimming pool
84240	75240	Local authority traffic wardens
49319	60219	Local authority transport department
84110	75110	Local government administration
60100	92201	Local radio station (broadcasting)
25720	28630	Lock (manufacture)
42910	45240	Lock construction
25720	28630	Lock for motor vehicle (manufacture)
25110	28110	Lock gates (manufacture)
25940	28740	Lock washer (manufacture)
31010	36120	Locker (manufacture)
13910	17600	Locknit fabric (manufacture)
46740	51540	Locks (wholesale)
25720	28630	Locksmiths (manufacture)
30200	35200	Locomotive (manufacture)
30200	35200	Locomotive parts and accessories (manufacture)
46690	51870	Locomotives (wholesale)
01250	01139	Locust bean growing
94990	91330	Lodge activities
87900	85312	Lodging house (local authority)
55900	55239	Lodging house (private)
28490	29430	Log decorticators (manufacture)
16100	20100	Log sawing (manufacture)
16100	20100	Log slicing, peeling or chipping (manufacture)
02400	02020	Log transport within the forest
01250	01139	Loganberry growing
46730	51530	Logged timber (wholesale)
02200	02010	Logging
02400	02020	Logging service activities
91020	92521	London Museum
16290	20520	Loofah articles (manufacture)
28940	29540	Loom (manufacture)
16290	20510	Loom made of wood (manufacture)
28940	29540	Loom winder (manufacture)
13921	17401	Loose cover for furniture (manufacture)
23140	26140	Loose glass fibre (manufacture)
17230	22220	Loose leaf binder (manufacture)
84230	75230	Lord Chancellor's Department
94910	91310	Lord's Day Observance Society
45190	50101	Lorries (retail)
45190	50102	Lorries (used) (retail)
45190	50102	Lorries (used) (wholesale)
45190	50101	Lorries (wholesale)
29100	34100	Lorry (manufacture)
28220	29220	Lorry loader (manufacture)
66210	67200	Loss adjuster
92000	92710	Lottery ticket sales
26400	32300	Loudspeaker (manufacture)

SIC 2007	SIC 2003	Activity
11050	15960	Low and non-alcoholic beer (manufacture)
11020	15932	Low and non-alcoholic wine based on concentrated grape must (manufacture)
11020	15931	Low and non-alcoholic wine from fresh grapes and grape juice (manufacture)
24100	27100	Low carbon ferro manganese (carbon 2% or less) (manufacture)
10860	15880	Low energy and energy-reduced foods (manufacture)
29202	34202	Low loader trailer (manufacture)
10860	15880	Low sodium foods (manufacture)
19100	23100	Low temperature carbonisation solid fuel (not ovoid or briquettes) (manufacture)
10860	15880	Low-sodium or sodium-free dietary salts (manufacture)
82990	74879	Loyalty programme administration
10822	15842	Lozenge (medicated) (manufacture)
10822	15842	Lozenge (not medicated) (manufacture)
46719	51519	Lubricants (wholesale)
19209	23209	Lubricating grease formulation outside refineries (manufacture)
19201	23201	Lubricating oil (at refinery) (manufacture)
20590	24660	Lubricating oil additive (manufacture)
19209	23209	Lubricating oil formulation outside refineries (manufacture)
46120	51120	Lubricating oils (commission agent)
46719	51519	Lubricating oils and greases (wholesale)
47300	50500	Lubricating products for motor vehicles
28131	29121	Lubricating pump (not for internal combustion engine) (manufacture)
28290	29240	Lubricator (manufacture)
15120	19200	Luggage and the like of leather (manufacture)
15120	19200	Luggage handbags made of paperboard (manufacture)
15120	19200	Luggage made of leather or leather substitute (manufacture)
30990	35500	Luggage trucks (hand propelled) (manufacture)
30200	35200	Luggage van for railway (manufacture)
15120	19200	Luggage, handbags made of plastic sheeting (manufacture)
15120	19200	Luggage, handbags made of textile materials (manufacture)
20120	24120	Luminophores (manufacture)
56101	55301	Luncheon bar (licensed)
56290	55510	Luncheon club
82990	74879	Luncheon voucher company
01110	01110	Lupin growing
20130	24130	Lyes (manufacture)
90030	92319	Lyric author
46460	51460	Lysine and glutamic acid and salts thereof (wholesale)
46180	51180	Lysine, glutamic acid and their salts (commission agent)
10730	15850	Macaroni (manufacture)
28930	29530	Macaroni, spaghetti or similar products machinery (manufacture)
46140	51140	Machine broker (commission agent)
13922	17402	Machine covers (manufacture)
25400	29600	Machine gun (manufacture)
16290	20510	Machine parts made of wood (manufacture)
13300	17300	Machine printing of textiles (manufacture)
33200	28520	Machine rigging (manufacture)

SIC 2007	SIC 2003	Activity
25730	28620	Machine tool interchangeable tools (manufacture)
28490	29430	Machine tool special attachments (excluding for metal working) (manufacture)
28410	29420	Machine tools (ultrasonic) (metal working) (manufacture)
46620	51810	Machine tools (wholesale)
46620	51810	Machine tools exporter (wholesale)
28490	29430	Machine tools for cork, bone, hard rubber, hard plastics or similar hard materials (manufacture)
28490	29430	Machine tools for working cold glass (manufacture)
28410	29420	Machine tools, for working metals (manufacture)
28410	29420	Machine tools, for working metal, plasma arc (manufacture)
28410	29420	Machine tools, for working metals, ultrasonic waves (manufacture)
28410	29420	Machine tools, for working metals, using a laser beam (manufacture)
28410	29420	Machine tools, for working metals, using a magnetic pulse (manufacture)
28490	29430	Machine tools, for working wood (manufacture)
46620	51810	Machine tools importer (wholesale)
77390	71340	Machine tools rental and operating leasing
46140	51140	Machinery (commission agent)
28290	29240	Machinery (not containing electrical connectors) n.e.c. parts (manufacture)
46900	51900	Machinery (undefined) (wholesale)
15120	19200	Machinery accessories made of leather (manufacture)
46690	51870	Machinery and apparatus for filtering or purifying gases (wholesale)
46660	51850	Machinery and equipment for offices (wholesale)
77299	71409	Machinery and equipment used by amateurs or as a hobby e.g. Home repair tools rental and leasing
71121	74205	Machinery and industrial plant design
13960	17542	Machinery belting (woven) (manufacture)
15120	19200	Machinery belting made of leather (manufacture)
28290	29240	Machinery for cleaning or drying bottles and for aerating beverages (manufacture)
77390	71340	Machinery for industrial use rental and operating leasing
28250	29230	Machinery for liquefying air or gas (manufacture)
28490	29430	Machinery for making wooden clogs, soles and heels for shoes (manufacture)
28921	29521	Machinery for treating minerals by screening, sorting, separating, washing, crushing (manufacture)
46690	51870	Machinery n.e.c., for use in trade, navigation and other services (wholesale)
46690	51870	Machinery n.e.c., for industrial use (except mining, construction, textile) (wholesale)
46690	51870	Machinery n.e.c., for industrial use (except mining, construction, textile) exporter (wholesale)
46690	51870	Machinery n.e.c., for industrial use (except mining, construction, textile) importer (wholesale)
46900	51900	Machinery stockist (undefined) (wholesale)
46690	51870	Machines and appliances for testing the mechanical properties of materials (wholesale)
28302	29320	Machines for cleaning, sorting or grading eggs, fruit (manufacture)
28990	29560	Machines for the assembly of electric or electronic lamps, tubes (valves) or bulbs (manufacture)
26200	30020	Machines for transcribing data media in coded form (manufacture)
28410	29420	Machining centre (metal working) (manufacture)
28290	33202	Machinists precision tools (manufacture)

M

SIC 2007	SIC 2003	Activity
14131	18221	Mackintoshes for men and boys (manufacture)
14132	18222	Mackintoshes for women and girls (manufacture)
62012	72400	Made-to-order software
13922	17402	Made-up filter cloth (manufacture)
13922	17402	Made-up goods of sailcloth (manufacture)
17120	21120	Magazine paper (manufacture)
18129	22220	Magazine printing (manufacture)
58142	22130	Magazine publishing
46499	51479	Magazines (wholesale)
77210	71401	Magazines rental
01490	01250	Maggot breeding
84230	75230	Magistrates' court
23200	26260	Magnesite chrome shape (manufacture)
24450	27450	Magnesium (manufacture)
08910	14300	Magnesium sulphates (natural kieserite) mining
26800	24650	Magnetic card (manufacture)
26200	30020	Magnetic card readers (manufacture)
26200	30020	Magnetic card storage units (manufacture)
26513	33202	Magnetic compass (manufacture)
26800	24650	Magnetic disc (unrecorded) (manufacture)
26200	30020	Magnetic disk drives (manufacture)
26200	30020	Magnetic flash drives (manufacture)
46690	51870	Magnetic lifting heads (wholesale)
26200	30020	Magnetic or optical readers (manufacture)
46510	51840	Magnetic or optical readers (wholesale)
26400	32300	Magnetic recording head (manufacture)
26600	33100	Magnetic resonance imaging (MRI) equipment (manufacture)
26200	30020	Magnetic storage devices for computers (manufacture)
26800	24650	Magnetic tape (unrecorded) (manufacture)
26400	32300	Magnetic tape recorders (manufacture)
29310	31610	Magneto (manufacture)
29310	31610	Magneto-dynamos (manufacture)
71122	74206	Magnetometric (subsurface) surveying activities
26110	32100	Magnetron (manufacture)
46520	51860	Magnetrons, klystrons and microwave tubes (wholesale)
26701	33402	Magnifying glass (manufacture)
24430	27430	Magnolia metal (manufacture)
53100	64110	Mail distribution and delivery
28230	30010	Mail handling machines (envelope stuffing, sealing, addressing; opening, sorting) (manufacture)
47910	52610	Mail order (retail)
45320	50300	Mail order sales of motor vehicle parts and accessories (retail)
82190	64110	Mailbox rental
18140	22230	Mailing finishing services such as customisation, envelope preparation (manufacture)
58120	22110	Mailing lists (in print) publishing
82190	74850	Mailing pre-sorting
26200	30020	Mainframe computers (manufacture)
96030	93030	Maintenance of graves
96030	93030	Maintenance of mausoleums
45200	50200	Maintenance of motor vehicles
68209	70209	Maisonettes letting
10611	15611	Maize (flaked) production (manufacture)
10611	15611	Maize flour and meal production (manufacture)
10620	15620	Maize starch (manufacture)
96020	93020	Make-up and beauty treatment

SIC 2007	SIC 2003	Activity
20420	24520	Make-up preparation (manufacture)
71111	74201	Making of architectural maquettes
24510	27510	Malleable castings (manufacture)
11060	15970	Malt and malt products (manufacture)
10890	15899	Malt extract (manufacture)
11050	15960	Malt liquors (manufacture)
10890	15899	Malted milk production (manufacture)
10620	15620	Maltose (manufacture)
26600	33100	Mammographs (manufacture)
62030	72300	Management and operation on a continuing basis of data processing facilities belonging to others
70229	74143	Management audits consultancy services
70229	74143	Management consultancy activities
85320	80220	Management training establishment
93110	92619	Managing and providing the staff for sports facilities
01230	01139	Mandarin growing
24450	27450	Manganese alloys (manufacture)
27200	31400	Manganese dioxide cells (manufacture)
07290	13200	Manganese mining and preparation
20120	24120	Manganese oxide (manufacture)
24450	27450	Manganese production and refining (manufacture)
24450	27450	Manganese wire made by drawing (manufacture)
01220	01139	Mango growing
01190	01110	Mangold growing
22230	25239	Manhole or access covers made of plastic (manufacture)
46180	51180	Manicure and pedicure preparations (commission agent)
20420	24520	Manicure and pedicure preparations (manufacture)
25710	28610	Manicure and pedicure sets (manufacture)
15120	19200	Manicure case made of leather (manufacture)
96020	93020	Manicurist
28110	29110	Manifold for industrial engine (manufacture)
20600	24700	Man-made fibre (not glass fibre) (manufacture)
13100	17150	Man-made fibre bulking (other than in man-made fibre producing establishments) (manufacture)
13100	17150	Man-made fibre crimping (other than in man-made fibre producing establishments) (manufacture)
13300	17300	Man-made fibre fabric bleaching, dyeing, printing or otherwise finishing (manufacture)
13100	17150	Man-made fibre texturing (other than in man-made fibre producing establishments) (manufacture)
13100	17150	Man-made fibre texturing, bulking, crimping in man-made fibre producing establishments (manufacture)
46180	51180	Man-made fibre waste (commission agent)
13200	17210	Man-made fibre weaving from yarns spun on the cotton system (manufacture)
13200	17220	Man-made fibre weaving of fabrics from yarns spun on the woollen system (manufacture)
13200	17230	Man-made fibre weaving of fabrics from yarns spun on worsted and semi-worsted systems (manufacture)
13300	17300	Man-made fibre yarn bleaching, dyeing or otherwise finishing (manufacture)
46120	51120	Man-made fibres and yarns (commission agent)
13100	17110	Man-made fibres spinning on the cotton system (manufacture)
13100	17120	Man-made fibres spinning on the woollen system (manufacture)
13100	17130	Man-made fibres spinning on the worsted and semi-worsted systems (manufacture)

M

SIC 2007	SIC 2003	Activity
13100	17110	Man-made fibres twisting on the cotton system (manufacture)
13100	17120	Man-made fibres twisting on the woollen system (manufacture)
13100	17130	Man-made fibres twisting on the worsted and semi-worsted systems (manufacture)
13100	17110	Man-made fibres warping on the cotton system (manufacture)
13100	17120	Man-made fibres warping on the woollen system (manufacture)
13100	17130	Man-made fibres warping on the worsted and semi-worsted systems (manufacture)
13100	17110	Man-made fibres winding on the cotton system (manufacture)
13100	17120	Man-made fibres winding on the woollen system (manufacture)
13100	17130	Man-made fibres winding on the worsted and semi-worsted systems (manufacture)
20600	24700	Man-made staple fibres, not carded, combed or otherwise processed for spinning (manufacture)
28940	29540	Man-made textile fibre or yarn producing machinery (manufacture)
20600	24700	Man-made tow (manufacture)
26511	33201	Manometers (electronic) (manufacture)
26513	33202	Manometers (non-electronic) (manufacture)
68209	70209	Mansions letting
30990	35500	Manually propelled trucks (manufacture)
24100	27100	Manufacture of hot-rolled rods of steel
25990	28750	Manufacture of metal combs (manufacture)
14200	18300	Manufacturing furrier (manufacture)
84130	75130	Manufacturing services administration and regulation (public sector)
46750	51550	Manure (wholesale)
28302	29320	Manure spreader (manufacture)
46610	51880	Manure spreaders, seeders (wholesale)
18129	22220	Manuscript book (manufacture)
58110	22110	Map and plan publishing
91011	92510	Map lending and storage
18129	22220	Map printing (manufacture)
47620	52470	Map seller (retail)
10810	15830	Maple syrup (manufacture)
46730	51530	Marble (wholesale)
23700	26700	Marble masonry working (manufacture)
08110	14110	Marble quarrying (rough trimming and sawing)
23190	26150	Marbles made of glass (manufacture)
10420	15430	Margarine (manufacture)
46330	51333	Margarine (wholesale)
42910	45240	Marina construction
93290	92629	Marinas
03110	05010	Marine and freshwater crustaceans and molluscs gathering
10410	15410	Marine animal crude oil and fat production (manufacture)
03210	05020	Marine aquaculture
52220	63220	Marine cargo lighterage
52220	63220	Marine cargo superintendent
71200	74300	Marine cargo surveyor
74909	74206	Marine consultant (other than environmental consultancy)
03110	05010	Marine crustacean and mollusc gathering
19201	23201	Marine diesel oil (manufacture)
84250	75250	Marine fireboat services
03110	05010	Marine fishing
03110	05010	Marine fishing vessels engaged in processing and preserving of fish
65120	66031	Marine insurance
71200	74300	Marine insurance survey activities
28110	29110	Marine non-propulsion engines (manufacture)
25300	28300	Marine or power boiler parts (manufacture)
20301	24301	Marine paint (manufacture)
46690	51870	Marine propulsion engines (wholesale)
52220	63220	Marine salvage
25990	28750	Marine screw propeller (manufacture)
46770	51570	Marine store waste (wholesale)
71200	74300	Marine surveyor
50200	61102	Marine tow out services
93290	92349	Marionette show
52290	63400	Maritime agent
84220	75220	Maritime search and rescue (military)
28230	36120	Marker boards (manufacture)
32990	36631	Marker pen (manufacture)
73200	74130	Market research agency
73200	74130	Market research consultant
73200	74130	Market research organisation
73200	74130	Market, social and economic research services
73110	74402	Marketing campaigns
70229	74149	Marketing management consultancy activities
20590	24660	Marking ink (manufacture)
08110	14120	Marl mining
10390	15330	Marmalade (manufacture)
13922	17402	Marquee (manufacture)
77210	71401	Marquee hire
88990	85321	Marriage and family guidance (charitable)
88990	85322	Marriage and family guidance (non-charitable)
96090	93059	Marriage bureau
10822	15842	Marshmallow (manufacture)
85510	92629	Martial arts instruction
10822	15842	Marzipan sweets (manufacture)
10310	15310	Mashed potatoes (dehydrated) production (manufacture)
26600	33100	Mask and respirator (not medical) (manufacture)
43999	45250	Mason (building)
86900	85140	Mass radiography service
32500	33100	Massage apparatus (manufacture)
96040	93040	Massage salons
24440	27440	Master alloys of copper (manufacture)
96090	93059	Master of Ceremonies
20301	24303	Mastics (manufacture)
46730	51530	Mastics and sealants (wholesale)
30120	35120	Masts and spars for pleasure boats (manufacture)
30110	35110	Masts and spars for ships (manufacture)
23140	26140	Mat made of glass fibre (manufacture)
20510	36639	Match (manufacture)
46499	51479	Matches (wholesale)
25730	28620	Matchet (manufacture)
01270	01139	Maté growing
46210	51210	Materials, residues and by-products used as animal feed (wholesale)
86900	85140	Maternity and child welfare services
86900	85140	Maternity clinic
86101	85112	Maternity hospital (private sector)

M

SIC 2007	SIC 2003	Activity	SIC 2007	SIC 2003	Activity
86101	85111	Maternity hospital (public sector)	10130	15139	Meat juices (manufacture)
26513	33202	Mathematical instrument (non-electronic) (manufacture)	10110	15112	Meat meal (ground meat) (manufacture)
			10130	15139	Meat pate (manufacture)
72190	73100	Mathematical research and experimental development	56103	55303	Meat pie shop
			10130	15139	Meat pies and puddings (manufacture)
59200	22140	Matrice for record production	46320	51320	Meat porter (wholesale)
13939	17519	Mats and matting made of coconut fibre (manufacture)	28930	29530	Meat processing machinery (manufacture)
13939	17519	Mats and matting made of coir (manufacture)	10110	15111	Meat production (fresh, chilled or frozen) in carcasses or cuts (manufacture)
13939	17519	Mats and matting made of sisal (manufacture)	10130	15139	Meat pudding (manufacture)
14200	18300	Mats and rugs made of fur (manufacture)	10130	15139	Meat rillettes (manufacture)
16290	20520	Mats made of cork (manufacture)	46320	51320	Meat salesman (wholesale)
13939	17519	Mats made of jute (manufacture)	71129	74204	Mechanical and electrical installation for buildings design activities
22190	25130	Mats made of rubber (manufacture)			
24440	27440	Mattes made of copper (manufacture)	28210	29210	Mechanical ash dischargers (manufacture)
24450	27450	Mattes of nickel production (manufacture)	38320	37100	Mechanical crushing of metal waste (cars, washing machines, etc.) with sorting and separation
16290	20520	Matting made of cane (manufacture)			
16290	20520	Matting made of rushes (manufacture)	38320	37100	Mechanical crushing of metal waste from used bikes
22230	25231	Matting made of woven plastic (manufacture)	25620	28520	Mechanical engineering (general) (manufacture)
25730	28620	Mattock (manufacture)	28220	29220	Mechanical industrial robots for lifting, handling (manufacture)
31030	36150	Mattress base (manufacture)			
23140	26140	Mattress made of glass fibre (manufacture)	15110	19100	Mechanical leather preparation (manufacture)
31030	36150	Mattress made of plastic foam (manufacture)	28220	29220	Mechanical lifting manipulators (manufacture)
31030	36150	Mattress made of sponge (manufacture)	28220	29220	Mechanical manipulators (manufacture)
31030	36150	Mattress support made of metal (manufacture)	28150	29140	Mechanical power transmission bearing housings (manufacture)
10720	15820	Matzos (manufacture)	28150	29140	Mechanical power transmission camshafts (manufacture)
10840	15870	Mayonnaise (manufacture)			
93120	92629	MCC	28150	29140	Mechanical power transmission chain (manufacture)
11030	15949	Mead (manufacture)	28150	29140	Mechanical power transmission cranks (manufacture)
10611	15611	Meal from grain (manufacture)	28150	29140	Mechanical power transmission crankshafts (manufacture)
10612	15612	Meal of dried leguminous vegetables production (manufacture)			
			28150	29140	Mechanical power transmission plain shaft bearings (manufacture)
56290	55520	Meals on wheels catering			
77390	71340	Measuring and controlling equipment rental and operating leasing	28150	29140	Mechanical power transmission plant (manufacture)
			28150	29140	Mechanical power transmission shafts (manufacture)
26511	33201	Measuring instruments and appliances (electronic) (manufacture)	38320	37100	Mechanical reduction of large iron pieces such as railway wagons into secondary raw materials
26513	33202	Measuring instruments and appliances (non-electronic) (manufacture)	28922	29522	Mechanical shovels (manufacture)
46690	51870	Measuring instruments and equipment (wholesale)	46630	51820	Mechanical shovels, shovel loaders, excavators, with 360 degree revolving superstructure (wholesale)
28290	33202	Measuring rods and tapes (manufacture)	28210	29210	Mechanical stokers (manufacture)
28290	33202	Measuring rule (manufacture)	17110	21110	Mechanical woodpulp (manufacture)
28290	33202	Measuring tape (manufacture)	71129	74204	Mechanical, industrial and systems engineering design projects
10110	15111	Meat (except poultry meat) processing and preserving (manufacture)			
			32500	33100	Mechano-therapy appliances (manufacture)
46320	51320	Meat (wholesale)	47789	52485	Medals (new) (retail)
10110	15112	Meat and bone meal from knackers (manufacture)	77291	71403	Media entertainment equipment renting and leasing
46390	51390	Meat and fish market porterage (wholesale)	26800	24650	Media for sound or video recording (unrecorded) (manufacture)
47220	52220	Meat and meat products (retail)			
46320	51320	Meat and meat products exporter (wholesale)	73120	74401	Media representation
46320	51320	Meat and meat products importer (wholesale)	74202	74813	Medical and biological photography
47220	52220	Meat and meat products in specialised stores (retail)	32500	33100	Medical and dental instruments and supplies (manufacture)
10130	15139	Meat and poultry meat processing (other than bacon and ham) (manufacture)	77299	71409	Medical and paramedical equipment (e.g. crutches) rental and leasing
10130	15139	Meat and poultry meat products (manufacture)	32500	33100	Medical appliances (manufacture)
10130	15139	Meat canning, cooking and preserving (manufacture)	94120	91120	Medical associations
10110	15111	Meat chilling or freezing for human consumption (manufacture)	86101	85111	Medical consultant (public sector)
			21200	24421	Medical diagnostic preparations, including pregnancy tests (manufacture)
47220	52220	Meat dealer (retail)			
10130	15139	Meat extract (manufacture)	46180	51180	Medical goods (commission agent)
46380	51380	Meat for domestic animals (wholesale)			

SIC 2007	SIC 2003	Activity	SIC 2007	SIC 2003	Activity
47749	52329	Medical goods (retail)	47710	52424	Men's outfitter (retail)
46460	51460	Medical goods (wholesale)	47710	52424	Men's wear dealer (retail)
86210	85120	Medical group practice	86101	85111	Mental disability (public sector)
21200	24421	Medical impregnated bandages, dressings, gauze and wadding (manufacture)	86220	85120	Mental health specialist (private practice)
			86101	85111	Mental health specialist (public sector)
32500	33100	Medical instrument (non-optical) (manufacture)	86101	85112	Mental hospital (private sector)
86900	85140	Medical laboratories	86101	85111	Mental hospital (public sector)
26600	33100	Medical laser equipment (manufacture)	85590	80429	Mentally disabled adult training
32500	33100	Medical nucleonic apparatus (manufacture)	28940	29540	Mercerising machinery for textiles (manufacture)
19201	23201	Medical paraffin (manufacture)	13300	17300	Mercerising of textiles and textile articles, including wearing apparel (manufacture)
78200	74500	Medical personnel (supply) (temporary employment agency)	13300	17300	Mercerising yarn (manufacture)
72190	73100	Medical research establishment not attached to hospital (other than biotechnological)	18140	22230	Merchandising display finishing (manufacture)
			46410	51410	Merchant converter (textiles) (wholesale)
32500	25130	Medical rubber dressings (manufacture)	50200	61102	Merchant Navy
32500	25130	Medical rubber goods (not dressings) (manufacture)	27200	31400	Mercuric dioxide cells (manufacture)
85421	80302	Medical school	27400	31500	Mercury vapour lamp (manufacture)
72190	73100	Medical sciences research and experimental development (other than biotechnological)	13100	17130	Merino yarn spinning (manufacture)
			22220	25220	Mesh bags made of plastic (manufacture)
32500	33202	Medical thermometers (manufacture)	53201	64120	Messenger licensed
46460	51460	Medical, surgical, dental and veterinary furniture (wholesale)	53201	64120	Messenger service licensed
			53202	64120	Messenger service unlicensed
32500	33100	Medical, surgical, dental or veterinary examination tables (manufacture)	53202	64120	Messenger unlicensed
			46770	51570	Metal and non-metal waste and scrap and materials for recycling (wholesale)
32500	33100	Medical, surgical, dental or veterinary operating tables (manufacture)	25990	28750	Metal badges and metal military insignia (manufacture)
21200	24421	Medicaments (manufacture)	46120	51120	Metal broker (not scrap) (commission agent)
46460	51460	Medicaments containing alkaloids or their derivatives (wholesale)	22290	25240	Metal coating of plastics
46460	51460	Medicaments containing hormones (wholesale)	28410	29420	Metal cutting machine (numerically controlled) (manufacture)
46180	51180	Medicaments containing hormones but (not antibiotics) (commission agent)	28410	29420	Metal cutting machine tool (manufacture)
			26513	33202	Metal detectors (non-electronic) (manufacture)
46180	51180	Medicaments containing penicillins or other antibiotics (commission agent)	25990	28750	Metal dinnerware bowls (manufacture)
46460	51460	Medicaments containing penicillins or other antibiotics (wholesale)	25990	28750	Metal dinnerware platters (manufacture)
			25910	28710	Metal drum reconditioning (manufacture)
10822	15842	Medicated confectionery (manufacture)	18130	22240	Metal etching (manufacture)
32500	24422	Medicated dressings (manufacture)	25610	28510	Metal finishing
21100	24410	Medicinal active substances to be used for their pharmacological properties (manufacture)	25500	28400	Metal forging, pressing, stamping and roll-forming (manufacture)
21100	24410	Medicinal feed additives (veterinary) (manufacture)	28410	29420	Metal forming machine (numerically controlled) (manufacture)
28131	29121	Medicinal pumps (manufacture)	31090	36110	Metal framed upholstery for seating (manufacture)
21200	24421	Medicine (manufacture)	25110	28110	Metal frameworks for blast furnaces (manufacture)
47730	52310	Medicine dealer (retail)	25110	28110	Metal frameworks for construction of bridges (manufacture)
16210	20200	Medium density fibreboard (MDF) and other fibreboard (manufacture)	25110	28110	Metal frameworks for construction of masts (manufacture)
19201	23201	Medium fuel oil (manufacture)			
26400	32300	Megaphone (manufacture)	25110	28110	Metal frameworks for construction of towers (manufacture)
20140	24140	Melamine (manufacture)			
20160	24160	Melamine resins (manufacture)	25110	28110	Metal frameworks for lifting and handling equipment (manufacture)
10420	15430	Melanges and similar spreads (manufacture)			
01130	01120	Melon growing	25990	28750	Metal goods for office use (manufacture)
28210	29210	Melting furnace (manufacture)	25990	28750	Metal hair curlers (manufacture)
94990	91330	Membership organisations n.e.c.	25990	28750	Metal hollow ware pots (manufacture)
26200	30020	Memory store for computers (manufacture)	25990	28750	Metal hollowware kettles (manufacture)
91040	92530	Menagerie	25500	28400	Metal objects production directly from metal powders by heat treatment (manufacture)
13300	17300	Mending of textile articles, including wearing apparel (manufacture)			
47710	52424	Men's bespoke tailor (retail)	46120	51120	Metal oxides, hydroxides and peroxides (commission agent)
47710	52424	Men's clothier and outfitter (retail)			
47710	52424	Men's clothing (retail)	46750	51550	Metal oxides, hydroxides and peroxides (wholesale)
46420	51429	Men's clothing (wholesale)			

SIC 2007	SIC 2003	Activity	SIC 2007	SIC 2003	Activity
20590	24660	Metal pickling substances (manufacture)	26513	33202	Meters for liquid supply (non-electronic) (manufacture)
25990	28750	Metal plates (manufacture)	26511	33201	Meters for petrol pumps (electronic) (manufacture)
20412	24512	Metal polish (manufacture)	26513	33202	Meters for petrol pumps (non-electronic) (manufacture)
20301	24301	Metal pre-treatment paint (manufacture)			
25990	28750	Metal road signs (manufacture)	26511	33201	Meters for water (electronic) (manufacture)
46690	51870	Metal rolling mills (wholesale)	26513	33202	Meters for water (non-electronic) (manufacture)
28910	29510	Metal rolling mills and rolls for such mills (manufacture)	06200	11100	Methane extraction from natural gas
24100	27100	Metal sand for sandblasting (manufacture)	20140	24140	Methanol (manufacture)
25110	28110	Metal skeletons for bridges (manufacture)	46750	51550	Methanol (wholesale)
25110	28110	Metal skeletons for masts (manufacture)	94910	91310	Methodist Church
25110	28110	Metal skeletons for towers (manufacture)	20140	15920	Methylated spirits (manufacture)
25990	28750	Metal spinning	32200	36300	Metronome (electronic or mechanical) (manufacture)
25610	28510	Metal spraying (manufacture)	49311	60213	Metropolitan area passenger railway transportation by underground, metro and similar systems
28290	29240	Metal spraying machine (manufacture)			
46720	51520	Metal stockholder (wholesale)	84240	75240	Metropolitan Police Commissioners Office
25110	28110	Metal structures and parts of structures (manufacture)	49319	60219	Metropolitan scheduled passenger land transport other than underground, metro and similar systems
20590	24660	Metal treatment chemical (manufacture)	23990	26829	Mica goods (manufacture)
25990	28750	Metal umbrella handles and frames (manufacture)	08990	14500	Mica mining and quarrying
25990	28750	Metal vacuum jugs and bottles (manufacture)	23990	26829	Mica slab and sheet processing (manufacture)
46120	51120	Metal waste and scrap (commission agent)	26200	30020	Mice, joysticks and trackballs (manufacture)
43320	45420	Metal window fixing	21100	24410	Microbiological cultures, toxins, etc. (manufacture)
28410	29420	Metal working machine tool (physical process) (manufacture)	26110	32100	Microchip (manufacture)
28410	29420	Metal working machine tool parts (manufacture)	46520	51860	Microchips (wholesale)
25920	28720	Metallic closures (manufacture)	26110	32100	Microcircuit (manufacture)
46120	51120	Metallic halogenates (commission agent)	26200	30020	Micro-computers (manufacture)
46750	51550	Metallic halogenates (wholesale)	26702	33403	Microfiche readers (manufacture)
20301	24301	Metallic paint (manufacture)	26702	33403	Microfilm equipment (manufacture)
15110	19100	Metallised leathers (manufacture)	26702	33403	Microfilm readers (manufacture)
13960	17549	Metallised yarn (manufacture)	74202	74813	Micro-filming activities
46120	51120	Metalloids (commission agent)	26513	33202	Micrometer (manufacture)
46750	51550	Metalloids (wholesale)	71122	74206	Micropalaeontogical analysis activities
19100	23100	Metallurgical coke (manufacture)	26400	32300	Microphone (manufacture)
71200	74300	Metallurgist (private practice)	26702	33402	Microphotography equipment (manufacture)
28910	29510	Metallurgy machinery (manufacture)	26110	32100	Microprocessors (manufacture)
46120	51120	Metals (commission agent)	26702	33402	Microprojection equipment (manufacture)
46720	51520	Metals (wholesale)	46690	51870	Microscopes (except optical) and diffraction equipment (wholesale)
46720	51520	Metals and metal ores exporter (wholesale)			
46720	51520	Metals and metal ores importer (wholesale)	26513	33202	Microscopes (other than optical) (manufacture)
46499	51440	Metalware for domestic use (wholesale)	26110	32100	Microwave components (manufacture)
26511	33201	Meteorological instruments (electronic) (manufacture)	27510	29710	Microwave ovens (manufacture)
26513	33202	Meteorological instruments (non-electronic) (manufacture)	46439	51439	Microwave ovens (wholesale)
			26110	32100	Microwave tube (manufacture)
26701	33201	Meteorological optical instruments (electronic) (manufacture)	85200	80100	Middle schools deemed primary
			86900	85140	Midwife (NHS)
26513	33202	Meteorological optical instruments (non-electronic) (manufacture)	86900	85140	Midwife (private)
			55900	55239	Migrant worker accommodation
22230	25239	Meter housing boxes made of plastic (manufacture)	84210	75210	Military aid missions accredited to foreign governments (public sector)
82990	74879	Meter reading on a fee or contract basis	86101	85111	Military base hospitals
28131	29121	Metering pump (manufacture)	25400	29600	Military carbine (manufacture)
26511	33201	Meters (other than for electricity and parking) (electronic) (manufacture)	14120	18210	Military clothing (manufacture)
			85421	80302	Military college
26513	33202	Meters (other than for electricity and parking) (non-electronic) (manufacture)	84220	75220	Military defence administration
			30400	29600	Military fighting tanks (manufacture)
26511	33201	Meters for electricity (electronic) (manufacture)	30400	29600	Military fighting vehicles (manufacture)
26513	33202	Meters for electricity (non-electronic) (manufacture)	86101	85111	Military hospital (public sector)
26511	33201	Meters for liquid supply (electronic) (manufacture)	84220	75220	Military logistics
			91020	92521	Military museums

M

SIC 2007	SIC 2003	Activity
84220	75220	Military ports
25400	29600	Military rifle (manufacture)
85320	80220	Military school
84230	75230	Military tribunals administration and operation
46330	51331	Milk (wholesale)
56103	55303	Milk bar
10860	15880	Milk based baby food (manufacture)
10821	15841	Milk chocolate (manufacture)
25920	28720	Milk churns made of aluminium (manufacture)
25910	28710	Milk churns made of iron or steel (manufacture)
10821	15841	Milk cocoa (manufacture)
49410	60249	Milk collection by tanker
28930	29530	Milk converting machinery (manufacture)
11070	15980	Milk drinks (flavoured) (manufacture)
10511	15511	Milk homogenising (manufacture)
10519	15519	Milk or cream in solid form (manufacture)
25990	28750	Milk pan (manufacture)
28930	29530	Milk pasteurisation plant (manufacture)
10519	15519	Milk powder (manufacture)
28930	29530	Milk processing machinery (manufacture)
10511	15511	Milk production (evaporated, condensed, etc.) (manufacture)
10519	15519	Milk products other than liquid milk and cream, butter, cheese n.e.c. (manufacture)
47990	52630	Milk roundsman (not farmer) (retail)
11070	15980	Milk shake base (manufacture)
10511	15511	Milk sterilising (manufacture)
10511	15511	Milk ultra heat treatment (manufacture)
28302	29320	Milking machine (manufacture)
46610	51880	Milking machines (wholesale)
47990	52630	Milkman (not farmer) (retail)
17120	21120	Mill board (manufacture)
23990	26821	Millboard made of asbestos (manufacture)
01110	01110	Millet growing
46420	51429	Millinery (wholesale)
47710	52423	Millinery dealer (retail)
46420	51429	Millinery importer (wholesale)
14190	18241	Millinery made of felt (manufacture)
25730	28620	Milling cutter (manufacture)
28930	29530	Milling machine (food processing) (manufacture)
28410	29420	Milling machine (metal cutting) (manufacture)
28940	29540	Milling machinery for textiles (manufacture)
23910	26810	Millstone and grindstone cutting (manufacture)
23910	26810	Millstones made of bonded abrasives (manufacture)
10390	15330	Mincemeat (manufacture)
25400	29600	Mine case and component (manufacture)
26511	31620	Mine detectors (electronic) (manufacture)
26513	33202	Mine detectors (non-electronic) (manufacture)
26513	31620	Mine detectors, pulse (signal) generators; metal detectors (manufacture)
43999	45250	Mine sinking
39000	90030	Minefield clearance
21100	24410	Mineral and pharmaceutical nutritional ingredients for food and feeding stuff (manufacture)
20120	24120	Mineral colours (manufacture)
28921	29521	Mineral cutter (manufacture)
28921	29521	Mineral dressing plant (manufacture)
23990	26829	Mineral insulating materials (manufacture)
23990	26829	Mineral insulation products (manufacture)

SIC 2007	SIC 2003	Activity
19209	23209	Mineral oil blending (manufacture)
06100	11100	Mineral oil extraction
19201	23201	Mineral oil refining (manufacture)
84130	75130	Mineral resource services administration and regulation (public sector)
71122	74206	Mineral surveyor
11070	15980	Mineral water bottling (manufacture)
46341	51341	Mineral water exporter (wholesale)
46341	51341	Mineral water importer (wholesale)
11070	15980	Mineral water production (manufacture)
46341	51341	Mineral waters (wholesale)
23990	26829	Mineral wool (manufacture)
28921	29521	Minerals treatment machinery (manufacture)
27400	31500	Miners' lamp (manufacture)
27120	31200	Miniature circuit breaker (manufacture)
29100	34100	Minibus (manufacture)
45111	50101	Minibuses with a weight not exceeding 3.5 tonnes (new) (retail)
45111	50101	Minibuses with a weight not exceeding 3.5 tonnes (new) (wholesale)
45112	50102	Minibuses with a weight not exceeding 3.5 tonnes (used) (retail)
45112	50102	Minibuses with a weight not exceeding 3.5 tonnes (used) (wholesale)
26200	30020	Mini-computers (manufacture)
09900	14500	Mining and quarrying of residual class 08990, support services provided on a fee or contract basis
71129	74204	Mining engineering design projects
77390	71340	Mining equipment rental and operating leasing
30200	35200	Mining locomotives and mining rail cars (manufacture)
28921	29521	Mining machinery (manufacture)
46630	51820	Mining machinery and equipment (wholesale)
07290	13200	Mining of non-ferrous metal ore
72190	73100	Mining research establishment
84130	75130	Mining services administration and regulation (public sector)
43120	45110	Mining site preparation and overburden removal
16100	20100	Mining timber (sawn) (manufacture)
28921	28620	Mining tool (bit) (manufacture)
28240	29410	Mining tool (powered portable) (manufacture)
84220	75220	Ministry of Defence (civilian personnel)
84220	75220	Ministry of Defence (forces personnel)
84220	75220	Ministry of Defence Headquarters
72190	73100	Ministry of Defence research and development (other than biotechnological)
10831	15861	Mint herb infusions maté (manufacture)
10840	15870	Mint sauce (manufacture)
16290	20510	Mirror frame made of wood (manufacture)
23120	26120	Mirror glass (manufacture)
32500	33100	Mirrors for medical use (manufacture)
23120	26120	Mirrors for motor vehicles (manufacture)
25400	29600	Missiles (guided weapons), except intercontinental ballistic missile (ICBM) (manufacture)
94910	91310	Missionary Society
26513	33202	Mitre (manufacture)
11030	15949	Mixed beverages containing fruit wines (manufacture)
47190	52120	Mixed business retailing both food and non food goods but non-food predominating (retail)
01500	01300	Mixed farming

SIC 2007	SIC 2003	Activity
20110	24110	Mixed industrial gases (manufacture)
28960	29560	Mixing machine for working rubber or plastic (manufacture)
29203	34203	Mobile bank (not self propelled) (manufacture)
29203	34203	Mobile canteen (not self propelled) (manufacture)
28220	29220	Mobile crane (manufacture)
56103	55304	Mobile food carts
68209	70209	Mobile home letting (residential)
29100	34100	Mobile library (not trailer) (manufacture)
28220	29220	Mobile lifting frames (manufacture)
26301	32202	Mobile telephone (manufacture)
61200	64200	Mobile telephone services
47421	52488	Mobile telephones (retail)
47421	52488	Mobile telephones for motor vehicles (retail)
33110	28520	Mobile welding repair of fabricated metal products (not of machinery) (manufacture)
29100	34100	Mobile x-ray unit (not trailer) (manufacture)
32409	36509	Model kit (manufacture)
93290	92349	Model railway installations
30110	35110	Model ship made by shipbuilder (manufacture)
46120	51120	Modelling pastes (commission agent)
20590	24660	Modelling pastes (manufacture)
32990	36639	Models for educational or exhibition purposes (manufacture)
32990	36639	Models for geographical use made of wax or plaster (manufacture)
32409	36509	Models for recreational use (manufacture)
22290	36639	Models for window display made of plastic (manufacture)
32990	36639	Models made of plaster (manufacture)
32990	36639	Models made of wax (manufacture)
26120	32100	Modem interface cards (manufacture)
26301	32201	Modems (manufacture)
25110	28110	Modular exhibition elements made of metal (manufacture)
30110	35110	Modules for oil platform (manufacture)
13100	17120	Mohair spinning on the woollen system (manufacture)
13100	17130	Mohair spinning on the worsted system (manufacture)
13200	17220	Mohair woollen weaving (manufacture)
13200	17230	Mohair worsted weaving (manufacture)
10910	15710	Molassed feeding stuff containing more than 30% molasses (manufacture)
10810	15830	Molasses (manufacture)
01610	01410	Mole catching by contractors
28921	29521	Moles for mining (manufacture)
14200	18300	Moleskin finishing (manufacture)
01490	01250	Mollusc farming, except aquatic molluscs
10200	15209	Mollusc preservation (other than by freezing) (manufacture)
10200	15201	Mollusc preservation by freezing (manufacture)
20200	24200	Molluscicides (manufacture)
47230	52230	Molluscs (retail)
03210	05020	Molluscs and other aquatic animals cultured in sea water
46380	51380	Molluscs distribution (wholesale)
24450	27450	Molybdenum (manufacture)
07290	13200	Molybdenum mining and preparation
94910	91310	Monastery
66120	67130	Money changer

SIC 2007	SIC 2003	Activity
64921	65221	Money lender
64191	65121	Money order activities
26110	31300	Monitor cables (manufacture)
80200	74602	Monitoring activities by mechanical or electrical protective devices
26309	32202	Monitoring equipment for radio and television (manufacture)
80200	74602	Monitoring of electronic security alarm systems
26200	30020	Monitors for computers (manufacture)
26400	32300	Monitors for videos (manufacture)
20140	24140	Mono and polycarboxyclic acids including acetic acid (manufacture)
32500	33401	Monocle (manufacture)
26701	33402	Monocular (manufacture)
20600	24700	Monofilament or strip (manufacture)
46120	51120	Monohydric alcohols (commission agent)
20140	24140	Monohydric alcohols (manufacture)
46750	51550	Monohydric alcohols (wholesale)
26110	32100	Monolithic integrated circuits (manufacture)
25110	28110	Monopod tower made of steel plate (manufacture)
28120	29121	Montejus (compressed air chamber elevators) (manufacture)
23700	26700	Monumental stonework (manufacture)
32910	36620	Mop (manufacture)
30910	35410	Moped (manufacture)
45400	50400	Moped sales (retail)
45400	50400	Moped sales (wholesale)
32910	36620	Mops for household use (manufacture)
13200	17210	Moquette (not woollen) weaving (manufacture)
13200	17220	Moquette woollen weaving (manufacture)
86220	85120	Morbid anatomy specialist (private practice)
86101	85111	Morbid anatomy specialist (public sector)
20120	24120	Mordant dye (manufacture)
25400	29600	Mortar (ordnance) (manufacture)
25400	29600	Mortar bomb (manufacture)
28923	29523	Mortar mixers (manufacture)
28923	29523	Mortar spreaders (manufacture)
23640	26640	Mortars (manufacture)
23640	26640	Mortars (powdered) (manufacture)
23200	26260	Mortars made of refractory (manufacture)
66190	67130	Mortgage agent
66190	67130	Mortgage broker activities
64929	65229	Mortgage corporation for agriculture
64922	65223	Mortgage finance companies' activities (other than banks and building societies)
28490	29430	Mortising machines (manufacture)
96030	93030	Mortuary
23310	26300	Mosaic cube (manufacture)
23190	26150	Mosaic cubes made of glass (manufacture)
23310	26300	Mosaic glazed tiles (manufacture)
94910	91310	Mosque
02300	02010	Moss collecting, cutting or gathering
23990	26829	Moss litter (manufacture)
46760	51560	Moss litter (wholesale)
71200	74300	MOT testing station
55100	55101	Motel (licensed with restaurant)
55100	55102	Motel (unlicensed with restaurant)
55100	55103	Motel without restaurant
28921	29521	Motion compensation equipment for oil drilling rigs (manufacture)

M

SIC 2007	SIC 2003	Activity
26513	33202	Motion detectors (non-electronic) (manufacture)
74203	74814	Motion picture developing
59131	92120	Motion picture distribution activities
59120	92119	Motion picture film laboratory activities
59111	92111	Motion picture production
77390	71340	Motion picture production equipment rental and leasing
59140	92130	Motion picture projection
59131	92120	Motion pictures distribution to other industries
59120	92119	Motion pictures post-production activities
52219	63210	Motive power depot (railway)
45320	50300	Motor accessories dealer (retail)
45310	50300	Motor accessories dealer (wholesale)
46120	51120	Motor and aviation spirit (commission agent)
49320	60220	Motor bike taxi service
30120	35120	Motor boats (manufacture)
49319	60219	Motor bus scheduled passenger transport
29100	34100	Motor car (manufacture)
29100	34100	Motor coach (manufacture)
49390	60231	Motor coach service
49390	60231	Motor coach with driver (private hire)
28120	29122	Motor for hydraulic equipment (manufacture)
28120	29122	Motor for pneumatic equipment (manufacture)
19201	23201	Motor fuel (manufacture)
27110	31100	Motor generator sets (manufacture)
45190	50102	Motor homes (used) (retail)
45190	50102	Motor homes (used) (wholesale)
65120	66031	Motor insurance
45200	50200	Motor repair depot
30910	35410	Motor scooter (manufacture)
19201	23201	Motor spirit (manufacture)
46711	51511	Motor spirit distribution (wholesale)
27120	31200	Motor starting and controlling gear (manufacture)
30910	35410	Motor tricycle and parts (manufacture)
45190	50102	Motor vehicle (used) exporter
22290	25240	Motor vehicle accessories, fittings and parts made of plastic (manufacture)
49410	60249	Motor vehicle collection
71121	74205	Motor vehicle design
29310	31610	Motor vehicle electrical generators (manufacture)
29310	31610	Motor vehicle electrical ignition wiring harnesses (manufacture)
29310	31610	Motor vehicle electrical power window and door systems (manufacture)
45200	50200	Motor vehicle painting and body repairing
45320	50300	Motor vehicle parts and accessories (retail)
45310	50300	Motor vehicle parts and accessories (wholesale)
29310	31610	Motor vehicle purchased gauges into instrument panels (manufacture)
29100	34100	Motor vehicle reconditioning by manufacturer (manufacture)
45200	50200	Motor vehicle servicing
45200	50200	Motor vehicle spraying
45111	50101	Motor vehicle with a weight not exceeding 3.5 tonnes (new) exporter
45111	50101	Motor vehicle with a weight not exceeding 3.5 tonnes (new) importer
45112	50102	Motor vehicle with a weight not exceeding 3.5 tonnes (used) importer
29100	34100	Motor vehicles (manufacture)
71200	74300	Motor vehicles certification
29100	34100	Motor vehicles for commercial use (manufacture)
45111	50101	Motor vehicles with a weight not exceeding 3.5 tonnes (new) (retail)
45111	50101	Motor vehicles with a weight not exceeding 3.5 tonnes (new) (wholesale)
45112	50102	Motor vehicles with a weight not exceeding 3.5 tonnes (used) (retail)
45112	50102	Motor vehicles with a weight not exceeding 3.5 tonnes (used) (wholesale)
30910	35410	Motorcycle (manufacture)
45400	50400	Motorcycle exporter (wholesale)
77390	71211	Motorcycle hire
45400	50400	Motorcycle importer (wholesale)
30910	35410	Motorcycle parts and accessories (manufacture)
45400	50400	Motorcycle parts and accessories (retail)
45400	50400	Motorcycle parts and accessories (wholesale)
77390	71211	Motorcycle rental
45400	50400	Motorcycle sales (retail)
45400	50400	Motorcycle sales (wholesale)
45190	50101	Motorhomes (retail)
45190	50101	Motorhomes (wholesale)
29100	34100	Motorised caravans (manufacture)
46690	51870	Motorised tanks and other armoured fighting vehicles (wholesale)
94990	91330	Motorists' organisation (not road patrol or touring service)
52219	50200	Motorists' organisation (road patrol)
79110	63301	Motorists' organisation touring department
30300	35300	Motors and engines of a kind typically found on aircraft (manufacture)
30300	35300	Motors for aircraft (manufacture)
42110	45230	Motorway and other dual carriageway construction
52219	63210	Motorway maintenance unit
56102	55302	Motorway services cafeteria (unlicensed)
47300	50500	Motorway services petrol filling station
25730	28620	Mould (engineers' small tools) (manufacture)
25730	29560	Mould for foundry (manufacture)
23200	26260	Mouldable refractory (manufacture)
27120	31200	Moulded case circuit breaker (manufacture)
22190	19300	Moulded rubber bottoms for footwear (manufacture)
16230	20300	Moulded skirting board made of wood (manufacture)
28930	29530	Moulders for bakery (manufacture)
28960	29560	Moulders for rubber or plastic (manufacture)
25730	29560	Moulding boxes for any material (manufacture)
20160	24160	Moulding compounds (plastics) (manufacture)
28960	29560	Moulding machine for rubber or plastic (manufacture)
28490	29430	Moulding machine for wood, etc. (manufacture)
25730	29560	Moulding machine for working rubber or plastics (manufacture)
28930	29530	Moulding machines for dairies (manufacture)
22190	25130	Mouldings for upholstery made of rubber (manufacture)
23200	26260	Mouldings made of magnesite (manufacture)
22290	25240	Mouldings made of plastic (manufacture)
16230	20300	Mouldings made of wood (manufacture)
23200	26260	Moulds made of silica (manufacture)
46690	51870	Moulds, moulding boxes for metal foundries, mould bases and moulding patterns (wholesale)

M

SIC 2007	SIC 2003	Activity
17290	21259	Mount cutting (manufacture)
93199	92629	Mountain guides
55202	55210	Mountain refuges
32300	36400	Mountaineering equipment (manufacture)
26110	32100	Mounted piezo-electric crystals (manufacture)
46520	51860	Mounted piezo-electric crystals (wholesale)
17290	21259	Mounting paper on linen (manufacture)
20520	24620	Moutant (manufacture)
32200	36300	Mouth blown signalling instruments (manufacture)
32200	36300	Mouth organ (manufacture)
26520	33500	Movements for clocks and watches (manufacture)
28220	29220	Moving walkways (manufacture)
28302	29320	Mowers for agricultural use (manufacture)
28302	29320	Mowers for lawns, parks and sports grounds (manufacture)
09100	11200	Mud logging services
28921	29521	Mudline suspension and tie-back equipment (manufacture)
23200	26260	Muffles (refractory product) (manufacture)
14200	18300	Muffs made of fur (manufacture)
01250	01139	Mulberry growing
01430	01220	Mule farming and breeding
23190	26150	Multicellular glass block (manufacture)
72190	73100	Multi-disciplinary research and development (not biotechnological or social sciences and humanities)
72200	73100	Multi-disciplinary research and development predominantly on social sciences and humanities
26200	30020	Multi-function office equipment (manufacture)
82190	74850	Multigraphing
17120	21120	Multi-layer paper obtained by compression (manufacture)
23120	26120	Multiple insulating glass (manufacture)
23120	26120	Multiple-walled insulating units of glass (manufacture)
26301	32201	Multiplexers for telephone exchanges (manufacture)
84130	75130	Multipurpose development project services administration (public sector)
17211	21211	Multi-wall paper sack (manufacture)
46760	51560	Mungo (carded or combed) (wholesale)
49319	60219	Municipal bus service
91020	92521	Museums of all kinds
91020	92521	Museums of furniture, costumes, ceramics, silverware
91020	92521	Museums of jewellery
02300	01120	Mushroom (wild) gathering
01130	01120	Mushroom growing (cultivated)
01300	01120	Mushroom spawn growing
47210	52210	Mushrooms (retail)
46310	51310	Mushrooms (wholesale)
59200	22140	Music (printed) publishing
32200	36300	Music box mechanisms (manufacture)
94990	91330	Music club
90030	92319	Music composer
90030	92319	Music copyist and transcriber (own account)
59200	72400	Music downloads (on-line publishing with provision of downloaded content)
90040	92320	Music hall
18130	22240	Music plate engraving (manufacture)
18129	22220	Music printing (manufacture)
47591	52450	Music shop (retail)
59200	22140	Music tape publishing
18201	22310	Music tape reproduction from master copies (manufacture)
85320	80220	Music teacher (own account)
32200	36300	Musical box (manufacture)
32200	36300	Musical instrument parts and accessories (manufacture)
77210	71401	Musical instrument rental
32200	36300	Musical instrument tuners (electronic) (manufacture)
46180	51180	Musical instruments (commission agent)
47591	52450	Musical instruments (retail)
46491	51475	Musical instruments (wholesale)
46491	51475	Musical instruments exporter (wholesale)
46491	51475	Musical instruments importer (wholesale)
32200	36300	Musical instruments including electronic (manufacture)
77299	71409	Musical instruments rental and leasing
47591	52450	Musical scores (retail)
90010	92311	Musicians
13300	17300	Muslin clipping (manufacture)
13300	17300	Muslin dressing (manufacture)
13300	17300	Muslin ending (manufacture)
13300	17300	Muslin finishing (manufacture)
13300	17300	Muslin gassing (manufacture)
13300	17300	Muslin mending (manufacture)
13200	17210	Muslin weaving (manufacture)
03110	05010	Mussel gathering
03220	05020	Mussel production, freshwater
03210	05020	Mussel production, marine
10840	15870	Mustard (manufacture)
10840	15870	Mustard flour and meal (manufacture)
10410	15410	Mustard oil production (manufacture)
28930	29530	Mustard processing machine (manufacture)
10410	15410	Mustard seed crushing (manufacture)
01110	01110	Mustard seed growing
66300	67121	Mutual funds management
32500	33100	Myograph (manufacture)
56290	55510	NAAFI canteen
56302	55404	NAAFI clubs
56290	55520	NAAFI headquarters
47110	52112	NAAFI shop with alcohol licence (retail)
47110	52113	NAAFI shop without alcohol licence (retail)
30300	35300	Nacelles for aircraft (manufacture)
25930	28730	Nail (not wire) (manufacture)
32910	36620	Nail brush (manufacture)
25710	28610	Nail file (manufacture)
20420	24520	Nail preparation (cosmetic) (manufacture)
28490	29430	Nailing machines (manufacture)
25930	28730	Nails made of steel wire (manufacture)
46740	51540	Nails, tacks, drawing pins and staples (wholesale)
25990	28750	Nameplates made of metal (manufacture)
22290	25240	Nameplates made of plastic (manufacture)
13990	17541	Napery lace (manufacture)
19201	23201	Naphtha (LDF) (manufacture)
20140	24140	Naphthalene (manufacture)
17220	21220	Napkin liners (manufacture)
17220	21220	Napkins made of paper (manufacture)
13960	17542	Narrow fabric (not elastic or elastomeric) (manufacture)

N

SIC 2007	SIC 2003	Activity
93290	92729	Narrow gauge railway (recreational)
24100	27100	Narrow slabs of semi-finished steel (manufacture)
13960	17542	Narrow woven fabrics, including weftless fabrics assembled by means of an adhesive (manufacture)
94990	91330	National council for civil liberties
72200	73200	National foundation for educational research
91020	92521	National galleries (Scotland)
91020	92521	National gallery
93120	92629	National greyhound racing club
85590	80429	National institute for adult continuing education
91011	92510	National library for the blind
91011	92510	National library of Scotland
91011	92510	National library of Wales
94120	91120	National maritime board
91020	92521	National maritime museum
72190	73100	National physical laboratory
91020	92521	National portrait gallery
64191	65121	National savings bank
88990	85321	National society for the prevention of cruelty to children
94990	91330	National trust (the)
91020	92521	National trust garden (property)
94990	91330	National union of students (not trading activities)
08910	14300	Native sulphur mining
46180	51180	Natural cork in plates, sheets, strips, crushed, granulated or ground (commission agent)
72190	73100	Natural environment research council
35220	40220	Natural gas booster/compression site
06200	11100	Natural gas condensates separation
35220	40220	Natural gas distribution
71122	74206	Natural gas exploration
06200	11100	Natural gas production well
35220	40220	Natural gas storage
91020	92521	Natural history museum
20530	24630	Natural material used in flavours or perfumes (manufacture)
03110	05010	Natural pearls gathering of
72190	73100	Natural sciences research and experimental development (other than biotechnological)
32990	36639	Natural sponge preparation (manufacture)
46120	51120	Natural uranium and plutonium and their compounds (commission agent)
24460	23300	Natural uranium production (manufacture)
96090	93059	Naturalisation agent
91040	92530	Nature reserves including wildlife preservation
26513	33202	Nautical instrument (non-electronic) (manufacture)
85320	80220	Nautical school
71121	74205	Naval architect
30110	35110	Naval dockyard (shipbuilding and repairing) (manufacture)
30110	35110	Naval ships of all types (manufacture)
52220	63220	Navigation activities
46690	51870	Navigation machinery (wholesale)
26511	33201	Navigational instruments and appliances (electronic) (manufacture)
26513	33202	Navigational instruments and appliances (non-electronic) (manufacture)
10410	15420	Neatsfoot oil (manufacture)
14190	18249	Necktie (manufacture)
14190	18249	Neckwear for women (manufacture)

SIC 2007	SIC 2003	Activity
01240	01139	Nectarine growing
28150	29140	Needle roller bearings (manufacture)
28140	29130	Needle valves (manufacture)
13990	17549	Needlefelt (other than carpet underlay) (manufacture)
13939	17519	Needlefelt carpet underlay (manufacture)
13939	17519	Needleloom and bonded fibre rugs (manufacture)
13939	17519	Needleloom and bonded fibre tiles (manufacture)
13939	17519	Needleloom carpet (manufacture)
13939	17519	Needleloom carpeting (manufacture)
13990	17549	Needleloom felt (other than carpet underlay) (manufacture)
13939	17519	Needleloom felt carpet underlay (manufacture)
47510	52410	Needles for sewing (retail)
28940	29540	Needles for sewing machines (manufacture)
25930	28730	Needles made of metal (manufacture)
32500	33100	Needles used in medicine (manufacture)
46410	51410	Needles, etc. for sewing (wholesale)
26702	33403	Negatoscopes (manufacture)
94110	91110	Negotiations of business and employer organisations
20200	24200	Nematocide (manufacture)
20110	24110	Neon (manufacture)
27400	31500	Neon tube (manufacture)
13910	17600	Net and window furnishing type fabrics (manufacture)
13910	17600	Net curtaining knitted or crocheted fabric (manufacture)
47530	52440	Net curtains (retail)
13990	17541	Net fabrics in the piece, in strips or in motifs (manufacture)
13940	17520	Nets for horticulture (manufacture)
13910	17600	Netted fabric (manufacture)
22290	25240	Netting made of plastic (not woven or knotted) (manufacture)
25930	28730	Netting made of steel wire (manufacture)
25930	28730	Netting made of wire (manufacture)
13940	17520	Netting products (manufacture)
26200	30020	Network interface (manufacture)
26120	32100	Network interface cards (manufacture)
86900	85140	Neuropath
20140	15920	Neutral spirits production (manufacture)
93110	92619	Newmarket Heath
63910	92400	News agency activities
63990	92400	News clipping service
20302	24302	News ink (manufacture)
18110	22210	Newspaper printing (manufacture)
58130	22120	Newspaper publishing
47620	52470	Newspapers (retail)
46499	51479	Newspapers (wholesale)
17120	21120	Newsprint (manufacture)
47890	52620	Newsvendor (retail)
27200	31400	NiCad batteries (manufacture)
24450	27450	Nickel (manufacture)
24450	27450	Nickel alloys (manufacture)
07290	13200	Nickel mining and preparation
24440	27440	Nickel silver (manufacture)
24450	27450	Nickel wire made by drawing (manufacture)
20200	24200	Nicotine preparation (manufacture)
01110	01110	Niger seed growing
25990	28750	Night safe (manufacture)

N

SIC 2007	SIC 2003	Activity
56301	55401	Night-clubs (licensed to sell alcohol)
14142	18232	Night-dresses for women and girls (manufacture)
32990	36639	Nightlight (manufacture)
14190	18249	Nightwear for infants (manufacture)
14141	18231	Nightwear for men and boys (manufacture)
14142	18232	Nightwear for women and girls (manufacture)
27200	31400	NiMH batteries (manufacture)
25730	28620	Nippers (manufacture)
46750	51550	Nitrate of soda importer (wholesale)
20150	24150	Nitrates and nitrites of potassium (manufacture)
46120	51120	Nitrates of potassium (commission agent)
46120	51120	Nitric acid, sulphonitric acid and ammonia (commission agent)
20150	24150	Nitric and sulphonitric acid (manufacture)
20510	24610	Nitro glycerine (manufacture)
13960	17549	Nitro-cellulose coated textile fabric (manufacture)
20110	24110	Nitrogen (manufacture)
20150	24150	Nitrogen products (manufacture)
20301	24301	Nitrogen resin type paint (manufacture)
20140	24140	Nitrogen-function organic compounds including amine (manufacture)
20150	24150	Nitrogenous straight fertiliser (manufacture)
20150	24150	Nitrogenous, phosphatic or potassic fertilisers (manufacture)
20110	24110	Nitrous oxide (manufacture)
24100	27100	Nodular pig iron (manufacture)
46760	51560	Noil (wholesale)
13100	17120	Noils (woollen industry) (manufacture)
74901	74206	Noise control consultancy activities
66190	67130	Nominee company
28230	30010	Non electronic calculators (manufacture)
46170	51170	Non-alcoholic beverages (commission agent)
47250	52250	Non-alcoholic beverages (retail)
11070	15980	Non-alcoholic flavoured and/or sweetened waters (manufacture)
24100	27100	Non-alloy pig iron (manufacture)
24100	27100	Non-alloy steel (manufacture)
10910	15710	Non-compound animal and cattle feed (not from grain milling or oilseed cakes and meal) (manufacture)
27330	25240	Non-current carrying plastic junction boxes (manufacture)
27330	25240	Non-current carrying plastic pole line fittings (manufacture)
27330	25240	Non-current carrying plastic switch covers (manufacture)
10890	15899	Non-dairy milk and cheese substitutes (manufacture)
10410	15410	Non-defatted flour production (manufacture)
28250	29230	Non-domestic cooling and ventilation equipment parts (manufacture)
28210	29720	Non-electric household heating equipment (permanently mounted) (manufacture)
28210	29210	Non-electric household-type furnaces (manufacture)
26513	33202	Non-electrical apparatus for testing physical and mechanical properties of materials (manufacture)
46470	51479	Non-electrical lamps and light fittings (wholesale)
27400	31500	Non-electrical lighting equipment (manufacture)
28290	29430	Non-electrical welding and soldering equipment (manufacture)
24540	27540	Non-ferrous metal foundry (manufacture)
09900	13200	Non-ferrous metal ore mining of class 07290, support services provided on a fee or contract basis
07290	13200	Non-ferrous metal ore quarrying
07290	13200	Non-ferrous metal ores mining and preparation
24450	27450	Non-ferrous other metals production (manufacture)
65202	66032	Non-life reinsurance
65202	66020	Non-life reinsurance related to pension funding
38320	37200	Non-metal waste and scrap recycling into secondary raw materials
23990	26829	Non-metallic mineral substances, articles thereof (manufacture)
84210	75210	Non-military aid programmes to developing countries (public sector)
24100	27100	Non-oriented electrical steel sheet (manufacture)
68209	70209	Non-residential buildings letting
28140	29130	Non-return, reflux and check valves (manufacture)
51102	62201	Non-scheduled air transport of passengers
49390	60239	Non-scheduled passenger transport n.e.c.
43341	45440	Non-specialised painting of metal structures (including ships)
47990	52630	Non-store auctions (retail)
47990	52630	Non-store commission agents (retail)
20160	24160	Non-vulcanisable thermoplastic elastomers (manufacture)
02300	02010	Non-wood products, gathering from the wild
13950	17530	Non-woven bonded fibre fabrics (manufacture)
22220	25220	Non-woven liners made of polyethylene (manufacture)
22220	25220	Non-woven sacks made of polyethylene (manufacture)
13950	17530	Non-wovens (manufacture)
13950	17530	Non-wovens and articles made from non-wovens (manufacture)
10730	15850	Noodle (manufacture)
69109	74119	Notary activities
69109	74119	Notary public
24420	27420	Notched bars made of aluminium (manufacture)
17230	22220	Notepad (manufacture)
22290	25240	Notice plates made of plastic (manufacture)
13990	17541	Nottingham lace (manufacture)
10822	15842	Nougat (manufacture)
22290	36639	Novelties made of plastic (manufacture)
15120	19200	Novelty goods made of leather (manufacture)
30300	35300	Nozzle for gas turbine aero engine (manufacture)
23200	26260	Nozzles made of refractory ceramic (manufacture)
25300	28300	Nuclear fired boiler (manufacture)
24460	23300	Nuclear fuel (except enrichment of uranium or thorium) (manufacture)
25300	28300	Nuclear fuel plant (manufacture)
24460	23300	Nuclear fuel processing (except enrichment of uranium or thorium) (manufacture)
32500	33100	Nuclear magnetic resonance apparatus (manufacture)
71200	74300	Nuclear plant certification
35110	40110	Nuclear power station
25300	28300	Nuclear reactor parts (manufacture)
25300	28300	Nuclear reactors (manufacture)
26511	33201	Nucleonic instrument (manufacture)
32500	33100	Nucleonic medical apparatus (manufacture)
86101	85112	Nuffield Hospital Trust
32990	36631	Numbering stamps (manufacture)
26511	33201	Numerical control and indication equipment for machine tools (electronic) (manufacture)

N

SIC 2007	SIC 2003	Activity
26513	33202	Numerical control and indication equipment for machine tools (non-electronic) (manufacture)
47789	52485	Numismatist (retail)
14120	18210	Nuns' clothing (manufacture)
86900	85140	Nurse (private)
01300	01120	Nursery (horticulture)
32300	36400	Nursery equipment (manufacture)
85100	80100	Nursery schools
13923	17403	Nursery square (manufacture)
14120	18210	Nurses' uniforms for men (manufacture)
14120	18210	Nurses' uniforms for women (manufacture)
78200	74500	Nursing agency (supplying nurses) (temporary employment agency)
87100	85140	Nursing care facilities
86900	85140	Nursing co-operative
86102	85113	Nursing home with medical care (under the direct supervision of medical doctors)
87100	85140	Nursing homes
94120	91120	Nursing society
25940	28740	Nut (manufacture)
10822	15842	Nut and bean confectionery (manufacture)
10390	15330	Nut foods and pastes (manufacture)
10822	15842	Nut preserving in sugar (manufacture)
10390	15330	Nut processing and preservation (except in sugar) (manufacture)
28930	29530	Nut processing machines and equipment (manufacture)
10390	15330	Nut shelling, grinding and preparing (manufacture)
01280	01139	Nutmeg, mace and cardamoms growing
10390	15330	Nuts preserved by freezing (manufacture)
02300	01139	Nuts, gathering from the wild
21100	24410	O-acetylsalicylic acids (manufacture)
10720	15820	Oat cake (manufacture)
10611	15611	Oat flour and meal (manufacture)
10611	15611	Oat grinding, rolling, crushing or flaking (manufacture)
10611	15611	Oats (manufacture)
01110	01110	Oats growing
26701	33402	Observation telescopes (manufacture)
88100	85322	Occupation and training centre for the mentally disordered (non-charitable)
88100	85321	Occupation and training centres for the mentally disordered (charitable)
86900	85140	Occupational therapist (private)
26511	33201	Oceanographic or hydrological instruments (electronic) (manufacture)
26513	33202	Oceanographic or hydrological instruments (non-electronic) (manufacture)
08910	14300	Ochre pit
20120	24120	Ochres (pigments) (manufacture)
20530	24630	Odoriferous products (manufacture)
47250	52250	Off licence (not public house) (retail)
10110	15111	Offal (edible) preparation (i.e. removal, freezing, packing, etc.) (manufacture)
46320	51320	Offal salesman (wholesale)
41201	45211	Office and shop construction
17219	21219	Office box files and similar articles (manufacture)
81210	74701	Office cleaning contractor
77390	71320	Office container renting
47410	52482	Office equipment (retail)
77330	71330	Office equipment hire
47599	52482	Office furniture (retail)
46650	51850	Office furniture (wholesale)
77330	71330	Office furniture hire
46140	51140	Office machinery (commission agent)
28230	30010	Office machinery (manufacture)
28230	30010	Office machinery and equipment (manufacture)
77330	71330	Office machinery and equipment leasing
77330	71330	Office machinery and equipment rental and operating leasing
99000	99000	Office of High Commissioner
32120	36220	Office or desk articles of base metals clad with precious metals (manufacture)
31010	36110	Office seating (manufacture)
47620	52470	Office supplies (retail)
22290	25240	Office supplies made of plastic (manufacture)
78200	74500	Office support personnel (supply) (temporary employment agency)
17219	21219	Office systems made of paper and board (manufacture)
56290	55510	Officers' messes
68209	70209	Offices letting
84230	75230	Official receiver
84230	75230	Official solicitor
45190	50101	Off-road motor vehicles with a weight exceeding 3.5 tonnes (new) (retail)
45190	50101	Off-road motor vehicles with a weight exceeding 3.5 tonnes (new) (wholesale)
45190	50102	Off-road motor vehicles with a weight exceeding 3.5 tonnes (used) (retail)
45190	50102	Off-road motor vehicles with a weight exceeding 3.5 tonnes (used) (wholesale)
45111	50101	Off-road motor vehicles with a weight not exceeding 3.5 tonnes (new) (retail)
45111	50101	Off-road motor vehicles with a weight not exceeding 3.5 tonnes (new) (wholesale)
45112	50102	Off-road motor vehicles with a weight not exceeding 3.5 tonnes (used) (retail)
45112	50102	Off-road motor vehicles with a weight not exceeding 3.5 tonnes (used) (wholesale)
28990	29560	Offset litho printing machine (manufacture)
18129	22220	Offset printing (manufacture)
46660	51850	Offset sheet fed printing machinery for offices (wholesale)
30110	35110	Offshore floating drilling rig (manufacture)
49500	60300	Offshore natural gas pipeline operation
42210	45213	Offshore oil pipeline laying
49500	60300	Offshore oil pipeline operation
52220	63220	Offshore positioning services
50200	61102	Offshore supply vessel services
30110	35110	Offshore support vessel (manufacture)
92000	92710	Off-track betting
26513	33202	Ohmmeter (manufacture)
10410	15420	Oil (edible) (manufacture)
20590	24660	Oil additive (manufacture)
09100	11200	Oil and gas exploration in connection with oil and gas extraction, including traditional prospecting
09100	11200	Oil and gas extraction service activities provided on a fee or contract basis
09100	11200	Oil and gas field fire fighting services
09100	11200	Oil and gas well casing, cementing, tubing and lining services
09100	11200	Oil and gas well cementing services

SIC 2007	SIC 2003	Activity	SIC 2007	SIC 2003	Activity
09100	11200	Oil and gas well coiled-tubing wellwork	87300	85312	Old persons' home (local authority)
09100	11200	Oil and gas well conductor driving services	87300	85311	Old persons' warden assisted dwellings (charitable)
09100	11200	Oil and gas well cutting, casing and abandonment services	87300	85312	Old persons' warden assisted dwellings (non-charitable)
09100	11200	Oil and gas well pipe refurbishment services	46210	51210	Oleaginous fruits (wholesale)
09100	11200	Oil and gas well sleeving repair services	20140	24140	Oleic acid (manufacture)
19209	23209	Oil based lubricating oils (manufacture)	20140	24140	Oleine (manufacture)
23990	26829	Oil based sealants (manufacture)	20301	24301	Oleo resinous paint (manufacture)
46210	51210	Oil cake (wholesale)	10410	15410	Oleo stearin (manufacture)
20411	24511	Oil dispersant (manufacture)	01260	01139	Olive growing
09100	11200	Oil extraction service activities	10410	15410	Olive oil (crude) production (manufacture)
77390	71340	Oil field equipment rental and operating leasing	10410	01139	Olive oil from self produced olives (if value of production exceeds that of growing) (manufacture)
28290	29240	Oil filter for motor vehicle (manufacture)	10410	15420	Olive oil refining (manufacture)
28210	29210	Oil fuel burner (manufacture)	10390	15330	Olive preserving in salt or brine (manufacture)
10410	15410	Oil kernel meal production (manufacture)	25990	28750	Omelette pan (manufacture)
47789	52489	Oil merchant (retail)	45200	50200	Omnibus repair depot
46719	51519	Oil merchant (wholesale)	49319	60219	Omnibus service
10410	15410	Oil nut meal production (manufacture)	74203	74814	One hour photo shop (not part of camera shop)
01260	01110	Oil palms growing	01130	01120	Onion growing
49500	60300	Oil pipeline terminal operating (for petroleum)	58190	72400	On-line advertising material publishing
30110	35110	Oil platform fabrication of steel plate (manufacture)	58110	72400	On-line book publishing
06100	11100	Oil platform operation	58190	72400	On-line catalogue publishing
30110	35110	Oil platform structural sections (manufacture)	58210	72400	On-line computer games publishing
43999	45250	Oil production platform (fixed concrete or composite steel/concrete) construction	58120	72400	On-line database publishing
			58120	72400	On-line directory publishing
06100	11100	Oil production well or platform operating	78109	74500	On-line employment placement agencies
19201	23201	Oil refinery (manufacture)	58190	72400	On-line forms publishing
28290	29240	Oil refining industry machinery (other than plant) (manufacture)	58190	72400	On-line greeting cards and postcards publishing
28290	29240	Oil seal (manufacture)	58142	72400	On-line journal (other than learned journals) and periodical publishing
10410	15410	Oil seed cake and meal (manufacture)	58141	72400	On-line learned journal publishing
10410	15410	Oil seed crushing (manufacture)	58120	72400	On-line mailing list publishing
46210	51210	Oil seeds (wholesale)	58130	72400	On-line newspaper publishing
01110	01110	Oil seeds growing	58190	72400	On-line publishing n.e.c.
06100	11100	Oil shale mine	58190	72400	On-line publishing of posters and reproductions of works of art
06100	11100	Oil shale retorting			
39000	90030	Oil spill clearance on land	58190	72400	On-line publishing of statistics and other information
39000	90030	Oil spill pollution control services	58190	72400	On-line publishing photos and engravings
39000	90030	Oil spills at sea containment, dispersion and clean up services	58290	72400	On-line software publishing (except computer games on-line publishing)
06100	11100	Oil stabilisation plant operation	20301	24301	Opacifiers and colours (manufacture)
25290	28210	Oil storage tank made of metal for domestic use exceeding 300 litres (manufacture)	21200	24421	Opacifying preparations for x-ray examinations (manufacture)
46120	51120	Oil trading (commodity broking) (commission agent)	91020	92521	Open air museums
10410	15410	Oilcakes and other residual products of oil production (manufacture)	24330	27330	Open sections made of steel formed on a roll mill (manufacture)
46410	51410	Oilcloth (wholesale)	85421	80302	Open University
28131	29121	Oil-cushion pumps (manufacture)	05102	10102	Opencast coal disposal point
50200	61102	Oil-rig transportation by towing or pushing	05102	10102	Opencast coal site
20590	24660	Oils and fats (chemically modified) (manufacture)	05102	10102	Opencast coal working
46750	51550	Oils and other products of distilling of high temperature coal tar, pitch and pitch tar (wholesale)	64304	65237	Open-ended investment companies
			90040	92320	Opera house
14131	18221	Oilskins for men and boys (manufacture)	90010	92311	Opera production
14132	18222	Oilskins for women and girls (manufacture)	32500	33100	Operating tables (manufacture)
23910	26810	Oilstones (bonded) (manufacture)	49319	60219	Operation of aerial cableways as part of urban, suburban or metropolitan transit
21200	24421	Ointment (manufacture)			
88100	85321	Old age and sick visiting (charitable)	93290	92341	Operation of dance floors
88100	85322	Old age and sick visiting (non-charitable)	35140	40130	Operation of electricity and transmission capacity exchanges for electric power
87300	85311	Old people's sheltered housing (charitable)			
87300	85312	Old people's sheltered housing (non-charitable)			

O

SIC 2007	SIC 2003	Activity	SIC 2007	SIC 2003	Activity
38220	90020	Operation of facilities for treatment of hazardous waste	26513	33202	Optical type measuring and checking appliances and instruments (non-electronic) (manufacture)
49319	60219	Operation of funicular railways as part of urban, suburban or metropolitan transit	26600	33100	Optometer (manufacture)
36000	41000	Operation of irrigation canals	46180	51180	Oral and dental hygiene preparations including denture fixative pastes and powders (commission agent)
38210	90020	Operation of landfills for the disposal of non-hazardous waste	21200	24421	Oral contraceptives (manufacture)
52212	63210	Operation of rail passenger facilities at railway stations	20420	24520	Oral hygiene preparations (manufacture)
93290	92629	Operation of recreational transport facilities e.g. marinas	86230	85130	Oral pathology
			01230	01139	Orange growing
93290	92729	Operation of ski hills	11070	15980	Orangeade production (manufacture)
49390	60239	Operation of teleferics, funiculars	30300	35300	Orbital stations (manufacture)
38110	90020	Operation of waste transfer facilities for non-hazardous waste	90010	92311	Orchestras
			25400	29600	Ordnance (manufacture)
38120	90020	Operation of waste transfer stations for hazardous waste	46120	51120	Ores (commission agent)
			46720	51520	Ores (wholesale)
81100	75140	Operational services of government owned or occupied buildings (public sector)	32200	36300	Organ tuning (manufacture)
			13200	17210	Organdie weaving (manufacture)
86900	85140	Ophthalmic clinic	20140	24140	Organic acids and their esters and halogenated, nitrosated and sulphonated derivatives (manufacture)
32500	33401	Ophthalmic eyeglasses (manufacture)			
86101	85112	Ophthalmic hospital (private sector)	20140	24140	Organic base chemicals (manufacture)
86101	85111	Ophthalmic hospital (public sector)	23910	26810	Organic bonded abrasives (manufacture)
32500	33100	Ophthalmic instrument (manufacture)	24100	27100	Organic coated steel sheet (manufacture)
46690	51870	Ophthalmic instruments (wholesale)	20301	24301	Organic composite solvents (manufacture)
32500	33401	Ophthalmic lenses ground to prescription, safety goggles (manufacture)	20140	24140	Organic compounds including wood distillation products (manufacture)
47789	52482	Optical and precision goods (retail)	20411	24511	Organic surface-active agents (manufacture)
20120	24120	Optical bleaching agent (manufacture)	66190	67130	Organisation and development of electronic money circulation
26200	30020	Optical CD-RW, CD-ROM, DVD-ROM, DVD-RW disk drives (manufacture)	99000	99000	Organisation for economic co-operation and development
26701	33402	Optical comparators (manufacture)			
26511	33201	Optical density measuring equipment (electronic) (manufacture)	99000	99000	Organisation of oil producing and exporting countries
26513	33202	Optical density measuring equipment (non-electronic) (manufacture)	85600	74149	Organisation of student exchange programmes
			90010	92311	Organist (own account)
26200	30020	Optical disk drives (manufacture)	46750	51550	Organo-sulphur and other organo-inorganic compounds (wholesale)
32500	33401	Optical element (mounted, (not photographic)) (manufacture)	20140	24140	Organo-sulphur compounds (manufacture)
32500	33401	Optical element (unmounted) (manufacture)	47789	52489	Oriental goods (retail)
23190	26150	Optical elements made of glass (not optically worked) (manufacture)	16210	20200	Oriented strand board (OSB) and other particle board (manufacture)
27310	31300	Optical fibre cables for coded data transmission (manufacture)	09900	14110	Ornamental and building stone quarrying support services provided on a fee or contract basis
46690	51870	Optical fibre cables made up of individually sheathed fibres (wholesale)	08110	14110	Ornamental and building stone, breaking and crushing
27310	33402	Optical fibres, optical fibre bundles and cables (manufacture)	13960	17542	Ornamental braids (manufacture)
26701	33402	Optical fire control equipment (manufacture)	23410	26210	Ornamental ceramic ware (manufacture)
23190	26150	Optical glass (manufacture)	23310	26300	Ornamental earthenware glazed tiles (manufacture)
46439	51479	Optical goods (wholesale)	03220	05020	Ornamental fish farming, freshwater
26701	33402	Optical gun sighting equipment (manufacture)	03210	05020	Ornamental fish farming, marine
26701	33402	Optical instruments and appliances (other than photographic goods) (manufacture)	01300	01120	Ornamental tree and shrub growing
			13960	17542	Ornamental trimmings (manufacture)
26701	33402	Optical machinist's precision tools (manufacture)	43390	45450	Ornamentation fitting work
26701	33402	Optical magnifying instruments (manufacture)	23130	26130	Ornaments made of glass (manufacture)
26800	24650	Optical media (manufacture)	22290	25240	Ornaments made of plastic (manufacture)
26701	33402	Optical microscope (manufacture)	32120	36220	Ornaments made of precious metal (manufacture)
26701	33401	Optical mirrors (manufacture)	16290	20510	Ornaments made of wood (manufacture)
26701	33402	Optical positioning equipment (manufacture)	32120	36220	Ornaments that are gold or silver plated (manufacture)
26701	33402	Optical projector (meteorological) (manufacture)			
26511	33201	Optical type measuring and checking appliances and instruments (electronic) (manufacture)	87900	85311	Orphanages (charitable)
			87900	85312	Orphanages (non-charitable)

SIC 2007	SIC 2003	Activity	SIC 2007	SIC 2003	Activity
86230	85130	Orthodontic activities	14120	18210	Overalls for girls (manufacture)
32500	33100	Orthopaedic appliances (not footwear) (manufacture)	14120	18210	Overalls for men and boys (manufacture)
47749	52329	Orthopaedic appliances (retail)	14120	18210	Overalls for women and girls (manufacture)
32500	33100	Orthopaedic footwear (manufacture)	15200	19300	Overboot (manufacture)
46460	51460	Orthopaedic goods (wholesale)	43120	45110	Overburden removal and other development of mineral properties and sites
86101	85112	Orthopaedic hospital (private sector)	13200	17220	Overcoating woollen weaving (manufacture)
86101	85111	Orthopaedic hospital (public sector)	14131	18221	Overcoats for men and boys (manufacture)
32500	33100	Orthopedic and prosthetic devices (manufacture)	14132	18222	Overcoats for women and girls (manufacture)
26511	33201	Oscilloscope (manufacture)	30300	35300	Overhaul and conversion of aircraft or aircraft engines (manufacture)
16290	20520	Osier articles (manufacture)	42220	45213	Overhead line construction
02200	02010	Osier growing	27320	31300	Overhead line fittings (manufacture)
16290	20520	Osier preparation (manufacture)	18130	22250	Overhead media foils and other forms of presentation (manufacture)
25110	28110	Ossature in metal for construction (manufacture)	18130	22250	Overhead projection foils production (manufacture)
86900	85140	Osteopath (not registered medical practitioner)	28220	29220	Overhead runway (manufacture)
86220	85120	Osteopath (registered medical practitioner)	26702	33403	Overhead transparency projectors (manufacture)
01490	01250	Ostriches raising and breeding	15200	19300	Overshoes made of rubber (manufacture)
85590	80429	Other adult and other education n.e.c.	19201	10103	Ovoid solid fuel production (manufacture)
24100	27100	Other alloy steel (manufacture)	88990	85321	Oxfam (not shops)
46760	51560	Other intermediate products exporter (wholesale)	20140	24140	Oxirane (ethylene oxide) (manufacture)
46760	51560	Other intermediate products importer (wholesale)	20110	24110	Oxygen (manufacture)
31090	36140	Other plastic furniture (manufacture)	32500	33100	Oxygen breathing equipment for medical use (manufacture)
56290	55510	Other ranks' messes	20130	24130	Oxygen compounds of non metals (excluding carbon dioxide) (manufacture)
01220	01139	Other tropical and subtropical fruit growing	32500	33100	Oxygen therapy apparatus (manufacture)
31090	36110	Ottoman (manufacture)	20140	24140	Oxygen-function compounds (dual or poly) (manufacture)
82200	74860	Outbound call centres	20140	24140	Oxygen-function compounds including aldehydes (manufacture)
27400	31500	Outdoor and road lighting (manufacture)	56101	55301	Oyster bar (licensed)
31090	36140	Outdoor furniture (non-upholstered) (manufacture)	03220	05020	Oyster cultivation, freshwater
31090	36140	Outdoor furniture made of metal (manufacture)	03210	05020	Oyster cultivation, marine
43999	45450	Outdoor private swimming pools	03220	05020	Oyster fishery, freshwater
31090	36110	Outdoor seating (manufacture)	03210	05020	Oyster fishery, marine
81299	90030	Outdoor sweeping and watering of parking lots, streets, paths, public spaces etc.	03220	05020	Oyster spat production, freshwater
14190	18249	Outerwear for infants (manufacture)	03210	05020	Oyster spat production, marine
14131	18221	Outerwear for men and boys (manufacture)	46380	51380	Oysters (wholesale)
14131	18221	Outerwear for men and boys made of stitched plastic (manufacture)	32500	33100	Ozone therapy apparatus (manufacture)
14132	18222	Outerwear for women and girls made of stitched plastic (manufacture)	26600	33100	Pacemaker (electro medical) (manufacture)
29201	34201	Outfitting of all types of motor vehicles (except caravans) (manufacture)	46460	51460	Pace-makers (wholesale)
29203	34203	Outfitting of caravans (manufacture)	79110	63301	Packaged tour sales
29202	34202	Outfitting of tankers and removal trailers for transport of goods (manufacture)	82920	74820	Packaging activities on a fee or contract basis
			28290	29240	Packaging machinery (manufacture)
29202	34202	Outfitting of trailers and semi-trailers for transport of goods (manufacture)	77390	71340	Packaging machinery leasing
			82920	74820	Packaging of solids
27330	31200	Outlet boxes for electrical wiring (manufacture)	47789	52489	Packaging products for food e.g. aluminium foil, plastics foil, bags, etc. (retail)
96090	93059	Outside porter	23490	26250	Packaging products made of ceramic (manufacture)
28930	29530	Oven (food processing) machine (manufacture)	22220	25220	Packaging products made of plastic (manufacture)
27510	29710	Ovens (electric) (manufacture)	52290	63400	Packer and shipper
28210	29210	Ovens for industrial use (except bakery) (manufacture)	17219	21219	Packing cases made of fibre board (manufacture)
46499	51479	Ovens, cookers, cooking plates, boiling rings, grills and roasters (non-electric) (wholesale)	16240	20400	Packing cases made of wood (manufacture)
46439	51439	Ovens, cookers, cooking plates, boiling rings, grills and roasters (electric) (wholesale)	22220	25220	Packing goods made of plastic (manufacture)
			28290	29240	Packing machinery (manufacture)
23130	26130	Ovenware made of glass (manufacture)	17120	21120	Packing made of cardboard (manufacture)
70229	74143	Overall planning, structuring and control of organisation consultancy services	23990	26821	Packing made of woven asbestos (manufacture)
14120	18210	Overalls for boys (manufacture)	82920	74820	Packing of medicaments into edible capsules
14120	18210	Overalls for domestic use (manufacture)			

P

95

SIC 2007	SIC 2003	Activity
17120	21120	Packing paper (manufacture)
46770	51570	Packing, repacking, storage and delivery of waste and scrap without transformation (wholesale)
52290	63400	Packing service incidental to transport
13100	17170	Padding for upholstery (manufacture)
25720	28630	Padlock (manufacture)
86230	85130	Paediatric dentistry
26301	32201	Pagers (manufacture)
61200	64200	Paging activities and maintenance
61200	64200	Paging services
25910	28710	Pails made of steel (manufacture)
20301	24301	Paint (not cement based) (manufacture)
47520	52460	Paint and varnish (retail)
46730	51530	Paint and varnish (wholesale)
32910	36620	Paint brush (manufacture)
32910	36620	Paint pads (manufacture)
20301	24301	Paint removers (manufacture)
28290	29240	Paint spraying machine (manufacture)
20301	24301	Paint with cement base (manufacture)
46730	51530	Paint, varnish and lacquer (wholesale)
24100	27100	Painted steel sheet (manufacture)
90030	92319	Painters (artistic)
94120	91120	Painters and other artists associations
43341	45440	Painting contractor
31090	36140	Painting of furniture (manufacture)
45200	50200	Painting of motor vehicles
33150	35110	Painting of ships
46130	51130	Paints and varnishes (commission agent)
72190	73100	Palaeontologist (consultant)
02200	02010	Pale fencing production
24410	27410	Palladium (manufacture)
28220	29220	Pallet hoist (manufacture)
77390	71219	Pallet rental
28220	29220	Pallet truck (manufacture)
28220	29220	Palletizer (manufacture)
25990	28750	Pallets made of metal (manufacture)
22290	25240	Pallets made of plastic (manufacture)
16240	20400	Pallets, box pallets and other load boards made of wood (manufacture)
46130	51130	Pallets, pallet boards and other load boards made of wood (commission agent)
10410	15410	Palm kernel crushing (manufacture)
10410	15420	Palm kernel oil refining (manufacture)
46330	51333	Palm oil (wholesale)
10410	15410	Palm oil production (manufacture)
10410	15420	Palm oil refining (manufacture)
96090	93059	Palmist
18129	22220	Pamphlet printing (manufacture)
58110	22110	Pamphlet publishing
10720	15820	Pancake making (manufacture)
45200	50200	Panel beating services
28490	29430	Panel forming machines (manufacture)
29320	34300	Panels for motor vehicle bodywork, made of metal or fibreglass (manufacture)
23990	26821	Panels made of asbestos (manufacture)
23610	26610	Panels made of concrete (manufacture)
23650	26650	Panels made of fibre-cement (manufacture)
23620	26620	Panels made of plaster (manufacture)
14142	18232	Panty (manufacture)

SIC 2007	SIC 2003	Activity
14310	17710	Pantyhose (manufacture)
01220	01139	Papaya growing
17120	21120	Paper (not sensitized) (manufacture)
47620	52470	Paper (retail)
17120	21120	Paper (uncut) for household use (manufacture)
17211	21211	Paper and paperboard (corrugated) (manufacture)
17240	21240	Paper and paperboard articles for interior decoration (manufacture)
17120	21120	Paper and paperboard coating, covering and impregnation (manufacture)
17120	21120	Paper and paperboard intended for further industrial processing (manufacture)
17120	21120	Paper and paperboard processing (manufacture)
28950	29550	Paper and paperboard production machinery (manufacture)
46690	51870	Paper and paperboard production machinery (wholesale)
28950	29550	Paper bag making machinery (manufacture)
46760	51560	Paper bags (wholesale)
46760	51560	Paper boards (wholesale)
25990	28750	Paper clips made of metal (manufacture)
17290	21259	Paper converting (unspecified) (manufacture)
17290	21259	Paper creping (manufacture)
17290	21259	Paper cut to size (not packaging products) (manufacture)
17290	21259	Paper embossing (manufacture)
25990	28750	Paper fasteners made of metal (manufacture)
46760	51560	Paper in bulk (wholesale)
17220	21220	Paper lace (manufacture)
23990	26821	Paper made of asbestos (manufacture)
17120	21120	Paper made of vegetable fibres for corrugated cardboard (manufacture)
28950	29550	Paper making machinery (manufacture)
46760	51560	Paper merchant (wholesale)
17290	36639	Paper novelties (manufacture)
17290	21259	Paper or paperboard cards for use on jacquard machines (manufacture)
17290	21259	Paper patterns (manufacture)
17290	21259	Paper perforating (manufacture)
17290	21259	Paper shavings (manufacture)
17240	21240	Paper staining (manufacture)
17290	21259	Paper transfer for embroidery, etc. (manufacture)
43330	45430	Paperhanging
17290	21259	Papier mache works (manufacture)
13922	17402	Parachute (manufacture)
46160	51160	Parachutes and rotochutes (commission agent)
19201	23201	Paraffin (manufacture)
47789	52489	Paraffin (retail)
46750	51550	Paraffin (wholesale)
19201	23201	Paraffin for medicinal use (manufacture)
19201	23201	Paraffin wax (manufacture)
28140	29130	Parallel slide valves (manufacture)
86900	85140	Para-medical practitioner activities
32990	36639	Parasol (manufacture)
10612	15612	Parboiled or converted rice (manufacture)
82920	74820	Parcel packing and gift wrapping on a fee or contract basis
53201	64120	Parcels delivery service (not post office) licensed
53202	64120	Parcels delivery service (not post office) unlicensed
53100	64110	Parcels, distribution and delivery of, by the post office

P

SIC 2007	SIC 2003	Activity
17120	21120	Parchment and imitation parchment paper (manufacture)
15110	19100	Parchment made of leather (manufacture)
28490	29430	Paring and slicing machines (manufacture)
93290	92729	Park (local authority or municipally owned)
81300	01410	Park laying out, planting and maintenance on a fee or contract basis
42110	45230	Parking lot markings painting
52219	63210	Parking lot operation
82990	74879	Parking meter coin collection services
52219	63210	Parking meter services
26520	33500	Parking meters (manufacture)
69109	74119	Parliamentary agent
43330	45430	Parquet floor laying (not by manufacturer)
16220	20300	Parquet flooring (manufacture)
01130	01120	Parsley growing
01130	01120	Parsnip growing
27900	31620	Particle accelerator (manufacture)
16210	20200	Particleboard agglomerated with non-mineral binding substances (manufacture)
25120	28120	Partitioning made of metal (manufacture)
31010	36120	Partitions (non-domestic) (manufacture)
74909	74879	Partnership agent
28490	31620	Parts and accessories for electroplating machinery (manufacture)
28490	29430	Parts and accessories for machine tools for working hard materials e.g. wood, stone (manufacture)
28490	29430	Parts and accessories for stationary drills, machines, riveters, sheet metal cutters (manufacture)
28490	29430	Parts and accessories for stationary machine tools for nailing, stapling, glueing etc. (manufacture)
23190	26150	Parts for electric lamps and electronic valves made of glass (manufacture)
29320	34300	Parts for motor vehicles (not electric) (manufacture)
28240	29410	Parts for portable hand held power tools (manufacture)
28150	29140	Parts of bearings, gearing and driving elements (manufacture)
14131	18221	Parts of coats for men and boys (manufacture)
14132	18222	Parts of coats for women and girls (manufacture)
14132	18222	Parts of ensembles for women and girls (manufacture)
14131	18221	Parts of jackets for men and boys (manufacture)
14132	18222	Parts of jackets for women and girls (manufacture)
28302	29530	Parts of milking machines and dairy machinery, n.e.c. (manufacturing)
14132	18222	Parts of skirts for women and girls (manufacture)
14131	18221	Parts of suits for men and boys (manufacture)
14132	18222	Parts of suits for women and girls (manufacture)
14131	18221	Parts of trousers for men and boys (manufacture)
14132	18222	Parts of trousers for women and girls (manufacture)
79110	63301	Passage agent
79110	63301	Passenger agent (not transport authority)
51102	62201	Passenger air transport (non-scheduled)
51101	62101	Passenger air transport (scheduled)
51101	62101	Passenger air transport over regular routes
51102	62201	Passenger aircraft rental services with crew (non-scheduled)
30110	35110	Passenger cargo liner (manufacture)
30200	35200	Passenger carriage for railways (manufacture)
29100	34100	Passenger cars (manufacture)

SIC 2007	SIC 2003	Activity
28220	29220	Passenger conveyor (manufacture)
50300	61201	Passenger ferry (river or estuary)
50100	61101	Passenger ferry between UK and international ports
50100	61101	Passenger ferry on domestic or coastal routes
50300	61201	Passenger ferry transport (inland waterway)
77390	71211	Passenger land transport equipment self drive rental (other than motor vehicles)
49319	60219	Passenger scheduled land transport (other than interurban railways or inter-city coach services)
50100	61101	Passenger shipping service (sea and coastal)
52220	63220	Passenger terminal services
49390	60239	Passenger transport by animal-drawn vehicles
49100	60101	Passenger transport by inter-city rail services
49100	60109	Passenger transport by inter-urban railways (other than inter-city services)
49390	60239	Passenger transport by man-drawn vehicles
30110	35110	Passenger vessel building (manufacture)
77341	71221	Passenger water transport equipment leasing (without operator)
74201	74812	Passport photography
18129	22220	Passport printing (manufacture)
10730	15850	Pasta products, canned or frozen (manufacture)
10850	15850	Pasta, cooked, stuffed or otherwise prepared
10730	15850	Pastas (manufacture)
32910	36620	Paste brush (manufacture)
20301	24301	Paste made of aluminium (manufacture)
23990	26821	Paste made of asbestos (manufacture)
32990	36631	Pastel (manufacture)
10511	15511	Pasteurized fresh liquid milk (manufacture)
10822	15842	Pastille (manufacture)
10130	15139	Pastrami and other salted, dried or smoked meats (manufacture)
47240	52240	Pastry (retail)
10720	15820	Pastry and buns (preserved) (manufacture)
32910	36620	Pastry brush (manufacture)
28930	29530	Pastry roller food preparation machinery (manufacture)
69109	74111	Patent agent
74909	74879	Patent broker
46719	51519	Patent fuel (wholesale)
19201	10103	Patent fuel production (manufacture)
15110	19100	Patent leather (manufacture)
46460	51460	Patent medicines (wholesale)
69109	74111	Patents preparation
10130	15139	Pâtés (manufacture)
86900	85140	Pathological laboratory
17290	21259	Pattern card (manufacture)
96090	93059	Pavement artist
23190	26150	Pavement light (manufacture)
23190	26150	Paving blocks made of glass (manufacture)
16230	20300	Paving blocks made of wood (manufacture)
42110	45230	Paving contractor
28923	29523	Paving machinery (manufacture)
23310	26300	Paving made of non-refractory ceramic (manufacture)
23610	26610	Paving slabs made of concrete (manufacture)
23700	26700	Paving stone (manufacture)
23310	26300	Paving tiles made of unglazed clay (manufacture)
64921	65221	Pawnbroker (principally lending money)
47799	52509	Pawnshops (principally dealing in second-hand goods) (retail)

P

SIC 2007	SIC 2003	Activity
84220	75220	Pay and personnel agency (armed forces)
66190	67130	Paying agent
69202	74122	Payroll bureau
56103	55303	Pea and pie vendor
01110	01120	Pea growing
28930	29530	Pea splitters (manufacture)
10612	15612	Pea splitting, milling or grinding (manufacture)
01240	01139	Peach growing
14190	18241	Peak cap (manufacture)
10390	15330	Peanut butter (manufacture)
01110	01110	Peanut growing
01240	01139	Pear growing
32120	36220	Pearl drilling (manufacture)
03110	05010	Pearl gathering
32120	36220	Pearl stringing (manufacture)
32120	36220	Pearls production (manufacture)
46719	51519	Peat (wholesale)
08920	10300	Peat agglomeration
19201	10300	Peat briquettes (manufacture)
08920	10300	Peat cutting and digging
08920	10300	Peat extraction
09900	10300	Peat extraction support services, provided on a fee or contract basis
23990	26829	Peat products (pots or for chemical use, etc.) (manufacture)
08920	10300	Peat, preparation to facilitate transport or storage
08920	10300	Peat, preparation to improve quality
08120	14210	Pebble dredging
65120	66031	Pecuniary loss insurance
25990	28750	Pedal bins made of metal (manufacture)
22290	25240	Pedal bins made of plastic (manufacture)
30920	35420	Pedals for bicycles (manufacture)
42110	45230	Pedestrian ways construction
96020	93020	Pedicure
26511	33201	Pedometers (electronic) (manufacture)
26513	33202	Pedometers (non-electronic) (manufacture)
10390	01410	Peeled or cut vegetables, mixed fresh salads, packaged (manufacture)
10390	51310	Peeled, cut fresh vegetables, mixed salads, packed (manufacture)
22290	25240	Pelmets made of plastic (manufacture)
10110	15113	Pelt from fellmongery (manufacture)
32990	36631	Pen nibs (manufacture)
32990	36631	Pencil (manufacture)
32990	36631	Pencil leads (manufacture)
28490	29430	Pencil-making machinery (manufacture)
28230	30010	Pencil sharpeners (manufacture)
28230	30010	Pencil-sharpening machines (manufacture)
47620	52470	Pencils (retail)
46499	51479	Pencils, crayons, leads, drawing charcoals, writing or drawing chalks and tailors chalk (wholesale)
19209	23209	Penetrating oil (manufacture)
32990	36631	Penholder (manufacture)
13922	17402	Pennants (manufacture)
47620	52470	Pens (retail)
32990	36631	Pens for writing or drawing (manufacture)
55900	55239	Pension (accommodation)
66290	67200	Pension consultancy services
66290	67200	Pension consultants (own account)

SIC 2007	SIC 2003	Activity
65300	66020	Pension fund (autonomous)
66300	67121	Pension fund management
65300	66020	Pension funding except compulsory social security
65300	66020	Pension funds and plans
84230	75230	Pensions appeal tribunal
25300	28300	Penstock made of steel (manufacture)
28140	29130	Penstock valves (manufacture)
75000	85200	People's dispensary for sick animals (not animal care units)
10840	15870	Pepper (ground) (manufacture)
01280	01120	Pepper growing
10840	15870	Pepper substitute (manufacture)
46170	51170	Peptic substances, mucilages and thickeners (commission agent)
20590	24660	Peptone derivatives (manufacture)
20590	24660	Peptones (manufacture)
46120	51120	Peptones/protein substances and derivatives (commission agent)
20140	24140	Peracetic acid (manufacture)
30920	36639	Perambulator (manufacture)
13922	17402	Perambulator awning (manufacture)
47789	52489	Perambulators (retail)
46499	51479	Perambulators (wholesale)
20510	24610	Perchlorate explosive (manufacture)
20140	24140	Perchloroethylene (manufacture)
27510	29710	Percolator (electric) (manufacture)
25990	28750	Percolators made of metal (non-electric) (manufacture)
20510	24610	Percussion cap (manufacture)
32200	36300	Percussion instrument (manufacture)
25500	28400	Perforated metal (manufacture)
71200	74300	Performance testing of complete machinery
94120	91120	Performers associations
85520	80429	Performing arts schools (except academic)
85421	80302	Performing arts schools providing tertiary education
90020	92311	Performing arts support activities
94120	91120	Performing Rights Society
20420	24520	Perfume (manufacture)
47750	52330	Perfume (retail)
46450	51450	Perfume (wholesale)
20530	24630	Perfume compounds (blended perfume concentrates) (manufacture)
20530	24630	Perfumery and flavour synthetic chemicals (manufacture)
46180	51180	Perfumery, cosmetic and toilet and bath preparations (commission agent)
46450	51450	Perfumes and cosmetics exporter (wholesale)
46450	51450	Perfumes and cosmetics importer (wholesale)
18129	22220	Periodical printing (manufacture)
58142	22130	Periodical publishing
91011	92510	Periodicals lending and storage
26200	30020	Peripheral equipment for computer uses including card punches and verifiers (manufacture)
26701	33402	Periscopes (manufacture)
10890	15810	Perishable food preparations of bread, pastry or cake (manufacture)
10390	15330	Perishable prepared fruit and vegetables (manufacture)
10390	15330	Perishable prepared salads; mixed salads (manufacture)
10390	15330	Perishable prepared tofu bean curd (manufacture)

P

SIC 2007	SIC 2003	Activity	SIC 2007	SIC 2003	Activity
10390	15330	Perishable prepared vegetables, packaged peeled or cut (manufacture)	46120	51120	Petroleum coke, bitumen and other residues of petroleum (commission agent)
28131	29121	Peristaltic pumps (manufacture)	46750	51550	Petroleum coke, bitumen and other residues of petroleum oils (wholesale)
03110	05010	Periwinkle gathering			
23440	31620	Permanent magnets (ceramic and ferrite) (manufacture)	28921	29521	Petroleum drilling equipment (manufacture)
			71122	74206	Petroleum exploration
25990	31620	Permanent magnets (metallic) (manufacture)	19201	23201	Petroleum feedstock (manufacture)
46690	51870	Permanent magnets and electro-magnetic couplings, clutches and brakes (wholesale)	19201	23201	Petroleum gas (manufacture)
			46120	51120	Petroleum gases and gaseous hydrocarbons (excluding natural gas) (commission agent)
29203	34203	Permanent residential caravan (manufacture)			
24100	27100	Permanent way material (except rails production) (manufacture)	71122	74206	Petroleum geologist
			19201	23201	Petroleum grease (at refinery) (manufacture)
20130	24130	Peroxides (inorganic) (manufacture)	19209	23209	Petroleum grease formulation outside refineries (manufacture)
11030	15941	Perry (manufacture)			
26200	30020	Personal computers and workstations (manufacture)	46120	51120	Petroleum jelly and paraffin wax (commission agent)
			19209	23209	Petroleum jelly formulation outside refineries (manufacture)
65120	66031	Personal injury insurance			
32990	28750	Personal safety devices of metal (manufacture)	19201	23201	Petroleum product (at refineries) (manufacture)
18129	22220	Personal stationery printing (manufacture)	46711	51511	Petroleum products distribution (wholesale)
30120	35120	Personal watercraft (manufacture)	19201	23201	Petroleum refining (manufacture)
30400	29600	Personnel carrier (armoured fighting vehicle) (manufacture)	19209	23209	Petroleum societies without refineries (manufacture)
			09100	11200	Petroleum test well drilling
78200	74500	Personnel provision (temporary employment agency)	09100	11200	Petroleum well drilling
01610	01410	Pest control in connection with agriculture	71122	74206	Petrophysical interpretation activities
81291	74703	Pest control services (except agricultural)	14142	18232	Petticoat (manufacture)
81291	74703	Pest destruction service (not especially for agriculture)	31090	36110	Pew (manufacture)
			24430	27430	Pewter (manufacture)
20130	24130	Pesticide inorganic chemicals (excluding formulated preparations) (manufacture)	25990	28750	Pewter ware (manufacture)
			26511	33201	Ph meters (electronic) (manufacture)
20140	24140	Pesticide organic chemicals (excluding formulated preparations) (manufacture)	26513	33202	Ph meters (non-electronic) (manufacture)
			26511	33201	Ph/gas blood analysers (electronic) (manufacture)
20200	24200	Pesticides and other agrochemical products (manufacture)	21100	24410	Pharmaceutical chemicals (manufacture)
			47730	52310	Pharmaceutical chemist (retail)
01490	01250	Pet animal breeding, other than fish	46460	51460	Pharmaceutical chemist (wholesale)
10920	15720	Pet animal feeds produced from slaughter waste (manufacture)	01280	01110	Pharmaceutical crops growing
			23190	26150	Pharmaceutical glassware (other than containers) (manufacture)
46380	51380	Pet animal food (wholesale)			
47760	52489	Pet animals (retail)	46180	51180	Pharmaceutical goods (commission agent)
01490	01250	Pet bird raising and breeding	46460	51460	Pharmaceutical goods exporter (wholesale)
10920	15720	Pet fish feeds (manufacture)	46460	51460	Pharmaceutical goods importer (wholesale)
47760	52489	Pet food (retail)	21200	24421	Pharmaceutical medicament products (manufacture)
10920	15720	Pet food including canned (manufacture)	32500	24422	Pharmaceutical non-medicament products (manufacture)
10920	15720	Pet foods (manufacture)			
26600	33100	Pet scanners (manufacture)	94120	91120	Pharmaceutical society
47760	52489	Pet shop (retail)	21200	24421	Pharmaceuticals for veterinary use (manufacture)
96090	93059	Pet sitting services	47730	52310	Pharmacy (retail)
13960	17542	Petersham ribbon (manufacture)	20140	24140	Phenol (manufacture)
28290	29240	Petro-chemical industry machinery (other than plant) (manufacture)	20160	24160	Phenolic resins (manufacture)
			46750	51550	Phenols, phenol-alcohols and derivatives of phenols (wholesale)
19201	23201	Petro-chemical industry products (manufacture)			
19201	23201	Petrol (manufacture)	46120	51120	Phenols, phenol-alcohols and phenol derivatives (commission agent)
47300	50500	Petrol filling station			
28110	29110	Petrol industrial engines (manufacture)	47789	52485	Philatelist (retail)
28131	29121	Petrol station pump (manufacture)	26120	32100	Phonecards and similar cards containing integrated circuits (smart cards) (manufacture)
09100	11200	Petroleum and natural gas mining support activities			
46711	51511	Petroleum and petroleum products exporter (wholesale)	18129	22220	Phonetic printing (manufacture)
			08910	14300	Phosphates (natural) mining
46711	51511	Petroleum and petroleum products importer (wholesale)	20150	24150	Phosphates of ammonium carbonates (manufacture)
			46120	51120	Phosphates of triammonium (commission agent)
19201	23201	Petroleum briquettes (manufacture)	20150	24150	Phosphates of triammonium carbonates (manufacture)
19201	23201	Petroleum coke (manufacture)			

P

SIC 2007	SIC 2003	Activity
20150	24150	Phosphatic straight fertiliser (manufacture)
46750	51550	Phosphides, carbides, hydrides, nitrides azides, silicides and borides (wholesale)
46120	51120	Phosphides, carbides, hydrides, nitrides, azides, silicides and borides (commission agent)
46120	51120	Phosphinates, phosphonates, phosphates and polyphosphates (commission agent)
46750	51550	Phosphinates, phosphonates, phosphates and polyphosphates (wholesale)
46750	51550	Phosphoric esters and esters of other inorganic acids and their derivatives (wholesale)
46120	51120	Phosphoric esters and esters of other inorganic acids, their salts and derivatives (commission agent)
20130	24130	Phosphorous compounds (excluding phosphatic fertiliser) (manufacture)
58190	22150	Photo and engraving publishing
26110	32100	Photo diode (manufacture)
26110	32100	Photo electric cell (manufacture)
26702	33201	Photo electric exposure meter (manufacture)
18130	22240	Photo engraving (manufacture)
18130	22250	Photo lithography (manufacture)
47789	52482	Photo pick-up services (retail)
26110	32100	Photo semi-conductor device (manufacture)
26110	32100	Photo-cathode valves or tubes (manufacture)
18129	22220	Photocopier printing (manufacture)
47410	52482	Photocopiers (retail)
82190	74850	Photocopying
46660	51850	Photo-copying apparatus (wholesale)
28230	30010	Photocopying machinery (manufacture)
28990	29560	Photo-engraving machine (manufacture)
27400	31500	Photoflash bulb (manufacture)
26511	33201	Photogrammetric equipment (electronic) (manufacture)
26513	33202	Photogrammetric equipment (non-electronic) (manufacture)
74203	74814	Photograph colouring
74203	74814	Photograph copying
74203	74814	Photograph developing
74203	74814	Photograph enlarging
74203	74814	Photograph finishing
17290	21259	Photograph mount (manufacture)
74203	74814	Photograph mounting
74203	74814	Photograph printing
17120	21120	Photographic base paper (manufacture)
20590	24640	Photographic chemicals (manufacture)
74203	74814	Photographic colour printing
20590	24640	Photographic developer (manufacture)
26702	33403	Photographic enlarger (manufacture)
46180	51180	Photographic equipment (commission agent)
26702	33403	Photographic equipment (manufacture)
77210	71401	Photographic equipment hire
20590	24640	Photographic film (sensitized) (manufacture)
22210	25210	Photographic film (unsensitized) (manufacture)
26702	33403	Photographic film instrument (manufacture)
20590	24640	Photographic film plate (sensitised) (manufacture)
59120	92119	Photographic film processing activities (for the motion picture and television industries)
46439	51476	Photographic flashbulbs and flashcubes (wholesale)
47789	52482	Photographic goods (retail)
46439	51476	Photographic goods (wholesale)
46439	51476	Photographic goods exporter (wholesale)

SIC 2007	SIC 2003	Activity
46439	51476	Photographic goods importer (wholesale)
26702	33201	Photographic light meters (manufacture)
96090	93051	Photographic machines (coin-operated)
46180	51180	Photographic paper (commission agent)
20590	24640	Photographic plates (manufacture)
46180	51180	Photographic plates and film and instant print film (commission agent)
74201	74812	Photographic studio
20590	24640	Photographic unexposed film (manufacture)
22210	25210	Photographic unsensitized film (manufacture)
74209	74819	Photographing of live events such as weddings, graduations, conventions, fashion shows, etc.
74209	74819	Photography for commercials, publishers or tourism purposes
85520	80429	Photography schools (except commercial)
28990	29560	Photogravure machine (manufacture)
18129	22220	Photogravure printing (manufacture)
74209	92400	Photojournalists
28990	29560	Photolitho machine (manufacture)
28990	33403	Photolithography equipment for the manufacture of semi-conductors (manufacturing)
26701	33201	Photometers (electronic) (manufacture)
26513	33202	Photometers (non-electronic) (manufacture)
74202	74813	Photomicrography
26110	32100	Photosensitive semi-conductor devices (manufacture)
18130	22240	Phototypesetting (manufacture)
20140	24140	Phthalic anhydride (manufacture)
93199	92629	Physical culture expert
46690	51870	Physical or chemical analysis instruments and apparatus (non-electronic) (wholesale)
26513	33202	Physical properties testing and inspection equipment (non-electronic) (manufacture)
72190	73100	Physical sciences research and experimental development
96040	93040	Physical well-being activities
71200	74300	Physicist
72190	73100	Physics research and experimental development
86220	85120	Physiologist
86900	85140	Physiotherapist (private)
86900	85140	Physiotherapy clinic
32200	36300	Piano (manufacture)
77210	71401	Piano hire
85520	80429	Piano teachers and other music instruction
95290	52740	Piano tuning
24340	27340	Piano wire made of steel (manufacture)
47591	52450	Pianofortes (retail)
10390	15330	Piccalilli production (manufacture)
25730	28620	Pick (manufacture)
26400	32300	Pick-up arm and cartridge for record player (manufacture)
28302	29320	Pick-up baler (manufacture)
02200	02010	Pickets (of wood) production
15120	19200	Picking bands made of leather (manufacture)
10390	15330	Pickle including beetroot and onion (manufacture)
10390	15330	Pickling of fruit and vegetables (manufacture)
46120	51120	Pickling preparations (commission agent)
20590	24660	Pickling preparations for metal treatment (manufacture)
74209	92400	Picture agency
16290	20510	Picture frame made of wood (manufacture)

SIC 2007	SIC 2003	Activity	SIC 2007	SIC 2003	Activity
17290	21259	Picture frame mount (manufacture)	13923	17403	Pillow case (manufacture)
47789	52489	Picture framing (retail)	21200	24421	Pills (medicinal) (manufacture)
18129	22220	Picture postcard (manufacture)	52220	63220	Pilotage activities
47789	52489	Picture postcards (retail)	13923	17403	Pin cushion (manufacture)
90030	92319	Picture restoring	25730	28620	Pincers (manufacture)
26301	32201	Picture transmitter (manufacture)	01220	01139	Pineapple growing
47510	52410	Piece goods (retail)	25710	28610	Pinking shears (manufacture)
46410	51410	Piece goods (wholesale)	25930	28730	Pins made of metal (manufacture)
13300	17300	Piece goods dyeing (manufacture)	32401	36501	Pin-tables (manufacture)
22190	25130	Piece goods made of unsupported rubber sheeting (manufacture)	24440	27440	Pipe and pipe fittings made of brass (manufacture)
24100	27100	Pieces roughly shaped by forging (manufacture)	24440	27440	Pipe blanks made of copper (manufacture)
52220	63220	Pier operation (not amusement)	15120	19200	Pipe case (not leather or plastic) (manufacture)
52220	63220	Pier owner or authority (not amusement)	23990	26821	Pipe covering sections made of asbestos (manufacture)
25500	28400	Piercing of base metal (manufacture)	25730	28620	Pipe cutters (manufacture)
10720	15820	Pies (other than meat) (manufacture)	24420	27420	Pipe fittings made of aluminium (manufacture)
26110	32100	Piezo electric crystal (manufacture)	24440	27440	Pipe fittings made of copper (manufacture)
46120	51120	Piezo-electric quartz (commission agent)	24510	27210	Pipe fittings of cast-iron (manufacture)
20140	24140	Piezo-electric quartz (manufacture)	28990	29560	Pipe making machinery (manufacture)
46750	51550	Piezo-electric quartz (wholesale)	32200	36300	Pipe organ (manufacture)
01460	01230	Pig farming	22190	36639	Pipe stems of hard rubber (manufacture)
46230	51230	Pig jobber (wholesale)	25300	28300	Pipe system construction for steam generators (manufacture)
01460	01230	Pig raising and breeding	12000	16000	Pipe tobacco (manufacture)
01110	01120	Pigeon pea growing	08120	14220	Pipeclay pit
20301	24301	Pigments (prepared) (manufacture)	71200	74300	Pipeline and ancillary equipment testing activities
46230	51230	Pigs (wholesale)	42210	45213	Pipeline construction
10720	15820	Pikelet making (manufacture)	42210	45213	Pipeline contracting
13931	17511	Pile carpet mat weaving (manufacture)	71129	74204	Pipeline design activities
13931	17511	Pile carpet weaving (manufacture)	49500	60300	Pipeline operator
13931	17512	Pile carpets of cotton, tufted (manufacture)	25110	28110	Pipeline supports (manufacture)
13931	17511	Pile carpets of cotton, woven (manufacture)	23320	26400	Pipes and conduits made of clay (manufacture)
13931	17512	Pile carpets of man-made fibres, tufted (manufacture)	23320	26400	Pipes and fittings made of ceramics (manufacture)
13931	17511	Pile carpets of man-made fibres, woven (manufacture)	23990	26829	Pipes and fittings made of pitch fibre (manufacture)
13931	17512	Pile carpets of wool, tufted (manufacture)	22210	25210	Pipes and fittings made of plastic (manufacture)
13931	17511	Pile carpets of wool, woven (manufacture)	29320	34300	Pipes for motor vehicles (manufacture)
43999	45250	Pile driving	24420	27420	Pipes made of aluminium (manufacture)
28923	29523	Pile driving equipment (manufacture)	23650	26650	Pipes made of asbestos cement (manufacture)
13300	17300	Pile fabric bleaching and finishing (manufacture)	23320	26400	Pipes made of clay (manufacture)
13910	17600	Pile knitted fabric (manufacture)	23610	26610	Pipes made of concrete (manufacture)
13931	17511	Pile rugs and mats of cotton, woven (manufacture)	24440	27440	Pipes made of copper (manufacture)
13931	17511	Pile rugs and mats of man-made fibres, woven (manufacture)	23650	26650	Pipes made of fibre-cement (manufacture)
13931	17511	Pile rugs and mats of wool, woven (manufacture)	23200	26260	Pipes made of refractory ceramic (manufacture)
13931	17512	Pile rugs, mats and tiles of cotton, tufted (manufacture)	22190	25130	Pipes made of rubber (manufacture)
13931	17512	Pile rugs, mats and tiles of man-made fibres, tufted (manufacture)	24200	27220	Pipes made of steel (manufacture)
13931	17512	Pile rugs, mats and tiles of wool, tufted (manufacture)	24510	27210	Pipes of cast-iron (manufacture)
28923	29523	Pile-drivers (manufacture)	23190	26150	Pipettes made of glass (manufacture)
28923	29523	Pile-extractors (manufacture)	13200	17210	Pique weaving (manufacture)
25920	28720	Pilfer-proof metal caps (manufacture)	28940	29540	Pirns (textile machinery accessory) (manufacture)
43999	45250	Piling (building)	01250	01139	Pistachio growing
43999	45250	Piling contractor (civil engineering)	25400	29600	Pistol (manufacture)
02200	02010	Pilings (of wood) production	28110	29110	Piston for industrial engine (manufacture)
25110	28110	Pilings (tubular welded) (manufacture)	28110	34300	Piston for motor vehicle engine (manufacture)
23200	26260	Pillars made of ceramic (manufacture)	28110	29110	Piston ring for industrial engine (manufacture)
30910	35410	Pillion seats for motorcycles (manufacture)	28110	34300	Piston ring for motor vehicle engine (manufacture)
13921	17401	Pillow (manufacture)	28110	29110	Piston rings for all internal combustion engines (manufacture)
			28110	29110	Pistons for all internal combustion engines (manufacture)

P

SIC 2007	SIC 2003	Activity
28921	29521	Pit bottom machinery (manufacture)
02200	02010	Pit props (of wood) production
46730	51530	Pit props (wholesale)
19100	24140	Pitch and pitch coke (manufacture)
32200	36300	Pitch pipes (manufacture)
46630	51820	Pit-head winding gear (wholesale)
10720	15820	Pizza (manufacture)
10850	15820	Pizza, frozen or otherwise preserved (manufacture)
94920	91320	Plaid Cymru
28150	29140	Plain bearing (manufacture)
28940	29540	Plaiting machinery for textiles (manufacture)
16290	20520	Plaiting material preparation (manufacture)
13300	17300	Plaiting of textiles (manufacture)
16290	20520	Plaits and products of plaiting materials (manufacture)
25990	28750	Plan chests made of metal (manufacture)
18129	22220	Plan printing (manufacture)
25730	28620	Plane (manufacture)
25730	28620	Planer tool (manufacture)
28923	29523	Planers for road surfacing (manufacture)
30300	35300	Planetary probes (manufacture)
28410	29420	Planing machine (metal cutting) (manufacture)
28490	29430	Planing machine for wood (not portable powered) (manufacture)
16100	20100	Plank (manufacture)
28490	29430	Plank glueing machines (manufacture)
77390	71340	Plant and equipment for industrial use hire
01300	01120	Plant growing for bulbs
01300	01120	Plant growing for cuttings and slips
01300	01120	Plant growing for roots
01300	01120	Plant growing for tubers
20200	24200	Plant growth regulators (manufacture)
46750	51550	Plant growth regulators (wholesale)
77320	71320	Plant hire for construction rental (without operator)
43999	45500	Plant hire for construction rental with operator
20200	24200	Plant hormone (manufacture)
10822	15842	Plant parts preserving in sugar (manufacture)
01300	01120	Plant propagation
25110	28110	Plant support of fabricated steelwork (manufacture)
28302	29320	Planter for agricultural use (manufacture)
81300	01410	Planting, laying out and maintenance of gardens, parks and green areas for sports installations
47760	52489	Plants (retail)
46220	51220	Plants (wholesale)
01160	01110	Plants bearing vegetable fibres, retting of
01300	01120	Plants for planting or ornamental purposes
01280	01110	Plants used chiefly in pharmacy or for insecticidal, fungicidal or similar purposes
23520	26530	Plaster (manufacture)
46730	51530	Plaster (wholesale)
23620	26620	Plaster articles for use in construction (manufacture)
32500	24422	Plaster bandages (manufacture)
32990	36639	Plaster cast (manufacture)
23520	26530	Plaster of Paris (manufacture)
23620	26620	Plaster products for construction purposes (manufacture)
23620	26620	Plasterboard (manufacture)
46730	51530	Plasterboards (wholesale)
43310	45410	Plastering contractor

SIC 2007	SIC 2003	Activity
32409	36509	Plastic bicycles and tricycles designed to be ridden (manufacture)
32910	36620	Plastic brush (complete) (manufacture)
24100	27100	Plastic coated steel sheet (manufacture)
13940	17520	Plastic coated twine, cordage rope and cables of textile fibres (manufacture)
25610	28510	Plastic coating of metals (manufacture)
96090	52740	Plastic coating services of identity cards, etc. (while you wait)
27330	25239	Plastic electrical conduit tubing (manufacture)
32409	36509	Plastic game (manufacture)
22290	25240	Plastic headgear (other than hard hats) made of plastic (manufacture)
46760	51550	Plastic materials in primary forms (wholesale)
27330	25240	Plastic non-current carrying face plates (manufacture)
28230	30010	Plastic office-type binding equipment or tape binding (manufacture)
46760	51560	Plastic packaging (wholesale)
28960	29560	Plastic product making machines (manufacture)
22210	25210	Plastic semi-manufactures (manufacture)
31090	36110	Plastic shell upholstery (manufacture)
18140	22230	Plastic wire binding and finishing of books and brochures (manufacture)
20160	24160	Plastics in primary forms (manufacture)
28960	29560	Plastics working machinery (manufacture)
23110	26110	Plate glass (manufacture)
65120	66031	Plate glass insurance
18130	22240	Plate making for printing (manufacture)
20412	24512	Plate polish (manufacture)
18130	22240	Plate processes direct to plate (also photopolymer plates) (manufacture)
18130	22240	Plate setting for letterpress processes (manufacture)
18130	22240	Plate setting for offset printing processes (manufacture)
27510	29710	Plate warmers for domestic use (electric) (manufacture)
27520	29720	Plate warmers for domestic use (non-electric) (manufacture)
24420	27420	Plates made of aluminium (manufacture)
23410	26210	Plates made of ceramic for domestic use (manufacture)
17220	21220	Plates made of paper (manufacture)
22210	25210	Plates made of plastic (semi-manufactures) (manufacture)
22290	25240	Plates made of plastic (tableware) (manufacture)
22190	25130	Plates made of rubber (semi-manufactures) (manufacture)
24430	27430	Plates, sheets, strip and foil made of lead (manufacture)
24430	27430	Plates, sheets, strip and foil made of tin (manufacture)
24430	27430	Plates, sheets, strip and foil made of zinc (manufacture)
33110	28220	Platework repair of central heating boilers and radiators (manufacture)
30110	35110	Platform for drilling rig (manufacture)
29202	34202	Platform trailer (motor drawn) (manufacture)
25610	28510	Plating (metal finishing) (manufacture)
24410	27410	Platinum (manufacture)
24410	27410	Platinum group metals (manufacture)
07290	13200	Platinum mining and preparation
32409	36509	Playballs made of rubber (manufacture)

P

SIC 2007	SIC 2003	Activity	SIC 2007	SIC 2003	Activity
32300	36400	Playground equipment (manufacture)	22110	25110	Pneumatic tyres (manufacture)
88910	85321	Playgroup (charitable)	15120	19200	Pochette made of leather (manufacture)
88910	85322	Playgroup (non-charitable)	15120	19200	Pocket book made of leather (manufacture)
32409	36509	Playing cards (manufacture)	25710	28610	Pocket knife (manufacture)
46499	51477	Playing cards (wholesale)	26520	33500	Pocket timer (manufacture)
90030	92319	Playwright	26520	33500	Pocket watch (manufacture)
30120	35120	Pleasure and sporting boats (manufacture)	13200	17210	Pocketing weaving (manufacture)
93290	92629	Pleasure boat hiring as an integral part of recreational facilities	90030	92319	Poet
			94990	91330	Poetry club
50100	61101	Pleasure boat rental with crew (e.g. for fishing cruises) (except for inland waterway service)	30200	35200	Point locks (manufacture)
			28230	30010	Point of sale unit (manufacture)
77210	71401	Pleasure boats rental	26200	30020	Point-of-sale (pos) computer terminals not mechanically operated (manufacture)
93290	92729	Pleasure ground			
93290	92729	Pleasure pier	26511	33201	Polarimeters (electronic) (manufacture)
42910	45240	Pleasure port construction	26513	33202	Polarimeters (non-electronic) (manufacture)
56290	55520	Pleasure steamer caterer	32500	33401	Polarising elements (manufacture)
17290	21259	Pleated paper (manufacture)	02200	02010	Pole fencing production
13300	17300	Pleating and similar work on textiles (manufacture)	02200	02010	Poles (of wood) production
			84240	75240	Police authorities
13300	17300	Pleating of textiles (manufacture)	88990	85322	Police court mission
25730	28620	Pliers (manufacture)	71200	75240	Police laboratories
28302	29320	Plough (manufacture)	84240	75240	Police records maintenance
28302	29320	Plough disc (manufacture)	70229	74143	Policy formulation consultancy services
46610	51880	Ploughs (wholesale)	20412	24512	Polish (manufacture)
26110	31200	Plug (electronic) (manufacture)	10612	15612	Polished rice (manufacture)
27330	31200	Plug and socket (electric) (manufacture)	20412	24512	Polishes and creams (manufacture)
28140	29130	Plug valves (manufacture)	20412	24512	Polishes and creams for leather (manufacture)
09100	11200	Plugging and abandoning oil and gas wells	20412	24512	Polishes and creams for wood (manufacture)
46439	51439	Plugs, sockets and other apparatus for protecting electrical circuits for domestic use (wholesale)	20412	24512	Polishes for coachwork (manufacture)
			20412	24512	Polishes for glass (manufacture)
46690	51870	Plugs, sockets and the like, for switching or protecting industrial electrical circuits (wholesale)	20412	24512	Polishes for metal (manufacture)
			25610	28510	Polishing (metal finishing) (manufacture)
01240	01139	Plum growing	13923	17403	Polishing cloths and pads, made of unprepared, non-bonded fibre fabric (manufacture)
46740	51540	Plumbers' merchant (wholesale)			
25990	28750	Plumbing and pipe fittings made of metal (not cast) (manufacture)	28990	29560	Polishing machine for glass (manufacture)
			32910	36620	Polishing mop (manufacture)
43220	45330	Plumbing contractor	20412	24512	Polishing paste and powder (manufacture)
46740	51540	Plumbing equipment and supplies (wholesale)	23910	26810	Polishing stones made of bonded abrasives (manufacture)
13200	17240	Plush silk (manufacture)			
13200	17220	Plush woollen weaving (manufacture)	94920	91320	Political organisations
13200	17230	Plush worsted weaving (manufacture)	71200	74300	Pollution measuring
20130	23300	Plutonium processing (manufacture)	84120	75120	Pollution standards, dissemination and information services (public sector)
65120	66031	Pluvium insurance			
16210	20200	Plywood (manufacture)	10130	15139	Polony (manufacture)
46730	51530	Plywood (wholesale)	20160	24160	Polyamide compounds (manufacture)
28490	29430	Plywood press (manufacture)	20600	24700	Polyamide man-made fibre (manufacture)
28220	29220	Pneumatic and hydraulic conveying plant (manufacture)	20160	24160	Polyamides (manufacture)
			46120	51120	Polyamides in primary forms (commission agent)
28220	29220	Pneumatic and hydraulic handling plant (manufacture)	46760	51550	Polyamides in primary forms (wholesale)
			46120	51120	Polycarbonates, alkyd and epoxide resins (commission agent)
46690	51870	Pneumatic and other continuous action elevators and conveyors for goods or materials (wholesale)			
			20520	24620	Polyester adhesive (manufacture)
28120	29122	Pneumatic equipment and systems for aircraft (manufacture)	20600	24700	Polyester man-made fibre (manufacture)
			20301	24301	Polyester paint (manufacture)
26513	33202	Pneumatic gauges (non-electronic) (manufacture)	20160	24160	Polyester resins (manufacture)
28240	29410	Pneumatic nailers (manufacture)	20160	24160	Polyesters (manufacture)
28120	29122	Pneumatic power engines and motors (manufacture)	20160	24160	Polyethers (manufacture)
28240	29410	Pneumatic power tools (portable) (manufacture)	46120	51120	Polyethers and polyesters (commission agent)
28240	29410	Pneumatic rivet guns (manufacture)	46760	51550	Polyethers, polyesters, polycarbonates, alkyd and epoxide resins (wholesale)
28960	29560	Pneumatic tyre making or retreading machines (manufacture)			

P

SIC 2007	SIC 2003	Activity
20160	24160	Polyethylene (manufacture)
26513	33202	Polygraph machines (non-electronic) (manufacture)
80100	74602	Polygraph services
20160	24160	Polymers (manufacture)
46760	51550	Polymers of ethylene and styrene in primary forms (wholesale)
46120	51120	Polymers of ethylene in primary forms (commission agent)
46120	51120	Polymers of propylene and other olefins in primary forms (commission agent)
46120	51120	Polymers of styrene in primary forms (commission agent)
46120	51120	Polymers of vinyl acetate, other vinyl esters and vinyl polymers in primary forms (commission agent)
46120	51120	Polymers of vinyl chloride and other halogenated olefins in primary forms (commission agent)
46750	51550	Polymers of vinyl chloride and other halogenated olefins in primary forms (wholesale)
20160	24160	Polypropylene (manufacture)
13200	17250	Polypropylene fabrics (manufacture)
20160	24160	Polystyrene (manufacture)
85421	80302	Polytechnics
20160	24160	Polytetrafluoroethylene (PTFE) (manufacture)
20520	24620	Polyurethane adhesive (manufacture)
13960	17549	Polyurethane coated textile fabrics (manufacture)
20301	24301	Polyurethane paint (manufacture)
20160	24160	Polyurethanes (manufacture)
20160	24160	Polyvinyl acetate (PVA) (manufacture)
20520	24620	Polyvinyl acetate and co-polymer adhesives (manufacture)
20160	24160	Polyvinyl chloride (PVC) (manufacture)
01240	01139	Pome fruit growing
01230	01139	Pomelo growing
10822	15842	Pomfret (pontefract) cakes (manufacture)
13960	17542	Pompons (manufacture)
30110	35110	Pontoons construction (manufacture)
93120	92629	Pony Club
96090	93059	Poodle clipping
90010	92311	Pop group
13200	17210	Poplin weaving (manufacture)
10130	15131	Pork (salted or pickled) (manufacture)
47220	52220	Pork butcher (retail)
46320	51320	Pork butcher (wholesale)
10130	15139	Pork pie (manufacture)
26513	33202	Porosimeter (manufacture)
52220	63220	Port Authority
84240	75240	Port guards administration and operation
52220	63220	Port of London Authority
25110	28110	Portable building metalwork (manufacture)
25730	28620	Portable forges (manufacture)
27400	31500	Portable lamp (electric) (manufacture)
28240	29410	Portable power tool parts (manufacture)
77320	71320	Portable road sign hire for construction
27510	29710	Portable space heaters (manufacture)
16230	20300	Portable wooden buildings (manufacture)
28220	29220	Portal and pedestal jib cranes (manufacture)
11050	15960	Porter brewing (manufacture)
96090	93059	Porters
66300	67121	Portfolio management services
23510	26510	Portland cement (manufacture)

SIC 2007	SIC 2003	Activity
74201	74812	Portrait photographer
74201	74812	Portrait photography
28120	29122	Positioner for pneumatic control equipment (manufacture)
28131	29121	Positive displacement pump (reciprocating) (manufacture)
46180	51180	Posphoaminolipids (commission agent)
46460	51460	Posphoaminolipids, amides and their salts and derivatives (wholesale)
53100	64110	Post activities
53100	64110	Post office regional headquarters
28490	29430	Post peeling machines (manufacture)
18140	22230	Post press services in support of printing activities (manufacture)
59120	92119	Post production film activities
30200	35200	Post van for railways (manufacture)
28230	30010	Postage franking machines (manufacture)
28230	30010	Postage meters (manufacture)
18129	22220	Postage stamp perforating (manufacture)
18129	22220	Postage stamp printing (manufacture)
64191	65121	Postal giro and postal savings bank activities
53100	64110	Postal headquarters
53100	64110	Postal sorting office
58190	22150	Postcard publishing
17230	21230	Postcards (plain) (manufacture)
53100	64110	Poste restante
73110	74402	Poster advertising
18130	22250	Poster aerographing (manufacture)
18129	22220	Poster printing (manufacture)
58190	22150	Poster publishing
18130	22240	Poster writing (manufacture)
85422	80303	Post-graduate college
01630	01410	Post-harvest crop activities
02200	02010	Posts (of wood) production
23610	26610	Posts made of precast concrete (manufacture)
85410	80301	Post-secondary non-tertiary vocational education
08910	14300	Potash mine
20150	24150	Potassic straight fertiliser (manufacture)
20130	24130	Potassium compounds (manufacture)
20150	24150	Potassium salts (manufacture)
08910	14300	Potassium salts (natural) mining
10310	15310	Potato chip production (frozen, raw, steamed or boiled) (manufacture)
10310	15310	Potato crisp (manufacture)
10910	15710	Potato dehydrating for animal feed (manufacture)
10310	15310	Potato flour and meal (manufacture)
01130	01110	Potato growing
28302	29320	Potato harvester and sorter (manufacture)
10310	15310	Potato peeling (industrial) (manufacture)
10310	15310	Potato processing and preserving (manufacture)
46310	51380	Potato products (wholesale)
10310	15310	Potato puff (manufacture)
10310	15310	Potato snacks production (manufacture)
10620	15620	Potato starch (manufacture)
10310	15310	Potato stick (manufacture)
10310	15310	Potato straw (manufacture)
10310	15310	Potatoes (prepared frozen) production (manufacture)
47210	52210	Potatoes (retail)
46310	51310	Potatoes (wholesale)

P

SIC 2007	SIC 2003	Activity
26511	33201	Potentiometric recorder (electronic) (manufacture)
26513	33202	Potentiometric recorder (non-electronic) (manufacture)
23490	26250	Pots made of ceramic (manufacture)
23130	26130	Pots made of glass (manufacture)
17220	21220	Pots made of paper (manufacture)
22220	25220	Pots made of plastic (not flower pots) (manufacture)
10130	15139	Potted meat. (manufacture)
10200	15209	Potted shrimp (manufacture)
08120	14220	Potters' clay mine or quarry
47599	52440	Pottery (retail)
46440	51440	Pottery (wholesale)
23410	26210	Pottery for domestic use (manufacture)
23410	26210	Pottery made of stone (manufacture)
28990	29560	Pottery making machinery (manufacture)
77299	71409	Pottery rental and leasing
20150	24150	Potting soil mixtures of natural soil, sand, clays and minerals (manufacture)
20150	24150	Potting soil with peat as main constituent (manufacture)
15120	19200	Pouch for tobacco (manufacture)
15120	19200	Pouch made of leather or leather substitute (manufacture)
17230	21230	Pouches containing an assortment of paper stationery (manufacture)
13921	17401	Pouffe (manufacture)
13200	17210	Poult weaving (manufacture)
10120	15120	Poultry (fresh, chilled or frozen) production (manufacture)
46320	51320	Poultry (wholesale)
47220	52220	Poultry and game (retail)
10130	15139	Poultry canning (manufacture)
01629	01429	Poultry caponising
10850	15139	Poultry dishes (manufacture)
10120	15120	Poultry dressing (manufacture)
01470	01240	Poultry farming
10120	15120	Poultry fat (edible) rendering (manufacture)
10910	15710	Poultry grit (manufacture)
01470	01240	Poultry hatcheries
16230	20300	Poultry house made of wood (manufacture)
28302	29320	Poultry keeping machinery (manufacture)
10120	15120	Poultry meat preparation (manufacture)
10120	15120	Poultry meat production and preserving (manufacture)
10120	15120	Poultry packing (manufacture)
10130	15139	Poultry potting (manufacture)
28930	29530	Poultry processing machines and equipment (manufacture)
01470	01240	Poultry raising and breeding
10120	15120	Poultry slaughtering (manufacture)
46210	51210	Poultry spice (wholesale)
46610	51880	Poultry-keeping machines (wholesale)
22290	36639	Powder compact made of plastic (manufacture)
24420	27420	Powder made of aluminium (manufacture)
24440	27440	Powder made of copper (manufacture)
20301	24301	Powder made of glass (manufacture)
25500	28400	Powder metallurgy (manufacture)
13990	36639	Powder puff (manufacture)
10890	15891	Powdered broth containing meat or vegetables or both (manufacture)

SIC 2007	SIC 2003	Activity
10890	15891	Powdered soup containing meat or vegetables or both (manufacture)
10810	15830	Powdered sugar (manufacture)
24430	27430	Powders and flakes made of lead (manufacture)
24430	27430	Powders and flakes made of tin (manufacture)
20590	24660	Powders and pastes used in soldering, brazing or welding (manufacture)
46180	51180	Powders for cosmetic or toilet use (commission agent)
30120	35120	Power boats of all types (manufacture)
30300	35300	Power control for aircraft (manufacture)
27110	31100	Power generators (manufacture)
42220	45213	Power line construction
35110	40110	Power station
25110	28110	Power structural steelwork (manufacture)
27110	31100	Power supply unit for electronic applications (manufacture)
27120	31200	Power switching equipment (manufacture)
28240	29410	Power tool (portable) (manufacture)
28240	29410	Power tools (portable electric) (manufacture)
28240	29410	Power-driven buffers (manufacture)
28220	29220	Power-driven cranes (manufacture)
28240	29410	Power-driven grinders (manufacture)
28240	29410	Power-driven impact wrenches (manufacture)
28220	29220	Power-driven lifting capstans (manufacture)
28220	29220	Power-driven lifting hoists (manufacture)
28220	29220	Power-driven lifting jacks (manufacture)
28220	29220	Power-driven lifting pulley tackle (manufacture)
28220	29220	Power-driven lifting winches (manufacture)
28220	29220	Power-driven loading and unloading derricks (manufacture)
28220	29220	Power-driven mobile lifting frames (manufacture)
28240	29410	Power-driven planers (manufacture)
28240	29410	Power-driven powder actuated nailers (manufacture)
28240	29410	Power-driven routers (manufacture)
28240	29410	Power-driven shears and nibblers (manufacture)
28240	29410	Power-driven staplers (manufacture)
28220	29220	Power-driven straddle carriers (manufacture)
28922	29522	Powered barrow (manufacture)
28921	29521	Powered roof support for mining (manufacture)
13923	17403	Pram blanket (outside knitting or weaving establishment) (manufacture)
23610	26610	Precast cement articles for use in construction (manufacture)
23610	26610	Precast concrete articles for use in construction (manufacture)
23610	26610	Precast concrete products (manufacture)
46730	51530	Precast concrete products (wholesale)
32120	36220	Precious and semi-precious stones in the worked state production (manufacture)
24410	27410	Precious metal foil laminates (manufacture)
07290	13200	Precious metal ores mining and preparation
24410	27410	Precious metals production (manufacture)
32120	36220	Precious stone cutting (manufacture)
46760	51560	Precious stones (wholesale)
26513	33202	Precision balance (non-electronic) (manufacture)
28150	29140	Precision chain (manufacture)
26513	33202	Precision drawing instrument (manufacture)
25940	28740	Precision screw (manufacture)
24200	27220	Precision tube made of steel (manufacture)

P

SIC 2007	SIC 2003	Activity	SIC 2007	SIC 2003	Activity
86101	85112	Pre-convalescent hospital (private sector)	28410	29420	Press (pneumatic) (metal forming) (manufacture)
86101	85111	Pre-convalescent hospital (public sector)	63990	92400	Press clipping service
25110	28110	Prefabricated building metalwork (manufacture)	13200	17220	Press cloth (manufacture)
46730	51530	Prefabricated buildings (wholesale)	63990	92400	Press cutting agency
23610	26610	Prefabricated buildings and components made of concrete (manufacture)	28490	29430	Press for chipboard (manufacture)
			28930	29530	Press for food and drink (manufacture)
25110	28110	Prefabricated buildings made of metal (manufacture)	28960	29560	Press machinery for working rubber or plastics (manufacture)
22230	25239	Prefabricated buildings made of plastic (manufacture)	74209	92400	Press photographers
46130	51130	Prefabricated buildings made of wood (commission agent)	25990	36639	Press stud (manufacture)
			25730	28620	Press tool (manufacture)
16230	20300	Prefabricated buildings made of wood (manufacture)	28930	29530	Press used to make wine, cider, fruit juices, etc. (manufacture)
16230	20300	Prefabricated buildings or elements thereof made of wood (manufacture)	17120	21120	Pressboard (manufacture)
42990	45213	Prefabricated constructions (civil engineering) assembly and erection	13990	17549	Pressed felt (not paper or roofing) (manufacture)
			13990	17549	Pressed wool felt (manufacture)
41201	45211	Prefabricated constructions (commercial) assembly and erection	28490	29430	Presses for the manufacture of particle board (manufacture)
41202	45212	Prefabricated constructions (domestic) assembly and erection	32990	36639	Press-fasteners (manufacture)
16230	20300	Prefabricated roof timbers (manufacture)	25990	36639	Press-fasteners made of metal (manufacture)
23610	26610	Prefabricated structural parts for construction, of cement, concrete, artificial stone (manufacture)	96010	93010	Pressing and valeting
			25500	28400	Pressing of base metal (manufacture)
10110	15112	Premier jus (manufacture)	96010	93010	Pressing of wearing apparel on a fee or contract basis
18130	22240	Preparation and linkage of digital data (manufacture)	17120	21120	Presspahn (manufacture)
85200	80100	Preparatory schools	32990	36639	Press-studs (manufacture)
20590	24660	Prepared additives for cement (manufacture)	25990	28750	Pressure cooker (manufacture)
46120	51120	Prepared binders for foundry moulds or cores (commission agent)	24540	27540	Pressure die casting of non-ferrous base metal (manufacture)
20590	24660	Prepared culture media for micro-organisms (manufacture)	28960	29560	Pressure forming machines for rubber or plastic (manufacture)
20301	24301	Prepared dyes (manufacture)	26511	33201	Pressure measuring and control instrument (electronic) (manufacture)
10910	15710	Prepared feeds for farm animals (manufacture)	26513	33202	Pressure measuring and control instrument (non-electronic) (manufacture)
46210	51210	Prepared feeds for farm animals (wholesale)	23610	26610	Pressure pipes made of pre-stressed concrete (manufacture)
10850	15139	Prepared meat dishes (manufacture)	28140	29130	Pressure reducing valves (manufacture)
46130	51130	Prepared pigments, opacifiers and colours (commission agent)	22290	25240	Pressure sensitive adhesive tape made of plastic (manufacture)
46120	51120	Prepared rubber accelerators (commission agent)	13990	17549	Pressure sensitive cloth-tape (manufacture)
32990	36631	Prepared typewriter ribbons (manufacture)	26511	33201	Pressure switch (electronic) (manufacture)
20412	24512	Prepared waxes (manufacture)	16100	20100	Pressure treatment of wood (manufacture)
96030	93030	Preparing the dead for burial or cremation	71200	74300	Pressurised containers certification
18130	22240	Pre-press data input electronic make-up (manufacture)	23610	26610	Pre-stressed concrete products (manufacture)
18130	22240	Pre-press data input optical character recognition, electronic make-up (manufacture)	10720	15820	Pretzels whether sweet or salted (manufacture)
			86101	85111	Preventoria providing hospital type care (public sector)
18130	22240	Pre-press data input scanning (manufacture)	85200	80100	Primary and pre-primary education
85100	80100	Pre-primary education	27200	31400	Primary battery (manufacture)
18201	22310	Pre-recorded tape, except master copies for records or audio material (manufacture)	27200	31400	Primary cells (manufacture)
94910	91310	Presbyterian Church	46439	51439	Primary cells and primary batteries for domestic use (wholesale)
94910	91310	Presbyterian Church of Wales	46690	51870	Primary cells and primary batteries for industrial use (wholesale)
91030	92522	Preservation society for historic houses			
10511	15511	Preserved cream (manufacture)	24440	27440	Primary copper (manufacture)
46310	51380	Preserved fruit (wholesale)	85200	80100	Primary education
47210	52270	Preserved fruit and vegetables (retail)	24100	27100	Primary products of stainless steel (manufacture)
10130	15139	Preserved meat (manufacture)	85200	80100	Primary schools
10720	15820	Preserved pastry goods and cakes (manufacture)	41201	45211	Primary, secondary and other schools construction
93210	92330	Preserved railway operation	27120	31200	Prime mover generator sets (manufacture)
20420	24520	Pre-shave lotion (manufacture)			
28410	29420	Press (hydraulic) (metal forming) (manufacture)			
28410	29420	Press (mechanical) (metal forming) (manufacture)			
28410	29420	Press (metal forming) (manufacture)			

SIC 2007	SIC 2003	Activity
25400	29600	Primer for cartridge (manufacture)
20301	24301	Primer paint (manufacture)
84220	75220	Princess Mary's RAFNS
13200	17210	Print cloth weaving (manufacture)
18130	22250	Print colouring (manufacture)
26110	32100	Printed circuit (manufacture)
46520	51860	Printed circuits (wholesale)
22230	25231	Printed felt base floorcovering (manufacture)
18121	21251	Printed labels (manufacture)
13960	17542	Printed labels of textile materials (manufacture)
17230	22220	Printed matter for accounting and technical use (manufacture)
58190	22150	Printed matter publishing
17211	21211	Printed paper bags (manufacture)
18140	22230	Printed paper or board finishing (manufacture)
18140	22230	Printed sheets finishing (manufacture)
26110	31300	Printer and monitor connectors (manufacture)
26110	31300	Printer cables (manufacture)
26200	30020	Printer ink cartridges (manufacture)
26200	30020	Printer ink cartridges (refilling)
26200	30020	Printer servers (manufacture)
26200	30020	Printers and plotters (manufacture)
22190	25130	Printers' blankets made of rubber (manufacture)
17230	21230	Printers' cards (manufacture)
18130	22250	Printers' designing (manufacture)
26200	30020	Printers for computers (manufacture)
20302	24302	Printers' varnish (manufacture)
13200	17210	Printers' weaving (manufacture)
18129	22220	Printing (undefined) (manufacture)
17230	21230	Printing and writing paper ready for use (manufacture)
46690	51870	Printing block and plates preparation and production machinery, equipment and apparatus (wholesale)
18129	22220	Printing by computer printers (manufacture)
18129	22220	Printing by duplication machines (manufacture)
18129	22220	Printing by embossers (manufacture)
18129	22220	Printing by quick printing (manufacture)
32990	36631	Printing devices (hand operated) (manufacture)
18129	22220	Printing directly onto ceramics (manufacture)
46120	51120	Printing ink (commission agent)
20302	24302	Printing ink (manufacture)
46750	51550	Printing ink (wholesale)
28990	29560	Printing machine or press (manufacture)
46690	51870	Printing machinery (wholesale)
28940	29540	Printing machinery for textiles (manufacture)
28230	30010	Printing machines (sheet fed office type offset) (manufacture)
18129	22220	Printing of banknotes
18129	22220	Printing of magazines appearing less than four times a week (manufacture)
18129	22220	Printing of other security papers (manufacture)
18110	22210	Printing of periodicals appearing at least four times a week (manufacture)
18129	22220	Printing of smart cards (manufacture)
59200	92119	Printing of sound tracks
18121	21251	Printing on labels or tags (manufacture)
18129	22220	Printing onto glass (manufacture)
18129	22220	Printing onto metal (manufacture)
18129	22220	Printing onto plastic (manufacture)
18129	21220	Printing onto sanitary goods and toilet requisites of paper (manufacture)
18129	22220	Printing onto textiles (manufacture)
18129	22220	Printing onto wood (manufacture)
17120	21120	Printing paper (manufacture)
18130	22240	Printing plate engraving (manufacture)
18130	22240	Printing roller engraving (manufacture)
32500	33401	Prisms (mounted, (not photographic)) (manufacture)
23190	26150	Prisms (pressed or moulded, unworked) (manufacture)
32500	33401	Prisms (unmounted) (manufacture)
84230	75230	Prison administration and operation
86101	85111	Prison hospitals
84230	75230	Prisons (excluding naval and military)
26301	32201	Private branch exchange (PBX) equipment (manufacture)
86220	85120	Private consultants clinics
80300	74601	Private detective
49320	60220	Private hire car with driver
86101	85112	Private hospital
80300	74601	Private investigator activities
55900	55239	Private lodging house
80100	74602	Private security activities
85590	80421	Private training providers
84230	75230	Probate registry
88990	85322	Probation and after care service
18130	22240	Process block making (manufacture)
26514	33302	Process control equipment (electric) (manufacture)
26512	33301	Process control equipment (electronic) (manufacture)
28140	29130	Process control valves (manufacture)
46690	51870	Process control valves, gate valves, globe valves and other valves (wholesale)
71129	74209	Process engineering contractor
18130	22240	Process engraving (manufacture)
25300	28300	Process heater (manufacture)
19201	23201	Process oil refining (manufacture)
19201	23201	Process oils (manufacture)
25300	28300	Process pipework (manufacture)
18130	22240	Process plate engraving (manufacture)
25300	28300	Process pressure sphere (manufacture)
25300	28300	Process pressure vessel (manufacture)
69109	74119	Process server
26520	33500	Process timers (manufacture)
23990	26821	Processed asbestos fibre (manufacture)
10512	15512	Processed cheese (manufacture)
46380	51380	Processed fruit (wholesale)
46320	51320	Processed meat and meat products (wholesale)
46380	51380	Processed vegetables (wholesale)
38320	37200	Processing cleaning, melting, grinding of plastic or rubber waste to granulates
38320	37200	Processing of food, beverage and tobacco waste and residual substances into secondary raw materials
38320	37200	Processing of used cooking oils and fats into secondary raw materials
46690	51870	Producer gas or water gas generators (wholesale)
35300	40300	Production and distribution of cooled air
46690	51870	Production line robot (wholesale)
70229	74149	Production management consultancy services (other than for construction)

P

SIC 2007	SIC 2003	Activity
38210	90020	Production of compost from organic waste
24100	27100	Production of granular iron
35300	40300	Production of ice for cooling purposes (manufacture)
35300	15980	Production of ice for food (manufacture)
59113	92111	Production of theatrical and non-theatrical television programmes
28921	29521	Production riser tensioners (manufacture)
28921	29521	Production riser tie-back equipment (manufacture)
85590	80429	Professional examination review courses
94120	91120	Professional organisations
22210	25210	Profile shapes of plastic materials (rods, tubes, etc.) (manufacture)
22190	25130	Profile shapes of rubber (manufacture)
24330	27330	Profiled steel sheet (manufacture)
70229	74143	Profit improvement programmes consultancy services
28940	29540	Programmer for textile machinery (manufacture)
62012	72220	Programming services
71129	74209	Project management
26702	33403	Projection screen (manufacture)
26702	33403	Projector (photographic or cinematographic) (manufacture)
26702	33403	Projector for cinema (manufacture)
27400	31500	Projector lamp (manufacture)
93199	92629	Promotion of sporting events
82190	74850	Proof reading
01629	01429	Propagation, growth and output of animals, promotion service activities
19201	23201	Propane (manufacture)
06200	11100	Propane extraction from natural gas
46711	51511	Propane gas (wholesale)
20510	24610	Propellant powder (manufacture)
46120	51120	Propellant powders and prepared explosives (commission agent)
30300	35300	Propeller for aircraft (manufacture)
30300	35300	Propeller rotor blades (manufacture)
29320	34300	Propeller shaft for motor vehicle (manufacture)
46499	51479	Propelling or sliding pencils (wholesale)
32990	36631	Propelling pencil (manufacture)
20510	24610	Propergol fuels and other propellant powders (manufacture)
68310	70310	Property consultant (own account)
41100	70110	Property developer
65120	66031	Property insurance
41100	70110	Property investment company
68209	70209	Property leasing (other than conference centres and exhibition halls)
68320	70320	Property management (as agents for owners)
94110	91110	Property owners' association
64305	65238	Property unit trusts
23200	26260	Props made of ceramic (manufacture)
28110	29110	Propulsion engine for marine use (manufacture)
20140	24140	Propylene (manufacture)
20140	24140	Propylene oxide (manufacture)
20160	24160	Propylene polymers (manufacture)
18140	22230	Prospectus finishing (manufacture)
18129	22220	Prospectus printing (manufacture)
46460	51460	Prosthesis and orthopaedic appliances (wholesale)
14120	18210	Protective clothing for industrial use (manufacture)
43341	45440	Protective coatings application work

SIC 2007	SIC 2003	Activity
15200	19300	Protective footwear made of plastic (manufacture)
32500	33401	Protective glasses (manufacture)
32990	18249	Protective gloves for industrial use (manufacture)
32990	18241	Protective headgear (manufacture)
32990	18241	Protective headgear for industrial use (manufacture)
10910	15710	Protein (synthetic) for animal feed (manufacture)
10910	15710	Protein concentrates (animal food) (manufacture)
20590	24660	Protein substances (manufacture)
94990	91330	Protest movement activities
26511	33201	Proton microscope (manufacture)
46210	51210	Provender (wholesale)
65110	66011	Provident fund (life)
65120	66031	Provident fund (non-life)
46170	51170	Provision exchange (commission agent)
87300	85140	Provision of residential care and treatment for the elderly and disabled by paramedical staff
84240	75240	Provision of supplies for domestic emergency use in case of peacetime disasters
46390	51390	Provisions (wholesale)
46180	51180	Provitamins, vitamins and their derivatives (commission agent)
21100	24410	Provitamins, vitamins and their derivatives (manufacture)
46460	51460	Provitamins, vitamins and their derivatives (wholesale)
25710	28610	Pruning knife (manufacture)
25730	28620	Pruning shears (manufacture)
86900	85140	Psychiatric clinic
86900	85140	Psychiatric day hospital
86101	85112	Psychiatric unit (private sector)
86220	85120	Psychiatrist (private practice)
32500	33100	Psychological testing apparatus (manufacture)
86900	85140	Psychologist
72200	73200	Psychology research and experimental development
32500	33100	Psychology testing apparatus (manufacture)
93290	92729	Psychometry
46690	51870	Public address equipment (wholesale)
26400	32300	Public address system (manufacture)
84120	75120	Public administration of education programmes
84120	75120	Public administration of environment programmes
84120	75120	Public administration of health programmes
84120	75120	Public administration of housing programmes
84120	75120	Public administration of recreation programmes
84120	75120	Public administration of research and development policies
84120	75120	Public administration of social services programmes
84120	75120	Public administration of sport programmes
71200	74300	Public analyst
96040	93040	Public baths
26400	32300	Public broadcasting equipment (manufacture)
84110	75110	Public debt services administration
84110	75110	Public fund services administration
86900	85140	Public health laboratory
56302	55402	Public houses (independent)
56302	55404	Public houses (managed)
56302	55403	Public houses (tenanted)
73200	74130	Public opinion polling
84240	75240	Public order and safety administration, regulation and operation
93290	92729	Public park

SIC 2007	SIC 2003	Activity	SIC 2007	SIC 2003	Activity
82990	74879	Public record searching	28410	29420	Punch presses (manufacture)
70210	74141	Public relations and communication	26200	30020	Punchcard readers (manufacture)
70210	74141	Public relations consultant (not advertising agency)	17120	21120	Punched card and punched paper tape stock (manufacture)
85310	80210	Public schools	28230	30010	Punched card machine (other than for computer use) (manufacture)
49319	60219	Public service vehicle operator
90010	92319	Public speaker	25730	28620	Punches for hand tools (interchangeable) (manufacture)
85590	80429	Public speaking training
82990	74850	Public stenography services	25730	28620	Punches for machine tools (interchangeable) (manufacture)
66190	67130	Public trust office
42990	45213	Public works contractor	28410	29420	Punching machine (metal forming) (manufacture)
74909	74879	Publicans' broker	22190	25130	Puncture repair outfit made of rubber (manufacture)
58190	22150	Publishers (other than of newspapers, books and periodicals)	16290	20520	Punnets made of cork, straw or plaiting materials (manufacture)
18140	22230	Publisher's case making (manufacture)	30120	35120	Punt (manufacture)
59200	22140	Publishing of music and sheet books	32409	36509	Puppet (manufacture)
58290	72210	Publishing of software for business	93290	92349	Puppet shows
58290	72210	Publishing of software for operating systems	61200	64200	Purchase of wireless access and network capacity
58190	22220	Publishing of stamps, banknotes, advertising material, catalogues and other printed matter n.e.c.	28290	29240	Purifying machinery (manufacture)
			15120	19200	Purses made of leather or leather substitute (manufacture)
58110	22110	Publishing on CD- ROM
10611	15611	Pudding mixture (manufacture)	30920	36639	Push chair (manufacture)
10612	15612	Puffed rice (manufacture)	32409	36509	Push toy on wheels (manufacture)
10612	15612	Puffed wheat (manufacture)	20301	24303	Putty (manufacture)
10110	15113	Pulled wool production (manufacture)	32409	36509	Puzzles (manufacture)
28150	29140	Pulley (manufacture)	46499	51477	Puzzles (wholesale)
28220	29220	Pulley block (manufacture)	20160	24160	PVC polyvinyl chloride as a raw material (manufacture)
16290	20510	Pulley made of wood (manufacture)
28220	29220	Pulley tackle (manufacture)	13990	17542	Pyjama cord (manufacture)
46690	51870	Pulley tackle and hoists (wholesale)	14141	18231	Pyjamas for men and boys (manufacture)
28150	29140	Pulley wheel (manufacture)	14142	18232	Pyjamas for women and girls (manufacture)
14390	17720	Pullovers (knitted or crocheted) (manufacture)	43999	45250	Pylon erection
17110	21110	Pulp for paper (manufacture)	23610	26610	Pylons made of precast concrete (manufacture)
28950	29550	Pulp making machinery (manufacture)	20140	24140	Pyridine base (manufacture)
17110	21110	Pulping recycled paper (manufacture)	08910	14300	Pyrites and pyrrhotite extraction and preparation
31010	36120	Pulpit (manufacture)	26513	33202	Pyrometer (non-electronic) (manufacture)
02100	02010	Pulpwood production	20510	24610	Pyrotechnics (manufacture)
26513	33202	Pulse (signal) generators (non-electronic) (manufacture)	74909	74149	Quality assurance consultancy activities
			71200	74300	Quality control
46310	51310	Pulses (wholesale)	74902	74203	Quantity surveying activities
28923	29523	Pulverising machinery (not for mines) (manufacture)	23320	26400	Quarry floor brick (manufacture)
28990	29560	Pulverising machinery for the chemical industry (manufacture)	23320	26400	Quarry tile made of clay (manufacture)
			26110	32100	Quartz crystal (manufacture)
08990	14500	Pumice extraction	08990	14500	Quartz mining and quarrying
23910	26810	Pumice stones (bonded) (manufacture)	21100	24410	Quaternary ammonium salts and fatty amines (manufacture)
28131	29121	Pump (not for hydraulic or for internal combustion engine) (manufacture)	46180	51180	Quaternary ammonium salts and hydroxides (commission agent)
28120	29121	Pump for hydraulic equipment (manufacture)
15120	19200	Pump leather (manufacture)	46460	51460	Quaternary ammonium salts and hydroxides (wholesale)
28131	29121	Pump parts (manufacture)
49500	60300	Pump stations operation	84220	75220	Queen Alexandra's RANC
09100	11200	Pumping of oil and gas wells	84220	75220	Queen Alexandra's RNNS
46690	51870	Pumping plant (wholesale)	69101	74112	Queen's counsel
28131	29121	Pumps and parts (non-electric) for oil, gas, petrol or water on motor vehicles (manufacture)	10390	15330	Quick freezing of fruit and vegetables (manufacture)
			46390	51390	Quick frozen foods (wholesale)
46630	51820	Pumps for concrete (wholesale)	23520	26520	Quicklime (manufacture)
46690	51870	Pumps for liquids (excluding motor vehicles) (wholesale)	13923	17403	Quilt (filled) (manufacture)
			13923	17403	Quilt fringing (manufacture)
28131	29121	Pumps for liquids whether or not fitted with measuring devices (manufacture)	13200	17210	Quilt weaving (manufacture)
			01240	01139	Quince growing
93290	92349	Punch and Judy show	20140	24140	Quinones (manufacture)

SIC 2007	SIC 2003	Activity
01490	01250	Rabbit and other fur animal raising
01490	01250	Rabbit breeding
01610	01410	Rabbit destroying and trapping on agricultural land
10110	15120	Rabbit meat preparation (manufacture)
01490	01250	Rabbit skin sorting
10110	15120	Rabbit slaughtering (manufacture)
93110	92619	Racecourse operation
93191	92621	Racehorse owner
93199	92629	Racehorse trainer
92000	92710	Racing pool
93199	92629	Racing stables
92000	92710	Racing tipster
49319	60219	Rack railway
32300	36400	Racket and racket frames (manufacture)
31010	36120	Racking (manufacture)
25110	28110	Racking systems for large scale heavy duty use in shops, workshops and warehouses (manufacture)
93120	92629	Racquet club
46520	51860	Radar apparatus (wholesale)
26511	33201	Radar equipment (manufacture)
61900	64200	Radar station operation
28131	29121	Radial flow pump (manufacture)
23200	26260	Radiant for gas and electric fire (manufacture)
25210	28220	Radiant panel (space heating equipment) (manufacture)
26513	33202	Radiation equipment and detection instruments (non-electronic) (manufacture)
26511	33201	Radiation measuring and detection instruments (electronic) (manufacture)
27510	29710	Radiator (electric) (manufacture)
29320	34300	Radiator for motor vehicle (manufacture)
25210	28220	Radiator for space heating equipment (manufacture)
29320	34300	Radiator grill for motor vehicle (manufacture)
25210	28220	Radiators and boilers for central heating (manufacture)
77291	71403	Radio (domestic) hire
25110	28110	Radio and television masts made of metal (manufacture)
26309	32202	Radio beacons (manufacture)
60100	92201	Radio broadcasting station
26400	32300	Radio cabinets made of wood (manufacture)
15120	19200	Radio cases made of leather (manufacture)
26309	32202	Radio communications equipment (manufacture)
52219	63210	Radio despatch offices for taxis, bicycle couriers etc.
47910	52610	Radio direct sales (retail)
77390	71340	Radio equipment for professional use, rental and operating leasing
26309	32202	Radio frequency booster stations (manufacture)
26511	33201	Radio navigational aid apparatus (manufacture)
46520	51860	Radio navigational aid apparatus (wholesale)
60100	64200	Radio programme transmission
26400	32300	Radio receiving set (manufacture)
61200	64200	Radio relay service
26511	33201	Radio remote control apparatus (manufacture)
46520	51860	Radio remote control apparatus (wholesale)
47430	52450	Radio sets and equipment (retail)
60100	64200	Radio station (telecommunications)
60100	92201	Radio studio
46439	51439	Radio, television and electrical household equipment n.e.c. exporter (wholesale)
46439	51439	Radio, television and electrical household goods n.e.c. importer (wholesale)
20130	23300	Radioactive compounds production (manufacture)
21200	23300	Radioactive in-vivo diagnostic substances (manufacture)
20130	24130	Radioactive isotopes (other than of uranium, thorium or plutonium) (manufacture)
20130	23300	Radioactive isotopes of uranium, thorium and plutonium (manufacture)
46750	51550	Radioactive residues (wholesale)
71200	74300	Radioactivity measuring
86900	85140	Radiographer (private)
71200	74300	Radiographic testing of welds and joints
86101	85111	Radiologist (public sector)
46439	51439	Radios and televisions (wholesale)
26309	32202	Radio-telephony apparatus (manufacture)
86101	85111	Radiotherapist (public sector)
86220	85120	Radiotherapy treatment centre
16290	20520	Raffia goods (manufacture)
16230	20300	Rafters (manufacture)
46770	51570	Rag and bone dealer (wholesale)
13300	17300	Rag bleaching (manufacture)
18130	22250	Rag book making (manufacture)
13300	17300	Rag dyeing (manufacture)
46770	51570	Rag merchant
08110	14110	Ragstone quarry
52211	63210	Rail freight terminals operation
30200	35200	Rail locomotives (manufacture)
49100	60109	Rail transport (inter-urban)
30200	35200	Railbrakes (manufacture)
25120	28120	Railings made of metal (manufacture)
16230	20300	Railings made of wood (manufacture)
77390	71211	Railroad passenger vehicles rental
24100	27100	Rails of iron, steel or cast iron (manufacture)
52290	63400	Railway agent (not transport authority)
30200	35200	Railway and tramway coaches (manufacture)
30200	35200	Railway and tramway locomotives and specialised parts (manufacture)
30200	35200	Railway and tramway rolling stock (manufacture)
30200	36110	Railway car seats (manufacture)
30200	35200	Railway coach (manufacture)
42120	45230	Railway construction
56101	55301	Railway dining car or buffet (licensed)
49100	60101	Railway dining cars operated as an integrated operation of railway companies
77390	71219	Railway freight vehicle hire
30200	35200	Railway locomotives and rolling stock (manufacture)
46690	51870	Railway or tramway coaches, vans and wagons (wholesale)
30200	35200	Railway or tramway not self-propelled passenger coaches, goods vans and other wagons (manufacture)
77390	71211	Railway passenger vehicle hire
52219	63210	Railway running shed
30200	35200	Railway self-propelled car (manufacture)
27900	31620	Railway signalling equipment (electric) (manufacture)
30200	35200	Railway signalling equipment (mechanical) (manufacture)
23610	26610	Railway sleepers made of pre-cast concrete (manufacture)
16100	20100	Railway sleepers made of wood (manufacture)

R

SIC 2007	SIC 2003	Activity
49100	60101	Railway sleeping cars operated as an integrated operation of railway companies
55900	55239	Railway sleeping cars when operated by separate units from the transport provider
52219	63210	Railway station operation
30200	35200	Railway test wagon (manufacture)
30200	35200	Railway track equipment (mechanical) (manufacture)
25990	28750	Railway track fixtures of assembled metal (manufacture)
42120	45213	Railway tunnelling contractor
42120	45213	Railway tunnel construction
30200	35200	Railway wagon (manufacture)
52290	63400	Railway wagon agent
30200	35200	Railway wagon axle box and axle lubricator (manufacture)
20510	24610	Rain rocket (manufacture)
37000	90010	Rain water collection and transportation by sewers, collectors, tanks and other means of transport
14131	18221	Raincoats for men and boys (manufacture)
14132	18222	Raincoats for women and girls (manufacture)
14132	18222	Raincoats for women and girls made of stitched plastic (manufacture)
14131	18221	Rainproof garments for men and boys (manufacture)
14132	18222	Rainproof garments for women and girls (manufacture)
28940	29540	Raising machinery for textiles (manufacture)
25730	28620	Rakes for garden use (manufacture)
16290	20510	Rakes made of wood (manufacture)
01160	01110	Ramie and other vegetable textile fibre growing
13200	17250	Ramie weaving (manufacture)
23200	26260	Ramming material made of refractory (manufacture)
86101	85111	Rampton Hospital
26701	33201	Range finder (optical) (manufacture)
26513	33202	Range finders (non-electronic) (manufacture)
26513	33202	Range finders (optical non-electronic) (manufacture)
26701	33201	Range finders (optical) (manufacture)
28921	29521	Ranging drum shearer for mining (manufacture)
01110	01110	Rape growing
10410	15410	Rape oil production (manufacture)
10410	15420	Rape oil refining (manufacture)
10410	15410	Rape seed crushing (manufacture)
20110	24110	Rare gases (manufacture)
13990	17541	Raschel lace (manufacture)
25730	28620	Rasp (manufacture)
01250	01139	Raspberry growing
81291	74703	Rat catcher (not especially for agriculture)
01610	01410	Rat destroying and trapping on agricultural land
10730	15850	Ravioli (manufacture)
10110	15112	Raw bones from knackers (manufacture)
13300	17300	Raw cotton bleaching, dyeing or otherwise finishing (manufacture)
46120	51120	Raw earth metals (commission agent)
01490	01250	Raw fur skin production
13300	17300	Raw silk dyeing (manufacture)
46210	51210	Raw wool (wholesale)
20600	24700	Rayon (manufacture)
27510	29710	Razor (electric) (manufacture)
25710	28610	Razor (not electric) (manufacture)
25710	28610	Razor blade (manufacture)
25710	28610	Razor blade blanks in strips (manufacture)

SIC 2007	SIC 2003	Activity
25710	28610	Razor set (manufacture)
28940	29540	Reaching-in machinery for textiles (manufacture)
25300	28300	Reactor column (manufacture)
27110	31100	Reactor shunt and limiting (manufacture)
25300	28300	Reactor vessel (manufacture)
32500	33401	Reading glasses (manufacture)
91011	92510	Reading room (library)
62011	72220	Ready-made interactive leisure and entertainment software development
58290	72210	Ready-made software publishing (except computer games publishing)
23630	26630	Ready-mixed concrete (manufacture)
23640	26640	Ready-mixed wet mortars (manufacture)
68310	70310	Real estate agencies
68100	70120	Real estate buying and selling
68310	70310	Real estate escrow agents activities
64306	65231	Real estate investment trusts
68320	70320	Real estate management on a fee or contract basis
68100	70120	Real estate owner
74209	74819	Real estate photography
41100	70110	Real estate project development
82990	74879	Real-time closed captioning
25730	28620	Reamer (manufacture)
13940	17520	Reaper twine (manufacture)
28922	29522	Rear digger (manufacture)
28922	29522	Rear digger unit (manufacture)
16100	20100	Rebated wood (manufacture)
26110	32100	Receiver or amplifier valves or tubes (manufacture)
26400	32300	Receivers for radio broadcasting (manufacture)
26400	32300	Receivers for television (manufacture)
77400	74879	Receiving royalties or licensing fees for the use of brand names
77400	74879	Receiving royalties or licensing fees for the use of franchise agreements
77400	74879	Receiving royalties or licensing fees for the use of mineral exploration and evaluation
77400	74879	Receiving royalties or licensing fees for the use of patented entities
77400	74879	Receiving royalties or licensing fees for the use of trademarks or service marks
46520	51860	Reception apparatus for radio telephony or telegraphy for professional use (wholesale)
26309	32202	Reception apparatus for radio-telephony or radio-telegraphy (manufacture)
28132	29122	Reciprocating compressor (manufacture)
46690	51870	Reciprocating displacement compressors (wholesale)
46690	51870	Reciprocating positive displacement pumps for liquids (wholesale)
38320	37100	Reclaiming metals out of photographic waste, e.g. Fixer solution or photographic films and paper
38320	37200	Reclaiming of chemicals from chemical waste
38320	37200	Reclaiming of rubber to produce secondary raw materials
26110	32300	Record cutters (manufacture)
91011	92510	Record lending and storage
26400	32300	Record player accessories (manufacture)
26400	32300	Record player cabinets made of wood (manufacture)
26400	32300	Record players (manufacture)
47430	52450	Record players (retail)
46431	51431	Record players (wholesale)
26400	32300	Record playing mechanism (manufacture)

R

SIC 2007	SIC 2003	Activity	SIC 2007	SIC 2003	Activity
46431	51431	Recorded audio and video tapes, CDs, DVDs and the equipment on which these are played (wholesale)	32500	33100	Reflectors used in medicine (manufacture)
32200	36300	Recorders made of plastic or wood (manufacture)	25300	28300	Reformer (manufacture)
59200	92201	Recording studio (radio)	26511	33201	Refractometers (electronic) (manufacture)
60200	92202	Recording studio (television)	26513	33202	Refractometers (non-electronic) (manufacture)
77220	71404	Records rental	23200	26260	Refractory articles containing magnesite, dolomite or chromite (manufacture)
47630	52450	Records, compact discs and tapes (retail)	23200	26260	Refractory blocks (manufacture)
13100	17120	Recovered wool (mungo and shoddy) (manufacture)	23200	26260	Refractory castable (manufacture)
30400	29600	Recovery vehicle (tracked military type) (manufacture)	23200	26260	Refractory cement (manufacture)
93290	92729	Recreational activities n.e.c.	23200	26260	Refractory ceramic goods (manufacture)
77210	71401	Recreational and sports goods renting and leasing	08120	14220	Refractory clays mining
14131	18221	Recreational clothing for men and boys (waterproofed) (manufacture)	23200	26260	Refractory concretes (manufacture)
14132	18222	Recreational clothing for women and girls (waterproof) (manufacture)	23200	26260	Refractory goods (manufacture)
			23200	26260	Refractory hollow ware (manufacture)
94990	91330	Recreational organisations	23200	26260	Refractory jointing cement (manufacture)
84120	75120	Recreational services administration (public sector)	23200	26260	Refractory mouldable (manufacture)
55300	55220	Recreational vehicle parks and trailer parks	23200	26260	Refractory ramming material (manufacture)
77120	71219	Recreational vehicles renting and leasing	23200	26260	Refractory tiles (manufacture)
14131	18221	Recreational weatherproof clothing for men and boys (manufacture)	56290	55510	Refreshment club
			56290	55520	Refreshment contracting
14132	18222	Recreational weatherproof clothing for women and girls (manufacture)	56102	55302	Refreshment room (unlicensed)
24200	27220	Rectangular hollow steel section (manufacture)	20110	24110	Refrigerant gases (manufacture)
11010	15910	Rectified spirits (manufacture)	49410	60249	Refrigerated haulage by road
26110	31100	Rectifier plant (electronic) (manufacture)	29100	34100	Refrigerated lorry (manufacture)
27900	32100	Rectifier, solid state (manufacture)	28250	29230	Refrigerated service cabinet (manufacture)
28290	29240	Rectifying plant (manufacture)	30200	35200	Refrigerated wagon for railway (manufacture)
27900	32100	Rectifying valve and tube (manufacture)	52102	63121	Refrigerated warehouses operation for air transport activities
17110	21110	Recycled fibre pulp (manufacture)	52103	63121	Refrigerated warehouses operation for land transport activities
88990	85321	Red Cross Society			
01250	01139	Redcurrant growing	52101	63121	Refrigerated warehouses operation for water transport activities
09100	11200	Redrilling oil and gas wells	46690	51870	Refrigerating and freezing equipment and heat pumps (commercial) (wholesale)
32409	36509	Reduced-size (scale) models (manufacture)			
32409	36509	Reduced-size scale model electrical trains (manufacture)	28250	29230	Refrigerating and freezing equipment parts (manufacture)
32409	36509	Reduced-size scale models construction sets (manufacture)	28250	29230	Refrigerating equipment (manufacture)
			28250	29230	Refrigerating or freezing equipment for industrial use (manufacture)
26702	33403	Reducer (photographic) (manufacture)	28250	29230	Refrigerator cabinets made of wood (manufacture)
96040	93040	Reducing and slimming salon activities	28250	29230	Refrigerator for commercial use (manufacture)
28140	29130	Reducing valves (manufacture)	27510	29710	Refrigerator for domestic use (electric) (manufacture)
28990	29560	Reduction gear for marine use (manufacture)	27520	29720	Refrigerators (gas) (manufacture)
28940	29540	Reed (textile machinery accessory) (manufacture)	46439	51439	Refrigerators and freezers for domestic use (wholesale)
16290	20520	Reed articles (manufacture)			
02300	02010	Reed collecting, cutting and gathering	88990	75210	Refugee and hunger relief programmes abroad
32200	36300	Reed for musical instrument (manufacture)	88990	85321	Refugee camp (charitable)
16290	20520	Reed preparation (manufacture)	88990	85322	Refugee camp (non-charitable)
13100	17150	Reeling and washing of silk (manufacture)	88990	85321	Refugee services (charitable)
28940	29540	Reeling machinery for textiles (manufacture)	88990	85322	Refugee services (non-charitable)
22220	25220	Reels made of printed polypropylene (manufacture)	38110	90020	Refuse collection by local authority cleansing department
16290	20510	Reels made of wood (manufacture)			
69109	74119	Referees legal activities	28290	29240	Refuse disposal plant (manufacture)
20140	24140	Refined coal tar (manufacture)	38210	90020	Refuse disposal plant or tip (local authority or municipally owned)
46711	51511	Refined petroleum products (wholesale)			
46120	51120	Refined sulphur (commission agent)	38210	90020	Refuse disposal service (not especially for agriculture)
46750	51550	Refined sulphur (wholesale)	38210	90020	Refuse disposal tip operator
19201	23201	Refinery gas (manufacture)	29100	34100	Refuse disposal vehicle (manufacture)
30120	35120	Refitting of pleasure craft (manufacture)	93290	92729	Regent's Park and Primrose Hill
30110	35110	Refitting of ships (manufacture)	84240	75240	Regional crime squad
			18129	22220	Register printing (manufacture)

SIC 2007	SIC 2003	Activity	SIC 2007	SIC 2003	Activity
75000	85200	Registered veterinarian	95290	52740	Repair and alteration of clothing
84230	75230	Registrars Office (Courts of Justice)	09100	11200	Repair and dismantling services of derricks
29320	34300	Registration plate for motor vehicle (manufacture)	33110	28520	Repair and maintenance for pipes and pipelines (manufacture)
51102	62201	Regular charter flights for passengers	42220	45213	Repair and maintenance of above-ground telecommunication lines
86101	85112	Rehabilitation hospital (private sector)			
86101	85111	Rehabilitation hospital (public sector)	33120	72500	Repair and maintenance of accounting machinery (manufacture)
84230	75230	Rehabilitation services for prisoners			
28210	29210	Re-heating furnace (manufacture)	33160	35300	Repair and maintenance of aero-engine parts and sub assemblies (manufacture)
43999	45250	Reinforced concrete engineer (civil engineering)			
23610	26610	Reinforced concrete products (manufacture)	33160	35300	Repair and maintenance of aero-space equipment (manufacture)
22190	25130	Reinforced hose made of rubber (manufacture)	33120	29320	Repair and maintenance of agricultural machinery (manufacture)
65202	66032	Re-insurance company (non-life)			
26309	32202	Relay link apparatus (manufacture)	33120	29310	Repair and maintenance of agricultural tractors (manufacture)
26309	32202	Relay transmitters (manufacture)			
26110	31200	Relays for electronic and telecommunications use (manufacture)	33160	35300	Repair and maintenance of air cushion vehicles (manufacture)
18140	22230	Relief stamping (manufacture)	33160	35300	Repair and maintenance of aircraft (manufacture)
28140	29130	Relief valves (manufacture)	33130	33201	Repair and maintenance of aircraft engine instruments (manufacture)
94910	91310	Religious funeral service activities			
47789	52489	Religious goods (retail)	33160	35300	Repair and maintenance of aircraft engines of all types (manufacture)
85590	80429	Religious instruction			
94910	91310	Religious organisations	33170	35500	Repair and maintenance of animal-drawn buggies and wagons (manufacture)
94910	91310	Religious retreat activities			
84120	75120	Religious services administration (public sector)	45200	50200	Repair and maintenance of auto electricals
58110	22110	Religious tract publishing	95110	30020	Repair and maintenance of automatic teller machines (ATMs)
10840	15870	Relish (manufacture)			
84230	75230	Remand centres	33130	33201	Repair and maintenance of automotive emissions testing equipment (manufacture)
45200	50200	REME workshop			
39000	90030	Remediation activities and other waste management services	33110	28300	Repair and maintenance of auxiliary plant for steam collectors and accumulators (manufacture)
24420	27420	Remelt ingots made of aluminium (manufacture)	33110	28300	Repair and maintenance of auxiliary plant for use with steam condensers (manufacture)
24100	27100	Remelting ferrous waste or scrap			
24100	27100	Remelting scrap ingots of iron	33110	28300	Repair and maintenance of auxiliary plant for use with steam economisers (manufacture)
41202	45212	Remodelling or renovating existing residential structures	33110	28300	Repair and maintenance of auxiliary plant for use with steam generators (manufacture)
49420	60241	Removal by road transport			
49420	60241	Removal contractor	33110	28300	Repair and maintenance of auxiliary plant for use with steam superheaters (manufacture)
17110	21110	Removal of ink from waste paper and subsequent manufacture of pulp (manufacture)			
			95110	30020	Repair and maintenance of bar code scanners
10110	15112	Rendering of lard and other edible fats of animal origin (manufacture)	33150	35120	Repair and maintenance of boats (manufacture)
			95120	32201	Repair and maintenance of carrier equipment modems
10519	15519	Rennet (not artificial) (manufacture)			
95290	52740	Renovating of hats	33120	72500	Repair and maintenance of cash registers (manufacture)
68320	70320	Rent collecting agencies			
81299	90030	Rental of lavatory cubicles	95120	52740	Repair and maintenance of cellular telephones
50300	61201	Rental of pleasure boats with crew for inland water transport	33190	26400	Repair and maintenance of ceramic pipes, etc. And systems thereof in industrial plants (manufacture)
50200	61102	Rental of vessels with crew for coastal freight water transport	33130	33500	Repair and maintenance of clocks in church towers and the like
50200	61102	Rental of vessels with crew for sea freight water transport	95120	32202	Repair and maintenance of commercial TV cameras
			45200	50200	Repair and maintenance of commercial vehicles
68201	70209	Renting and operating of housing association real estate	95120	32202	Repair and maintenance of commercial video cameras
68209	70209	Renting and operating of self-owned or leased real estate	33120	29240	Repair and maintenance of commercial-type general purpose machinery (manufacture)
43999	45500	Renting of cranes with operator	95120	32201	Repair and maintenance of communications transmission bridges
93290	92729	Renting of leisure and pleasure equipment as an integral part of recreational facilities	95120	32201	Repair and maintenance of communications transmission modems
43999	45500	Renting of other building equipment with operator	95120	32201	Repair and maintenance of communications transmission routers
43991	45250	Renting of scaffolds and work platforms with erection and dismantling	33120	29122	Repair and maintenance of compressors (manufacture)

SIC 2007	SIC 2003	Activity	SIC 2007	SIC 2003	Activity
95110	30020	Repair and maintenance of computer monitors	33120	29230	Repair and maintenance of industrial refrigeration equipment, air purifying equipment (manufacture)
95110	30020	Repair and maintenance of computer projectors	33130	33500	Repair and maintenance of industrial time measuring instruments and apparatus (manufacture)
95110	30020	Repair and maintenance of computer servers			
95110	72500	Repair and maintenance of computing machinery	33120	29230	Repair and maintenance of industrial type air conditioning (manufacture)
33120	29522	Repair and maintenance of construction machinery (earth moving type) (manufacture)	95110	30020	Repair and maintenance of internal and external computer modems
33120	29523	Repair and maintenance of construction machinery (except earth moving type) (manufacture)	33170	35430	Repair and maintenance of invalid carriages (manufacture)
33110	34202	Repair and maintenance of containers for freight (manufacture)	33130	33100	Repair and maintenance of irradiation apparatus (manufacture)
95110	30020	Repair and maintenance of dedicated computer terminals	95110	30020	Repair and maintenance of keyboards
95110	30020	Repair and maintenance of desktop computers	95110	30020	Repair and maintenance of laptop computers
43220	45330	Repair and maintenance of domestic air conditioning	33120	29220	Repair and maintenance of lifting and handling equipment (manufacture)
43220	45330	Repair and maintenance of domestic boilers	33120	29220	Repair and maintenance of lifting, handling equipment, elevators, moving walkways etc. (manufacture)
95220	52740	Repair and maintenance of domestic cookers			
95220	52740	Repair and maintenance of domestic ovens	33120	29220	Repair and maintenance of lifts and escalators (not in buildings or civil engineering) (manufacture)
95220	52740	Repair and maintenance of domestic ranges			
33120	29522	Repair and maintenance of earth-moving and excavating equipment (manufacture)	33120	29560	Repair and maintenance of machinery for bookbinding (manufacture)
33140	31500	Repair and maintenance of electric lighting equipment (manufacture)	33120	29530	Repair and maintenance of machinery for food, beverage and tobacco processing (manufacture)
33140	31620	Repair and maintenance of electrical signalling equipment (manufacture)	33120	29510	Repair and maintenance of machinery for metallurgy (manufacture)
33140	31200	Repair and maintenance of electricity distribution and control apparatus (manufacture)	33120	29521	Repair and maintenance of machinery for mining (manufacture)
35130	33202	Repair and maintenance of electricity meters	33120	29550	Repair and maintenance of machinery for paper and paperboard production (manufacture)
33130	33100	Repair and maintenance of electrocardiographs (manufacture)	33120	29560	Repair and maintenance of machinery for printing (manufacture)
33130	33100	Repair and maintenance of electromedical endoscopic equipment (manufacture)	33120	29540	Repair and maintenance of machinery for textile, apparel and leather production (manufacture)
33130	33201	Repair and maintenance of electronic equipment for measuring, checking, testing, etc. (manufacture)	33120	29560	Repair and maintenance of machinery for working rubber or plastics (manufacture)
43290	29220	Repair and maintenance of elevators and escalators	95110	30020	Repair and maintenance of magnetic disk drives
33120	29110	Repair and maintenance of engines and turbines (except aircraft, vehicle and cycle) (manufacture)	95110	30020	Repair and maintenance of magnetic flash drives and other storage devices
33120	29523	Repair and maintenance of equipment for concrete crushing and screening and roadworks (manufacture)	33130	33100	Repair and maintenance of magnetic resonance imaging equipment (manufacture)
95120	32201	Repair and maintenance of fax machines	33120	29110	Repair and maintenance of marine engines (manufacture)
33110	29600	Repair and maintenance of firearms and ordnance (manufacture)	33130	33201	Repair and maintenance of materials' properties testing and inspection equipment (manufacture)
33120	29122	Repair and maintenance of fluid power machinery (compressors) (manufacture)	33130	33100	Repair and maintenance of medical and surgical equipment and apparatus (manufacture)
33120	29121	Repair and maintenance of fluid power machinery (pumps) (manufacture)	33140	33100	Repair and maintenance of medical equipment, electrical (manufacture)
33120	29320	Repair and maintenance of forestry and logging machinery (manufacture)	33130	33100	Repair and maintenance of medical ultrasound equipment (manufacture)
33120	29210	Repair and maintenance of furnaces and furnace burners (manufacture)	33120	29420	Repair and maintenance of metal cutting or metal forming machine tools and accessories (manufacture)
35220	33202	Repair and maintenance of gas meters			
35220	33202	Repair and maintenance of gas meters (non-electronic) (manufacture)	33120	29420	Repair and maintenance of metal working machine tools (manufacture)
33120	29110	Repair and maintenance of gas turbines (manufacture)	33130	33201	Repair and maintenance of meteorological instruments (manufacture)
33120	29240	Repair and maintenance of general purpose machinery n.e.c. (manufacture)	95110	30020	Repair and maintenance of mice, joysticks, and trackball accessories
33190	26150	Repair and maintenance of glass tubes, etc. And systems thereof in industrial plants (manufacture)	33190	26810	Repair and maintenance of millstones, grindstones, polishing stones and the like (manufacture)
95110	30020	Repair and maintenance of hand-held computers (PDA's)	45400	50400	Repair and maintenance of motor cycles
33130	33100	Repair and maintenance of hearing aids (manufacture)	33140	31100	Repair and maintenance of motor generator sets (manufacture)
33160	35300	Repair and maintenance of helicopters (manufacture)			
33120	29210	Repair and maintenance of industrial process furnaces (manufacture)			

R

SIC 2007	SIC 2003	Activity	SIC 2007	SIC 2003	Activity
33110	28220	Repair and maintenance of non-domestic central heating boilers (manufacture)	33140	31200	Repair and maintenance of relays and industrial controls (manufacture)
33120	29230	Repair and maintenance of non-domestic cooling and ventilating equipment (manufacture)	33120	72500	Repair and maintenance of reprographic machinery
33120	29560	Repair and maintenance of non-domestic machinery for drying wood, paper pulp, etc. (manufacture)	33140	31620	Repair and maintenance of road and other non-domestic exterior lighting equipment (manufacture)
33140	33202	Repair and maintenance of non-electronic measuring, checking, testing etc. equipment (manufacture)	52219	63210	Repair and maintenance of rolling stock (minor)
33110	28300	Repair and maintenance of nuclear reactors (manufacture)	95110	30020	Repair and maintenance of scanners
33120	72500	Repair and maintenance of office machinery (other than computers)	33110	35500	Repair and maintenance of shopping carts (manufacture)
43220	45330	Repair and maintenance of office, shop and computer centre air conditioning	95110	30020	Repair and maintenance of smart card readers
33120	29521	Repair and maintenance of oil and gas extraction machinery (manufacture)	33160	35300	Repair and maintenance of spacecraft (manufacture)
95110	30020	Repair and maintenance of optical disk drives (CD-RW, CD-ROM, DVD-ROM, DVD-RW)	33120	29560	Repair and maintenance of special purpose machinery n.e.c. (manufacture)
33130	33402	Repair and maintenance of optical precision instruments (manufacture)	33110	29600	Repair and maintenance of sporting and recreational guns (manufacture)
33120	29430	Repair and maintenance of other machine tools (except metal working) (manufacture)	33110	28300	Repair and maintenance of steam generators (manufacture)
33120	29420	Repair and maintenance of other machine tools (metal working) (manufacture)	33110	28300	Repair and maintenance of steam or other vapour generators (manufacture)
33190	25130	Repair and maintenance of other rubber products (excluding tyres) (manufacture)	33120	29110	Repair and maintenance of steam turbines (manufacture)
33130	33100	Repair and maintenance of pacemakers (manufacture)	33130	33201	Repair and maintenance of surveying instruments (manufacture)
33120	29550	Repair and maintenance of papermaking machinery (manufacture)	33140	31200	Repair and maintenance of switchgear and switchboard apparatus (manufacture)
33110	28300	Repair and maintenance of parts for marine or power boilers (manufacture)	33110	28210	Repair and maintenance of tanks, reservoirs and containers of metal (manufacture)
33120	72500	Repair and maintenance of photocopy machines (manufacture)	33120	29130	Repair and maintenance of taps (manufacture)
33120	29560	Repair and maintenance of plastic and rubber working machinery (manufacture)	95120	32300	Repair and maintenance of telephone answering machines
33190	25210	Repair and maintenance of plastic tubes etc. And systems thereof in industrial plants (manufacture)	95120	32201	Repair and maintenance of telephone sets
33150	35120	Repair and maintenance of pleasure and sporting craft (manufacture)	95120	32201	Repair and maintenance of telex machines and other line telephony or telegraphy equipment
95110	30020	Repair and maintenance of point-of-sale (pos) terminals	45200	50200	Repair and maintenance of trailers and semi-trailers
33140	31400	Repair and maintenance of primary and storage batteries (manufacture)	33170	35200	Repair and maintenance of tramway rolling stock (manufacture)
95110	30020	Repair and maintenance of printers	33170	35200	Repair and maintenance of transmissions and other parts for locomotives (manufacture)
33130	29710	Repair and maintenance of professional electric appliances (manufacture)	95120	32202	Repair and maintenance of two-way radios
33130	33403	Repair and maintenance of professional photographic and cinematographic equipment (manufacture)	42220	45213	Repair and maintenance of underground communication lines
95120	32300	Repair and maintenance of professional radio, television, sound and video equipment	33120	29130	Repair and maintenance of valves (manufacture)
33120	29121	Repair and maintenance of pumps (manufacture)	33120	29240	Repair and maintenance of vending machines (manufacture)
33130	33201	Repair and maintenance of radiation detection and monitoring instruments (manufacture)	95110	30020	Repair and maintenance of virtual reality helmets
95120	32202	Repair and maintenance of radio and television transmitters	33110	29600	Repair and maintenance of weapons and weapon systems (manufacture)
95120	32201	Repair and maintenance of radio telephony apparatus	33120	29240	Repair and maintenance of weighing equipment (manufacture)
33170	35200	Repair and maintenance of railway cars (manufacture)	33140	31200	Repair and maintenance of wiring devices for wiring electrical circuits (manufacture)
33120	29110	Repair and maintenance of railway diesel engines (manufacture)	33190	20510	Repair and maintenance of wooden products n.e.c. (manufacture)
33170	35200	Repair and maintenance of railway locomotives (manufacture)	33150	35110	Repair and maintenance or alteration of ships (manufacture)
33170	35200	Repair and maintenance of railway rolling stock (major) (manufacture)	33190	36300	Repair and rebuilding of organs (in factory) (manufacture)
			95290	36300	Repair and reconditioning of musical instruments (except organs and historical musical instruments)
			95240	36110	Repair and restoration of chairs and seats
			95240	36140	Repair and restoration of furniture n.e.c.
			95240	36150	Repair and restoration of mattresses
			95240	36130	Repair and restoration of other kitchen furniture

R

SIC 2007	SIC 2003	Activity	SIC 2007	SIC 2003	Activity
95240	36120	Repair and restoration of other office and shop furniture	95220	29320	Repair of lawnmowers
90030	92319	Repair and restoration of works of art	33110	28630	Repair of locks and hinges (manufacture)
33140	31100	Repair and rewiring of armatures (manufacture)	95230	52710	Repair of luggage
33150	35120	Repair and routine maintenance performed on boats by floating dry docks (manufacture)	33110	28110	Repair of metal structures (manufacture)
			33130	33402	Repair of microscopes (manufacture)
33150	35110	Repair and routine maintenance performed on ships by floating dry docks (manufacture)	33130	31620	Repair of mine detectors (manufacture)
			95120	52740	Repair of mobile telephones
45200	50200	Repair and servicing in garages, of motor vehicles	45200	50200	Repair of motor cars (except roadside assistance)
24200	27220	Repair clamps and collars, made of iron or steel (manufacture)	45200	50200	Repair of motor vehicle parts
			45200	50200	Repair of motor vehicle seats
22190	25130	Repair materials made of rubber (manufacture)	45200	50200	Repair of motor vehicle windows
43290	28120	Repair of automated and revolving doors in buildings and civil engineering works (manufacture)	45200	50200	Repair of motor vehicle windscreens
			45200	50200	Repair of motor vehicles (except roadside assistance)
95290	52740	Repair of bicycles	13940	17520	Repair of nets and ropework (manufacture)
33130	33402	Repair of binoculars (manufacture)	33130	33302	Repair of non-electronic industrial process control equipment (manufacture)
95290	52740	Repair of books			
95230	52710	Repair of boots	95290	52740	Repair of non-professional photographic equipment
33120	72500	Repair of calculators (manufacture)			
33190	17402	Repair of camping goods made of canvas (manufacture)	33120	29410	Repair of other power-driven hand-tools (manufacture)
45200	50200	Repair of car bodies	33130	33403	Repair of photographic equipment (manufacture)
45200	50200	Repair of car electrical systems	33190	36300	Repair of pianos (in factory) (manufacture)
45200	50200	Repair of car electronics	33190	36501	Repair of pinball machines and other coin-operated games (manufacture)
33170	34203	Repair of caravans			
95210	52720	Repair of CD players	33190	25240	Repair of plexiglas plane windows (manufacture)
95250	52730	Repair of clock cases and housings of all materials	33130	33401	Repair of prisms and lenses (manufacture)
95250	52730	Repair of clock movements and chronometers	95210	52720	Repair of radios
95250	52730	Repair of clocks	95220	52720	Repair of refrigerators
95220	52720	Repair of clothes dryers	33190	17520	Repair of rigging (manufacture)
95110	30020	Repair of computers and peripheral equipment	95220	52720	Repair of room air conditioners
95210	52720	Repair of consumer electronics:	33190	17520	Repair of ropes (manufacture)
95120	32201	Repair of cordless telephones	33190	17402	Repair of sails (manufacture)
33110	28610	Repair of cutlery (manufacture)	42210	45213	Repair of sewer systems
95290	52740	Repair of cycles	95230	52710	Repair of shoes
95210	52720	Repair of domestic audio and video equipment	95220	52720	Repair of snow and leaf blowers
95220	52720	Repair of domestic electrical appliances	95290	52740	Repair of sporting and camping equipment (except tents)
95290	52740	Repair of domestic lighting articles			
95210	52720	Repair of dvd players	33110	28710	Repair of steel shipping drums (manufacture)
33130	33301	Repair of electronic industrial process control equipment (manufacture)	95220	52720	Repair of stoves
			33190	17402	Repair of tarpaulins (manufacture)
33130	33402	Repair of electronic optical equipment (manufacture)	33130	33402	Repair of telescopes (manufacture)
33130	31100	Repair of electronic transformers (solid state), coils, chokes, and other inductors (manufacture)	95210	52720	Repair of televisions
			95290	17402	Repair of tents
33130	32100	Repair of electronic valves and tubes and other electronic components (manufacture)	33110	28620	Repair of tools (manufacture)
			95290	52740	Repair of toys
33190	17520	Repair of fishing nets (manufacture)	33120	72500	Repair of typewriters (manufacture)
95230	52710	Repair of footwear and leather goods	22110	25120	Repair of tyres and inner tubes by specialists (manufacture)
45200	50200	Repair of fuel injection systems for motor vehicles			
95220	52720	Repair of garden edger	45200	50200	Repair of tyres and tubes (fitting or replacement)
95220	52720	Repair of garden trimmers	95290	52740	Repair of umbrellas
33120	29140	Repair of gearing and driving elements (manufacture)	33120	29130	Repair of valves for machinery (manufacture)
			95210	52720	Repair of video cassette recorders (VCR)
95230	52710	Repair of handbags	95220	52720	Repair of washing machines
33130	32300	Repair of heads (pickup, recording, read/write, etc.), phonograph needles (manufacturing)	95250	52730	Repair of watch cases and housings of all materials
			95250	52730	Repair of watches and clocks
95290	52740	Repair of household textile articles	33190	20400	Repair or reconditioning of wooden pallets, shipping drums or barrels (manufacture)
95210	52720	Repair of household-type video cameras			
95250	52730	Repair of jewellery	45200	50200	Repair to bodywork of motor vehicles
33130	29240	Repair of laboratory distilling apparatus, centrifuges, ultrasonic cleaning machinery (manufacture)	33140	31100	Repair, maintenance and rewinding of electric motors, generators and transformers (manufacture)

R

SIC 2007	SIC 2003	Activity	SIC 2007	SIC 2003	Activity
90010	92311	Repertory company	87200	85312	Residential care (social) in mental disability facilities (non-charitable)
52101	63129	Repository for water transport activities	87200	85311	Residential care (social) in psychiatric convalescent homes (charitable)
82990	74879	Repossession services			
18130	22240	Reproduction and composing (manufacture)	87200	85312	Residential care (social) in psychiatric convalescent homes (non-charitable)
58190	22150	Reproduction of works of art publishing			
82190	74850	Reprographic activities (other than printing)	87200	85140	Residential care activities (paramedical) for mental health
18130	22250	Reprographic dummies production (manufacture)	87200	85140	Residential care activities (paramedical) for mental disability
18130	22250	Reprographic lay-outs production (manufacture)			
18130	22250	Reprographic products production (manufacture)	87200	85140	Residential care activities (paramedical) for substance abuse
18130	22250	Reprographic sketch production (manufacture)			
15110	19100	Reptile leather (manufacture)	87200	85311	Residential care activities (social) for mental health (charitable)
01700	01500	Reptile skin production (from hunting)			
01490	01250	Reptile skin production from ranching operation	87200	85312	Residential care activities (social) for mental health (non-charitable)
72190	73100	Research and development consultants (other than biotechnological)	87200	85311	Residential care activities (social) for learning difficulties (charitable)
72110	73100	Research and experimental development on bioinformatics	87200	85312	Residential care activities (social) for learning difficulties (non-charitable)
72110	73100	Research and experimental development on cell and tissue culture and engineering	87200	85311	Residential care activities (social) for substance abuse (charitable)
72110	73100	Research and experimental development on DNA/RNA	87200	85312	Residential care activities (social) for substance abuse (non-charitable)
72110	73100	Research and experimental development on gene and RNA vectors	87300	85311	Residential care activities for the elderly and disabled (charitable)
72110	73100	Research and experimental development on nanobiotechnology	87300	85312	Residential care activities for the elderly and disabled (non-charitable)
72110	73100	Research and experimental development on process biotechnology techniques	87300	85311	Residential care home for epileptics (charitable)
			87300	85312	Residential care home for epileptics (non-charitable)
72110	73100	Research and experimental development on proteins and other molecules	87300	85311	Residential care home for disabled children (charitable)
72190	73100	Research association (other than biotechnological)	87300	85312	Residential care home for disabled children (non-charitable)
72190	73100	Research chemist (private practice)			
72190	73100	Research institution (other than biotechnological)	87200	85311	Residential care home for the mentally disabled (charitable)
72190	73100	Research laboratory (other than biotechnological)	87200	85312	Residential care home for the mentally disabled (non-charitable)
30110	35110	Research vessel (manufacture)			
28120	29122	Reservoir (hydraulic) (manufacture)	87200	85311	Residential care home for the mentally ill (charitable)
28120	29122	Reservoir (pneumatic) (manufacture)	87200	85312	Residential care home for the mentally ill (non-charitable)
81222	74704	Reservoir and tank cleaning			
42210	45240	Reservoir construction	87200	85112	Residential care in alcoholism or drug addiction treatment facilities (private sector)
23650	26650	Reservoirs made of fibre-cement (manufacture)			
25290	28210	Reservoirs made of metal exceeding 300 litres (manufacture)	87200	85111	Residential care in alcoholism or drug addiction treatment facilities (public sector)
22230	25239	Reservoirs made of plastic (manufacture)	87200	85112	Residential care in rehabilitation centres (private sector)
46690	51870	Reservoirs, tanks and containers of metal (not for central heating) (300 litres or more) (wholesale)	87200	85111	Residential care in rehabilitation health centres (public sector)
46740	51540	Reservoirs, tanks and containers of metal for central heating (300 litres or more) (wholesale)	68209	70209	Residential chambers letting
			68209	70209	Residential mobile home sites operation
87200	85140	Residential care (paramedical) in group homes for the emotionally disturbed (charitable)	87900	85311	Residential nurseries (charitable)
			87900	85312	Residential nurseries (non-charitable)
87200	85140	Residential care (paramedical) in mental health halfway houses	87100	85140	Residential nursing care facilities
87200	85140	Residential care (paramedical) in mental disability facilities	87100	85113	Residential nursing care facilities (not directly supervised by medical doctors)
87200	85140	Residential care (paramedical) in psychiatric convalescent homes	81100	70320	Residents' property management
			22290	25240	Resin bonded glass fibre mouldings (excluding for motor vehicles) (manufacture)
87200	85311	Residential care (social) in group homes for the emotionally disturbed (charitable)	02300	02010	Resin gathering
87200	85312	Residential care (social) in group homes for the emotionally disturbed (non charitable)	20530	24630	Resinoids (manufacture)
			20160	24160	Resins for paint (manufacture)
87200	85311	Residential care (social) in mental health halfway houses (charitable)	20160	24160	Resins made of urea formaldehyde (manufacture)
87200	85312	Residential care (social) in mental health halfway houses (non-charitable)	26511	33201	Resistance checking instruments (electronic) (manufacture)
87200	85311	Residential care (social) in mental disability facilities (charitable)			

R

SIC 2007	SIC 2003	Activity
26110	32100	Resistor (electronic) (manufacture)
27900	32100	Resistors including rheostats and potentiometers (manufacture)
20520	24620	Resorcinol formaldehyde adhesive (manufacture)
32500	33100	Respirator and mask for medical use (manufacture)
87100	85140	Rest homes with nursing care
87300	85311	Rest homes without nursing care (charitable)
87300	85312	Rest homes without nursing care (non-charitable)
56101	55301	Restaurant (licensed)
56102	55302	Restaurant (unlicensed)
74203	74814	Restoration, copying and retouching of photographs
33190	36300	Restoring of organs and other historical musical instruments (manufacture)
84230	75230	Restrictive Practices Court
32500	33100	Resuscitation equipment (manufacture)
47799	52630	Retail auction house, second-hand goods other than antiques (except internet auctions) (retail)
47791	52630	Retail sale of antiques including antique books in retail auction houses (except internet auctions)
65300	66020	Retirement incomes provision
65300	66020	Retirement plans
46120	51120	Retort carbon (commission agent)
43999	45250	Retort setting
23200	26260	Retorts made of fireclay, silica and siliceous (manufacture)
23200	26260	Retorts made of graphite (manufacture)
23200	26260	Retorts made of refractory ceramic (manufacture)
18129	22220	Review printing (manufacture)
58142	22130	Review publishing
46690	51870	Revolution and production counters (wholesale)
26511	33201	Revolution counters (electronic) (manufacture)
26513	33202	Revolution counters (non-electronic) (manufacture)
25400	29600	Revolver (manufacture)
46690	51870	Revolvers, pistols and other firearms and similar devices (wholesale)
25120	28120	Revolving doors made of metal (manufacture)
90010	92311	Revue company
33140	31100	Rewind electric motor (manufacture)
27110	31100	Rewinding of armatures on a factory basis (manufacture)
24410	27410	Rhodium (manufacture)
32990	36631	Ribbon (inked) (manufacture)
13960	17542	Ribbon made of textile (manufacture)
10612	15612	Rice cleaning (manufacture)
10612	15612	Rice flaking (manufacture)
10612	15612	Rice flour production (manufacture)
01120	01110	Rice growing
28930	29530	Rice hullers (manufacture)
10612	15612	Rice husking (manufacture)
10612	15612	Rice milling (manufacture)
10890	15899	Rice pudding (canned) (manufacture)
10612	15612	Rice rolling (manufacture)
10620	15620	Rice starch (manufacture)
01610	01410	Rice transplanting
01610	01410	Rice transplanting, on a fee or contract basis
93290	92729	Richmond Park
13922	17402	Rick cloths and covers (manufacture)
32990	18241	Riding caps (manufacture)
15120	36639	Riding crops (manufacture)
85510	92629	Riding school
93199	92629	Riding stables
93110	92629	Rifle butts
25930	28730	Rigging for ships (manufacture)
28940	29540	Rigging machinery for textiles (manufacture)
26800	24650	Rigid magnetic disk (manufacture)
22210	25210	Rigid plastic foam (manufacture)
22290	25240	Ring binders and folders made of plastic (manufacture)
25930	28740	Ring spring (manufacture)
28940	29540	Ring traveller for textile machinery (manufacture)
22190	25130	Rings and washers made of rubber (manufacture)
23990	26821	Rings made of asbestos (manufacture)
32130	36610	Rings, bracelets, necklaces made from base metals plated with precious metals (manufacture)
28922	29522	Ripper (manufacture)
28921	29521	Riser connector apparatus (manufacture)
36000	41000	River management
42910	45240	River work construction
25940	28740	Rivet (manufacture)
22290	25240	Road cones made of plastic (manufacture)
42110	45230	Road construction and repair
19201	23201	Road coverings derived from crude petroleum or bituminous minerals (manufacture)
49410	60249	Road haulage contracting for general hire or reward
49410	60249	Road haulage contractor
08120	14210	Road metal production (crushed and processed)
28921	29521	Road ripper for mining (manufacture)
28923	29523	Road roller (manufacture)
46630	51820	Road rollers (wholesale)
22290	25240	Road signs made of plastic (manufacture)
42110	45230	Road surface markings painting
29100	34100	Road tanker (not trailer) (manufacture)
29202	34202	Road tractor trailer (manufacture)
29100	34100	Road tractor unit (manufacture)
77120	71219	Road trailer hire
29320	34300	Road wheels for motor vehicle (manufacture)
52219	63210	Roads operation
71200	74300	Road-safety testing of motor vehicles
52219	50200	Roadside assistance for motor vehicles
08120	14210	Roadstone (coated) production
46120	51120	Roasted iron pyrites (commission agent)
46750	51550	Roasted iron pyrites (wholesale)
27510	29710	Roasters (electric) (manufacture)
10890	15899	Roasting of nuts (manufacture)
14120	18210	Robes for academic, legal and ecclesiastical use (manufacture)
28220	29220	Robots designed for lifting and handling in industry (manufacture)
28990	29560	Robots for multiple industrial uses (manufacture)
32300	36400	Rock climbing equipment (manufacture)
28921	29521	Rock cutting machinery (manufacture)
28240	29410	Rock drill (portable) (manufacture)
25730	28620	Rock drilling and earth boring interchangeable tools (e.g. Augers, boring bits, drills) (manufacture)
28921	29521	Rock drilling machinery (manufacture)
43120	45110	Rock removal
08930	14400	Rock salt production
28921	29521	Rocker shovel for mining (manufacture)
30300	35300	Rocket (aerospace) (manufacture)
25400	29600	Rocket launch systems (manufacture)

SIC 2007	SIC 2003	Activity
30300	35300	Rocket motor (manufacture)
23410	26210	Rockingham ware (manufacture)
23990	26829	Rockwool (manufacture)
81291	74703	Rodent destroying (not agricultural)
01610	01410	Rodent destroying and trapping on agricultural land
20200	24200	Rodenticide (manufacture)
93290	92349	Rodeos
24420	27420	Rods made of aluminium (manufacture)
24440	27440	Rods made of brass (manufacture)
24440	27440	Rods made of copper (manufacture)
23190	26150	Rods made of glass (manufacture)
22190	25130	Rods made of rubber (manufacture)
24100	27100	Rods made of steel (in coils) (manufacture)
24100	27100	Rods of stainless steel (manufacture)
25930	28730	Rods or wires (coated/covered with flux) for gas welding, soldering or brazing (manufacture)
10200	15209	Roe production (manufacture)
28960	29560	Roll mill for rubber or plastic (manufacture)
23110	26110	Rolled glass (manufacture)
24420	27420	Rolled products made of aluminium (manufacture)
24440	27440	Rolled products made of copper (manufacture)
28940	29540	Roller (textile machinery accessory) (manufacture)
28150	29140	Roller bearing (manufacture)
13922	17402	Roller blinds made of canvas (manufacture)
16290	20510	Roller blinds made of wood (manufacture)
22190	25130	Roller coverings made of rubber (manufacture)
22190	25130	Roller covers made of rubber (manufacture)
28302	29320	Roller for agricultural use (manufacture)
32990	36631	Roller pens and refills (manufacture)
13300	17300	Roller printing of textiles (manufacture)
32300	36400	Roller skates (manufacture)
93110	92611	Roller skating rink
15120	19200	Roller skin (cut) made of leather (manufacture)
13923	17403	Roller towel (manufacture)
32910	36620	Rollers for paint (manufacture)
24420	27420	Rolling ingots and slabs made of aluminium (manufacture)
28910	29510	Rolling mill for metals (manufacture)
16290	20510	Rolling pins made of wood
30200	35200	Rolling stock (manufacture)
32200	36300	Rolls for automatic mechanical instruments (manufacture)
25500	28400	Rolls made of alloy or steel forgings (manufacture)
28910	29510	Rolls made of iron or steel (manufacture)
25990	28750	Rolls made of metal for cable, hose, etc. (manufacture)
13921	17401	Roman blinds (manufacture)
94910	91310	Roman Catholic Church
43910	45220	Roof covering
43910	45220	Roof covering erection
43290	45320	Roof insulation contractor
22230	25239	Roof lights made of plastic (manufacture)
16230	20300	Roof struts (manufacture)
28921	29521	Roof support (hydraulic) (manufacture)
25110	28110	Roof trusses made of metal (manufacture)
23610	26610	Roof units made of precast concrete (manufacture)
25940	28740	Roofbolts, plates and accessories (manufacture)
43910	45220	Roofing contractor
23320	26400	Roofing tiles made of ceramic (manufacture)
23610	26610	Roofing tiles made of precast concrete (manufacture)
23320	26400	Roofing tiles made of unglazed clay (manufacture)
31090	36140	Room divide system (manufacture)
25120	28120	Room partitions made of metal, for floor attachment (manufacture)
55900	55239	Rooming and boarding houses
28302	29320	Root crop harvesting and sorting machinery (manufacture)
01130	01120	Root vegetable growing
28922	29522	Rooter (not agricultural) (manufacture)
01130	01110	Roots and tubers growing
01130	01110	Roots and tubers with a high starch or inulin content
46410	51410	Rope (new) (wholesale)
13940	17520	Rope and cables of textile fibres (manufacture)
23990	26821	Rope lagging made of asbestos (manufacture)
13940	17520	Rope made of cotton (manufacture)
22190	25130	Rope made of rubber (manufacture)
25930	28730	Rope made of wire (manufacture)
28990	29560	Rope making machines (manufacture)
13940	17520	Rope or cable fitted with metal rings (manufacture)
13940	17520	Rope products (manufacture)
13940	17520	Rope slings (manufacture)
13940	17520	Rope walk (manufacture)
13940	17520	Rope, cord and line made of sisal (manufacture)
13940	17520	Rope, cord or line made of jute (manufacture)
13940	17520	Rope, cord or line made of manila (manufacture)
13940	17520	Rope, cord or line made of man-made fibre (manufacture)
20140	24140	Rosin size (manufacture)
94990	91330	Rotary clubs
46690	51870	Rotary displacement compressors (single or multi shaft) (wholesale)
28131	29121	Rotary piston lobe-type pumps (manufacture)
46690	51870	Rotary positive displacement pumps for liquids (wholesale)
28921	29521	Rotary tables (manufacture)
28132	29122	Rotating compressors (manufacture)
30300	35300	Rotor blades for aircraft (manufacture)
28220	29220	Rough terrain industrial trucks (manufacture)
08110	14110	Rough trimming and sawing of building and monumental stone
94990	91330	Round Table
28990	36639	Roundabout for fairground (manufacture)
24100	27100	Rounds of semi-finished steel for seamless tube production (manufacture)
02200	02010	Roundwood production (untreated)
02200	02010	Roundwood production for forest-based manufacturing industries
26301	32201	Routers for telecommunications (manufacture)
28940	29540	Roving frames (manufacture)
30120	35120	Rowing boat (manufacture)
93120	92629	Rowing club
94120	91120	Royal Academy of Arts
85320	80220	Royal Academy of Dramatic Art
94120	91120	Royal Aeronautical Society
94120	91120	Royal Agricultural Society of England
84220	75220	Royal Air Force Establishments (civilian personnel)
84220	75220	Royal Air Force Establishments (service personnel)
94990	91330	Royal Automobile Club headquarters
52219	50200	Royal Automobile Club road patrols

R

SIC 2007	SIC 2003	Activity
79110	63301	Royal Automobile Club touring department
91040	92530	Royal Botanical Gardens
94120	91120	Royal College of Midwives
94120	91120	Royal College of Nursing
94120	91120	Royal College of Physicians
94120	91120	Royal College of Surgeons
50200	61102	Royal Fleet Auxiliary
94120	91120	Royal Geographical Society
94990	91330	Royal Horticultural Society
94120	91120	Royal Institute of Chartered Surveyors
94120	91120	Royal Institute of Public Health
84220	75220	Royal Marines
88990	85321	Royal Masonic Benevolent Institute
32110	36210	Royal Mint (manufacture)
84220	75220	Royal Navy establishments (civilian personnel)
84220	75220	Royal Navy establishments (service personnel)
72190	73100	Royal Observatory
93290	92729	Royal Park
94990	91330	Royal Scottish Automobile Club
91020	92521	Royal Scottish Museum (Edinburgh)
94120	91120	Royal Society
94120	91120	Royal Society for Health
94990	91330	Royal Society for the Prevention of Accidents
94990	91330	Royal Society for the Prevention of Cruelty to Animals (not animal hospitals or homes)
94120	91120	Royal Society of Medicine
94120	91120	Royal Statistical Society
94120	91120	Royal United Services Institution
22190	25130	Rubber (vulcanized, unvulcanized or hardened) (manufacture)
46760	51550	Rubber (wholesale)
20590	24660	Rubber accelerators (manufacture)
20520	24620	Rubber based glues and adhesives (manufacture)
20301	24301	Rubber based paint (chlorinated) (manufacture)
13940	17520	Rubber coated twine, cordage rope and cables of textile fibres (manufacture)
25610	28510	Rubber coating of metals (manufacture)
22190	25130	Rubber compounds (manufacture)
32500	25130	Rubber gloves (medical) (manufacture)
46690	51870	Rubber or plastics working machinery (wholesale)
20590	24660	Rubber processing chemicals (manufacture)
22190	25130	Rubber products (manufacture)
22190	25130	Rubber thread (uncovered) (manufacture)
13960	17542	Rubber thread and cord, textile covered, coated or sheathed with rubber or plastics (manufacture)
13960	17542	Rubber thread or cord covered with textile material (manufacture)
01290	01110	Rubber trees growing for harvesting of latex
22110	25110	Rubber tyres for furniture and other uses (manufacture)
22110	25110	Rubber tyres for mobile machinery (manufacture)
22110	25110	Rubber tyres for toys (manufacture)
28960	29560	Rubber working machinery (manufacture)
22190	25130	Rubberised fabrics (manufacture)
22190	25130	Rubberised hair (manufacture)
22190	25130	Rubberised textile fabric (manufacture)
13300	17300	Rubberising of purchased garments (manufacture)
20170	24170	Rubber-like gums (balata, etc.) (manufacture)
08990	14500	Rubbing stone mine, pit or quarry
38110	90020	Rubbish collection

SIC 2007	SIC 2003	Activity
47510	52410	Rug making materials (retail)
96010	93010	Rug shampooing
13931	17512	Rug tufting (manufacture)
13931	17511	Rug weaving (not travelling rug) (manufacture)
93199	92629	Rugby League
93199	92629	Rugby Union
13939	17519	Rugs (other than woven or tufted) (manufacture)
47530	52481	Rugs (retail)
46470	51479	Rugs (wholesale)
13939	17519	Rugs made of coir (manufacture)
13939	17519	Rugs made of rag (manufacture)
14200	18300	Rugs made of sheepskin (manufacture)
14200	18300	Rugs made of skins (manufacture)
28290	33202	Rule (measuring) (manufacture)
22290	25240	Rulers made of plastic (manufacture)
28990	29560	Ruling machinery for printing (manufacture)
11010	15910	Rum distilling (manufacture)
29320	34300	Running gear for motor vehicles (manufacture)
16290	20520	Rush matting (manufacture)
10720	15820	Rusk making (manufacture)
96040	93040	Russian baths
31090	36140	Rustic furniture (manufacture)
10611	15611	Rye (manufacture)
10611	15611	Rye flaking (manufacture)
10611	15611	Rye flour and meal (manufacture)
01110	01110	Rye growing
10611	15611	Rye milling (manufacture)
10611	15611	Rye rolling (manufacture)
20140	24140	Saccharin tablet (manufacture)
22220	25220	Sachets made of plastic (manufacture)
13922	17402	Sacks (woven) (manufacture)
46410	51410	Sacks and bags (wholesale)
13922	17402	Sacks and bags made of noil (manufacture)
17211	21211	Sacks and bags made of paper (manufacture)
46160	51160	Sacks and bags used for packing of goods (commission agent)
13922	17402	Sacks for coal (manufacture)
13922	17402	Sacks made of canvas (manufacture)
13922	17402	Sacks made of jute (manufacture)
17120	21120	Sacks made of kraft paper (manufacture)
22220	25220	Sacks made of non-woven polyethylene (manufacture)
17211	21211	Sacks made of paper (manufacture)
22220	25220	Sacks made of plastic (manufacture)
25990	28750	Sacrificial anodes made of zinc, magnesium or other non-ferrous metal (manufacture)
15120	19200	Saddle horse (manufacture)
15120	19200	Saddlery (manufacture)
47722	52432	Saddlery (retail)
15120	19200	Saddlery and harness made of leather (manufacture)
46499	51479	Saddlery and leather goods (wholesale)
30910	35410	Saddles for motorcycles (manufacture)
30920	35420	Saddles for pedal cycles (manufacture)
25990	28750	Safe (manufacture)
52103	63129	Safe deposit company
30300	35300	Safety belt or harness for aircraft crew or passengers (manufacture)
29320	34300	Safety belts for cars (manufacture)
15200	19300	Safety boots (manufacture)

S

SIC 2007	SIC 2003	Activity
28301	29310	Safety frame for tractors (manufacture)
20510	24610	Safety fuse (manufacture)
46120	51120	Safety fuses, detonating fuses, caps, igniters and electric detonators (commission agent)
23120	26120	Safety glass (manufacture)
32990	28750	Safety headgear made of metal (manufacture)
32990	25240	Safety helmets made of plastic (manufacture)
25930	28730	Safety pins (manufacture)
25710	28610	Safety razors (manufacture)
28140	29130	Safety valves (manufacture)
01110	01110	Safflower seed growing
23200	26260	Saggar (manufacture)
10612	15612	Sago grinding or milling (manufacture)
32300	36400	Sailboards (manufacture)
30120	35120	Sailboat building (manufacture)
46499	51479	Sailboats for pleasure or sports (wholesale)
13200	17250	Sailcloth weaving (manufacture)
30120	35120	Sailing boats less than 100 gross tons (manufacture)
14131	18221	Sailing clothing for men and boys (weatherproof) (manufacture)
14132	18222	Sailing clothing for women and girls (weatherproof) (manufacture)
85530	80410	Sailing schools not issuing commercial certificates and permits
30300	35300	Sailplane (manufacture)
13922	17402	Sails (manufacture)
46499	51479	Sails (wholesale)
46180	51180	Sails for boats, sailboards or landcraft (commission agents)
01190	01110	Sainfoin growing
11030	15949	Sake (manufacture)
10840	15870	Salad cream (manufacture)
10840	15870	Salad dressing (manufacture)
10130	15139	Salami (manufacture)
47810	52620	Sale of food beverages and tobacco via stalls and markets (retail)
35230	40220	Sale of gas to the user through mains
96030	93030	Sale of graves
47890	52620	Sale via stalls and markets of books (retail)
47890	52620	Sale via stalls and markets of carpets and rugs (retail)
47820	52620	Sale via stalls and markets of clothing
47890	52620	Sale via stalls and markets of consumer electronics (retail)
47820	52620	Sale via stalls and markets of footwear (retail)
47890	52620	Sale via stalls and markets of games and toys (retail)
47890	52620	Sale via stalls and markets of household appliances (retail)
47890	52620	Sale via stalls and markets of music and video recordings (retail)
47820	52620	Sale via stalls and markets of textiles (retail)
78109	74500	Sales management recruitment consultant
46180	51180	Salicylic acids, o-acetylsalicylic acid and their salts and esters (commission agent)
21100	24410	Salicylic medicaments (manufacture)
21100	24410	Salines (manufacture)
03220	05020	Salmon and trout fishery (hatchery), freshwater
03210	05020	Salmon and trout fishery (hatchery), marine
03110	05010	Salmon netting
08930	14400	Salt crushing, purification and refining by the producer
08930	14400	Salt extraction
08930	14400	Salt, extraction of, from underground including by dissolving and pumping
08930	14400	Salt mine
08930	14400	Salt preparation (not at salt mine or brine pit)
08930	14400	Salt production
08930	14400	Salt production by evaporation of sea water or other saline waters
09900	14400	Salt production support services provided on a fee or contract basis
08930	14400	Salt works
10840	14400	Salt, processing of into food-grade salt, iodised salt (manufacture)
10720	15820	Salted crackers (manufacture)
46120	51120	Salts of oxometallic or perometallic acids (commission agent)
52220	63220	Salvage activities supporting water transport activities
30110	35110	Salvage vessel (manufacture)
94910	91310	Salvation army
96090	93059	Salvation army emigration department
87900	85311	Salvation army shelter (charitable)
23410	26210	Samian ware (manufacture)
18140	22230	Sample card finishing (manufacture)
15120	19200	Sample case made of leather (manufacture)
18140	22230	Sample mounting in support of printing activities (manufacture)
86101	85112	Sanatoria providing hospital type care (private sector)
86101	85111	Sanatoria providing hospital type care (public sector)
46730	51530	Sand (wholesale)
46730	51530	Sand and gravel merchant (wholesale)
43999	45250	Sand blasting for building exteriors
28290	29240	Sand blasting machines (manufacture)
25610	28510	Sand blasting of metals (manufacture)
24540	27540	Sand casting of non-ferrous base metal (manufacture)
08120	14210	Sand dredging
08120	14210	Sand extraction and dredging for industrial use
28990	29560	Sand handling, mixing, treatment or reclamation plant for foundries (manufacture)
08120	14210	Sand pit
08120	14210	Sand quarry
15200	19300	Sandals (manufacture)
43390	45450	Sandblasting of buildings
28490	29430	Sanding and polishing machines for wood (not portable) (manufacture)
81299	90030	Sanding or salting of highways, etc.
28240	29410	Sanding tool (powered portable) (manufacture)
23910	26810	Sandpaper (manufacture)
08110	14110	Sandstone mine, pit or quarry
56103	55303	Sandwich bar
10720	15820	Sandwich cake baking (manufacture)
24330	28110	Sandwich panels of coated steel sheet (manufacture)
10840	15870	Sandwich spread (manufacture)
10890	15512	Sandwiches with cheese filling (manufacture)
10890	15899	Sandwiches with egg filling (manufacture)
10890	15209	Sandwiches with fish, crustacean and mollusc filling (manufacture)
10890	15139	Sandwiches with ham or bacon filling (manufacture)
10890	15139	Sandwiches with meat or poultry filling, other than ham or bacon (manufacture)

S

SIC 2007	SIC 2003	Activity
10890	15330	Sandwiches with vegetable filling (manufacture)
13300	17300	Sanforising of textiles and textile articles, including wearing apparel (manufacture)
20412	24512	Sanitary cleanser (manufacture)
43220	45330	Sanitary engineering for buildings
47520	52460	Sanitary equipment (retail)
23420	26220	Sanitary fixtures made of ceramic (manufacture)
46740	51540	Sanitary installation connections, rubber pipes (wholesale)
46740	51540	Sanitary installation equipment (wholesale)
46740	51540	Sanitary installation taps, t-pieces (wholesale)
46740	51540	Sanitary installation tubes, pipes, fittings (wholesale)
46730	51530	Sanitary porcelain (wholesale)
28140	29130	Sanitary taps (manufacture)
17220	21220	Sanitary towels made of paper (manufacture)
17220	17549	Sanitary towels made of textile wadding (manufacture)
28140	29130	Sanitary valves (manufacture)
46730	51530	Sanitary ware (wholesale)
25990	28750	Sanitary ware and fittings made of metal (manufacture)
23420	26220	Sanitary ware made of ceramics (manufacture)
23420	26220	Sanitary ware made of fireclay (manufacture)
22230	25239	Sanitary ware made of plastic (manufacture)
23420	26220	Sanitary ware made of vitreous china (manufacture)
13940	17520	Sash line (manufacture)
15120	19200	Satchels made of leather (manufacture)
13200	17210	Sateen weaving (manufacture)
61300	64200	Satellite circuit rental services
51220	62300	Satellite launching
46520	51860	Satellite navigation (wholesale)
26309	32202	Satellite relay (manufacture)
61900	64200	Satellite terminal stations
61900	64200	Satellite tracking
30300	35300	Satellites (manufacture)
13200	17240	Satin weaving (manufacture)
17120	21120	Saturated and impregnated base paper (manufacture)
10840	15870	Sauce (manufacture)
25990	28750	Saucepan (manufacture)
25990	28750	Saucepans made of aluminium (manufacture)
96040	93040	Saunas
47520	52460	Saunas (retail)
16230	20300	Saunas made from wood (manufacture)
10720	15820	Sausage filler made of cereal (manufacture)
10130	15139	Sausage meat (manufacture)
10130	15139	Sausage rolls (manufacture)
46320	51320	Sausage skins (wholesale)
10110	15112	Sausage skins and casings (natural) (manufacture)
10130	15139	Sausages (manufacture)
10130	15139	Saveloys (manufacture)
64191	65121	Savings bank
28490	28620	Saw blades for machines including wood cutting (manufacture)
46770	51570	Sawdust (wholesale)
28410	29420	Sawing machine (metal cutting) (manufacture)
28490	29430	Sawing machine for wood (manufacture)
28490	29430	Sawing machines for stone, ceramics, asbestos-cement or similar materials (manufacture)
02200	02010	Sawlog production

SIC 2007	SIC 2003	Activity
16100	20100	Sawmilling (manufacture)
16100	20100	Sawn fencing (manufacture)
25730	28620	Saws (hand tools) (manufacture)
28240	29410	Saws (powered portable) (manufacture)
25730	28620	Saws and sawblades (manufacture)
46740	51540	Saws and sawblades (wholesale)
46740	51540	Saws, screwdrivers and similar hand tools (wholesale)
25110	28110	Scaffolding (manufacture)
77320	71320	Scaffolding hire (without staff)
43991	45250	Scaffolding hiring and erecting
24200	27220	Scaffolding tubes made of steel (manufacture)
43991	45250	Scaffolds and work platform erecting and dismantling
46499	51477	Scale models (wholesale)
28290	29240	Scales (platform) (manufacture)
28290	29240	Scales for domestic use (manufacture)
28290	29240	Scales for postal use (manufacture)
28290	29240	Scales for shop use (manufacture)
46120	51120	Scandium, yttrium and mercury (commission agent)
28990	29560	Scanner (printing machinery) (manufacture)
26200	30020	Scanners for computer use (manufacture)
14190	18249	Scarf (lace or knitted) (manufacture)
14190	18249	Scarf (not lace or knitted) (manufacture)
90030	92319	Scenario writer
46690	51870	Scene lighting, road lighting and other lighting (not domestic) (wholesale)
90020	92311	Scene shifters and lighting engineers
77210	71401	Scenery rental
51102	62201	Scenic and sightseeing flights
90030	92319	Scenic artist
32990	36639	Scent sprays (manufacture)
46750	51550	Scents (wholesale)
49390	60211	Scheduled long-distance bus services
13990	17541	Schiffli embroidery (manufacture)
85600	93059	Scholastic agent
85600	93059	School agent
49390	60219	School bus service
56290	55510	School canteen
56290	55520	School canteen (run by catering contractor)
84240	75240	School crossing patrols
86900	85140	School dental nurse
55900	55239	School dormitories
85310	80210	School examination board
86900	85140	School health service
86900	85140	School medical clinic
86210	85120	School medical officer
85320	80220	School of arts and crafts
85410	80301	School of languages
85530	80410	School of motoring
85320	80220	School of speech and drama
74201	74812	School photography
17230	22220	School stationery (manufacture)
22290	25240	School supplies made of plastic (manufacture)
91020	92521	Science museums
72190	73100	Science research council
26513	33202	Scientific laboratory equipment (non-electrical or non-optical) (manufacture)
77390	71340	Scientific machinery rental and operating leasing

SIC 2007	SIC 2003	Activity
32990	36639	Scientific models for educational and exhibition purposes (manufacture)
94120	91120	Scientific organisation
32500	33100	Scintigraphy apparatus (manufacture)
26600	33100	Scintillation scanners (manufacture)
28220	29220	Scissor lift (manufacture)
25710	28610	Scissors (manufacture)
10720	15820	Scone baking (manufacture)
32409	36509	Scooters for children (manufacture)
86900	85140	Scottish Ambulance Service
94920	91320	Scottish Nationalist Party
41202	45212	Scottish Special Housing Association (building work)
25990	28750	Scourers made of metal (manufacture)
28940	29540	Scouring machinery for textiles (manufacture)
25990	28750	Scouring pads made of metal (manufacture)
20412	24512	Scouring paste or powder coated paper (manufacture)
20412	24512	Scouring pastes (manufacture)
20412	24512	Scouring powder (manufacture)
94990	91330	Scout Association
46770	51570	Scrap (wholesale)
46770	51570	Scrap iron (wholesale)
46770	51570	Scrap leather (wholesale)
46770	51570	Scrap merchant (general dealer) (wholesale)
46770	51570	Scrap metal (wholesale)
38320	37100	Scrap metal recycling into new raw materials (except remelting of ferrous waste and scrap)
28922	29522	Scraper (earth moving equipment) (manufacture)
25730	28620	Scraper (hand tool) (manufacture)
46630	51820	Scrapers (wholesale)
15110	18300	Scraping of fur skins and hides with the hair on (manufacture)
43999	45250	Screed laying
26702	33403	Screen for cinema (manufacture)
18129	22220	Screen printing (except silk screen printing on textiles and apparel) (manufacture)
18129	22220	Screen printing (manufacture)
18129	22220	Screen printing of logos (manufacture)
13300	17300	Screen printing of textiles (manufacture)
18129	22220	Screen printing on glass or pottery (manufacture)
20302	24302	Screen process ink (manufacture)
28290	29240	Screening plant (not for mines) (manufacture)
28290	29240	Screening plant for effluent treatment (manufacture)
31010	36120	Screens (non-domestic) (manufacture)
16290	20520	Screens made of plaiting materials (manufacture)
25920	28720	Screw caps made of metal (manufacture)
28132	29122	Screw compressor (manufacture)
25940	28740	Screw machine products (manufacture)
28131	29121	Screw pump (manufacture)
25730	28620	Screwdrivers (manufacture)
46740	51540	Screwdrivers (wholesale)
28410	29420	Screwing machines (metal cutting) (manufacture)
25940	28740	Screws (self tapping) (manufacture)
22290	25240	Screws made of plastic (manufacture)
25940	28740	Screws of all types made of metal (manufacture)
13200	17250	Scrim weaving from flax, hemp, ramie and man-made fibres processed on the flax system (manufacture)
69109	74119	Scrivenery
28250	29230	Scrubber for air conditioning equipment (manufacture)
32910	36620	Scrubbing brush (manufacture)
46499	51479	Scuba diving breathing equipment (wholesale)
90030	92319	Sculptors
13100	17140	Scutching of flax (manufacture)
25730	28620	Scythe (manufacture)
50100	61101	Sea ferry (passenger)
52290	63400	Sea freight forwarder activities
30110	35110	Sea going luxury yachts of 100 gross tons or more (manufacture)
01700	01500	Sea mammal catching
08930	14400	Sea salt production
94990	91330	Sea scout association
03110	05010	Sea urchin hunting
36000	41000	Sea water desalination
47230	52230	Seafood and seafood products (retail)
28930	29530	Sea-food processing machines and equipment (manufacture)
01700	01500	Seal catching
23990	26829	Sealants (bituminous) (manufacture)
23990	26829	Sealants (oil based) (manufacture)
20301	24303	Sealants (other than bituminous or oil based) (manufacture)
28290	29240	Sealing machinery (manufacture)
09100	11200	Sealing of oil and gas wells
46499	51479	Sealing or numbering stamps (wholesale)
32990	36631	Sealing stamps (manufacture)
32990	36631	Seals for use with sealing wax (manufacture)
22190	25130	Seals made of rubber (manufacture)
24520	27520	Seamless pipes of steel by centrifugal casting (manufacture)
24520	27520	Seamless tubes of steel by centrifugal casting (manufacture)
27400	31500	Search light (manufacture)
10840	15870	Seasoning (manufacture)
03110	05010	Sea-squirt hunting
31090	36110	Seating (manufacture)
31010	36110	Seating for office or school (manufacture)
31090	36110	Seating made of metal (not for road vehicle or aircraft) (manufacture)
30300	36110	Seats for aircraft (manufacture)
31010	36110	Seats for cinema (manufacture)
29320	36110	Seats for motor vehicles (manufacture)
30110	36110	Seats for ships and floating structures (manufacture)
31010	36110	Seats for theatre (manufacture)
32990	36639	Seat-sticks (manufacture)
03110	05010	Seaweed collecting, cutting and gathering (uncultivated)
10200	15209	Seaweed processing (manufacture)
25730	28620	Secateurs (manufacture)
27200	31400	Secondary battery (manufacture)
24440	27440	Secondary copper (manufacture)
85310	80210	Secondary level education
85310	80210	Secondary modern schools
85310	80210	Secondary schools
47799	52509	Second-hand books (retail)
47799	52509	Second-hand clothing (retail)
47799	52509	Second-hand furniture (retail)
47799	52509	Second-hand general goods (retail)
85320	80220	Secretarial college
28250	29230	Sectional coldroom (manufacture)

S

123

SIC 2007	SIC 2003	Activity	SIC 2007	SIC 2003	Activity
24330	27330	Sections cold formed from flat steel products (manufacture)	01640	01139	Seed post harvest processing (fruit, nuts, beverage and spice crops)
30110	35110	Sections for ships and floating structures (manufacture)	01640	01120	Seed post harvest processing (vegetables, horticultural specialties and other nursery products)
24420	27420	Sections made of aluminium (manufacture)	46210	51210	Seed potatoes (wholesale)
24440	27440	Sections made of brass (manufacture)	01640	01110	Seed processing for propagation (cereals and other crops)
24440	27440	Sections made of copper (manufacture)			
24100	27100	Sections of stainless steel (manufacture)	01640	01410	Seed processing for propagation (fee or contract basis)
66120	67122	Securities broking	01640	01139	Seed processing for propagation (fruit, nuts, beverage and spice crops)
64991	65233	Securities dealer on own account			
66120	67122	Securities dealing on behalf of others	01640	01120	Seed processing for propagation (vegetables, horticultural specialties and other nursery products)
66110	67110	Securities exchanges administration			
64999	65239	Securitization activities	01190	01120	Seed production for flowers, fruit or vegetables (not for oil)
80100	74602	Security activities (not government)			
26301	31620	Security alarms and systems (manufacture)	22290	25240	Seed trays made of plastic (manufacture)
66120	67122	Security and commodity contracts dealing activities	28302	29320	Seeders (manufacture)
74909	74602	Security consultancy for industrial, household and public services	47760	52489	Seeds (retail)
			46210	51210	Seeds (wholesale)
80100	74602	Security delivery of prisoners	23910	26810	Segment bonded abrasive (manufacture)
80100	74602	Security guard services	71122	74206	Seismic surveying for petroleum
82920	74820	Security packaging of pharmaceutical preparations	26511	33201	Seismometers (electronic) (manufacture)
17120	21120	Security paper (manufacture)	26513	33202	Seismometers (non-electronic) (manufacture)
18129	22220	Security printing (manufacture)	30200	35200	Self-propelled railway car (manufacture)
25110	28110	Security screens made of metal (manufacture)	10611	15611	Self-raising and patent flour (manufacture)
80100	74602	Security shredding of information on any media	22290	25240	Self-adhesive tapes made of plastic (manufacture)
80200	74602	Security systems service activities	17230	21230	Self-copy paper ready for use (manufacture)
80100	74602	Security transport of valuables and money	30200	35200	Self-propelled railway or tramway coaches, vans, maintenance or service vehicles (manufacture)
28290	29240	Sedimentation plant for effluent treatment (manufacture)			
			56101	55301	Self-service restaurant (licensed)
10410	15410	Seed and nut crushing (manufacture)	56102	55302	Self-service restaurant (unlicensed)
28302	29320	Seed cleaner or pre-cleaner for agricultural use (manufacture)	17110	21110	Semi-bleached paper pulp made by chemical dissolving (manufacture)
01640	01110	Seed cleaning and treating (cereals and other crops)	17110	21110	Semi-bleached paper pulp made by mechanical processes (manufacture)
01640	01410	Seed cleaning and treating (fee or contract basis)			
01640	01139	Seed cleaning and treating (fruit, nuts, beverage and spice crops)	17110	21110	Semi-bleached paper pulp made by non-dissolving processes (manufacture)
01640	01120	Seed cleaning and treating (vegetables, horticultural specialties and other nursery products)	17110	21110	Semi-bleached paper pulp made by semi-chemical processes (manufacture)
			17110	21110	Semi-chemical woodpulp (manufacture)
28302	29320	Seed cleaning, sorting or grading machines (manufacture)	19100	23100	Semi-coke (manufacture)
20200	24200	Seed dressing (manufacture)	26110	32100	Semi-conductor (not power) (manufacture)
01640	01110	Seed drying (cereals and other crops)	26110	31100	Semi-conductor control equipment (converters) (manufacture)
01640	01410	Seed drying (fee or contract basis)			
01640	01139	Seed drying (fruit, nuts, beverage and spice crops)	46520	51860	Semi-conductor devices (wholesale)
01640	01120	Seed drying (vegetables, horticultural specialties and other nursery products)	26110	24660	Semi-finished dice or wafers, semiconductor (manufacture)
			22190	25130	Semi-finished products made of rubber (manufacture)
01640	01110	Seed genetically modified, treatment of these (cereals and other crops)			
01640	01410	Seed genetically modified, treatment of these (fee or contract basis)	24100	27100	Semi-finished products of stainless steel (manufacture)
			24440	27440	Semi-manufactures made of copper (manufacture)
01640	01139	Seed genetically modified, treatment of these (fruit, nuts, beverage and spice crops)	24450	27450	Semi-manufacturing of chrome (manufacture)
			24450	27450	Semi-manufacturing of manganese (manufacture)
01640	01120	Seed genetically modified, treatment of these (vegetables, horticultural and other nursery products)	24450	27450	Semi-manufacturing of nickel (manufacture)
			85320	80220	Seminary
01640	01110	Seed grading (cereals and other crops)	08990	14500	Semi-precious stones extraction
01640	01410	Seed grading (fee or contract basis)	29202	34202	Semi-trailers (manufacture)
01640	01139	Seed grading (fruit, nuts, beverage and spice crops)	45190	50101	Semi-trailers (retail)
01640	01120	Seed grading (vegetables, horticultural specialties and other nursery products)	45190	50102	Semi-trailers (used) (retail)
			45190	50102	Semi-trailers (used) (wholesale)
01640	01110	Seed post harvest processing (cereals and other crops)	45190	50101	Semi-trailers (wholesale)
01640	01410	Seed post harvest processing (fee or contract basis)			

SIC 2007	SIC 2003	Activity	SIC 2007	SIC 2003	Activity
13100	17130	Semi-worsted yarn spinning (manufacture)	25990	28750	Shackle (manufacture)
10611	15611	Semolina milling (manufacture)	23190	26150	Shades made of glass (manufacture)
20590	24640	Sensitized cloth (manufacture)	28150	29140	Shaft bearings (manufacture)
20590	24640	Sensitized emulsions for photographic use (manufacture)	28150	29140	Shaft couplings (manufacture)
			43999	45250	Shaft drilling (civil engineering)
20590	24640	Sensitized paper (manufacture)	43999	45250	Shaft sinking
26514	33302	Sensor for electric process control equipment (manufacture)	19201	23201	Shale oil refining (manufacture)
			01130	01120	Shallot growing
09100	11200	Separation terminal operation (natural gas)	20420	24520	Shampoo (manufacture)
28930	29530	Separators for the grain milling industry (manufacture)	46180	51180	Shampoos, hair lacquers and permanent waving or straightening preparations (commission agent)
37000	90010	Septic tanks emptying and cleaning	11070	15980	Shandy (manufacture)
21200	24421	Sera (manufacture)	24100	27100	Shapes of stainless steel (manufacture)
56290	55510	Sergeants' messes	28410	29420	Shaping machine (metal cutting) (manufacture)
21200	24421	Serum albumin (manufacture)	66120	67122	Share dealer on behalf of others
26200	30020	Servers and network servers (manufacture)	64991	65233	Share dealer on own account
52230	63230	Service activities incidental to air transportation	61100	64200	Shared business telephone network services (wired telecommunications)
52230	62300	Service activities incidental to space transportation			
68209	70209	Service flat letting	23910	26810	Sharpening stones made of bonded abrasives (manufacture)
28302	29320	Servicing of agricultural machinery (manufacture)	25730	28620	Shave hook (manufacture)
95220	52720	Servicing of clothes dryers	27510	29710	Shaver (electric) (manufacture)
95220	52720	Servicing of domestic electrical appliances	46439	51439	Shavers and hair clippers with self-contained motors (wholesale)
95220	52720	Servicing of garden edger			
95220	52720	Servicing of garden trimmers	32910	36620	Shaving brush (manufacture)
95220	29320	Servicing of lawnmowers	20420	24520	Shaving cream (brushless) (manufacture)
45200	50200	Servicing of motor vehicles	20420	24520	Shaving cream (manufacture)
95220	52720	Servicing of refrigerators	20420	24520	Shaving preparations (manufacture)
95220	52720	Servicing of room air conditioners	46180	51180	Shaving preparations, personal deodorants and antiperspirants (commission agent)
95220	52720	Servicing of snow and leaf blowers			
95220	52720	Servicing of stoves	20420	24511	Shaving soap (manufacture)
95220	52720	Servicing of washing machines	14190	18249	Shawls (manufacture)
17220	21220	Serviettes made of paper (manufacture)	10410	15420	Shea butter (manufacture)
25990	28750	Serving dishes made of base metal (manufacture)	10410	15410	Shea nut crushing (manufacture)
10410	15420	Sesame oil refining (manufacture)	15110	18300	Shearing and plucking of fur skins and hides with the hair on (manufacture)
10410	15410	Sesame seed crushing (manufacture)			
01110	01110	Sesame seed growing	28410	29420	Shearing machine (metal forming) (manufacture)
08110	14110	Sett quarry	28940	29540	Shearing machinery for textiles (manufacture)
31090	36110	Settee (manufacture)	25730	28620	Shears for agricultural or horticultural use (manufacture)
28290	29240	Settlement plant for water treatment (manufacture)			
37000	90010	Sewage farm	25730	28620	Shears for garden use (manufacture)
71122	74206	Sewage treatment consultancy activities	22190	25130	Sheath contraceptives made of rubber (manufacture)
28290	29240	Sewage treatment plant (manufacture)	25710	28610	Sheath knife (manufacture)
37000	90010	Sewage works	25110	28110	Sheds made of metal (manufacture)
42210	45213	Sewerage construction	16230	20300	Sheds made of wood (manufacture)
37000	90010	Sewerage system maintenance and operation	46230	51230	Sheep (wholesale)
28940	29540	Sewing machine (manufacture)	01629	01429	Sheep agisting (grazing)
28940	29540	Sewing machine heads (manufacture)	32500	24422	Sheep and cattle dressings (manufacture)
28990	29560	Sewing machinery for bookbinding (manufacture)	10110	15113	Sheep and lambskin pulling (manufacture)
47540	52450	Sewing machines (retail)	20200	24200	Sheep dip (manufacture)
46640	51830	Sewing machines (wholesale)	01450	01220	Sheep farming
25930	28730	Sewing needles (manufacture)	01450	01220	Sheep milk (raw) production
47510	52410	Sewing thread (retail)	13940	17520	Sheep net (manufacture)
13100	17160	Sewing thread made of cotton (manufacture)	01450	01220	Sheep raising and breeding
16290	20510	Sewing thread reels and similar articles of turned wood (manufacture)	01629	01429	Sheep shearing on a fee or contract basis
			25730	28620	Sheep shears (not power) (manufacture)
46410	51410	Sewing thread, etc. (wholesale)	15110	19100	Sheep skin preparation (manufacture)
22190	36639	Sex articles of rubber (manufacture)	46730	51530	Sheet glass merchant (wholesale)
26511	33201	Sextant (electronic) (manufacture)	13923	17403	Sheet hemming (textiles) (manufacture)
26513	33202	Sextant (non-electronic) (manufacture)	28410	29420	Sheet metal forming machine (manufacture)

S

SIC 2007	SIC 2003	Activity
18129	22220	Sheet metal printing (manufacture)
25620	28520	Sheet metal working (manufacture)
47591	52450	Sheet music (retail)
18129	22220	Sheet music printing (manufacture)
24100	27100	Sheet piling production (manufacture)
23990	26821	Sheeting made of non-woven asbestos/rubber composite (manufacture)
22230	25239	Sheeting made of plastic for roofing and cladding (manufacture)
13200	17250	Sheeting weaving from flax and man-made fibres processed on the flax system (manufacture)
13200	17210	Sheeting weaving from yarn spun on the cotton system (manufacture)
23990	26821	Sheets and sheeting made of woven asbestos (manufacture)
24420	27420	Sheets made of aluminium (manufacture)
24440	27440	Sheets made of brass (manufacture)
22210	25210	Sheets made of cellophane (manufacture)
23610	26610	Sheets made of concrete (manufacture)
24440	27440	Sheets made of copper (manufacture)
16290	20520	Sheets made of cork (manufacture)
23650	26650	Sheets made of fibre cement (manufacture)
23140	26140	Sheets made of glass fibre (manufacture)
22210	25210	Sheets made of laminated thermosetting plastic (manufacture)
23620	26620	Sheets made of plaster (manufacture)
22210	25210	Sheets made of plastic (manufacture)
22210	25210	Sheets made of polyethylene (manufacture)
22210	25210	Sheets made of polypropylene (manufacture)
22210	25210	Sheets made of polyvinyl chloride (PVC) (manufacture)
22190	25130	Sheets made of rubber (manufacture)
47510	52410	Sheets made of textiles (retail)
24330	27330	Sheets of square corrugated steel (manufacture)
25300	28300	Shell boiler (manufacture)
25400	29600	Shell case (manufacture)
28930	29530	Shell fish processing machines and equipment (manufacture)
20301	24301	Shellac varnish (manufacture)
47230	52230	Shellfish (retail)
46380	51380	Shellfish (wholesale)
10200	15201	Shellfish freezing (manufacture)
10200	15209	Shellfish preserving (not freezing) (manufacture)
87900	85311	Shelter (the charity)
31010	36120	Shelving (non-domestic) (manufacture)
84230	75230	Sheriff's Court (Scotland)
69109	74119	Sheriff's officer
08120	14210	Shingle dredging
16230	20300	Shingles and shakes (manufacture)
81291	74703	Ship disinfecting and exterminating activities
81291	74703	Ship fumigating and scrubbing
77342	71229	Ship hire for freight (without crew)
77341	71221	Ship hire for passengers (without crew)
85530	80410	Ship licence tuition (not commercial certificates)
64922	65223	Ship mortgage finance company
25990	28750	Ship propellers (manufacture)
25990	28750	Ship propeller blades of metal (manufacture)
77341	71221	Ship rental for passengers (without operator)
71200	74300	Ship surveyor
30110	35110	Shipbuilding (manufacture)
22230	25239	Shiplap cladding made of plastic (manufacture)
52290	63400	Shipping agent or broker
85530	80410	Shipping schools not issuing commercial certificates and permits
46140	51140	Ships (commission agent)
46690	51870	Ships and boats for the carriage of passengers or goods (wholesale)
46690	51870	Ships and boats propellers and blades (wholesale)
20301	24301	Ship's bottom composition (manufacture)
71200	74300	Ships certification
46900	51900	Ships chandler (wholesale)
13940	17520	Ship's fenders (manufacture)
27320	31300	Ship's wiring (manufacture)
94990	91330	Shire Horse Association
96010	93010	Shirt and collar pressing
14190	18249	Shirt front (manufacture)
14190	18249	Shirt neckband (manufacture)
13200	17250	Shirting weaving from flax and man made fibre processed on the flax system (manufacture)
13200	17210	Shirting weaving from yarn spun on the cotton system (manufacture)
14120	18210	Shirts for industrial use (manufacture)
14141	18231	Shirts for men and boys (manufacture)
14142	18232	Shirts for women and girls (manufacture)
29320	34300	Shock absorber for motor vehicle (manufacture)
30200	35200	Shock absorbers; bodies; etc. of railway or tramway locomotives or of rolling stock (manufacture)
46760	51560	Shoddy (carded or combed) (wholesale)
32910	36620	Shoe brush (manufacture)
20412	24512	Shoe dye (manufacture)
13990	17542	Shoe laces (braided) (manufacture)
22290	25240	Shoe lasts made of plastic (manufacture)
16290	19300	Shoe parts made of wood (manufacture)
20412	24512	Shoe polish (manufacture)
96090	93059	Shoe shiners
13960	17542	Shoe trimmings made of textile materials (manufacture)
15200	19300	Shoes (manufacture)
47721	52431	Shoes (retail)
46420	51423	Shoes (wholesale)
15200	19300	Shoes made of wood (manufacture)
16290	20510	Shoetrees made of wood (manufacture)
93120	92629	Shooting clubs
93290	92349	Shooting galleries
28990	36639	Shooting galleries (manufacture)
43320	45420	Shop fitter
31010	36120	Shop fixtures for display and storage of goods (manufacture)
25120	28120	Shop fronts and entrances made of aluminium (manufacture)
16230	20300	Shop fronts made of wood (manufacture)
68209	70209	Shop letting
15120	19200	Shopping bags made of plastic (designed for prolonged use) (manufacture)
15120	19200	Shopping bags made of stitched plastic (manufacture)
30990	35500	Shopping carts (hand-propelled) (manufacture)
52220	63220	Shore base (sea transport)
82190	74850	Shorthand writing
28230	30010	Shorthand writing machines (manufacture)
49200	60109	Shortline freight railways

S

SIC 2007	SIC 2003	Activity
14131	18221	Shorts for men and boys (manufacture)
25610	28510	Shot peening of metals (manufacture)
43390	45450	Shotblasting of buildings
28922	29522	Shovel loaders (manufacture)
25730	28620	Shovels (manufacture)
31010	36120	Show case (manufacture)
18129	22220	Showcard (manufacture)
22230	25239	Shower baths made of plastic (manufacture)
25990	28750	Shower cabinets made of metal (manufacture)
73110	74402	Showroom design
38320	37100	Shredding of metal waste, end-of-life vehicles
10200	15209	Shrimp preserving (not freezing) (manufacture)
03220	05020	Shrimp production (post-larvae), freshwater
03210	05020	Shrimp production (post-larvae), marine
03110	05010	Shrimping
46380	51380	Shrimps (wholesale)
28940	29540	Shrinking machines for textiles (manufacture)
13300	17300	Shrinking of textiles and textile articles, including wearing apparel (manufacture)
32500	17403	Shroud and cerement (manufacture)
13922	17402	Shutes made of canvas (manufacture)
25110	28110	Shuttering made of steel (manufacture)
16230	20300	Shuttering made of wood (manufacture)
25120	28120	Shutters made of metal (manufacture)
22230	25239	Shutters made of plastic (manufacture)
16230	20300	Shutters made of wood (manufacture)
28940	29540	Shuttle (textile machinery accessory) (manufacture)
28940	29540	Shuttle changing machinery (manufacture)
25730	28620	Sickles (manufacture)
28220	29220	Side loader (manufacture)
31090	36140	Sideboard (manufacture)
30910	35410	Sidecars for motorcycles (manufacture)
28930	29530	Sieving belts (manufacture)
28930	29530	Sifters (manufacture)
26701	33402	Sight telescopes (manufacture)
49390	60231	Sightseeing buses
43210	45340	Sign (electric) erection and maintenance
25990	28750	Sign plates made of metal (manufacture)
30200	35200	Signal box equipment (manufacture)
27900	31620	Signal generator (manufacture)
20510	24610	Signal rocket (manufacture)
30200	35200	Signalling equipment for airports, inland waterways, ports, roads and tramways (manufacture)
30200	31620	Signalling equipment for railways (electro mechanical) manufacture
27900	31620	Signalling equipment for road traffic (electric) (manufacture)
46120	51120	Signalling flares, rain rockets, fog signals and other pyrotechnic articles (commission agent)
23190	26150	Signalling glassware (manufacture)
30200	35200	Signalling, safety and traffic control equipment for railways (manufacture)
30200	31620	Signalling, safety or traffic control equipment (electrical) (manufacture)
22290	25240	Signs made of plastic (manufacture)
16290	20510	Signs made of wood (manufacture)
73110	74402	Signwriting
28302	29320	Silage making machinery (manufacture)
29320	34300	Silencer for motor vehicle (manufacture)
08990	14500	Silica stone extraction or quarrying
46750	51550	Silicates, borates, perborates and other salts of inorganic acids or peroxoacids (wholesale)
08990	14500	Siliceous fossil meals mining
24100	27100	Silico manganese steel (manufacture)
46120	51120	Silicon and sulphur dioxide (commission agent)
46750	51550	Silicon and sulphur dioxide (wholesale)
20160	24160	Silicones (manufacture)
46120	51120	Silicones in primary forms (commission agent)
46760	51550	Silicones in primary forms (wholesale)
13300	17300	Silk bleaching (manufacture)
13100	17150	Silk creping (manufacture)
13300	17300	Silk dyeing (manufacture)
13300	17300	Silk finishing and weighting (manufacture)
13300	17300	Silk printing (manufacture)
13300	17300	Silk screen-printing on wearing apparel (manufacture)
13100	17150	Silk throwing (manufacture)
13100	17150	Silk twisting (manufacture)
13200	17240	Silk type fabrics (manufacture)
13100	17150	Silk warping (manufacture)
13100	17150	Silk waste noil spinning (manufacture)
13100	17150	Silk winding (manufacture)
01490	01250	Silk worm cocoon production
01490	01250	Silk worm raising
13200	17240	Silk woven cloth (manufacture)
13100	17150	Silk yarn (manufacture)
46410	51410	Silk yarn and fabrics (wholesale)
25290	28210	Silos made of steel exceeding 300 litres (manufacture)
16230	20300	Silos made of wood (manufacture)
20160	24160	Siloxanes (manufacture)
24410	27410	Silver (manufacture)
32120	36220	Silver burnishing (manufacture)
07290	13200	Silver mining and preparation
07290	13200	Silver ore and concentrate extraction and preparation
27200	31400	Silver oxide cells (manufacture)
24410	27410	Silver rolled onto base metals production (manufacture)
32120	36220	Silversmiths' work (manufacture)
47789	52489	Silverware (retail)
02100	02010	Silviculture and other forestry activities
27900	31620	Simulator (battle) (manufacture)
27900	31620	Simulator (driving) (manufacture)
27900	31620	Simulator (other training (except flying trainers)) (manufacture)
28940	29540	Singeing machinery for textiles (manufacture)
90010	92311	Singer (own account)
10910	15710	Single feeds, unmixed, for farm animals (manufacture)
24340	27340	Single strand wire made of steel (manufacture)
20600	24700	Single yarn of man-made continuous fibres, including high tenacity and textured yarn (manufacture)
28921	29521	Sinking machine for mining (manufacture)
23420	26220	Sinks made of ceramic fireclay (manufacture)
23650	26650	Sinks made of fibre cement (manufacture)
25990	28750	Sinks made of metal (other than cast iron) (manufacture)
22230	25239	Sinks made of plastic (manufacture)
25500	28400	Sintering of metals (manufacture)

S

SIC 2007	SIC 2003	Activity	SIC 2007	SIC 2003	Activity
29310	31610	Sirens for motor vehicles (manufacture)	16290	20510	Slats for the manufacture of pencils (manufacture)
01160	01110	Sisal growing and other textile fibre of the genus agave growing	10110	15111	Slaughterhouse (manufacture)
			28930	29530	Slaughterhouse machinery (manufacture)
43120	45110	Site preparation	10110	15111	Slaughterhouses killing, dressing or packing meat (manufacture)
85310	80210	Sixth form colleges			
28990	29560	Size reduction equipment for the chemical industry (manufacture)	10120	15120	Slaughterhouses killing, dressing or packing poultry (manufacture)
28990	29560	Size separation equipment for the chemical industry (manufacture)	30990	35500	Sledges (hand-propelled) (manufacture)
			24100	27100	Sleepers (cross ties) made of iron or steel (hot rolled) (manufacture)
28940	29540	Sizing machinery for textiles (manufacture)			
32300	36400	Skateboards (manufacture)	46730	51530	Sleepers (wholesale)
29202	34202	Skeletal trailer (motor drawn) (manufacture)	16100	20100	Sleepers made of wood (manufacture)
18130	22250	Sketches for presentation (manufacture)	13923	17403	Sleeping bag (manufacture)
49390	60239	Ski and cable lifts operation	46180	51180	Sleeping bags (commission agent)
32300	36400	Ski bindings and poles (manufacture)	24200	27220	Sleeves made of steel (manufacture)
85510	92629	Ski instructor (own account)	28930	29530	Slicers for bakeries (manufacture)
28220	29220	Ski lifts (manufacture)	74203	74814	Slide and negative duplicating
14190	18249	Ski suits (manufacture)	32990	36639	Slide fasteners (manufacture)
20412	24512	Ski wax (manufacture)	25990	36639	Slide fasteners made of metal (manufacture)
32300	36400	Ski-boots (manufacture)	22290	36639	Slide fasteners made of plastic (manufacture)
25930	28740	Skid chain (manufacture)	74203	74814	Slide mounting
30120	35120	Skiff building (manufacture)	26702	33403	Slide projector (manufacture)
32300	36400	Skiing equipment (manufacture)	26513	33202	Slide rules (manufacture)
20420	24520	Skin care preparations (manufacture)	15200	19300	Slipper soles (manufacture)
10110	15111	Skin drying (manufacture)	15200	19300	Slippers (manufacture)
10110	15111	Skin pickling (manufacture)	14142	18232	Slips (manufacture)
10110	15111	Skin production from slaughterhouses (manufacture)	28950	29550	Slitting machine for paper (manufacture)
10110	15111	Skin sorting (manufacture)	28940	29540	Slitting machinery for textiles (manufacture)
46240	51249	Skins (wholesale)	25730	28620	Slitting saw (manufacture)
38110	90020	Skip hire (waste transportation)	28940	29540	Sliver can (textile machinery accessory) (manufacture)
28921	29521	Skip plant for mining (manufacture)	13100	17120	Sliver dyeing of wool (manufacture)
15120	19200	Skips made of leather (manufacture)	13910	17600	Sliver knitted fabric (manufacture)
22230	25239	Skirting boards made of plastic (manufacture)	01240	01139	Sloe growing
16100	20100	Skirting boards made of unmoulded wood (manufacture)	77390	71340	Slot machine rental
			28410	29420	Slotting machine (metal working) (manufacture)
14132	18222	Skirts for women and girls (dressmade) (manufacture)	28110	29110	Slow speed diesel engine for marine use (manufacture)
32300	36400	Skis (manufacture)	13300	17300	Slub dyeing (manufacture)
77210	71401	Skis renting and leasing	13100	17120	Slubbing dyeing of wool (manufacture)
93110	92629	Skittle alley	30110	35110	Sludge vessel (manufacture)
25120	28120	Skylights made of metal (manufacture)	24420	27420	Slugs made of aluminium (manufacture)
24440	27440	Slabs made of brass (manufacture)	25110	28110	Sluice gates made of steel (manufacture)
24440	27440	Slabs made of copper (manufacture)	49500	60300	Slurry transport via pipelines
16290	20520	Slabs made of cork (manufacture)	25400	29600	Small arms (manufacture)
23140	26140	Slabs made of glass fibre (manufacture)	25990	28750	Small metal hand-operated kitchen appliances and accessories (manufacture)
14132	18222	Slacks for women and girls (manufacture)			
23510	26510	Slag cement (manufacture)	13300	17300	Small ware bleaching (manufacture)
23990	26829	Slag wool (manufacture)	13300	17300	Small ware dyeing (manufacture)
23520	26520	Slaked lime (manufacture)	13990	17542	Small ware made of textiles (manufacture)
46730	51530	Slate (wholesale)	86101	85111	Smallpox hospital (public sector)
08110	14130	Slate mine or quarry	26200	30020	Smart card readers (manufacture)
23700	26700	Slate polishing (manufacture)	26120	32100	Smart cards (manufacture)
09900	14130	Slate quarrying support services provided on a fee or contract basis	28210	29210	Smelting furnace (manufacture)
			26511	33201	Smoke detection equipment (electronic) (manufacture)
23700	26700	Slate slab and sheet cutting and preparation (manufacture)	10130	15139	Smoked meat (other than bacon and ham) (manufacture)
46730	51530	Slate slabs (wholesale)	10200	15209	Smoked salmon, trout and herring production (manufacture)
23700	26700	Slate working (manufacture)			
46730	51479	Slates and boards (wholesale)	32990	36639	Smokers' requisites (manufacture)
32990	36631	Slates for writing (manufacture)	47260	52260	Smokers' requisites (retail)

SIC 2007	SIC 2003	Activity
32990	36639	Smoking pipes (manufacture)
46499	51479	Smoking pipes and cigarette and cigar holders (wholesale)
46439	51439	Smoothing irons (electric) (wholesale)
27510	29710	Smoothing irons (manufacture)
56103	55303	Snack bar
10720	15820	Snack products of puffed or extruded farinaceous or proteinaceous materials (manufacture)
10720	15820	Snack products whether sweet or salted (manufacture)
01490	01250	Snail farming
32990	36639	Snap fasteners (manufacture)
93120	92629	Snooker club
81299	90030	Snow and ice clearing of highways, etc.
28923	29523	Snow blowers (manufacture)
29100	34100	Snow mobiles (manufacture)
28923	29523	Snow ploughs (manufacture)
46499	51479	Snow skis, ice skates and roller skates (wholesale)
46690	51870	Snow-ploughs and blowers (wholesale)
12000	16000	Snuff (manufacture)
20411	24511	Soap (manufacture)
46450	51450	Soap (wholesale)
46180	51180	Soap and organic surface-active products and preparations for use as soap (commission agent)
20411	24511	Soap chips (manufacture)
20411	24511	Soap flakes (manufacture)
28990	29560	Soap making machinery (manufacture)
20411	24511	Soap or detergent coated felt, paper and wadding (manufacture)
20411	24511	Soap powder (manufacture)
94990	91330	Social club (not licensed to sell alcohol)
56301	55401	Social clubs (licensed to sell alcohol)
94920	91320	Social Democratic and Labour Party
86220	85120	Social medicine specialist (private practice)
86101	85111	Social medicine specialist (public sector)
72200	73200	Social Science Research Council
72200	73200	Social sciences research and experimental development
88990	85322	Social Services Department
88990	85321	Social welfare society (charitable)
88990	85321	Social work activities for immigrants (charitable)
88990	85322	Social work activities for immigrants (non-charitable)
87900	85311	Social work activities with accommodation (charitable)
87900	85312	Social work activities with accommodation (non-charitable)
88990	85321	Social work activities without accommodation (charitable)
88990	85322	Social work activities without accommodation (non-charitable)
88990	85321	Social worker (charitable)
88990	85322	Social worker (non-charitable)
94120	91120	Society of Apothecaries
94120	91120	Society of Arts
94910	91310	Society of Friends
72200	73200	Sociology research and experimental development
25730	28620	Socket set (manufacture)
27330	31200	Sockets (electric) (manufacture)
14310	17710	Socks and stockings (knitted) (manufacture)
14310	17710	Socks for children (manufacture)
14310	17710	Socks for men (manufacture)

SIC 2007	SIC 2003	Activity
14310	17710	Socks for women and girls (manufacture)
23990	26821	Socks made of asbestos (manufacture)
15200	19300	Socks made of leather (manufacture)
46750	51550	Soda (wholesale)
11070	15980	Soda water (manufacture)
20130	24130	Sodium and sodium compounds (manufacture)
46120	51120	Sodium nitrate (commission agent)
46750	51550	Sodium nitrate (wholesale)
27400	31500	Sodium vapour lamp (manufacture)
31090	36110	Sofa (manufacture)
31090	36110	Sofa sets (manufacture)
31090	36110	Sofabeds (manufacture)
11070	15980	Soft drinks (manufacture)
10519	15519	Soft drinks (milk based) (manufacture)
47250	52250	Soft drinks (retail)
46341	51341	Soft drinks (wholesale)
46341	51341	Soft drinks exporter (wholesale)
46341	51341	Soft drinks importer (wholesale)
28930	29530	Soft drinks machinery (manufacture)
13921	17401	Soft furnishing articles, made-up, of knitted or crocheted fabrics (manufacture)
13921	17401	Soft furnishing articles, made-up, of textile material (manufacture)
13921	17401	Soft furnishings (manufacture)
47599	52440	Soft furnishings (retail)
46180	51180	Soft goods (commission agent)
32409	36509	Soft toys (manufacture)
46510	51840	Software (non customised) (wholesale)
47410	52482	Software (non-customised) (retail)
58290	72210	Software (ready-made) publishing (except computer games publishing)
62020	72220	Software consultancy
69109	74111	Software copyright consultancy activities
62090	72220	Software disaster recovery services
62012	72220	Software house
62090	72220	Software installation services
18203	22330	Software reproduction from master copies (manufacture)
62012	72220	Software systems maintenance services
20200	24200	Soil fumigant (manufacture)
26110	32100	Solar panels (photovoltaic cell type) (manufacture)
28210	29720	Solar panels, domestic (other than photovoltaic cell type) (manufacture)
28210	29720	Solar panels, non-domestic (other than photovoltaic cell type) (manufacture)
96040	93040	Solariums
24450	27450	Solder (manufacture)
27900	29430	Soldering irons (electrical, hand-held) (manufacture)
27900	29430	Soldering machines (electric) (manufacture)
28290	29430	Soldering machines (gas) (manufacture)
15110	19100	Sole leather preparation (manufacture)
24100	27100	Sole plates (hot rolled) (manufacture)
26110	32100	Solenoids for electronic apparatus (manufacture)
69102	74113	Solicitor (own account)
19201	10103	Solid fuel briquettes production (manufacture)
46719	51519	Solid fuels (wholesale)
19201	10103	Solid fuels production (manufacture)
24420	27420	Solid sections made of aluminium (manufacture)
19201	10103	Solid smokeless ovoids and briquettes preparation (manufacture)

S

SIC 2007	SIC 2003	Activity
27900	31100	Solid state battery chargers (manufacture)
26110	32100	Solid state circuit (manufacture)
27900	31100	Solid state fuel cells (manufacture)
27900	31100	Solid state inverters (manufacture)
10832	15862	Soluble coffee (manufacture)
10620	15620	Soluble starch (manufacture)
10831	15861	Soluble tea (manufacture)
20120	24120	Solvent dye (manufacture)
28290	29240	Solvent extraction equipment for the chemical industry (manufacture)
28290	29240	Solvent recovery equipment for the chemical industry (manufacture)
26511	33201	Sonar (manufacture)
90030	92319	Song writer
10520	15520	Sorbet production (manufacture)
01110	01110	Sorghum growing
38320	37200	Sorting and pelleting of plastics to produce secondary raw material for tubes, flower pots, pallets
38320	37200	Sorting and pelleting of plastics to produce secondary raw materials
28923	29523	Sorting, grinding and mixing machinery for earth, stones and other mineral substances (manufacture)
23990	26829	Sound absorbing materials (manufacture)
77210	71401	Sound equipment rental
26400	32300	Sound heads (manufacture)
23990	26829	Sound insulating materials (manufacture)
26120	32100	Sound interface cards (manufacture)
29310	31610	Sound or visual signalling equipment for cycles and motor vehicles (manufacture)
26400	32300	Sound recording and reproducing equipment (manufacture)
59200	22140	Sound recording publishing
18201	22310	Sound recording reproduction (manufacture)
59200	92119	Sound recording studios
61100	64200	Sound transmission via cables, broadcasting, relay or satellite
10890	15891	Soup containing meat or vegetables or both (manufacture)
01240	01139	Sour cherry growing
47789	52489	Souvenirs (retail)
10410	15410	Soya bean crushing (manufacture)
10612	15612	Soya bean grinding (manufacture)
10612	15612	Soya bean milling (manufacture)
10410	15410	Soya bean oil (crude) (manufacture)
10410	15420	Soya bean oil refining (manufacture)
10612	15612	Soya flour and meal (manufacture)
01110	01110	Soya growing
27520	29720	Space heaters (gas) (manufacture)
27520	29720	Space heaters (oil) (manufacture)
27510	29710	Space heaters for domestic use (electric) (manufacture)
46439	51439	Space heating and soil heating apparatus (electric) (wholesale)
30300	35300	Space shuttles (manufacture)
51220	62300	Space transport
51220	62300	Space transport of freight
51220	62300	Space transport of passengers
51220	62300	Space vehicle launching
30300	35300	Spacecraft (manufacture)

SIC 2007	SIC 2003	Activity
25730	28620	Spade (manufacture)
10730	15850	Spaghetti (manufacture)
10730	15850	Spaghetti canning (manufacture)
25730	28620	Spanner (manufacture)
28410	29420	Spark erosion machines (metal working) (manufacture)
46690	51870	Spark ignition and compression ignition engines (except motor vehicle and outboard) (wholesale)
28110	29110	Spark ignition engines for industrial use (manufacture)
29310	31610	Sparking plug (manufacture)
11020	15931	Sparkling wine (manufacture)
96040	93040	Spas
15200	19300	Spats made of leather (manufacture)
26400	32300	Speaker systems (manufacture)
90010	92319	Speakers (after dinner etc.)
86101	85112	Special hospital (private sector)
86101	85111	Special hospital (public sector)
29203	34203	Special purpose caravans (manufacture)
46690	51870	Special purpose machinery n.e.c. (wholesale)
17120	21120	Special purpose paper (manufacture)
30200	35200	Special purpose railway wagon (manufacture)
85200	80100	Special schools at primary and pre-primary level
85310	80210	Special schools at secondary level
46499	51479	Special sports footwear such as ski boots (wholesale)
28220	29220	Special steelworks crane (manufacture)
81222	74704	Specialised cleaning of tanks and reservoirs
39000	90030	Specialised pollution control activities
86220	85120	Specialist (not employed full time by a hospital)
86220	85120	Specialist medical consultant (private practice)
86220	85120	Specialist medical consultation and treatment
86220	85120	Specialist physician and surgeon (private practice)
15120	19200	Spectacle cases made of leather (manufacture)
32500	33401	Spectacle frames (manufacture)
23190	26150	Spectacle glass (manufacture)
32500	33401	Spectacle lens (manufacture)
32500	33401	Spectacle mounts (manufacture)
32500	33401	Spectacles (manufacture)
47782	52487	Spectacles (retail)
26511	33201	Spectrofluorimeter (electronic) (manufacture)
26513	33202	Spectrofluorimeter (non-electronic) (manufacture)
26511	33201	Spectrograph (electronic) (manufacture)
26513	33202	Spectrograph (non-electronic) (manufacture)
26511	33201	Spectrometers (electronic) (manufacture)
26513	33202	Spectrometers (non-electronic) (manufacture)
26511	33201	Spectrophotometer (manufacture)
26511	33201	Spectrum analysers (manufacture)
86900	85140	Speech therapist (NHS)
86900	85140	Speech therapist (own account)
28150	29140	Speed changers (manufacture)
46690	51870	Speed indicators and tachometers (wholesale)
85590	80429	Speed reading instruction
26511	33201	Speedometers (electronic) (manufacture)
26513	33202	Speedometers (non-electronic) (manufacture)
93199	92629	Speedway racing
46720	51520	Spelter (wholesale)
38120	40110	Spent (irradiated) fuel elements (cartridges) of nuclear reactors
20130	23300	Spent nuclear fuel re-processing (manufacture)

S

SIC 2007	SIC 2003	Activity
86900	85140	Sperm banks
10410	15420	Sperm oil refining (manufacture)
28150	29140	Spherical roller bearings (manufacture)
24100	27100	Spheroidal graphite pig iron (manufacture)
10840	15870	Spice (ground) (manufacture)
10840	15870	Spice (purifying) (manufacture)
46370	51370	Spice (wholesale)
46170	51170	Spice broker (commission agent)
01280	01139	Spice crops growing
24100	27100	Spiegeleisen (manufacture)
16290	20510	Spills made of wood (manufacture)
27510	29710	Spin dryer
01130	01120	Spinach growing
28940	29540	Spindles (textile machinery accessories) (manufacture)
28940	29540	Spindles and spindle flyers (manufacture)
13100	17160	Spinning and manufacture of yarn for weaving, for the trade or for further processing (manufacture)
13100	17160	Spinning and manufacture of yarn or sewing thread, for the trade or further processing (manufacture)
28410	29420	Spinning lathes (metal working) (manufacture)
28940	29540	Spinning machinery for textiles (manufacture)
28940	29540	Spinning machines (manufacture)
13100	17110	Spinning on the cotton system (manufacture)
13100	17140	Spinning on the flax system (manufacture)
13100	17120	Spinning on the woollen system (manufacture)
13100	17130	Spinning on the worsted and semi-worsted systems (manufacture)
16290	20510	Spinning wheels made of wood (manufacture)
18140	22230	Spiral binding and finishing of books and brochures (manufacture)
11010	15910	Spirit distilling and compounding (manufacture)
26513	33202	Spirit level (manufacture)
20140	24140	Spirit of turpentine (manufacture)
46342	51342	Spirits (wholesale)
94910	91310	Spiritualist church
96090	93059	Spiritualists' activities
32500	33100	Splints (manufacture)
02200	02010	Split pole production
16100	02010	Split poles, pickets and other products (manufacture)
28490	29430	Splitting or cleaving machines for stone, ceramics, asbestos-cement or similar (manufacture)
28490	29430	Splitting, stamping and fragmenting machines (manufacture)
02200	02010	Splitwood production
25730	28620	Spoke shave (manufacture)
32990	36639	Sponge bleaching (manufacture)
13990	36639	Sponge dressing (manufacture)
03110	05010	Sponge gathering
46210	51210	Sponge importer (wholesale)
24100	27100	Sponge iron (manufacture)
32990	36639	Sponge trimming (manufacture)
22190	25130	Sponges made of rubber (manufacture)
28940	29540	Spool (textile machinery accessory) (manufacture)
28940	29540	Spooling machinery for carpet making (manufacture)
17290	21259	Spools made of paper (manufacture)
16290	20510	Spools made of wood (manufacture)
25710	28610	Spoons made of metal (manufacture)
22290	25240	Spoons made of plastic (manufacture)
16290	20510	Spoons made of wood (manufacture)

SIC 2007	SIC 2003	Activity
14190	18249	Sporran (manufacture)
85510	92629	Sport and game schools
42990	45230	Sport facilities construction
32300	36400	Sport fishing requisites (manufacture)
46160	51160	Sport nets (commission agent)
25400	29600	Sporting carbine (manufacture)
46180	51180	Sporting goods (commission agent)
25400	29600	Sporting gun (manufacture)
25400	29600	Sporting rifle (manufacture)
32300	36400	Sports and outdoor and indoor games, articles and equipment (manufacture)
85510	92629	Sports and recreation education
42990	45230	Sports and recreation grounds, laying out
93110	92629	Sports arenas
15120	19200	Sports bags made of fabric (manufacture)
15120	19200	Sports bags made of leather (manufacture)
46420	51429	Sports clothes (wholesale)
93120	92629	Sports club
32300	36400	Sports equipment made of plastic (manufacture)
77210	71401	Sports equipment rental
93110	92619	Sports facilities operation
15200	19300	Sports footwear (manufacture)
32300	36400	Sports gloves (specialist) (manufacture)
32300	36400	Sports goods (manufacture)
47640	52485	Sports goods (retail)
46499	51479	Sports goods (wholesale)
15120	19200	Sports goods carrier (manufacture)
65120	66031	Sports insurance
93199	92629	Sports leagues and regulating bodies
13940	17520	Sports nets (manufacture)
47640	52485	Sports outfitter (retail)
93199	92629	Sports referees
93199	92629	Sportsmen and sportswomen
14131	18221	Sportswear for men and boys (manufacture)
14132	18222	Sportswear for women and girls (manufacture)
27400	31500	Spotlight (manufacture)
28940	29540	Spotting table (manufacture)
28290	29240	Spray guns (manufacture)
29100	34100	Spraying lorry (manufacture)
28302	29320	Spraying machine for agricultural use (manufacture)
28290	29240	Spraying machinery parts (manufacture)
31090	36140	Spraying of furniture (manufacture)
28290	29240	Spring balance (manufacture)
25930	28740	Spring presswork (manufacture)
29320	34300	Spring suspension for motor vehicles (manufacture)
25930	28740	Spring washer (manufacture)
31030	36150	Spring wire mattress (manufacture)
25930	28740	Springs (manufacture)
25930	28740	Springs for upholstery (manufacture)
25930	28740	Springs made of steel for upholstery (manufacture)
28290	29240	Sprinklers for fire extinguishing (manufacture)
28150	29140	Sprocket chain (manufacture)
09100	11200	Spudding in for oil wells
93120	92629	Squash club
11070	15980	Squash drink (manufacture)
32300	36400	Squash racket (manufacture)
32910	36620	Squeegees (manufacture)
28490	29430	Squeeze presses (manufacture)

S

SIC 2007	SIC 2003	Activity
86900	85140	St Andrew's ambulance brigade
93290	92729	St James's park
86900	85140	St John's ambulance brigade
28990	29560	Stabiliser for ship (manufacture)
20590	24660	Stabilisers and extenders for PVC processing (manufacture)
20590	24660	Stabilisers for rubber or plastics (manufacture)
26110	32100	Stabilising valve (manufacture)
28220	29220	Stacking machines (manufacture)
42990	45230	Stadium construction
93110	92629	Stadium operation
56290	55520	Staff canteen (run by catering contractor)
27400	31500	Stage lighting (manufacture)
90010	92311	Stage productions
90020	92311	Stage set designers and builders
93110	92629	Staging of sports events by organizations with their own facilities
20301	24301	Stain (manufacture)
23120	26120	Stained glass (manufacture)
24100	27100	Stainless steel (manufacture)
25720	28630	Stair rods made of metal (manufacture)
16230	20300	Stair rods made of wood (manufacture)
16230	20300	Staircase made of wood (manufacture)
02200	02010	Stakes (of wood) production
47789	52485	Stamp dealer (retail)
18129	22220	Stamp embossed paper (manufacture)
17290	21259	Stamp hinges (manufacture)
25500	28400	Stamping of base metal (manufacture)
28410	29420	Stamping or pressing machine tools (manufacture)
47789	52485	Stamps (retail)
32990	36631	Stamps made of rubber (manufacture)
50200	61102	Standby vessel services
20600	24700	Staple fibre of acetate, synthetic or viscose production (manufacture)
28230	30010	Staple removers (manufacture)
28230	30010	Staplers (manufacture)
25930	28730	Staples (not wire) (manufacture)
25990	28750	Staples for office use (manufacture)
20600	24700	Staples of man-made fibre (manufacture)
28490	29430	Stapling machines (machine tools) (manufacture)
28490	29430	Stapling machines for industrial use (manufacture)
28230	30010	Stapling machines for office use (manufacture)
10620	15620	Starch (manufacture)
46380	51380	Starch (wholesale)
20520	24620	Starch based adhesives (manufacture)
46750	51550	Starch derivatives (wholesale)
29310	31610	Starter motor for vehicle (manufacture)
68100	70120	Static caravan sales
46690	51870	Static converters (wholesale)
25110	28110	Static drilling derricks (manufacture)
29100	34100	Station wagon (manufacture)
28490	29430	Stationary machines for nailing, glueing or otherwise assembling wood, cork bone, etc. (manufacture)
28490	29430	Stationary rotary or rotary percussion drills, filing machines, sheet metal cutters (manufacture)
46499	51479	Stationers' sundries (wholesale)
46180	51180	Stationery (commission agent)
47620	52470	Stationery (retail)
46499	51479	Stationery (wholesale)

SIC 2007	SIC 2003	Activity
17230	21230	Stationery paper (manufacture)
84110	75110	Statistical services (public sector)
23690	26660	Statuary made of concrete, plaster, cement or artificial stone (manufacture)
25990	28750	Statuettes and other ornaments made of base metal (manufacture)
46150	51150	Statuettes and other ornaments made of wood (commission agent)
23410	26210	Statuettes made of ceramic (manufacture)
22290	25240	Statuettes made of plastic (manufacture)
16290	20510	Statuettes made of wood (manufacture)
28490	29430	Stave and cask croze cutting machines (manufacture)
28490	29430	Stave jointing, planing, bending machinery (manufacture)
16240	20400	Staves made of wood (manufacture)
56101	55301	Steak houses (licensed)
25300	28300	Steam accumulator (manufacture)
28110	29110	Steam and other vapour turbine parts (manufacture)
46690	51870	Steam and sand blasting machinery and appliances (excluding agricultural) (wholesale)
96040	93040	Steam baths
25300	28300	Steam boiler (manufacture)
43999	45250	Steam cleaning for building exteriors
28290	29240	Steam cleaning machines (manufacture)
43390	45450	Steam cleaning of buildings
25300	28300	Steam collector (manufacture)
28110	29110	Steam engine (manufacture)
25300	28300	Steam generator (manufacture)
25300	28300	Steam generator parts (manufacture)
46690	51870	Steam generators (wholesale)
35300	40300	Steam production, collection and distribution
28120	29121	Steam pulsators (pulsometers) (manufacture)
30200	35200	Steam rail locomotives (manufacture)
28110	29110	Steam turbine (not marine or for electricity generation) (manufacture)
28110	29110	Steam turbine for marine use (manufacture)
28110	29110	Steam turbines and other vapour turbines (manufacture)
28940	29540	Steaming machinery for textiles (manufacture)
13300	17300	Steaming of textiles and textile articles, including wearing apparel (manufacture)
20140	24140	Stearic acid (manufacture)
20590	24660	Stearin (manufacture)
08990	14500	Steatite (talc) mining and quarrying
46720	51520	Steel (wholesale)
43999	45250	Steel bending
43999	45250	Steel elements (not self-manufactured) erection
24520	27520	Steel founders (manufacture)
24100	27100	Steel in primary form, from ore or scrap, production (manufacture)
24100	27100	Steel making pig iron (manufacture)
23200	26260	Steel moulder's composition (manufacture)
24100	27100	Steel powders (manufacture)
24100	27100	Steel sheet (not finally annealed) (manufacture)
24100	27100	Steel shot (manufacture)
46720	51520	Steel stockholder (wholesale)
25990	28750	Steel wool for domestic use (manufacture)
28210	29210	Steelmaking furnace (manufacture)
43999	45250	Steelwork erection (building)
43999	45250	Steelwork erection (civil engineering)

SIC 2007	SIC 2003	Activity	SIC 2007	SIC 2003	Activity
25110	28110	Steelwork for agricultural buildings (manufacture)	28220	29220	Stillage truck (manufacture)
25110	28110	Steelwork for bridges (manufacture)	23200	26260	Stilts made of ceramic (manufacture)
25110	28110	Steelwork for buildings (manufacture)	84230	75230	Stipendiary magistrates
25110	28110	Steelwork for commercial buildings (manufacture)	46640	51830	Stitch bonding machinery (wholesale)
25110	28110	Steelwork for distribution depots (manufacture)	28990	29560	Stitching machine for bookbinding (manufacture)
25110	28110	Steelwork for docks (manufacture)	18140	22230	Stitching of books, brochures, etc. (manufacture)
25110	28110	Steelwork for domestic buildings (manufacture)	18129	22220	Stochastic printing (manufacture)
25110	28110	Steelwork for exhibition centres (manufacture)	82990	74879	Stock control activities
25110	28110	Steelwork for factory buildings (manufacture)	66110	67110	Stock exchange activities
25110	28110	Steelwork for glass roofs (manufacture)	64991	65233	Stock exchange money broker activities
25110	28110	Steelwork for harbours (manufacture)	59120	92119	Stock footage, film library activities
25110	28110	Steelwork for hospital buildings (manufacture)	49410	60249	Stock haulage by road
25110	28110	Steelwork for jetties (manufacture)	63910	92400	Stock market reporting service
25110	28110	Steelwork for school buildings (manufacture)	66110	67110	Stock or commodity options exchanges administration
25110	28110	Steelwork for tunnels (manufacture)	28950	29550	Stock preparation plant for paper and board (manufacture)
25110	28110	Steelwork for viaducts (manufacture)	14310	17710	Stockinette goods (manufacture)
43999	45250	Steeplejacking	14310	17710	Stockings (manufacture)
29320	34300	Steering box for motor vehicle (manufacture)	14310	17710	Stockings for women and girls (manufacture)
30920	35430	Steering column and gear for powered invalid carriage (manufacture)	14200	18300	Stoles made of fur (manufacture)
29320	34300	Steering column for motor vehicle (manufacture)	23610	26610	Stone articles for use in construction (manufacture)
29320	34300	Steering equipment components for motor vehicles (manufacture)	43999	45250	Stone carving
28990	29560	Steering gear for marine use (manufacture)	08120	14210	Stone chippings production
29320	34300	Steering wheels for motor vehicle (manufacture)	08120	14210	Stone dust production
23130	26130	Stemmed drinking vessels made of glass (manufacture)	01240	01139	Stone fruit growing
			23700	26700	Stone furniture (manufacture)
17120	21120	Stencil basepaper (manufacture)	43999	45250	Stone setting
17230	21230	Stencil duplicating (manufacture)	43999	45250	Stone walling
28230	30010	Stencil duplicating machines (manufacture)	23700	26700	Stone working (manufacture)
17120	21120	Stencil paper in large sheets (manufacture)	43999	45250	Stonemasonry (building)
17120	21120	Stencil paper in rolls (manufacture)	46730	51530	Stones (wholesale)
28230	30010	Stenography machines (manufacture)	23910	26810	Stones for sharpening or polishing (manufacture)
28940	29540	Stentering machinery for textiles (manufacture)	23410	26210	Stoneware for domestic use (manufacture)
16290	20510	Step ladders made of wood (manufacture)	43390	45450	Stonework cleaning and renovation
25990	28750	Steps made of metal (manufacture)	26520	33500	Stop watch (manufacture)
16290	20510	Steps made of wood (manufacture)	28140	29130	Stopcocks for domestic use (manufacture)
26400	32300	Stereo systems (manufacture)	16290	20520	Stopper insets made of cork (manufacture)
18130	22240	Stereotyping (manufacture)	16290	20520	Stoppers made of cork (manufacture)
28930	29530	Sterilisation equipment for food and drink (manufacture)	23130	26130	Stoppers made of glass (manufacture)
81222	74704	Sterilisation of objects or premises (e.g. operating theatres)	25920	28720	Stoppers made of metal (manufacture)
			31090	36140	Storage cabinets for domestic use (manufacture)
10110	15112	Sterilised bone flour (not for fertilisers) (manufacture)	52102	63129	Storage facilities n.e.c. for air transport activities
10511	15511	Sterilised cream (manufacture)	52103	63129	Storage facilities n.e.c. for land transport activities
46460	51460	Sterilisers for medical, surgical or laboratory use (wholesale)	52101	63129	Storage facilities n.e.c. for water transport activities
32500	33100	Sterilising equipment for medical use (manufacture)	27510	29710	Storage heaters (manufacture)
32500	33100	Sterilizers (manufacture)	52102	63129	Storage of goods in foreign trade zones for air transport activities
28990	29560	Stern gear (manufacture)	52103	63129	Storage of goods in foreign trade zones for land transport activities
46180	51180	Steroids used primarily as hormones (commission agent)	52101	63129	Storage of goods in foreign trade zones for water transport activities
46460	51460	Steroids used primarily as hormones (wholesale)	25290	28210	Storage tanks made of heavy steel plate exceeding 300 litres (manufacture)
24450	27450	Sterro metal (manufacture)	22230	25239	Storage tanks made of plastic (manufacture)
52241	63110	Stevedoring	52102	63122	Storage tanks operation for air transport activities
25990	28750	Stewpans (manufacture)	52103	63122	Storage tanks operation for land transport activities
32500	24422	Sticking plaster (surgical) (manufacture)	52101	63122	Storage tanks operation for water transport activities
13960	17549	Stiffened textile fabrics (manufacture)	80100	74602	Store detective activities
25990	28750	Stillage made of metal (manufacture)	13300	17300	Storing yarn (manufacture)
16240	20400	Stillage made of wood (manufacture)			

S

SIC 2007	SIC 2003	Activity
11050	15960	Stout brewing (manufacture)
27520	29720	Stove (gas) (manufacture)
27520	29720	Stove (oil) (manufacture)
27520	29720	Stove (solid fuel) (manufacture)
81223	74705	Stove cleaning
25610	28510	Stove painting (manufacture)
27520	29720	Stoves for domestic use (non-electric) (manufacture)
28921	29521	Stowing machine for mining (manufacture)
28220	29220	Straddle carrier (manufacture)
46690	51870	Straddle carriers and works trucks fitted with a crane (wholesale)
10390	15330	Strained fruit (manufacture)
10390	15330	Strained vegetables (manufacture)
13960	17549	Straining cloth (manufacture)
25930	28730	Stranded un-insulated wire made of copper (manufacture)
46690	51870	Stranded wires, cables, plaited bands, slings and the like (not electrically insulated) (wholesale)
19209	23209	Strap paste for transmission belts (manufacture)
15120	19200	Straps made of leather (manufacture)
70229	74143	Strategic business plan consultancy services
46210	51210	Straw (wholesale)
46420	51429	Straw and felt hats (wholesale)
16290	20520	Straw articles (manufacture)
14190	18241	Straw hat blocking (manufacture)
14190	18241	Straw hats (manufacture)
01250	01139	Strawberry growing
17120	21120	Strawboard (manufacture)
17120	21120	Strawpaper (manufacture)
49319	60219	Street car (scheduled passenger transport)
81299	90030	Street cleaning and watering
42110	45230	Street construction
25990	28750	Street furniture (manufacture)
27400	31500	Street lighting fixtures (manufacture)
90010	92311	Street musician or singer
80100	74602	Street patrol
74209	74819	Street photographer
29100	34100	Street sweepers (manufacture)
29100	34100	Street sweeping lorry (manufacture)
71200	74300	Strength and failure testing
13921	17401	Stretch covers for furniture (manufacture)
13922	17402	Stretchers made of canvas (manufacture)
25990	28750	Striking of medals (manufacture)
13940	17520	String (manufacture)
13940	17520	String bag (manufacture)
23990	26821	String made of asbestos (manufacture)
32200	36300	Stringed instruments (manufacture)
32200	36300	Strings for musical instruments (manufacture)
16220	20300	Strip flooring made of hardwood (manufacture)
28210	29210	Strip processing line furnace (manufacture)
24420	27420	Strips made of aluminium (manufacture)
24440	27440	Strips made of brass (manufacture)
24440	27440	Strips made of copper (manufacture)
22210	25210	Strips made of plastic (manufacture)
22190	25130	Strips made of rubber (manufacture)
26513	33202	Stroboscope (manufacture)
46690	51870	Stroboscopes (wholesale)
25990	28750	Strong box (manufacture)
25990	28750	Strong room (manufacture)

SIC 2007	SIC 2003	Activity
25990	28750	Strong room door (manufacture)
15120	19200	Strops made of leather (manufacture)
24100	27100	Structural steel (manufacture)
43999	45250	Structural steelwork erection (building)
43999	45250	Structural steelwork erection (civil engineering)
25110	28110	Structural steelwork for buildings (manufacture)
23610	26610	Structural wall panels made of precast concrete (manufacture)
25110	28110	Structures for buildings made of aluminium (manufacture)
25110	28110	Structures for civil engineering made of aluminium (manufacture)
43310	45410	Stucco application in buildings
01430	01220	Stud farming
25930	28740	Stud link chain (manufacture)
01629	01429	Stud services on a fee or contract basis
94990	91330	Student associations
94910	91310	Student Christian Movement
55900	55239	Student house accommodation
55900	55239	Student residences
94990	91330	Student Union
31090	36110	Studio couch (manufacture)
85421	80302	Study leading to a one year post graduate certificate of education (PGCE)
32409	36509	Stuffed toy (manufacture)
10840	15870	Stuffing (manufacture)
32990	36631	Stylographic pen (manufacture)
26400	32300	Stylus for record player (manufacture)
20140	24140	Styrene (manufacture)
20160	24160	Styrene polymers (manufacture)
53100	64110	Sub post office (principally devoted to post office business)
30110	35110	Submarine (manufacture)
27320	31300	Submarine cable (manufacture)
28131	29121	Submersible motor pump (manufacture)
26301	32201	Subscriber apparatus (telephone) (manufacture)
27110	31100	Substation transformers for electric power distribution (manufacture)
71122	74206	Sub-surface surveying activities
43999	45240	Subsurface work
49311	60213	Suburban area passenger railway transportation by underground, metro and similar systems
49319	60219	Suburban scheduled passenger land transport other than by underground, metro and similar systems
42120	45213	Subway construction
22190	25130	Suction and discharge hose made of rubber (manufacture)
10110	15112	Suet (manufacture)
46360	51360	Sugar (wholesale)
46360	51360	Sugar and chocolate and sugar confectionery exporter (wholesale)
46360	51360	Sugar and chocolate and sugar confectionery importer (wholesale)
01130	01110	Sugar beet growing
28302	29320	Sugar beet harvester (manufacture)
01130	01110	Sugar beet seed production
01140	01110	Sugar cane growing
10822	15842	Sugar confectionery (manufacture)
47240	52240	Sugar confectionery (retail)
28930	29530	Sugar confectionery making machinery (manufacture)

SIC 2007	SIC 2003	Activity	SIC 2007	SIC 2003	Activity
46180	51180	Sugar ethers, sugar esters and their salts and chemically pure sugar (commission agent)	47110	52113	Superstore (selling mainly foodstuffs) without alcohol licence (retail)
28930	29530	Sugar making and refining machinery (manufacture)	56102	55302	Supper bar or room (unlicensed)
10810	15830	Sugar milling (manufacture)	78200	74500	Supply and provision of personnel (temporary employment agency)
10810	15830	Sugar refining (manufacture)	42120	45213	Supply line (third rail) for railway construction
10810	15830	Sugar substitutes refining and production (manufacture)	90020	92320	Support activities to performing arts e.g. stage set-up, costume and lighting design etc.
10810	15830	Sugar sucrose and sugar substitutes from cane, manufacture or refining (manufacture)	84220	75220	Support for defence plans and exercises for civilians
10810	15830	Sugar sucrose and sugar substitutes from maple, manufacture or refining (manufacture)	84230	75230	Supreme court of judicature
10810	15830	Sugar sucrose and sugar substitutes from palm, manufacture or refining (manufacture)	20590	24660	Surface active chemicals (excluding finished detergents and scouring powder) (manufacture)
10810	15830	Sugar syrups (manufacture)	20411	24511	Surface active preparations (manufacture)
25720	28630	Suitcase fittings made of metal (manufacture)	05102	10102	Surface mining of hard coal
15120	19200	Suitcases made of leather or leather substitute (manufacture)	28410	29420	Surface tempering machines (manufacture)
16240	20400	Suitcases made of wood (manufacture)	26511	33201	Surface tension instruments (electronic) (manufacture)
13200	17220	Suiting woollen weaving (manufacture)	26513	33202	Surface tension instruments (non-electronic) (manufacture)
13200	17230	Suiting worsted weaving (manufacture)	42110	45230	Surface work on elevated highways, bridges and in tunnels
14131	18221	Suits for men and boys (manufacture)			
14132	18222	Suits for women and girls (dressmade) (manufacture)	42110	45230	Surface work on streets, roads, highways, bridges or tunnels
17110	21110	Sulphate and soda woodpulp (manufacture)	32300	36400	Surfboard (manufacture)
19100	23100	Sulphate of ammonia from coke ovens (manufacture)	28990	29560	Surge damper for hydraulic equipment (manufacture)
35210	40210	Sulphate of ammonia from gas works	27900	31200	Surge suppressors (manufacture)
46750	51550	Sulphides, sulphites and sulphates (wholesale)	86220	85120	Surgeon (private practice)
17110	21110	Sulphite woodpulp (manufacture)	86101	85111	Surgeon (public sector)
17120	21120	Sulphite wrapping paper (manufacture)	86210	85120	Surgery (doctor's)
46180	51180	Sulphonamides (commission agent)	46460	51460	Surgical and dental instruments and appliances (wholesale)
21100	24410	Sulphonamides (manufacture)	47749	52329	Surgical appliances (retail)
46460	51460	Sulphonamides (wholesale)	32500	24422	Surgical bandage (manufacture)
20130	24130	Sulphur (manufacture)	32500	33100	Surgical belts (manufacture)
46750	51550	Sulphur (wholesale)	32500	33100	Surgical boot (manufacture)
20120	24120	Sulphur dye (manufacture)	32500	33100	Surgical corset (manufacture)
20130	24130	Sulphuric acid (manufacture)	32500	17403	Surgical drapes and sterile string and tissue (manufacture)
13922	17402	Sun blinds (manufacture)			
32990	36639	Sun car (manufacture)	32500	24422	Surgical dressing (manufacture)
46499	51479	Sun umbrellas and garden umbrellas (wholesale)	32500	33100	Surgical equipment (manufacture)
20420	24520	Sunburn prevention and sun tan preparations (manufacture)	32500	24422	Surgical gauze (manufacture)
			32500	25130	Surgical goods made of rubber (manufacture)
10410	15420	Sunflower oil refining (manufacture)	32500	24422	Surgical gut string (manufacture)
10410	15410	Sunflower seed crushing (manufacture)	32500	33100	Surgical hosiery (manufacture)
01110	01110	Sunflower seed growing	32500	33100	Surgical implants (manufacture)
01110	01110	Sunflower seed production	32500	33100	Surgical instrument (manufacture)
10410	15410	Sunflower-seed oil production (manufacture)	32500	24422	Surgical lint (manufacture)
23190	26150	Sunglass blank (manufacture)	32500	24422	Surgical sutures (manufacture)
32500	33401	Sunglasses (manufacture)	32500	33100	Surgical truss (manufacture)
13100	17170	Sunn hemp (manufacture)	32500	24422	Surgical wadding (manufacture)
32990	36639	Sunshade (manufacture)	14120	18210	Surplice (manufacture)
32990	36639	Sun-umbrellas (manufacture)	80100	74602	Surveillance activities
65300	66020	Superannuation fund (autonomous)	73200	74130	Survey analysis and other social and economic intelligence services
25300	28300	Super-heaters (manufacture)			
47110	52112	Supermarket (selling mainly foodstuffs) with alcohol licence (retail)	73200	74130	Survey design services
			71122	74206	Surveying activities (industrial and engineering)
47110	52113	Supermarket (selling mainly foodstuffs) without alcohol licence (retail)	26511	33201	Surveying instruments (electronic) (manufacture)
20150	24150	Superphosphate (manufacture)	26513	33202	Surveying instruments (non-electronic) (manufacture)
23510	26510	Superphosphate cements (manufacture)			
47110	52112	Superstore (selling mainly foodstuffs) with alcohol licence (retail)	26513	33202	Surveying instruments (optical non-electronic) (manufacture)

S

SIC 2007	SIC 2003	Activity
26701	33201	Surveying instruments (optical) (manufacture)
46690	51870	Surveying, hydrographic, oceanographic, hydrological and meteorological instruments (wholesale)
71122	74206	Surveyor (other than valuer)
68310	70310	Surveyor and valuer (real estate)
85590	80429	Survival training
14142	18232	Suspender (manufacture)
14142	18232	Suspender belt (manufacture)
28220	29220	Suspension railway (manufacture)
29320	34300	Suspension shock absorbers for motor vehicle (manufacture)
29320	34300	Suspension springs for motor vehicle (manufacture)
28410	29420	Swaging machine (metal forming) (manufacture)
64999	65239	Swaps, options and other hedging arrangements
01130	01120	Swede growing
10720	15820	Sweet crackers (manufacture)
01130	01120	Sweet marjoram growing
01130	01110	Sweet potato growing
10519	15519	Sweetened skimmed whey production (manufacture)
47240	52240	Sweets (retail)
10822	15842	Sweets (sugar confectionery) (manufacture)
46499	51479	Swimming and paddling pools (wholesale)
93110	92619	Swimming baths
93120	92629	Swimming clubs
85510	92629	Swimming instruction
81299	74709	Swimming pool cleaning and maintenance activities
93110	92619	Swimming pools
41201	45230	Swimming pools construction
14190	18249	Swimwear (manufacture)
01460	01230	Swine farming
01460	01230	Swine raising and breeding
28990	36639	Swings (fairground equipment) (manufacture)
32300	36400	Swings for playgrounds (manufacture)
13990	17541	Swiss embroidery (manufacture)
27330	31200	Switch (electric) (manufacture)
27330	31200	Switch boxes for electrical wiring (manufacture)
93210	92330	Switchback (fairground)
26301	32201	Switchboard for telecommunications (manufacture)
46690	51870	Switches (wholesale)
26110	31200	Switches and transducers for electronic applications (manufacture)
27120	31200	Switchgear (power) (manufacture)
52219	60109	Switching and shunting
52219	60101	Switching and shunting on railways
26301	32201	Switching equipment for telegraph and telex (manufacture)
25710	28750	Sword (manufacture)
46690	51870	Swords, cutlasses, bayonets, lances and similar arms (wholesale)
71200	74300	Sworn timber measurer
71200	74300	Sworn weigher
94910	91310	Synagogue
20140	24140	Synthetic alcohol (manufacture)
46120	51120	Synthetic and organic colouring matter, colouring lakes and preparations (commission agent)
10910	15710	Synthetic animal feed protein (manufacture)
20140	24140	Synthetic aromatic products (manufacture)
20411	24511	Synthetic detergent (manufacture)
20120	24120	Synthetic dyestuffs (manufacture)
20140	24140	Synthetic ethyl alcohol (manufacture)
20600	24700	Synthetic fibre (manufacture)
17110	21110	Synthetic fibre woodpulp (manufacture)
20140	24140	Synthetic glycerol (manufacture)
20120	24120	Synthetic iron oxide (manufacture)
20140	24140	Synthetic or reconstructed precious or semi-precious stones (manufacture)
46120	51120	Synthetic or reconstructed precious or semi-precious unworked stones (commission agent)
46760	51550	Synthetic or reconstructed precious or semi-precious unworked stones (wholesale)
46750	51550	Synthetic organic colouring matter and colouring lakes and preparations based on them (wholesale)
20120	24120	Synthetic organic pigment (manufacture)
32120	36220	Synthetic precious and semi-precious stones (manufacture)
46750	51550	Synthetic resin (wholesale)
20520	24620	Synthetic resin adhesive (manufacture)
20160	24160	Synthetic resin adhesive (unformulated) (manufacture)
20160	24160	Synthetic resins (manufacture)
46120	51120	Synthetic rubber (commission agent)
20170	24170	Synthetic rubber (manufacture)
46760	51550	Synthetic rubber (wholesale)
20170	24170	Synthetic rubber and natural rubber mixtures (manufacture)
46120	51120	Synthetic, organic or inorganic tanning extracts and preparations (commission agent)
23130	26130	Syphons made of glass (manufacture)
32500	33100	Syringes (manufacture)
46460	51460	Syringes, needles, catheters and cannulae (wholesale)
10810	15830	Syrup (sugar) (manufacture)
62012	72220	System maintenance and support services
62020	72220	System software acceptance testing consultancy services
62012	72220	Systems analysis (computer)
62020	72220	Systems and technical consultancy services
10390	15330	Table jelly (manufacture)
13923	17403	Table linen (manufacture)
47510	52410	Table linen (retail)
13923	17403	Table mats made of textiles (manufacture)
32409	36509	Table or parlour games (manufacture)
13923	17403	Table runner (manufacture)
32300	36400	Table tennis ball (manufacture)
32300	36400	Table tennis equipment (manufacture)
11070	15980	Table water (manufacture)
47510	52410	Table-cloths (retail)
17220	21220	Table-cloths made of paper (manufacture)
32401	36501	Tables for casino games (manufacture)
31090	36140	Tables for domestic use (manufacture)
31010	36120	Tables for office or school (manufacture)
26400	32300	Tables for turn-tables (manufacture)
25730	29560	Tableting and pelleting press for the chemical industry (manufacture)
46150	51150	Tableware and kitchenware made of wood (commission agent)
25990	28750	Tableware made of base metal (manufacture)
23410	26210	Tableware made of ceramic (manufacture)
23130	26130	Tableware made of glass (manufacture)
23130	26130	Tableware made of lead crystal (manufacture)

SIC 2007	SIC 2003	Activity
22290	25240	Tableware made of plastic (manufacture)
32120	36220	Tableware made of precious metal (manufacture)
77299	71409	Tableware rental and leasing
17290	21259	Tabulating machine cards (manufacture)
28230	30010	Tabulating machines (manufacture)
63110	72300	Tabulating service
26511	33201	Tachometer (electronic) (manufacture)
26513	33202	Tachometer (non-electronic) (manufacture)
25930	28730	Tack made of metal (not wire) (manufacture)
13200	17240	Taffeta made of silk (manufacture)
13200	17220	Taffeta woollen weaving (manufacture)
19201	23201	Tail gas (manufacture)
28220	29220	Tailboard lift (manufacture)
14131	18221	Tailored outerwear for men and boys (manufacture)
14132	18222	Tailored outerwear for women and girls (manufacture)
14132	18222	Tailored skirts (manufacture)
32990	36631	Tailors' chalk (manufacture)
32990	36639	Tailors' dummy (not plastic) (manufacture)
22290	36639	Tailors' dummy made of plastic (manufacture)
14190	18249	Tailors' pad (manufacture)
25710	28610	Tailors' shears (manufacture)
46770	51570	Tailors' trimmings (wholesale)
56103	55303	Take away food shop
56103	55304	Take-out eating places
08990	14500	Talc mine or quarry
20420	24520	Talcum powder (manufacture)
46760	51560	Tallow (wholesale)
26513	33202	Tally counters (non-electronic) (manufacture)
46630	51820	Tamping machines (wholesale)
17220	21220	Tampons made of paper and cellulose wadding (manufacture)
17220	17549	Tampons made of textile wadding (manufacture)
30920	35420	Tandems (manufacture)
01230	01139	Tangerine growing
30110	35110	Tanker (ship) (manufacture)
29202	34202	Tanker trailer (motor drawn) (manufacture)
30200	35200	Tanker wagon for railway (manufacture)
30400	29600	Tanks (tracked armoured fighting vehicles) (manufacture)
30400	29600	Tanks and other fighting vehicles (manufacture)
25290	28210	Tanks made of galvanised steel exceeding 300 litres (manufacture)
23190	26150	Tanks made of glass (manufacture)
25290	28210	Tanks made of metal exceeding 300 litres (manufacture)
25290	28210	Tanks made of metal for domestic storage exceeding 300 litres (manufacture)
22230	25239	Tanks made of plastic (open and closed) (manufacture)
46690	51870	Tanks, casks, drums, cans, boxes and similar containers (excluding for gas) (wholesale)
20120	24120	Tanning agents (synthetic) (manufacture)
28990	31620	Tanning beds (manufacture)
15110	19100	Tanning leather (manufacture)
15110	18300	Tanning of fur skins and hides with the hair on (manufacture)
15110	19100	Tanning of hairless skins and hides (manufacture)
46750	51550	Tanning preparations (wholesale)
24450	27450	Tantalum (manufacture)
07290	13200	Tantalum mining and preparation
28140	29130	Tap parts (manufacture)
13300	17300	Tape bleaching (manufacture)
28230	30010	Tape dispensers (manufacture)
28290	33202	Tape measure (manufacture)
28230	30010	Tape office-type binding equipment (manufacture)
26400	32300	Tape player and recorder (audio and visual) (manufacture)
18201	22310	Tape pre-recording, except master copies for records or audio material (manufacture)
26200	30020	Tape reader for computers (manufacture)
26400	32300	Tape recorder cabinets made of wood (manufacture)
47430	52450	Tape recorders (retail)
26200	30020	Tape streamers and other magnetic tape storage units (manufacture)
59200	92201	Taped radio programming production
28150	29140	Tapered roller bearings (manufacture)
32990	36639	Tapers and the like (manufacture)
91011	92510	Tapes (music and video) lending and storage
46431	51431	Tapes (recorded) (wholesale)
23990	26821	Tapes made of asbestos (manufacture)
13960	17542	Tapes made of non-elastic and non-elastomeric textile materials (manufacture)
13200	17210	Tapestry (not woollen or worsted) (manufacture)
47510	52410	Tapestry making materials (retail)
13200	17220	Tapestry woollen weaving (manufacture)
10620	15620	Tapioca (manufacture)
10620	15620	Tapioca grinding (manufacture)
28140	29130	Taps (manufacture)
22230	25239	Taps and valves made of plastic (manufacture)
28140	29130	Taps for domestic use (manufacture)
20140	24140	Tar acids (manufacture)
28923	29523	Tar laying plant (manufacture)
28923	29523	Tar processing plant (manufacture)
42110	45230	Tar spraying contractor (civil engineering)
42110	45230	Tarmacadam laying contracting
28923	29523	Tarmacadam laying plant (manufacture)
28923	29523	Tarmacadam processing plant (manufacture)
08120	14210	Tarmacadam (coated) production
13922	17402	Tarpaulins (manufacture)
47510	52410	Tarpaulins (retail)
46410	51410	Tarpaulins (wholesale)
46160	51160	Tarpaulins, awnings, sunblinds and circus tents (commission agent)
01130	01120	Tarragon growing
10710	15810	Tarts (manufacture)
13960	17542	Tassels made of textile material (manufacture)
91020	92521	Tate Gallery
96090	93059	Tattooist
56302	55402	Taverns (independent)
56302	55404	Taverns (managed)
56302	55403	Taverns (tenanted)
69203	74123	Tax consultancy
84110	75110	Tax violation investigation services
84110	75110	Taxation schemes
18129	22220	Taxation stamps printing (manufacture)
29100	34100	Taxi (manufacture)
49320	60220	Taxi cab service
32990	36639	Taxidermy activities (manufacture)

T

137

SIC 2007	SIC 2003	Activity
26511	33201	Taximeters (electronic) (manufacture)
26513	33202	Taximeters (non-electronic) (manufacture)
46690	51870	Taximeters (wholesale)
10831	15861	Tea (packing into tea bags) (manufacture)
46370	51370	Tea (wholesale)
47290	52270	Tea and coffee grocer (retail)
10831	15861	Tea and maté blending (manufacture)
56103	55304	Tea bar
10831	15861	Tea blending (manufacture)
16240	20400	Tea chests made of wood (manufacture)
46170	51170	Tea exchange (commission agent)
10831	15861	Tea extract and essence (manufacture)
56102	55302	Tea garden (unlicensed)
01270	01139	Tea growing
47290	52270	Tea merchant (retail)
46370	51370	Tea merchant (wholesale)
10831	15861	Tea or maté based extracts and preparations (manufacture)
28930	29530	Tea processing machinery and plant (manufacture)
56102	55302	Tea room or shop (unlicensed)
25990	28750	Tea sets made of base metal (manufacture)
13923	17403	Tea towels (manufacture)
52102	63129	Tea warehouse for air transport activities
52103	63129	Tea warehouse for land transport activities
52101	63129	Tea warehouse for water transport activities
85590	80429	Teacher n.e.c.
85510	92629	Teachers of sport
94120	91120	Teachers' Registration Council
32990	36639	Teaching aids (electronic) (manufacture)
78200	74500	Teaching personnel (supply) (temporary employment agency)
25990	28750	Teapots made of base metal (manufacture)
23410	26210	Teapots made of ceramic (manufacture)
32120	36220	Teapots made of precious metal (manufacture)
02300	02010	Teasel growing
28940	29540	Teasel rod (textile machinery accessory) (manufacture)
22190	25130	Teats made of rubber (manufacture)
71200	74300	Technical and non-destructive testing services
90030	92319	Technical and training manual authors
85320	80429	Technical and vocational adult education (excl. cultural, sports, recreation education and the like)
85320	80220	Technical and vocational education
85320	80220	Technical and vocational secondary education
84210	75210	Technical assistance and training programmes abroad (public sector)
71200	74300	Technical automobile inspection activities
23440	26240	Technical ceramic products (manufacture)
85320	80220	Technical college
71200	74300	Technical inspection services of buildings
94120	91120	Technical organisations
10410	15410	Technical tallow (manufacture)
71200	74300	Technical testing of bridges and other engineering structures
71200	74300	Technical testing of lifting and handling equipment
32409	36509	Technical toy (manufacture)
19201	23201	Technical white oil (manufacture)
91020	92521	Technological museums
46520	51860	Telecommunication instruments and apparatus (wholesale)
61100	64200	Telecommunication network maintenance (wired telecommunications)
71122	74206	Telecommunications consultancy activities
46520	51860	Telecommunications equipment (wholesale)
47429	52482	Telecommunications equipment other than mobile telephones (retail)
46520	51860	Telecommunications machinery, equipment and materials for professional use (wholesale)
61900	64200	Telecommunications resellers
61300	64200	Telecommunications satellite relay station
27320	31300	Telecommunications wire (manufacture)
61100	64200	Teleconferencing services (wired telecommunications)
28220	29220	Teleferics (manufacture)
61100	64200	Telegram service
26301	32201	Telegraph apparatus (manufacture)
61100	64200	Telegraph communication
16100	20100	Telegraph poles (manufacture)
26511	33201	Telemetering instruments (manufacture)
26511	33201	Telemetric equipment (manufacture)
26301	32201	Telephone (manufacture)
46520	51860	Telephone and communications equipment (wholesale)
61900	64200	Telephone and internet access in public facilities
26301	32300	Telephone answering machines (manufacture)
26301	32201	Telephone apparatus (manufacture)
63990	74879	Telephone based information services
58120	22110	Telephone books (in print) publishing
25110	28110	Telephone booths made of metal (manufacture)
81222	74704	Telephone cleaning and sterilising service
61100	64200	Telephone communication (wired telecommunications)
47910	52610	Telephone direct sales (retail)
61100	64200	Telephone exchange
26301	32201	Telephone exchange equipment (manufacture)
26301	32201	Telephone exchanges (manufacture)
26301	32201	Telephone handset (manufacture)
77390	71340	Telephone hire (other than by public telephone undertakings)
61100	64200	Telephone service (wired telecommunications)
61100	64200	Telephone service operation of 0898 numbers
81222	74704	Telephone sterilising
26301	32201	Teleprinter (manufacture)
26701	33402	Telescope (manufacture)
26701	33402	Telescopic sights (manufacture)
61100	64200	Teletext and other electronic message and information services (wired telecommunications)
77291	71403	Television (domestic) hire
60200	92202	Television broadcasting station
26400	32300	Television cabinets made of wood (manufacture)
26309	32202	Television camera (manufacture)
26702	33403	Television camera lens (manufacture)
26110	32100	Television camera tubes (manufacture)
46439	51439	Television cameras for domestic use (wholesale)
46520	51860	Television cameras for professional use (wholesale)
60200	92202	Television channel programme creation from purchased and/or self produced programme components
47910	52610	Television direct sales (retail)
59133	92202	Television distribution rights acquisition
77390	71340	Television equipment (not domestic) rental and operating leasing

T

SIC 2007	SIC 2003	Activity	SIC 2007	SIC 2003	Activity
47430	52450	Television goods (retail)	26511	33201	Test benches (electronic) (manufacture)
26400	32300	Television monitors and displays (manufacture)	26513	33202	Test benches (non-electronic) (manufacture)
26110	32100	Television picture tube (manufacture)	43130	45120	Test boring for construction
59120	92202	Television post-production activities	09100	11200	Test boring incidental to oil and gas extraction
59133	92120	Television programme distribution activities	43130	45120	Test drilling for construction
59113	92202	Television programme production activities	23190	26150	Test tube (manufacture)
60200	64200	Television programmes transmission	30200	35200	Test wagons for railway (manufacture)
26309	32300	Television receiver (manufacture)	71200	74300	Testing and measuring of environmental indicators: air and water pollution
60200	64200	Television relay service			
26110	32100	Television scan coil (manufacture)	26511	33201	Testing instruments and appliances (electronic) (manufacture)
60200	92202	Television service			
47430	52450	Television sets and equipment (retail)	26513	33202	Testing instruments and appliances (non-electronic) (manufacture)
59113	92202	Television studio			
26309	32202	Television transmitter (manufacture)	26511	33201	Testing machines and equipment (electronic) (manufacture)
26301	32201	Telewriter (manufacture)			
26301	32201	Telex machine (manufacture)	71200	74300	Testing of calculations for building elements
61100	64200	Telex service (wired telecommunications)	71200	74300	Testing of composition and purity of minerals
88990	85321	Temperance association	71200	74300	Testing of physical characteristics and performance of materials
56102	55302	Temperance buffet			
26513	33202	Temperature measuring and control instrument (non-electronic) (manufacture)	71200	74300	Testing or analysing laboratory
			20140	24140	Tetrachloroethylene (manufacture)
28140	29130	Temperature regulators (manufacture)	13300	17300	Textile binding and mending (manufacture)
94910	91310	Temple (for worship)	13300	17300	Textile bleaching (manufacture)
87900	85311	Temporary accommodation for the homeless (charitable)	13300	17300	Textile calendering (manufacture)
			20590	24660	Textile chemical auxiliaries (manufacture)
87900	85312	Temporary accommodation for the homeless (non-charitable)	46410	51410	Textile converter (wholesale)
			28940	29540	Textile dressing or impregnating machinery (manufacture)
78200	74500	Temporary employment agency activities			
87900	85311	Temporary homeless shelters (charitable)	13300	17300	Textile dyeing (manufacture)
87900	85312	Temporary homeless shelters (non-charitable)	13300	17300	Textile embossing (manufacture)
30200	35200	Tenders (manufacture)	13300	17300	Textile ending and mending (manufacture)
22190	25130	Tennis ball core (manufacture)	28940	29540	Textile fabric making machinery (manufacture)
32300	36400	Tennis balls (finished) (manufacture)	28940	29540	Textile fibre preparation machinery (manufacture)
93120	92629	Tennis club	46640	51830	Textile fibre preparation machinery (wholesale)
93110	92619	Tennis court	46760	51560	Textile fibres (wholesale)
42990	45230	Tennis courts construction	13300	17300	Textile finishing (manufacture)
32300	36400	Tennis racket (manufacture)	46640	51830	Textile industry machinery (wholesale)
25730	28620	Tension strapping tool (manufacture)	13300	17300	Textile lacquering (manufacture)
25990	28750	Tensional steel strapping (manufacture)	28940	29540	Textile machinery (manufacture)
16290	20510	Tent poles made of wood (manufacture)	13940	17520	Textile material cordage (manufacture)
13922	17402	Tents (manufacture)	74100	74872	Textile or wallpaper printing designing
46499	51479	Tents (wholesale)	52290	63400	Textile packing incidental to transport
46180	51180	Tents and other camping goods (commission agent)	13923	17403	Textile part of electric blankets (manufacture)
26301	32201	Terminal equipment for telegraphic and data communications (manufacture)	01160	01110	Textile plants growing
			28940	29560	Textile printing machinery (manufacture)
52230	63230	Terminal facilities operation (air transport)	46110	51110	Textile raw materials (commission agent)
52213	63210	Terminal facilities operation (land transport)	20411	24511	Textile soap (manufacture)
52220	63220	Terminal facilities operation (water transport)	20411	24511	Textile softeners (manufacture)
26110	31200	Terminals for electronic apparatus (manufacture)	46640	51830	Textile spinning, doubling or twisting machinery (wholesale)
28230	30010	Terminals for issuing of tickets and reservations (manufacture)	28940	29540	Textile unreeling, folding, or pinking machinery (manufacture)
24100	27100	Terneplate (manufacture)	17220	17549	Textile wadding and articles of wadding (manufacture)
23410	26210	Terracotta ware (manufacture)			
43330	45430	Terrazzo work (building)	17240	21240	Textile wall coverings (manufacture)
84220	75220	Territorial Army	17240	21240	Textile wallpaper (manufacture)
13910	17600	Terry fabrics (manufacture)	46770	51570	Textile waste (wholesale)
13200	17210	Terry towelling (manufacture)	46640	51830	Textile weaving machinery (wholesale)
85320	80220	Tertiary college	13960	17542	Textile wicks (manufacture)
23310	26300	Tessellated glazed pavement tiles (manufacture)	46640	51830	Textile winding or reeling machinery (wholesale)
23310	26300	Tesserae made of earthenware (manufacture)			

T

SIC 2007	SIC 2003	Activity
13960	17549	Textile yarn impregnated, coated or sheathed with rubber or plastic (manufacture)
28940	29540	Textile yarn preparation machinery (manufacture)
46160	51160	Textiles (commission agent)
47510	52410	Textiles (retail)
46410	51410	Textiles (wholesale)
20590	24660	Textiles and leather finishing materials (manufacture)
46410	51410	Textiles exporter (wholesale)
46410	51410	Textiles importer (wholesale)
77299	71409	Textiles rental and leasing
20600	24700	Textured single yarn (manufacture)
28940	29540	Texturing and softening machinery for textiles (manufacture)
43910	45220	Thatching
93191	92621	The seeking of sponsorship, appearance money and prize money for horse racing
58110	22110	The Stationery Office
90040	92320	Theatre halls operation
79909	92320	Theatre ticket agency
74909	74879	Theatrical agency
14190	18249	Theatrical costumes (manufacture)
77299	71409	Theatrical costumes leasing
77210	71401	Theatrical costumes rental
90010	92311	Theatrical presentations (live production)
77299	71409	Theatrical scenery leasing
90010	92311	Theatrical touring company
93210	92330	Theme park operation
26511	33201	Theodolite (electronic) (manufacture)
26513	33202	Theodolite (non-electronic) (manufacture)
85421	80302	Theological college specialising in higher education course
94910	91310	Theosophical Society
46690	51870	Therapeutic instruments and appliances (wholesale)
23140	26140	Thermal and sound insulating material made of glass fibre (manufacture)
26110	32100	Thermionic valves or tubes (manufacture)
28990	29560	Thermo forming machines (manufacture)
18129	22220	Thermocopier printing (manufacture)
26600	33100	Thermographs (manufacture)
26511	33201	Thermometer (electronic) (manufacture)
26513	33202	Thermometer (non-electronic) (manufacture)
20160	24160	Thermoplastic resins (manufacture)
20160	24160	Thermosetting resins (manufacture)
26511	33201	Thermostat (electronic) (manufacture)
26513	33202	Thermostat (non-electronic) (manufacture)
20301	24301	Thinners for paint and varnish (manufacture)
20160	24160	Thiourea resins (manufacture)
07210	12000	Thorium ores mining
46410	51410	Thread (wholesale)
26701	33402	Thread counters (manufacture)
13100	17160	Thread for sewing and embroidery, made of cotton (manufacture)
28940	29540	Thread guide (textile machinery accessory) (manufacture)
13100	17160	Thread made of hemp (manufacture)
13100	17160	Thread made of jute (manufacture)
13100	17160	Thread made of linen (manufacture)
13100	17160	Thread made of silk (manufacture)
28410	29420	Thread rollers or machines for working wires (manufacture)

SIC 2007	SIC 2003	Activity
28410	29420	Thread rolling machines (manufacture)
46740	51540	Threaded and non-threaded fasteners (wholesale)
25940	28740	Threaded fasteners (manufacture)
25730	28620	Threading die (manufacture)
28410	29420	Threading machine (metal cutting) (manufacture)
25730	28620	Threading tap (manufacture)
46610	51880	Threshers (wholesale)
01610	01410	Threshing by contractor
28302	29320	Threshing machine (manufacture)
26110	32100	Thyratron (manufacture)
27900	32100	Thyristor (manufacture)
92000	92710	Tic tac person
79909	92349	Ticket agencies for other entertainment activities
79909	92320	Ticket agencies for theatre
79110	63301	Ticket agencies for travel
17290	21259	Ticket cutting and punching (manufacture)
28230	30010	Ticket issuing machine (manufacture)
77330	71330	Ticket machine hire
18129	22220	Ticket printing (manufacture)
28230	30010	Ticket punch (manufacture)
79909	92629	Ticket sales activities for sports events
79909	92729	Ticket sales for other recreational activities
13200	17210	Ticking weaving (manufacture)
32130	36610	Tie pin (not of, or clad in, precious metal, or of precious/semi precious stones) (manufacture)
13200	17240	Tie silk (manufacture)
47710	52424	Ties (retail)
46420	51429	Ties (wholesale)
14190	18249	Ties made of silk (manufacture)
14310	17710	Tights (manufacture)
28990	29560	Tile making machine (not plastic working) (manufacture)
22230	25239	Tiles (other than floor tiles made of plastic) (manufacture)
46730	51530	Tiles (wholesale)
23320	26400	Tiles and construction products, made of baked clay (manufacture)
47520	52460	Tiles for wall or floor made of ceramic (retail)
43330	45430	Tiles laying or fitting
22230	25231	Tiles made of asphalt thermoplastic (manufacture)
23310	26300	Tiles made of ceramics (manufacture)
23610	26610	Tiles made of concrete (manufacture)
16290	20520	Tiles made of cork (manufacture)
13939	17519	Tiles made of felt (manufacture)
23650	26650	Tiles made of fibre cement (manufacture)
23190	26150	Tiles made of glass (manufacture)
23310	26300	Tiles made of glazed earthenware (manufacture)
13939	17519	Tiles made of needleloom carpet (manufacture)
23620	26620	Tiles made of plaster (manufacture)
23700	26700	Tiles made of slate (manufacture)
13931	17512	Tiles made of tufted carpet (manufacture)
22230	25231	Tiles made of vinyl asbestos (manufacture)
23990	26821	Tiles made of woven asbestos (manufacture)
43330	45430	Tiling contractor (floors and walls)
22190	25130	Tiling made of rubber (manufacture)
18140	22230	Tillot and seal making (manufacture)
46130	51130	Timber (commission agent)
46130	51130	Timber broker (commission agent)
02400	02020	Timber evaluation

SIC 2007	SIC 2003	Activity
02200	02010	Timber felling
02100	02010	Timber growing
46730	51530	Timber importer (wholesale)
71200	74300	Timber measurer
46730	51530	Timber merchant (wholesale)
46730	51530	Timber yard (wholesale)
26520	33500	Time clock (manufacture)
26520	33500	Time lock (manufacture)
26520	33500	Time recorder (manufacture)
63110	72300	Time sharing services (computer)
26520	33500	Time switch (manufacture)
26520	33500	Time/date stamps (manufacture)
93199	92629	Timekeepers of sport
26520	33500	Timer for industrial use (manufacture)
79909	63309	Time-share exchange services
68209	70209	Timeshare operations (real estate)
18129	22220	Timetable printing (manufacture)
58190	22150	Timetable publishing
22190	25130	Timing belt for motor vehicles (manufacture)
24430	27430	Tin (manufacture)
24430	27430	Tin foil (manufacture)
07290	13200	Tin mining and preparation
27510	29710	Tin openers (electric) (manufacture)
07290	13200	Tin ore and concentrate extraction and preparation
18129	22220	Tin printing (manufacture)
24430	27430	Tin wire made by drawing (manufacture)
25730	28620	Tinman's snips (manufacture)
10890	15891	Tinned broth (manufacture)
10130	15131	Tinned ham (manufacture)
10130	15139	Tinned meat (other than tinned ham) (manufacture)
10890	15891	Tinned soup (manufacture)
24100	27100	Tinplate (manufacture)
46720	51520	Tinplate (wholesale)
25920	28720	Tins for food products (manufacture)
23110	26110	Tinted glass (manufacture)
25730	28620	Tip for cutting tool (manufacture)
28220	29220	Tipper (manufacture)
29320	34300	Tipping gear and parts thereof for motor vehicles (not hydraulic) (manufacture)
16290	20520	Tips made of cork (manufacture)
23140	26140	Tissue made of glass fibre (manufacture)
17120	21120	Tissue paper (uncut) (manufacture)
24450	27450	Titanium (manufacture)
20120	24120	Titanium dioxide (manufacture)
46750	51550	Titanium oxide (wholesale)
18129	22220	Title document printing (manufacture)
25990	28750	Toast racks made of base metal (manufacture)
27510	29710	Toaster (electric) (manufacture)
96090	93059	Toastmaster
46170	51170	Tobacco (commission agent)
47260	52260	Tobacco (retail)
46210	51250	Tobacco (unmanufactured) (wholesale)
46170	51170	Tobacco broker (commission agent)
46210	51250	Tobacco exporter (unmanufactured) (wholesale)
12000	16000	Tobacco for use in pipes and rolled cigarettes (manufacture)
01150	01110	Tobacco growing
46210	51250	Tobacco importer (unmanufactured) (wholesale)
01630	01110	Tobacco leaves, drying and preparation

SIC 2007	SIC 2003	Activity
46350	51350	Tobacco merchant (wholesale)
46690	51870	Tobacco preparation and making-up machinery (wholesale)
28930	29530	Tobacco processing machinery (manufacture)
12000	16000	Tobacco products (manufacture)
46350	51350	Tobacco products exporter (wholesale)
46350	51350	Tobacco products importer (wholesale)
46110	51110	Tobacco refuse (commission agent)
12000	16000	Tobacco stemming and redrying (manufacture)
12000	16000	Tobacco substitute products (manufacture)
12000	16000	Tobacco, homogenised or reconstituted (manufacture)
47260	52260	Tobacconist (retail)
46350	51350	Tobacconist (wholesale)
46350	51350	Tobacconists' sundriesman (wholesale)
22290	19300	Toe puff (manufacture)
10822	15842	Toffee (manufacture)
23410	26210	Toilet articles made of ceramic (manufacture)
22290	25240	Toilet articles made of plastic (manufacture)
32120	36220	Toilet articles of base metals clad with precious metals (manufacture)
32910	36620	Toilet brush (manufacture)
15120	19200	Toilet case (fitted) made of leather (manufacture)
47750	52330	Toilet goods (retail)
13923	17403	Toilet linen (manufacture)
17220	21220	Toilet paper (cut to size) (manufacture)
17120	21120	Toilet paper (uncut) (manufacture)
20420	24520	Toilet preparations (manufacture)
46450	51450	Toilet preparations (wholesale)
20420	24511	Toilet soap (manufacture)
20420	24520	Toilet water (manufacture)
20420	24520	Toiletries (manufacture)
46730	51530	Toilets (wholesale)
52219	63210	Toll bridge, road or tunnel
20140	24140	Toluene (manufacture)
01130	01120	Tomato growing
26600	33100	Tomographs (manufacture)
26400	32300	Tone arms (manufacture)
20120	24120	Toner (pigment) (manufacture)
28230	30010	Toner cartridges (manufacture)
20590	24640	Toner for photographic use (manufacture)
16100	20100	Tongued wood (manufacture)
11070	15980	Tonic water production (manufacture)
11020	15932	Tonic wine production (manufacture)
28302	29320	Tool bar for agricultural use (manufacture)
16290	20510	Tool handles made of wood (manufacture)
28490	29430	Tool holder (manufacture)
15120	19200	Toolbags made of leather (manufacture)
47520	52460	Tools (not machine tools) (retail)
25730	28620	Tools (precision) (manufacture)
46740	51540	Tools (wholesale)
25730	28620	Tools except power driven hand tools (manufacture)
77320	71320	Tools for construction hire (without operator)
77390	71340	Tools for mechanics or engineers hire
16290	20510	Tools made of wood (manufacture)
27510	29710	Tooth brush (electric) (manufacture)
32910	36620	Tooth brush (not electric) (manufacture)
46499	51479	Tooth brushes (wholesale)
20420	24520	Tooth powder (manufacture)

SIC 2007	SIC 2003	Activity
20420	24520	Toothpaste (manufacture)
32990	36639	Toothpicks made of bone (manufacture)
43120	45110	Top soil stripping work
13100	17130	Topmaking (wool) (manufacture)
46760	51560	Tops (wholesale)
27400	31500	Torch (manufacture)
25400	29600	Torpedo (manufacture)
25930	28740	Torsion bar spring (manufacture)
92000	92710	Totalisator
23120	26120	Toughened glass (manufacture)
79120	63302	Tour operator activities
29203	34203	Touring caravan (manufacture)
94990	91330	Touring clubs
70229	74149	Tourism development consultancy services
79909	63309	Tourism promotion activities
84130	75130	Tourism services administration and regulation (public sector)
79909	63309	Tourist assistance activities n.e.c.
79909	63309	Tourist board or information service
79901	63303	Tourist guide activities
85320	80220	Tourist guide instruction
13100	17140	Tow of flax (manufacture)
13940	17520	Tow yarn made of hard fibres (manufacture)
13923	17403	Towel (manufacture)
96010	93010	Towel hire
27510	29710	Towel rail (electric) (manufacture)
96010	93010	Towel supply company
13200	17210	Towelling weaving (manufacture)
47510	52410	Towels (retail)
17220	21220	Towels made of paper (manufacture)
25110	28110	Tower made of steel (manufacture)
16230	20300	Tower made of wood (manufacture)
52219	50200	Towing and road side assistance
52219	63210	Towing away of vehicles
13940	17520	Towing rope (manufacture)
52220	61209	Towing services for distressed freight vessels in inland waters
52220	61102	Towing services for distressed freight vessels on sea and coastal waters
52220	61201	Towing services for distressed passenger vessels in inland waters
52220	61101	Towing services for distressed passenger vessels on sea and coastal waters
71112	74202	Town and city planning
96090	93059	Town crier
35220	40220	Town gas distribution
35210	40210	Town gas production
49319	60219	Town-to-airport transport by bus
49311	60213	Town-to-airport transport by rail
49319	60219	Town-to-station transport by bus
49311	60213	Town-to-station transport by rail
38220	90020	Toxic waste treatment service
32409	36509	Toy animal (manufacture)
32409	36509	Toy balloon (manufacture)
32409	36509	Toy car circuit (electric) (manufacture)
32409	36509	Toy cars (electric) (manufacture)
32409	36509	Toy cars (pedal) (manufacture)
32409	36509	Toy furniture (manufacture)
32409	36509	Toy guns (not operated by compressed air) (manufacture)
32409	36509	Toy musical instruments including electronic (manufacture)
32409	36509	Toy perambulators and pushchairs (manufacture)
32409	36509	Toy push cart (manufacture)
32409	36509	Toy trains (electric) (manufacture)
46499	51477	Toy trains and accessories (wholesale)
32409	36509	Toy wheelbarrow (manufacture)
32409	36509	Toys (mechanical) (manufacture)
47650	52485	Toys (retail)
46499	51477	Toys (wholesale)
32409	36509	Toys and games (electronic) with fixed (non replaceable software) (manufacture)
46499	51477	Toys and games exporter (wholesale)
46499	51477	Toys and games importer (wholesale)
32409	36509	Toys and games made of paper (manufacture)
32409	36509	Toys and games made of wood (manufacture)
32401	36501	Toys for professional and arcade use (manufacture)
32409	36509	Toys made of cardboard (manufacture)
32409	36509	Toys made of metal (manufacture)
32409	36509	Toys made of plastic (manufacture)
32409	36509	Toys made of rubber (manufacture)
13300	17300	Tracing cloth (textile finishing) (manufacture)
13960	17549	Tracing cloth (woven) (manufacture)
13960	17549	Tracing cloth weaving (manufacture)
17120	21120	Tracing paper (manufacture)
28923	29523	Track laying and other tractors used in construction (manufacture)
28921	29521	Track laying tractors and other tractors used in mining (manufacture)
29320	34300	Track rods for motor vehicles (manufacture)
14190	18249	Tracksuits (manufacture)
27200	31400	Traction battery (rechargeable) (manufacture)
27110	31100	Traction motors with or without associated control equipment (manufacture)
32500	33100	Traction or suspension devices for medical beds (manufacture)
28301	29310	Tractor (half track) (manufacture)
28301	29310	Tractor (pedestrian controlled) (manufacture)
28301	29310	Tractor (skidded unit) (manufacture)
28301	29310	Tractor (wheeled) (manufacture)
28301	29310	Tractor for agricultural use (manufacture)
28301	29310	Tractor for forestry use (manufacture)
77310	71310	Tractor hire for agriculture (without driver)
28302	29320	Tractor hoe (manufacture)
28301	29310	Tractor parts for wheeled and half-track tractors (manufacture)
28302	29320	Tractor plough (manufacture)
28922	29522	Tractor shovel (manufacture)
28922	29522	Tractor winch (manufacture)
46610	51880	Tractors (wholesale)
46610	51880	Tractors for agricultural use (wholesale)
29100	34100	Tractors for semi-trailers (manufacture)
46690	51870	Tractors of a type used on railway station platforms (wholesale)
46610	51880	Tractors used in forestry (wholesale)
94110	91110	Trade association
18140	22230	Trade binding (manufacture)
82301	74873	Trade centre
18129	22220	Trade journal printing (manufacture)
58142	22130	Trade journal publishing
69109	74111	Trade mark agent

T

SIC 2007	SIC 2003	Activity	SIC 2007	SIC 2003	Activity
94200	91200	Trade unions	42120	45213	Transmission line construction
82301	74873	Trades exhibition organiser	60100	92201	Transmission of aural programming via over-the-air broadcasts, cable or satellite
94200	91200	Trades union congress			
25730	28620	Tradesmen's knife (manufacture)	13960	17542	Transmission or conveyor belts or belting, made of textiles (manufacture)
82990	74879	Trading stamp activities			
18129	22220	Trading stamp printing (manufacture)	27330	31200	Transmission pole and line hardware (manufacture)
30200	31620	Traffic control equipment for roads and inland waterways (manufacture)	28150	29140	Transmission shafts (manufacture)
			22190	25130	Transmission v-belts made of rubber (manufacture)
71129	74204	Traffic engineering design projects	26309	32202	Transmitter-receivers (manufacture)
29310	31610	Traffic indicators for motor vehicles (manufacture)	26309	32300	Transmitting and receiving antenna (manufacture)
84240	75240	Traffic regulation administration and operation	86900	85140	Transplant organ banks
84240	75240	Traffic wardens	28302	29320	Transplanter for agricultural use (manufacture)
29202	34202	Trailer (motor drawn) (manufacture)	26309	32202	Transponders (manufacture)
45190	50101	Trailers (new) (wholesale)	50200	61102	Transport by towing or pushing of barges (except inland waterway)
45190	50101	Trailers (retail)			
45190	50102	Trailers (used) (retail)	52290	63400	Transport documents issue and procurement
45190	50102	Trailers (used) (wholesale)	46690	51870	Transport equipment (except motor vehicles, motorcycles and bicycles) (wholesale)
29202	34202	Trailers and semi-trailers (manufacture)			
28302	29320	Trailers and semi-trailers for agricultural use (manufacture)	81299	74709	Transport equipment cleaning (non-specialised)
			65120	66031	Transport insurance
77120	71219	Trailers and semi-trailers rental	50200	61102	Transport of freight over seas and coastal waters (whether scheduled or not)
81299	74709	Train cleaning (non-specialised)			
81291	74703	Train disinfecting and exterminating activities	50400	61209	Transport of freight via canals
32990	36639	Trainer (electronic training equipment) (manufacture)	50400	61209	Transport of freight via lakes
			50400	61209	Transport of freight via ports
93199	92629	Trainer (racehorse or greyhound)	50400	61209	Transport of freight via rivers
59111	92111	Training film production	49500	60300	Transport of gases slurry via pipelines
96090	93059	Training of pet animals	50100	61101	Transport of passengers over seas and coastal waters
93199	92629	Training stables	50100	61101	Transport of passengers over water (except for inland waterway service)
59112	92111	Training video production			
49319	60219	Tramway (scheduled passenger transport)	50300	61201	Transport of passengers over water (inland waterway service)
42120	45230	Tramways construction			
74300	74850	Transcribing services from tapes, discs, etc.	50300	61201	Transport of passengers via canals
82190	74850	Transcription of documents, and other secretarial services	50300	61201	Transport of passengers via inside harbours
			50300	61201	Transport of passengers via lakes
28960	29560	Transfer moulding press for rubber or plastics (manufacture)	50300	61201	Transport of passengers via ports
			50300	61201	Transport of passengers via rivers
18129	22220	Transfer printing (manufacture)	49410	60249	Transport of waste and waste materials
27110	31100	Transformer (generator, transmission system and distribution) (manufacture)	49500	60300	Transport of water via pipelines
			52290	63400	Transport operations arranging or carrying out by air
26110	31100	Transformer for electronic apparatus (manufacture)	52290	63400	Transport operations arranging or carrying out by road
27110	31100	Transformer for industrial use (manufacture)			
19201	23201	Transformer oil (at refineries) (manufacture)	52290	63400	Transport operations arranging or carrying out by sea
19209	23209	Transformer oil formulation outside refineries (manufacture)	84130	75130	Transport services administration and regulation (public sector)
46690	51870	Transformers (wholesale)	49500	60300	Transport via pipeline
32500	33100	Transfusion apparatus (manufacture)	28220	29220	Transporter (manufacture)
32500	33100	Transfusion pods (manufacture)	38110	90020	Trash collection
26110	32100	Transistors (manufacture)	27510	29710	Trash compactor for domestic use (electric) (manufacture)
74300	74850	Translation activities			
27110	31100	Transmission and distribution voltage regulators (manufacture)	46499	51478	Travel accessories (wholesale)
			47722	52432	Travel accessories made of leather and leather substitutes (retail)
26309	32202	Transmission apparatus for radio-broadcasting (manufacture)			
			79110	63301	Travel agency activities
22190	25130	Transmission belting made of rubber (manufacture)	46499	51478	Travel and fancy goods exporter (wholesale)
22290	25240	Transmission belts made of plastic (manufacture)	46499	51478	Travel and fancy goods importer (wholesale)
28150	29140	Transmission chain (manufacture)	47722	52432	Travel goods (retail)
26301	32201	Transmission equipment for telephone and telegraph (manufacture)	15120	19200	Travel goods made of leather or leather substitute (manufacture)
			65120	66031	Travel insurance
46520	51860	Transmission kit for radio-telephony and telegraphy, radio or television broadcasting (wholesale)	26520	33500	Travelling clock (manufacture)

SIC 2007	SIC 2003	Activity
28220	29220	Travelling crane (manufacture)
29100	34100	Travelling libraries, banks, etc. (not trailers) (manufacture)
13923	17403	Travelling rug making-up (outside weaving establishment) (manufacture)
13923	17403	Travelling rugs made of wool (manufacture)
90010	92349	Travelling show
16240	20400	Travelling trunks made of wood (manufacture)
26110	32100	Travelling wave tube (manufacture)
28990	29560	Trawl door (manufacture)
30110	35110	Trawler (manufacture)
25990	28750	Trays made of base metal (manufacture)
17220	21220	Trays made of paper (manufacture)
22290	25240	Trays made of plastic (manufacture)
16290	20510	Trays made of wood (manufacture)
10810	15830	Treacle (manufacture)
25730	28620	Treadle operated and other non power-operated tools (manufacture)
25610	28510	Treatment and coating of metals (manufacture)
38220	90020	Treatment and disposal of foul liquids (e.g. leachate)
38220	90020	Treatment and disposal of hazardous waste
38210	90020	Treatment and disposal of non-hazardous waste
38220	23300	Treatment and disposal of nuclear waste
38220	90020	Treatment and disposal of radioactive waste from hospitals, etc.
38220	90020	Treatment and disposal of toxic live or dead animals and other contaminated waste
38220	23300	Treatment and disposal of transition radioactive waste
37000	90010	Treatment of human waste water by means of physical, chemical and biological processes
38210	90020	Treatment of organic waste for disposal
37000	90010	Treatment of waste water (human, industrial, from swimming pools etc.) by various processes
36000	41000	Treatment of water for industrial and other purposes
38220	23300	Treatment, disposal and storage of radioactive nuclear waste
01290	01110	Tree growing for extraction of sap
01300	01120	Tree nurseries
02100	02010	Tree nursery (not fruit or ornamental trees)
81300	01410	Tree pruning, replanting, on a fee or contract basis (except as an agricultural service activity)
02200	02010	Treefelling (own account)
25990	28750	Trellis work made of metal (manufacture)
16230	20300	Trellis work made of wood (manufacture)
43120	45110	Trench digging
28922	29522	Trencher (manufacture)
84230	75230	Tribunals
96020	93020	Trichologist
30920	35420	Tricycles (including delivery tricycles) (manufacture)
30920	35420	Tricycles and parts (manufacture)
32409	36509	Tricycles for children (manufacture)
18140	22230	Trimming of books, brochures, etc. (manufacture)
14200	18300	Trimmings made of fur (manufacture)
15110	19100	Trimmings made of leather (manufacture)
20510	24610	Trinitrotoluene (TNT) (manufacture)
52220	63220	Trinity House
47220	52220	Tripe (retail)
10110	15112	Tripe dressing (manufacture)
49390	60239	Trishaw (cycle rickshaw) taxi service
29100	34100	Trolley bus (manufacture)

SIC 2007	SIC 2003	Activity
49319	60219	Trolley bus (scheduled passenger transport)
30400	29600	Troop carrier (armoured) (manufacture)
14190	18241	Tropical helmet (manufacture)
93120	92629	Trotting Club
23650	26650	Troughs made of fibre cement (manufacture)
14132	18222	Trouser suits for women and girls (manufacture)
14120	18210	Trousers for industrial use (manufacture)
14131	18221	Trousers for men and boys (manufacture)
14132	18222	Trousers for women and girls (manufacture)
25730	28620	Trowel (not garden) (manufacture)
29100	34100	Truck (commercial vehicle) (manufacture)
49410	60249	Truck rental (with driver)
77120	71219	Truck rental (without driver)
77120	71219	Trucks and other heavy vehicles exceeding 3.5 tonnes renting and leasing
01160	01110	True hemp growing
28490	29430	Trueing and grinding machines for cold working of glass (manufacture)
01130	01120	Truffles growing
02300	01120	Truffles, gathering from the wild
25400	29600	Truncheons and night sticks (manufacture)
15120	19200	Trunk handles made of leather (manufacture)
15120	19200	Trunks made of leather (manufacture)
16230	20300	Truss rafter (manufacture)
84110	75110	Trust territory programme administration (public sector)
66190	67130	Trustee, fiduciary and custody services on a fee or contract basis
66120	67122	Trustees
69109	74119	Trusts preparation
14141	18231	T-shirts for men and boys (manufacture)
26110	32100	Tube (electronic) (manufacture)
24440	27440	Tube blanks made of copper (manufacture)
22220	25220	Tube containers made of plastic (manufacture)
28120	29122	Tube coupling and equipment for pneumatics (manufacture)
24420	27420	Tube fittings made of aluminium (manufacture)
24440	27440	Tube fittings made of copper (manufacture)
23190	26150	Tube fittings made of glass for electric lights (manufacture)
17290	21259	Tube fittings made of paper (manufacture)
24200	27220	Tube fittings made of steel (manufacture)
24200	27220	Tube hollows made of steel (manufacture)
28910	29510	Tube mill plant (manufacture)
24440	27440	Tube shells made of copper (manufacture)
86101	85111	Tuberculosis hospital (public sector)
01130	01110	Tuberous vegetable growing
28990	29560	Tubes (valves) or bulbs producing machinery (manufacture)
24510	27210	Tubes and fittings made of cast iron (manufacture)
25920	28720	Tubes made of aluminium for packaging (collapsible) (manufacture)
24440	27440	Tubes made of brass (manufacture)
13922	17402	Tubes made of canvas for ventilating purposes (manufacture)
17290	21259	Tubes made of cardboard (not for packing) (manufacture)
24510	27210	Tubes made of centrifugally cast steel (manufacture)
24200	27220	Tubes made of cold drawn steel (manufacture)
23610	26610	Tubes made of concrete (manufacture)
24440	27440	Tubes made of copper (manufacture)

T

SIC 2007	SIC 2003	Activity	SIC 2007	SIC 2003	Activity
23650	26650	Tubes made of fibre cement (manufacture)	52219	63210	Tunnels operation
23190	26150	Tubes made of glass (manufacture)	16240	20400	Tuns made of wood (manufacture)
25920	28720	Tubes made of metal (collapsible) (manufacture)	09100	11200	Turbine drilling services
22210	25210	Tubes made of plastic (manufacture)	27110	31100	Turbine for electricity generation (manufacture)
23200	26260	Tubes made of refractory ceramic (manufacture)	77390	71340	Turbine rental and operating leasing
22190	25130	Tubes made of rubber (manufacture)	28110	29110	Turbine-generator sets (manufacture)
24200	27220	Tubes made of seamless steel (manufacture)	28110	29110	Turbines and parts thereof (manufacture)
24200	27220	Tubes made of steel (manufacture)	27110	31100	Turbo alternator (manufacture)
24430	27430	Tubes, pipes, tube and pipe fittings made of lead (manufacture)	46690	51870	Turbo-compressors (wholesale)
24430	27430	Tubes, pipes, tube and pipe fittings made of tin (manufacture)	30300	35300	Turbo-jets and parts for aircraft (manufacture)
			30300	35300	Turbo-propellers and parts for aircraft (manufacture)
24430	27430	Tubes, pipes, tube and pipe fittings made of zinc (manufacture)	92000	92710	Turf accountant
			92000	92710	Turf commission agency
23190	26150	Tubing made of glass (manufacture)	01300	01120	Turf for transplanting
22210	25210	Tubing made of plastic (manufacture)	01470	01240	Turkey farming
22190	25130	Tubing made of rubber (manufacture)	01470	01240	Turkey raising and breeding
24200	27220	Tubing made of steel (manufacture)	96040	93040	Turkish baths
22220	25220	Tubs made of plastic (manufacture)	10822	15842	Turkish delight (manufacture)
16240	20400	Tubs made of wood (manufacture)	16290	20510	Turned wood products (manufacture)
25920	28720	Tubular containers made of aluminium (manufacture)	28410	29420	Turning machine (metal cutting) (manufacture)
23130	26130	Tubular containers made of glass (manufacture)	28490	29430	Turning, engraving, carving, etc. machines for stone, ceramics, and similar (manufacture)
25920	28720	Tubular containers made of metal (manufacture)	01130	01120	Turnip growing
30110	35110	Tubular modules for oil rigs (manufacture)	26400	32300	Turn-tables (record decks) (manufacture)
25940	28740	Tubular rivets (manufacture)	03110	05010	Turtle hunting
13931	17512	Tufted carpet (manufacture)	13200	17220	Tweed (manufacture)
13990	17549	Tufted fabrics (other than household textiles) (manufacture)	13200	17210	Twill weaving (manufacture)
13923	17403	Tufting blankets (manufacture)	14390	17720	Twin sets (knitted) (manufacture)
13923	17403	Tufting household textiles (manufacture)	13940	17520	Twine (manufacture)
46640	51830	Tufting machinery (wholesale)	46410	51410	Twine (wholesale)
28940	29540	Tufting machinery for carpet making (manufacture)	13100	17170	Twine made of paper (manufacture)
30110	35110	Tug (manufacture)	13960	17542	Twist cord (fabric) (manufacture)
52220	63220	Tug boat service for inland waterways	25730	28620	Twist drill (manufacture)
52220	63220	Tug boat service for offshore installations	28940	29540	Twisting machinery for textiles (manufacture)
52220	63220	Tug boat service for sea barge or off-shore well	24430	27430	Type metal (manufacture)
52220	63220	Tug boat service for sea barges on domestic coastal routes	28990	29560	Type setting machine (manufacture)
			18130	22240	Typesetting and phototypesetting (manufacture)
52220	63220	Tug lessee or owner for inland waterways service	46690	51870	Type-setting machinery, equipment and apparatus (wholesale)
52220	63220	Tug owner or lessee for in port service or salvage	28230	30010	Typewriter (manufacture)
46690	51870	Tugs and pusher craft (wholesale)	15120	19200	Typewriter case made of leather (manufacture)
85320	63220	Tuition for ships' licences for commercial certificates and permits	77330	71330	Typewriter rental and operating leasing
46210	51210	Tulip bulbs (wholesale)	32990	36631	Typewriter ribbons (manufacture)
13990	17541	Tulles and other net fabrics (manufacture)	46499	51479	Typewriter ribbons (wholesale)
13990	17541	Tulles in motifs (manufacture)	47410	52482	Typewriters (retail)
13990	17541	Tulles in one piece (manufacture)	46660	51850	Typewriters (wholesale)
13990	17541	Tulles in strips (manufacture)	82190	74850	Typing, word processing and desk top publication service
27510	29710	Tumble dryer for domestic use (manufacture)	28990	29560	Tyre alignment and balancing equipment (except wheel balancing) (manufacture)
23130	26130	Tumblers made of glass (manufacture)			
26400	32300	Tuner (audio separate) (manufacture)	13100	17110	Tyre cord (cotton system) (manufacture)
26400	32300	Tuner for radio and television (other than audio separates) (manufacture)	13960	17549	Tyre cord fabric made of high-tenacity man-made yarn (manufacture)
10410	15410	Tung oil extraction (manufacture)	13960	17549	Tyre cord fabric of high-tenacity man-made yarn (manufacture)
24450	27450	Tungsten (manufacture)	45320	50300	Tyre dealer (retail)
03110	05010	Tunicate hunting	13960	17549	Tyre fabric woven from yarn spun on the cotton system (manufacture)
32200	36300	Tuning fork (manufacture)			
95290	52740	Tuning of musical instruments	22110	25110	Tyre flaps (manufacture)
23200	26260	Tunnel oven refractory (manufacture)	22110	25120	Tyre rebuilding and retreading (manufacture)
28921	29521	Tunnelling machine for mining (manufacture)			

SIC 2007	SIC 2003	Activity
22190	25130	Tyre repair materials and kits (manufacture)
22110	25120	Tyre retreading (manufacture)
22110	25110	Tyres (manufacture)
22110	25110	Tyres for aircraft (manufacture)
22110	25110	Tyres for cars or vans (manufacture)
22110	25110	Tyres for commercial vehicles (manufacture)
22110	25110	Tyres for cycles (manufacture)
22110	25110	Tyres for industrial use (manufacture)
22110	25110	Tyres for motorcycles and mopeds (manufacture)
22110	25110	Tyres for scooters (manufacture)
22110	25110	Tyres for tractors (manufacture)
22110	25110	Tyres made of rubber (manufacture)
22110	25110	Tyres made of solid rubber (manufacture)
27900	31620	Ultrasonic cleaning machines (except laboratory and dental) (manufacture)
26600	33100	Ultrasonic diagnostic equipment (manufacture)
26511	33201	Ultrasonic sounding instruments (electronic) (manufacture)
26513	33202	Ultrasonic sounding instruments (non-electronic) (manufacture)
46460	51460	Ultra-violet and infra-red apparatus for medical use (wholesale)
27400	31500	Ultra-violet lamps (manufacture)
32990	36639	Umbrella (manufacture)
13960	17542	Umbrella trimmings made of textile material (manufacture)
47789	52489	Umbrellas (retail)
46420	51429	Umbrellas (wholesale)
17110	21110	Unbleached paper pulp made by chemical dissolving (manufacture)
17110	21110	Unbleached paper pulp made by mechanical processes (manufacture)
17110	21110	Unbleached paper pulp made by non-dissolving processes (manufacture)
17110	21110	Unbleached paper pulp made by semi-chemical processes (manufacture)
10890	15899	Uncooked pizza (manufacture)
11010	15920	Undenatured ethyl alcohol (manufacture)
14190	18249	Underclothing for infants (manufacture)
49311	60213	Underground railways (scheduled passenger transport)
81299	74709	Underground train cleaning (non-specialised)
14142	18232	Underskirt (manufacture)
96030	93030	Undertaking
74202	74813	Underwater photography services
22190	25130	Underwater swimming suit made of rubber (manufacture)
14141	18231	Underwear for men and boys (manufacture)
14142	18232	Underwear for women and girls (manufacture)
17220	21220	Underwear made of paper (manufacture)
65110	66011	Underwriter (life insurance)
65120	66031	Underwriter (non-life insurance)
64999	65239	Underwriter (stock and share issues)
66220	67200	Underwriting brokers
98100	96000	Undifferentiated goods producing activities of private households for own use
98200	97000	Undifferentiated services producing activities of private households for own use
20590	24640	Unexposed materials (manufacture)
23320	26400	Unglazed building brick (manufacture)
14190	18241	Uniform hats and caps (manufacture)

SIC 2007	SIC 2003	Activity
32990	18241	Uniform helmets (manufacture)
14120	18210	Uniforms for men and boys (manufacture)
14120	18210	Uniforms for women and girls (manufacture)
25930	28730	Uninsulated metal cable or insulated cable not useable as a conductor of electricity (manufacture)
27900	31620	Uninterruptible power supplies (UPS) (manufacture)
13200	17250	Union cloth (cotton/linen) (manufacture)
94920	91320	Unionist parties
28410	29420	Unit construction and transfer machine (metal working) (manufacture)
31090	36140	Unit furniture (non-upholstered) (manufacture)
31090	36110	Unit seating for domestic use (upholstered) (manufacture)
64302	65232	Unit trust activities
94910	91310	Unitarian Church
99000	99000	United Nations and affiliated organisations (not United Nations Association)
94990	91330	United Nations associations
94910	91310	United Reform Church
94910	91310	United Society for Christian Literature
27110	31100	Universal ac/dc motors (manufacture)
29320	34300	Universal joints for motor vehicles (manufacture)
85421	80302	Universities' Central Council on Admissions
85421	80302	University
56290	55510	University canteen
85421	80302	University college
56290	55510	University dining halls
55900	55239	University halls of residence
85421	80302	University medical or dental school
53202	64120	Unlicensed carriers
13940	17520	Unloading cushions (manufacture)
46180	51180	Unrecorded media for sound recording or similar recording of other phenomena (commission agent)
24420	27420	Unwrought aluminium (manufacture)
24440	27440	Unwrought brass (manufacture)
24440	27440	Unwrought bronze (manufacture)
24440	27440	Unwrought cadmium copper (manufacture)
24440	27440	Unwrought copper (manufacture)
24440	27440	Unwrought cupro-nickel (manufacture)
24440	27440	Unwrought delta metal (manufacture)
24440	27440	Unwrought German silver (manufacture)
24440	27440	Unwrought gun metal (manufacture)
24440	27440	Unwrought manganese bronze (manufacture)
24440	27440	Unwrought naval brass (manufacture)
24440	27440	Unwrought red metal (manufacture)
31030	36150	Upholstered base for mattress (manufacture)
31090	36140	Upholstered furniture (other than chairs and seats) (manufacture)
31090	36110	Upholsterer (not repair and maintenance) (manufacture)
95240	36110	Upholsterer (repair and restoration)
13960	17542	Upholsterers' trimmings (textile material) (manufacture)
46770	51570	Upholsterers' trimmings (wholesale)
13100	17170	Upholstery hair fibre and filling (manufacture)
15110	19100	Upholstery leather preparation (manufacture)
31090	36110	Upholstery of chairs and seats (manufacture)
15110	19100	Upper leather (manufacture)
20130	23300	Uranium (enriched) (manufacture)
07210	12000	Uranium and thorium ore concentration

U

SIC 2007	SIC 2003	Activity
09900	12000	Uranium and thorium ore mining support services provided on a fee or contract basis
24460	23300	Uranium metal production from pitchblende or other ores (manufacture)
07210	12000	Uranium ore mining
24460	23300	Uranium smelting and refining (manufacture)
49311	60213	Urban area passenger railway transportation by underground, metro and similar systems
49319	60219	Urban area scheduled passenger land transport other than by underground, metro and similar systems
42220	45213	Urban communication and powerlines construction
42210	45213	Urban pipelines construction
71112	74202	Urban planning activities
20140	24140	Urea (not for use as fertiliser) (manufacture)
20150	24150	Urea for use as fertiliser (manufacture)
46120	51120	Urea, thiourea and melamine resins in primary forms (commission agent)
46750	51550	Urea, thiourea and melamine resins in primary forms (wholesale)
20140	24140	Ureines (manufacture)
23420	26220	Urinals made of ceramic, fireclay, etc. (manufacture)
32500	33100	Urine bottle holders and other accessories for medical beds (manufacture)
21200	24421	Urological reagents (manufacture)
86220	85120	Urologist (private practice)
86101	85111	Urologist (public sector)
26110	31300	USB cables (manufacture)
16100	20100	V-jointed wood (manufacture)
21200	24421	Vaccine (manufacture)
27510	29710	Vacuum cleaners for domestic use (manufacture)
28990	29560	Vacuum cleaners for industrial and commercial use (manufacture)
32990	36639	Vacuum flask (complete) (manufacture)
23130	26130	Vacuum flask inners (manufacture)
28990	29560	Vacuum forming machine (manufacture)
32990	36639	Vacuum jar (manufacture)
28131	29121	Vacuum pump (manufacture)
46690	51870	Vacuum pumps (wholesale)
16100	20100	Vacuum treatment of wood (manufacture)
32990	36639	Vacuum vessels for personal or household use (manufacture)
13923	17403	Valances (manufacture)
96090	93059	Valet car parkers
96010	93010	Valet service
74909	74879	Valuer (any trade except real estate)
68310	70310	Valuer (real estate)
28140	29130	Valve actuators (electrical (other than electric motors)) (manufacture)
28120	29130	Valve actuators (hydraulic and pneumatic) (manufacture)
28140	29130	Valve parts (manufacture)
26110	32100	Valves (electronic) (manufacture)
22290	25240	Valves for aerosols (moulded components) (manufacture)
28120	29130	Valves for hydraulic equipment (manufacture)
28140	29130	Valves for industrial use (manufacture)
28110	34300	Valves for motor vehicle engines (manufacture)
28120	29130	Valves for pneumatic control equipment (manufacture)
28140	29130	Valves for tyres (manufacture)
29100	34100	Van (manufacture)
77120	71219	Van hire (exceeding 3.5 tonnes without driver)
77110	71100	Van rental (self drive not exceeding 3.5 tonnes)
24450	27450	Vanadium (manufacture)
07290	13200	Vanadium mining and preparation
28131	29121	Vane pump (manufacture)
01280	01139	Vanilla growing
19201	23201	Vaporising oil (manufacture)
25300	28300	Vapour generators (manufacture)
74909	74879	Variety agency
90010	92311	Variety artiste (own account)
20301	24301	Varnish (manufacture)
20301	24301	Varnish removers (manufacture)
47520	52460	Varnishes (retail)
46730	51530	Varnishes (wholesale)
18130	22250	Varnishing (manufacture)
32500	33100	Vascular prostheses (manufacture)
19201	23201	Vaseline (at refinery) (manufacture)
23410	26210	Vases made of ceramic (manufacture)
23690	26660	Vases made of concrete, plaster, cement or artificial stone (manufacture)
23130	26130	Vases made of glassware (manufacture)
22290	25240	Vases made of plastic (manufacture)
20120	24120	Vat dye (manufacture)
16240	20400	Vats made of wood (manufacture)
20590	24660	Vegetable-based bio diesel (manufacture)
20590	24660	Vegetable-based biodiesel (manufacture)
10390	15330	Vegetable dehydrating for human consumption (manufacture)
10850	15330	Vegetable dishes (manufacture)
13100	17170	Vegetable down (manufacture)
28930	29530	Vegetable fats or oils extraction or preparation machinery (manufacture)
17110	21110	Vegetable fibre pulp (manufacture)
01130	01120	Vegetable growing (except potatoes)
02300	02010	Vegetable hair gathering
28302	29320	Vegetable harvesting and sorting machinery (manufacture)
10320	15320	Vegetable juice (manufacture)
02100	02010	Vegetable materials used for plaiting growing
10612	15612	Vegetable milling (manufacture)
10410	15420	Vegetable oil processing: blowing, boiling, dehydration, hydrogenation (manufacture)
10410	15420	Vegetable oil refining (manufacture)
46120	51120	Vegetable or resin product derivatives (commission agent)
46750	51550	Vegetable or resin product derivatives (wholesale)
10390	15330	Vegetable pickling (manufacture)
01290	01110	Vegetable plaiting material growing
10390	15330	Vegetable preparation and preserving (manufacture)
28930	29530	Vegetable processing machines and equipment (manufacture)
10390	15330	Vegetable quick freezing (manufacture)
46170	51170	Vegetable saps and extracts (commission agent)
01130	01120	Vegetable seed growing
20120	24120	Vegetable tanning and dyeing extracts (manufacture)
47210	52210	Vegetables (retail)
46310	51310	Vegetables (unprocessed) (wholesale)
46310	51310	Vegetables (wholesale)
47290	52270	Vegetarian foods (retail)

V

SIC 2007	SIC 2003	Activity
64921	65221	Vehicle fuel credit card services
27400	31500	Vehicle lamps (bulb and sealed beam unit) (manufacture)
82990	75130	Vehicle licence issuing on a fee or contract basis
26513	33202	Vehicle motors testing and regulating apparatus (non-electronic) (manufacture)
30990	35500	Vehicles drawn by animals (manufacture)
13990	17541	Veiling (not silk) (manufacture)
13200	17240	Veiling made of silk (manufacture)
15110	19100	Vellum (manufacture)
26511	33201	Velocity measuring instruments (electronic) (manufacture)
26513	33202	Velocity measuring instruments (non-electronic) (manufacture)
13200	17210	Velvet (manufacture)
13300	17300	Velvet cutting and shearing (manufacture)
13300	17300	Velvet dyeing (manufacture)
13200	17210	Velveteen (manufacture)
13300	17300	Velveteen cutting and shearing (manufacture)
13300	17300	Velveteen dyeing (manufacture)
28290	29240	Vending machine (manufacture)
47990	52630	Vending machine sales (retail)
46690	51870	Vending machines (wholesale)
16210	20200	Veneer (manufacture)
16100	20100	Veneer log sawing (manufacture)
28490	29430	Veneer press (manufacture)
28490	29430	Veneer shearing machines (manufacture)
16210	20200	Veneer sheet (manufacture)
28490	29430	Veneer splicing machines (manufacture)
28250	29230	Ventilating fans (manufacture)
27510	29710	Ventilating or recycling hoods (manufacture)
28250	29230	Ventilating unit (manufacture)
81222	74704	Ventilation ducts cleaning
46740	51540	Ventilation equipment (wholesale)
28250	29710	Ventilation equipment for domestic use (manufacture)
28250	29230	Ventilation equipment for non-domestic use (manufacture)
90010	92311	Ventriloquist
64303	65235	Venture and development capital companies and funds activities
10730	15850	Vermicelli (manufacture)
81291	74703	Vermin destroying (not agricultural)
01610	01410	Vermin destroying and trapping on agricultural land
11040	15950	Vermouth (manufacture)
25300	28300	Vertical boiler (manufacture)
10831	15861	Vervain herb infusions (manufacture)
52220	63220	Vessel laying up and storage services
52220	63220	Vessel registration services
46499	51479	Vessels for pleasure or sports (wholesale)
10200	15209	Vessels only engaged in processing and preserving fish (other than by freezing) (manufacture)
10200	15201	Vessels only engaged in processing and preserving fish by freezing (manufacture)
14120	18210	Vestments for clerical use (manufacture)
14141	18231	Vests for men and boys (manufacture)
14142	18232	Vests for women and girls (manufacture)
75000	85200	Veterinary activities
75000	85200	Veterinary assistants or other auxiliary veterinary personnel
21100	24410	Veterinary biologicals (manufacture)

SIC 2007	SIC 2003	Activity
46460	51460	Veterinary drugs (wholesale)
32500	33100	Veterinary equipment (manufacture)
21100	24410	Veterinary feed additives (medicinal) (manufacture)
75000	85200	Veterinary laboratory
21200	24421	Veterinary medicines (manufacture)
21200	24421	Veterinary pharmaceuticals (manufacture)
75000	85200	Veterinary surgery
42130	45213	Viaduct construction
23130	26130	Vial (manufacture)
64999	65239	Viatical settlement company activities
25730	28620	Vices (manufacture)
91020	92521	Victoria and Albert Museum
26309	32202	Video conferencing equipment (manufacture)
18202	22320	Video disc reproduction (manufacture)
47430	52450	Video equipment (retail)
26400	36509	Video game consoles (manufacture)
47410	52482	Video game consoles (retail)
46499	51477	Video games (wholesale)
32409	36509	Video games machines (manufacture)
46499	51477	Video games of a kind used with a television receiver (wholesale)
26120	32100	Video interface cards (manufacture)
59120	92119	Video post-production activities
59112	92111	Video producer (own account)
59112	92111	Video production
26400	32300	Video projector (manufacture)
77291	71403	Video recorder/player (domestic) hire
47430	52450	Video recorders (retail)
26400	32300	Video recording or reproducing apparatus including camcorders (manufacture)
26309	32202	Video signalling equipment (manufacture)
59112	92111	Video studios
59132	92120	Video tape and dvd distribution rights acquisition
59140	92130	Video tape projection
18202	22320	Video tape recordings reproduction (manufacture)
77220	71405	Video tape rental
47630	52450	Video tapes (retail)
59132	92120	Video tapes distribution to other industries
74209	74819	Videoing of live events such as weddings, graduations, conventions, fashion shows, etc.
46431	51431	Videos (recorded) (wholesale)
47110	52112	Village general store (selling mainly foodstuffs) with alcohol licence (retail)
47110	52113	Village general store (selling mainly foodstuffs) without alcohol licence (retail)
28930	29530	Vinegar processing machinery (manufacture)
10840	15870	Vinegars (malt, spirit, wine, acetic acid) (manufacture)
01210	01131	Vineyards
20140	24140	Vinyl acetate (manufacture)
20160	24160	Vinyl acetate polymers (manufacture)
20160	24160	Vinyl chloride polymers (manufacture)
22230	25231	Vinyl floor covering (homogeneous and printed) (manufacture)
20301	24301	Vinyl paint (manufacture)
17240	21240	Vinyl-coated wall coverings (manufacture)
17240	21240	Vinyl-coated wallpaper (manufacture)
32200	36300	Viola (manufacture)
32200	36300	Violin (manufacture)
15120	19200	Violin case (not wooden) (manufacture)

SIC 2007	SIC 2003	Activity
16290	20510	Violin cases made of wood (manufacture)
92000	92710	Virtual gambling web site operation
26200	30020	Virtual reality helmets (manufacture)
26511	33201	Viscometers (electronic) (manufacture)
26513	33202	Viscometers (non-electronic) (manufacture)
28230	30010	Visible record computer (tabulator) (manufacture)
79909	63309	Visitor assistance services
55201	55231	Visitor flats and bungalows provided in holiday centres and holiday villages
26200	30020	Visual display unit for computer (manufacture)
20301	24301	Vitreous enamel frits (manufacture)
20301	24301	Vitrifiable enamels (manufacture)
46130	51130	Vitrifiable enamels and glazes, englobes, liquid lustres and glass frit (commission agent)
23910	26810	Vitrified bonded abrasives (manufacture)
85410	80301	Vocational education at post-secondary non-tertial level
88100	85321	Vocational rehabilitation (charitable)
88100	85322	Vocational rehabilitation (non-charitable)
11010	15910	Vodka distilling (manufacture)
13200	17210	Voile weaving (manufacture)
61900	64200	VOIP (voice over internet protocol) provision
26511	33201	Voltage checking instruments (manufacture)
27120	31200	Voltage limiters (manufacture)
29310	31610	Voltage regulators for vehicles (manufacture)
26511	33201	Voltmeter (electronic) (manufacture)
26513	33202	Voltmeter (non-electronic) (manufacture)
23190	26150	Volumetric glassware (manufacture)
28230	30010	Voting machines (manufacture)
46120	51120	Vulcanized and unvulcanized rubber and articles thereof (commission agent)
22210	25210	Vulcanized fibre (manufacture)
28960	29560	Vulcanizing machines for working rubber and plastics (manufacture)
17220	17549	Wadding made from yarn spun on the cotton system (manufacture)
10720	15820	Wafer biscuits (manufacture)
27510	29710	Waffle irons (manufacture)
10710	15810	Waffles (manufacture)
30200	35200	Wagon and locomotive frames (manufacture)
13922	17402	Wagon cover (manufacture)
16100	20100	Wagon timber (sawn) (manufacture)
14390	17720	Waistcoats and other similar articles (knitted) (manufacture)
14131	18221	Waistcoats for men and boys (manufacture)
28922	29522	Walking draglines (manufacture)
32990	36639	Walking sticks (manufacture)
46420	51429	Walking sticks and seat sticks (wholesale)
16290	36639	Walking sticks made of wood (manufacture)
46730	51530	Wall boards (wholesale)
43330	45430	Wall covering
22230	25239	Wall coverings made of plastic (manufacture)
25990	28750	Wall mountings made of metal (manufacture)
23610	26610	Wall panels for structural use made of precast concrete (manufacture)
23310	26300	Wall tiles (glazed) (manufacture)
23310	26300	Wall tiles made of unglazed clay (manufacture)
31090	36140	Wall unit (manufacture)
91020	92521	Wallace collection
47722	52432	Wallets (retail)
17230	21230	Wallets containing an assortment of paper stationery (manufacture)
15120	19200	Wallets made of leather or leather substitute (manufacture)
47530	52489	Wallpaper (retail)
46730	51440	Wallpaper (wholesale)
17240	21240	Wallpaper and lining paper (manufacture)
17120	21120	Wallpaper base (manufacture)
43330	45430	Wallpaper hanging
01250	01139	Walnut growing
01700	01500	Walrus catching
25400	29600	War ammunition (manufacture)
94990	91330	War veterans' associations
31090	36140	Wardrobes (manufacture)
31090	36140	Wardrobes made of metal (manufacture)
52102	63129	Warehouse (general) operation for air transport activities
52103	63129	Warehouse (general) operation for land transport activities
52101	63129	Warehouse (general) operation for water transport activities
27520	29720	Warm air generator (non-electric) (manufacture)
13100	17120	Warp dressing (woollen) (manufacture)
13910	17600	Warp knitted fabric (manufacture)
13300	17300	Warp knitted fabric dyeing and finishing (manufacture)
13910	17600	Warp knitting (manufacture)
13100	17130	Warp sizing and dressing (worsted) (manufacture)
10620	15620	Warp starch (manufacture)
28940	29540	Warpers for preparing textile yarns (manufacture)
28940	29540	Warping machinery (manufacture)
30110	35110	Warship (manufacture)
25990	28750	Wash basins made of metal (manufacture)
22230	25239	Wash basins made of plastic (manufacture)
23420	26220	Wash basins or sinks made of ceramic, fireclay, etc. (manufacture)
46730	51530	Washbasins (wholesale)
15120	19200	Washers made of leather (manufacture)
25940	28740	Washers made of metal (manufacture)
22190	25130	Washers made of rubber (manufacture)
96010	93010	Washing and cleaning of fur products
96010	93010	Washing and dry cleaning of clothing
96010	93010	Washing and dry cleaning of textile products
28940	29540	Washing machines (laundry) (manufacture)
28940	29540	Washing machines (textile) (non-domestic) (manufacture)
27510	29710	Washing machines for domestic use (manufacture)
20411	24511	Washing powders and preparations in solid or liquid form (manufacture)
46440	51440	Washing products (e.g. washing powder) (wholesale)
46770	51570	Waste (wholesale)
46180	51180	Waste and scrap (commission agent)
46770	51570	Waste and scrap exporter (wholesale)
46770	51570	Waste and scrap importer (wholesale)
38110	90020	Waste collection
38110	90020	Waste collection centre
13100	17110	Waste cotton yarn (manufacture)
38210	90020	Waste disposal
27510	29710	Waste disposers (manufacture)
23130	26130	Waste glass resulting from glass container production (manufacture)

W

SIC 2007	SIC 2003	Activity
23190	26150	Waste glass resulting from glass product production (other than glass container) (manufacture)
25300	28300	Waste heat boiler (manufacture)
38210	90020	Waste incineration
46770	51570	Waste paper (wholesale)
46770	51570	Waste rubber (wholesale)
46770	51570	Waste string (wholesale)
13300	17300	Waste textile dyeing (manufacture)
38210	90020	Waste treatment by composting of plant materials
46760	51550	Waste, parings and scrap of plastic (wholesale)
46120	51120	Waste, parings and scrap of rubber (commission agent)
26520	33500	Watch (manufacture)
46499	51479	Watch and clock movements (wholesale)
15120	19200	Watch bands of non-metallic material such as fabric, leather or plastic (manufacture)
26520	33500	Watch case (manufacture)
23190	26150	Watch glass (manufacture)
15120	19200	Watch straps (non-metallic) (manufacture)
15120	19200	Watch straps made of leather or leather substitute (manufacture)
32130	33500	Watch straps, bands and bracelets made of non-precious metal (manufacture)
32120	33500	Watchbands, wristbands and watch straps, of precious metal (manufacture)
47770	52484	Watches and clocks (retail)
46480	51479	Watches and clocks (wholesale)
26520	33500	Watchmakers' jewels (manufacture)
80100	74602	Watchman activities
36000	41000	Water authority (headquarters and water supply)
11070	15980	Water bottling (manufacture)
22290	25240	Water butts made of plastic (manufacture)
25400	29600	Water cannon (manufacture)
23420	26220	Water closet bowls made of ceramic, fireclay, etc. (manufacture)
36000	41000	Water collection, purification and distribution
36000	41000	Water company
36000	41000	Water conservation
71122	74206	Water divining and other scientific prospecting activities
77342	71229	Water freight transport equipment rental (without operator)
27520	29720	Water heaters (gas) (manufacture)
27510	29710	Water heaters for domestic use (electric) (manufacture)
27520	29720	Water heaters for domestic use (non-electric) (manufacture)
10520	15520	Water ices (manufacture)
42210	45213	Water main and line construction
71129	74204	Water management projects design
71200	74300	Water measuring related to cleanness
26513	33202	Water meters (non-electronic) (manufacture)
77341	71221	Water passenger transport equipment rental (without operator)
42910	45240	Water project construction
22290	25240	Water sensitive adhesive tapes made of plastic (manufacture)
28290	29240	Water softening plant (manufacture)
32300	36400	Water sports equipment (manufacture)
22230	25239	Water stop and bar made of plastic (manufacture)
50100	61101	Water taxis operation (except for inland waterway service)
50300	61201	Water taxis operation (inland waterway service)
25110	28110	Water tower made of steel plate (manufacture)
52220	63220	Water transport (supporting activities)
50400	61209	Water transport of freight inside harbours and docks
20590	24660	Water treatment chemicals (manufacture)
28290	29240	Water treatment plant (manufacture)
42210	45240	Water treatment plant construction
25300	28300	Water tube boilers (manufacture)
42210	45250	Water well drilling
22190	25130	Waterbed mattresses of rubber (manufacture)
50200	61102	Waterborne freight transport (except for inland waterway service)
01130	01120	Watercress growing
22290	25240	Watering cans made of plastic (manufacture)
01130	01120	Watermelon growing
17120	21120	Waterproof paper (manufacture)
13922	17402	Waterproofed covers made of canvas (manufacture)
43999	45220	Waterproofing of buildings
13300	17300	Waterproofing of purchased garments (manufacture)
46499	51479	Water-skis, surf-boards, sail-boards and other water-sport equipment (wholesale)
42910	45240	Waterway construction
52220	63220	Waterway locks operation
28110	29110	Water-wheels (manufacture)
28110	29110	Waterwheels and regulators and parts thereof (manufacture)
32300	36400	Waterwings made of rubber (manufacture)
26511	33201	Watt meter (electronic) (manufacture)
26513	33202	Watt meter (non-electronic) (manufacture)
27900	31620	Wave form generator (manufacture)
20420	24520	Waving and hair straightening preparations (manufacture)
20412	24512	Wax (manufacture)
93290	92349	Waxworks
74909	63400	Way-bills issue and procurement
22230	25239	WC seat and cover units made of plastic (manufacture)
25400	29600	Weapons (manufacture)
47789	52489	Weapons (retail)
14200	18300	Wearing apparel and clothing accessories made of fur (manufacture)
77299	71409	Wearing apparel rental and leasing
74909	74206	Weather forecasting activities
65120	66031	Weather insurance
14120	18210	Weather protective industrial clothing (manufacture)
16100	20100	Weatherboard (manufacture)
22230	25239	Weatherboarding made of plastic (manufacture)
14131	18221	Weatherproof jackets for men and boys (manufacture)
14132	18222	Weatherproof jackets for women and girls (manufacture)
14190	18249	Weatherproof outerwear for infants (manufacture)
14131	18221	Weatherproof outerwear for men and boys (manufacture)
14132	18222	Weatherproof outerwear for women and girls (manufacture)
14190	18249	Weatherproof skiing clothing (manufacture)
13200	17220	Weaving (woollen) (manufacture)
13200	17230	Weaving (worsted) (manufacture)
28940	29540	Weaving machinery (looms) (manufacture)
28940	29540	Weaving machines (manufacture)

W

SIC 2007	SIC 2003	Activity
13200	17210	Weaving of cotton and man-made fibres (manufacture)
13922	17402	Web equipment making-up (manufacture)
63110	72300	Web hosting
62012	72220	Web page design
63120	72400	Web search portals
13960	17542	Webbing made of non-elastic and non-elastomeric (manufacture)
13960	17542	Webbing weaving (manufacture)
56210	55520	Wedding catering
74209	74819	Wedding photography
26701	33402	Wedge (optical) (manufacture)
20200	24200	Weed killer (manufacture)
13910	17600	Weft knitted fabric (manufacture)
13300	17300	Weft knitted fabric dyeing and finishing (manufacture)
28290	29240	Weighbridge (manufacture)
52219	63210	Weighbridge services
28290	29240	Weighing machine (manufacture)
96090	93059	Weighing machine operation (coin operated)
46690	51870	Weighing machines and scales for commercial use (wholesale)
28290	29240	Weights for weighing machine (manufacture)
24100	27100	Welded sections of iron and steel (manufacture)
24200	27220	Welded tubes (manufacture)
27900	31620	Welding electrode (manufacture)
27900	29430	Welding machines (electric) (manufacture)
28290	29430	Welding machines (gas) (manufacture)
25930	28730	Welding rods (coated/covered with flux) made of steel (manufacture)
28290	29430	Welding torch (manufacture)
24340	27340	Welding wire (uncoated) made of steel (manufacture)
88990	85321	Welfare and guidance activities for children and adolescents (charitable)
88990	85322	Welfare and guidance activities for children and adolescents (non-charitable)
88990	85321	Welfare service (charitable)
88990	85322	Welfare service (non-charitable)
23190	26150	Well and bulkhead glass (manufacture)
28921	29521	Well drilling equipment (manufacture)
09100	11200	Well logging
43999	45250	Well sinking (except gas or oil)
28921	29521	Wellhead running tools (manufacture)
28921	29521	Wellheads (manufacture)
15200	19300	Wellington boot (manufacture)
09100	11200	Well-perforating services
94910	91310	Wesleyan reform union
27200	31400	Wet cell batteries (manufacture)
10620	15620	Wet corn milling (manufacture)
47230	52230	Wet fish (retail)
46380	51380	Wet fish dealer (wholesale)
22190	25130	Wet suits of rubber (manufacture)
46330	51333	Whale oil (wholesale)
10410	15410	Whale oil production (manufacture)
10410	15420	Whale oil refining (manufacture)
32990	36639	Whalebone cutting and splitting (manufacture)
30110	35110	Whaler (manufacture)
10110	15111	Whales processing on land or on specialised vessels (manufacture)

SIC 2007	SIC 2003	Activity
03110	05010	Whaling
52220	63220	Wharfinger
10611	15611	Wheat flake (manufacture)
01110	01110	Wheat growing
10611	15611	Wheat milling (manufacture)
10611	15611	Wheat offal (manufacture)
10611	15611	Wheat pellets (manufacture)
10620	15620	Wheat starch (manufacture)
26511	33201	Wheel balancing machine (electronic) (manufacture)
26513	33202	Wheel balancing machine (non-electronic) (manufacture)
30920	35430	Wheel chair (manufacture)
77390	71219	Wheelbarrow hire
28220	35500	Wheelbarrows made of metal (manufacture)
28220	35500	Wheelbarrows made of plastic (manufacture)
28220	35500	Wheelbarrows made of wood (manufacture)
30920	35430	Wheelchair (manufacture)
33170	35430	Wheelchair repair and maintenance (manufacture)
32409	36509	Wheeled toys designed to be ridden (manufacture)
46499	51477	Wheeled toys designed to be ridden by children (wholesale)
46610	51880	Wheeled tractor (wholesale)
29320	34300	Wheels and hubs for motor vehicles (manufacture)
30910	35410	Wheels for motorcycles (manufacture)
30920	35420	Wheels for pedal cycles (manufacture)
10519	15519	Whey (sweetened skimmed) production (manufacture)
13300	52740	'While-you-wait' printing of textile articles (manufacture)
08110	14110	Whinstone quarry
15120	36639	Whips (manufacture)
46499	51479	Whips and riding crops (wholesale)
11010	15910	Whisky (manufacture)
11010	15910	Whisky blending (manufacture)
11010	15910	Whisky distilling (manufacture)
32200	36300	Whistles (manufacture)
28230	36120	White boards (manufacture)
10822	15842	White chocolate (manufacture)
01250	01139	White currant growing
20130	24130	White lead (not in paste form) (manufacture)
20301	24301	White lead in paste form (manufacture)
24450	27450	White metal (manufacture)
08930	14400	White salt production
19201	23201	White spirit (manufacture)
10810	15830	White sugar (manufacture)
32910	36620	Whitewash brush (manufacture)
08110	14120	Whiting and prepared chalk production
46390	51390	Wholesale grocer (wholesale)
46900	51900	Wholesale of a variety of goods without any particular specialisation (wholesale)
16290	20520	Wicker baskets (manufacture)
31090	36140	Wicker furniture (manufacture)
16290	20520	Wickerwork (manufacture)
46499	51479	Wickerwork (wholesale)
47599	52440	Wickerwork goods (retail)
13960	17542	Wicks for lamps, stoves or candles (manufacture)
19201	23201	Wide cut gasoline (manufacture)
24100	27100	Wide slabs made of semi-finished steel (manufacture)
32990	36639	Wig (manufacture)

W

SIC 2007	SIC 2003	Activity	SIC 2007	SIC 2003	Activity
91040	92530	Wildlife preservation services	03110	05010	Winkle gathering
02100	02010	Willow growing	28302	29320	Winnower for agricultural use (manufacture)
69109	74119	Wills preparation	28930	29530	Winnowers for milling (manufacture)
13931	17511	Wilton carpet (manufacture)	93110	92619	Winter sport arenas and stadiums
13200	17210	Winceyette weaving (manufacture)	93120	92629	Winter sport clubs
28220	29220	Winch (manufacture)	16240	20400	Wire and cable drums made of wood (manufacture)
46690	51870	Winches and capstans (wholesale)	32910	36620	Wire brush (manufacture)
46630	51820	Winches specially designed for use underground (wholesale)	25930	28730	Wire cable made of steel (manufacture)
			28990	29560	Wire coiling machine (manufacture)
35110	40110	Wind farms	25930	28730	Wire fabric made of steel (manufacture)
32200	36300	Wind instrument (manufacture)	46690	51870	Wire for industrial use (wholesale)
28110	29110	Wind turbines (manufacture)	24420	27420	Wire made of aluminium (manufacture)
28220	29220	Winding device (manufacture)	24440	27440	Wire made of copper (uninsulated) (manufacture)
28921	29521	Winding machine for mining (manufacture)	46740	51540	Wire netting (wholesale)
28940	29540	Winding machine for textiles (manufacture)	24410	27410	Wire of precious metals, made by drawing (manufacture)
27320	31300	Winding wire and strip (manufacture)	25930	28730	Wire products (manufacture)
28220	29220	Windlass (manufacture)	25930	28730	Wire products made of uninsulated copper (manufacture)
47599	52440	Window blinds (retail)			
22230	25239	Window blinds and accessories made of plastic (manufacture)	24440	27440	Wire rods made of copper (manufacture)
81221	74702	Window cleaning	24100	27100	Wire rods made of steel (manufacture)
13940	17520	Window cord (manufacture)	28990	29560	Wire rope making machine (manufacture)
73110	74402	Window dressing	25930	28730	Wire strands made of aluminium (manufacture)
25720	28630	Window fittings made of metal (manufacture)	28490	29430	Wire weaving machine (manufacture)
23650	26650	Window frames made of fibre-cement (manufacture)	46690	51870	Wire, switches and other installation equipment for industrial use (wholesale)
25120	28120	Window frames made of metal (manufacture)	24420	27420	Wirebar made of aluminium (manufacture)
22230	25239	Window frames made of plastic (manufacture)	23110	26110	Wired glass (manufacture)
16230	20300	Window frames made of wood (manufacture)	61200	64200	Wireless telecommunications activities
13910	17600	Window furnishing knitted fabric (manufacture)	46470	51439	Wires and switches for domestic lighting use (wholesale)
23110	26110	Window glass (not cut to size) (manufacture)			
23120	26120	Window glass cut to size (manufacture)	27330	31200	Wiring accessories (manufacture)
18129	22220	Window ticket (manufacture)	27330	31200	Wiring devices (manufacture)
29320	34300	Window winding gear for motor vehicles (not electric) (manufacture)	29310	31610	Wiring sets (manufacture)
45200	50200	Windscreen replacement services	08990	14500	Witherite mine
29310	31610	Windscreen wipers (manufacture)	02200	02010	Withy growing
29320	34300	Windscreen wipers for motor vehicles (non-electric) (manufacture)	24450	27450	Wolfram (manufacture)
			47710	52423	Women's bespoke tailor (retail)
23120	26120	Windscreens made of glass (manufacture)	47710	52423	Women's clothier and outfitter (retail)
93290	92729	Windsor Great Park	47710	52423	Women's clothing (retail)
11020	15931	Wine (of own manufacture) blending, purification and bottling (manufacture)	46420	51429	Women's clothing (wholesale)
11020	51342	Wine (purchased in bulk) blending, purification and bottling (manufacture)	47710	52423	Women's clothing accessories (retail)
			47710	52423	Women's hats (retail)
47250	52250	Wine and spirit merchant (retail)	47710	52423	Women's hosiery (retail)
46342	51342	Wine and spirit merchant (wholesale)	47710	52423	Women's outfitter (retail)
47250	52250	Wine and spirits (retail)	84220	75220	Women's Royal Air Force
11020	15932	Wine based on concentrated grape must (manufacture)	84220	75220	Women's Royal Army Corps
			84220	75220	Women's Royal Naval Service
46342	51342	Wine importer (wholesale)	88990	85321	Women's Royal Voluntary Service
28930	29530	Wine making machinery (manufacture)	47710	52423	Women's tailor (retail)
20590	24660	Wine making preparations (excluding yeast) (manufacture)	47710	52423	Women's wear (retail)
			46130	51130	Wood (commission agent)
11020	15931	Wine production from fresh grapes and grape juice (manufacture)	46730	51530	Wood (wholesale)
11020	01131	Wine production from self produced grapes	18130	22240	Wood blocks for printing (manufacture)
16290	20510	Wine racks made of wood (manufacture)	25730	28620	Wood boring bit (manufacture)
46690	51870	Wine, cider, fruit beverage production machinery (wholesale)	16290	20510	Wood carving (manufacture)
			46750	51550	Wood charcoal (not fuel) (wholesale)
28930	29530	Wine, cider, fruit juice, etc. press (manufacture)	46120	51120	Wood charcoal for fuel (commission agent)
30300	35300	Wings for aircraft (manufacture)	16100	20100	Wood chip (manufacture)

W

SIC 2007	SIC 2003	Activity
16210	20200	Wood chipboard agglomerated with non-mineral binding substances (manufacture)
25730	28620	Wood chisel (manufacture)
16100	20100	Wood creosoting, impregnation, preservation, spraying, varnishing and drying (manufacture)
16100	20100	Wood drying (manufacture)
90030	92319	Wood engraver (artistic)
16100	20100	Wood flour (manufacture)
02200	02010	Wood gathering and production, for energy
16100	20100	Wood grooving, milling, planing, sawing etc. (manufacture)
16100	20100	Wood impregnation (manufacture)
46730	51530	Wood in the rough (wholesale)
02200	02010	Wood in the rough production (untreated)
02200	02010	Wood logging, etc. within forestry site
16100	20100	Wood machining (manufacture)
16290	20510	Wood marquetry (manufacture)
16100	20100	Wood mill (manufacture)
16100	20100	Wood particles (manufacture)
16230	20300	Wood partitions (manufacture)
16100	20100	Wood planing (manufacture)
16100	20100	Wood preservation (manufacture)
46730	51530	Wood products of primary processing (wholesale)
17110	21110	Wood pulp (manufacture)
17290	21259	Wood pulp vessel (manufacture)
16100	20100	Wood sawing (manufacture)
28490	29430	Wood sculpturing and engraving machines (manufacture)
16100	20100	Wood shavings (manufacture)
16100	20100	Wood spraying (manufacture)
20301	24301	Wood stain (manufacture)
20140	24140	Wood tar chemicals (manufacture)
16100	20100	Wood varnishing (manufacture)
16210	20200	Wood veneers (manufacture)
16100	20100	Wood wool (manufacture)
81291	74703	Wood worm preventative treatment service
46730	51530	Wood, construction materials and sanitary equipment exporter (wholesale)
46730	51530	Wood, construction materials and sanitary equipment importer (wholesale)
16290	20510	Wooden articles n.e.c. (manufacture)
16230	20300	Wooden beams for the construction industry (manufacture)
28490	29430	Wooden button making machinery (manufacture)
16290	20510	Wooden cases for jewellery (manufacture)
16230	20300	Wooden goods intended to be used primarily in the construction industry (manufacture)
16220	20300	Wooden parquet floor blocks assembled into panels (manufacture)
16220	20300	Wooden parquet floor strips assembled into panels (manufacture)
47599	52440	Wooden ware (retail)
46499	51479	Wooden ware (wholesale)
46760	51560	Woodpulp and paper making materials (wholesale)
81291	74703	Wood rot preventative treatment service
32200	36300	Woodwind instruments (manufacture)
28490	29430	Woodworking machinery (manufacture)
10110	15113	Wool (fellmongery) (manufacture)
01450	01220	Wool (raw) production
46110	51110	Wool broker (commission agent)
28940	29540	Wool carbonisers (manufacture)

SIC 2007	SIC 2003	Activity
13100	17120	Wool carbonising (manufacture)
13100	17120	Wool carding (manufacture)
13100	17120	Wool cleaning (manufacture)
13100	17120	Wool condensing (manufacture)
13300	17300	Wool dyeing (loose) (manufacture)
46110	51110	Wool exchange (commission agent)
13100	17120	Wool extracting (manufacture)
14190	18241	Wool felt and fur felt hood and capeline (manufacture)
46110	51110	Wool grease including lanolin (commission agent)
46760	51560	Wool merchant (wholesale)
13100	17120	Wool opening and willeying (manufacture)
13300	17300	Wool printing (manufacture)
13100	17120	Wool recovery (manufacture)
28940	29540	Wool scourers (manufacture)
13100	17120	Wool scouring (manufacture)
13100	17120	Wool sorting (manufacture)
13100	17130	Wool topmaking (manufacture)
52102	63129	Wool warehouse for air transport activities
52103	63129	Wool warehouse for land transport activities
52101	63129	Wool warehouse for water transport activities
13300	17300	Woollen and worsted fabric bleaching, dyeing, or otherwise finishing (manufacture)
13200	17220	Woollen cloth weaving (manufacture)
47510	52410	Woollen draper (retail)
46410	51410	Woollen flock (wholesale)
46770	51570	Woollen rag (wholesale)
13100	17120	Woollen rag carbonising (manufacture)
13100	17120	Woollen rag carding (manufacture)
13100	17120	Woollen rag garnetting (manufacture)
13100	17120	Woollen rag grinding or pulling (manufacture)
13100	17120	Woollen waste breaking (manufacture)
13100	17120	Woollen waste garnetting (manufacture)
13100	17120	Woollen waste grinding (manufacture)
13100	17120	Woollen waste opening and willeying (manufacture)
13100	17120	Woollen yarn carding (manufacture)
13100	17120	Woollen yarn condensing (manufacture)
13100	17120	Woollen yarn reeling (manufacture)
13100	17120	Woollen yarn sizing (manufacture)
13100	17120	Woollen yarn spinning (manufacture)
13100	17120	Woollen yarn twisting (manufacture)
13100	17120	Woollen yarn warping (manufacture)
13100	17120	Woollen yarn winding (manufacture)
46410	51410	Woollens (wholesale)
13200	17220	Woollen-type fabrics (manufacture)
13100	17120	Woollen-type fibre preparation and spinning (manufacture)
13200	17220	Woollen-type weaving (manufacture)
13100	17120	Woollen-type yarns (manufacture)
28230	30010	Word processing machines (manufacture)
77330	71330	Word processing machines rental and operating leasing
31090	36140	Work benches made of wood (manufacture)
28490	29430	Work holders (engineers' small tools) (manufacture)
28490	29430	Work holders for machine tools (manufacture)
77320	71320	Work platform rental without erection and dismantling
96010	93010	Work uniforms rental from laundries
43999	45250	Work with specialist access requirements necessitating climbing skills and related equipment

W

SIC 2007	SIC 2003	Activity	SIC 2007	SIC 2003	Activity
32120	36220	Worked pearls (manufacture)	17230	22220	Writing paper pads (manufacture)
23990	26829	Worked peat, articles thereof (manufacture)	28230	30010	Xerographic copying machines (manufacture)
85590	80429	Workers' Educational Association	74202	74813	X-ray and other speciality photography activities
55900	55239	Workers' hostels	26600	33100	X-ray apparatus for industrial use (manufacture)
70221	74142	Working capital and liquidity management consultancy services	26600	33100	X-ray apparatus for medical use (manufacture)
94990	91330	Working men's club (not licensed to sell alcohol)	26600	33100	X-ray diffraction or fluorescence apparatus (manufacture)
56301	55401	Working men's clubs (licensed to sell alcohol)	26600	33100	X-ray or alpha, beta or gamma radiation apparatus (manufacture)
47781	52486	Works of art (retail)	26600	33100	X-ray tubes (manufacture)
85320	80220	Works school (if separately identifiable)	46520	51860	X-ray, alpha, beta or gamma radiation apparatus (wholesale)
28220	29220	Works trucks (manufacture)	20140	24140	Xylene (manufacture)
46690	51870	Works trucks (wholesale)	47789	52489	Yacht chandler (retail)
30200	35200	Workshop wagon for railways (manufacture)	93120	92629	Yacht club
46160	51160	Workwear (commission agent)	30120	35120	Yacht building (manufacture)
14120	18210	Workwear (manufacture)	46499	51479	Yachts (wholesale)
14120	18210	Workwear clothing for women (manufacture)	01130	01110	Yam growing
99000	99000	World bank	46160	51160	Yarn (commission agent)
01490	01250	Worm (other than marine) farming	13100	17110	Yarn (core spun on the cotton system) (manufacture)
03210	05020	Worm farms, marine	22190	25130	Yarn (rubberised) (manufacture)
13100	17130	Worsted carding (manufacture)	46410	51410	Yarn (wholesale)
13100	17130	Worsted doubling (manufacture)	13300	17300	Yarn bleaching, dyeing or otherwise finishing (manufacture)
13100	17130	Worsted spinning and twisting (manufacture)	13300	17300	Yarn finishing (manufacture)
13100	17130	Worsted waste grinding (manufacture)	13300	17300	Yarn gassing (manufacture)
13100	17130	Worsted yarn reeling (manufacture)	23990	26821	Yarn made of asbestos (manufacture)
13100	17130	Worsted yarn warping (manufacture)	23140	26140	Yarn made of glass fibre (manufacture)
13100	17130	Worsted yarn winding (manufacture)	13100	17170	Yarn made of paper (manufacture)
13200	17230	Worsted-type fabrics (manufacture)	13300	17300	Yarn mercerising (manufacture)
13100	17130	Worsted-type fibres preparation and spinning (manufacture)	13100	17110	Yarn of cotton (manufacture)
13200	17230	Worsted-type weaving (manufacture)	13300	17300	Yarn polishing (manufacture)
13100	17130	Worsted-type yarns (manufacture)	13300	17300	Yarn storing (manufacture)
13200	17250	Woven cloth made of polypropylene (manufacture)	46380	51380	Yeast (wholesale)
13960	17542	Woven conveyor belting (manufacture)	10890	15899	Yeast and vegetable extract (manufacture)
13200	17250	Woven elastic over 30 cm wide (manufacture)	10890	15899	Yeast preparation (manufacture)
13200	17250	Woven elastomeric over 30 cm wide (manufacture)	46170	51170	Yeasts and prepared baking powders (commission agent)
13300	17300	Woven fabric bleaching, dyeing, printing or otherwise finishing (manufacture)	07210	12000	Yellowcake production
31090	36140	Woven fibre furniture (manufacture)	24460	23300	Yellowcake to uranium tetrafluoride and hexafluoride conversion (manufacture)
13960	17542	Woven machinery belting (manufacture)	55900	55239	YMCA hostel
13200	17210	Woven pile cotton-type fabrics (manufacture)	85510	93059	Yoga instruction
13960	17542	Woven trimmings (manufacture)	10519	15519	Yoghurt (manufacture)
17120	21120	Wrapping and packaging paper including coated (manufacture)	46330	51331	Yoghurt (wholesale)
24420	27420	Wrapping foil made of aluminium (manufacture)	84230	75230	Young offender centres
28290	29240	Wrapping machinery (manufacture)	94920	91320	Young people's auxiliaries associated with a political party
17290	21259	Wrapping paper cut and packed in ready to use sheets or rolls (manufacture)	94990	91330	Young persons associations
52220	63220	Wreck raising	94990	91330	Youth centre
25730	28620	Wrecking bars (manufacture)	94990	91330	Youth club
25730	28620	Wrench (manufacture)	55202	55210	Youth hostel
93120	92629	Wrestling clubs	94990	91330	YMCA (not hostel)
26520	33500	Wrist watch (manufacture)		17210	Zephyr weaving (manufacture)
69109	74119	Writer to the signet	32500	33100	Zimmer frames and other walking aids (manufacture)
94120	91120	Writers associations	24430	27430	Zinc (manufacture)
17230	21230	Writing compendiums (manufacture)	46720	51520	Zinc (wholesale)
46499	51479	Writing implement sets (wholesale)	07290	13200	Zinc mining and preparation
20590	24660	Writing ink (manufacture)	20120	24120	Zinc oxide (manufacture)
32990	36631	Writing instrument sets (manufacture)			
25620	28520	Writing on metal by laser beam (manufacture)			
17120	21120	Writing paper (manufacture)			

X

Y

Z

SIC 2007	SIC 2003	Activity	SIC 2007	SIC 2003	Activity
46750	51550	Zinc oxide and peroxide (wholesale)	25990	36639	Zip fasteners made of metal (manufacture)
20301	24301	Zinc paint (manufacture)	22290	36639	Zip fasteners made of plastic (manufacture)
24430	27430	Zinc wire made by drawing (manufacture)	24450	27450	Zirconium (manufacture)
94990	91330	Zionist organisation	91040	92530	Zoological gardens

Z

SIC 2007	SIC 2003	Activity		SIC 2007	SIC 2003	Activity

Numerical Index

Part 2

Numerical Index

UK SIC 2007	UK SIC 2003	Activity
01110		**Growing of cereals (except rice), leguminous crops and oil seeds**
	01110	Barley growing
	01120	Bean growing
	01120	Broad bean growing
	01110	Castor bean growing
	01110	Cereal grains growing, except rice
	01120	Chick pea growing
	01110	Colza growing
	01110	Cottonseed growing
	01120	Cow pea growing
	01110	Dried leguminous vegetables growing
	01110	Grain maize growing
	01110	Groundnut growing
	01120	Leguminous crops growing
	01110	Lentil growing
	01110	Linseed growing
	01110	Lupin growing
	01110	Millet growing
	01110	Mustard seed growing
	01110	Niger seed growing
	01110	Oats growing
	01110	Oil seeds growing
	01120	Pea growing
	01110	Peanut growing
	01120	Pigeon pea growing
	01110	Rape growing
	01110	Rye growing
	01110	Safflower seed growing
	01110	Sesame seed growing
	01110	Sorghum growing
	01110	Soya growing
	01110	Sunflower seed growing
	01110	Sunflower seed production
	01110	Wheat growing
01120		**Growing of rice**
	01110	Rice growing
01130		**Growing of vegetables and melons, roots and tubers**
	01120	Alliaceous vegetable growing
	01120	Artichoke growing
	01120	Asparagus growing
	01120	Aubergine (egg-plant) growing
	01120	Beetroot growing
	01120	Broccoli growing
	01120	Brussel sprout growing
	01120	Bulb growing
	01120	Cabbage growing
	01120	Cantaloupe growing

SIC 2007	SIC 2003	Activity
	01120	Capers growing
	01120	Carrot growing
	01110	Cassava growing
	01120	Cauliflower growing
	01120	Chervil growing
	01120	Chicory growing
	01120	Courgette growing
	01120	Cress growing
	01120	Cucumber growing
	01120	Egg plant growing
	01120	Fennel growing
	01120	Fruit bearing vegetables growing
	01120	Garlic growing
	01120	Gherkin growing
	01120	Herbs (culinary) growing
	01120	Jerusalem artichoke growing
	01120	Leafy or stem vegetables growing
	01120	Leek growing
	01120	Lettuce growing
	01120	Melon growing
	01120	Mushroom growing (cultivated)
	01120	Onion growing
	01120	Parsley growing
	01120	Parsnip growing
	01110	Potato growing
	01120	Root vegetable growing
	01110	Roots and tubers growing
	01110	Roots and tubers with a high starch or inulin content
	01120	Shallot growing
	01120	Spinach growing
	01110	Sugar beet growing
	01110	Sugar beet seed production
	01120	Swede growing
	01120	Sweet marjoram growing
	01110	Sweet potato growing
	01120	Tarragon growing
	01120	Tomato growing
	01120	Truffles growing
	01110	Tuberous vegetable growing
	01120	Turnip growing
	01120	Vegetable growing (except potatoes)
	01120	Vegetable seed growing
	01120	Watercress growing
	01120	Watermelon growing
	01110	Yam growing
01140		**Growing of sugar cane**
	01110	Sugar cane growing
01150		**Growing of tobacco**
	01110	Tobacco growing

UK SIC 2007	UK SIC 2003	Activity	SIC 2007	SIC 2003	Activity
01160		**Growing of fibre crops**		01139	Cherry growing
	01110	Abaca and other vegetable textile fibre growing		01139	Cider apple growing
	01110	Cotton growing		01139	Nectarine growing
	01110	Fibre crop growing		01139	Peach growing
	01110	Flax growing		01139	Pear growing
	01110	Jute textile bast fibre growing		01139	Plum growing
	01110	Kenaf and other textile bast fibre growing		01139	Pome fruit growing
	01110	Plants bearing vegetable fibres, retting of		01139	Quince growing
	01110	Ramie and other vegetable textile fibre growing		01139	Sloe growing
	01110	Sisal growing and other textile fibre of the genus agave growing		01139	Sour cherry growing
	01110	Textile plants growing		01139	Stone fruit growing
	01110	True hemp growing	**01250**		**Growing of other tree and bush fruits and nuts**
01190		**Growing of other non-perennial crops**		01139	Almond growing
	01110	Alfalfa growing		01139	Blackberry (cultivated) growing
	01110	Beet seed growing		01139	Blackcurrant growing
	01110	Clover growing		01139	Blueberry growing
	01120	Cut flowers and flower bud production		01139	Cashew nut growing
	01120	Dried flower production		01139	Chestnut growing
	01120	Flower growing		01139	Currant growing
	01120	Flower seed growing		01139	Edible nuts growing
	01110	Fodder maize and other grass growing		01120	Fruit seed growing
	01110	Fodder root growing		01139	Gooseberry growing
	01110	Forage kale and similar forage products growing		01139	Growing of berries
	01110	Forage plants seed production including grasses		01139	Hazelnut growing
	01110	Forage production		01139	Kiwi fruit growing
	01110	Mangold growing		01139	Locust bean growing
	01110	Sainfoin growing		01139	Loganberry growing
	01120	Seed production for flowers, fruit or vegetables (not for oil)		01139	Mulberry growing
				01139	Pistachio growing
01210		**Growing of grapes**		01139	Raspberry growing
	01131	Grape production		01139	Redcurrant growing
	01131	Vineyards		01139	Strawberry growing
				01139	Walnut growing
01220		**Growing of tropical and subtropical fruits**		01139	White currant growing
	01139	Avocado growing	**01260**		**Growing of oleaginous fruits**
	01139	Bananas and plantain growing		01139	Coconut growing
	01139	Dates growing		01110	Growing of oleaginous fruits other than olives
	01139	Fig growing		01110	Oil palms growing
	01139	Mango growing		01139	Olive growing
	01139	Other tropical and subtropical fruit growing			
	01139	Papaya growing	**01270**		**Growing of beverage crops**
	01139	Pineapple growing		01139	Beverage crop growing other than wine grapes
				01139	Cocoa growing
01230		**Growing of citrus fruits**		01139	Coffee growing
	01139	Clementine growing		01139	Maté growing
	01139	Grapefruit growing		01139	Tea growing
	01139	Growing of citrus fruits			
	01139	Lemon growing	**01280**		**Growing of spices, aromatic, drug and pharmaceutical crops**
	01139	Lime growing		01139	Anise growing
	01139	Mandarin growing		01139	Aromatic crops growing
	01139	Orange growing		01139	Badian growing
	01139	Pomelo growing		01139	Basil growing
	01139	Tangerine growing		01139	Bay growing
				01139	Chilli growing
01240		**Growing of pome fruits and stone fruits**		01139	Chillies and peppers capsicum sop. Growing
	01139	Apple growing		01139	Cinnamon growing
	01139	Apricot growing			

UK SIC 2007	UK SIC 2003	Activity	SIC 2007	SIC 2003	Activity
	01139	Clove growing		01220	Sheep farming
	01139	Coriander growing		01220	Sheep milk (raw) production
	01110	Drug and narcotic crops growing		01220	Sheep raising and breeding
	01139	Ginger growing		01220	Wool (raw) production
	01110	Hop cones growing			
	01139	Nutmeg, mace and cardamoms growing	01460		**Raising of swine/pigs**
	01120	Pepper growing		01230	Pig farming
	01110	Pharmaceutical crops growing		01230	Pig raising and breeding
	01110	Plants used chiefly in pharmacy or for insecticidal, fungicidal or similar purposes		01230	Swine farming
				01230	Swine raising and breeding
	01139	Spice crops growing			
	01139	Vanilla growing	01470		**Raising of poultry**
				01240	Chicken farm (battery rearing)
01290		**Growing of other perennial crops**		01240	Chicken raising and breeding
	02010	Christmas tree growing		01240	Duck farming
	01110	Rubber trees growing for harvesting of latex		01240	Duck raising and breeding
	01110	Tree growing for extraction of sap		01240	Egg hatchery
	01110	Vegetable plaiting material growing		01240	Egg production
				01240	Egg production from poultry
01300		**Plant propagation**		01240	Geese farming
	01120	Mushroom spawn growing		01240	Goose raising and breeding
	01120	Nursery (horticulture)		01240	Guinea fowl production
	01120	Ornamental tree and shrub growing		01240	Guinea fowl raising and breeding
	01120	Plant growing for bulbs		01240	Poultry farming
	01120	Plant growing for cuttings and slips		01240	Poultry hatcheries
	01120	Plant growing for roots		01240	Poultry raising and breeding
	01120	Plant growing for tubers		01240	Turkey farming
	01120	Plant propagation		01240	Turkey raising and breeding
	01120	Plants for planting or ornamental purposes			
	01120	Tree nurseries	01490		**Raising of other animals**
	01120	Turf for transplanting		01250	Angora rabbit breeding
				01250	Animal rearing for medical research
01410		**Raising of dairy cattle**		01250	Bee keeping
	01210	Buffalo milk, raw		01250	Bird raising, other than poultry
	01210	Cows' milk (raw) production		01250	Bird skin production from ranching operation
	01210	Dairy farming		01250	Cat and dog raising and breeding
				01250	Emu raising and breeding
01420		**Raising of other cattle and buffaloes**		01220	Fine or coarse animal hair, not including sheep or goats wool
	01210	Bovine semen production		01250	Fur animal raising
	01210	Buffalo raising and breeding for meat		01250	Fur farming
	01210	Cattle farming		01250	Fur skin production from ranching operation
	01210	Cattle raising and breeding for meat		01250	Game bird farming
				01250	Hamster raising and breeding
01430		**Raising of horses and other equines**		01250	Honey and beeswax production
	01220	Animal hair (not carded or combed) including horsehair production		01250	Honey processing and packing
	01220	Ass farming and breeding		01250	Insect raising
	01220	Hinny farming and breeding		01250	Maggot breeding
	01220	Horse farming and breeding		01250	Mollusc farming, except aquatic molluscs
	01220	Mule farming and breeding		01250	Ostriches raising and breeding
	01220	Stud farming		01250	Pet animal breeding, other than fish
				01250	Pet bird raising and breeding
01440		**Raising of camels and camelids**		01250	Rabbit and other fur animal raising
	01250	Camelid raising and breeding		01250	Rabbit breeding
	01250	Camels (dromedary) raising and breeding		01250	Rabbit skin sorting
				01250	Raw fur skin production
01450		**Raising of sheep and goats**		01250	Reptile skin production from ranching operation
	01220	Goat farming		01250	Silk worm cocoon production
	01220	Goat milk (raw) production		01250	Silk worm raising
	01220	Goat raising and breeding			

UK SIC 2007	UK SIC 2003	Activity	SIC 2007	SIC 2003	Activity
	01250	Snail farming		01429	Stud services on a fee or contract basis
	01250	Worm (other than marine) farming			
			01630		**Post-harvest crop activities**
01500		**Mixed farming**		01139	Cocoa beans, peeling and preparation
	01300	Crop growing in combination with farming of livestock		01410	Cooling and bulk packing of crops for primary market
	01300	Mixed farming		01410	Cotton ginning
				01410	Crop drying and disinfecting for primary market
01610		**Support activities for crop production**		01410	Crop preparation for primary market (cleaning, trimming, grading, etc.)
	01410	Agricultural contracting		01410	Crop wax covering, polishing and wrapping for primary market
	01410	Agricultural land maintenance in good agricultural and environmental condition		01410	Fruit packing, for primary market
	01410	Agricultural machinery and equipment rental with operator		01410	Fruit waxing
	01410	Beet thinning on a fee or contract basis		01410	Grain drying
	01410	Crop and grass drying plant operation by contractor		01410	Post-harvest crop activities
	01410	Crop establishing for subsequent crop production, on a fee or contract basis		01110	Tobacco leaves, drying and preparation
	01410	Crop harvesting and preparation	**01640**		**Seed processing for propagation**
	01410	Crop production support activities		01110	Seed cleaning and treating (cereals and other crops)
	01410	Crop spraying on a fee or contract basis		01410	Seed cleaning and treating (fee or contract basis)
	01410	Crop treatment on a fee or contract basis		01139	Seed cleaning and treating (fruit, nuts, beverage and spice crops)
	01410	Fencing by agricultural contractor		01120	Seed cleaning and treating (vegetables, horticultural specialties and other nursery products)
	01410	Field preparation on a fee or contract basis		01110	Seed drying (cereals and other crops)
	01410	Fruit tree and vine trimming, on a fee or contract basis		01410	Seed drying (fee or contract basis)
	01410	Irrigation systems operation on a fee or contract basis		01139	Seed drying (fruit, nuts, beverage and spice crops)
	01410	Mole catching by contractors		01120	Seed drying (vegetables, horticultural specialties and other nursery products)
	01410	Pest control in connection with agriculture		01110	Seed genetically modified, treatment of these (cereals and other crops)
	01410	Rabbit destroying and trapping on agricultural land		01410	Seed genetically modified, treatment of these (fee or contract basis)
	01410	Rat destroying and trapping on agricultural land		01139	Seed genetically modified, treatment of these (fruit, nuts, beverage and spice crops)
	01410	Rice transplanting		01120	Seed genetically modified, treatment of these (vegetables, horticultural and other nursery products)
	01410	Rice transplanting, on a fee or contract basis			
	01410	Rodent destroying and trapping on agricultural land		01110	Seed grading (cereals and other crops)
	01410	Threshing by contractor		01410	Seed grading (fee or contract basis)
	01410	Vermin destroying and trapping on agricultural land		01139	Seed grading (fruit, nuts, beverage and spice crops)
				01120	Seed grading (vegetables, horticultural specialties and other nursery products)
01621		**Farm animal boarding and care**		01110	Seed post harvest processing (cereals and other crops)
	01421	Farm animal boarding and care (except pets)		01410	Seed post harvest processing (fee or contract basis)
	01421	Farm animal pound		01139	Seed post harvest processing (fruit, nuts, beverage and spice crops)
01629		**Support activities for animal production (other than farm animal boarding and care) n.e.c.**		01120	Seed post harvest processing (vegetables, horticultural specialties and other nursery products)
	01429	Agistment services on a fee or contract basis		01110	Seed processing for propagation (cereals and other crops)
	01429	Animal production support activities other than farm animal boarding and care		01410	Seed processing for propagation (fee or contract basis)
	01429	Animal propagation, growth and output – activities to promote these, on a fee or contract basis		01139	Seed processing for propagation (fruit, nuts, beverage and spice crops)
	01429	Animal rearing for production of serum		01120	Seed processing for propagation (vegetables, horticultural specialties and other nursery products)
	01429	Artificial insemination activities on a fee or contract basis			
	01429	Coop cleaning	**01700**		**Hunting, trapping and related service activities**
	01429	Droving services		01500	Animal hunting and trapping
	28520	Farriers, on a fee or contract basis		01500	Bird skin production (from hunting)
	01429	Grazing			
	01429	Herd testing services			
	01429	Poultry caponising			
	01429	Propagation, growth and output of animals, promotion service activities			
	01429	Sheep agisting (grazing)			
	01429	Sheep shearing on a fee or contract basis			

UK SIC 2007	UK SIC 2003	Activity	SIC 2007	SIC 2003	Activity
	01500	Furskin production (from hunting)	**02300**		**Gathering of wild growing non-wood products**
	01500	Game propagation		02010	Acorn gathering
	01500	Hunting or trapping of animals for food		02010	Balata gathering
	01500	Hunting or trapping of animals for fur		02010	Balsam gathering
	01500	Hunting or trapping of animals for pets		01139	Berries, gathering from the wild
	01500	Hunting or trapping of animals for skin		02010	Cork, gathering from the wild
	01500	Hunting or trapping of animals for use in research		02010	Eel grass gathering
	01500	Hunting or trapping of animals for use in zoos		02010	Fern collecting, cutting, gathering
	01500	Hunting, trapping and related service activities		02010	Growing materials, gathering from the wild
	01500	Reptile skin production (from hunting)		02010	Horse-chestnut gathering
	01500	Sea mammal catching		02010	Lac gathering
	01500	Seal catching		02010	Lichen gathering
	01500	Walrus catching		02010	Moss collecting, cutting or gathering
				01120	Mushroom (wild) gathering
02100		**Silviculture and other forestry activities**		02010	Non-wood products, gathering from the wild
	02010	Coppice and pulpwood growing		01139	Nuts, gathering from the wild
	02010	Forest enterprises		02010	Reed collecting, cutting and gathering
	02010	Forest tree nursery operation		02010	Resin gathering
	02010	Forests and timber tract planting, replanting, transplanting, thinning and conservation		02010	Teasel growing
	02010	Fuel wood production		01120	Truffles, gathering from the wild
	02010	Furze collecting, cutting or gathering		02010	Vegetable hair gathering
	01120	Growing of forest tree seeds			
	02010	Heath collecting, cutting or gathering	**02400**		**Support services to forestry**
	02010	Hoopwood production		74149	Forest management consulting services
	02010	Pulpwood production		02020	Forest pest control
	02010	Silviculture and other forestry activities		02020	Forestry fire protection
	02010	Timber growing		02020	Forestry inventories
	02010	Tree nursery (not fruit or ornamental trees)		02020	Forestry service activities
	02010	Vegetable materials used for plaiting growing		02020	Forestry support services
	02010	Willow growing		02020	Log transport within the forest
				02020	Logging service activities
02200		**Logging**		02020	Timber evaluation
	02010	Charcoal production in the forest using traditional methods			
	02010	Forest harvesting residues, gathering of these for energy	**03110**		**Marine fishing**
	02010	Logging		05010	Algae gathering
	02010	Osier growing		05010	Cockle gathering
	02010	Pale fencing production		05010	Coral gathering
	02010	Pickets (of wood) production		05010	Fishing by line, except for recreation or sport
	02010	Pilings (of wood) production		05010	Fishing for shellfish
	02010	Pit props (of wood) production		05010	Fishing in ocean, sea, coastal or inland waters
	02010	Pole fencing production		05010	Fishing on a commercial basis in ocean and coastal waters
	02010	Poles (of wood) production		05010	Fishing service activities
	02010	Posts (of wood) production		05010	Kelp collecting, cutting and gathering (uncultivated)
	02010	Roundwood production (untreated)		05010	Marine and freshwater crustaceans and molluscs gathering
	02010	Roundwood production for forest-based manufacturing industries		05010	Marine crustacean and mollusc gathering
	02010	Sawlog production		05010	Marine fishing
	02010	Split pole production		05010	Marine fishing vessels engaged in processing and preserving of fish
	02010	Splitwood production		05010	Mussel gathering
	02010	Stakes (of wood) production		05010	Natural pearls gathering of
	02010	Timber felling		05010	Pearl gathering
	02010	Treefelling (own account)		05010	Periwinkle gathering
	02010	Withy growing		05010	Salmon netting
	02010	Wood gathering and production, for energy		05010	Sea urchin hunting
	02010	Wood in the rough production (untreated)		05010	Sea-squirt hunting
	02010	Wood logging, etc. within forestry site		05010	Seaweed collecting, cutting and gathering (uncultivated)
				05010	Shrimping

UK SIC 2007	UK SIC 2003	Activity	SIC 2007	SIC 2003	Activity
	05010	Sponge gathering		05020	Oyster spat production, freshwater
	05010	Tunicate hunting		05020	Salmon and trout fishery (hatchery), freshwater
	05010	Turtle hunting		05020	Shrimp production (post-larvae), freshwater
	05010	Whaling			
	05010	Winkle gathering	**05101**		**Mining of hard coal from deep coal mines (underground mining)**
03120		**Freshwater fishing**		10101	Coal cleaning, sizing, grading and pulverising (hard, deep mined)
	05010	Fishing on a commercial basis in inland waters		10101	Coal mine (deep or drift)
	05010	Freshwater aquatic animal taking		10101	Coal preparation (deep mined)
	05010	Freshwater crustacean and mollusc taking		10101	Coal washing (deep mined)
	05010	Freshwater fishing		10101	Deep coal mines
	05010	Freshwater materials gathering		10101	Hard coal mining (underground)
03210		**Marine aquaculture**	**05102**		**Mining of hard coal from open cast coal working (surface mining)**
	05020	Aquaculture in salt water filled tanks or reservoirs		10102	Coal cleaning, sizing, grading and pulverising (hard, opencast)
	05020	Aquaculture in sea or brackish waters		10102	Coal contractor (opencast)
	01250	Bait digging		10102	Coal preparation (opencast)
	01250	Bait production		10102	Coal recovery from dumps, tips etc.
	05020	Bivalves cultured in sea water		10102	Coal recovery of from culm banks
	05020	Crustaceans cultured in sea water		10102	Coal site (opencast)
	05020	Edible seaweed growing		10102	Coal washing (opencast)
	05020	Fingerling production, marine		10102	Hard coal recovery from tips
	05020	Fish breeding, marine		10102	Opencast coal disposal point
	05020	Fish farming, marine		10102	Opencast coal site
	05020	Fish fry production, marine		10102	Opencast coal working
	05020	Fish hatcheries and farms service activities, marine		10102	Surface mining of hard coal
	05020	Fish hatcheries, marine	**05200**		**Mining of lignite**
	05020	Laver gathering (cultivated)		10200	Lignite (brown coal) mining including mining through liquefaction methods
	05020	Laver growing		10200	Lignite mining
	05020	Lobsterling production, marine		10200	Lignite washing, dehydrating, pulverising etc. To improve quality or facilitate transport or storage
	05020	Marine aquaculture			
	05020	Molluscs and other aquatic animals cultured in sea water	**06100**		**Extraction of crude petroleum**
	05020	Mussel production, marine		11100	Bituminous or oil shale and sand extraction
	05020	Ornamental fish farming, marine		11100	Crude oil extraction
	05020	Oyster cultivation, marine		11100	Crude oils obtained by decantation processes
	05020	Oyster fishery, marine		11100	Crude oils obtained by dehydration processes
	05020	Oyster spat production, marine		11100	Crude oils obtained by desalting processes
	05020	Salmon and trout fishery (hatchery), marine		11100	Crude oils obtained by stabilisation processes
	05020	Shrimp production (post-larvae), marine		11100	Crude petroleum extraction
	05020	Worm farms, marine		11100	Crude petroleum production
				11100	Crude petroleum production from bituminous shale and sand
03220		**Freshwater aquaculture**		11100	Mineral oil extraction
	05020	Aquaculture, freshwater		11100	Oil platform operation
	05020	Fingerling production, freshwater		11100	Oil production well or platform operating
	01250	Fish breeding, freshwater		11100	Oil shale mine
	05020	Fish farming in fresh water, including farming of freshwater ornamental fish		11100	Oil shale retorting
	05020	Fish fry production, freshwater		11100	Oil stabilisation plant operation
	05020	Fish hatcheries and farms service activities, freshwater	**06200**		**Extraction of natural gas**
	05020	Fish hatcheries, freshwater		11100	Butane extraction from natural gas
	05020	Freshwater crustaceans, bivalves, other molluscs and other aquatic animals, culture of		11100	Condensate extraction
	05020	Frog farming		11100	Crude gaseous hydrocarbon production (natural gas)
	05020	Lobsterling production, freshwater		11100	Ethane extraction from natural gas
	05020	Mussel production, freshwater		11100	Gas desulphurisation
	05020	Ornamental fish farming, freshwater			
	05020	Oyster cultivation, freshwater			
	05020	Oyster fishery, freshwater			

UK SIC 2007	UK SIC 2003	Activity	SIC 2007	SIC 2003	Activity
	11100	Gas extraction (natural gas)		14110	Crushing and breaking of stone
	11100	Hydrocarbon liquids mining, by liquefaction or pyrolysis		14110	Flagstone quarry
	11100	Liquid hydrocarbon fractions draining and separation		14110	Freestone mine or quarry
				14110	Granite quarrying (rough trimming and sawing)
	11100	Methane extraction from natural gas		14120	Gypsum mine or quarry
	11100	Natural gas condensates separation		14110	Igneous rock quarry
	11100	Natural gas production well		14120	Limestone including dolomite mine or quarry
	11100	Propane extraction from natural gas		14120	Limestone quarrying, crushing and breaking for constructional purposes
07100		**Mining of iron ores**		14110	Marble quarrying (rough trimming and sawing)
	13100	Haematite quarry		14120	Marl mining
	13100	Iron ore beneficiation and agglomeration		14110	Ornamental and building stone, breaking and crushing
	13100	Iron ore calcining		14110	Ragstone quarry
	13100	Iron ore crushing		14110	Rough trimming and sawing of building and monumental stone
	13100	Iron ore mine or quarry			
	13100	Iron ore preparation		14110	Sandstone mine, pit or quarry
	13100	Iron ore sintering		14110	Sett quarry
	13100	Iron ore washing		14130	Slate mine or quarry
				14110	Whinstone quarry
07210		**Mining of uranium and thorium ores**		14120	Whiting and prepared chalk production
	12000	Thorium ores mining			
	12000	Uranium and thorium ore concentration	**08120**		**Operation of gravel and sand pits; mining of clays and kaolin**
	12000	Uranium ore mining		14220	Ball clay extraction (mine or opencast working)
	12000	Yellowcake production		14220	China clay (ground) production
				14220	China clay pit
07290		**Mining of other non-ferrous metal ores**		14220	China stone mine
	13200	Aluminium ore (bauxite) mining or preparation		14220	Clay extraction for brick, pipe and tile production
	13200	Chrome ore mining and preparation		14220	Clay mining
	13200	Cobalt mining and preparation		14220	Clay quarrying
	13200	Copper mining and preparation		14210	Coated roadstone production
	13200	Copper ore and concentrate extraction and preparation		14210	Coated tarmacadam production
				14220	Fireclay mine or quarry
	13200	Gold mining and preparation		14210	Flint bed, pit or quarry
	13200	Lead mining and preparation		14210	Flint grit production
	13200	Lead ore and concentrate extraction and preparation		14220	Fuller's earth pit
	13200	Manganese mining and preparation		14210	Gravel and sand breaking and crushing
	13200	Mining of non-ferrous metal ore		14210	Gravel pit or quarry
	13200	Molybdenum mining and preparation		14220	Kaolin mining
	13200	Nickel mining and preparation		14210	Pebble dredging
	13200	Non-ferrous metal ore quarrying		14220	Pipeclay pit
	13200	Non-ferrous metal ores mining and preparation		14220	Potters' clay mine or quarry
	13200	Platinum mining and preparation		14220	Refractory clays mining
	13200	Precious metal ores mining and preparation		14210	Road metal production (crushed and processed)
	13200	Silver mining and preparation		14210	Roadstone (coated) production
	13200	Silver ore and concentrate extraction and preparation		14210	Sand dredging
				14210	Sand extraction and dredging for industrial use
	13200	Tantalum mining and preparation		14210	Sand pit
	13200	Tin mining and preparation		14210	Sand quarry
	13200	Tin ore and concentrate extraction and preparation		14210	Shingle dredging
	13200	Vanadium mining and preparation		14210	Stone chippings production
	13200	Zinc mining and preparation		14210	Stone dust production
				14210	Tarmacadam (coated) production
08110		**Quarrying of ornamental and building stone, limestone, gypsum, chalk and slate**			
			08910		**Mining of chemical and fertiliser minerals**
	14110	Alabaster mine		14300	Alum mine
	14120	Anhydrite mine or quarry		14300	Barium sulphate (natural) mining
	14110	Basalt mine		14300	Barytes mine
	14110	Blackstone quarry		14300	Borates (natural) mining
	14110	Blue pennant stone quarry			
	14120	Chalk pit or quarry			

UK SIC 2007	UK SIC 2003	Activity	SIC 2007	SIC 2003	Activity
	14300	Carbonate (barytes and witherite) mining		14500	Rubbing stone mine, pit or quarry
	14300	Celestine pit		14500	Semi-precious stones extraction
	14300	Chalk (ground) production		14500	Silica stone extraction or quarrying
	14300	Chemical minerals mining		14500	Siliceous fossil meals mining
	14300	Earth colours and fluorspar mining		14500	Steatite (talc) mining and quarrying
	14300	Fertiliser minerals mining		14500	Talc mine or quarry
	14300	Fluorspar mining		14500	Witherite mine
	14300	Guano mining			
	14300	Iron pyrites extraction (not for iron production)	09100		**Support activities for petroleum and natural gas extraction**
	14300	Jet mine		11200	Derrick erection in situ, repairing and dismantling
	14300	Magnesium sulphates (natural kieserite) mining		11200	Directional drilling services
	14300	Native sulphur mining		11200	Diving services incidental to oil and gas exploration
	14300	Ochre pit		11200	Downhole-fishing services
	14300	Phosphates (natural) mining		11200	Downhole-milling services
	14300	Potash mine		11200	Draining and pumping services incidental to oil and gas extraction, on a fee or contract basis
	14300	Potassium salts (natural) mining		11200	Drilling contractor for offshore oil or gas well
	14300	Pyrites and pyrrhotite extraction and preparation		11200	Drilling services to oil and gas extraction wells
08920		**Extraction of peat**		11200	Floating drilling rig operation for petroleum or natural gas exploration or production
	10300	Peat agglomeration		11200	Fluid-displacement services
	10300	Peat cutting and digging		11200	Fracture/stimulation services
	10300	Peat extraction		11200	Gas extraction service activities
	10300	Peat, preparation to facilitate transport or storage		11200	Gravel packing services
	10300	Peat, preparation to improve quality		11200	Horizontal drilling services
				11200	Hot-tap operation services
08930		**Extraction of salt**		11200	Hyperbaric welding services
	14400	Brine pit		11100	Liquefaction and regasification of natural gas for transport
	14400	Brine production		11200	Mud logging services
	14400	Rock salt production		11200	Oil and gas exploration in connection with oil and gas extraction, including traditional prospecting
	14400	Salt crushing, purification and refining by the producer		11200	Oil and gas extraction service activities provided on a fee or contract basis
	14400	Salt extraction		11200	Oil and gas field fire fighting services
	14400	Salt mine		11200	Oil and gas well casing, cementing, tubing and lining services
	14400	Salt preparation (not at salt mine or brine pit)		11200	Oil and gas well cementing services
	14400	Salt production		11200	Oil and gas well coiled-tubing wellwork
	14400	Salt production by evaporation of sea water or other saline waters		11200	Oil and gas well conductor driving services
	14400	Salt works		11200	Oil and gas well cutting, casing and abandonment services
	14400	Salt, extraction of, from underground including by dissolving and pumping		11200	Oil and gas well pipe refurbishment services
	14400	Sea salt production		11200	Oil and gas well sleeving repair services
	14400	White salt production		11200	Oil extraction service activities
08990		**Other mining and quarrying n.e.c.**		11200	Petroleum and natural gas mining support activities
	14500	Abrasive materials mining and quarrying		11200	Petroleum test well drilling
	14500	Asbestos mining and quarrying		11200	Petroleum well drilling
	14500	Asphalt (natural) mining and quarrying		11200	Plugging and abandoning oil and gas wells
	14500	Asphaltites and asphaltic rock mining and quarrying		11200	Pumping of oil and gas wells
	14500	Bitumen mining and quarrying		11200	Redrilling oil and gas wells
	14500	Chert quarry		11200	Repair and dismantling services of derricks
	14500	Diatomite bed		11200	Sealing of oil and gas wells
	14500	Emery extraction		11200	Separation terminal operation (natural gas)
	14500	Feldspar mining and quarrying		11200	Spudding in for oil wells
	14500	French chalk production		11200	Test boring incidental to oil and gas extraction
	14500	Ganister extraction		11200	Turbine drilling services
	14500	Gem stones mining and quarrying		11200	Well logging
	14500	Graphite (natural) mining		11200	Well-perforating services
	14500	Mica mining and quarrying			
	14500	Pumice extraction			
	14500	Quartz mining and quarrying			

UK SIC 2007	UK SIC 2003	Activity	SIC 2007	SIC 2003	Activity
09900		**Support activities for other mining and quarrying**		15111	Hide pickling (manufacture)
	14300	Chemicals and fertiliser minerals mining support services provided on a fee or contract basis		15111	Hides and skins production from abattoirs (manufacture)
	14220	Clay and kaolin mining support services provided on a fee or contract basis		15111	Hides and skins production from knackers (manufacture)
	14210	Gravel and sand pit support services provided on a fee or contract basis		15111	Hides and skins production from slaughterhouses (manufacture)
	10102	Hard coal mining (opencast) support services, provided on a fee or contract basis		15112	Hooves from knackers production (manufacture)
	10101	Hard coal mining (underground) support services, provided on a fee or contract basis		15112	Lard from knackers (manufacture)
				15112	Lard refining (manufacture)
	13100	Iron ore mining support services provided on a fee or contract basis		15111	Meat (except poultry meat) processing and preserving (manufacture)
	10200	Lignite mining support services provided on a fee or contract basis		15112	Meat and bone meal from knackers (manufacture)
	14120	Limestone, gypsum and chalk quarrying support services provided on a fee or contract basis		15111	Meat chilling or freezing for human consumption (manufacture)
				15112	Meat meal (ground meat) (manufacture)
	14500	Mining and quarrying of residual class 08990, support services provided on a fee or contract basis		15111	Meat production (fresh, chilled or frozen) in carcasses or cuts (manufacture)
	13200	Non-ferrous metal ore mining of class 07290, support services provided on a fee or contract basis		15111	Offal (edible) preparation (i.e. removal, freezing, packing, etc.) (manufacture)
	14110	Ornamental and building stone quarrying support services provided on a fee or contract basis		15113	Pelt from fellmongery (manufacture)
				15112	Premier jus (manufacture)
	10300	Peat extraction support services, provided on a fee or contract basis		15113	Pulled wool production (manufacture)
				15120	Rabbit meat preparation (manufacture)
	14400	Salt production support services provided on a fee or contract basis		15120	Rabbit slaughtering (manufacture)
	14130	Slate quarrying support services provided on a fee or contract basis		15112	Raw bones from knackers (manufacture)
				15112	Rendering of lard and other edible fats of animal origin (manufacture)
	12000	Uranium and thorium ore mining support services provided on a fee or contract basis		15112	Sausage skins and casings (natural) (manufacture)
				15113	Sheep and lambskin pulling (manufacture)
10110		**Processing and preserving of meat**		15111	Skin drying (manufacture)
	15111	Abattoir (manufacture)		15111	Skin pickling (manufacture)
	15112	Animal grease (manufacture)		15111	Skin production from slaughterhouses (manufacture)
	15112	Animal offal (inedible) production (manufacture)		15111	Skin sorting (manufacture)
	15112	Animal offal processing (manufacture)		15111	Slaughterhouse (manufacture)
	15112	Bile processing by knackers (manufacture)		15111	Slaughterhouses killing, dressing or packing meat (manufacture)
	15112	Bladder processing (manufacture)		15112	Sterilised bone flour (not for fertilisers) (manufacture)
	15112	Bone boiling by knackers (manufacture)		15112	Suet (manufacture)
	15112	Bone crushing by knackers (manufacture)		15112	Tripe dressing (manufacture)
	15112	Bone degreasing by knackers (manufacture)		15111	Whales processing on land or on specialised vessels (manufacture)
	15112	Bone flour from knackers (manufacture)			
	15112	Bone meal (manufacture)		15113	Wool (fellmongery) (manufacture)
	15112	Bone meal from knackers (manufacture)			
	15112	Bone scraping by knackers (manufacture)	**10120**		**Processing and preserving of poultry meat**
	15112	Bone sorting by knackers (manufacture)		15120	Chicken cuts (fresh, chilled or frozen) (manufacture)
	15111	Bovine hides and skins production from knackers (manufacture)		15120	Down production (manufacture)
	15112	Bristles from knackers (manufacture)		15120	Duck (fresh, chilled or frozen) slaughter and dressing (manufacture)
	15112	Casings for sausages (manufacture)			
	15113	De-woolling (manufacture)		15120	Feather production (manufacture)
	15112	Dripping (manufacture)		15120	Game bird (fresh, chilled or frozen) dressing or preparation (manufacture)
	15112	Edible fats of animal origin rendering (manufacture)		15120	Goose (fresh, chilled or frozen) slaughter and dressing (manufacture)
	15112	Edible offal (processed) production (manufacture)			
	15112	Edible tallow production (manufacture)		15120	Poultry (fresh, chilled or frozen) production (manufacture)
	15112	Fat recovery from knackers (manufacture)		15120	Poultry dressing (manufacture)
	15113	Fellmongery (manufacture)		15120	Poultry fat (edible) rendering (manufacture)
	15112	Flours and meals of meat (manufacture)		15120	Poultry meat preparation (manufacture)
	15112	Hair (animal by-product) from knackers (manufacture)		15120	Poultry meat production and preserving (manufacture)
	15111	Hide degreasing (manufacture)			

UK SIC 2007	UK SIC 2003	Activity
	15120	Poultry packing (manufacture)
	15120	Poultry slaughtering (manufacture)
	15120	Slaughterhouses killing, dressing or packing poultry (manufacture)
10130		**Production of meat and poultry meat products**
	15139	Andouillettes (manufacture)
	15131	Bacon curing (manufacture)
	15131	Bacon production (manufacture)
	15131	Bacon smoking (manufacture)
	15139	Beef extract (manufacture)
	15139	Beef paste (manufacture)
	15139	Beef pickling (manufacture)
	15139	Black pudding (manufacture)
	15139	Blood pudding (manufacture)
	15131	Boiled ham production (manufacture)
	15139	Bolognas (manufacture)
	15139	Brawn (manufacture)
	15139	Calves' foot jelly (manufacture)
	15139	Chicken paste (manufacture)
	15139	Cooked and preserved meat (manufacture)
	15139	Dried, salted or smoked meat (manufacture)
	15139	Forcemeat (manufacture)
	15139	Galantines (manufacture)
	15139	Haggis (manufacture)
	15131	Ham boiling (manufacture)
	15131	Ham cooking or preparing in bulk (manufacture)
	15131	Ham curing (manufacture)
	15131	Ham production (manufacture)
	15131	Ham smoking (manufacture)
	15139	Meat and poultry meat processing (other than bacon and ham) (manufacture)
	15139	Meat and poultry meat products (manufacture)
	15139	Meat canning, cooking and preserving (manufacture)
	15139	Meat extract (manufacture)
	15139	Meat juices (manufacture)
	15139	Meat pate (manufacture)
	15139	Meat pies and puddings (manufacture)
	15139	Meat pudding (manufacture)
	15139	Meat rillettes (manufacture)
	15139	Pastrami and other salted, dried or smoked meats (manufacture)
	15139	Pâtés (manufacture)
	15139	Polony (manufacture)
	15131	Pork (salted or pickled) (manufacture)
	15139	Pork pie (manufacture)
	15139	Potted meat. (manufacture)
	15139	Poultry canning (manufacture)
	15139	Poultry potting (manufacture)
	15139	Preserved meat (manufacture)
	15139	Salami (manufacture)
	15139	Sausage meat (manufacture)
	15139	Sausage rolls (manufacture)
	15139	Sausages (manufacture)
	15139	Saveloys (manufacture)
	15139	Smoked meat (other than bacon and ham) (manufacture)
	15131	Tinned ham (manufacture)
	15139	Tinned meat (other than tinned ham) (manufacture)

SIC 2007	SIC 2003	Activity
10200		**Processing and preserving of fish, crustaceans and molluscs**
	15209	Caviar (manufacture)
	15209	Caviar substitute (manufacture)
	15209	Crustacean and mollusc canning (manufacture)
	15209	Crustacean and mollusc preservation by drying (manufacture)
	15209	Crustacean and mollusc products (manufacture)
	15209	Crustacean and mollusc salting (manufacture)
	15201	Crustacean freezing (manufacture)
	15209	Crustacean preservation (other than by freezing) (manufacture)
	15209	Fish and other aquatic animal meals and solubles unfit for human consumption (manufacture)
	15209	Fish cakes (manufacture)
	15209	Fish canning (manufacture)
	15209	Fish curing (other than by distributors) (manufacture)
	15209	Fish drying (manufacture)
	15209	Fish fillet production (manufacture)
	15209	Fish meal (manufacture)
	15209	Fish paste (manufacture)
	15209	Fish preservation (other than by freezing) (manufacture)
	15201	Fish preservation by freezing (manufacture)
	15209	Fish processing (not freezing) (manufacture)
	15209	Fish products (manufacture)
	15209	Fish salting (manufacture)
	15209	Fish, crustacean and mollusc cooking (manufacture)
	15209	Fish, crustacean and mollusc preparation and preservation, by immersing in brine (manufacture)
	15209	Fish, crustacean and mollusc smoking (manufacture)
	15209	Inedible flours, meal and pellets of fish, crustaceans and molluscs production (manufacture)
	15209	Kipper (manufacture)
	15209	Mollusc preservation (other than by freezing) (manufacture)
	15201	Mollusc preservation by freezing (manufacture)
	15209	Potted shrimp (manufacture)
	15209	Roe production (manufacture)
	15209	Seaweed processing (manufacture)
	15201	Shellfish freezing (manufacture)
	15209	Shellfish preserving (not freezing) (manufacture)
	15209	Shrimp preserving (not freezing) (manufacture)
	15209	Smoked salmon, trout and herring production (manufacture)
	15209	Vessels only engaged in processing and preserving fish (other than by freezing) (manufacture)
	15201	Vessels only engaged in processing and preserving fish by freezing (manufacture)
10310		**Processing and preserving of potatoes**
	15310	Mashed potatoes (dehydrated) production (manufacture)
	15310	Potato chip production (frozen, raw, steamed or boiled) (manufacture)
	15310	Potato crisp (manufacture)
	15310	Potato flour and meal (manufacture)
	15310	Potato peeling (industrial) (manufacture)
	15310	Potato processing and preserving (manufacture)
	15310	Potato puff (manufacture)
	15310	Potato snacks production (manufacture)

UK SIC 2007	UK SIC 2003	Activity	SIC 2007	SIC 2003	Activity
	15310	Potato stick (manufacture)		15330	Perishable prepared vegetables, packaged peeled or cut (manufacture)
	15310	Potato straw (manufacture)		15330	Piccalilli production (manufacture)
	15310	Potatoes (prepared frozen) production (manufacture)		15330	Pickle including beetroot and onion (manufacture)
				15330	Pickling of fruit and vegetables (manufacture)
10320		**Manufacture of fruit and vegetable juice**		15330	Quick freezing of fruit and vegetables (manufacture)
	15320	Fruit and vegetable concentrates (manufacture)		15330	Strained fruit (manufacture)
	15320	Fruit juice (manufacture)		15330	Strained vegetables (manufacture)
	15320	Vegetable juice (manufacture)		15330	Table jelly (manufacture)
				15330	Vegetable dehydrating for human consumption (manufacture)
10390		**Other processing and preserving of fruit and vegetables**		15330	Vegetable pickling (manufacture)
	15330	Banana ripening and conditioning (manufacture)		15330	Vegetable preparation and preserving (manufacture)
	15330	Canning of fruit and vegetables (except fruit juices and potatoes) (manufacture)		15330	Vegetable quick freezing (manufacture)
	15330	Chutney (manufacture)	**10410**		**Manufacture of oils and fats**
	15330	Coconut flakes including desiccated but (not sugared) (manufacture)		15410	Animal fat and oil production (non-edible) (manufacture)
	15330	Dehydrating fruit for human consumption (manufacture)		15420	Animal oil refining (manufacture)
	15330	Dehydrating of vegetables for human consumption (manufacture)		15410	Benniseed crushing (manufacture)
	15330	Dried fruit (except field dried) (manufacture)		15420	Benniseed oil refining (manufacture)
	15330	Dried fruit cleaning (manufacture)		15410	Bone oil (manufacture)
	15330	Dried vegetables (except field dried) (manufacture)		15420	Castor oil processing (manufacture)
	15330	Flaked coconut including desiccated but (not sugared) (manufacture)		15410	Castor seed crushing (manufacture)
	15330	Fruit freezing (manufacture)		15420	Coconut oil refining (manufacture)
	15330	Fruit jelly (preserve) (manufacture)		15420	Cod liver oil refining (manufacture)
	15330	Fruit or vegetable food products (manufacture)		15420	Cola oil refining (manufacture)
	15330	Fruit pickling (manufacture)		15410	Colza oil production (manufacture)
	15330	Fruit preserving (manufacture)		15410	Copra (coconut) crushing (manufacture)
	15330	Fruit processing and preserving (except in sugar) (manufacture)		15410	Cotton linters production (manufacture)
	15330	Fruit pulp (manufacture)		15410	Cotton seed crushing including delinting or cleaning (manufacture)
	15330	Fruit, nuts or vegetables preserved by immersing in oil (manufacture)		15410	Cotton seed oil production (manufacture)
	15330	Fruit, nuts or vegetables preserved by immersing in vinegar (manufacture)		15420	Cotton seed oil refining (manufacture)
	15330	Gherkin pickling (manufacture)		15410	Crude vegetable oil production (manufacture)
	15330	Heat treatment of fruit and vegetables (manufacture)		15410	Fat of marine animals production (manufacture)
	15330	Homogenised fruit and vegetables (manufacture)		15410	Fish and marine mammal oil extraction (manufacture)
	15330	Jam (manufacture)		15410	Fish liver oil (unrefined) production (manufacture)
	15330	Jelly (table) (manufacture)		15420	Fish liver oil refining (manufacture)
	15330	Jelly powder (manufacture)		15410	Fish oil (crude) production (manufacture)
	15330	Marmalade (manufacture)		15420	Gingelly oil refining (manufacture)
	15330	Mincemeat (manufacture)		15410	Gingelly seed crushing (manufacture)
	15330	Nut foods and pastes (manufacture)		15410	Groundnut crushing (manufacture)
	15330	Nut processing and preservation (except in sugar) (manufacture)		15420	Groundnut oil refining (manufacture)
	15330	Nut shelling, grinding and preparing (manufacture)		15420	Herring oil refining (manufacture)
	15330	Nuts preserved by freezing (manufacture)		15410	Kapok seed crushing (manufacture)
	15330	Olive preserving in salt or brine (manufacture)		15420	Kapok seed oil refining (manufacture)
	15330	Peanut butter (manufacture)		15410	Kernel crushing (manufacture)
	01410	Peeled or cut vegetables, mixed fresh salads, packaged (manufacture)		15410	Lanolin recovery (manufacture)
	51310	Peeled, cut fresh vegetables, mixed salads, packed (manufacture)		15410	Lard oil (manufacture)
				15410	Linseed crushing (manufacture)
	15330	Perishable prepared fruit and vegetables (manufacture)		15410	Linseed oil production (manufacture)
				15420	Linseed oil refining (manufacture)
	15330	Perishable prepared salads; mixed salads (manufacture)		15410	Marine animal crude oil and fat production (manufacture)
				15410	Mustard oil production (manufacture)
	15330	Perishable prepared tofu bean curd (manufacture)		15410	Mustard seed crushing (manufacture)
				15420	Neatsfoot oil (manufacture)
				15410	Non-defatted flour production (manufacture)

UK SIC 2007	UK SIC 2003	Activity	SIC 2007	SIC 2003	Activity
	15420	Oil (edible) (manufacture)		15511	Homogenised milk production (manufacture)
	15410	Oil kernel meal production (manufacture)		15511	Milk homogenising (manufacture)
	15410	Oil nut meal production (manufacture)		15511	Milk production (evaporated, condensed, etc.) (manufacture)
	15410	Oil seed cake and meal (manufacture)		15511	Milk sterilising (manufacture)
	15410	Oil seed crushing (manufacture)		15511	Milk ultra heat treatment (manufacture)
	15410	Oilcakes and other residual products of oil production (manufacture)		15511	Pasteurized fresh liquid milk (manufacture)
	15410	Oleo stearin (manufacture)		15511	Preserved cream (manufacture)
	15410	Olive oil (crude) production (manufacture)		15511	Sterilised cream (manufacture)
	01139	Olive oil from self produced olives (if value of production exceeds that of growing) (manufacture)	**10512**		**Butter and cheese production**
	15420	Olive oil refining (manufacture)		15512	Butter blending (manufacture)
	15410	Palm kernel crushing (manufacture)		15512	Butter milk (manufacture)
	15420	Palm kernel oil refining (manufacture)		15512	Butter oil (manufacture)
	15410	Palm oil production (manufacture)		15512	Butter production (manufacture)
	15420	Palm oil refining (manufacture)		15512	Butterfat (manufacture)
	15410	Rape oil production (manufacture)		15512	Cheese (manufacture)
	15420	Rape oil refining (manufacture)		15512	Curd production (manufacture)
	15410	Rape seed crushing (manufacture)		15512	Dairy preparation of cheese and butter (manufacture)
	15410	Seed and nut crushing (manufacture)		15512	Processed cheese (manufacture)
	15420	Sesame oil refining (manufacture)			
	15410	Sesame seed crushing (manufacture)	**10519**		**Manufacture of milk products (other than liquid milk and cream, butter, cheese) n.e.c**
	15420	Shea butter (manufacture)		15519	Casein production (manufacture)
	15410	Shea nut crushing (manufacture)		15519	Concentrated dried milk (manufacture)
	15410	Soya bean crushing (manufacture)		15519	Dairy preparation of milk products n.e.c. (manufacture)
	15410	Soya bean oil (crude) (manufacture)		15519	Desserts with a milk base (manufacture)
	15420	Soya bean oil refining (manufacture)		15519	Dried milk (manufacture)
	15420	Sperm oil refining (manufacture)		15519	Junket powder (manufacture)
	15420	Sunflower oil refining (manufacture)		15519	Lactose production (manufacture)
	15410	Sunflower seed crushing (manufacture)		15519	Milk or cream in solid form (manufacture)
	15410	Sunflower-seed oil production (manufacture)		15519	Milk powder (manufacture)
	15410	Technical tallow (manufacture)		15519	Milk products other than liquid milk and cream, butter, cheese n.e.c. (manufacture)
	15410	Tung oil extraction (manufacture)		15519	Rennet (not artificial) (manufacture)
	15420	Vegetable oil processing: blowing, boiling, dehydration, hydrogenation (manufacture)		15519	Soft drinks (milk based) (manufacture)
	15420	Vegetable oil refining (manufacture)		15519	Sweetened skimmed whey production (manufacture)
	15410	Whale oil production (manufacture)		15519	Whey (sweetened skimmed) production (manufacture)
	15420	Whale oil refining (manufacture)		15519	Yoghurt (manufacture)
10420		**Manufacture of margarine and similar edible fats**	**10520**		**Manufacture of ice cream**
	15430	Cooking fat (compound) (manufacture)		15520	Ice cream (manufacture)
	15430	Fats (edible) (manufacture)		15520	Ice cream powder (manufacture)
	15430	Margarine (manufacture)		15520	Sorbet production (manufacture)
	15430	Melanges and similar spreads (manufacture)		15520	Water ices (manufacture)
10511		**Liquid milk and cream production**	**10611**		**Grain milling**
	15511	Clotted cream (manufacture)		15611	Barley meal production (manufacture)
	15511	Condensed milk (manufacture)		15611	Barley milling (manufacture)
	15511	Cream (sterilised) (manufacture)		15611	Barley processing (blocked, flaked, puffed or pearled) (manufacture)
	15511	Cream from fresh homogenized liquid milk (manufacture)		15611	Bran (manufacture)
	15511	Cream from fresh pasteurized liquid milk (manufacture)		15611	Cake mixture (manufacture)
	15511	Cream production (manufacture)		15611	Cereal grains, flour, groats, meal or pellets (manufacture)
	15511	Creamery (not farm or retail shop) (manufacture)		15611	Corn or other cereal grains (manufacture)
	15511	Dairy preparation of milk and cream (manufacture)		15611	Cornflour (manufacture)
	15511	Double cream (manufacture)		15611	Flaked maize (manufacture)
	15511	Evaporated milk (manufacture)			
	15511	Heat treatment of milk (manufacture)			

UK SIC 2007	UK SIC 2003	Activity	SIC 2007	SIC 2003	Activity
	15611	Flour (manufacture)		15612	Rice flour production (manufacture)
	15611	Flour milling (manufacture)		15612	Rice husking (manufacture)
	15611	Flour mixes and prepared blended flour and dough for biscuits (manufacture)		15612	Rice milling (manufacture)
	15611	Flour mixes and prepared blended flour and dough for bread (manufacture)		15612	Rice rolling (manufacture)
	15611	Flour mixes and prepared blended flour and dough for cakes (manufacture)		15612	Sago grinding or milling (manufacture)
	15611	Flour mixes and prepared blended flour and dough for pancakes (manufacture)		15612	Soya bean grinding (manufacture)
	15611	Flour of cereal grains production (manufacture)		15612	Soya bean milling (manufacture)
	15611	Grain milling (manufacture)		15612	Soya flour and meal (manufacture)
	15611	Grist milling (manufacture)		15612	Vegetable milling (manufacture)
	15611	Groats production (manufacture)	**10620**		**Manufacture of starches and starch products**
	15611	Maize (flaked) production (manufacture)		15620	Arrowroot (manufacture)
	15611	Maize flour and meal production (manufacture)		15620	Corn oil (manufacture)
	15611	Meal from grain (manufacture)		15620	Dextrin (manufacture)
	15611	Oat flour and meal (manufacture)		15620	Dextrose (manufacture)
	15611	Oat grinding, rolling, crushing or flaking (manufacture)		15620	Glucose (manufacture)
	15611	Oats (manufacture)		15620	Glucose syrup (manufacture)
	15611	Pudding mixture (manufacture)		15620	Gluten (manufacture)
	15611	Rye (manufacture)		15620	Inulin (manufacture)
	15611	Rye flaking (manufacture)		15620	Laundry starch (manufacture)
	15611	Rye flour and meal (manufacture)		15620	Maize starch (manufacture)
	15611	Rye milling (manufacture)		15620	Maltose (manufacture)
	15611	Rye rolling (manufacture)		15620	Potato starch (manufacture)
	15611	Self-raising and patent flour (manufacture)		15620	Rice starch (manufacture)
	15611	Semolina milling (manufacture)		15620	Soluble starch (manufacture)
	15611	Wheat flake (manufacture)		15620	Starch (manufacture)
	15611	Wheat milling (manufacture)		15620	Tapioca (manufacture)
	15611	Wheat offal (manufacture)		15620	Tapioca grinding (manufacture)
	15611	Wheat pellets (manufacture)		15620	Warp starch (manufacture)
				15620	Wet corn milling (manufacture)
				15620	Wheat starch (manufacture)
10612		**Manufacture of breakfast cereals and cereals-based foods**	**10710**		**Manufacture of bread; manufacture of fresh pastry goods and cakes**
	15612	Almond grinding (manufacture)		15810	Bakery (baking main activity) (manufacture)
	15612	Bean grinding (manufacture)		15810	Bread and flour confectionery baking (manufacture)
	15612	Bean milling (manufacture)		15810	Bread baking including rolls (manufacture)
	15612	Bean splitting (manufacture)		15810	Bread rolls (manufacture)
	15612	Breakfast cereal (cooked) (manufacture)		15810	Cakes (manufacture)
	15612	Breakfast cereal (uncooked) (manufacture)		15810	Crumpet making (manufacture)
	15612	Cereal breakfast foods (manufacture)		15810	Fancy pastry (manufacture)
	15612	Chicory root drying (manufacture)		15810	Flour confectionery (manufacture)
	15612	Cornflake (manufacture)		15810	Fresh pastry (manufacture)
	15612	Edible nut flour or meal production (manufacture)		15810	Fruit cake baking (manufacture)
	15612	Flour of dried leguminous vegetables production (manufacture)		15810	Fruit loaf baking (manufacture)
	15612	Flour or meal of roots or tubers (manufacture)		15810	Fruit pie making (manufacture)
	15612	Glazed rice (manufacture)		15810	Tarts (manufacture)
	15612	Lentil splitting, grinding or milling (manufacture)		15810	Waffles (manufacture)
	15612	Meal of dried leguminous vegetables production (manufacture)	**10720**		**Manufacture of rusks and biscuits; manufacture of preserved pastry goods and cakes**
	15612	Parboiled or converted rice (manufacture)		15820	Biscuits (manufacture)
	15612	Pea splitting, milling or grinding (manufacture)		15820	Cakes (preserved) (manufacture)
	15612	Polished rice (manufacture)		15820	Cereal for sausage filler (manufacture)
	15612	Puffed rice (manufacture)		15820	Cookies (manufacture)
	15612	Puffed wheat (manufacture)		15820	Crispbread (manufacture)
	15612	Rice cleaning (manufacture)		15820	Dry bakery products (manufacture)
	15612	Rice flaking (manufacture)		15820	Matzos (manufacture)
				15820	Oat cake (manufacture)

UK SIC 2007	UK SIC 2003	Activity	SIC 2007	SIC 2003	Activity
	15820	Pancake making (manufacture)		15841	Chocolate couverture (manufacture)
	15820	Pastry and buns (preserved) (manufacture)		15841	Cocoa (manufacture)
	15820	Pies (other than meat) (manufacture)		15841	Cocoa bean roasting and dressing (manufacture)
	15820	Pikelet making (manufacture)		15841	Cocoa butter (manufacture)
	15820	Pizza (manufacture)		15841	Cocoa fat (manufacture)
	15820	Preserved pastry goods and cakes (manufacture)		15841	Cocoa oil (manufacture)
	15820	Pretzels whether sweet or salted (manufacture)		15841	Cocoa powder (manufacture)
	15820	Rusk making (manufacture)		15841	Cocoa products (manufacture)
	15820	Salted crackers (manufacture)		15841	Confectionery made of chocolate (manufacture)
	15820	Sandwich cake baking (manufacture)		15841	Drinking chocolate (manufacture)
	15820	Sausage filler made of cereal (manufacture)		15841	Milk chocolate (manufacture)
	15820	Scone baking (manufacture)		15841	Milk cocoa (manufacture)
	15820	Snack products of puffed or extruded farinaceous or proteinaceous materials (manufacture)	**10822**		**Manufacture of sugar confectionery**
	15820	Snack products whether sweet or salted (manufacture)		15842	Boiled sweet (manufacture)
	15820	Sweet crackers (manufacture)		15842	Butterscotch (manufacture)
	15820	Wafer biscuits (manufacture)		15842	Cachous (manufacture)
				15842	Candied peel (manufacture)
10730		**Manufacture of macaroni, noodles, couscous and similar farinaceous products**		15842	Caramel sweets (manufacture)
	15850	Couscous (manufacture)		15842	Chewing gum (manufacture)
	15850	Farinaceous products (manufacture)		15842	Clear gum confectionery (manufacture)
	15850	Macaroni (manufacture)		15842	Confectioner's novelty (manufacture)
	15850	Noodle (manufacture)		15842	Confectionery (medicated) (manufacture)
	15850	Pasta products, canned or frozen (manufacture)		15842	Confectionery made of sugar (manufacture)
	15850	Pastas (manufacture)		15842	Crystallised fruit (manufacture)
	15850	Ravioli (manufacture)		15842	Fondant (manufacture)
	15850	Spaghetti (manufacture)		15842	Fruit peel preserving in sugar (manufacture)
	15850	Spaghetti canning (manufacture)		15842	Fruit preserving in sugar (manufacture)
	15850	Vermicelli (manufacture)		15842	Jujube (manufacture)
				15842	Liquorice (manufacture)
10810		**Manufacture of sugar**		15842	Lozenge (medicated) (manufacture)
	15830	Beet pulp (manufacture)		15842	Lozenge (not medicated) (manufacture)
	15830	Beet sugar (manufacture)		15842	Marshmallow (manufacture)
	15830	Castor sugar (manufacture)		15842	Marzipan sweets (manufacture)
	15830	Icing sugar (manufacture)		15842	Medicated confectionery (manufacture)
	15830	Invert sugar (manufacture)		15842	Nougat (manufacture)
	15830	Liquid sugar (manufacture)		15842	Nut and bean confectionery (manufacture)
	15830	Maple syrup (manufacture)		15842	Nut preserving in sugar (manufacture)
	15830	Molasses (manufacture)		15842	Pastille (manufacture)
	15830	Powdered sugar (manufacture)		15842	Plant parts preserving in sugar (manufacture)
	15830	Sugar milling (manufacture)		15842	Pomfret (pontefract) cakes (manufacture)
	15830	Sugar refining (manufacture)		15842	Sugar confectionery (manufacture)
	15830	Sugar substitutes refining and production (manufacture)		15842	Sweets (sugar confectionery) (manufacture)
	15830	Sugar sucrose and sugar substitutes from cane, manufacture or refining (manufacture)		15842	Toffee (manufacture)
	15830	Sugar sucrose and sugar substitutes from maple, manufacture or refining (manufacture)		15842	Turkish delight (manufacture)
	15830	Sugar sucrose and sugar substitutes from palm, manufacture or refining (manufacture)		15842	White chocolate (manufacture)
	15830	Sugar syrups (manufacture)	**10831**		**Tea processing**
	15830	Syrup (sugar) (manufacture)		15861	Chamomile herb infusions (manufacture)
	15830	Treacle (manufacture)		15861	Herb tea (manufacture)
	15830	White sugar (manufacture)		15861	Mint herb infusions maté (manufacture)
				15861	Soluble tea (manufacture)
10821		**Manufacture of cocoa, and chocolate confectionery**		15861	Tea (packing into tea bags) (manufacture)
	15841	Chocolate (manufacture)		15861	Tea and maté blending (manufacture)
	15841	Chocolate confectionery (manufacture)		15861	Tea blending (manufacture)
				15861	Tea extract and essence (manufacture)
				15861	Tea or maté based extracts and preparations (manufacture)
				15861	Vervain herb infusions (manufacture)

UK SIC 2007	UK SIC 2003	Activity	SIC 2007	SIC 2003	Activity
10832		**Production of coffee and coffee substitutes**	10860		**Manufacture of homogenised food preparations and dietetic food**
	15862	Coffee (manufacture)		15880	Baby foods (manufacture)
	15862	Coffee and chicory essence and extract (manufacture)		15880	Baby foods (milk based) (manufacture)
	15862	Coffee bags (manufacture)		15880	Diabetic food (manufacture)
	15862	Coffee blending (manufacture)		15880	Dietary foods for special medical purposes (manufacture)
	15862	Coffee essence and extract (manufacture)		15880	Dietetic food (excluding milk based) (manufacture)
	15862	Coffee extracts and concentrates (manufacture)		15880	Dietetic food with a milk base (manufacture)
	15862	Coffee grinding and roasting (manufacture)		15880	Follow-up milk (manufacture)
	15862	Coffee processing (manufacture)		15880	Follow-up milk for infants (manufacture)
	15862	Coffee products (manufacture)		15880	Food for particular nutritional uses (manufacture)
	15862	Coffee roasting (manufacture)		15880	Foods for persons suffering from carbohydrate metabolism disorders (manufacture)
	15862	Coffee substitutes (manufacture)		15880	Foods to meet the expenditure of intense muscular effort, especially for sportsmen (manufacture)
	15862	Dandelion coffee (manufacture)			
	15862	De-caffeinated coffee (manufacture)		15880	Gluten-free foods (manufacture)
	15862	Ground coffee (manufacture)		15880	Homogenised food preparations (manufacture)
	15862	Instant coffee (manufacture)		15880	Infant food (milk based) (manufacture)
	15862	Liquid coffee (manufacture)		15880	Infant food (other than milk based) (manufacture)
	15862	Soluble coffee (manufacture)		15880	Infant formulae (manufacture)
				15880	Invalid food (milk based) (manufacture)
10840		**Manufacture of condiments and seasonings**		15880	Invalid food (other than milk based) (manufacture)
	15870	Catsup (manufacture)		15880	Low energy and energy-reduced foods (manufacture)
	15870	Condiments (manufacture)		15880	Low sodium foods (manufacture)
	15870	Curry powder (manufacture)		15880	Low-sodium or sodium-free dietary salts (manufacture)
	15870	Dried herbs (except field dried) (manufacture)		15880	Milk based baby food (manufacture)
	15870	Gravy (manufacture)			
	15870	Ketchup (manufacture)	10890		**Manufacture of other food products n.e.c.**
	15870	Mayonnaise (manufacture)		15899	Apple pomace and pectin (manufacture)
	15870	Mint sauce (manufacture)		15899	Artificial concentrates (manufacture)
	15870	Mustard (manufacture)		15620	Artificial honey (manufacture)
	15870	Mustard flour and meal (manufacture)		15899	Bakers' yeast from distillery (manufacture)
	15870	Pepper (ground) (manufacture)		15899	Baking powder (manufacture)
	15870	Pepper substitute (manufacture)		15899	Blancmange powder (manufacture)
	15870	Relish (manufacture)		15891	Broth containing meat or vegetables or both (manufacture)
	15870	Salad cream (manufacture)		15891	Canned broth containing meat or vegetables or both (manufacture)
	15870	Salad dressing (manufacture)			
	14400	Salt, processing of into food-grade salt, iodised salt (manufacture)		15891	Canned soup containing meat or vegetables or both (manufacture)
	15870	Sandwich spread (manufacture)		15620	Caramel (not sweets) (manufacture)
	15870	Sauce (manufacture)		15899	Cider pectin (manufacture)
	15870	Seasoning (manufacture)		15899	Custard powder (manufacture)
	15870	Spice (ground) (manufacture)		15899	Dried egg (manufacture)
	15870	Spice (purifying) (manufacture)		15899	Egg drying (manufacture)
	15870	Stuffing (manufacture)		15899	Egg pickling (manufacture)
	15870	Vinegars (malt, spirit, wine, acetic acid) (manufacture)		15899	Egg products and egg albumin (manufacture)
				15899	Egg substitute (manufacture)
10850		**Manufacture of prepared meals and dishes**		15899	Eggs (powdered or reconstituted) (manufacture)
	15330	Bottling of fruit and vegetables (manufacture)		15139	Extracts and juices of meat, fish, crustaceans or molluscs (manufacture)
	15139	Chicken ready to eat meals (manufacture)			
	15209	Fish dish (prepared) production (manufacture)		15899	Food products enriched with vitamins or proteins (manufacture)
	15209	Fish dishes, including fish and chips (manufacture)			
	15209	Fish fingers (manufacture)		15899	Food supplements (manufacture)
	15139	Frozen meals based on meat (manufacture)		15899	Hop extracts (manufacture)
	15850	Pasta, cooked, stuffed or otherwise prepared		15899	Malt extract (manufacture)
	15820	Pizza, frozen or otherwise preserved (manufacture)		15899	Malted milk production (manufacture)
	15139	Poultry dishes (manufacture)		15899	Non-dairy milk and cheese substitutes (manufacture)
	15139	Prepared meat dishes (manufacture)			
	15330	Vegetable dishes (manufacture)			

UK SIC 2007	UK SIC 2003	Activity	SIC 2007	SIC 2003	Activity
	15810	Perishable food preparations of bread, pastry or cake (manufacture)	**11010**		**Distilling, rectifying and blending of spirits**
	15891	Powdered broth containing meat or vegetables or both (manufacture)		15910	Alcoholic distilled potable beverage (manufacture)
	15891	Powdered soup containing meat or vegetables or both (manufacture)		15910	Aperitif (spirit based) (manufacture)
	15899	Rice pudding (canned) (manufacture)		15910	Blending and bottling of whisky (manufacture)
	15899	Roasting of nuts (manufacture)		15910	Blending of distilled spirits (manufacture)
	15512	Sandwiches with cheese filling (manufacture)		51342	Blending of distilled spirits by wholesalers (manufacture)
	15899	Sandwiches with egg filling (manufacture)		15910	Brandy (manufacture)
	15209	Sandwiches with fish, crustacean and mollusc filling (manufacture)		15910	Cherry brandy (manufacture)
	15139	Sandwiches with ham or bacon filling (manufacture)		15910	Drinks mixed with distilled alcoholic beverages (manufacture)
	15139	Sandwiches with meat or poultry filling, other than ham or bacon (manufacture)		15910	Gin (manufacture)
	15330	Sandwiches with vegetable filling (manufacture)		15910	Liqueurs (manufacture)
	15891	Soup containing meat or vegetables or both (manufacture)		15910	Rectified spirits (manufacture)
	15891	Tinned broth (manufacture)		15910	Rum distilling (manufacture)
	15891	Tinned soup (manufacture)		15910	Spirit distilling and compounding (manufacture)
	15899	Uncooked pizza (manufacture)		15920	Undenatured ethyl alcohol (manufacture)
	15899	Yeast and vegetable extract (manufacture)		15910	Vodka distilling (manufacture)
	15899	Yeast preparation (manufacture)		15910	Whisky (manufacture)
				15910	Whisky blending (manufacture)
10910		**Manufacture of prepared feeds for farm animals**		15910	Whisky distilling (manufacture)
	15710	Alfalfa (Lucerne) meal and pellets (manufacture)	**11020**		**Manufacture of wine from grape**
	15710	Animal compound feed (manufacture)		51342	Buying wine in bulk with blending, purification and bottling of wine (manufacture)
	15710	Animal feed supplement (manufacture)		15932	Low and non-alcoholic wine based on concentrated grape must (manufacture)
	15710	Chicken food (manufacture)		15931	Low and non-alcoholic wine from fresh grapes and grape juice (manufacture)
	15710	Compound animal feed (manufacture)		15931	Sparkling wine (manufacture)
	15710	Concentrated animal feed and feed supplements (manufacture)		15932	Tonic wine production (manufacture)
	15710	Dairy concentrate (animal feed) (manufacture)		15931	Wine (of own manufacture) blending, purification and bottling (manufacture)
	15710	Farm animal feeds produced from slaughter waste (manufacture)		51342	Wine (purchased in bulk) blending, purification and bottling (manufacture)
	15710	Feed supplements for animals (manufacture)		15932	Wine based on concentrated grape must (manufacture)
	15710	Molassed feeding stuff containing more than 30% molasses (manufacture)		15931	Wine production from fresh grapes and grape juice (manufacture)
	15710	Non-compound animal and cattle feed (not from grain milling or oilseed cakes and meal) (manufacture)		01131	Wine production from self produced grapes
	15710	Potato dehydrating for animal feed (manufacture)	**11030**		**Manufacture of cider and other fruit wines**
	15710	Poultry grit (manufacture)		15949	Apple wine making
	15710	Prepared feeds for farm animals (manufacture)		15949	Beverages made of fermented fruit n.e.c. (manufacture)
	15710	Protein (synthetic) for animal feed (manufacture)		15941	Cider (alcoholic) (manufacture)
	15710	Protein concentrates (animal food) (manufacture)		15941	Cider perry (manufacture)
	15710	Single feeds, unmixed, for farm animals (manufacture)		15949	Fruit wines other than cider and perry (manufacture)
	15710	Synthetic animal feed protein (manufacture)		15949	Mead (manufacture)
				15949	Mixed beverages containing fruit wines (manufacture)
10920		**Manufacture of prepared pet foods**		15941	Perry (manufacture)
	15720	Bird food (manufacture)		15949	Sake (manufacture)
	15720	Cat food (manufacture)	**11040**		**Manufacture of other non-distilled fermented beverages**
	15720	Dog biscuit (manufacture)		15950	Beverages (non-distilled, fermented) (manufacture)
	15720	Dog food (manufacture)		15950	Vermouth (manufacture)
	15720	Pet animal feeds produced from slaughter waste (manufacture)	**11050**		**Manufacture of beer**
	15720	Pet fish feeds (manufacture)		15960	Ale brewing (manufacture)
	15720	Pet food including canned (manufacture)		15960	Beer (non-alcoholic) (manufacture)
	15720	Pet foods (manufacture)			

UK SIC 2007	UK SIC 2003	Activity	SIC 2007	SIC 2003	Activity
	15960	Beer brewing (manufacture)		16000	Tobacco, homogenised or reconstituted (manufacture)
	15960	Black beer brewing (manufacture)			
	15960	Brewery (beer and other brewing products) (manufacture)	13100		**Preparation and spinning of textile fibres**
	15960	Lager brewing (manufacture)		17120	Alpaca and mohair spinning on the woollen system (manufacture)
	15960	Low and non-alcoholic beer (manufacture)		17130	Alpaca and mohair spinning on the worsted system (manufacture)
	15960	Malt liquors (manufacture)		17170	Bast fibres preparation and spinning (manufacture)
	15960	Porter brewing (manufacture)		17130	Blending of fibres on the worsted system (manufacture)
	15960	Stout brewing (manufacture)		17130	Botany spinning (manufacture)
11060		**Manufacture of malt**		17120	Camel hair spinning on the woollen system (manufacture)
	15970	Barley malting (manufacture)		17130	Camel hair spinning on the worsted system (manufacture)
	15970	Malt and malt products (manufacture)		17120	Carded sliver preparation for textiles industry (manufacture)
11070		**Manufacture of soft drinks; production of mineral waters and other bottled waters**		17110	Carpet pile yarn spun on the cotton system (manufacture)
	15980	Aerated water (manufacture)		17120	Carpet pile yarn spun on the woollen system (manufacture)
	15980	Carbonated soft drink (manufacture)		17130	Carpet pile yarn spun on the worsted and semi-worsted systems (manufacture)
	15980	Cider (non-alcoholic) (manufacture)		17120	Coffin cloth (manufacture)
	15980	Cola production (manufacture)		17120	Combers' shoddy for woollen industry (manufacture)
	15980	Cordial (non-alcoholic) (manufacture)		17130	Combing and slubbing of wool on a commission basis (manufacture)
	15980	Cream soda (manufacture)		17110	Core spun yarn spun on the cotton system (manufacture)
	15980	Energy drinks (manufacture)		17110	Cotton carding (manufacture)
	15980	Fruit cordial (manufacture)		17110	Cotton combing (manufacture)
	15980	Fruit drinks (non-alcoholic) (manufacture)		17110	Cotton doubling (manufacture)
	15980	Fruit squash (manufacture)		17110	Cotton drawing (manufacture)
	15980	Fruit syrup (manufacture)		17110	Cotton lap, sliver, rovings and other intermediate bobbin (manufacture)
	15980	Ginger beer (manufacture)		17110	Cotton opening (manufacture)
	15980	Hop bitters (manufacture)		17110	Cotton reeling (manufacture)
	15980	Lemonade powder (manufacture)		17110	Cotton sorting (manufacture)
	15980	Lemonade production (manufacture)		17110	Cotton spinning (manufacture)
	15980	Milk drinks (flavoured) (manufacture)		17160	Cotton thread mill (manufacture)
	15980	Milk shake base (manufacture)		17110	Cotton warp (manufacture)
	15980	Mineral water bottling (manufacture)		17110	Cotton waste spinning (manufacture)
	15980	Mineral water production (manufacture)		17110	Cotton yarn (manufacture)
	15980	Non-alcoholic flavoured and/or sweetened waters (manufacture)		17110	Cotton yarn doubling (manufacture)
	15980	Orangeade production (manufacture)		17110	Cotton yarn twisting (manufacture)
	15980	Shandy (manufacture)		17110	Cotton yarn warping (manufacture)
	15980	Soda water (manufacture)		17110	Cotton yarn winding (manufacture)
	15980	Soft drinks (manufacture)		17110	Cotton-type yarn (manufacture)
	15980	Squash drink (manufacture)		17120	Degreasing and carbonising of wool (manufacture)
	15980	Table water (manufacture)		17170	Down of vegetable origin (manufacture)
	15980	Tonic water production (manufacture)		17140	Dressed line made of flax (manufacture)
	15980	Water bottling (manufacture)		17120	Dyeing of wool fleece (manufacture)
12000		**Manufacture of tobacco products**		17160	Embroidery cotton (manufacture)
	16000	Cheroot (manufacture)		17120	Fingering wool (manufacture)
	16000	Chewing tobacco (manufacture)		17110	Fishing net yarn made of cotton (manufacture)
	16000	Cigar (manufacture)		17140	Flax carding (manufacture)
	16000	Cigarette (manufacture)		17140	Flax deseeding (manufacture)
	16000	Cigarillo (manufacture)		17140	Flax dressing (manufacture)
	16000	Fine cut tobacco (manufacture)		17140	Flax hackling (manufacture)
	16000	Pipe tobacco (manufacture)		17140	Flax preparing (manufacture)
	16000	Snuff (manufacture)		17140	Flax roughing (manufacture)
	16000	Tobacco for use in pipes and rolled cigarettes (manufacture)			
	16000	Tobacco products (manufacture)			
	16000	Tobacco stemming and redrying (manufacture)			
	16000	Tobacco substitute products (manufacture)			

UK SIC 2007	UK SIC 2003	Activity	SIC 2007	SIC 2003	Activity
	17140	Flax sorting (manufacture)		17130	Mohair spinning on the worsted system (manufacture)
	17140	Flax spinning (manufacture)		17120	Noils (woollen industry) (manufacture)
	17140	Flax tow (manufacture)		17170	Padding for upholstery (manufacture)
	17140	Flax type yarns (manufacture)		17120	Recovered wool (mungo and shoddy) (manufacture)
	17140	Flax-type fibre preparation and spinning (manufacture)		17150	Reeling and washing of silk (manufacture)
	17170	Hair dressing for upholsterers (manufacture)		17140	Scutching of flax (manufacture)
	17170	Hemp carding (manufacture)		17130	Semi-worsted yarn spinning (manufacture)
	17170	Hemp dressing (manufacture)		17160	Sewing thread made of cotton (manufacture)
	17170	Hemp sorting (manufacture)		17150	Silk creping (manufacture)
	17170	Hemp spinning (manufacture)		17150	Silk throwing (manufacture)
	17170	Horsehair curling (manufacture)		17150	Silk twisting (manufacture)
	17170	Horsehair dressing (manufacture)		17150	Silk warping (manufacture)
	17170	Horsehair hackling (manufacture)		17150	Silk waste noil spinning (manufacture)
	17170	Horsehair sorting (manufacture)		17150	Silk winding (manufacture)
	17170	Horsehair teasing (manufacture)		17150	Silk yarn (manufacture)
	17170	Jute sorting (manufacture)		17120	Sliver dyeing of wool (manufacture)
	17170	Jute spinning (manufacture)		17120	Slubbing dyeing of wool (manufacture)
	17170	Jute tow (manufacture)		17160	Spinning and manufacture of yarn for weaving, for the trade or for further processing (manufacture)
	17170	Jute winding (manufacture)		17160	Spinning and manufacture of yarn or sewing thread, for the trade or further processing (manufacture)
	17170	Jute yarn (manufacture)			
	17170	Kapok willowing (manufacture)		17110	Spinning on the cotton system (manufacture)
	17120	Knitting yarn made of wool (manufacture)		17140	Spinning on the flax system (manufacture)
	17130	Knitting yarn made of worsted (manufacture)		17120	Spinning on the woollen system (manufacture)
	17110	Knitting yarns (cotton) (manufacture)		17130	Spinning on the worsted and semi-worsted systems (manufacture)
	17110	Knitting yarns (man made fibres) (manufacture)			
	17140	Line yarn made of flax (manufacture)		17170	Sunn hemp (manufacture)
	17150	Man-made fibre bulking (other than in man-made fibre producing establishments) (manufacture)		17160	Thread for sewing and embroidery, made of cotton (manufacture)
	17150	Man-made fibre crimping (other than in man-made fibre producing establishments) (manufacture)		17160	Thread made of hemp (manufacture)
				17160	Thread made of jute (manufacture)
	17150	Man-made fibre texturing (other than in man-made fibre producing establishments) (manufacture)		17160	Thread made of linen (manufacture)
				17160	Thread made of silk (manufacture)
	17150	Man-made fibre texturing, bulking, crimping in man-made fibre producing establishments (manufacture)		17130	Topmaking (wool) (manufacture)
				17140	Tow of flax (manufacture)
	17110	Man-made fibres spinning on the cotton system (manufacture)		17170	Twine made of paper (manufacture)
	17120	Man-made fibres spinning on the woollen system (manufacture)		17110	Tyre cord (cotton system) (manufacture)
				17170	Upholstery hair fibre and filling (manufacture)
	17130	Man-made fibres spinning on the worsted and semi-worsted systems (manufacture)		17170	Vegetable down (manufacture)
	17110	Man-made fibres twisting on the cotton system (manufacture)		17120	Warp dressing (woollen) (manufacture)
				17130	Warp sizing and dressing (worsted) (manufacture)
	17120	Man-made fibres twisting on the woollen system (manufacture)		17110	Waste cotton yarn (manufacture)
				17120	Wool carbonising (manufacture)
	17130	Man-made fibres twisting on the worsted and semi-worsted systems (manufacture)		17120	Wool carding (manufacture)
				17120	Wool cleaning (manufacture)
	17110	Man-made fibres warping on the cotton system (manufacture)		17120	Wool condensing (manufacture)
				17120	Wool extracting (manufacture)
	17120	Man-made fibres warping on the woollen system (manufacture)		17120	Wool opening and willeying (manufacture)
				17120	Wool recovery (manufacture)
	17130	Man-made fibres warping on the worsted and semi-worsted systems (manufacture)		17120	Wool scouring (manufacture)
				17120	Wool sorting (manufacture)
	17110	Man-made fibres winding on the cotton system (manufacture)		17130	Wool topmaking (manufacture)
				17120	Woollen rag carbonising (manufacture)
	17120	Man-made fibres winding on the woollen system (manufacture)		17120	Woollen rag carding (manufacture)
				17120	Woollen rag garnetting (manufacture)
	17130	Man-made fibres winding on the worsted and semi-worsted systems (manufacture)		17120	Woollen rag grinding or pulling (manufacture)
	17130	Merino yarn spinning (manufacture)		17120	Woollen waste breaking (manufacture)
	17120	Mohair spinning on the woollen system (manufacture)		17120	Woollen waste garnetting (manufacture)

UK SIC 2007	UK SIC 2003	Activity	SIC 2007	SIC 2003	Activity
	17120	Woollen waste grinding (manufacture)		17210	Cloth made of cotton and similar man made fibres (manufacture)
	17120	Woollen waste opening and willeying (manufacture)		17220	Coating woollen weaving (manufacture)
	17120	Woollen yarn carding (manufacture)		17230	Coating worsted weaving (manufacture)
	17120	Woollen yarn condensing (manufacture)		17210	Corduroy weaving from yarn spun on the cotton system (manufacture)
	17120	Woollen yarn reeling (manufacture)			
	17120	Woollen yarn sizing (manufacture)		17210	Corset cloth weaving (from yarn spun on the cotton system) (manufacture)
	17120	Woollen yarn spinning (manufacture)			
	17120	Woollen yarn twisting (manufacture)		17210	Cotton weaving (manufacture)
	17120	Woollen yarn warping (manufacture)		17210	Cotton-type fabrics (manufacture)
	17120	Woollen yarn winding (manufacture)		17240	Crepe weaving (manufacture)
	17120	Woollen-type fibre preparation and spinning (manufacture)		17210	Cretonne weaving (manufacture)
				17210	Damask weaving (not woollen or worsted) (manufacture)
	17120	Woollen-type yarns (manufacture)			
	17130	Worsted carding (manufacture)		17220	Damask woollen weaving (manufacture)
	17130	Worsted doubling (manufacture)		17230	Damask worsted weaving (manufacture)
	17130	Worsted spinning and twisting (manufacture)		17210	Denim weaving (manufacture)
	17130	Worsted waste grinding (manufacture)		17210	Downproof cloth weaving (manufacture)
	17130	Worsted yarn reeling (manufacture)		17210	Dress fabric (woven (not wool)) (manufacture)
	17130	Worsted yarn warping (manufacture)		17220	Dress goods woollen weaving (manufacture)
	17130	Worsted yarn winding (manufacture)		17230	Dress goods worsted weaving (manufacture)
	17130	Worsted-type fibres preparation and spinning (manufacture)		17210	Drill weaving from yarn spun on the cotton system (manufacture)
	17130	Worsted-type yarns (manufacture)		17210	Duck weaving (manufacture)
	17110	Yarn (core spun on the cotton system) (manufacture)		17220	Flag fabric (woollen) (manufacture)
				17220	Flannel (manufacture)
	17170	Yarn made of paper (manufacture)		17210	Flannelette weaving (manufacture)
	17110	Yarn of cotton (manufacture)		17250	Flax woven cloth (manufacture)
				17210	Flock made of cotton (manufacture)
13200		**Weaving of textiles**		17220	Frieze cloth (manufacture)
	17220	Alpaca woollen weaving (manufacture)		17210	Furnishing fabric (woven (not wool or worsted)) (manufacture)
	17230	Alpaca worsted weaving (manufacture)			
	17210	Apparel cloth woven from yarns spun on the cotton system (manufacture)		17230	Furnishing fabric worsted weaving (manufacture)
				17220	Furnishing fabrics woollen weaving (manufacture)
	17250	Awning cloth weaving (manufacture)		17210	Fustian weaving (manufacture)
	17250	Bagging cloth (manufacture)		17210	Gabardine (cotton) weaving (manufacture)
	17210	Bandage cloth weaving (manufacture)		17210	Gauze weaving (manufacture)
	17250	Bast fibres and special yarns weaving (manufacture)		17210	Gingham weaving (manufacture)
	17210	Bedford cord (not worsted) weaving (manufacture)		17250	Glass fibre woven fabric (manufacture)
				17210	Haircord weaving (manufacture)
	17230	Bedford cord worsted weaving (manufacture)		17250	Hemp weaving (manufacture)
	17210	Belting duck weaving (manufacture)		17250	Hessian (manufacture)
	17210	Book cloth weaving (manufacture)		17210	Imitation fur (woven long pile fabrics) (manufacture)
	17210	Brocade weaving (manufacture)			
	17210	Bunting made of cotton (manufacture)		17220	Imitation fur of long pile fabrics made on the woollen system (manufacture)
	17220	Bunting woollen weaving (manufacture)			
	17210	Calico weaving (manufacture)		17210	Interlining weaving from yarn spun on the cotton system (manufacture)
	17210	Cambric weaving (manufacture)			
	17220	Camel hair woollen weaving (manufacture)		17210	Jeans cloth weaving (manufacture)
	17230	Camel hair worsted weaving (manufacture)		17250	Jute cloth (manufacture)
	17250	Canvas weaving (manufacture)		17250	Jute sacking (manufacture)
	17250	Carbon and aramid thread weaving (manufacture)		17250	Jute weaving (manufacture)
				17210	Leno fabric weaving (manufacture)
	17210	Casement cloth weaving (manufacture)		17250	Linen and union cloth (manufacture)
	17220	Cashmere woollen weaving (manufacture)		17250	Linen weaving (manufacture)
	17230	Cashmere worsted weaving (manufacture)		17210	Man-made fibre weaving from yarns spun on the cotton system (manufacture)
	17210	Cellular cloth weaving from yarn spun on the cotton system (manufacture)			
				17220	Man-made fibre weaving of fabrics from yarns spun on the woollen system (manufacture)
	17210	Chenille (manufacture)			
	17210	Chiffon weaving from yarn spun on the cotton system (manufacture)		17230	Man-made fibre weaving of fabrics from yarns spun on worsted and semi-worsted systems (manufacture)
	17210	Chintz weaving from yarn spun on the cotton system (manufacture)		17220	Mohair woollen weaving (manufacture)

UK SIC 2007	UK SIC 2003	Activity
	17230	Mohair worsted weaving (manufacture)
	17210	Moquette (not woollen) weaving (manufacture)
	17220	Moquette woollen weaving (manufacture)
	17210	Muslin weaving (manufacture)
	17210	Organdie weaving (manufacture)
	17220	Overcoating woollen weaving (manufacture)
	17210	Pique weaving (manufacture)
	17240	Plush silk (manufacture)
	17220	Plush woollen weaving (manufacture)
	17230	Plush worsted weaving (manufacture)
	17210	Pocketing weaving (manufacture)
	17250	Polypropylene fabrics (manufacture)
	17210	Poplin weaving (manufacture)
	17210	Poult weaving (manufacture)
	17220	Press cloth (manufacture)
	17210	Print cloth weaving (manufacture)
	17210	Printers' weaving (manufacture)
	17210	Quilt weaving (manufacture)
	17250	Ramie weaving (manufacture)
	17250	Sailcloth weaving (manufacture)
	17210	Sateen weaving (manufacture)
	17240	Satin weaving (manufacture)
	17250	Scrim weaving from flax, hemp, ramie and man-made fibres processed on the flax system (manufacture)
	17250	Sheeting weaving from flax and man-made fibres processed on the flax system (manufacture)
	17210	Sheeting weaving from yarn spun on the cotton system (manufacture)
	17250	Shirting weaving from flax and man made fibre processed on the flax system (manufacture)
	17210	Shirting weaving from yarn spun on the cotton system (manufacture)
	17240	Silk type fabrics (manufacture)
	17240	Silk woven cloth (manufacture)
	17220	Suiting woollen weaving (manufacture)
	17230	Suiting worsted weaving (manufacture)
	17240	Taffeta made of silk (manufacture)
	17220	Taffeta woollen weaving (manufacture)
	17210	Tapestry (not woollen or worsted) (manufacture)
	17220	Tapestry woollen weaving (manufacture)
	17210	Terry towelling (manufacture)
	17210	Ticking weaving (manufacture)
	17240	Tie silk (manufacture)
	17210	Towelling weaving (manufacture)
	17220	Tweed (manufacture)
	17210	Twill weaving (manufacture)
	17250	Union cloth (cotton/linen) (manufacture)
	17240	Veiling made of silk (manufacture)
	17210	Velvet (manufacture)
	17210	Velveteen (manufacture)
	17210	Voile weaving (manufacture)
	17220	Weaving (woollen) (manufacture)
	17230	Weaving (worsted) (manufacture)
	17210	Weaving of cotton and man-made fibres (manufacture)
	17210	Winceyette weaving (manufacture)
	17220	Woollen cloth weaving (manufacture)
	17220	Woollen-type fabrics (manufacture)
	17220	Woollen-type weaving (manufacture)

SIC 2007	SIC 2003	Activity
	17230	Worsted-type fabrics (manufacture)
	17230	Worsted-type weaving (manufacture)
	17250	Woven cloth made of polypropylene (manufacture)
	17250	Woven elastic over 30 cm wide (manufacture)
	17250	Woven elastomeric over 30 cm wide (manufacture)
	17210	Woven pile cotton-type fabrics (manufacture)
	17210	Zephyr weaving (manufacture)
13300		**Finishing of textiles**
	17300	Binding and mending of textiles (manufacture)
	17300	Bleach works (manufacture)
	17300	Bleaching and dyeing of fabrics (manufacture)
	17300	Bleaching and dyeing of textile articles, including wearing apparel (manufacture)
	17300	Bleaching and dyeing of textile fibres (manufacture)
	17300	Bleaching and dyeing of yarns (manufacture)
	17300	Bleaching of jeans (manufacture)
	17300	Bleaching of wool cop, hank, warp, etc. (manufacture)
	17300	Block printing of textiles (manufacture)
	17300	Bonding of fabric to fabric (manufacture)
	17300	Calendering of textiles (manufacture)
	17300	Calico printing (manufacture)
	17300	Chintz glazing (manufacture)
	17300	Cloth beetling (manufacture)
	17300	Cloth crease resisting treatment (manufacture)
	17300	Cloth degreasing (manufacture)
	17300	Cloth dressing (manufacture)
	17300	Cloth dyeing (manufacture)
	17300	Cloth embossing (manufacture)
	17300	Cloth ending and mending (manufacture)
	17300	Cloth finishing (manufacture)
	17300	Cloth fireproofing (manufacture)
	17300	Cloth mercerising (manufacture)
	17300	Cloth piece goods printing (manufacture)
	17300	Cloth proofing (manufacture)
	17300	Cloth rot proofing (manufacture)
	17300	Cloth shrinking (manufacture)
	17300	Cloth waterproofing
	17300	Coating of purchased garments (manufacture)
	17300	Commission mending of textiles (manufacture)
	17300	Contract cutting of textiles (not self-owned) (manufacture)
	17300	Cotton cord and velveteen finishing (manufacture)
	17300	Cotton fabric bleaching, dyeing or otherwise finishing (manufacture)
	17300	Cotton waste bleaching, dying or otherwise finishing (manufacture)
	17300	Cotton yarn gassing (manufacture)
	17300	Cotton yarn polishing (manufacture)
	17300	Cotton yarn printing (manufacture)
	17300	Dressing of textiles and textile articles, including wearing apparel (manufacture)
	17300	Drying of textiles and textile articles, including wearing apparel (manufacture)
	17300	Dyework (manufacture)
	17300	Embroidery on made up textile goods
	17300	Ending and mending of textiles (manufacture)
	17300	Finishing of leather wearing apparel (manufacture)
	17300	Finishing of wearing apparel n.e.c. (manufacture)

UK SIC 2007	UK SIC 2003	Activity	SIC 2007	SIC 2003	Activity
	17300	Flannel ending (manufacture)		17300	Pleating of textiles (manufacture)
	17300	Flannel filling (manufacture)		17300	Rag bleaching (manufacture)
	17300	Flannel finishing (manufacture)		17300	Rag dyeing (manufacture)
	17300	Flannel preparing (manufacture)		17300	Raw cotton bleaching, dyeing or otherwise finishing (manufacture)
	17300	Flannel scouring (manufacture)			
	17300	Flannel shrinking (manufacture)		17300	Raw silk dyeing (manufacture)
	17300	Flannelette raising and finishing (manufacture)		17300	Roller printing of textiles (manufacture)
	17300	Flax yarn bleaching, dyeing or otherwise finishing (manufacture)		17300	Rubberising of purchased garments (manufacture)
	17300	Foam backed fabric finishing (manufacture)		17300	Sanforising of textiles and textile articles, including wearing apparel (manufacture)
	17300	Foam backing (single textile material) (manufacture)		17300	Screen printing of textiles (manufacture)
	17300	Foam backing (texture material sandwich) (manufacture)		17300	Shrinking of textiles and textile articles, including wearing apparel (manufacture)
	17300	Fulling mill (manufacture)		17300	Silk bleaching (manufacture)
	17300	Gassing yarn (manufacture)		17300	Silk dyeing (manufacture)
	17300	Hair dyeing (textile) (manufacture)		17300	Silk finishing and weighting (manufacture)
	17300	Hand block printing of textiles (manufacture)		17300	Silk printing (manufacture)
	17300	Hose bleaching, dyeing or otherwise finishing (manufacture)		17300	Silk screen-printing on wearing apparel (manufacture)
	17300	Hosiery finishing (manufacture)		17300	Slub dyeing (manufacture)
	17300	Hosiery printing (manufacture)		17300	Small ware bleaching (manufacture)
	17300	Hosiery scouring (manufacture)		17300	Small ware dyeing (manufacture)
	17300	Hosiery shrinking (manufacture)		17300	Steaming of textiles and textile articles, including wearing apparel (manufacture)
	17300	Hosiery trimming		17300	Storing yarn (manufacture)
	17300	Impregnating purchased garments (manufacture)		17300	Tape bleaching (manufacture)
	17300	Jute calendering (manufacture)		17300	Textile binding and mending (manufacture)
	17300	Jute fabrics bleaching, dyeing or otherwise finishing (manufacture)		17300	Textile bleaching (manufacture)
	17300	Knitted goods finishing (manufacture)		17300	Textile calendering (manufacture)
	17300	Knitted goods printing (manufacture)		17300	Textile dyeing (manufacture)
	17300	Knitted goods scouring (manufacture)		17300	Textile embossing (manufacture)
	17300	Knitted goods shrinking (manufacture)		17300	Textile ending and mending (manufacture)
	17300	Knitted goods trimming (manufacture)		17300	Textile finishing (manufacture)
	17300	Lace bleaching, dyeing and dressing (on commission) (manufacture)		17300	Textile lacquering (manufacture)
	17300	Laminating of textile material (manufacture)		17300	Tracing cloth (textile finishing) (manufacture)
	17300	Lapping of textiles (manufacture)		17300	Velvet cutting and shearing (manufacture)
	17300	Linen bleaching (manufacture)		17300	Velvet dyeing (manufacture)
	17300	Linen dyeing (manufacture)		17300	Velveteen cutting and shearing (manufacture)
	17300	Linen printing (manufacture)		17300	Velveteen dyeing (manufacture)
	17300	Machine printing of textiles (manufacture)		17300	Warp knitted fabric dyeing and finishing (manufacture)
	17300	Man-made fibre fabric bleaching, dyeing, printing or otherwise finishing (manufacture)		17300	Waste textile dyeing (manufacture)
	17300	Man-made fibre yarn bleaching, dyeing or otherwise finishing (manufacture)		17300	Waterproofing of purchased garments (manufacture)
	17300	Mending of textile articles, including wearing apparel (manufacture)		17300	Weft knitted fabric dyeing and finishing (manufacture)
	17300	Mercerising of textiles and textile articles, including wearing apparel (manufacture)		52740	'While-you-wait' printing of textile articles (manufacture)
	17300	Mercerising yarn (manufacture)		17300	Wool dyeing (loose) (manufacture)
	17300	Muslin clipping (manufacture)		17300	Wool printing (manufacture)
	17300	Muslin dressing (manufacture)		17300	Woollen and worsted fabric bleaching, dyeing, or otherwise finishing (manufacture)
	17300	Muslin ending (manufacture)		17300	Woven fabric bleaching, dyeing, printing or otherwise finishing (manufacture)
	17300	Muslin finishing (manufacture)			
	17300	Muslin gassing (manufacture)		17300	Yarn bleaching, dyeing or otherwise finishing (manufacture)
	17300	Muslin mending (manufacture)		17300	Yarn finishing (manufacture)
	17300	Piece goods dyeing (manufacture)		17300	Yarn gassing (manufacture)
	17300	Pile fabric bleaching and finishing (manufacture)		17300	Yarn mercerising (manufacture)
	17300	Plaiting of textiles (manufacture)		17300	Yarn polishing (manufacture)
	17300	Pleating and similar work on textiles (manufacture)		17300	Yarn storing (manufacture)

UK SIC 2007	UK SIC 2003	Activity	SIC 2007	SIC 2003	Activity
13910		**Manufacture of knitted and crocheted fabrics**		17402	Escape chute for aircraft (manufacture)
	17600	Crocheted fabric (manufacture)		17402	Filter cloth (made-up) (manufacture)
	17600	Crocheted fabrics (manufacture)		17402	Flag making up (manufacture)
	17600	Elastic and elastomeric fabric (manufacture)		17402	Flexible ventilating ducting made of textiles (manufacture)
	17600	Elastic or elastomeric knitted or netted fabric more than 30 cm wide (manufacture)		17402	Haversack (manufacture)
	17600	Imitation fur obtained by knitting (manufacture)		17402	Lifejacket made of canvas (manufacture)
	17600	Knitted fabrics (manufacture)		17402	Machine covers (manufacture)
	17600	Knitted or netted elastic over 30 cm wide (manufacture)		17402	Made-up filter cloth (manufacture)
	17600	Locknit fabric (manufacture)		17402	Made-up goods of sailcloth (manufacture)
	17600	Net and window furnishing type fabrics (manufacture)		17402	Marquee (manufacture)
	17600	Net curtaining knitted or crocheted fabric (manufacture)		17402	Parachute (manufacture)
	17600	Netted fabric (manufacture)		17402	Pennants (manufacture)
	17600	Pile knitted fabric (manufacture)		17402	Perambulator awning (manufacture)
	17600	Sliver knitted fabric (manufacture)		17402	Rick cloths and covers (manufacture)
	17600	Terry fabrics (manufacture)		17402	Roller blinds made of canvas (manufacture)
	17600	Warp knitted fabric (manufacture)		17402	Sacks (woven) (manufacture)
	17600	Warp knitting (manufacture)		17402	Sacks and bags made of noil (manufacture)
	17600	Weft knitted fabric (manufacture)		17402	Sacks for coal (manufacture)
	17600	Window furnishing knitted fabric (manufacture)		17402	Sacks made of canvas (manufacture)
				17402	Sacks made of jute (manufacture)
13921		**Manufacture of soft furnishings**		17402	Sails (manufacture)
	17401	Austrian blinds (manufacture)		17402	Shutes made of canvas (manufacture)
	17401	Blinds (soft furnishings) (manufacture)		17402	Stretchers made of canvas (manufacture)
	17401	Bolster (manufacture)		17402	Sun blinds (manufacture)
	17401	Curtains (made-up) (manufacture)		17402	Tarpaulins (manufacture)
	17401	Cushion covers (manufacture)		17402	Tents (manufacture)
	17401	Cushions (manufacture)		17402	Tubes made of canvas for ventilating purposes (manufacture)
	17401	Festoon blinds (manufacture)		17402	Wagon cover (manufacture)
	17401	Furnishing articles (made-up) (manufacture)		17402	Waterproofed covers made of canvas (manufacture)
	17401	Furniture covers (manufacture)		17402	Web equipment making-up (manufacture)
	17401	Lace curtains (manufacture)			
	17401	Loose cover for furniture (manufacture)	**13923**		**Manufacture of household textiles (other than soft furnishings of 13.42/1)**
	17401	Pillow (manufacture)		17403	Art needlework (manufacture)
	17401	Pouffe (manufacture)		17403	Baby napkins made of towelling (manufacture)
	17401	Roman blinds (manufacture)		17403	Bath towel (manufacture)
	17401	Soft furnishing articles, made-up, of knitted or crocheted fabrics (manufacture)		17403	Bed linen (manufacture)
	17401	Soft furnishing articles, made-up, of textile material (manufacture)		17403	Bedspread (manufacture)
	17401	Soft furnishings (manufacture)		17403	Bedspreads made of lace (manufacture)
	17401	Stretch covers for furniture (manufacture)		17403	Blanket making up, outside weaving or knitting establishment (manufacture)
				17403	Blankets including travelling rugs (manufacture)
13922		**Manufacture of canvas goods, sacks etc.**		17403	Blankets made of cotton and man-made fibres (manufacture)
	17402	Awnings (manufacture)		17403	Blankets made of wool (manufacture)
	17402	Bags made of canvas (manufacture)		17403	Bolster case (manufacture)
	17402	Bags made of canvas or cotton cloth (manufacture)		17403	Cleaning cloth (not of bonded fibre fabric) (manufacture)
	17402	Banner (making up) (manufacture)		17403	Cot blanket (manufacture)
	17402	Blinds made of canvas (manufacture)		17403	Cot quilt (manufacture)
	17402	Brattice cloth (manufacture)		17403	Cotton patch quilt (manufacture)
	17402	Buckets made of canvas (manufacture)		17403	Cotton, silk, etc. embroidering (except lace and apparel) (manufacture)
	17402	Bunting (making-up) (manufacture)		17403	Counterpane (manufacture)
	17402	Camping goods (manufacture)		17403	Dish-cloths and similar articles (manufacture)
	17402	Canvas goods (manufacture)		17403	Doilies made of textiles (manufacture)
	17402	Car loose covers (manufacture)		17403	Duchesse set (manufacture)
	17402	Cart cover made of canvas (manufacture)		17403	Dust cloths (manufacture)
	17402	Covers made of waterproofed canvas (manufacture)			

UK SIC 2007	UK SIC 2003	Activity	SIC 2007	SIC 2003	Activity
	17403	Dust sheet (manufacture)		17512	Pile rugs, mats and tiles of man-made fibres, tufted (manufacture)
	17403	Duster (cleaning cloth (not of bonded fibre fabric)) (manufacture)		17512	Pile rugs, mats and tiles of wool, tufted (manufacture)
	17403	Duvet (manufacture)		17512	Rug tufting (manufacture)
	17403	Eiderdowns (manufacture)		17511	Rug weaving (not travelling rug) (manufacture)
	17403	Embroidering on made-up textile goods (manufacture)		17512	Tiles made of tufted carpet (manufacture)
	17403	Hand towel (manufacture)		17512	Tufted carpet (manufacture)
	17403	Hand-woven tapestries (manufacture)		17511	Wilton carpet (manufacture)
	17403	Household textile made-up articles (manufacture)	**13939**		**Manufacture of carpets and rugs (other than woven or tufted) n.e.c.**
	17403	Household textiles (manufacture)		17519	Bonded fibre carpets (manufacture)
	17403	Kitchen linen (manufacture)		17519	Carpets (other than woven or tufted) (manufacture)
	17403	Lace tablecloth (manufacture)		17519	Carpets and rugs, other than woven or tufted (manufacture)
	17403	Lavatory seat cover (manufacture)		17519	Carpets made of jute (manufacture)
	17403	Nursery square (manufacture)		17519	Floor rugs made of jute (manufacture)
	17403	Pillow case (manufacture)		17519	Knotted carpets (manufacture)
	17403	Pin cushion (manufacture)		17519	Mats and matting made of coconut fibre (manufacture)
	17403	Polishing cloths and pads, made of unprepared, non-bonded fibre fabric (manufacture)		17519	Mats and matting made of coir (manufacture)
	17403	Pram blanket (outside knitting or weaving establishment) (manufacture)		17519	Mats and matting made of sisal (manufacture)
	17403	Quilt (filled) (manufacture)		17519	Mats made of jute (manufacture)
	17403	Quilt fringing (manufacture)		17519	Needlefelt carpet underlay (manufacture)
	17403	Roller towel (manufacture)		17519	Needleloom and bonded fibre rugs (manufacture)
	17403	Sheet hemming (textiles) (manufacture)		17519	Needleloom and bonded fibre tiles (manufacture)
	17403	Sleeping bag (manufacture)		17519	Needleloom carpet (manufacture)
	17403	Table linen (manufacture)		17519	Needleloom carpeting (manufacture)
	17403	Table mats made of textiles (manufacture)		17519	Needleloom felt carpet underlay (manufacture)
	17403	Table runner (manufacture)		17519	Rugs (other than woven or tufted) (manufacture)
	17403	Tea towels (manufacture)		17519	Rugs made of coir (manufacture)
	17403	Textile part of electric blankets (manufacture)		17519	Rugs made of rag (manufacture)
	17403	Toilet linen (manufacture)		17519	Tiles made of felt (manufacture)
	17403	Towel (manufacture)		17519	Tiles made of needleloom carpet (manufacture)
	17403	Travelling rug making-up (outside weaving establishment) (manufacture)	**13940**		**Manufacture of cordage, rope, twine and netting**
	17403	Travelling rugs made of wool (manufacture)		17520	Agricultural twine (manufacture)
	17403	Tufting blankets (manufacture)		17520	Baler twine (manufacture)
	17403	Tufting household textiles (manufacture)		17520	Binder twine (manufacture)
	17403	Valances (manufacture)		17520	Cable made of textile materials (manufacture)
13931		**Manufacture of woven or tufted carpets and rugs**		17520	Cargo sling (manufacture)
	17511	Axminster carpet (manufacture)		17520	Combination rope (manufacture)
	17511	Brussels carpet (manufacture)		17520	Cordage made of textile material (manufacture)
	17511	Carpet weaving (manufacture)		17520	Fibre core for wire rope (manufacture)
	17511	Pile carpet mat weaving (manufacture)		17520	Fishing line (manufacture)
	17511	Pile carpet weaving (manufacture)		17520	Fishing net (manufacture)
	17512	Pile carpets of cotton, tufted (manufacture)		17520	Fishing net mending (manufacture)
	17511	Pile carpets of cotton, woven (manufacture)		17520	Garden and horticultural net (manufacture)
	17512	Pile carpets of man-made fibres, tufted (manufacture)		17520	Hammocks (manufacture)
	17511	Pile carpets of man-made fibres, woven (manufacture)		17520	Hemp rope, cord or line (manufacture)
	17512	Pile carpets of wool, tufted (manufacture)		17520	Knotted netting of twine, cordage or rope (manufacture)
	17511	Pile carpets of wool, woven (manufacture)		17520	Line yarn made of hard fibre (manufacture)
	17511	Pile rugs and mats of cotton, woven (manufacture)		17520	Loading slings (manufacture)
	17511	Pile rugs and mats of man-made fibres, woven (manufacture)		17520	Nets for horticulture (manufacture)
	17511	Pile rugs and mats of wool, woven (manufacture)		17520	Netting products (manufacture)
	17512	Pile rugs, mats and tiles of cotton, tufted (manufacture)		17520	Plastic coated twine, cordage rope and cables of textile fibres (manufacture)
				17520	Reaper twine (manufacture)

UK SIC 2007	UK SIC 2003	Activity	SIC 2007	SIC 2003	Activity
	17520	Repair of nets and ropework (manufacture)		17549	Fabrics coated with gum or amylaceous substances (manufacture)
	17520	Rope and cables of textile fibres (manufacture)		17549	Fabrics impregnated, coated, covered or laminated with plastics (manufacture)
	17520	Rope made of cotton (manufacture)			
	17520	Rope or cable fitted with metal rings (manufacture)		17549	Filter cloth weaving (manufacture)
	17520	Rope products (manufacture)		17542	Frilling (manufacture)
	17520	Rope slings (manufacture)		17542	Fringe (textile material) (manufacture)
	17520	Rope walk (manufacture)		17542	Galloon ribbon (manufacture)
	17520	Rope, cord and line made of sisal (manufacture)		17549	Gas mantles and tubular gas mantle fabric (manufacture)
	17520	Rope, cord or line made of jute (manufacture)		17542	Gimp (manufacture)
	17520	Rope, cord or line made of manila (manufacture)		17542	Haberdashery (narrow fabrics) (manufacture)
	17520	Rope, cord or line made of man-made fibre (manufacture)		17542	Hat bands (manufacture)
				17542	Hosepiping made of textiles (manufacture)
	17520	Rubber coated twine, cordage rope and cables of textile fibres (manufacture)		17549	Incandescent mantle (manufacture)
				17542	Labels (manufacture)
	17520	Sash line (manufacture)		17542	Labels made of textiles (manufacture)
	17520	Sheep net (manufacture)		17542	Labels made of woven fabric (manufacture)
	17520	Ship's fenders (manufacture)		17542	Ladder tape (textile material) (manufacture)
	17520	Sports nets (manufacture)		17549	Leathercloth (manufacture)
	17520	String (manufacture)		17549	Leathercloth made of polyvinyl chloride (manufacture)
	17520	String bag (manufacture)			
	17520	Textile material cordage (manufacture)		17549	Linen buckram weaving (manufacture)
	17520	Tow yarn made of hard fibres (manufacture)		17542	Machinery belting (woven) (manufacture)
	17520	Towing rope (manufacture)		17549	Metallised yarn (manufacture)
	17520	Twine (manufacture)		17542	Narrow fabric (not elastic or elastomeric) (manufacture)
	17520	Unloading cushions (manufacture)			
	17520	Window cord (manufacture)		17542	Narrow woven fabrics, including weftless fabrics assembled by means of an adhesive (manufacture)
13950		**Manufacture of non-wovens and articles made from non-wovens, except apparel**		17549	Nitro-cellulose coated textile fabric (manufacture)
	17530	Articles made from non-wovens (manufacture)		17542	Ornamental braids (manufacture)
	17530	Bonded fibre fabric (manufacture)		17542	Ornamental trimmings (manufacture)
	17530	Cleaning cloth (non-woven) (manufacture)		17542	Petersham ribbon (manufacture)
	17530	Non-woven bonded fibre fabrics (manufacture)		17549	Polyurethane coated textile fabrics (manufacture)
	17530	Non-wovens (manufacture)		17542	Pompons (manufacture)
	17530	Non-wovens and articles made from non-wovens (manufacture)		17542	Printed labels of textile materials (manufacture)
				17542	Ribbon made of textile (manufacture)
13960		**Manufacture of other technical and industrial textiles**		17542	Rubber thread and cord, textile covered, coated or sheathed with rubber or plastics (manufacture)
	17549	Artists' canvases (manufacture)		17542	Rubber thread or cord covered with textile material (manufacture)
	17549	Artists' tracing cloth (manufacture)			
	17542	Automotive trimmings (manufacture)		17542	Shoe trimmings made of textile materials (manufacture)
	17542	Badges (textile) (manufacture)			
	17549	Bolting cloth (manufacture)		17549	Stiffened textile fabrics (manufacture)
	17542	Braid made of elastic (manufacture)		17549	Straining cloth (manufacture)
	17542	Braid made of elastomeric (manufacture)		17542	Tapes made of non-elastic and non-elastomeric textile materials (manufacture)
	17542	Braid made of non-elastic (manufacture)			
	17542	Braid made of textile material (manufacture)		17542	Tassels made of textile material (manufacture)
	17549	Buckram (manufacture)		17542	Textile wicks (manufacture)
	17549	Buckram and similar stiffened textile fabrics (manufacture)		17549	Textile yarn impregnated, coated or sheathed with rubber or plastic (manufacture)
	17549	Canvas prepared for use by painters (manufacture)		17549	Tracing cloth (woven) (manufacture)
	17542	Carding of trimmings (manufacture)		17549	Tracing cloth weaving (manufacture)
	17542	Coach trimming (manufacture)		17542	Transmission or conveyor belts or belting, made of textiles (manufacture)
	17542	Coffin frilling (manufacture)			
	17542	Conveyor belting (woven) (manufacture)		17542	Twist cord (fabric) (manufacture)
	17542	Elastic fabric (not more than 30 cm wide) (manufacture)		17549	Tyre cord fabric made of high-tenacity man-made yarn (manufacture)
				17549	Tyre cord fabric of high-tenacity man-made yarn (manufacture)
	17542	Elastomeric fabric (not more than 30 cm wide) (manufacture)			
				17549	Tyre fabric woven from yarn spun on the cotton system (manufacture)

UK SIC 2007	UK SIC 2003	Activity	SIC 2007	SIC 2003	Activity
	17542	Umbrella trimmings made of textile material (manufacture)		17549	Pressure sensitive cloth-tape (manufacture)
	17542	Upholsterers' trimmings (textile material) (manufacture)		17542	Pyjama cord (manufacture)
	17542	Webbing made of non-elastic and non-elastomeric (manufacture)		17541	Raschel lace (manufacture)
				17541	Schiffli embroidery (manufacture)
	17542	Webbing weaving (manufacture)		17542	Shoe laces (braided) (manufacture)
	17542	Wicks for lamps, stoves or candles (manufacture)		17542	Small ware made of textiles (manufacture)
	17542	Woven conveyor belting (manufacture)		36639	Sponge dressing (manufacture)
	17542	Woven machinery belting (manufacture)		17541	Swiss embroidery (manufacture)
	17542	Woven trimmings (manufacture)		17549	Tufted fabrics (other than household textiles) (manufacture)
13990		**Manufacture of other textiles n.e.c.**		17541	Tulles and other net fabrics (manufacture)
	17549	Baize (manufacture)		17541	Tulles in motifs (manufacture)
	17542	Banding (woven) (manufacture)		17541	Tulles in one piece (manufacture)
	17542	Bias binding (manufacture)		17541	Tulles in strips (manufacture)
	17549	Billiard table cloth (manufacture)		17541	Veiling (not silk) (manufacture)
	17542	Binding (woven) (manufacture)			
	17542	Boot lace (manufacture)	**14110**		**Manufacture of leather clothes**
	36639	Button carding (manufacture)		18100	Imitation leather clothes for men and boys (manufacture)
	36639	Button covering (manufacture)		18100	Imitation leather clothes for women and girls (manufacture)
	17542	Cord made of elastic (manufacture)		18100	Industrial leather welders aprons (manufacture)
	17542	Cord made of elastomeric material (manufacture)		18100	Leather clothes (manufacture)
	17542	Corset lace (manufacture)		18100	Leather garments for men and boys (manufacture)
	17542	Curtain loop (manufacture)		18100	Leather garments for women and girls (manufacture)
	17542	Dress binding (manufacture)		18100	Leather industrial work accessories (manufacture)
	17542	Dressing gown cord and girdle (manufacture)		18100	Leather wearing apparel (manufacture)
	17541	Embroidery lace (manufacture)			
	17549	Felt (manufacture)	**14120**		**Manufacture of workwear**
	17541	Furnishing lace (manufacture)		18210	Aprons for domestic use (manufacture)
	17541	Lace (manufacture)		18210	Aprons for industrial use (manufacture)
	17541	Lace and embroidery in the piece, in strips or in motifs (manufacture)		18210	Battledress for men (manufacture)
	17541	Lace bleaching (not on commission) (manufacture)		18210	Battledress for women (manufacture)
	17541	Lace clipping (manufacture)		18210	Boiler suit (manufacture)
	17541	Lace drawing (manufacture)		18210	Cassock (manufacture)
	17541	Lace dressing (manufacture)		18210	Chefs' clothing (manufacture)
	17541	Lace dyeing (not on commission) (manufacture)		18210	Clerical vestment (manufacture)
	17541	Lace edging (manufacture)		18210	Dungarees (manufacture)
	17541	Lace embroidery (manufacture)		18210	Gowns for academic, legal or ecclesiastical use (manufacture)
	17541	Lace ending (manufacture)		18210	Industrial clothing (manufacture)
	17541	Lace finishing (manufacture)		18210	Livery (manufacture)
	17541	Lace flouncing (manufacture)		18210	Military clothing (manufacture)
	17541	Lace mending (manufacture)		18210	Nuns' clothing (manufacture)
	17541	Lace net (manufacture)		18210	Nurses' uniforms for men (manufacture)
	17541	Lace scalloping (manufacture)		18210	Nurses' uniforms for women (manufacture)
	17541	Lace trimming (manufacture)		18210	Overalls for boys (manufacture)
	17542	Laces for boots and shoes (manufacture)		18210	Overalls for domestic use (manufacture)
	17541	Leavers lace (manufacture)		18210	Overalls for girls (manufacture)
	17541	Napery lace (manufacture)		18210	Overalls for men and boys (manufacture)
	17549	Needlefelt (other than carpet underlay) (manufacture)		18210	Overalls for women and girls (manufacture)
	17549	Needleloom felt (other than carpet underlay) (manufacture)		18210	Protective clothing for industrial use (manufacture)
	17541	Net fabrics in the piece, in strips or in motifs (manufacture)		18210	Robes for academic, legal and ecclesiastical use (manufacture)
	17541	Nottingham lace (manufacture)		18210	Shirts for industrial use (manufacture)
	36639	Powder puff (manufacture)		18210	Surplice (manufacture)
	17549	Pressed felt (not paper or roofing) (manufacture)		18210	Trousers for industrial use (manufacture)
	17549	Pressed wool felt (manufacture)		18210	Uniforms for men and boys (manufacture)
				18210	Uniforms for women and girls (manufacture)
				18210	Vestments for clerical use (manufacture)

UK SIC 2007	UK SIC 2003	Activity	SIC 2007	SIC 2003	Activity
	18210	Weather protective industrial clothing (manufacture)		18222	Dressmaking (manufacture)
	18210	Workwear (manufacture)		18222	Ensembles (manufacture)
	18210	Workwear clothing for women (manufacture)		18222	Fashion (manufacture)
				18222	Gowns (manufacture)
14131		**Manufacture of men's outerwear, other than leather clothes and workwear**		18222	Housecoats (manufacture)
	18221	Anoraks for men and boys (manufacture)		18222	Jackets for women and girls (manufacture)
	18221	Blazers for men and boys (manufacture)		18222	Jeans for women and girls (manufacture)
	18221	Breeches (manufacture)		18222	Knitted dress and jacket ensemble (manufacture)
	18221	Climbing clothing for men and boys (weatherproof) (manufacture)		18222	Knitted dresses for women and girls (manufacture)
	18221	Cloaks for men and boys (manufacture)		18222	Knitted outerwear for women and girls (manufacture)
	18221	Coats for men and boys (manufacture)		18222	Knitted skirts (manufacture)
	18221	Custom tailored outerwear for men and boys (manufacture)		18222	Knitted suits for women and girls (manufacture)
	18221	Custom tailoring for men and boys (manufacture)		18222	Light outerwear for women and girls (manufacture)
	18221	Jackets for men and boys (manufacture)		18222	Lightweight jackets for women and girls (manufacture)
	18221	Jeans for men and boys (manufacture)		18222	Mackintoshes for women and girls (manufacture)
	18221	Knitted outerwear for men and boys (manufacture)		18222	Oilskins for women and girls (manufacture)
	18221	Mackintoshes for men and boys (manufacture)		18222	Outerwear for women and girls made of stitched plastic (manufacture)
	18221	Oilskins for men and boys (manufacture)		18222	Overcoats for women and girls (manufacture)
	18221	Outerwear for men and boys (manufacture)		18222	Parts of coats for women and girls (manufacture)
	18221	Outerwear for men and boys made of stitched plastic (manufacture)		18222	Parts of ensembles for women and girls (manufacture)
	18221	Overcoats for men and boys (manufacture)		18222	Parts of jackets for women and girls (manufacture)
	18221	Parts of coats for men and boys (manufacture)		18222	Parts of skirts for women and girls (manufacture)
	18221	Parts of jackets for men and boys (manufacture)		18222	Parts of suits for women and girls (manufacture)
	18221	Parts of suits for men and boys (manufacture)		18222	Parts of trousers for women and girls (manufacture)
	18221	Parts of trousers for men and boys (manufacture)		18222	Raincoats for women and girls (manufacture)
	18221	Raincoats for men and boys (manufacture)		18222	Raincoats for women and girls made of stitched plastic (manufacture)
	18221	Rainproof garments for men and boys (manufacture)		18222	Rainproof garments for women and girls (manufacture)
	18221	Recreational clothing for men and boys (waterproofed) (manufacture)		18222	Recreational clothing for women and girls (waterproof) (manufacture)
	18221	Recreational weatherproof clothing for men and boys (manufacture)		18222	Recreational weatherproof clothing for women and girls (manufacture)
	18221	Sailing clothing for men and boys (weatherproof) (manufacture)		18222	Sailing clothing for women and girls (weatherproof) (manufacture)
	18221	Shorts for men and boys (manufacture)		18222	Skirts for women and girls (dressmade) (manufacture)
	18221	Sportswear for men and boys (manufacture)		18222	Slacks for women and girls (manufacture)
	18221	Suits for men and boys (manufacture)		18222	Sportswear for women and girls (manufacture)
	18221	Tailored outerwear for men and boys (manufacture)		18222	Suits for women and girls (dressmade) (manufacture)
	18221	Trousers for men and boys (manufacture)		18222	Tailored outerwear for women and girls (manufacture)
	18221	Waistcoats for men and boys (manufacture)		18222	Tailored skirts (manufacture)
	18221	Weatherproof jackets for men and boys (manufacture)		18222	Trouser suits for women and girls (manufacture)
	18221	Weatherproof outerwear for men and boys (manufacture)		18222	Trousers for women and girls (manufacture)
				18222	Weatherproof jackets for women and girls (manufacture)
14132		**Manufacture of women's outerwear, other than leather clothes and workwear**		18222	Weatherproof outerwear for women and girls (manufacture)
	18222	Anoraks for women and girls (manufacture)			
	18222	Blazers for women and girls (manufacture)	**14141**		**Manufacture of men's underwear**
	18222	Climbing clothing for women and girls (weatherproof) (manufacture)		18231	Briefs for men and boys (manufacture)
	18222	Cloaks for women and girls (manufacture)		18231	Dressing gowns for men and boys (manufacture)
	18222	Coats for women and girls (manufacture)		18231	Knitted nightwear for men and boys (manufacture)
	18222	Costumes for women and girls (manufacture)		18231	Knitted underwear for men and boys (manufacture)
	18222	Custom tailoring for women and girls (manufacture)		18231	Knitted vests for men and boys (manufacture)
	18222	Divided lightweight skirt (dress made) (manufacture)		18231	Nightwear for men and boys (manufacture)
	18222	Dress and jacket knitted ensemble (manufacture)		18231	Pyjamas for men and boys (manufacture)
	18222	Dresses for women and girls (manufacture)			

UK SIC 2007	UK SIC 2003	Activity	SIC 2007	SIC 2003	Activity
	18231	Shirts for men and boys (manufacture)		18249	Dress gloves made of fabric (manufacture)
	18231	T-shirts for men and boys (manufacture)		18249	Dress shield (manufacture)
	18231	Underwear for men and boys (manufacture)		18241	Felt hat bleaching and dying (manufacture)
	18231	Vests for men and boys (manufacture)		18241	Felt hat body making (manufacture)
				18241	Felt hat finishing (manufacture)
14142		**Manufacture of women's underwear**		18249	Footwear made of textile fabric with applied soles (manufacture)
	18232	Blouses for women and girls (manufacture)			
	18232	Brassiere (manufacture)		17710	Footwear made of textiles without applied soles (manufacture)
	18232	Briefs for women and girls (manufacture)			
	18232	Corselet (manufacture)		18249	Garter (manufacture)
	18232	Corset (manufacture)		18249	Gloves (other than knitted) (manufacture)
	18232	Corset belt (manufacture)		18249	Gloves for children (manufacture)
	18232	Dressing gown for women and girls (manufacture)		18249	Gloves made of cloth (manufacture)
	18232	Foundation garment (manufacture)		18249	Gloves made of fur (manufacture)
	18232	Knickers (manufacture)		18249	Gloves made of leather (not sports) (manufacture)
	18232	Knitted nightwear for women and girls (manufacture)		18249	Gloves made of textiles for household use (manufacture)
	18232	Knitted underwear for women and girls (manufacture)			
				18249	Hair nets (manufacture)
	18232	Knitted vests for women and girls (manufacture)		18249	Hair nets made of lace (manufacture)
	18232	Lingerie (manufacture)		18249	Handkerchief folding (manufacture)
	18232	Night-dresses for women and girls (manufacture)		18249	Handkerchief hemming (manufacture)
	18232	Nightwear for women and girls (manufacture)		18249	Handkerchief made of textile material (manufacture)
	18232	Panty (manufacture)		18241	Hat lining (manufacture)
	18232	Petticoat (manufacture)		18241	Hat pad (manufacture)
	18232	Pyjamas for women and girls (manufacture)		18241	Hat shape (manufacture)
	18232	Shirts for women and girls (manufacture)		18241	Hats made of cloth (manufacture)
	18232	Slips (manufacture)		18241	Hats made of felt (manufacture)
	18232	Suspender (manufacture)		18241	Hats made of fur (manufacture)
	18232	Suspender belt (manufacture)		18241	Hats made of fur fabric (manufacture)
	18232	Underskirt (manufacture)		18241	Hats made of silk (manufacture)
	18232	Underwear for women and girls (manufacture)		18241	Hats made of wool (manufacture)
	18232	Vests for women and girls (manufacture)		18241	Headgear made of furskins (manufacture)
				18249	Headsquare (manufacture)
14190		**Manufacture of other wearing apparel and accessories**		18241	Knitted bonnet (manufacture)
				18249	Knitted bootees (manufacture)
	18249	Athletic clothing (manufacture)		18249	Knitted gloves (manufacture)
	18249	Babies garments (manufacture)		18249	Knitted mittens and mitts (manufacture)
	18249	Baby clothing (manufacture)		18249	Knitted scarf (manufacture)
	18249	Baby linen (manufacture)		18249	Knitted shawl (manufacture)
	18241	Balaclava helmet (manufacture)		18249	Knitted swimwear (manufacture)
	18249	Beachwear for women and girls (manufacture)		18249	Knitted swimwear for infants (manufacture)
	18241	Beret (knitted) (manufacture)		18249	Knitted ties (manufacture)
	18241	Beret (not knitted) (manufacture)		18249	Knitted underclothing for infants (manufacture)
	18249	Braces (not made of leather or leather substitute) (manufacture)		18249	Ladies fan (manufacture)
				18241	Linings for hats (manufacture)
	18249	Braces made of leather (manufacture)		18241	Millinery made of felt (manufacture)
	18241	Buckram shape (manufacture)		18249	Necktie (manufacture)
	18241	Capeline felt (manufacture)		18249	Neckwear for women (manufacture)
	18241	Caps made of cloth (manufacture)		18249	Nightwear for infants (manufacture)
	18249	Clothing accessories (manufacture)		18249	Outerwear for infants (manufacture)
	18249	Clothing pad (manufacture)		18241	Peak cap (manufacture)
	18249	Collars for men and boys (manufacture)		18249	Scarf (lace or knitted) (manufacture)
	18249	Cravats (manufacture)		18249	Scarf (not lace or knitted) (manufacture)
	18249	Cuffs for men and boys (manufacture)		18249	Shawls (manufacture)
	18243	Cut, make and trim on a fee or contract basis (manufacture)		18249	Shirt front (manufacture)
				18249	Shirt neckband (manufacture)
	18249	Disposable clothing (manufacture)		18249	Ski suits (manufacture)
	18249	Dress belts (not made of leather or leather substitute) (manufacture)		18249	Sporran (manufacture)
				18241	Straw hat blocking (manufacture)

UK SIC 2007	UK SIC 2003	Activity	SIC 2007	SIC 2003	Activity
	18241	Straw hats (manufacture)		17720	Jerseys (knitted) (manufacture)
	18249	Swimwear (manufacture)		17720	Jumpers (knitted) (manufacture)
	18249	Tailors' pad (manufacture)		17720	Knitwear (manufacture)
	18249	Theatrical costumes (manufacture)		17720	Pullovers (knitted or crocheted) (manufacture)
	18249	Ties made of silk (manufacture)		17720	Twin sets (knitted) (manufacture)
	18249	Tracksuits (manufacture)		17720	Waistcoats and other similar articles (knitted) (manufacture)
	18241	Tropical helmet (manufacture)			
	18249	Underclothing for infants (manufacture)			
	18241	Uniform hats and caps (manufacture)	**15110**		**Tanning and dressing of leather; dressing and dyeing of fur**
	18249	Weatherproof outerwear for infants (manufacture)		19100	Box and willow calf leather (manufacture)
	18249	Weatherproof skiing clothing (manufacture)		19100	Buckskin (manufacture)
	18241	Wool felt and fur felt hood and capeline (manufacture)		19100	Cattle hide leather (manufacture)
				19100	Chamois leather (manufacture)
14200		**Manufacture of articles of fur**		19100	Chrome tanning (manufacture)
	18300	Apparel made of fur (manufacture)		19100	Combing leather (manufacture)
	18300	Artificial fur and articles thereof (manufacture)		19100	Composition leather (manufacture)
	18300	Capes made of fur (manufacture)		18300	Currying of fur skins and hides with the hair on (manufacture)
	18300	Clothing made of sheepskin (manufacture)		18300	Dressing and dying of furskins and hides with the hair on (manufacture)
	18300	Cravats made of fur (manufacture)		19100	Dressing of leather (manufacture)
	18300	Fur skin assemblies including dropped fur skins, plates, mats and strips (manufacture)		18300	Dyed lamb including beaver lamb (manufacture)
	18300	Furrier (manufacture)		19100	Footwear leather preparation (manufacture)
	18300	Furskin articles (manufacture)		18300	Fur dressing (manufacture)
	18300	Furskin assemblies (manufacture)		18300	Fur dressing and dyeing (manufacture)
	18300	Furskin pouffes, unstuffed (manufacture)		19100	Gill leather (manufacture)
	18300	Garments made of rabbit fur (manufacture)		19100	Glace kid (manufacture)
	18300	Hatters' fur (manufacture)		19100	Glove leather preparation (manufacture)
	18300	Industrial polishing cloths made of fur (manufacture)		19100	Goldbeaters' skin or bung (manufacture)
	18300	Lambskin clothing (manufacture)		19100	Harness and saddlery leather preparation (manufacture)
	18300	Manufacturing furrier (manufacture)		19100	Hat and cap leather preparation (manufacture)
	18300	Mats and rugs made of fur (manufacture)		19100	Hydraulic leather (manufacture)
	18300	Moleskin finishing (manufacture)		19100	Leather dressing (manufacture)
	18300	Muffs made of fur (manufacture)		19100	Leather dying (manufacture)
	18300	Rugs made of sheepskin (manufacture)		19100	Leather enamelling (manufacture)
	18300	Rugs made of skins (manufacture)		19100	Leather gilding (manufacture)
	18300	Stoles made of fur (manufacture)		19100	Leather proofing (manufacture)
	18300	Trimmings made of fur (manufacture)		19100	Leather tanning and dressing (manufacture)
	18300	Wearing apparel and clothing accessories made of fur (manufacture)		19100	Leather trimmings (manufacture)
				19100	Mechanical leather preparation (manufacture)
14310		**Manufacture of knitted and crocheted hosiery**		19100	Metallised leathers (manufacture)
	17710	Bedsock (manufacture)		19100	Parchment made of leather (manufacture)
	17710	Fancy hosiery (manufacture)		19100	Patent leather (manufacture)
	17710	Hosiery (knitted and crocheted) (manufacture)		19100	Reptile leather (manufacture)
	17710	Hosiery blank (manufacture)		18300	Scraping of fur skins and hides with the hair on (manufacture)
	17710	Pantyhose (manufacture)		18300	Shearing and plucking of fur skins and hides with the hair on (manufacture)
	17710	Socks and stockings (knitted) (manufacture)		19100	Sheep skin preparation (manufacture)
	17710	Socks for children (manufacture)		19100	Sole leather preparation (manufacture)
	17710	Socks for men (manufacture)		19100	Tanning leather (manufacture)
	17710	Socks for women and girls (manufacture)		18300	Tanning of fur skins and hides with the hair on (manufacture)
	17710	Stockinette goods (manufacture)		19100	Tanning of hairless skins and hides (manufacture)
	17710	Stockings (manufacture)		19100	Trimmings made of leather (manufacture)
	17710	Stockings for women and girls (manufacture)		19100	Upholstery leather preparation (manufacture)
	17710	Tights (manufacture)		19100	Upper leather (manufacture)
				19100	Vellum (manufacture)
14390		**Manufacture of other knitted and crocheted apparel**			
	17720	Cardigans (knitted) (manufacture)			
	17720	Crocheted articles (manufacture)			

UK SIC 2007	UK SIC 2003	Activity	SIC 2007	SIC 2003	Activity
15120		**Manufacture of luggage, handbags and the like, saddlery and harness**		19200	Luggage, handbags made of plastic sheeting (manufacture)
	19200	Army accoutrement made of leather (manufacture)		19200	Luggage, handbags made of textile materials (manufacture)
	19200	Art leather work (manufacture)		19200	Machinery accessories made of leather (manufacture)
	19200	Attaché case made of leather or leather substitute (manufacture)		19200	Machinery belting made of leather (manufacture)
	19200	Bags made of leather or leather substitute (manufacture)		19200	Manicure case made of leather (manufacture)
	19200	Belts made of leather or leather substitute (manufacture)		19200	Novelty goods made of leather (manufacture)
	19200	Billfolds made of leather (manufacture)		19200	Picking bands made of leather (manufacture)
	19200	Bridle cutting (manufacture)		19200	Pipe case (not leather or plastic) (manufacture)
	19200	Brown saddlery (manufacture)		19200	Pochette made of leather (manufacture)
	19200	Brush case (not of leather or plastics) (manufacture)		19200	Pocket book made of leather (manufacture)
	19200	Buff and mop made of leather (manufacture)		19200	Pouch for tobacco (manufacture)
	19200	Buffalo pickers made of leather (manufacture)		19200	Pouch made of leather or leather substitute (manufacture)
	19200	Cases for cutlery (not wooden) (manufacture)		19200	Pump leather (manufacture)
	19200	Cases for jewellery (not wooden) (manufacture)		19200	Purses made of leather or leather substitute (manufacture)
	19200	Cases for musical instruments (not wooden) (manufacture)		19200	Radio cases made of leather (manufacture)
	19200	Cases made of leather for cutlery, instruments, etc. (manufacture)		36639	Riding crops (manufacture)
	19200	Check strap made of leather (manufacture)		19200	Roller skin (cut) made of leather (manufacture)
	19200	Container for typewriter, radio, etc. Made of leather (manufacture)		19200	Saddle horse (manufacture)
	19200	Conveyor bands made of leather (manufacture)		19200	Saddlery (manufacture)
	19200	Cutlery case (not leather or plastics) (manufacture)		19200	Saddlery and harness made of leather (manufacture)
	19200	Cycle bags made of leather (manufacture)		19200	Sample case made of leather (manufacture)
	19200	Dog lead made of leather (manufacture)		19200	Satchels made of leather (manufacture)
	19200	Dressing case made of leather or leather substitute (manufacture)		19200	Shopping bags made of plastic (designed for prolonged use) (manufacture)
	19200	Driving belts made of leather (manufacture)		19200	Shopping bags made of stitched plastic (manufacture)
	19200	Elevator bands made of leather (manufacture)		19200	Skips made of leather (manufacture)
	19200	Fancy leather goods (manufacture)		19200	Spectacle cases made of leather (manufacture)
	19200	Gas meter diaphragm made of leather (manufacture)		19200	Sports bags made of fabric (manufacture)
	19200	Gun cases made of leather (manufacture)		19200	Sports bags made of leather (manufacture)
	19200	Hand luggage made of leather (manufacture)		19200	Sports goods carrier (manufacture)
	19200	Handbags and the like of composition leather (manufacture)		19200	Straps made of leather (manufacture)
	19200	Handbags made of leather or leather substitutes (manufacture)		19200	Strops made of leather (manufacture)
	19200	Harness (manufacture)		19200	Suitcases made of leather or leather substitute (manufacture)
	19200	Harness front and rosette made of leather (manufacture)		19200	Toilet case (fitted) made of leather (manufacture)
	19200	Hat box made of leather or leather substitute (manufacture)		19200	Toolbags made of leather (manufacture)
	19200	Horse collars made of leather (manufacture)		19200	Travel goods made of leather or leather substitute (manufacture)
	36639	Horse whips (manufacture)		19200	Trunk handles made of leather (manufacture)
	19200	Industrial leather (manufacture)		19200	Trunks made of leather (manufacture)
	19200	Jewel case (not wood or metal) (manufacture)		19200	Typewriter case made of leather (manufacture)
	19200	Key tags and cases made of leather (manufacture)		19200	Violin case (not wooden) (manufacture)
	19200	Ladies handbags made of leather (manufacture)		19200	Wallets made of leather or leather substitute (manufacture)
	19200	Leather articles for use in machinery or mechanical appliances (manufacture)		19200	Washers made of leather (manufacture)
	19200	Leather belting for use in machinery (manufacture)		19200	Watch bands of non-metallic material such as fabric, leather or plastic (manufacture)
	19200	Leather goods (not industrial) (manufacture)		19200	Watch straps (non-metallic) (manufacture)
	19200	Leather shoe-laces (manufacture)		19200	Watch straps made of leather or leather substitute (manufacture)
	19200	Luggage and the like of leather (manufacture)		36639	Whips (manufacture)
	19200	Luggage handbags made of paperboard (manufacture)	**15200**		**Manufacture of footwear**
	19200	Luggage made of leather or leather substitute (manufacture)		19300	Athletic footwear (manufacture)
				19300	Ballet shoe (manufacture)

UK SIC 2007	UK SIC 2003	Activity	SIC 2007	SIC 2003	Activity
	19300	Beach footwear (manufacture)		20100	V-jointed wood (manufacture)
	19300	Boot (manufacture)		20100	Wagon timber (sawn) (manufacture)
	19300	Boot closing (manufacture)		20100	Weatherboard (manufacture)
	19300	Boot stiffener (manufacture)		20100	Wood chip (manufacture)
	19300	Boot upper (manufacture)		20100	Wood creosoting, impregnation, preservation, spraying, varnishing and drying (manufacture)
	19300	Clog (manufacture)		20100	Wood drying (manufacture)
	19300	Cut soles for footwear (manufacture)		20100	Wood flour (manufacture)
	19300	Footwear (manufacture)		20100	Wood grooving, milling, planing, sawing etc. (manufacture)
	19300	Gaiters made of leather (manufacture)		20100	Wood impregnation (manufacture)
	19300	Galoshes made of rubber (manufacture)		20100	Wood machining (manufacture)
	19300	Heels made of leather (manufacture)		20100	Wood mill (manufacture)
	19300	House shoe (manufacture)		20100	Wood particles (manufacture)
	19300	Insoles made of leather (manufacture)		20100	Wood planing (manufacture)
	19300	Leather fillings (manufacture)		20100	Wood preservation (manufacture)
	19300	Leggings and gaiters made of cloth (manufacture)		20100	Wood sawing (manufacture)
	19300	Leggings made of leather (manufacture)		20100	Wood shavings (manufacture)
	19300	Overboot (manufacture)		20100	Wood spraying (manufacture)
	19300	Overshoes made of rubber (manufacture)		20100	Wood varnishing (manufacture)
	19300	Protective footwear made of plastic (manufacture)		20100	Wood wool (manufacture)
	19300	Safety boots (manufacture)			
	19300	Sandals (manufacture)	**16210**		**Manufacture of veneer sheets and wood-based panels**
	19300	Shoes (manufacture)		20200	Battenboard (manufacture)
	19300	Shoes made of wood (manufacture)		20200	Blockboard (manufacture)
	19300	Slipper soles (manufacture)		20200	Building boards made of fibre (manufacture)
	19300	Slippers (manufacture)		20200	Building boards made of wood waste (manufacture)
	19300	Socks made of leather (manufacture)		20200	Cellular wood panel (manufacture)
	19300	Spats made of leather (manufacture)		20200	Chipboard agglomerated with non-mineral binding substances (manufacture)
	19300	Sports footwear (manufacture)		20200	Densified wood (manufacture)
	19300	Wellington boot (manufacture)		20200	Fibre board (manufacture)
				20200	Fibre building board (manufacture)
16100		**Sawmilling and planing of wood**		20200	Glue laminated wood (manufacture)
	20100	Bargeboard (manufacture)		20200	Hardboard (manufacture)
	20100	Beaded wood (manufacture)		20200	Improved wood (manufacture)
	20100	Bent timber (manufacture)		20200	Laminated veneer wood (manufacture)
	20100	Boxboard (manufacture)		20200	Laminated wood products (manufacture)
	20100	Chamfered wood (manufacture)		20200	Laminboard (manufacture)
	20100	Drying of timber (manufacture)		20200	Medium density fibreboard (MDF) and other fibreboard (manufacture)
	20100	Flooring made of wood (not parquet flooring) (manufacture)		20200	Oriented strand board (OSB) and other particle board (manufacture)
	20100	Flooring made of wood (unassembled) (manufacture)		20200	Particleboard agglomerated with non-mineral binding substances (manufacture)
	20100	Grooved wood (manufacture)		20200	Plywood (manufacture)
	20100	Immersion treatment of wood (manufacture)		20200	Veneer (manufacture)
	20100	Log sawing (manufacture)		20200	Veneer sheet (manufacture)
	20100	Log slicing, peeling or chipping (manufacture)		20200	Wood chipboard agglomerated with non-mineral binding substances (manufacture)
	20100	Mining timber (sawn) (manufacture)		20200	Wood veneers (manufacture)
	20100	Plank (manufacture)			
	20100	Pressure treatment of wood (manufacture)	**16220**		**Manufacture of assembled parquet floors**
	20100	Railway sleepers made of wood (manufacture)		20300	Hardwood flooring strip (manufacture)
	20100	Rebated wood (manufacture)		20300	Parquet flooring (manufacture)
	20100	Sawmilling (manufacture)		20300	Strip flooring made of hardwood (manufacture)
	20100	Sawn fencing (manufacture)		20300	Wooden parquet floor blocks assembled into panels (manufacture)
	20100	Skirting boards made of unmoulded wood (manufacture)		20300	Wooden parquet floor strips assembled into panels (manufacture)
	20100	Sleepers made of wood (manufacture)			
	02010	Split poles, pickets and other products (manufacture)			
	20100	Telegraph poles (manufacture)			
	20100	Tongued wood (manufacture)			
	20100	Vacuum treatment of wood (manufacture)			
	20100	Veneer log sawing (manufacture)			

UK SIC 2007	UK SIC 2003	Activity
16230		**Manufacture of other builders' carpentry and joinery**
	20300	Bannister rails made of wood (manufacture)
	20300	Beading made of wood (manufacture)
	20300	Beadings and mouldings made of wood (manufacture)
	20300	Beams (manufacture)
	20300	Blinds made of wood (excluding shop blinds) (manufacture)
	20300	Bridges made of wood (manufacture)
	20300	Builders' carpentry and joinery made of wood (manufacture)
	20300	Builders' woodwork such as window frames etc. (manufacture)
	20300	Display stand (manufacture)
	20300	Door frames made of wood (manufacture)
	20300	Doors made of wood (manufacture)
	20300	Exhibition stand (manufacture)
	20300	Fencing made of wood (assembled) (manufacture)
	20300	Garage made of wood (manufacture)
	20300	Garden frames made of wood (manufacture)
	20300	Gate made of wood (manufacture)
	20300	Glue-laminated and metal connected prefabricated wooden roof trusses (manufacture)
	20300	Greenhouse made of wood (manufacture)
	20300	Huts made of wood (manufacture)
	20300	Industrialised building component made of timber (manufacture)
	20300	Moulded skirting board made of wood (manufacture)
	20300	Mouldings made of wood (manufacture)
	20300	Paving blocks made of wood (manufacture)
	20300	Portable wooden buildings (manufacture)
	20300	Poultry house made of wood (manufacture)
	20300	Prefabricated buildings made of wood (manufacture)
	20300	Prefabricated buildings or elements thereof made of wood (manufacture)
	20300	Prefabricated roof timbers (manufacture)
	20300	Rafters (manufacture)
	20300	Railings made of wood (manufacture)
	20300	Roof struts (manufacture)
	20300	Saunas made from wood (manufacture)
	20300	Sheds made of wood (manufacture)
	20300	Shingles and shakes (manufacture)
	20300	Shop fronts made of wood (manufacture)
	20300	Shuttering made of wood (manufacture)
	20300	Shutters made of wood (manufacture)
	20300	Silos made of wood (manufacture)
	20300	Stair rods made of wood (manufacture)
	20300	Staircase made of wood (manufacture)
	20300	Tower made of wood (manufacture)
	20300	Trellis work made of wood (manufacture)
	20300	Truss rafter (manufacture)
	20300	Window frames made of wood (manufacture)
	20300	Wood partitions (manufacture)
	20300	Wooden beams for the construction industry (manufacture)
	20300	Wooden goods intended to be used primarily in the construction industry (manufacture)

SIC 2007	SIC 2003	Activity
16240		**Manufacture of wooden containers**
	20400	Barrels made of wood (manufacture)
	20400	Box pallet (manufacture)
	20400	Boxes made of wood (manufacture)
	20400	Boxes made of wood (wirebound) (manufacture)
	20400	Buckets made of wood (manufacture)
	20400	Bungs made of wood (manufacture)
	20400	Cable drums made of wood (manufacture)
	20400	Cask heads made of wood (manufacture)
	20400	Casks made of wood (manufacture)
	20400	Chests made of wood (manufacture)
	20400	Churns made of wood (manufacture)
	20400	Cigar box made of wood (manufacture)
	20400	Cock made of wood (manufacture)
	20400	Collapsible box made of wood (manufacture)
	20400	Containers made of wood (manufacture)
	20400	Cooper's products (manufacture)
	20400	Cooper's products reconditioning (manufacture)
	20400	Cooper's wood (manufacture)
	20400	Crates made of wood (manufacture)
	20400	Drums and similar packings made of wood (manufacture)
	20400	Egg box made of wood (manufacture)
	20400	Fish boxes made of wood (manufacture)
	20400	Hoops made of wood (manufacture)
	20400	Kegs made of wood (manufacture)
	20400	Packing cases made of wood (manufacture)
	20400	Pallets, box pallets and other load boards made of wood (manufacture)
	20400	Staves made of wood (manufacture)
	20400	Stillage made of wood (manufacture)
	20400	Suitcases made of wood (manufacture)
	20400	Tea chests made of wood (manufacture)
	20400	Travelling trunks made of wood (manufacture)
	20400	Tubs made of wood (manufacture)
	20400	Tuns made of wood (manufacture)
	20400	Vats made of wood (manufacture)
	20400	Wire and cable drums made of wood (manufacture)
16290		**Manufacture of other products of wood; manufacture of articles of cork, straw and plaiting materials**
	20520	Agglomerated cork (manufacture)
	20520	Bamboo preparation (manufacture)
	20520	Baskets made of materials (other than plastic) (manufacture)
	20520	Basketware (manufacture)
	20510	Beads made of wood (manufacture)
	20510	Beehives made of wood (manufacture)
	36639	Blocks for the manufacture of smoking pipes (manufacture)
	20510	Bobbins made of wood (not textile accessory) (manufacture)
	20510	Boot or shoe lasts and trees made of wood (manufacture)
	20510	Breadboards made of wood (manufacture)
	20510	Broom handles made of wood (manufacture)
	20510	Brush back made of wood (manufacture)
	20510	Brush head made of wood (manufacture)
	20510	Brush top made of wood (manufacture)
	20510	Brush wood ware (manufacture)

UK SIC 2007	UK SIC 2003	Activity	SIC 2007	SIC 2003	Activity
	20520	Buoyancy apparatus made of cork (except cork life preservers) (manufacture)		20510	Mirror frame made of wood (manufacture)
	20520	Cane preparation (manufacture)		20510	Ornaments made of wood (manufacture)
	20520	Cane splitting and weaving (manufacture)		20520	Osier articles (manufacture)
	20520	Cane working (manufacture)		20520	Osier preparation (manufacture)
	20510	Cases made of wood, for musical instruments (manufacture)		20510	Picture frame made of wood (manufacture)
	20510	Caskets (except burial caskets) and cases made of wood (manufacture)		20520	Plaiting material preparation (manufacture)
	20510	Clothes hangers made of wood (manufacture)		20520	Plaits and products of plaiting materials (manufacture)
	20510	Clothes horse made of wood (manufacture)		20510	Pulley made of wood (manufacture)
	20510	Clothes pegs made of wood (manufacture)		20520	Punnets made of cork, straw or plaiting materials (manufacture)
	20510	Coat and hat racks made of wood (manufacture)		20520	Raffia goods (manufacture)
	20510	Coat hangers made of wood (manufacture)		20510	Rakes made of wood (manufacture)
	20520	Composition cork (manufacture)		20520	Reed articles (manufacture)
	20510	Cooking utensils made of wood (manufacture)		20520	Reed preparation (manufacture)
	20510	Cops made of wood (manufacture)		20510	Reels made of wood (manufacture)
	20520	Cork products (except cork life preservers) (manufacture)		20510	Roller blinds made of wood (manufacture)
	20510	Cutlery case made of wood (manufacture)		20510	Rolling pins made of wood
	20510	Dishes made of wood (manufacture)		20520	Rush matting (manufacture)
	20510	Domestic woodware (manufacture)		20520	Screens made of plaiting materials (manufacture)
	20510	Dowel pin (manufacture)		20510	Sewing thread reels and similar articles of turned wood (manufacture)
	20510	Drawing instruments case made of wood (not containing instruments) (manufacture)		20520	Sheets made of cork (manufacture)
	20510	Egg cup made of wood (manufacture)		19300	Shoe parts made of wood (manufacture)
	20520	Envelopes for bottles made of straw (manufacture)		20510	Shoetrees made of wood (manufacture)
	20520	Fenders made of cork (manufacture)		20510	Signs made of wood (manufacture)
	20510	Fire logs and pellets, of pressed wood, or of coffee or soybean grounds and the like (manufacture)		20520	Slabs made of cork (manufacture)
	20520	Floor coverings of natural cork (manufacture)		20510	Slats for the manufacture of pencils (manufacture)
	20510	Foundry moulding pattern made of wood (manufacture)		20510	Spills made of wood (manufacture)
	20510	Frames for artists canvases (manufacture)		20510	Spinning wheels made of wood (manufacture)
	20510	Fruit bowls made of wood (manufacture)		20510	Spools made of wood (manufacture)
	20510	Grids made of wood (manufacture)		20510	Spoons made of wood (manufacture)
	20510	Gunstock made of wood (manufacture)		20510	Statuettes made of wood (manufacture)
	20510	Handicraft articles made of wood (manufacture)		20510	Step ladders made of wood (manufacture)
	20510	Handles and bodies for brooms made of wood (manufacture)		20510	Steps made of wood (manufacture)
	20510	Handles and bodies for brushes made of wood (manufacture)		20520	Stopper insets made of cork (manufacture)
	20510	Handles and bodies for tools made of wood (manufacture)		20520	Stoppers made of cork (manufacture)
	36639	Handles for canes, umbrellas and similar (manufacture)		20520	Straw articles (manufacture)
	20510	Handles made of wood (manufacture)		20510	Tent poles made of wood (manufacture)
	20510	Household utensils made of wood (manufacture)		20520	Tiles made of cork (manufacture)
	20510	Hurdles made of wood (manufacture)		20520	Tips made of cork (manufacture)
	20510	Inlaid wood (manufacture)		20510	Tool handles made of wood (manufacture)
	20510	Instrument case made of wood (manufacture)		20510	Tools made of wood (manufacture)
	20520	Insulating materials made of cork (manufacture)		20510	Trays made of wood (manufacture)
	20510	Kitchenware made of wood (manufacture)		20510	Turned wood products (manufacture)
	20510	Ladders made of wood (manufacture)		20510	Violin cases made of wood (manufacture)
	20510	Lasts made of wood (manufacture)		36639	Walking sticks made of wood (manufacture)
	20520	Loofah articles (manufacture)		20520	Wicker baskets (manufacture)
	20510	Loom made of wood (manufacture)		20520	Wickerwork (manufacture)
	20510	Machine parts made of wood (manufacture)		20510	Wine racks made of wood (manufacture)
	20520	Mats made of cork (manufacture)		20510	Wood carving (manufacture)
	20520	Matting made of cane (manufacture)		20510	Wood marquetry (manufacture)
	20520	Matting made of rushes (manufacture)		20510	Wooden articles n.e.c. (manufacture)
				20510	Wooden cases for jewellery (manufacture)

17110　　Manufacture of pulp

SIC 2007	SIC 2003	Activity
	21110	Bleached paper pulp made by chemical dissolving (manufacture)
	21110	Bleached paper pulp made by mechanical processes (manufacture)

UK SIC 2007	UK SIC 2003	Activity	SIC 2007	SIC 2003	Activity
	21110	Bleached paper pulp made by non-dissolving processes (manufacture)		21120	Filter paper stock (manufacture)
	21110	Bleached paper pulp made by semi-chemical processes (manufacture)		21120	Flong paperboard (manufacture)
	21110	Chemical woodpulp (manufacture)		21120	Fluting paper (manufacture)
	21110	Cotton-linters pulp (manufacture)		21120	Folding boxboard (manufacture)
	21110	Dissolving chemical wood pulp (manufacture)		21120	Glassine paper (manufacture)
	21110	Mechanical woodpulp (manufacture)		21120	Greaseproof paper (manufacture)
	21110	Pulp for paper (manufacture)		21120	Grey board (manufacture)
	21110	Pulping recycled paper (manufacture)		21120	Hand made paper (manufacture)
	21110	Recycled fibre pulp (manufacture)		21120	Hygienic paper (uncut) (manufacture)
	21110	Removal of ink from waste paper and subsequent manufacture of pulp (manufacture)		21120	Industrial paper (manufacture)
	21110	Semi-bleached paper pulp made by chemical dissolving (manufacture)		21120	Kraft wrapping and packaging paper (manufacture)
	21110	Semi-bleached paper pulp made by mechanical processes (manufacture)		21120	Laminates and foils laminated with paper or paperboard (manufacture)
	21110	Semi-bleached paper pulp made by non-dissolving processes (manufacture)		21120	Magazine paper (manufacture)
	21110	Semi-bleached paper pulp made by semi-chemical processes (manufacture)		21120	Mill board (manufacture)
	21110	Semi-chemical woodpulp (manufacture)		21120	Multi-layer paper obtained by compression (manufacture)
	21110	Sulphate and soda woodpulp (manufacture)		21120	Newsprint (manufacture)
	21110	Sulphite woodpulp (manufacture)		21120	Packing made of cardboard (manufacture)
	21110	Synthetic fibre woodpulp (manufacture)		21120	Packing paper (manufacture)
	21110	Unbleached paper pulp made by chemical dissolving (manufacture)		21120	Paper (not sensitized) (manufacture)
	21110	Unbleached paper pulp made by mechanical processes (manufacture)		21120	Paper (uncut) for household use (manufacture)
	21110	Unbleached paper pulp made by non-dissolving processes (manufacture)		21120	Paper and paperboard coating, covering and impregnation (manufacture)
	21110	Unbleached paper pulp made by semi-chemical processes (manufacture)		21120	Paper and paperboard intended for further industrial processing (manufacture)
	21110	Vegetable fibre pulp (manufacture)		21120	Paper and paperboard processing (manufacture)
	21110	Wood pulp (manufacture)		21120	Paper made of vegetable fibres for corrugated cardboard (manufacture)
17120		**Manufacture of paper and paperboard**		21120	Parchment and imitation parchment paper (manufacture)
	21120	Abrasive base paper (manufacture)		21120	Photographic base paper (manufacture)
	21120	Bank note paper (manufacture)		21120	Pressboard (manufacture)
	21120	Base paper for printing and writing paper (manufacture)		21120	Presspahn (manufacture)
	21120	Bible paper (manufacture)		21120	Printing paper (manufacture)
	21120	Bituminised building board (manufacture)		21120	Punched card and punched paper tape stock (manufacture)
	21120	Blotting paper (manufacture)		21120	Sacks made of kraft paper (manufacture)
	21120	Boot and shoe board (manufacture)		21120	Saturated and impregnated base paper (manufacture)
	21120	Bristol board (manufacture)		21120	Security paper (manufacture)
	21120	Building boards made of paper (manufacture)		21120	Special purpose paper (manufacture)
	21120	Carbon paper in large sheets (manufacture)		21120	Stencil basepaper (manufacture)
	21120	Carbon paper in rolls (manufacture)		21120	Stencil paper in large sheets (manufacture)
	21120	Carbonising base paper (manufacture)		21120	Stencil paper in rolls (manufacture)
	21120	Cardboard (manufacture)		21120	Strawboard (manufacture)
	21120	Case making materials (manufacture)		21120	Strawpaper (manufacture)
	21120	Cellulose fibre webs (manufacture)		21120	Sulphite wrapping paper (manufacture)
	21120	Cellulose wadding (manufacture)		21120	Tissue paper (uncut) (manufacture)
	21120	Cigarette paper (uncut in rolls) (manufacture)		21120	Toilet paper (uncut) (manufacture)
	21120	Creped paper (manufacture)		21120	Tracing paper (manufacture)
	21120	Crinkled paper (manufacture)		21120	Wallpaper base (manufacture)
	21120	Drawing paper (manufacture)		21120	Waterproof paper (manufacture)
	21120	Electrical paper (manufacture)		21120	Wrapping and packaging paper including coated (manufacture)
	21120	Fancy paper (manufacture)		21120	Writing paper (manufacture)
	21120	Feltboard including felt paper (manufacture)	**17211**		**Manufacture of corrugated paper and paperboard; manufacture of sacks and bags of paper**
				21211	Bags made of paper (manufacture)

UK SIC 2007	UK SIC 2003	Activity
	21211	Corrugated paper (manufacture)
	21211	Corrugated paper board (manufacture)
	21211	Multi-wall paper sack (manufacture)
	21211	Paper and paperboard (corrugated) (manufacture)
	21211	Printed paper bags (manufacture)
	21211	Sacks and bags made of paper (manufacture)
	21211	Sacks made of paper (manufacture)
17219		**Manufacture of paper and paperboard containers other than sacks and bags**
	21213	Boxes made of corrugated cardboard (manufacture)
	21213	Boxes made of corrugated paper (manufacture)
	21214	Boxes made of non-corrugated cardboard (manufacture)
	21214	Boxes made of non-corrugated paper (manufacture)
	21214	Boxes made of rigid board (manufacture)
	21213	Boxes made of rigid corrugated board (manufacture)
	21215	Cartons and similar containers for carrying liquids (unwaxed) (manufacture)
	21215	Cartons and similar containers for carrying liquids (waxed) (manufacture)
	21213	Cartons made of corrugated board (manufacture)
	21213	Cartons made of corrugated paper (manufacture)
	21214	Cartons made of non-corrugated board (manufacture)
	21214	Cartons made of non-corrugated paper (manufacture)
	21213	Cases made of corrugated cardboard (manufacture)
	21213	Cases made of corrugated fibreboard (manufacture)
	21213	Cases made of corrugated paper (manufacture)
	21214	Cases made of non-corrugated cardboard (manufacture)
	21214	Cases made of non-corrugated paper (manufacture)
	21219	Cigarette packets (manufacture)
	21219	Containers and canisters made of cardboard n.e.c. (manufacture)
	21219	Containers made of corrugated paper or paperboard n.e.c. (manufacture)
	21219	Containers made of paper and paperboard n.e.c. (manufacture)
	21219	Containers made of solid board n.e.c. (manufacture)
	21213	Corrugated packing case (manufacture)
	21219	Cylinders made of board (open ended for posting documents) (manufacture)
	21219	Folding boxes made of board (manufacture)
	21219	Folding paperboard containers (manufacture)
	21219	Letter file (manufacture)
	21219	Office box files and similar articles (manufacture)
	21219	Office systems made of paper and board (manufacture)
	21219	Packing cases made of fibre board (manufacture)
17220		**Manufacture of household and sanitary goods and of toilet requisites**
	21220	Beakers made of paper (manufacture)
	21220	Cake board (manufacture)
	21220	Cellulose wadding products (manufacture)
	21220	Cleansing tissues (manufacture)
	21220	Cups made of paper (manufacture)
	21220	Dishes made of paper (manufacture)
	21220	Disposable baby napkins made of paper or cellulose wadding (manufacture)

SIC 2007	SIC 2003	Activity
	21220	Disposable bed linen made of paper or cellulose wadding (manufacture)
	21220	Doilies made of paper (manufacture)
	21220	Handkerchief made of paper (manufacture)
	21220	Household and personal hygiene paper (manufacture)
	21220	Household cellulose wadding paper products (manufacture)
	21220	Kitchen cloth made of paper (manufacture)
	21220	Kitchen towels made of paper (manufacture)
	21220	Napkin liners (manufacture)
	21220	Napkins made of paper (manufacture)
	21220	Paper lace (manufacture)
	21220	Plates made of paper (manufacture)
	21220	Pots made of paper (manufacture)
	21220	Sanitary towels made of paper (manufacture)
	17549	Sanitary towels made of textile wadding (manufacture)
	21220	Serviettes made of paper (manufacture)
	21220	Table-cloths made of paper (manufacture)
	21220	Tampons made of paper and cellulose wadding (manufacture)
	17549	Tampons made of textile wadding (manufacture)
	17549	Textile wadding and articles of wadding (manufacture)
	21220	Toilet paper (cut to size) (manufacture)
	21220	Towels made of paper (manufacture)
	21220	Trays made of paper (manufacture)
	21220	Underwear made of paper (manufacture)
	17549	Wadding made from yarn spun on the cotton system (manufacture)
17230		**Manufacture of paper stationery**
	22220	Account books (manufacture)
	21230	Adhesive paper ready for use (manufacture)
	21230	Boxed stationery (manufacture)
	21230	Carbon paper ready for use (manufacture)
	21230	Carbon paper stencil ready for use (manufacture)
	21230	Carbonless copy paper ready for use (manufacture)
	21230	Card cutting for index cards (manufacture)
	22220	Commercial notebooks (manufacture)
	22220	Commercial stationery (manufacture)
	22220	Commercial stationery binders (manufacture)
	22220	Commercial stationery business forms (manufacture)
	22220	Commercial stationery registers (manufacture)
	21230	Computer print-out paper (manufacture)
	21230	Continuous stationery (manufacture)
	21230	Duplicating paper (cut to size) (manufacture)
	21230	Duplicator stencils ready for use (manufacture)
	22220	Educational notebooks (manufacture)
	22220	Educational stationery (manufacture)
	22220	Educational stationery binders (manufacture)
	22220	Educational stationery business forms (manufacture)
	22220	Educational stationery registers
	21230	Envelopes and letter-cards (manufacture)
	21230	Gummed paper ready for use (manufacture)
	21230	Index card (manufacture)
	21230	Letter card (manufacture)
	22220	Loose leaf binder (manufacture)
	22220	Notepad (manufacture)

UK SIC 2007	UK SIC 2003	Activity	SIC 2007	SIC 2003	Activity
	21230	Postcards (plain) (manufacture)		21259	Paper cut to size (not packaging products) (manufacture)
	21230	Pouches containing an assortment of paper stationery (manufacture)		21259	Paper embossing (manufacture)
	22220	Printed matter for accounting and technical use (manufacture)		36639	Paper novelties (manufacture)
	21230	Printers' cards (manufacture)		21259	Paper or paperboard cards for use on jacquard machines (manufacture)
	21230	Printing and writing paper ready for use (manufacture)		21259	Paper patterns (manufacture)
	22220	School stationery (manufacture)		21259	Paper perforating (manufacture)
	21230	Self-copy paper ready for use (manufacture)		21259	Paper shavings (manufacture)
	21230	Stationery paper (manufacture)		21259	Paper transfer for embroidery, etc. (manufacture)
	21230	Stencil duplicating (manufacture)		21259	Papier mache works (manufacture)
	21230	Wallets containing an assortment of paper stationery (manufacture)		21259	Pattern card (manufacture)
	21230	Writing compendiums (manufacture)		21259	Photograph mount (manufacture)
	22220	Writing paper pads (manufacture)		21259	Picture frame mount (manufacture)
				21259	Pleated paper (manufacture)
17240		**Manufacture of wallpaper**		21259	Spools made of paper (manufacture)
	21240	Fabric wallcoverings (manufacture)		21259	Stamp hinges (manufacture)
	21240	Lincrusta (manufacture)		21259	Tabulating machine cards (manufacture)
	21240	Paper and paperboard articles for interior decoration (manufacture)		21259	Ticket cutting and punching (manufacture)
	21240	Paper staining (manufacture)		21259	Tube fittings made of paper (manufacture)
	21240	Textile wall coverings (manufacture)		21259	Tubes made of cardboard (not for packing) (manufacture)
	21240	Textile wallpaper (manufacture)		21259	Wood pulp vessel (manufacture)
	21240	Vinyl-coated wall coverings (manufacture)		21259	Wrapping paper cut and packed in ready to use sheets or rolls (manufacture)
	21240	Vinyl-coated wallpaper (manufacture)			
	21240	Wallpaper and lining paper (manufacture)	**18110**		**Printing of newspapers**
				22210	Newspaper printing (manufacture)
17290		**Manufacture of other articles of paper and paperboard**		22210	Printing of periodicals appearing at least four times a week (manufacture)
	21259	Articles made of paper and paperboard n.e.c. (manufacture)	**18121**		**Manufacture of printed labels**
	21259	Blinds made of paper (manufacture)		21251	Flexographic printing on labels or tags (manufacture)
	21259	Bobbins made of paper and paperboard (manufacture)		21251	Gravure printing on labels or tags (manufacture)
	21259	Bobbins, spools and cops made of paper and paperboard (manufacture)		21251	Labels (printed) made of gummed paper (manufacture)
	21259	Bonbon paper (manufacture)		21251	Labels (printed) made of paper (manufacture)
	36639	Christmas cracker (manufacture)		21251	Lithographic printing on labels or tags (manufacture)
	36639	Christmas decorations made of paper (manufacture)		21251	Printed labels (manufacture)
	21259	Cigarette paper in booklets (manufacture)		21251	Printing on labels or tags (manufacture)
	21259	Cigarette tube (manufacture)			
	36639	Confetti paper (manufacture)	**18129**		**Printing (other than printing of newspapers and printing on labels and tags) n.e.c.**
	21259	Cop paper (manufacture)		22220	Advertising catalogue printing (manufacture)
	21259	Cop tube (manufacture)		22220	Advertising printed matter printing (manufacture)
	21259	Cylinder made of hardened paper (manufacture)		22220	Album printing (manufacture)
	21259	Discs made of cardboard (manufacture)		22220	Almanac printing (manufacture)
	21259	Egg boxes made of paper (manufacture)		22220	Amusement guide periodical printing (manufacture)
	21259	Egg trays and other moulded pulp packaging products (manufacture)		22220	Atlas printing (manufacture)
	21259	Filter paper and paperboard (cut to size) (manufacture)		22220	Bank note printing (manufacture)
	21259	Flexible paper packaging (manufacture)		22220	Book printing (manufacture)
	36639	Hats made of paper (manufacture)		22220	Braille printing (manufacture)
	21252	Labels (unprinted) made of gummed paper (manufacture)		22220	Brochure printing (manufacture)
	21252	Labels (unprinted) made of paper (manufacture)		22220	Business form printing (manufacture)
	21259	Mount cutting (manufacture)		22220	Calendar printing (manufacture)
	21259	Mounting paper on linen (manufacture)		22220	Chart printing (manufacture)
	21259	Paper converting (unspecified) (manufacture)		22220	Cheque book printing (manufacture)
	21259	Paper creping (manufacture)		22220	Christmas card printing (manufacture)
				22220	Collotype printing (manufacture)
				22220	Commercial printed matter printing (manufacture)
				22220	Copper plate printing (manufacture)

UK SIC 2007	UK SIC 2003	Activity
	22220	Decal printing (manufacture)
	22220	Diary printing (manufacture)
	22220	Directory printing (manufacture)
	22220	Documents of title printing (manufacture)
	22220	Fashion printing (manufacture)
	22220	Flexographic printing (manufacture)
	22220	General printing (manufacture)
	22220	Greeting card printing (manufacture)
	22220	Hexachrome printing (manufacture)
	22220	Job printing (manufacture)
	22220	Letterpress printing (manufacture)
	22220	Lithographic printing (manufacture)
	22220	Magazine printing (manufacture)
	22220	Manuscript book (manufacture)
	22220	Map printing (manufacture)
	22220	Music printing (manufacture)
	22220	Offset printing (manufacture)
	22220	Pamphlet printing (manufacture)
	22220	Passport printing (manufacture)
	22220	Periodical printing (manufacture)
	22220	Personal stationery printing (manufacture)
	22220	Phonetic printing (manufacture)
	22220	Photocopier printing (manufacture)
	22220	Photogravure printing (manufacture)
	22220	Picture postcard (manufacture)
	22220	Plan printing (manufacture)
	22220	Postage stamp perforating (manufacture)
	22220	Postage stamp printing (manufacture)
	22220	Poster printing (manufacture)
	22220	Printing (undefined) (manufacture)
	22220	Printing by computer printers (manufacture)
	22220	Printing by duplication machines (manufacture)
	22220	Printing by embossers (manufacture)
	22220	Printing by quick printing (manufacture)
	22220	Printing directly onto ceramics (manufacture)
	22220	Printing of banknotes
	22220	Printing of magazines appearing less than four times a week (manufacture)
	22220	Printing of other security papers (manufacture)
	22220	Printing of smart cards (manufacture)
	22220	Printing onto glass (manufacture)
	22220	Printing onto metal (manufacture)
	22220	Printing onto plastic (manufacture)
	21220	Printing onto sanitary goods and toilet requisites of paper (manufacture)
	22220	Printing onto textiles (manufacture)
	22220	Printing onto wood (manufacture)
	22220	Prospectus printing (manufacture)
	22220	Register printing (manufacture)
	22220	Review printing (manufacture)
	22220	Screen printing (except silk screen printing on textiles and apparel) (manufacture)
	22220	Screen printing (manufacture)
	22220	Screen printing of logos (manufacture)
	22220	Screen printing on glass or pottery (manufacture)
	22220	Security printing (manufacture)
	22220	Sheet metal printing (manufacture)
	22220	Sheet music printing (manufacture)
	22220	Showcard (manufacture)

SIC 2007	SIC 2003	Activity
	22220	Stamp embossed paper (manufacture)
	22220	Stochastic printing (manufacture)
	22220	Taxation stamps printing (manufacture)
	22220	Thermocopier printing (manufacture)
	22220	Ticket printing (manufacture)
	22220	Timetable printing (manufacture)
	22220	Tin printing (manufacture)
	22220	Title document printing (manufacture)
	22220	Trade journal printing (manufacture)
	22220	Trading stamp printing (manufacture)
	22220	Transfer printing (manufacture)
	22220	Window ticket (manufacture)
18130		**Pre-press and pre-media services**
	22250	Aerographing (manufacture)
	22250	Blocking (printing)
	22250	Ceramic transfer litho engraving (manufacture)
	22240	Composition for printing (manufacture)
	22240	Computer to plate CTP processing of plates for relief printing (manufacture)
	22240	Computer to plate CTP processing of plates for relief stamping
	22240	Cylinder engraving for gravure printing (manufacture)
	22240	Cylinder etching for gravure printing (manufacture)
	22240	Data files preparation for multi-media printing on CD-ROM (manufacture)
	22240	Data files preparation for multi-media printing on internet applications (manufacture)
	22240	Data files preparation for multi-media printing on paper (manufacture)
	22250	Die sinking of stationery (manufacture)
	22250	Die stamping of stationery (manufacture)
	22240	Digital imposition (manufacture)
	22250	Dummies for presentation (manufacture)
	22240	Electronic makeup (manufacture)
	22240	Electrotyping (manufacture)
	22250	Embossing (manufacture)
	22240	Engraving for printing (manufacture)
	22240	Etching for printing (manufacture)
	22240	Image setting for letterpress processes (manufacture)
	22240	Image setting for offset printing processes (manufacture)
	22250	Insetting (manufacture)
	22250	Laminating (manufacture)
	22250	Layouts for presentation (manufacture)
	22240	Litho plate making (unsensitized) (manufacture)
	22240	Lithographic stones and wood blocks (manufacture)
	22250	Lithography (manufacture)
	22240	Metal etching (manufacture)
	22240	Music plate engraving (manufacture)
	22250	Overhead media foils and other forms of presentation (manufacture)
	22250	Overhead projection foils production (manufacture)
	22240	Photo engraving (manufacture)
	22250	Photo lithography (manufacture)
	22240	Phototypesetting (manufacture)
	22240	Plate making for printing (manufacture)
	22240	Plate processes direct to plate (also photopolymer plates) (manufacture)
	22240	Plate setting for letterpress processes (manufacture)

UK SIC 2007	UK SIC 2003	Activity	SIC 2007	SIC 2003	Activity
	22240	Plate setting for offset printing processes (manufacture)		22230	Post press services in support of printing activities (manufacture)
	22250	Poster aerographing (manufacture)		22230	Printed paper or board finishing (manufacture)
	22240	Poster writing (manufacture)		22230	Printed sheets finishing (manufacture)
	22240	Preparation and linkage of digital data (manufacture)		22230	Prospectus finishing (manufacture)
	22240	Pre-press data input electronic make-up (manufacture)		22230	Publisher's case making (manufacture)
	22240	Pre-press data input optical character recognition, electronic make-up (manufacture)		22230	Relief stamping (manufacture)
				22230	Sample card finishing (manufacture)
	22240	Pre-press data input scanning (manufacture)		22230	Sample mounting in support of printing activities (manufacture)
	22250	Print colouring (manufacture)			
	22250	Printers' designing (manufacture)		22230	Spiral binding and finishing of books and brochures (manufacture)
	22240	Printing plate engraving (manufacture)			
	22240	Printing roller engraving (manufacture)		22230	Stitching of books, brochures, etc. (manufacture)
	22240	Process block making (manufacture)		22230	Tillot and seal making (manufacture)
	22240	Process engraving (manufacture)		22230	Trade binding (manufacture)
	22240	Process plate engraving (manufacture)		22230	Trimming of books, brochures, etc. (manufacture)
	22250	Rag book making (manufacture)	**18201**		**Reproduction of sound recording**
	22240	Reproduction and composing (manufacture)		22310	Audio tape recording, except master copies for records or audio material (manufacture)
	22250	Reprographic dummies production (manufacture)			
	22250	Reprographic lay-outs production (manufacture)		22310	Compact disc reproduction from master copies (manufacture)
	22250	Reprographic products production (manufacture)		22310	Gramophone record reproduction from master copies (manufacture)
	22250	Reprographic sketch production (manufacture)			
	22250	Sketches for presentation (manufacture)		22310	Gramophone records (except master copies) including blanks for cutting) (manufacture)
	22240	Stereotyping (manufacture)			
	22240	Typesetting and phototypesetting (manufacture)		22310	Music tape reproduction from master copies (manufacture)
	22250	Varnishing (manufacture)		22310	Pre-recorded tape, except master copies for records or audio material (manufacture)
	22240	Wood blocks for printing (manufacture)			
18140		**Binding and related services**		22310	Sound recording reproduction (manufacture)
	22230	Adhesive binding of books, brochures, etc. (manufacture)		22310	Tape pre-recording, except master copies for records or audio material (manufacture)
	22230	Advertising mailing literature finishing (manufacture)	**18202**		**Reproduction of video recording**
				22320	Video disc reproduction (manufacture)
	22230	Assembling of books, brochures, etc. (manufacture)		22320	Video tape recordings reproduction (manufacture)
	22230	Binding and finishing of books, brochures, magazines, catalogues etc. (manufacture)	**18203**		**Reproduction of computer media**
				22330	Computer media reproduction (manufacture)
	22230	Binding and related services (manufacture)		22330	Software reproduction from master copies (manufacture)
	22230	Bookbinding (manufacture)			
	22230	Braille copying (manufacture)	**19100**		**Manufacture of coke oven products**
	22230	Business forms finishing (manufacture)		23100	Agglomeration of coke (manufacture)
	22230	Calendar finishing (manufacture)		23100	Ammoniacal liquor from coke ovens (manufacture)
	22230	Calico printers' engraving (manufacture)		23100	Ammonium sulphate from coke ovens (manufacture)
	22230	Card embossing (manufacture)		23100	Coal carbonisation (manufacture)
	22230	Collating of books, brochures, etc. (manufacture)		23100	Coke oven gas (manufacture)
	22230	Cutting, cover laying, gluing, collating books, brochures, magazines, catalogues etc. (manufacture)		23100	Coke oven products (manufacture)
				23100	Coke production (manufacture)
	22230	Die sinking or stamping finishing activities (manufacture)		23100	Crude benzole from coke ovens (manufacture)
				23100	Crude coal tar from coke ovens (manufacture)
	22230	Finishing services for CD-ROMS (manufacture)		23100	Crude coal tar production (manufacture)
	22230	Gilding (printing service) (manufacture)		23100	Foundry coke (manufacture)
	22230	Glueing of books, brochures, etc. (manufacture)		23100	Hard coke (manufacture)
	22230	Gold blocking (manufacture)		23100	Hard coke breeze (manufacture)
	22230	Gold stamping (manufacture)		23100	Lignite tars production (manufacture)
	22230	Heraldic chasing and seal engraving (manufacture)		23100	Low temperature carbonisation solid fuel (not ovoid or briquettes) (manufacture)
	22230	Heraldic engraving (manufacture)			
	22230	Mailing finishing services such as customisation, envelope preparation (manufacture)		23100	Metallurgical coke (manufacture)
				24140	Pitch and pitch coke (manufacture)
	22230	Merchandising display finishing (manufacture)			
	22230	Plastic wire binding and finishing of books and brochures (manufacture)			

UK SIC 2007	UK SIC 2003	Activity	SIC 2007	SIC 2003	Activity
	23100	Semi-coke (manufacture)		23201	Petroleum refining (manufacture)
	23100	Sulphate of ammonia from coke ovens (manufacture)		23201	Process oil refining (manufacture)
				23201	Process oils (manufacture)
19201		**Mineral oil refining**		23201	Propane (manufacture)
	23201	Aviation spirit (manufacture)		23201	Refinery gas (manufacture)
	23201	Aviation turbine fuel (manufacture)		23201	Road coverings derived from crude petroleum or bituminous minerals (manufacture)
	23201	Biofuels from blending of alcohols with petroleum, e.g. Gasohol (manufacture)		23201	Shale oil refining (manufacture)
	23201	Bitumen (manufacture)		10103	Solid fuel briquettes production (manufacture)
	10103	Briquette solid fuel production (manufacture)		10103	Solid fuels production (manufacture)
	23201	Burning oil (manufacture)		10103	Solid smokeless ovoids and briquettes preparation (manufacture)
	23201	Butane (manufacture)		23201	Tail gas (manufacture)
	23201	Chemical feedstock (manufacture)		23201	Technical white oil (manufacture)
	10103	Coal tar (crude) from manufactured fuel plants (manufacture)		23201	Transformer oil (at refineries) (manufacture)
	23201	Coke petroleum (manufacture)		23201	Vaporising oil (manufacture)
	23201	Crude oil refining (manufacture)		23201	Vaseline (at refinery) (manufacture)
	23201	Crude petroleum jelly (at refinery) (manufacture)		23201	White spirit (manufacture)
	23201	Derv (manufacture)		23201	Wide cut gasoline (manufacture)
	23201	Diesel oil (manufacture)			
	23201	Ethane production by refining (manufacture)	**19209**		**Other treatment of petroleum products (excluding mineral oil refining/petrochemicals manufacture)**
	23201	Fuel heavy fuel oil (manufacture)		23209	Cutting oil (manufacture)
	23201	Fuel oil (manufacture)		23209	Grease formulation outside refineries (manufacture)
	23201	Gas oil (manufacture)		23209	Hydraulic oil formulation outside refineries (manufacture)
	23201	Gasoline motor fuel (manufacture)		23209	Insulating oil formulation outside refineries (manufacture)
	23201	Greases (at refinery) (manufacture)		23209	Lubricating grease formulation outside refineries (manufacture)
	10103	Hard coal agglomeration (manufacture)		23209	Lubricating oil formulation outside refineries (manufacture)
	10103	Hard-coal briquettes (manufacture)		23209	Mineral oil blending (manufacture)
	23201	Industrial benzole (manufacture)		23209	Oil based lubricating oils (manufacture)
	23201	Industrial spirit from petroleum (manufacture)		23209	Penetrating oil (manufacture)
	23201	Insulating oil (at refineries) (manufacture)		23209	Petroleum grease formulation outside refineries (manufacture)
	23201	Kerosene (manufacture)		23209	Petroleum jelly formulation outside refineries (manufacture)
	23201	Light fuel oil (manufacture)		23209	Petroleum societies without refineries (manufacture)
	10200	Lignite fuel briquettes (manufacture)		23209	Strap paste for transmission belts (manufacture)
	23201	Liquid butane gas (manufacture)		23209	Transformer oil formulation outside refineries (manufacture)
	23201	Liquid propane gas (manufacture)			
	23201	Lubricating oil (at refinery) (manufacture)	**20110**		**Manufacture of industrial gases**
	23201	Marine diesel oil (manufacture)		24110	Acetylene (manufacture)
	23201	Medical paraffin (manufacture)		24110	Argon (manufacture)
	23201	Medium fuel oil (manufacture)		24110	Carbon dioxide (manufacture)
	23201	Mineral oil refining (manufacture)		24110	Compressed industrial gases (manufacture)
	23201	Motor fuel (manufacture)		24110	Elemental gases (manufacture)
	23201	Motor spirit (manufacture)		24110	Hydrogen (manufacture)
	23201	Naphtha (LDF) (manufacture)		24110	Industrial gases (manufacture)
	23201	Oil refinery (manufacture)		24110	Inert gases such as carbon dioxide (manufacture)
	10103	Ovoid solid fuel production (manufacture)		24110	Isolating gases (manufacture)
	23201	Paraffin (manufacture)		24110	Liquefied or compressed industrial gases (manufacture)
	23201	Paraffin for medicinal use (manufacture)		24110	Liquefied or compressed industrial or medical refrigerant gases (manufacture)
	23201	Paraffin wax (manufacture)		24110	Liquefied or compressed medical gases (manufacture)
	10103	Patent fuel production (manufacture)		24110	Liquid or compressed air (manufacture)
	10300	Peat briquettes (manufacture)			
	23201	Petro-chemical industry products (manufacture)			
	23201	Petrol (manufacture)			
	23201	Petroleum briquettes (manufacture)			
	23201	Petroleum coke (manufacture)			
	23201	Petroleum feedstock (manufacture)			
	23201	Petroleum gas (manufacture)			
	23201	Petroleum grease (at refinery) (manufacture)			
	23201	Petroleum product (at refineries) (manufacture)			

UK SIC 2007	UK SIC 2003	Activity	SIC 2007	SIC 2003	Activity
	24110	Mixed industrial gases (manufacture)		24130	Carbon disulphide (manufacture)
	24110	Neon (manufacture)		24130	Chemical elements (except metals) (manufacture)
	24110	Nitrogen (manufacture)		24130	Chlorine and chloride (manufacture)
	24110	Nitrous oxide (manufacture)		24130	Chromium compounds (excluding prepared pigments) (manufacture)
	24110	Oxygen (manufacture)		24130	Distilled water (manufacture)
	24110	Rare gases (manufacture)		23300	Enriched thorium (manufacture)
	24110	Refrigerant gases (manufacture)		23300	Enriched uranium production (manufacture)
20120		**Manufacture of dyes and pigments**		24130	Flocculating agents (chemical) (manufacture)
	24120	Acid dye (manufacture)		24130	Fluorine, hydrofluoric acid and fluorides (manufacture)
	24120	Alizarin dye (manufacture)		23300	Fuel elements for nuclear reactors production (manufacture)
	24120	Aniline dye (manufacture)		24130	Halogens and halides (inorganic) (manufacture)
	24120	Azoic dye (manufacture)		24130	Hydrochloric acid (manufacture)
	24120	Basic dye (manufacture)		24130	Hydrogen peroxide (manufacture)
	24120	Chromium pigment (manufacture)		24130	Hydrosulphite (manufacture)
	24120	Colour lake (manufacture)		24130	Inorganic acid (manufacture)
	24120	Colours for food and cosmetics (manufacture)		24130	Inorganic bases (manufacture)
	24120	Colours in dry, liquid or paste form (manufacture)		24130	Inorganic chemical (manufacture)
	24120	Crushed pigment colours (manufacture)		24130	Inorganic compounds (manufacture)
	24120	Direct dye (manufacture)		24130	Iodine and iodides (manufacture)
	24120	Disperse dye (manufacture)		23300	Iron pyrites roasting (manufacture)
	24120	Dye (manufacture)		24130	Lyes (manufacture)
	24120	Dyes and pigments from any source in basic or concentrated forms (manufacture)		24130	Oxygen compounds of non metals (excluding carbon dioxide) (manufacture)
	24120	Dyes for food, drink and cosmetics (manufacture)		24130	Peroxides (inorganic) (manufacture)
	24120	Dyes modified for dying acrylic fibres (manufacture)		24130	Pesticide inorganic chemicals (excluding formulated preparations) (manufacture)
	24120	Fluorescent brightening agent (manufacture)		24130	Phosphorous compounds (excluding phosphatic fertiliser) (manufacture)
	24120	Jewellers' rouge (manufacture)		23300	Plutonium processing (manufacture)
	24120	Lake (pigment) (manufacture)		24130	Potassium compounds (manufacture)
	24120	Laundry blue (manufacture)		23300	Radioactive compounds production (manufacture)
	24120	Luminophores (manufacture)		24130	Radioactive isotopes (other than of uranium, thorium or plutonium) (manufacture)
	24120	Manganese oxide (manufacture)		23300	Radioactive isotopes of uranium, thorium and plutonium (manufacture)
	24120	Mineral colours (manufacture)		24130	Sodium and sodium compounds (manufacture)
	24120	Mordant dye (manufacture)		23300	Spent nuclear fuel re-processing (manufacture)
	24120	Ochres (pigments) (manufacture)		24130	Sulphur (manufacture)
	24120	Optical bleaching agent (manufacture)		24130	Sulphuric acid (manufacture)
	24120	Solvent dye (manufacture)		23300	Uranium (enriched) (manufacture)
	24120	Sulphur dye (manufacture)		24130	White lead (not in paste form) (manufacture)
	24120	Synthetic dyestuffs (manufacture)			
	24120	Synthetic iron oxide (manufacture)	**20140**		**Manufacture of other organic basic chemicals**
	24120	Synthetic organic pigment (manufacture)		24140	Acetic acid (manufacture)
	24120	Tanning agents (synthetic) (manufacture)		24140	Acetone (manufacture)
	24120	Titanium dioxide (manufacture)		24140	Acid (organic) (manufacture)
	24120	Toner (pigment) (manufacture)		24140	Acrylonitrile (manufacture)
	24120	Vat dye (manufacture)		24140	Activated and unactivated charcoal (other than wood charcoal) (manufacture)
	24120	Vegetable tanning and dyeing extracts (manufacture)		24140	Activated earths (manufacture)
	24120	Zinc oxide (manufacture)		24140	Acyclic (fatty) alcohols (manufacture)
20130		**Manufacture of other inorganic basic chemicals**		24140	Acyclic hydrocarbons (saturated and unsaturated) (manufacture)
	24130	Acid (inorganic) (manufacture)		24140	Aldehyde (manufacture)
	24130	Alkali (manufacture)		24140	Amines (manufacture)
	24130	Aluminium compounds (except bauxite and abrasives) (manufacture)		24140	Anthracene (manufacture)
	24130	Alums (manufacture)		24140	Aromatic hydrocarbons (manufacture)
	24130	Bromine and bromides (manufacture)		24140	Benzene (manufacture)
	24130	Calcium and calcium compounds (manufacture)			
	24130	Calcium carbide (manufacture)			
	24130	Carbon (manufacture)			
	24130	Carbon black (manufacture)			

UK SIC 2007	UK SIC 2003	Activity	SIC 2007	SIC 2003	Activity
	24140	Butadiene (manufacture)		24140	Organo-sulphur compounds (manufacture)
	24140	Carboxylic acid (manufacture)		24140	Oxirane (ethylene oxide) (manufacture)
	24140	Charcoal (other than wood charcoal) (manufacture)		24140	Oxygen-function compounds (dual or poly) (manufacture)
	24140	Charcoal burning (manufacture)		24140	Oxygen-function compounds including aldehydes (manufacture)
	24140	Citric acid (manufacture)		24140	Peracetic acid (manufacture)
	24140	Coal tar (refined) (manufacture)		24140	Perchloroethylene (manufacture)
	24140	Coal tar distillation (manufacture)		24140	Pesticide organic chemicals (excluding formulated preparations) (manufacture)
	24140	Coal tar naphtha (manufacture)		24140	Phenol (manufacture)
	24140	Creosote (manufacture)		24140	Phthalic anhydride (manufacture)
	24140	Cresylic acid (manufacture)		24140	Piezo-electric quartz (manufacture)
	24140	Cumene (manufacture)		24140	Propylene (manufacture)
	24140	Cyclic alcohols (manufacture)		24140	Propylene oxide (manufacture)
	24140	Cyclic hydrocarbons (saturated and unsaturated) (manufacture)		24140	Pyridine base (manufacture)
	24140	Cyclohexane (manufacture)		24140	Quinones (manufacture)
	15920	Denatured ethyl alcohol (manufacture)		24140	Refined coal tar (manufacture)
	24140	Diethyl phenylamine diamine sulphate (chlorine tablets) (manufacture)		24140	Rosin size (manufacture)
	15920	Distillery draft production (manufacture)		24140	Saccharin tablet (manufacture)
	24140	Enzymes and other organic compounds (manufacture)		24140	Spirit of turpentine (manufacture)
	24140	Epoxides (manufacture)		24140	Stearic acid (manufacture)
	24140	Esters (but (not polyesters)) (manufacture)		24140	Styrene (manufacture)
	24140	Esters of methacrylic acid (manufacture)		24140	Synthetic alcohol (manufacture)
	24140	Ethane diol (excluding anti-freeze mixtures) (manufacture)		24140	Synthetic aromatic products (manufacture)
	24140	Ethanol (synthetic) (manufacture)		24140	Synthetic ethyl alcohol (manufacture)
	15920	Ethyl alcohol (non-potable) obtained by fermentation (manufacture)		24140	Synthetic glycerol (manufacture)
	24140	Ethylene (manufacture)		24140	Synthetic or reconstructed precious or semi-precious stones (manufacture)
	24140	Ethylene glycol (excluding anti-freeze mixtures) (manufacture)		24140	Tar acids (manufacture)
	24140	Fat splitting and distilling (manufacture)		24140	Tetrachloroethylene (manufacture)
	24140	Fatty acid (manufacture)		24140	Toluene (manufacture)
	24140	Formaldehyde (manufacture)		24140	Urea (not for use as fertiliser) (manufacture)
	24140	Halogenated derivatives of hydrocarbon (manufacture)		24140	Ureines (manufacture)
	24140	Heterocyclic compounds (manufacture)		24140	Vinyl acetate (manufacture)
	24140	Hydrocarbon derivatives (sulphated, nitrated or nitrosated) (manufacture)		24140	Wood tar chemicals (manufacture)
	24140	Hydrocarbons (not fuels) (manufacture)		24140	Xylene (manufacture)
	24140	Ketones (manufacture)			
	24140	Lactones (coumarin, methylcoumarins and ethylcoumarins) (manufacture)	**20150**		**Manufacture of fertilisers and nitrogen compounds**
	24140	Melamine (manufacture)		24150	Ammonia (manufacture)
	24140	Methanol (manufacture)		24150	Ammonium chloride (manufacture)
	15920	Methylated spirits (manufacture)		24150	Ammonium compounds (excluding ammonium nitrate, sulphate and phosphate) (manufacture)
	24140	Mono and polycarboxyclic acids including acetic acid (manufacture)		24150	Ammonium nitrate (not for explosives) (manufacture)
	24140	Monohydric alcohols (manufacture)		24150	Ammonium phosphate (manufacture)
	24140	Naphthalene (manufacture)		24150	Ammonium sulphate (manufacture)
	15920	Neutral spirits production (manufacture)		24150	Artificial manure (manufacture)
	24140	Nitrogen-function organic compounds including amine (manufacture)		24150	Basic slag (ground) (manufacture)
	24140	Oleic acid (manufacture)		24150	Compound fertiliser (manufacture)
	24140	Oleine (manufacture)		24150	Crude natural phosphates (manufacture)
	24140	Organic acids and their esters and halogenated, nitrosated and sulphonated derivatives (manufacture)		24150	Fertiliser (manufacture)
	24140	Organic base chemicals (manufacture)		24150	Lawn sand (manufacture)
	24140	Organic compounds including wood distillation products (manufacture)		24150	Lime (ammonium nitrate) (manufacture)
				24150	Nitrates and nitrites of potassium (manufacture)
				24150	Nitric and sulphonitric acid (manufacture)
				24150	Nitrogen products (manufacture)
				24150	Nitrogenous straight fertiliser (manufacture)
				24150	Nitrogenous, phosphatic or potassic fertilisers (manufacture)

UK SIC 2007	UK SIC 2003	Activity	SIC 2007	SIC 2003	Activity
	24150	Phosphates of ammonium carbonates (manufacture)		24160	Siloxanes (manufacture)
	24150	Phosphates of triammonium carbonates (manufacture)		24160	Styrene polymers (manufacture)
	24150	Phosphatic straight fertiliser (manufacture)		24160	Synthetic resin adhesive (unformulated) (manufacture)
	24150	Potassic straight fertiliser (manufacture)		24160	Synthetic resins (manufacture)
	24150	Potassium salts (manufacture)		24160	Thermoplastic resins (manufacture)
	24150	Potting soil mixtures of natural soil, sand, clays and minerals (manufacture)		24160	Thermosetting resins (manufacture)
	24150	Potting soil with peat as main constituent (manufacture)		24160	Thiourea resins (manufacture)
	24150	Superphosphate (manufacture)		24160	Vinyl acetate polymers (manufacture)
	24150	Urea for use as fertiliser (manufacture)		24160	Vinyl chloride polymers (manufacture)
20160		**Manufacture of plastics in primary forms**	**20170**		**Manufacture of synthetic rubber in primary forms**
	24160	Acrylic resins (manufacture)		24170	Factice (manufacture)
	24160	Acrylics (manufacture)		24170	Rubber-like gums (balata, etc.) (manufacture)
	24160	Acrylonitrile butadiene styrene (abs) polymers (manufacture)		24170	Synthetic rubber (manufacture)
	24160	Alginates (manufacture)		24170	Synthetic rubber and natural rubber mixtures (manufacture)
	24160	Alkyd resins (manufacture)			
	24160	Aminoplastic resins (manufacture)	**20200**		**Manufacture of pesticides and other agrochemical products**
	24160	Casein resins (manufacture)		24200	Acaricide (manufacture)
	24160	Cellulose (manufacture)		24200	Agro-chemical products n.e.c. (manufacture)
	24160	Cellulose acetate (manufacture)		24200	Anti-sprouting products (manufacture)
	24160	Cellulose ester and ether ester (manufacture)		24200	Biocides (manufacture)
	24160	Cellulose nitrate (manufacture)		24200	Cattle dip (manufacture)
	24160	Condensation, polycondensation and polyaddition products (plastic material) (manufacture)		24200	Disinfectant (manufacture)
	24160	Co-polymer plastics (manufacture)		24200	Disinfectants for agricultural and other use (manufacture)
	24160	Cresylic resins (manufacture)		24200	Fly paper (manufacture)
	24160	Dispersions of synthetic resin (manufacture)		24200	Formulated pesticide (manufacture)
	24160	Emulsions of synthetic resin (manufacture)		24200	Fruit dropping compound (manufacture)
	24160	Epoxide resins (manufacture)		24200	Fruit setting compound (manufacture)
	24160	Ethylene polymers (manufacture)		24200	Fumigating block (manufacture)
	24160	Extrusion compounds (plastics) (manufacture)		24200	Fungicide (manufacture)
	24160	Ion-exchangers based on polymers (manufacture)		24200	Herbicide (manufacture)
	24160	Melamine resins (manufacture)		24200	Insecticide (manufacture)
	24160	Moulding compounds (plastics) (manufacture)		24200	Molluscicides (manufacture)
	24160	Non-vulcanisable thermoplastic elastomers (manufacture)		24200	Nematocide (manufacture)
	24160	Phenolic resins (manufacture)		24200	Nicotine preparation (manufacture)
	24160	Plastics in primary forms (manufacture)		24200	Pesticides and other agrochemical products (manufacture)
	24160	Polyamide compounds (manufacture)		24200	Plant growth regulators (manufacture)
	24160	Polyamides (manufacture)		24200	Plant hormone (manufacture)
	24160	Polyester resins (manufacture)		24200	Rodenticide (manufacture)
	24160	Polyesters (manufacture)		24200	Seed dressing (manufacture)
	24160	Polyethers (manufacture)		24200	Sheep dip (manufacture)
	24160	Polyethylene (manufacture)		24200	Soil fumigant (manufacture)
	24160	Polymers (manufacture)		24200	Weed killer (manufacture)
	24160	Polypropylene (manufacture)			
	24160	Polystyrene (manufacture)	**20301**		**Manufacture of paints, varnishes and similar coatings, mastics and sealants**
	24160	Polytetrafluoroethylene (PTFE) (manufacture)		24301	Acrylic paints (manufacture)
	24160	Polyurethanes (manufacture)		24301	Alkyd (manufacture)
	24160	Polyvinyl acetate (PVA) (manufacture)		24301	Aluminium paint (manufacture)
	24160	Polyvinyl chloride (PVC) (manufacture)		24301	Aluminium paste (manufacture)
	24160	Propylene polymers (manufacture)		24301	Anti-corrosive paint (manufacture)
	24160	PVC polyvinyl chloride as a raw material (manufacture)		24301	Artists' colours (manufacture)
	24160	Resins for paint (manufacture)		24301	Bituminous paint (manufacture)
	24160	Resins made of urea formaldehyde (manufacture)		24303	Caulking compounds and similar non-refractory filling or surfacing preparations (manufacture)
	24160	Silicones (manufacture)			

UK SIC 2007	UK SIC 2003	Activity
	24301	Cellulose paint (manufacture)
	24301	Cellulose varnish (manufacture)
	24301	Cement based paint (manufacture)
	24301	Ceramic colours (manufacture)
	24301	Ceramic glaze (manufacture)
	24301	Chlorinated rubber based paint (manufacture)
	24301	Distemper (manufacture)
	24301	Electrocoats paint (manufacture)
	24301	Emulsion paint (manufacture)
	24301	Enamel (manufacture)
	24301	Epoxy paint (manufacture)
	24303	Filling and sealing compounds for painters (manufacture)
	24301	French polish (manufacture)
	24301	Glass powder (manufacture)
	24301	Glazes and engobes and similar preparations (manufacture)
	24301	Lacquer (manufacture)
	24301	Lead paint (manufacture)
	24301	Liquid lustres (manufacture)
	24301	Marine paint (manufacture)
	24303	Mastics (manufacture)
	24301	Metal pre-treatment paint (manufacture)
	24301	Metallic paint (manufacture)
	24301	Nitrogen resin type paint (manufacture)
	24301	Oleo resinous paint (manufacture)
	24301	Opacifiers and colours (manufacture)
	24301	Organic composite solvents (manufacture)
	24301	Paint (not cement based) (manufacture)
	24301	Paint removers (manufacture)
	24301	Paint with cement base (manufacture)
	24301	Paste made of aluminium (manufacture)
	24301	Pigments (prepared) (manufacture)
	24301	Polyester paint (manufacture)
	24301	Polyurethane paint (manufacture)
	24301	Powder made of glass (manufacture)
	24301	Prepared dyes (manufacture)
	24301	Primer paint (manufacture)
	24303	Putty (manufacture)
	24301	Rubber based paint (chlorinated) (manufacture)
	24303	Sealants (other than bituminous or oil based) (manufacture)
	24301	Shellac varnish (manufacture)
	24301	Ship's bottom composition (manufacture)
	24301	Stain (manufacture)
	24301	Thinners for paint and varnish (manufacture)
	24301	Varnish (manufacture)
	24301	Varnish removers (manufacture)
	24301	Vinyl paint (manufacture)
	24301	Vitreous enamel frits (manufacture)
	24301	Vitrifiable enamels (manufacture)
	24301	White lead in paste form (manufacture)
	24301	Wood stain (manufacture)
	24301	Zinc paint (manufacture)
20302		**Manufacture of printing ink**
	24302	Flexographic ink (manufacture)
	24302	Gravure ink (manufacture)
	24302	Letterpress ink (manufacture)

SIC 2007	SIC 2003	Activity
	24302	Lithographic ink (manufacture)
	24302	News ink (manufacture)
	24302	Printers' varnish (manufacture)
	24302	Printing ink (manufacture)
	24302	Screen process ink (manufacture)
20411		**Manufacture of soap and detergents**
	24511	Abrasive soap (manufacture)
	24511	Carpet soap (manufacture)
	24511	Crude glycerol (manufacture)
	24511	Detergent (soapless, formulated) (manufacture)
	24511	Detergent (synthetic) (manufacture)
	24511	Dish-washing preparations (manufacture)
	24511	Dog soap (manufacture)
	24511	Glycerol (manufacture)
	24511	Hard soap (manufacture)
	24511	Industrial soap (manufacture)
	24511	Liquid soap (manufacture)
	24511	Oil dispersant (manufacture)
	24511	Organic surface-active agents (manufacture)
	24511	Soap (manufacture)
	24511	Soap chips (manufacture)
	24511	Soap flakes (manufacture)
	24511	Soap or detergent coated felt, paper and wadding (manufacture)
	24511	Soap powder (manufacture)
	24511	Surface active preparations (manufacture)
	24511	Synthetic detergent (manufacture)
	24511	Textile soap (manufacture)
	24511	Textile softeners (manufacture)
	24511	Washing powders and preparations in solid or liquid form (manufacture)
20412		**Manufacture of cleaning and polishing preparations**
	24512	Artificial waxes (manufacture)
	24512	Car polish (manufacture)
	24512	Cleaning and polishing preparations (manufacture)
	24512	Cleaning powder (other than detergents and scouring powder) (manufacture)
	24512	Deodoriser for household use (manufacture)
	24512	Deodorisers (manufacture)
	24512	Floor cleanser (manufacture)
	24512	Floor polish (manufacture)
	24512	Floor seal (manufacture)
	24512	Furniture polish (manufacture)
	24512	Glass polish (manufacture)
	24512	Impregnated cleaning and polishing cloth (manufacture)
	24512	Metal polish (manufacture)
	24512	Plate polish (manufacture)
	24512	Polish (manufacture)
	24512	Polishes and creams (manufacture)
	24512	Polishes and creams for leather (manufacture)
	24512	Polishes and creams for wood (manufacture)
	24512	Polishes for coachwork (manufacture)
	24512	Polishes for glass (manufacture)
	24512	Polishes for metal (manufacture)
	24512	Polishing paste and powder (manufacture)
	24512	Prepared waxes (manufacture)

UK SIC 2007	UK SIC 2003	Activity	SIC 2007	SIC 2003	Activity
	24512	Sanitary cleanser (manufacture)		24610	Chlorate explosive (manufacture)
	24512	Scouring paste or powder coated paper (manufacture)		24610	Cordite (manufacture)
	24512	Scouring pastes (manufacture)		24610	Detonating fuse (manufacture)
	24512	Scouring powder (manufacture)		24610	Detonator (manufacture)
	24512	Shoe dye (manufacture)		24610	Dynamite (manufacture)
	24512	Shoe polish (manufacture)		24610	Explosive signalling flares (manufacture)
	24512	Ski wax (manufacture)		24610	Explosives (manufacture)
	24512	Wax (manufacture)		24610	Firework (manufacture)
				24610	Fog signal (manufacture)
20420		**Manufacture of perfumes and toilet preparations**		24610	Fuse for explosives (manufacture)
	24520	After shave lotion (manufacture)		24610	Gelignite (manufacture)
	24520	Anti-perspirant (manufacture)		24610	Guncotton (manufacture)
	24520	Bath preparations (manufacture)		24610	Gunpowder (manufacture)
	24520	Bath salts (manufacture)		24610	Incendiary composition (manufacture)
	24520	Beauty and make-up preparations (manufacture)		36639	Match (manufacture)
	24520	Brushless shaving cream (manufacture)		24610	Nitro glycerine (manufacture)
	24520	Colognes (manufacture)		24610	Perchlorate explosive (manufacture)
	24511	Cosmetic soap (manufacture)		24610	Percussion cap (manufacture)
	24520	Cosmetics (manufacture)		24610	Propellant powder (manufacture)
	24520	Dental cleansing preparation (manufacture)		24610	Propergol fuels and other propellant powders (manufacture)
	24520	Dentifrices (manufacture)		24610	Pyrotechnics (manufacture)
	24520	Denture fixative preparations (manufacture)		24610	Rain rocket (manufacture)
	24520	Deodorant (manufacture)		24610	Safety fuse (manufacture)
	24520	Depilatory (manufacture)		24610	Signal rocket (manufacture)
	24520	Face powder or cream (manufacture)		24610	Trinitrotoluene (TNT) (manufacture)
	24520	Hair lacquers (manufacture)			
	24520	Hair preparations (manufacture)	**20520**		**Manufacture of glues**
	24520	Hand cream (manufacture)		24620	Acrylic adhesives (manufacture)
	24520	Lipstick (manufacture)		24620	Adhesive (formulated) (manufacture)
	24520	Make-up preparation (manufacture)		24620	Adhesive coating (manufacture)
	24520	Manicure and pedicure preparations (manufacture)		24620	Adhesive made of urea formaldehyde (manufacture)
	24520	Nail preparation (cosmetic) (manufacture)		24620	Adhesive paste (manufacture)
	24520	Oral hygiene preparations (manufacture)		24620	Anaerobic adhesive (manufacture)
	24520	Perfume (manufacture)		24620	Bone glue (manufacture)
	24520	Pre-shave lotion (manufacture)		24620	Casein based adhesive (manufacture)
	24520	Shampoo (manufacture)		24620	Cellulose based adhesive (manufacture)
	24520	Shaving cream (brushless) (manufacture)		24620	Cross linking adhesive (manufacture)
	24520	Shaving cream (manufacture)		24620	Cyanoacrylate adhesive (manufacture)
	24520	Shaving preparations (manufacture)		24620	Decorators' size (manufacture)
	24511	Shaving soap (manufacture)		24620	Dextrin based adhesive (manufacture)
	24520	Skin care preparations (manufacture)		24620	Emulsion adhesive (manufacture)
	24520	Sunburn prevention and sun tan preparations (manufacture)		24620	Epoxide adhesive (manufacture)
	24520	Talcum powder (manufacture)		24620	Glue (manufacture)
	24520	Toilet preparations (manufacture)		24620	Gum (manufacture)
	24511	Toilet soap (manufacture)		24620	Hot melt adhesive (manufacture)
	24520	Toilet water (manufacture)		24620	Industrial adhesives (manufacture)
	24520	Toiletries (manufacture)		24620	Moutant (manufacture)
	24520	Tooth powder (manufacture)		24620	Polyester adhesive (manufacture)
	24520	Toothpaste (manufacture)		24620	Polyurethane adhesive (manufacture)
	24520	Waving and hair straightening preparations (manufacture)		24620	Polyvinyl acetate and co-polymer adhesives (manufacture)
				24620	Resorcinol formaldehyde adhesive (manufacture)
20510		**Manufacture of explosives**		24620	Rubber based glues and adhesives (manufacture)
	24610	Ammonium nitrate for explosives (manufacture)		24620	Starch based adhesives (manufacture)
	24610	Amorce (manufacture)		24620	Synthetic resin adhesive (manufacture)
	24610	Black powder (manufacture)			
	24610	Blasting powder	**20530**		**Manufacture of essential oils**
				24630	Aromatic distilled waters (manufacture)

UK SIC 2007	UK SIC 2003	Activity
	24630	Compound flavour (blended flavour concentrates) (manufacture)
	24630	Essential oils and essence (other than turpentine) (manufacture)
	24630	Extracts of aromatic products (manufacture)
	24630	Natural material used in flavours or perfumes (manufacture)
	24630	Odoriferous products (manufacture)
	24630	Perfume compounds (blended perfume concentrates) (manufacture)
	24630	Perfumery and flavour synthetic chemicals (manufacture)
	24630	Resinoids (manufacture)
20590		**Manufacture of other chemical products n.e.c.**
	24660	Activated carbon (manufacture)
	24660	Anti-freeze mixtures (excluding pure ethyl glycol) (manufacture)
	24660	Anti-knock compounds (manufacture)
	24660	Anti-rust preparations (manufacture)
	24660	Brewing preparations (excluding yeast) (manufacture)
	24660	Catalysts (manufacture)
	24660	Chemicals specially prepared for laboratory use (manufacture)
	24640	Cinematographic sensitized film (manufacture)
	24640	Clearing agents for photographic use (manufacture)
	24660	Composite diagnostic or laboratory reagents (manufacture)
	24660	Compound plasticisers for rubber or plastics (manufacture)
	24660	De-icing fluid (manufacture)
	24660	Dental wax (manufacture)
	24660	Desiccants (chemical) (manufacture)
	24660	Doped compounds for use in electronics (manufacture)
	24660	Drawing ink (manufacture)
	24660	Drilling mud (manufacture)
	24660	Duplicating ink (manufacture)
	24660	Finings (manufacture)
	24660	Fire extinguishing chemicals (manufacture)
	24640	Fixer for photographic use (manufacture)
	24660	Flux (manufacture)
	24660	Foundry bonding clays (manufacture)
	24660	Foundry core binder (manufacture)
	24660	Foundry facing (manufacture)
	24660	Foundry preparation (manufacture)
	24660	Fuel additive (manufacture)
	24660	Fusel oil (manufacture)
	24620	Gelatine (manufacture)
	24620	Gelatine derivatives (manufacture)
	24660	Heat treatment salts (manufacture)
	24660	Hydraulic brake fluid (less than 70% petroleum oil) (manufacture)
	24660	Hydraulic transmission liquids (manufacture)
	24660	Indian ink (manufacture)
	24660	Industrial catalyst (manufacture)
	24660	Industrial cleaning preparation (manufacture)
	24660	Ink for impregnating ink pads (manufacture)
	24640	Instant print film (manufacture)
	24640	Intensifier for photographic use (manufacture)
	24620	Isinglass (manufacture)

SIC 2007	SIC 2003	Activity
	24640	Litho plate making (sensitized) (manufacture)
	24660	Lubricating oil additive (manufacture)
	24660	Marking ink (manufacture)
	24660	Metal pickling substances (manufacture)
	24660	Metal treatment chemical (manufacture)
	24660	Modelling pastes (manufacture)
	24660	Oil additive (manufacture)
	24660	Oils and fats (chemically modified) (manufacture)
	24660	Peptone derivatives (manufacture)
	24660	Peptones (manufacture)
	24640	Photographic chemicals (manufacture)
	24640	Photographic developer (manufacture)
	24640	Photographic film (sensitized) (manufacture)
	24640	Photographic film plate (sensitised) (manufacture)
	24640	Photographic plates (manufacture)
	24640	Photographic unexposed film (manufacture)
	24660	Pickling preparations for metal treatment (manufacture)
	24660	Powders and pastes used in soldering, brazing or welding (manufacture)
	24660	Prepared additives for cement (manufacture)
	24660	Prepared culture media for micro-organisms (manufacture)
	24660	Protein substances (manufacture)
	24660	Rubber accelerators (manufacture)
	24660	Rubber processing chemicals (manufacture)
	24640	Sensitized cloth (manufacture)
	24640	Sensitized emulsions for photographic use (manufacture)
	24640	Sensitized paper (manufacture)
	24660	Stabilisers and extenders for PVC processing (manufacture)
	24660	Stabilisers for rubber or plastics (manufacture)
	24660	Stearin (manufacture)
	24660	Surface active chemicals (excluding finished detergents and scouring powder) (manufacture)
	24660	Textile chemical auxiliaries (manufacture)
	24660	Textiles and leather finishing materials (manufacture)
	24640	Toner for photographic use (manufacture)
	24640	Unexposed materials (manufacture)
	24660	Vegetable-based bio diesel (manufacture)
	24660	Vegetable-based biodiesel (manufacture)
	24660	Water treatment chemicals (manufacture)
	24660	Wine making preparations (excluding yeast) (manufacture)
	24660	Writing ink (manufacture)
20600		**Manufacture of man-made fibres**
	24700	Continuous filament yarn of man-made fibres (manufacture)
	24700	Fibrillated yarn (manufacture)
	24700	Filament tow (manufacture)
	24700	High tenacity yarn made of viscose rayon (manufacture)
	24700	Man-made fibre (not glass fibre) (manufacture)
	24700	Man-made staple fibres, not carded, combed or otherwise processed for spinning (manufacture)
	24700	Man-made tow (manufacture)
	24700	Monofilament or strip (manufacture)
	24700	Polyamide man-made fibre (manufacture)
	24700	Polyester man-made fibre (manufacture)

UK SIC 2007	UK SIC 2003	Activity	SIC 2007	SIC 2003	Activity
	24700	Rayon (manufacture)		24421	Immunoglobin (manufacture)
	24700	Single yarn of man-made continuous fibres, including high tenacity and textured yarn (manufacture)		24421	Medical diagnostic preparations, including pregnancy tests (manufacture)
	24700	Staple fibre of acetate, synthetic or viscose production (manufacture)		24421	Medical impregnated bandages, dressings, gauze and wadding (manufacture)
	24700	Staples of man-made fibre (manufacture)		24421	Medicaments (manufacture)
	24700	Synthetic fibre (manufacture)		24421	Medicine (manufacture)
	24700	Textured single yarn (manufacture)		24421	Ointment (manufacture)
				24421	Opacifying preparations for x-ray examinations (manufacture)
21100		**Manufacture of basic pharmaceutical products**		24421	Oral contraceptives (manufacture)
	24410	Antibiotics (manufacture)		24421	Pharmaceutical medicament products (manufacture)
	24410	Blood processing (manufacture)		24421	Pharmaceuticals for veterinary use (manufacture)
	24410	Chemically pure sugars (manufacture)		24421	Pills (medicinal) (manufacture)
	24410	Cyclamates (manufacture)		23300	Radioactive in-vivo diagnostic substances (manufacture)
	24410	Fatty amines and quaternary ammonium salts (manufacture)		24421	Sera (manufacture)
	24410	Gland extracts (manufacture)		24421	Serum albumin (manufacture)
	24410	Gland processing (manufacture)		24421	Urological reagents (manufacture)
	24410	Glycosides and their salts, ethers, esters and other derivatives (manufacture)		24421	Vaccine (manufacture)
	24410	Hormone (not plant hormone) (manufacture)		24421	Veterinary medicines (manufacture)
	24410	Lactones (other than coumarin, methylcoumarins and ethylcoumarins) (manufacture)		24421	Veterinary pharmaceuticals (manufacture)
	24410	Medicinal active substances to be used for their pharmacological properties (manufacture)	**22110**		**Manufacture of rubber tyres and tubes; retreading and rebuilding of rubber tyres**
	24410	Medicinal feed additives (veterinary) (manufacture)		25110	Camel back strips for retreading tyres (manufacture)
	24410	Microbiological cultures, toxins, etc. (manufacture)		25110	Inner tube for tyre (manufacture)
	24410	Mineral and pharmaceutical nutritional ingredients for food and feeding stuff (manufacture)		25110	Interchangeable tyre flaps for retreading tyres (manufacture)
	24410	O-acetylsalicylic acids (manufacture)		25110	Interchangeable tyre treads for retreading tyres (manufacture)
	24410	Pharmaceutical chemicals (manufacture)		25110	Pneumatic tyres (manufacture)
	24410	Provitamins, vitamins and their derivatives (manufacture)		25120	Repair of tyres and inner tubes by specialists (manufacture)
	24410	Quaternary ammonium salts and fatty amines (manufacture)		25110	Rubber tyres for furniture and other uses (manufacture)
	24410	Salicylic medicaments (manufacture)		25110	Rubber tyres for mobile machinery (manufacture)
	24410	Salines (manufacture)		25110	Rubber tyres for toys (manufacture)
	24410	Sulphonamides (manufacture)		25110	Tyre flaps (manufacture)
	24410	Veterinary biologicals (manufacture)		25120	Tyre rebuilding and retreading (manufacture)
	24410	Veterinary feed additives (medicinal) (manufacture)		25120	Tyre retreading (manufacture)
				25110	Tyres (manufacture)
21200		**Manufacture of pharmaceutical preparations**		25110	Tyres for aircraft (manufacture)
	24421	Anaesthetics (manufacture)		25110	Tyres for cars or vans (manufacture)
	24421	Analgesics (manufacture)		25110	Tyres for commercial vehicles (manufacture)
	24421	Anti-infectives (manufacture)		25110	Tyres for cycles (manufacture)
	24421	Antiseptics (manufacture)		25110	Tyres for industrial use (manufacture)
	24421	Antisera and other blood fractions (manufacture)		25110	Tyres for motorcycles and mopeds (manufacture)
	24421	Biotech pharmaceuticals (manufacture)		25110	Tyres for scooters (manufacture)
	24421	Blood-grouping reagents (manufacture)		25110	Tyres for tractors (manufacture)
	24421	Botanical products for pharmaceutical use (manufacture)		25110	Tyres made of rubber (manufacture)
	24421	Chemical contraceptive products (manufacture)		25110	Tyres made of solid rubber (manufacture)
	24421	Diagnostic reagents (manufacture)	**22190**		**Manufacture of other rubber products**
	24421	Drug (medicinal) (manufacture)		25130	Adhesive repair material made of rubber (manufacture)
	24421	Embrocation (manufacture)		25130	Adhesive tape of rubberised textile (manufacture)
	24421	Enema preparations (manufacture)		25130	Apparel made of rubber (if only sealed together, not sewn) (manufacture)
	24421	Gammaglobulin (manufacture)		25130	Apparel made of sealed rubber (manufacture)
	24421	Homeopathic preparations (manufacture)		25130	Aprons (manufacture)
	24421	Hormonal contraceptive medicaments (manufacture)			
	24421	Human plasma extract (manufacture)			

UK SIC 2007	UK SIC 2003	Activity	SIC 2007	SIC 2003	Activity
	25130	Armoured hose made of rubber (manufacture)		25130	Mouldings for upholstery made of rubber (manufacture)
	25130	Balata belting (manufacture)		25130	Piece goods made of unsupported rubber sheeting (manufacture)
	25130	Balata goods (excluding belting) (manufacture)			
	25130	Ball core (rubber) (manufacture)		36639	Pipe stems of hard rubber (manufacture)
	25130	Balloons, rubber (except pilot and sounding balloons, dirigibles and hot-air balloons) (manufacture)		25130	Pipes made of rubber (manufacture)
				25130	Plates made of rubber (semi-manufactures) (manufacture)
	25130	Bands made of elastic (manufacture)		25130	Printers' blankets made of rubber (manufacture)
	25130	Bands made of rubber (manufacture)		25130	Profile shapes of rubber (manufacture)
	25130	Bathing caps of rubber (manufacture)		25130	Puncture repair outfit made of rubber (manufacture)
	25130	Bellows made of rubber (manufacture)		25130	Reinforced hose made of rubber (manufacture)
	25130	Belting for domestic appliances made of rubber (manufacture)		25130	Repair materials made of rubber (manufacture)
	25130	Belting made of rubber (manufacture)		25130	Rings and washers made of rubber (manufacture)
	19300	Bootee (rubber protective) (manufacture)		25130	Rods made of rubber (manufacture)
	36620	Brushes of rubber (manufacture)		25130	Roller coverings made of rubber (manufacture)
	25130	Buckets made of rubberised fabric (manufacture)		25130	Roller covers made of rubber (manufacture)
	25130	Carpet underlay made of rubber (manufacture)		25130	Rope made of rubber (manufacture)
	25130	Cellular rubber products (manufacture)		25130	Rubber (vulcanized, unvulcanized or hardened) (manufacture)
	36639	Combs of hard rubber (manufacture)			
	25130	Conveyor belts made of rubber (manufacture)		25130	Rubber compounds (manufacture)
	25130	Cushioning for upholstery made of rubber (manufacture)		25130	Rubber products (manufacture)
				25130	Rubber thread (uncovered) (manufacture)
	25130	Delivery hose made of rubber (manufacture)		25130	Rubberised fabrics (manufacture)
	25130	Diving suit made of rubber (manufacture)		25130	Rubberised hair (manufacture)
	25130	Ebonite, vulcanite or hard rubber goods (manufacture)		25130	Rubberised textile fabric (manufacture)
				25130	Seals made of rubber (manufacture)
	25130	Elevator belting made of rubber (manufacture)		25130	Semi-finished products made of rubber (manufacture)
	25130	Eraser rubber (manufacture)			
	25130	Expansion joints made of rubber (manufacture)		36639	Sex articles of rubber (manufacture)
	25130	Fan belts for motor vehicles (manufacture)		25130	Sheath contraceptives made of rubber (manufacture)
	25130	Felting made of rubber (manufacture)		25130	Sheets made of rubber (manufacture)
	25130	Fittings made of rubber (manufacture)		25130	Sponges made of rubber (manufacture)
	25130	Floor coverings made of rubber (manufacture)		25130	Strips made of rubber (manufacture)
	25130	Flooring made of rubber (manufacture)		25130	Suction and discharge hose made of rubber (manufacture)
	25130	Fluid seals made of rubber (manufacture)			
	25130	Foam rubber (manufacture)		25130	Teats made of rubber (manufacture)
	25130	Garden hose made of rubber (manufacture)		25130	Tennis ball core (manufacture)
	25130	Gloves and gauntlets of unstitched rubber (manufacture)		25130	Tiling made of rubber (manufacture)
				25130	Timing belt for motor vehicles (manufacture)
	25130	Golf ball core (manufacture)		25130	Transmission belting made of rubber (manufacture)
	25130	Groundsheet made of rubber (manufacture)		25130	Transmission v-belts made of rubber (manufacture)
	25130	Grout packers (manufacture)		25130	Tubes made of rubber (manufacture)
	25130	Gutta percha goods (manufacture)		25130	Tubing made of rubber (manufacture)
	36639	Hair pins of hard rubber (manufacture)		25130	Tyre repair materials and kits (manufacture)
	36639	Hair rollers and similar of hard rubber (manufacture)		25130	Underwater swimming suit made of rubber (manufacture)
	19300	Heel and sole made of rubber (manufacture)			
	25130	Hose made of rubber (manufacture)		25130	Washers made of rubber (manufacture)
	25130	Hot water bottles made of rubber (manufacture)		25130	Waterbed mattresses of rubber (manufacture)
	25130	Hydraulic hose made of rubber (manufacture)		25130	Wet suits of rubber (manufacture)
	25130	Hygienic articles made of rubber (manufacture)		25130	Yarn (rubberised) (manufacture)
	25130	Industrial belting made of rubber (manufacture)	**22210**		**Manufacture of plastic plates, sheets, tubes and profiles**
	25130	Inflatable cushion made of rubber (manufacture)			
	25130	Inflatable mattress made of rubber (manufacture)		25210	Belting made of plastic (manufacture)
	25130	Insulating cloth tape (manufacture)		25210	Blocks made of plastic (manufacture)
	25130	Insulating material made of rubber (manufacture)		25210	Carpet underlay made of plastic (manufacture)
	25130	Latex foam (manufacture)		25210	Conveyor belts made of plastic (manufacture)
	25130	Mats made of rubber (manufacture)		25210	Decorative unsupported polyvinyl chloride film and sheet (manufacture)
	19300	Moulded rubber bottoms for footwear (manufacture)			
				25210	Drainpipes and fittings made of plastic (manufacture)

UK SIC 2007	UK SIC 2003	Activity	SIC 2007	SIC 2003	Activity
	25210	Film and sheet of decorated unsupported polyvinyl chloride (manufacture)		25220	Egg boxes made of plastic (manufacture)
	25210	Film made of cellophane (manufacture)		25220	Jam pot covers made of plastic (manufacture)
	25210	Film made of plastic (manufacture)		25220	Jars made of plastic (manufacture)
	25210	Film made of polyethylene (manufacture)		25220	Jerry can made of plastic (manufacture)
	25210	Film made of polypropylene (manufacture)		25220	Kegs made of plastic (manufacture)
	25210	Film made of polythene (manufacture)		25220	Liner made of non-woven polyethylene (manufacture)
	25210	Film made of polyvinyl chloride (PVC) (manufacture)		25220	Mesh bags made of plastic (manufacture)
	25210	Flexible plastic foam (manufacture)		25220	Non-woven liners made of polyethylene (manufacture)
	25210	Foil made of plastic (manufacture)			
	25210	Hose and pipe fittings made of plastic (manufacture)		25220	Non-woven sacks made of polyethylene (manufacture)
	25210	Hose made of plastic (manufacture)		25220	Packaging products made of plastic (manufacture)
	25210	Laminate made wholly of plastics and/or transparent regenerated cellulose film (manufacture)		25220	Packing goods made of plastic (manufacture)
				25220	Pots made of plastic (not flower pots) (manufacture)
	25210	Laminated plastic film (manufacture)		25220	Reels made of printed polypropylene (manufacture)
	25210	Laminated thermosetting plastics sheet (manufacture)		25220	Sachets made of plastic (manufacture)
	25210	Laminates made of plastic (manufacture)		25220	Sacks made of non-woven polyethylene (manufacture)
	25210	Photographic film (unsensitized) (manufacture)			
	25210	Photographic unsensitized film (manufacture)		25220	Sacks made of plastic (manufacture)
	25210	Pipes and fittings made of plastic (manufacture)		25220	Tube containers made of plastic (manufacture)
	25210	Plastic semi-manufactures (manufacture)		25220	Tubs made of plastic (manufacture)
	25210	Plates made of plastic (semi-manufactures) (manufacture)	**22230**		**Manufacture of builders' ware of plastic**
	25210	Profile shapes of plastic materials (rods, tubes, etc.) (manufacture)		25239	Architrave made of plastic (manufacture)
				25239	Artificial stone made of plastic (manufacture)
	25210	Rigid plastic foam (manufacture)		25239	Baths made of fibre glass (manufacture)
	25210	Sheets made of cellophane (manufacture)		25239	Baths made of plastic (manufacture)
	25210	Sheets made of laminated thermosetting plastic (manufacture)		25239	Blinds made of plastic (manufacture)
				25239	Builders' ware made of plastic (manufacture)
	25210	Sheets made of plastic (manufacture)		25239	Building products made of plastic (manufacture)
	25210	Sheets made of polyethylene (manufacture)		25239	Ceiling coverings made of plastic (manufacture)
	25210	Sheets made of polypropylene (manufacture)		25239	Ceiling tiles made of plastic (manufacture)
	25210	Sheets made of polyvinyl chloride (PVC) (manufacture)		25239	Cistern floats made of plastic (manufacture)
				25239	Cisterns made of plastic (manufacture)
	25210	Strips made of plastic (manufacture)		25239	Coving made of plastic (manufacture)
	25210	Tubes made of plastic (manufacture)		25239	Dome lights made of plastic (manufacture)
	25210	Tubing made of plastic (manufacture)		25239	Door frames made of plastic (manufacture)
	25210	Vulcanized fibre (manufacture)		25239	Door furniture for buildings (handles, hinges, knobs, etc.) Made of plastic (manufacture)
22220		**Manufacture of plastic packing goods**		25239	Doors made of plastic (manufacture)
	25220	Bags made of plastic (not designed for prolonged use) (manufacture)		25239	Double glazing made of plastic (manufacture)
				25239	Ducting made of plastic (manufacture)
	25220	Bags made of plastic for packaging (manufacture)		25239	Fencing made of plastic (manufacture)
	25220	Bags made of polyethylene (manufacture)		36639	Floor coverings (hard surface) (manufacture)
	25220	Bags made of transparent regenerated cellulose film (manufacture)		25231	Floor coverings made of plastic (manufacture)
				25231	Floor coverings made of printed vinyl (manufacture)
	25220	Barrels made of plastic (manufacture)		25231	Floor coverings made of supported vinyl (manufacture)
	25220	Bin liners made of plastic (manufacture)			
	25220	Bottle crates made of plastic (manufacture)		25239	Flushing cisterns made of plastic (manufacture)
	25220	Bottles made of plastic (manufacture)		25239	Frames made of plastic (manufacture)
	25220	Boxes made of plastic (manufacture)		25239	Gutter and fittings made of plastic (manufacture)
	25220	Canisters made of plastic (manufacture)		25239	Insulating (heat and sound) sheet, tiles, blocks and granules made of plastic (manufacture)
	25220	Caps and closures made of plastic (manufacture)			
	25220	Caps for bottles made of plastic (manufacture)		25239	Intermediate bulk containers (other than drums made of plastic) (manufacture)
	25220	Carboy made of plastic (manufacture)			
	25220	Cases made of plastic (manufacture)		25239	Lavatory pans made of plastic (manufacture)
	25220	Closures made of plastic (manufacture)		36639	Linoleum (manufacture)
	25220	Container made of plastic for closed transit (manufacture)		25239	Manhole or access covers made of plastic (manufacture)
	25220	Drums (containers) made of plastic (manufacture)		25231	Matting made of woven plastic (manufacture)

UK SIC 2007	UK SIC 2003	Activity	SIC 2007	SIC 2003	Activity
	25239	Meter housing boxes made of plastic (manufacture)		36639	Combs of plastic (manufacture)
	25239	Prefabricated buildings made of plastic (manufacture)		25240	Conveyer belts made of plastic (manufacture)
	25231	Printed felt base floorcovering (manufacture)		25240	Cooking utensils made of plastic (manufacture)
	25239	Reservoirs made of plastic (manufacture)		25240	Cups made of plastic (manufacture)
	25239	Roof lights made of plastic (manufacture)		25240	Curtain hooks, rings and runners made of plastic (manufacture)
	25239	Sanitary ware made of plastic (manufacture)		25240	Curtain rail, rollers and fittings made of plastic (manufacture)
	25239	Sheeting made of plastic for roofing and cladding (manufacture)		25240	Cutlery made of plastic (manufacture)
	25239	Shiplap cladding made of plastic (manufacture)		25240	Dishes made of plastic (manufacture)
	25239	Shower baths made of plastic (manufacture)		25240	Doilies made of plastic (manufacture)
	25239	Shutters made of plastic (manufacture)		25240	Dustbins made of plastic (manufacture)
	25239	Sinks made of plastic (manufacture)		25240	Dustpans made of plastic (manufacture)
	25239	Skirting boards made of plastic (manufacture)		25240	Egg cups made of plastic (manufacture)
	25239	Storage tanks made of plastic (manufacture)		25240	Filtration elements made of plastic (manufacture)
	25239	Tanks made of plastic (open and closed) (manufacture)		25240	Flower pots and tubs made of plastic (manufacture)
	25239	Taps and valves made of plastic (manufacture)		19300	Footwear parts and accessories made of plastic (manufacture)
	25239	Tiles (other than floor tiles made of plastic) (manufacture)		25240	Forks made of plastic (manufacture)
	25231	Tiles made of asphalt thermoplastic (manufacture)		25240	Funnels made of plastic (manufacture)
	25231	Tiles made of vinyl asbestos (manufacture)		25240	Furniture fittings made of plastic (manufacture)
	25231	Vinyl floor covering (homogeneous and printed) (manufacture)		25240	Gloves made of unstitched plastic (manufacture)
	25239	Wall coverings made of plastic (manufacture)		36639	Hair comb made of plastic (manufacture)
	25239	Wash basins made of plastic (manufacture)		36639	Hair curler made of plastic (manufacture)
	25239	Water stop and bar made of plastic (manufacture)		25240	Handles for furniture made of plastic (manufacture)
	25239	WC seat and cover units made of plastic (manufacture)		25240	Headgear made of plastic (manufacture)
	25239	Weatherboarding made of plastic (manufacture)		25240	Heat sensitive adhesive tape made of plastic (manufacture)
	25239	Window blinds and accessories made of plastic (manufacture)		25240	Hollow ware made of plastic (manufacture)
	25239	Window frames made of plastic (manufacture)		25240	Household utensils made of plastic (manufacture)
22290		**Manufacture of other plastic products**		25240	Illuminated street furniture made of plastic (manufacture)
	25240	Adhesive labels of plastic or cellulose (manufacture)		25240	Inflatable air bed made of plastic (manufacture)
	25240	Advertising material made of plastic (manufacture)		25240	Inflatable plastic products (excluding playballs) (manufacture)
	25240	Air beds made of inflatable plastic (manufacture)		25240	Insulated plastic fittings (manufacture)
	25240	Aircraft parts and accessories made of plastic (manufacture)		25240	Kitchenware made of plastic (manufacture)
	25240	Apparel made of plastic (if only sealed together, not sewn) (manufacture)		25240	Knitting needles made of plastic (manufacture)
	25240	Awnings made of plastic (manufacture)		25240	Knives made of plastic (manufacture)
	25240	Baby baths made of plastic (manufacture)		25240	Knobs for furniture made of plastic (manufacture)
	25240	Baskets made of plastic (manufacture)		25240	Labels (not self-adhesive) made of plastic (manufacture)
	25240	Bathing caps made of plastic (manufacture)		25240	Lampshades, reflectors, covers and diffusers made of plastic (manufacture)
	25240	Bins made of plastic (manufacture)		25240	Light fittings made of plastic (manufacture)
	25240	Bowls made of plastic (manufacture)		25240	Lighting fittings parts made of plastic (manufacture)
	25240	Buckets made of plastic (manufacture)		25240	Metal coating of plastics
	36639	Button and button moulds made of plastic (manufacture)		36639	Models for window display made of plastic (manufacture)
	36639	Buttons and button bases (not metal or glass) (manufacture)		25240	Motor vehicle accessories, fittings and parts made of plastic (manufacture)
	25240	Cabinet components made of plastic (manufacture)		25240	Mouldings made of plastic (manufacture)
	25240	Cellulose adhesive tape (manufacture)		25240	Nameplates made of plastic (manufacture)
	25240	Chains made of plastic (manufacture)		25240	Netting made of plastic (not woven or knotted) (manufacture)
	36639	Cigar holder (manufacture)		25240	Notice plates made of plastic (manufacture)
	36639	Cigarette holder (manufacture)		36639	Novelties made of plastic (manufacture)
	25240	Clothes pegs made of plastic (manufacture)		25240	Office supplies made of plastic (manufacture)
	25240	Coat hangers made of plastic (manufacture)		25240	Ornaments made of plastic (manufacture)
	25240	Colanders made of plastic (manufacture)		25240	Pallets made of plastic (manufacture)
				25240	Pedal bins made of plastic (manufacture)

UK SIC 2007	UK SIC 2003	Activity	SIC 2007	SIC 2003	Activity
	25240	Pelmets made of plastic (manufacture)		26120	Mirrors for motor vehicles (manufacture)
	25240	Plastic headgear (other than hard hats) made of plastic (manufacture)		26120	Multiple insulating glass (manufacture)
	25240	Plates made of plastic (tableware) (manufacture)		26120	Multiple-walled insulating units of glass (manufacture)
	36639	Powder compact made of plastic (manufacture)		26120	Safety glass (manufacture)
	25240	Pressure sensitive adhesive tape made of plastic (manufacture)		26120	Stained glass (manufacture)
	25240	Resin bonded glass fibre mouldings (excluding for motor vehicles) (manufacture)		26120	Toughened glass (manufacture)
	25240	Ring binders and folders made of plastic (manufacture)		26120	Window glass cut to size (manufacture)
				26120	Windscreens made of glass (manufacture)
	25240	Road cones made of plastic (manufacture)	**23130**		**Manufacture of hollow glass**
	25240	Road signs made of plastic (manufacture)		26130	Bottle stoppers made of glass (manufacture)
	25240	Rulers made of plastic (manufacture)		26130	Bottles made of glass or crystal (manufacture)
	25240	School supplies made of plastic (manufacture)		26130	Bowls made of glass (manufacture)
	25240	Screws made of plastic (manufacture)		26130	Bulbs for vacuum flask inners (manufacture)
	25240	Seed trays made of plastic (manufacture)		26130	Carboys made of glass (manufacture)
	25240	Self-adhesive tapes made of plastic (manufacture)		26130	Containers made of glass or crystal (manufacture)
	25240	Shoe lasts made of plastic (manufacture)		26130	Containers made of tubular glass (manufacture)
	25240	Signs made of plastic (manufacture)		26130	Crystal articles (manufacture)
	36639	Slide fasteners made of plastic (manufacture)		26130	Culinary glassware (manufacture)
	25240	Spoons made of plastic (manufacture)		26130	Drinking glass (manufacture)
	25240	Statuettes made of plastic (manufacture)		26130	Glass inners for vacuum flasks and other vacuum vessels (manufacture)
	25240	Tableware made of plastic (manufacture)		26130	Glassware for domestic use (manufacture)
	36639	Tailors' dummy made of plastic (manufacture)		26130	Heat resisting glassware for cooking purposes (manufacture)
	19300	Toe puff (manufacture)			
	25240	Toilet articles made of plastic (manufacture)		26130	Hollow glass (manufacture)
	25240	Transmission belts made of plastic (manufacture)		26130	Jars made of glass (manufacture)
	25240	Trays made of plastic (manufacture)		26130	Kitchenware made of glass (manufacture)
	25240	Valves for aerosols (moulded components) (manufacture)		26130	Ornaments made of glass (manufacture)
				26130	Ovenware made of glass (manufacture)
	25240	Vases made of plastic (manufacture)		26130	Pots made of glass (manufacture)
	25240	Water butts made of plastic (manufacture)		26130	Stemmed drinking vessels made of glass (manufacture)
	25240	Water sensitive adhesive tapes made of plastic (manufacture)			
				26130	Stoppers made of glass (manufacture)
	25240	Watering cans made of plastic (manufacture)		26130	Syphons made of glass (manufacture)
	36639	Zip fasteners made of plastic (manufacture)		26130	Tableware made of glass (manufacture)
				26130	Tableware made of lead crystal (manufacture)
23110		**Manufacture of flat glass**		26130	Tubular containers made of glass (manufacture)
	26110	Antique glass (manufacture)		26130	Tumblers made of glass (manufacture)
	26110	Blown glass (manufacture)		26130	Vacuum flask inners (manufacture)
	26110	Cast glass (manufacture)		26130	Vases made of glassware (manufacture)
	26110	Coloured glass (manufacture)		26130	Vial (manufacture)
	26110	Drawn sheet glass (manufacture)		26130	Waste glass resulting from glass container production (manufacture)
	26110	Figured glass (manufacture)			
	26110	Flat glass (manufacture)			
	26110	Float glass (manufacture)	**23140**		**Manufacture of glass fibres**
	26110	Plate glass (manufacture)		26140	Boards made of glass fibre (manufacture)
	26110	Rolled glass (manufacture)		26140	Chopped roving and strand made of glass fibre (manufacture)
	26110	Tinted glass (manufacture)			
	26110	Window glass (not cut to size) (manufacture)		26140	Doubled glass fibre (manufacture)
	26110	Wired glass (manufacture)		26140	Doubled yarn made of glass fibre (manufacture)
				26140	Felt made of glass fibre (manufacture)
23120		**Shaping and processing of flat glass**		26140	Flock made of glass fibre (manufacture)
	26120	Glass mirrors (manufacture)		26140	Glass fibre spinning and doubling (manufacture)
	26120	Glass mirrors for motor vehicles (not further assembled) (manufacture)		26140	Glass fibres (manufacture)
				26140	Glass wool (manufacture)
	26120	Glass shaping and processing (manufacture)		26140	Insulating material made of glass fibre (manufacture)
	26120	Laminated glass (manufacture)		26140	Loose glass fibre (manufacture)
	26120	Mirror glass (manufacture)		26140	Mat made of glass fibre (manufacture)

UK SIC 2007	UK SIC 2003	Activity	SIC 2007	SIC 2003	Activity
	26140	Mattress made of glass fibre (manufacture)		26150	Lens made of coloured glass for rail and road signals (not optically worked) (manufacture)
	26140	Sheets made of glass fibre (manufacture)		26150	Marbles made of glass (manufacture)
	26140	Slabs made of glass fibre (manufacture)		26150	Mosaic cubes made of glass (manufacture)
	26140	Thermal and sound insulating material made of glass fibre (manufacture)		26150	Multicellular glass block (manufacture)
	26140	Tissue made of glass fibre (manufacture)		26150	Optical elements made of glass (not optically worked) (manufacture)
	26140	Yarn made of glass fibre (manufacture)		26150	Optical glass (manufacture)
23190		**Manufacture and processing of other glass, including technical glassware**		26150	Parts for electric lamps and electronic valves made of glass (manufacture)
	26150	Absorption drums made of glass (manufacture)		26150	Pavement light (manufacture)
	26150	Accumulator cell cases made of glass (manufacture)		26150	Paving blocks made of glass (manufacture)
	26150	Ampoules made of glass (hygienic and pharmaceutical) (manufacture)		26150	Pharmaceutical glassware (other than containers) (manufacture)
	26150	Architectural glass (manufacture)		26150	Pipettes made of glass (manufacture)
	26150	Ballotini (manufacture)		26150	Prisms (pressed or moulded, unworked) (manufacture)
	26150	Bars made of glass (manufacture)		26150	Rods made of glass (manufacture)
	26150	Basement lights made of glass (manufacture)		26150	Shades made of glass (manufacture)
	26150	Beads made of glass (manufacture)		26150	Signalling glassware (manufacture)
	26150	Blanks for corrective spectacle lens (manufacture)		26150	Spectacle glass (manufacture)
	26150	Bricks made of glass (manufacture)		26150	Sunglass blank (manufacture)
	26150	Bulbs made of glass (manufacture)		26150	Tanks made of glass (manufacture)
	26150	Burettes made of glass (manufacture)		26150	Test tube (manufacture)
	26150	Catseye reflector (manufacture)		26150	Tiles made of glass (manufacture)
	26150	Clock and watch glass (manufacture)		26150	Tube fittings made of glass for electric lights (manufacture)
	26150	Container made of glass tubing (hygienic and pharmaceutical) (manufacture)		26150	Tubes made of glass (manufacture)
	26150	Desiccator made of glass (manufacture)		26150	Tubing made of glass (manufacture)
	26150	Electrical insulators made of glass (manufacture)		26150	Volumetric glassware (manufacture)
	26150	Enamel glass in the mass (manufacture)		26150	Waste glass resulting from glass product production (other than glass container) (manufacture)
	26150	Envelopes made of glass (manufacture)		26150	Watch glass (manufacture)
	26150	Envelopes made of glass for light bulbs and electronic valves (manufacture)		26150	Well and bulkhead glass (manufacture)
	26150	Fancy articles and goods made of glass (manufacture)	**23200**		**Manufacture of refractory products**
	26150	Figurines made of glass (manufacture)		26260	Bats made of ceramic (manufacture)
	26150	Gauge glass (manufacture)		26260	Blocks made of graphite (manufacture)
	26150	Glass ball (manufacture)		26260	Boiler block (manufacture)
	26150	Glass ball, bar, rod and tube for processing (manufacture)		26260	Bricks and blocks made of refractory ceramic (manufacture)
	26150	Glass in the mass (manufacture)		26260	Bricks and mouldings made of magnesite (manufacture)
	26150	Glassware (hygienic and pharmaceutical) (manufacture)		26260	Bricks made for refractory insulating (manufacture)
	26150	Glassware for laboratory, hygienic or pharmaceutical use (manufacture)		26260	Bricks made of alumina (manufacture)
	26150	Glassware for technical use (manufacture)		26260	Bricks made of bauxite (manufacture)
	26150	Glassware used in imitation jewellery (manufacture)		26260	Bricks made of chrome (manufacture)
	26150	Globes made of glass (manufacture)		26260	Bricks made of chromite (manufacture)
	26150	Graduated glassware (manufacture)		26260	Bricks made of dolomite (manufacture)
	26150	Hygienic glassware (other than containers) (manufacture)		26260	Bricks made of gannister (manufacture)
	26150	Illuminated glassware (manufacture)		26260	Bricks made of high alumina (manufacture)
	26150	Industrial glassware (not container) (manufacture)		26260	Bricks made of magnesite (manufacture)
	26150	Insulated fittings made of glass (manufacture)		26260	Bricks made of magnesite chrome (manufacture)
	26150	Insulators made of glass (manufacture)		26260	Bricks made of refractory (manufacture)
	26150	Jewellery made of glass (manufacture)		26260	Bricks made of silica (manufacture)
	26150	Lamp chimneys made of glass (manufacture)		26260	Bricks made of siliceous (manufacture)
	26150	Lamps made of glass (manufacture)		26260	Bricks made of sillimanite (manufacture)
	26150	Leaded light (manufacture)		26260	Casting pot (manufacture)
	26150	Lens (pressed or moulded, unworked, (not coloured glass for traffic signals)) (manufacture)		26260	Cement made of dolomite (manufacture)
				26260	Cement made of fireclay (manufacture)
				26260	Cement made of high alumina (manufacture)

UK SIC 2007	UK SIC 2003	Activity	SIC 2007	SIC 2003	Activity
	26260	Cement made of silica and siliceous (manufacture)		26300	Encaustic tile (manufacture)
	26260	Cement refractory jointing (manufacture)		26300	Flags made of clay (manufacture)
	26260	Chrome magnesite shape (manufacture)		26300	Flags made of non-refractory ceramic (manufacture)
	26260	Chromite articles (manufacture)		26300	Glazed fireplace brick (manufacture)
	26260	Crucibles made of fireclay (manufacture)		26300	Glazed tile (manufacture)
	26260	Crucibles made of fireclay or graphite (manufacture)		26300	Glazed tiles for fireplaces (manufacture)
	26260	Crucibles made of graphite (manufacture)		26300	Hearth or wall tiles made of non-refractory ceramic (manufacture)
	26260	Crucibles made of refractory ceramic (manufacture)		26300	Hearth tile made of clay (unglazed) (manufacture)
	26260	Firebrick and shape (manufacture)		26300	Mosaic cube (manufacture)
	26260	Furnace block and pot (manufacture)		26300	Mosaic glazed tiles (manufacture)
	26260	Gas mantle ring and rod (manufacture)		26300	Ornamental earthenware glazed tiles (manufacture)
	26260	Gas retort and kiln lining (manufacture)		26300	Paving made of non-refractory ceramic (manufacture)
	26260	Heat insulating ceramic goods made of siliceous fossil meals (manufacture)		26300	Paving tiles made of unglazed clay (manufacture)
	26260	Hollow ware made of refractory (manufacture)		26300	Tessellated glazed pavement tiles (manufacture)
	26260	Intermediate goods of mined or quarried non-metallic minerals e.g. Sand, gravel, clay (manufacture)		26300	Tesserae made of earthenware (manufacture)
				26300	Tiles made of ceramics (manufacture)
	26260	Kiln furniture (manufacture)		26300	Tiles made of glazed earthenware (manufacture)
	26260	Kiln lining (manufacture)		26300	Wall tiles (glazed) (manufacture)
	26260	Magnesite chrome shape (manufacture)		26300	Wall tiles made of unglazed clay (manufacture)
	26260	Mortars made of refractory (manufacture)			
	26260	Mouldable refractory (manufacture)	**23320**		**Manufacture of bricks, tiles and construction products, in baked clay**
	26260	Mouldings made of magnesite (manufacture)		26400	Block flooring made of clay (manufacture)
	26260	Moulds made of silica (manufacture)		26400	Blue brick (manufacture)
	26260	Muffles (refractory product) (manufacture)		26400	Bricks made of ceramic (manufacture)
	26260	Nozzles made of refractory ceramic (manufacture)		26400	Bricks made of clay (manufacture)
	26260	Pillars made of ceramic (manufacture)		26400	Building materials made of clay (non-refractory) (manufacture)
	26260	Pipes made of refractory ceramic (manufacture)		26400	Cable conduit made of clay (manufacture)
	26260	Props made of ceramic (manufacture)		26400	Chimney liners made of clay (manufacture)
	26260	Radiant for gas and electric fire (manufacture)		26400	Chimney pots made of ceramic (manufacture)
	26260	Ramming material made of refractory (manufacture)		26400	Chimney pots made of clay (manufacture)
	26260	Refractory articles containing magnesite, dolomite or chromite (manufacture)		26400	Conduits made of ceramic (manufacture)
	26260	Refractory blocks (manufacture)		26400	Conduits made of clay (manufacture)
	26260	Refractory castable (manufacture)		26400	Drainpipes and fittings made of clay (manufacture)
	26260	Refractory cement (manufacture)		26400	Engineering brick (manufacture)
	26260	Refractory ceramic goods (manufacture)		26400	Floor and quarry tiles made of unglazed clay (manufacture)
	26260	Refractory concretes (manufacture)		26400	Flue tiles made of clay (manufacture)
	26260	Refractory goods (manufacture)		26400	Hollow partition made of clay (manufacture)
	26260	Refractory hollow ware (manufacture)		26400	Pipes and conduits made of clay (manufacture)
	26260	Refractory jointing cement (manufacture)		26400	Pipes and fittings made of ceramics (manufacture)
	26260	Refractory mouldable (manufacture)		26400	Pipes made of clay (manufacture)
	26260	Refractory ramming material (manufacture)		26400	Quarry floor brick (manufacture)
	26260	Refractory tiles (manufacture)		26400	Quarry tile made of clay (manufacture)
	26260	Retorts made of fireclay, silica and siliceous (manufacture)		26400	Roofing tiles made of ceramic (manufacture)
	26260	Retorts made of graphite (manufacture)		26400	Roofing tiles made of unglazed clay (manufacture)
	26260	Retorts made of refractory ceramic (manufacture)		26400	Tiles and construction products, made of baked clay (manufacture)
	26260	Saggar (manufacture)		26400	Unglazed building brick (manufacture)
	26260	Steel moulder's composition (manufacture)			
	26260	Stilts made of ceramic (manufacture)	**23410**		**Manufacture of ceramic household and ornamental articles**
	26260	Tubes made of refractory ceramic (manufacture)		26210	Art pottery (manufacture)
	26260	Tunnel oven refractory (manufacture)		26210	Brown stone pottery (manufacture)
				26210	Ceramic ware for domestic use (manufacture)
23310		**Manufacture of ceramic tiles and flags**		26210	Cups and saucers made of china or porcelain (manufacture)
	26300	Biscuit tile (manufacture)		26210	Earthenware for domestic use (manufacture)
	26300	Decorative tile made of glazed earthenware (manufacture)			
	26300	Enamelled tile (glazed) (manufacture)			

UK SIC 2007	UK SIC 2003	Activity	SIC 2007	SIC 2003	Activity
	26210	Jet ware (pottery) (manufacture)		26510	Clinkers and hydraulic cement (manufacture)
	26210	Kitchenware made of ceramics (manufacture)		26510	Keene's cement (manufacture)
	26210	Ornamental ceramic ware (manufacture)		26510	Portland cement (manufacture)
	26210	Plates made of ceramic for domestic use (manufacture)		26510	Slag cement (manufacture)
	26210	Pottery for domestic use (manufacture)		26510	Superphosphate cements (manufacture)
	26210	Pottery made of stone (manufacture)	**23520**		**Manufacture of lime and plaster**
	26210	Rockingham ware (manufacture)		26520	Agricultural lime processing (manufacture)
	26210	Samian ware (manufacture)		26530	Anhydrite plaster (manufacture)
	26210	Statuettes made of ceramic (manufacture)		26520	Blue lias lime kiln (manufacture)
	26210	Stoneware for domestic use (manufacture)		26530	Building plaster (manufacture)
	26210	Tableware made of ceramic (manufacture)		26520	Calcined dolomite (manufacture)
	26210	Teapots made of ceramic (manufacture)		26530	Calcined sulphate plaster (manufacture)
	26210	Terracotta ware (manufacture)		26530	Gypsum plaster (manufacture)
	26210	Toilet articles made of ceramic (manufacture)		26520	Hydrated lime (manufacture)
	26210	Vases made of ceramic (manufacture)		26520	Hydraulic lime (manufacture)
23420		**Manufacture of ceramic sanitary fixtures**		26520	Lime (manufacture)
	26220	Baths made of ceramic (manufacture)		26530	Plaster (manufacture)
	26220	Bidets made of ceramic, fireclay, etc. (manufacture)		26530	Plaster of Paris (manufacture)
	26220	Sanitary fixtures made of ceramic (manufacture)		26520	Quicklime (manufacture)
	26220	Sanitary ware made of ceramics (manufacture)		26520	Slaked lime (manufacture)
	26220	Sanitary ware made of fireclay (manufacture)	**23610**		**Manufacture of concrete products for construction purposes**
	26220	Sanitary ware made of vitreous china (manufacture)		26610	Blocks made of breeze (manufacture)
	26220	Sinks made of ceramic fireclay (manufacture)		26610	Blocks made of concrete (manufacture)
	26220	Urinals made of ceramic, fireclay, etc. (manufacture)		26610	Boards made of concrete (manufacture)
	26220	Wash basins or sinks made of ceramic, fireclay, etc. (manufacture)		26610	Bricks made of concrete (manufacture)
	26220	Water closet bowls made of ceramic, fireclay, etc. (manufacture)		26610	Bricks made of sand lime (manufacture)
23430		**Manufacture of ceramic insulators and insulating fittings**		26610	Cast concrete products (manufacture)
	26230	Electrical ceramic fittings (manufacture)		26610	Cast stone units made of precast concrete (manufacture)
	26230	Electrical insulating components made of ceramic (manufacture)		26610	Cement articles for use in construction (manufacture)
	26230	Insulated ceramic fittings (manufacture)		26610	Cement products (manufacture)
	26230	Insulators made of ceramic (manufacture)		26610	Cement wood products (manufacture)
23440		**Manufacture of other technical ceramic products**		26610	Cladding wall panels made of precast concrete (manufacture)
	26240	Bacteria bed tile (manufacture)		26610	Flagstones made of precast concrete (manufacture)
	26240	Ceramic chemical products (manufacture)		26610	Floor and wall tiles made of concrete and terrazzo (manufacture)
	26240	Industrial non-refractory ceramic products (manufacture)		26610	Floor units made of precast concrete (manufacture)
	26240	Laboratory non-refractory ceramic products (manufacture)		26610	Gullies made of concrete (manufacture)
	31620	Permanent magnets (ceramic and ferrite) (manufacture)		26610	Kerbs and edging made of pre-cast concrete (manufacture)
	26240	Technical ceramic products (manufacture)		26610	Panels made of concrete (manufacture)
23490		**Manufacture of other ceramic products**		26610	Paving slabs made of concrete (manufacture)
	26250	Agricultural ceramic ware (manufacture)		26610	Pipes made of concrete (manufacture)
	26250	Flower pots made of clay (manufacture)		26610	Posts made of precast concrete (manufacture)
	26250	Jars made of ceramic (manufacture)		26610	Precast cement articles for use in construction (manufacture)
	26250	Packaging products made of ceramic (manufacture)		26610	Precast concrete articles for use in construction (manufacture)
	26250	Pots made of ceramic (manufacture)		26610	Precast concrete products (manufacture)
23510		**Manufacture of cement**		26610	Prefabricated buildings and components made of concrete (manufacture)
	26510	Aluminous cement (manufacture)		26610	Prefabricated structural parts for construction, of cement, concrete, artificial stone (manufacture)
	26510	Calcareous cement (manufacture)		26610	Pressure pipes made of pre-stressed concrete (manufacture)
	26510	Cement (manufacture)		26610	Pre-stressed concrete products (manufacture)

UK SIC 2007	UK SIC 2003	Activity	SIC 2007	SIC 2003	Activity
	26610	Pylons made of precast concrete (manufacture)		26650	Troughs made of fibre cement (manufacture)
	26610	Railway sleepers made of pre-cast concrete (manufacture)		26650	Tubes made of fibre cement (manufacture)
	26610	Reinforced concrete products (manufacture)		26650	Window frames made of fibre-cement (manufacture)
	26610	Roof units made of precast concrete (manufacture)	23690		**Manufacture of other articles of concrete, plaster and cement**
	26610	Roofing tiles made of precast concrete (manufacture)		26660	Bas-relief and haut-relief made of concrete, plaster, cement or artificial stone (manufacture)
	26610	Sheets made of concrete (manufacture)		26660	Cement products n.e.c. (manufacture)
	26610	Stone articles for use in construction (manufacture)		26660	Concrete articles n.e.c. (manufacture)
	26610	Structural wall panels made of precast concrete (manufacture)		26660	Flower pots made of concrete, plaster, cement or artificial stone (manufacture)
	26610	Tiles made of concrete (manufacture)		26660	Furniture made of concrete, plaster, cement or artificial stone (manufacture)
	26610	Tubes made of concrete (manufacture)		26660	Statuary made of concrete, plaster, cement or artificial stone (manufacture)
	26610	Wall panels for structural use made of precast concrete (manufacture)		26660	Vases made of concrete, plaster, cement or artificial stone (manufacture)
23620		**Manufacture of plaster products for construction purposes**	23700		**Cutting, shaping and finishing of stone**
	26620	Boards made of plaster (manufacture)		26700	Alabaster bowl cutting (manufacture)
	26620	Gypsum plaster products (manufacture)		26700	Cutting, shaping and finishing of stone for use as roofing (manufacture)
	26620	Panels made of plaster (manufacture)		26700	Cutting, shaping and finishing of stone for use in cemeteries (manufacture)
	26620	Plaster articles for use in construction (manufacture)		26700	Cutting, shaping and finishing of stone for use in construction (manufacture)
	26620	Plaster products for construction purposes (manufacture)		26700	Cutting, shaping and finishing of stone for use on roads (manufacture)
	26620	Plasterboard (manufacture)		26700	Decorated building stone (manufacture)
	26620	Sheets made of plaster (manufacture)		26700	Dolomite (ground) (manufacture)
	26620	Tiles made of plaster (manufacture)		26700	Funerary stonework (manufacture)
23630		**Manufacture of ready-mixed concrete**		26700	Granite working (manufacture)
	26630	Concrete dry mix (manufacture)		26700	Kerbstone (not concrete) (manufacture)
	26630	Ready-mixed concrete (manufacture)		26700	Limestone (ground) (manufacture)
23640		**Manufacture of mortars**		26700	Limestone working (manufacture)
	26640	Mortars (manufacture)		26700	Litho stone working (manufacture)
	26640	Mortars (powdered) (manufacture)		26700	Marble masonry working (manufacture)
	26640	Ready-mixed wet mortars (manufacture)		26700	Monumental stonework (manufacture)
23650		**Manufacture of fibre cement**		26700	Paving stone (manufacture)
	26650	Asbestos cement products (manufacture)		26700	Slate polishing (manufacture)
	26650	Basins made of fibre-cement (manufacture)		26700	Slate slab and sheet cutting and preparation (manufacture)
	26650	Building boards made of asbestos (manufacture)		26700	Slate working (manufacture)
	26650	Building boards made of asbestos cement (manufacture)		26700	Stone furniture (manufacture)
	26650	Building materials made of vegetable substances (manufacture)		26700	Stone working (manufacture)
	26650	Building materials of vegetable materials agglomerated with cement, plaster etc. (manufacture)		26700	Tiles made of slate (manufacture)
	26650	Cellulose fibre-cement articles (manufacture)	23910		**Production of abrasive products**
	26650	Corrugated sheets made of fibre-cement (manufacture)		26810	Abrasive bonded disc, wheel and segment (manufacture)
	26650	Fibre cement (manufacture)		26810	Abrasive cloth (manufacture)
	26650	Furniture made of fibre-cement (manufacture)		26810	Abrasive grain (manufacture)
	26650	Garages made of asbestos cement and concrete (manufacture)		26810	Abrasive grain of aluminium oxide (manufacture)
	26650	Jars made of fibre-cement (manufacture)		26810	Abrasive grain of artificial corundum (manufacture)
	26650	Panels made of fibre-cement (manufacture)		26810	Abrasive grain of boron carbide (manufacture)
	26650	Pipes made of asbestos cement (manufacture)		26810	Abrasive grain of silicon carbide (manufacture)
	26650	Pipes made of fibre-cement (manufacture)		26810	Abrasive paper (manufacture)
	26650	Reservoirs made of fibre-cement (manufacture)		26810	Abrasive wheel (bonded) (manufacture)
	26650	Sheets made of fibre cement (manufacture)		26810	Agglomerated abrasives (manufacture)
	26650	Sinks made of fibre cement (manufacture)		26810	Bonded abrasives (manufacture)
	26650	Tiles made of fibre cement (manufacture)		26810	Coated abrasives (manufacture)

UK SIC 2007	UK SIC 2003	Activity
	26810	Diamond impregnated disc and wheel (manufacture)
	26810	Discs made of bonded abrasives (manufacture)
	26810	Emery cloth (manufacture)
	26810	Emery paper (manufacture)
	26810	Emery wheel (manufacture)
	26810	Flint cloth (manufacture)
	26810	Flint paper (manufacture)
	26810	Garnet abrasives (manufacture)
	26810	Glass paper (manufacture)
	26810	Grinding paste (manufacture)
	26810	Grindstones made of bonded abrasives (manufacture)
	26810	Hones (bonded) (manufacture)
	26810	Millstone and grindstone cutting (manufacture)
	26810	Millstones made of bonded abrasives (manufacture)
	26810	Oilstones (bonded) (manufacture)
	26810	Organic bonded abrasives (manufacture)
	26810	Polishing stones made of bonded abrasives (manufacture)
	26810	Pumice stones (bonded) (manufacture)
	26810	Sandpaper (manufacture)
	26810	Segment bonded abrasive (manufacture)
	26810	Sharpening stones made of bonded abrasives (manufacture)
	26810	Stones for sharpening or polishing (manufacture)
	26810	Vitrified bonded abrasives (manufacture)
23990		**Manufacture of other non-metallic mineral products n.e.c.**
	26829	Artificial corundum (manufacture)
	26821	Asbestos carding (manufacture)
	26821	Asbestos felting (manufacture)
	26821	Asbestos mixing (manufacture)
	26821	Asbestos moulding (manufacture)
	26821	Asbestos spinning (manufacture)
	26821	Asbestos weaving (manufacture)
	26829	Asphalt (manufacture)
	26829	Asphalt or similar materials (e.g. Asphalt-based adhesives), articles thereof (manufacture)
	26829	Bitumen and flax felts for roofing and damp-proof courses (manufacture)
	26829	Bituminous sealants (manufacture)
	26821	Boards made of asbestos (manufacture)
	26821	Boiler packing made of asbestos (manufacture)
	26821	Brake linings made of asbestos (manufacture)
	26829	Carbon products (except carbon paper and electrical carbon) (manufacture)
	26821	Carded asbestos fibre (manufacture)
	26821	Clothing made of asbestos (manufacture)
	26821	Clutch linings made of asbestos (manufacture)
	26829	Coal tar pitch, articles thereof (manufacture)
	26821	Composition asbestos (manufacture)
	26821	Cord made of asbestos (manufacture)
	26821	Engine packing made of asbestos (manufacture)
	26829	Exfoliated vermiculite (manufacture)
	26829	Expanded clay (manufacture)
	26829	Expanded vermiculite (manufacture)
	26821	Fabric made of asbestos (manufacture)
	26821	Felt made of asbestos (manufacture)
	26829	Foamed slag (manufacture)

SIC 2007	SIC 2003	Activity
	26821	Footwear made of asbestos (manufacture)
	26829	Friction material and unmounted articles thereof, with a mineral or cellulose base (manufacture)
	26829	Friction material made of non-metallic mineral (manufacture)
	26821	Gaskets made of asbestos (manufacture)
	26829	Graphite (manufacture)
	26829	Graphite products (other than block and crucible) (manufacture)
	26821	Headgear made of asbestos (manufacture)
	26829	Heat insulating materials (other than asbestos) (manufacture)
	26821	Insulation made of asbestos (manufacture)
	26821	Joints made of asbestos (manufacture)
	26821	Lagging rope made of asbestos (manufacture)
	26829	Mica goods (manufacture)
	26829	Mica slab and sheet processing (manufacture)
	26821	Millboard made of asbestos (manufacture)
	26829	Mineral insulating materials (manufacture)
	26829	Mineral insulation products (manufacture)
	26829	Mineral wool (manufacture)
	26829	Moss litter (manufacture)
	26829	Non-metallic mineral substances, articles thereof (manufacture)
	26829	Oil based sealants (manufacture)
	26821	Packing made of woven asbestos (manufacture)
	26821	Panels made of asbestos (manufacture)
	26821	Paper made of asbestos (manufacture)
	26821	Paste made of asbestos (manufacture)
	26829	Peat products (pots or for chemical use, etc.) (manufacture)
	26821	Pipe covering sections made of asbestos (manufacture)
	26829	Pipes and fittings made of pitch fibre (manufacture)
	26821	Processed asbestos fibre (manufacture)
	26821	Rings made of asbestos (manufacture)
	26829	Rockwool (manufacture)
	26821	Rope lagging made of asbestos (manufacture)
	26829	Sealants (bituminous) (manufacture)
	26829	Sealants (oil based) (manufacture)
	26821	Sheeting made of non-woven asbestos/rubber composite (manufacture)
	26821	Sheets and sheeting made of woven asbestos (manufacture)
	26829	Slag wool (manufacture)
	26821	Socks made of asbestos (manufacture)
	26829	Sound absorbing materials (manufacture)
	26829	Sound insulating materials (manufacture)
	26821	String made of asbestos (manufacture)
	26821	Tapes made of asbestos (manufacture)
	26821	Tiles made of woven asbestos (manufacture)
	26829	Worked peat, articles thereof (manufacture)
	26821	Yarn made of asbestos (manufacture)
24100		**Manufacture of basic iron and steel and of ferro-alloys**
	27100	Alloy bearing steel (manufacture)
	27100	Alloy pig iron (manufacture)
	27100	Alloy tool steel (manufacture)
	27100	Angles of stainless steel (manufacture)
	27100	Bars of stainless steel (manufacture)

UK SIC 2007	UK SIC 2003	Activity	SIC 2007	SIC 2003	Activity
	27100	Billets made of semi-finished steel (manufacture)		27100	High carbon ferro-manganese (carbon over 2%) (manufacture)
	27100	Blackplate (manufacture)		27100	High speed tool steel (manufacture)
	27100	Blooms made of semi-finished steel (manufacture)		27100	Hollow drill bars made of steel (manufacture)
	27100	Boron steel (manufacture)		27100	Hot dip metal coated sheet steel (manufacture)
	27100	Bulb flats made of steel (manufacture)		27100	Hot dip zinc coated sheet steel (manufacture)
	27100	Cold reduced steel slit strip ≥600 mm (manufacture)		27100	Hot rolled steel angles (l-sections) (manufacture)
	27100	Cold reduced wide steel strip ≥600 mm (cold reduced coil) (manufacture)		27100	Hot rolled steel base plates (manufacture)
	27100	Cold rolled narrow steel strip ≥600 mm (manufacture)		27100	Hot rolled steel beams (manufacture)
	27100	Cold rolled steel plate (manufacture)		27100	Hot rolled steel bearing piling (manufacture)
	27100	Cold rolled steel sheet (manufacture)		27100	Hot rolled steel channels (manufacture)
	27100	Cold rolled steel slit strip ≥600 mm (manufacture)		27100	Hot rolled steel columns (manufacture)
	27100	Cold rolled wide steel strip ≥600 mm (cold rolled wide coil) (manufacture)		27100	Hot rolled steel cut lengths (manufacture)
	27100	Continuous cast products made of steel (manufacture)		27100	Hot rolled steel fish plates (manufacture)
	27100	Corrugated sheets made of steel (manufacture)		27100	Hot rolled steel flat bars (manufacture)
	27100	Directly reduced iron (manufacture)		27100	Hot rolled steel heavy sections (manufacture)
	27100	Electrical sheet steel (not finally annealed) (manufacture)		27100	Hot rolled steel hexagonal bars (manufacture)
	27100	Electrical steel (manufacture)		27100	Hot rolled steel h-sections (manufacture)
	27100	Electro zinc coated sheet steel (manufacture)		27100	Hot rolled steel i-sections (manufacture)
	27100	Electrolytic chromium/chromium oxide coated steel (manufacture)		27100	Hot rolled steel joists (manufacture)
	27100	Electrolytically metal coated sheet steel (manufacture)		27100	Hot rolled steel light sections (manufacture)
	27100	Engineering steel (manufacture)		27100	Hot rolled steel narrow strip (manufacture)
	27100	Ferro alloys (high carbon ferro manganese) (manufacture)		27100	Hot rolled steel plate (manufacture)
	27100	Ferro aluminium (manufacture)		27100	Hot rolled steel quarto plate (manufacture)
	27100	Ferro chromium (manufacture)		27100	Hot rolled steel railway materials (manufacture)
	27100	Ferro molybdenum (manufacture)		27100	Hot rolled steel reversing mill plate (manufacture)
	27100	Ferro nickel (manufacture)		27100	Hot rolled steel round bars (manufacture)
	27100	Ferro niobium (manufacture)		27100	Hot rolled steel sections for mining frames (manufacture)
	27100	Ferro phosphorus (manufacture)		27100	Hot rolled steel sheet (manufacture)
	27100	Ferro titanium (manufacture)		27100	Hot rolled steel sheet piling (manufacture)
	27100	Ferro tungsten (manufacture)		27100	Hot rolled steel sole plates (manufacture)
	27100	Ferro vanadium (manufacture)		27100	Hot rolled steel special bars (manufacture)
	27100	Ferro zirconium (manufacture)		27100	Hot rolled steel special sections (manufacture)
	27100	Ferro-alloys (except high carbon ferro-manganese production) (manufacture)		27100	Hot rolled steel square bars (manufacture)
	27100	Ferrosilicon (manufacture)		27100	Hot rolled steel t-sections (manufacture)
	27100	Ferrosilicon chromium (manufacture)		27100	Hot rolled steel u-sections (manufacture)
	27100	Ferrosilicon manganese (manufacture)		27100	Hot rolled steel wide strip (hot rolled wide coil) (manufacture)
	27100	Ferrosilicon titanium (manufacture)		27100	Ingots of stainless steel (manufacture)
	27100	Ferrosilicon tungsten (manufacture)		27100	Ingots of steel (manufacture)
	27100	Ferrous products production by reduction of iron ore (manufacture)		27100	Iron (manufacture)
	27100	Fish plates (hot rolled) (manufacture)		27100	Iron of exceptional purity production by electrolysis or other chemical processes (manufacture)
	27100	Fish plates and sole plates (non-rolled) (manufacture)		27100	Iron powder production (manufacture)
	27100	Flat rolled steel products in coils or straight lengths ≥600 mm (manufacture)		27100	Iron shot (manufacture)
	27100	Forged bars (manufacture)		27100	Lead coated steel sheet (manufacture)
	27100	Forged rail accessories (manufacture)		27100	Liquid steel (manufacture)
	27100	Forged sections (manufacture)		27100	Low carbon ferro manganese (carbon 2% or less) (manufacture)
	27100	Forged semi-finished products (manufacture)		27100	Manufacture of hot-rolled rods of steel
	27100	Foundry pig iron (manufacture)		27100	Metal sand for sandblasting (manufacture)
	27100	Galvanised sheet steel (manufacture)		27100	Narrow slabs of semi-finished steel (manufacture)
	27100	Grain oriented electrical sheet steel (manufacture)		27100	Nodular pig iron (manufacture)
				27100	Non-alloy pig iron (manufacture)
				27100	Non-alloy steel (manufacture)
				27100	Non-oriented electrical steel sheet (manufacture)
				27100	Organic coated steel sheet (manufacture)
				27100	Other alloy steel (manufacture)
				27100	Painted steel sheet (manufacture)

UK SIC 2007	UK SIC 2003	Activity	SIC 2007	SIC 2003	Activity
	27100	Permanent way material (except rails production) (manufacture)		27220	Repair clamps and collars, made of iron or steel (manufacture)
	27100	Pieces roughly shaped by forging (manufacture)		27220	Scaffolding tubes made of steel (manufacture)
	27100	Plastic coated steel sheet (manufacture)		27220	Sleeves made of steel (manufacture)
	27100	Primary products of stainless steel (manufacture)		27220	Tube fittings made of steel (manufacture)
	27100	Production of granular iron		27220	Tube hollows made of steel (manufacture)
	27100	Rails of iron, steel or cast iron (manufacture)		27220	Tubes made of cold drawn steel (manufacture)
	27100	Remelting ferrous waste or scrap		27220	Tubes made of seamless steel (manufacture)
	27100	Remelting scrap ingots of iron		27220	Tubes made of steel (manufacture)
	27100	Rods made of steel (in coils) (manufacture)		27220	Tubing made of steel (manufacture)
	27100	Rods of stainless steel (manufacture)		27220	Welded tubes (manufacture)
	27100	Rounds of semi-finished steel for seamless tube production (manufacture)	**24310**		**Cold drawing of bars**
	27100	Sections of stainless steel (manufacture)		27310	Bright steel bars (manufacture)
	27100	Semi-finished products of stainless steel (manufacture)		27310	Cold drawn steel bars (manufacture)
	27100	Shapes of stainless steel (manufacture)		27310	Cold drawn steel sections (manufacture)
	27100	Sheet piling production (manufacture)		27310	Cold finished steel bars (manufacture)
	27100	Silico manganese steel (manufacture)	**24320**		**Cold rolling of narrow strip**
	27100	Sleepers (cross ties) made of iron or steel (hot rolled) (manufacture)		27320	Cold reduced steel slit strip <600 mm (manufacture)
	27100	Sole plates (hot rolled) (manufacture)		27320	Cold rolled narrow steel strip <600 mm (manufacture)
	27100	Spheroidal graphite pig iron (manufacture)		27320	Flat rolled steel products in coils or straight lengths <600 mm (manufacture)
	27100	Spiegeleisen (manufacture)			
	27100	Sponge iron (manufacture)	**24330**		**Cold forming or folding**
	27100	Stainless steel (manufacture)		27330	Cold formed steel angles (manufacture)
	27100	Steel in primary form, from ore or scrap, production (manufacture)		27330	Cold formed steel channels (manufacture)
	27100	Steel making pig iron (manufacture)		27330	Cold formed steel sections (manufacture)
	27100	Steel powders (manufacture)		27330	Cold-folded ribbed sheets and sandwich panels (manufacture)
	27100	Steel sheet (not finally annealed) (manufacture)		27330	Cold-formed ribbed sheets and sandwich panels (manufacture)
	27100	Steel shot (manufacture)		27330	Open sections made of steel formed on a roll mill (manufacture)
	27100	Structural steel (manufacture)		27330	Profiled steel sheet (manufacture)
	27100	Terneplate (manufacture)		28110	Sandwich panels of coated steel sheet (manufacture)
	27100	Tinplate (manufacture)		27330	Sections cold formed from flat steel products (manufacture)
	27100	Welded sections of iron and steel (manufacture)		27330	Sheets of square corrugated steel (manufacture)
	27100	Wide slabs made of semi-finished steel (manufacture)	**24340**		**Cold drawing of wire**
	27100	Wire rods made of steel (manufacture)		27340	Cold drawing or stretching of steel wire (manufacture)
24200		**Manufacture of tubes, pipes, hollow profiles and related fittings, of steel**		27340	Filament wire made of steel (manufacture)
	27220	Bends made of steel (manufacture)		27340	Piano wire made of steel (manufacture)
	27220	Butt welding fittings made of steel (manufacture)		27340	Single strand wire made of steel (manufacture)
	27220	Casing made of steel (manufacture)		27340	Welding wire (uncoated) made of steel (manufacture)
	27220	Circular hollow sections made of steel (manufacture)			
	27220	Couplings and flange adapters made of iron or steel (manufacture)	**24410**		**Precious metals production**
	27220	Drill pipe made of steel (manufacture)		27410	Gold (manufacture)
	27220	Elbows made of steel (manufacture)		27410	Gold and silver bullion (manufacture)
	27220	Electrical conduit tube made of steel (manufacture)		27410	Gold rolled onto base metals or silver production (manufacture)
	27220	Fittings made of steel (manufacture)		27410	Iridium (manufacture)
	27220	Flanges made of steel (manufacture)		27410	Palladium (manufacture)
	27220	Gas pipes made of steel (manufacture)		27410	Platinum (manufacture)
	27220	Hollow bars made of steel (manufacture)		27410	Platinum group metals (manufacture)
	27220	Hot finished steel tube (manufacture)		27410	Precious metal foil laminates (manufacture)
	27220	Line pipe made of steel (manufacture)		27410	Precious metals production (manufacture)
	27220	Pipes made of steel (manufacture)		27410	Rhodium (manufacture)
	27220	Precision tube made of steel (manufacture)			
	27220	Rectangular hollow steel section (manufacture)			

UK SIC 2007	UK SIC 2003	Activity	SIC 2007	SIC 2003	Activity
	27410	Silver (manufacture)		27420	Sections made of aluminium (manufacture)
	27410	Silver rolled onto base metals production (manufacture)		27420	Sheets made of aluminium (manufacture)
	27410	Wire of precious metals, made by drawing (manufacture)		27420	Slugs made of aluminium (manufacture)
				27420	Solid sections made of aluminium (manufacture)
				27420	Strips made of aluminium (manufacture)
24420		**Aluminium production**		27420	Tube fittings made of aluminium (manufacture)
	27420	Aluminium alloys production (manufacture)		27420	Unwrought aluminium (manufacture)
	27420	Aluminium foil laminates made from aluminium foil as primary component (manufacture)		27420	Wire made of aluminium (manufacture)
	27420	Aluminium from alumina production (manufacture)		27420	Wirebar made of aluminium (manufacture)
	27420	Aluminium hardener (manufacture)		27420	Wrapping foil made of aluminium (manufacture)
	27420	Aluminium oxide (alumina) production (manufacture)	**24430**		**Lead, zinc and tin production**
	27420	Aluminium refining (manufacture)		27430	Bars, rods, profiles and wire made of lead (manufacture)
	27420	Aluminium semi-manufactures production (manufacture)		27430	Bars, rods, profiles and wire made of tin (manufacture)
	27420	Aluminium smelting (manufacture)		27430	Bars, rods, profiles and wire made of zinc (manufacture)
	27420	Aluminium wire made by drawing (manufacture)		27430	Britannia metal (manufacture)
	27420	Angles made of aluminium (manufacture)		27430	Dust, powder and flakes made of zinc (manufacture)
	27420	Bars made of aluminium (manufacture)		27430	Lead (manufacture)
	27420	Billets made of aluminium (manufacture)		27430	Lead wire made by drawing (manufacture)
	27420	Blanks made of aluminium (manufacture)		27430	Magnolia metal (manufacture)
	27420	Circles made of aluminium (manufacture)		27430	Pewter (manufacture)
	27420	Continuous cast rod made of aluminium (manufacture)		27430	Plates, sheets, strip and foil made of lead (manufacture)
	27420	Cooking foil made of aluminium (manufacture)		27430	Plates, sheets, strip and foil made of tin (manufacture)
	27420	Corrugated plate, sheet or strip made of aluminium (manufacture)		27430	Plates, sheets, strip and foil made of zinc (manufacture)
	27420	Deoxidiser made of aluminium (manufacture)		27430	Powders and flakes made of lead (manufacture)
	27420	Discs made of aluminium (manufacture)		27430	Powders and flakes made of tin (manufacture)
	27420	Drawn products made of aluminium (manufacture)		27430	Tin (manufacture)
	27420	Extruded sections made of aluminium (manufacture)		27430	Tin foil (manufacture)
	27420	Extruded tubes made of aluminium (manufacture)		27430	Tin wire made by drawing (manufacture)
	27420	Extrusion ingots made of aluminium (manufacture)		27430	Tubes, pipes, tube and pipe fittings made of lead (manufacture)
	27420	Extrusions made of aluminium (manufacture)		27430	Tubes, pipes, tube and pipe fittings made of tin (manufacture)
	27420	Flake made of aluminium (manufacture)		27430	Tubes, pipes, tube and pipe fittings made of zinc (manufacture)
	27420	Foil laminate made of aluminium (manufacture)		27430	Type metal (manufacture)
	27420	Foil made of aluminium (decorated, embossed or cut to size) (manufacture)		27430	Zinc (manufacture)
	27420	Foil made of aluminium (not put up as a packaging product) (manufacture)		27430	Zinc wire made by drawing (manufacture)
	27420	Foil packaging goods, made of aluminium (manufacture)	**24440**		**Copper production**
	27420	Foil stock made of aluminium (manufacture)		27440	Bars made of brass (manufacture)
	27420	Forging bars made of aluminium (manufacture)		27440	Bars made of copper (manufacture)
	27420	Foundry alloy made of aluminium (manufacture)		27440	Billets made of brass (manufacture)
	27420	Foundry ingot made of aluminium (manufacture)		27440	Billets made of copper (manufacture)
	27420	Hollow sections made of aluminium (manufacture)		27440	Blister copper (manufacture)
	27420	Kitchen foil made of aluminium (manufacture)		27440	Blooms made of copper (manufacture)
	27420	Laminates of aluminium foil with other materials (manufacture)		27440	Brass powder (manufacture)
	27420	Notched bars made of aluminium (manufacture)		27440	Circles made of brass (manufacture)
	27420	Pipe fittings made of aluminium (manufacture)		27440	Circles made of copper (manufacture)
	27420	Pipes made of aluminium (manufacture)		27440	Coils made of copper (manufacture)
	27420	Plates made of aluminium (manufacture)		27440	Continuous cast rod made of copper (manufacture)
	27420	Powder made of aluminium (manufacture)		27440	Copper refining (manufacture)
	27420	Remelt ingots made of aluminium (manufacture)		27440	Copper smelting (manufacture)
	27420	Rods made of aluminium (manufacture)		27440	Discs made of brass (manufacture)
	27420	Rolled products made of aluminium (manufacture)		27440	Discs made of copper (manufacture)
	27420	Rolling ingots and slabs made of aluminium (manufacture)			

UK SIC 2007	UK SIC 2003	Activity
	27440	Drawn products made of copper (manufacture)
	27440	Electrolytic copper (manufacture)
	27440	Extruded products made of copper (manufacture)
	27440	Fire refined copper (manufacture)
	27440	Flake made of copper (manufacture)
	27440	Foil made of brass (manufacture)
	27440	Foil made of copper (manufacture)
	27440	Fuse wire (manufacture)
	27440	Ingots made of brass (manufacture)
	27440	Ingots made of copper (manufacture)
	27440	Master alloys of copper (manufacture)
	27440	Mattes made of copper (manufacture)
	27440	Nickel silver (manufacture)
	27440	Pipe and pipe fittings made of brass (manufacture)
	27440	Pipe blanks made of copper (manufacture)
	27440	Pipe fittings made of copper (manufacture)
	27440	Pipes made of copper (manufacture)
	27440	Powder made of copper (manufacture)
	27440	Primary copper (manufacture)
	27440	Rods made of brass (manufacture)
	27440	Rods made of copper (manufacture)
	27440	Rolled products made of copper (manufacture)
	27440	Secondary copper (manufacture)
	27440	Sections made of brass (manufacture)
	27440	Sections made of copper (manufacture)
	27440	Semi-manufactures made of copper (manufacture)
	27440	Sheets made of brass (manufacture)
	27440	Sheets made of copper (manufacture)
	27440	Slabs made of brass (manufacture)
	27440	Slabs made of copper (manufacture)
	27440	Strips made of brass (manufacture)
	27440	Strips made of copper (manufacture)
	27440	Tube blanks made of copper (manufacture)
	27440	Tube fittings made of copper (manufacture)
	27440	Tube shells made of copper (manufacture)
	27440	Tubes made of brass (manufacture)
	27440	Tubes made of copper (manufacture)
	27440	Unwrought brass (manufacture)
	27440	Unwrought bronze (manufacture)
	27440	Unwrought cadmium copper (manufacture)
	27440	Unwrought copper (manufacture)
	27440	Unwrought cupro-nickel (manufacture)
	27440	Unwrought delta metal (manufacture)
	27440	Unwrought German silver (manufacture)
	27440	Unwrought gun metal (manufacture)
	27440	Unwrought manganese bronze (manufacture)
	27440	Unwrought naval brass (manufacture)
	27440	Unwrought red metal (manufacture)
	27440	Wire made of copper (uninsulated) (manufacture)
	27440	Wire rods made of copper (manufacture)
24450		**Other non-ferrous metal production**
	27450	Antifriction metal (manufacture)
	27450	Antimony (manufacture)
	27450	Arsenic (manufacture)
	27450	Beryllium (manufacture)
	27450	Bismuth (manufacture)
	27450	Cadmium (manufacture)
	27450	Chrome alloys (manufacture)

SIC 2007	SIC 2003	Activity
	27450	Chrome production and refining (manufacture)
	27450	Chrome wire made by drawing (manufacture)
	27450	Chromium (manufacture)
	27450	Cobalt (manufacture)
	27450	Continuous cast rods of other base non-ferrous metals (manufacture)
	27450	Germanium (manufacture)
	27450	Magnesium (manufacture)
	27450	Manganese alloys (manufacture)
	27450	Manganese production and refining (manufacture)
	27450	Manganese wire made by drawing (manufacture)
	27450	Mattes of nickel production (manufacture)
	27450	Molybdenum (manufacture)
	27450	Nickel (manufacture)
	27450	Nickel alloys (manufacture)
	27450	Nickel wire made by drawing (manufacture)
	27450	Non-ferrous other metals production (manufacture)
	27450	Semi-manufacturing of chrome (manufacture)
	27450	Semi-manufacturing of manganese (manufacture)
	27450	Semi-manufacturing of nickel (manufacture)
	27450	Solder (manufacture)
	27450	Sterro metal (manufacture)
	27450	Tantalum (manufacture)
	27450	Titanium (manufacture)
	27450	Tungsten (manufacture)
	27450	Vanadium (manufacture)
	27450	White metal (manufacture)
	27450	Wolfram (manufacture)
	27450	Zirconium (manufacture)
24460		**Processing of nuclear fuel**
	23300	Natural uranium production (manufacture)
	23300	Nuclear fuel (except enrichment of uranium or thorium) (manufacture)
	23300	Nuclear fuel processing (except enrichment of uranium or thorium) (manufacture)
	23300	Uranium metal production from pitchblende or other ores (manufacture)
	23300	Uranium smelting and refining (manufacture)
	23300	Yellowcake to uranium tetrafluoride and hexafluoride conversion (manufacture)
24510		**Casting of iron**
	27510	Casting of ferrous metal (manufacture)
	27510	Casting of ferrous patterns (manufacture)
	27510	Casting of grey iron (manufacture)
	27510	Casting of iron (manufacture)
	27510	Casting of iron products (finished or semi-finished) (manufacture)
	27510	Casting of spheroidal graphite iron (manufacture)
	27510	Ferrous metal foundry (manufacture)
	27210	Fittings for tubes made of cast iron (manufacture)
	27210	Fittings for tubes made of cast steel (manufacture)
	27210	Hollow profiles of cast-iron (manufacture)
	27510	Iron foundry (manufacture)
	27510	Malleable castings (manufacture)
	27210	Pipe fittings of cast-iron (manufacture)
	27210	Pipes of cast-iron (manufacture)
	27210	Tubes and fittings made of cast iron (manufacture)
	27210	Tubes made of centrifugally cast steel (manufacture)

UK SIC 2007	UK SIC 2003	Activity	SIC 2007	SIC 2003	Activity
24520		**Casting of steel**		28110	Metal frameworks for lifting and handling equipment (manufacture)
	27520	Casting of steel (manufacture)		28110	Metal skeletons for bridges (manufacture)
	27520	Casting of steel products (finished or semi-finished) (manufacture)		28110	Metal skeletons for masts (manufacture)
	27520	Seamless pipes of steel by centrifugal casting (manufacture)		28110	Metal skeletons for towers (manufacture)
	27520	Seamless tubes of steel by centrifugal casting (manufacture)		28110	Metal structures and parts of structures (manufacture)
	27520	Steel founders (manufacture)		28110	Modular exhibition elements made of metal (manufacture)
24530		**Casting of light metals**		28110	Monopod tower made of steel plate (manufacture)
	27530	Casting of aluminium (manufacture)		28110	Ossature in metal for construction (manufacture)
	27530	Casting of aluminium products (manufacture)		28110	Pilings (tubular welded) (manufacture)
	27530	Casting of beryllium products (manufacture)		28110	Pipeline supports (manufacture)
	27530	Casting of light metal products (manufacture)		28110	Plant support of fabricated steelwork (manufacture)
	27530	Casting of light metals (manufacture)		28110	Portable building metalwork (manufacture)
	27530	Casting of magnesium products (manufacture)		28110	Power structural steelwork (manufacture)
	27530	Casting of scandium products (manufacture)		28110	Prefabricated building metalwork (manufacture)
	27530	Casting of titanium products (manufacture)		28110	Prefabricated buildings made of metal (manufacture)
	27530	Casting of yttrium products (manufacture)		28110	Racking systems for large scale heavy duty use in shops, workshops and warehouses (manufacture)
	27530	Die casting of aluminium (manufacture)		28110	Radio and television masts made of metal (manufacture)
24540		**Casting of other non-ferrous metals**		28110	Roof trusses made of metal (manufacture)
	27540	Bell founding (manufacture)		28110	Scaffolding (manufacture)
	27540	Britannia metal founding (manufacture)		28110	Security screens made of metal (manufacture)
	27540	Bronze founding (manufacture)		28110	Sheds made of metal (manufacture)
	27540	Casting of heavy metal and precious metal (manufacture)		28110	Shuttering made of steel (manufacture)
	27540	Casting of non-ferrous base metal (manufacture)		28110	Sluice gates made of steel (manufacture)
	27540	Casting of ornamental brass (manufacture)		28110	Static drilling derricks (manufacture)
	27540	Casting of other non-ferrous metals (manufacture)		28110	Steelwork for agricultural buildings (manufacture)
	27540	Die casting of copper or copper alloy (manufacture)		28110	Steelwork for bridges (manufacture)
	27540	Die casting of non-ferrous base metal (manufacture)		28110	Steelwork for buildings (manufacture)
	27540	Founding of non-ferrous base metal (manufacture)		28110	Steelwork for commercial buildings (manufacture)
	27540	Gravity casting of non-ferrous base metal (manufacture)		28110	Steelwork for distribution depots (manufacture)
	27540	Non-ferrous metal foundry (manufacture)		28110	Steelwork for docks (manufacture)
	27540	Pressure die casting of non-ferrous base metal (manufacture)		28110	Steelwork for domestic buildings (manufacture)
	27540	Sand casting of non-ferrous base metal (manufacture)		28110	Steelwork for exhibition centres (manufacture)
25110		**Manufacture of metal structures and parts of structures**		28110	Steelwork for factory buildings (manufacture)
	28110	Bus shelters made of metal (manufacture)		28110	Steelwork for glass roofs (manufacture)
	28110	Chimneys made of steel (manufacture)		28110	Steelwork for harbours (manufacture)
	28110	Column (fabricated structural steelwork) (manufacture)		28110	Steelwork for hospital buildings (manufacture)
	28110	Construction site huts made of metal (manufacture)		28110	Steelwork for jetties (manufacture)
	28110	Fabricated structural steelwork for buildings (manufacture)		28110	Steelwork for school buildings (manufacture)
	28110	Flooring systems made of metal (manufacture)		28110	Steelwork for tunnels (manufacture)
	28110	Glasshouses with metal frame (manufacture)		28110	Steelwork for viaducts (manufacture)
	28110	Industrial ossature in metal (manufacture)		28110	Structural steelwork for buildings (manufacture)
	28110	Lock gates (manufacture)		28110	Structures for buildings made of aluminium (manufacture)
	28110	Metal frameworks for blast furnaces (manufacture)		28110	Structures for civil engineering made of aluminium (manufacture)
	28110	Metal frameworks for construction of bridges (manufacture)		28110	Telephone booths made of metal (manufacture)
	28110	Metal frameworks for construction of masts (manufacture)		28110	Tower made of steel (manufacture)
	28110	Metal frameworks for construction of towers (manufacture)		28110	Water tower made of steel plate (manufacture)
			25120		**Manufacture of doors and windows of metal**
				28120	Builders' carpentry and joinery made of metal (manufacture)
				28120	Casements made of metal (manufacture)
				28120	Curtain walling made of metal (manufacture)
				28120	Door frames made of metal (manufacture)

UK SIC 2007	UK SIC 2003	Activity
	28120	Doors (other than safe doors) made of metal (manufacture)
	28120	Double glazing made of metal (manufacture)
	28120	Garden frames made of metal (manufacture)
	28120	Gates made of metal (manufacture)
	28120	Glazing bars (manufacture)
	28120	Grilles made of metal (not cast) (manufacture)
	28120	Partitioning made of metal (manufacture)
	28120	Railings made of metal (manufacture)
	28120	Revolving doors made of metal (manufacture)
	28120	Room partitions made of metal, for floor attachment (manufacture)
	28120	Shop fronts and entrances made of aluminium (manufacture)
	28120	Shutters made of metal (manufacture)
	28120	Skylights made of metal (manufacture)
	28120	Window frames made of metal (manufacture)
25210		**Manufacture of central heating radiators and boilers**
	28220	Boiler for central heating (manufacture)
	28220	Central heating boiler parts (manufacture)
	28220	Radiant panel (space heating equipment) (manufacture)
	28220	Radiator for space heating equipment (manufacture)
	28220	Radiators and boilers for central heating (manufacture)
25290		**Manufacture of other tanks, reservoirs and containers of metal**
	28210	Bunkers made of heavy steel plate exceeding 300 litres (manufacture)
	28210	Cistern made of metal exceeding 300 litres (manufacture)
	28210	Containers for compressed or liquefied gases made of metal (manufacture)
	28210	Containers made of metal of a capacity exceeding 300 litres (manufacture)
	28210	Expansion tank made of metal exceeding 300 litres (manufacture)
	28210	Fuel bunkers made of metal exceeding 300 litres (manufacture)
	28210	Oil storage tank made of metal for domestic use exceeding 300 litres (manufacture)
	28210	Reservoirs made of metal exceeding 300 litres (manufacture)
	28210	Silos made of steel exceeding 300 litres (manufacture)
	28210	Storage tanks made of heavy steel plate exceeding 300 litres (manufacture)
	28210	Tanks made of galvanised steel exceeding 300 litres (manufacture)
	28210	Tanks made of metal exceeding 300 litres (manufacture)
	28210	Tanks made of metal for domestic storage exceeding 300 litres (manufacture)
25300		**Manufacture of steam generators, except central heating hot water boilers**
	28300	Auxiliary plant for use with steam generators (manufacture)
	28300	Boiler (nuclear powered) (manufacture)
	28300	Boiler drum (manufacture)
	28300	Boiler feed water heater (manufacture)

SIC 2007	SIC 2003	Activity
	28300	Boiler for marine applications (manufacture)
	28300	Boiler fuel economiser (manufacture)
	28300	Boiler fuel handling plant (manufacture)
	28300	Boiler house plant (manufacture)
	28300	Boilers and associated equipment and parts (manufacture)
	28300	Column (process plant) (manufacture)
	28300	Condenser (steam) (manufacture)
	28300	Condenser (vapour) (manufacture)
	28300	Cracker for process plant (manufacture)
	28300	Duct of heavy steel plate (manufacture)
	28300	Economic boiler (manufacture)
	28300	Economisers (manufacture)
	28300	Evaporator (manufacture)
	28300	Firing plant for boilers, etc. (manufacture)
	28300	Heat exchanger for process plant (manufacture)
	28300	Industrial air heater for boilers (manufacture)
	28300	Marine or power boiler parts (manufacture)
	28300	Nuclear fired boiler (manufacture)
	28300	Nuclear fuel plant (manufacture)
	28300	Nuclear reactor parts (manufacture)
	28300	Nuclear reactors (manufacture)
	28300	Penstock made of steel (manufacture)
	28300	Pipe system construction for steam generators (manufacture)
	28300	Process heater (manufacture)
	28300	Process pipework (manufacture)
	28300	Process pressure sphere (manufacture)
	28300	Process pressure vessel (manufacture)
	28300	Reactor column (manufacture)
	28300	Reactor vessel (manufacture)
	28300	Reformer (manufacture)
	28300	Shell boiler (manufacture)
	28300	Steam accumulator (manufacture)
	28300	Steam boiler (manufacture)
	28300	Steam collector (manufacture)
	28300	Steam generator (manufacture)
	28300	Steam generator parts (manufacture)
	28300	Super-heaters (manufacture)
	28300	Vapour generators (manufacture)
	28300	Vertical boiler (manufacture)
	28300	Waste heat boiler (manufacture)
	28300	Water tube boilers (manufacture)
25400		**Manufacture of weapons and ammunition**
	29600	Air gun (manufacture)
	29600	Air pistol (manufacture)
	29600	Air rifle (manufacture)
	29600	Aircraft bomb (manufacture)
	29600	Ammunition (manufacture)
	29600	Arms (manufacture)
	29600	Artillery (manufacture)
	29600	Artillery ammunition (manufacture)
	29600	Automatic gun (manufacture)
	29600	Ballistic missile, except intercontinental ballistic missile (ICBM) (manufacture)
	29600	Bomb fuse (manufacture)
	29600	Carbine (manufacture)
	29600	Cartridge case (manufacture)

UK SIC 2007	UK SIC 2003	Activity	SIC 2007	SIC 2003	Activity
	29600	Cartridge primer (manufacture)		28400	Heavy forging (manufacture)
	29600	Cartridges for riveting guns (manufacture)		28400	Hot pressing of ferrous metals (manufacture)
	29600	Conventional missiles (manufacture)		28400	Hot stamping of ferrous metals (manufacture)
	29600	Directed energy weapons (manufacture)		28400	Metal forging, pressing, stamping and roll-forming (manufacture)
	29600	Firearms for hunting, sporting or protective use (manufacture)		28400	Metal objects production directly from metal powders by heat treatment (manufacture)
	29600	Flame throwers (manufacture)		28400	Perforated metal (manufacture)
	29600	Fuse for shells and bombs (manufacture)		28400	Piercing of base metal (manufacture)
	29600	Gas guns (manufacture)		28400	Powder metallurgy (manufacture)
	29600	Grenade (manufacture)		28400	Pressing of base metal (manufacture)
	29600	Guided weapon airborne delivery system, not intercontinental ballistic missile (ICBM) (manufacture)		28400	Rolls made of alloy or steel forgings (manufacture)
	29600	Guided weapon warheads (manufacture)		28400	Sintering of metals (manufacture)
	29600	Guided weapon, except intercontinental ballistic missile (ICBM) (manufacture)		28400	Stamping of base metal (manufacture)
	29600	Gun (manufacture)	**25610**		**Treatment and coating of metals**
	29600	Gun carriage mounting or platform (manufacture)		28510	Anodising (manufacture)
	29600	Howitzer (manufacture)		28510	Case hardening (manufacture)
	29600	Machine gun (manufacture)		28510	Chrome plating (manufacture)
	29600	Military carbine (manufacture)		28510	Chromium plating (manufacture)
	29600	Military rifle (manufacture)		28510	Concrete coating of metals (manufacture)
	29600	Mine case and component (manufacture)		28510	Electroplating (manufacture)
	29600	Missiles (guided weapons), except intercontinental ballistic missile (ICBM) (manufacture)		28510	Enamelling of metals including vitreous enamelling (manufacture)
	29600	Mortar (ordnance) (manufacture)		28510	Galvanising (manufacture)
	29600	Mortar bomb (manufacture)		28510	Gilding of metals (manufacture)
	29600	Ordnance (manufacture)		28510	Grinding (metal finishing) (manufacture)
	29600	Pistol (manufacture)		28510	Hot dip coating (metal finishing) (manufacture)
	29600	Primer for cartridge (manufacture)		28510	Japanning (metal finishing) (manufacture)
	29600	Revolver (manufacture)		28510	Lacquering (metal finishing) (manufacture)
	29600	Rocket launch systems (manufacture)		28510	Metal finishing
	29600	Shell case (manufacture)		28510	Metal spraying (manufacture)
	29600	Small arms (manufacture)		28510	Plastic coating of metals (manufacture)
	29600	Sporting carbine (manufacture)		28510	Plating (metal finishing) (manufacture)
	29600	Sporting gun (manufacture)		28510	Polishing (metal finishing) (manufacture)
	29600	Sporting rifle (manufacture)		28510	Rubber coating of metals (manufacture)
	29600	Torpedo (manufacture)		28510	Sand blasting of metals (manufacture)
	29600	Truncheons and night sticks (manufacture)		28510	Shot peening of metals (manufacture)
	29600	War ammunition (manufacture)		28510	Stove painting (manufacture)
	29600	Water cannon (manufacture)		28510	Treatment and coating of metals (manufacture)
	29600	Weapons (manufacture)	**25620**		**Machining**
25500		**Forging, pressing, stamping and roll-forming of metal; powder metallurgy**		28520	Blacksmith (not including farriers) (manufacture)
	28400	Alloy and steel forging roll (manufacture)		28520	Cutting of metals by laser beam (manufacture)
	28400	Closed die forging (manufacture)		28520	General engineering (manufacture)
	28400	Cold pressing of base metals (manufacture)		28520	Mechanical engineering (general) (manufacture)
	28400	Cold stamping of base metals (manufacture)		28520	Sheet metal working (manufacture)
	28400	Die forging of ferrous metals (manufacture)		28520	Writing on metal by laser beam (manufacture)
	28400	Drop forging of ferrous metals (manufacture)	**25710**		**Manufacture of cutlery**
	28400	Drop stamping of base non-ferrous metals (manufacture)		28750	Bayonet (manufacture)
	28400	Drop stamping of ferrous metals (manufacture)		28610	Butter knife (manufacture)
	28400	Engineers' stampings and pressings of base non-ferrous metals (manufacture)		28610	Choppers (manufacture)
	28400	Engineers' stampings and pressings of ferrous metals (manufacture)		28610	Cleavers (manufacture)
	28400	Forging (manufacture)		28750	Cutlasses (manufacture)
	28400	Hammer forging of steel (manufacture)		28610	Cutlery (electro plated nickel silver) (manufacture)
				28610	Cutlery (manufacture)
				28610	Cutlery for domestic use (manufacture)
				28610	Electro plated nickel silver cutlery (manufacture)
				28610	Fish eater (manufacture)

UK SIC 2007	UK SIC 2003	Activity	SIC 2007	SIC 2003	Activity
	28610	Fork (cutlery) (manufacture)		28620	Builders' knife (manufacture)
	28610	Hair clippers (manufacture)		28620	Carpenter's drill (manufacture)
	28610	Kitchen knife (manufacture)		28620	Case opener (manufacture)
	28610	Knife (cutlery) (manufacture)		28620	Chainsaw blades (manufacture)
	28610	Knife with folding blade (manufacture)		28620	Circular sawblades (manufacture)
	28610	Ladle (manufacture)		28620	Circular saws for all materials (manufacture)
	28610	Manicure and pedicure sets (manufacture)		28620	Clamp (manufacture)
	28610	Nail file (manufacture)		28620	Cold chisel (manufacture)
	28610	Pinking shears (manufacture)		28620	Cramp (manufacture)
	28610	Pocket knife (manufacture)		28620	Cutting blades for machines or mechanical appliances (manufacture)
	28610	Pruning knife (manufacture)			
	28610	Razor (not electric) (manufacture)		28620	Diamond tipped tool (manufacture)
	28610	Razor blade (manufacture)		28620	Die (press tool) (manufacture)
	28610	Razor blade blanks in strips (manufacture)		28620	Die for machine tools (manufacture)
	28610	Razor set (manufacture)		28620	Die pellet (manufacture)
	28610	Safety razors (manufacture)		28620	Draw knife (manufacture)
	28610	Scissors (manufacture)		28620	Drill tools (interchangeable) (manufacture)
	28610	Sheath knife (manufacture)		28620	File (hand tool) (manufacture)
	28610	Spoons made of metal (manufacture)		28620	Forges (manufacture)
	28750	Sword (manufacture)		28620	Garden fork (manufacture)
	28610	Tailors' shears (manufacture)		28620	Garden shears (manufacture)
				28620	Garden trowel (manufacture)
25720		**Manufacture of locks and hinges**		28620	Glass cutter (manufacture)
	28630	Adjustable seat mechanisms (manufacture)		28620	Gouge (wood frame) (manufacture)
	28630	Bicycle locks with or without keys (manufacture)		28620	Hacksaw blades (manufacture)
	28630	Carpet fittings made of metal (manufacture)		28620	Hammer (manufacture)
	28630	Curtain rail and runners made of metal (manufacture)		28620	Hand tools (manufacture)
				28620	Handsaw (manufacture)
	28630	Door and window catches (manufacture)		28620	Hard metal tipped tools (manufacture)
	28630	Door fittings made of metal (manufacture)		28620	Hatchet (manufacture)
	28630	Door hardware for buildings, furniture and vehicles (manufacture)		28620	Hoe (manufacture)
				28620	Interchangeable tools for dies (manufacture)
	28630	Furniture fittings made of metal (manufacture)		28620	Knives for horticultural use (manufacture)
	28630	Hinge (manufacture)		28620	Knives for industrial use (manufacture)
	28630	Key (manufacture)		28620	Knives for machines (manufacture)
	28630	Key blank (manufacture)		28620	Knives for tradesmen (manufacture)
	28630	Latch (manufacture)		28620	Lapping tools (manufacture)
	28630	Lock (manufacture)		28620	Lathe tool (manufacture)
	28630	Lock for motor vehicle (manufacture)		28620	Machine tool interchangeable tools (manufacture)
	28630	Locksmiths (manufacture)		28620	Matchet (manufacture)
	28630	Padlock (manufacture)		28620	Mattock (manufacture)
	28630	Stair rods made of metal (manufacture)		28620	Milling cutter (manufacture)
	28630	Suitcase fittings made of metal (manufacture)		28620	Mould (engineers' small tools) (manufacture)
	28630	Window fittings made of metal (manufacture)		29560	Mould for foundry (manufacture)
				29560	Moulding boxes for any material (manufacture)
25730		**Manufacture of tools**		29560	Moulding machine for working rubber or plastics (manufacture)
	28620	Adze (manufacture)			
	28620	Agricultural hand tools, not power-driven (manufacture)		28620	Nippers (manufacture)
				28620	Pick (manufacture)
	28620	Agricultural knife (manufacture)		28620	Pincers (manufacture)
	28620	Anvils (manufacture)		28620	Pipe cutters (manufacture)
	28620	Auger and auger bit (manufacture)		28620	Plane (manufacture)
	28620	Axe (manufacture)		28620	Planer tool (manufacture)
	28620	Bandsaw blades (manufacture)		28620	Pliers (manufacture)
	28620	Bench vice (manufacture)		28620	Portable forges (manufacture)
	28620	Bit stock drill (manufacture)		28620	Press tool (manufacture)
	28620	Blacksmiths' tools (manufacture)		28620	Pruning shears (manufacture)
	28620	Blow lamp (manufacture)		28620	Punches for hand tools (interchangeable) (manufacture)
	28620	Bolt cropper (manufacture)			
	28620	Brazing lamp (manufacture)			

UK SIC 2007	UK SIC 2003	Activity
	28620	Punches for machine tools (interchangeable) (manufacture)
	28620	Rakes for garden use (manufacture)
	28620	Rasp (manufacture)
	28620	Reamer (manufacture)
	28620	Rock drilling and earth boring interchangeable tools (e.g. Augers, boring bits, drills) (manufacture)
	28620	Saws (hand tools) (manufacture)
	28620	Saws and sawblades (manufacture)
	28620	Scraper (hand tool) (manufacture)
	28620	Screwdrivers (manufacture)
	28620	Scythe (manufacture)
	28620	Secateurs (manufacture)
	28620	Shave hook (manufacture)
	28620	Shears for agricultural or horticultural use (manufacture)
	28620	Shears for garden use (manufacture)
	28620	Sheep shears (not power) (manufacture)
	28620	Shovels (manufacture)
	28620	Sickles (manufacture)
	28620	Slitting saw (manufacture)
	28620	Socket set (manufacture)
	28620	Spade (manufacture)
	28620	Spanner (manufacture)
	28620	Spoke shave (manufacture)
	29560	Tableting and pelleting press for the chemical industry (manufacture)
	28620	Tension strapping tool (manufacture)
	28620	Threading die (manufacture)
	28620	Threading tap (manufacture)
	28620	Tinman's snips (manufacture)
	28620	Tip for cutting tool (manufacture)
	28620	Tools (precision) (manufacture)
	28620	Tools except power driven hand tools (manufacture)
	28620	Tradesmen's knife (manufacture)
	28620	Treadle operated and other non power-operated tools (manufacture)
	28620	Trowel (not garden) (manufacture)
	28620	Twist drill (manufacture)
	28620	Vices (manufacture)
	28620	Wood boring bit (manufacture)
	28620	Wood chisel (manufacture)
	28620	Wrecking bars (manufacture)
	28620	Wrench (manufacture)
25910		**Manufacture of steel drums and similar containers**
	28710	Barrels made of iron or steel (manufacture)
	28710	Boxes and other containers made of iron or steel, of capacity not exceeding 300 litres (manufacture)
	28710	Boxes made of iron or steel (manufacture)
	28710	Buckets made of steel (manufacture)
	28710	Cans and boxes made of iron or steel (manufacture)
	28710	Cans made of blackplate (manufacture)
	28710	Cans made of steel (manufacture)
	28710	Casks made of iron or steel (manufacture)
	28710	Churns made of iron or steel (manufacture)
	28710	Drums made of iron or steel (manufacture)
	28710	Kegs made of iron or steel (manufacture)
	28710	Metal drum reconditioning (manufacture)

SIC 2007	SIC 2003	Activity
	28710	Milk churns made of iron or steel (manufacture)
	28710	Pails made of steel (manufacture)
25920		**Manufacture of light metal packaging**
	28720	Aerosol cans made of metal (manufacture)
	28720	Barrels made of aluminium (manufacture)
	28720	Bottle tops made of metal (manufacture)
	28720	Boxes made of aluminium (manufacture)
	28720	Boxes made of metal (collapsible) (manufacture)
	28720	Cans and boxes made of aluminium (manufacture)
	28720	Cans and boxes made of tin (manufacture)
	28720	Cans for food products (manufacture)
	28720	Cans made of aluminium (manufacture)
	28720	Capsules made of metal (manufacture)
	28720	Casks made of aluminium (manufacture)
	28720	Churns made of aluminium (manufacture)
	28720	Closures made of metal (manufacture)
	28720	Containers made of foil (manufacture)
	28720	Crown cork (manufacture)
	28720	Drums made of aluminium (manufacture)
	28720	Foil containers made of aluminium (manufacture)
	28720	Kegs made of aluminium (manufacture)
	28720	Light metal packaging (manufacture)
	28720	Metallic closures (manufacture)
	28720	Milk churns made of aluminium (manufacture)
	28720	Pilfer-proof metal caps (manufacture)
	28720	Screw caps made of metal (manufacture)
	28720	Stoppers made of metal (manufacture)
	28720	Tins for food products (manufacture)
	28720	Tubes made of aluminium for packaging (collapsible) (manufacture)
	28720	Tubes made of metal (collapsible) (manufacture)
	28720	Tubular containers made of aluminium (manufacture)
	28720	Tubular containers made of metal (manufacture)
25930		**Manufacture of wire products, chain and springs**
	28730	Bands made of plaited metal (manufacture)
	28730	Bands made of uninsulated plaited copper (manufacture)
	28730	Bands, slings, etc. Made of uninsulated plaited iron or steel (manufacture)
	28730	Barbed wire made of steel (manufacture)
	28730	Cable made of uninsulated aluminium (manufacture)
	28730	Cable sheathing made of aluminium (manufacture)
	28730	Cable strands made of aluminium (manufacture)
	28740	Chain (manufacture)
	28740	Chain (non-precision) (manufacture)
	28740	Chain (not articulated transmission) (manufacture)
	28730	Cloth made of wire (manufacture)
	28730	Coated electrodes for electric arc-welding (manufacture)
	28730	Coated or cored wire (manufacture)
	28740	Coil springs (not for motor vehicle suspension) (manufacture)
	28730	Conductor cable made of steel reinforced aluminium (manufacture)
	28730	Drawing pin (manufacture)
	28730	Expanded metal (manufacture)
	28730	Fencing made of steel wire (manufacture)

UK SIC 2007	UK SIC 2003	Activity	SIC 2007	SIC 2003	Activity
	28740	Fish plates for arches made of steel (manufacture)		28740	Screw machine products (manufacture)
	28740	Flange jointing sets (manufacture)		28740	Screws (self tapping) (manufacture)
	28730	Grills made of wire (manufacture)		28740	Screws of all types made of metal (manufacture)
	28730	Heavy wire (manufacture)		28740	Threaded fasteners (manufacture)
	28740	Helical springs (manufacture)		28740	Tubular rivets (manufacture)
	28730	Knitting needles made of metal (manufacture)		28740	Washers made of metal (manufacture)
	28740	Leaf springs (manufacture)			
	28740	Leaves for springs (manufacture)	**25990**		**Manufacture of other fabricated metal products n.e.c.**
	28740	Link chain (welded) (manufacture)		28750	Anchor (manufacture)
	28730	Nail (not wire) (manufacture)		28750	Armoured doors (manufacture)
	28730	Nails made of steel wire (manufacture)		28750	Art metal work (manufacture)
	28730	Needles made of metal (manufacture)		28750	Badges made of metal (manufacture)
	28730	Netting made of steel wire (manufacture)		28750	Bag clasp (manufacture)
	28730	Netting made of wire (manufacture)		28750	Bag frame (manufacture)
	28730	Pins made of metal (manufacture)		28750	Baking dish, pan and tin (manufacture)
	28730	Rigging for ships (manufacture)		28750	Base metal articles (manufacture)
	28740	Ring spring (manufacture)		28750	Basins made of metal (manufacture)
	28730	Rods or wires (coated/covered with flux) for gas welding, soldering or brazing (manufacture)		28750	Baths made of metal (manufacture)
				28750	Bells for pedal cycles (manufacture)
	28730	Rope made of wire (manufacture)		28750	Bells made of base metals (manufacture)
	28730	Safety pins (manufacture)		28750	Bins made of metal (manufacture)
	28730	Sewing needles (manufacture)		28750	Brackets made of base metal (manufacture)
	28740	Skid chain (manufacture)		28750	Buckets made of metal (manufacture)
	28740	Spring presswork (manufacture)		28750	Buckles made of metal (manufacture)
	28740	Spring washer (manufacture)		28750	Building components of zinc e.g. Gutters, roof capping (manufacture)
	28740	Springs (manufacture)			
	28740	Springs for upholstery (manufacture)		28750	Butter dishes made of metal (manufacture)
	28740	Springs made of steel for upholstery (manufacture)		36639	Buttons made of metal (manufacture)
	28730	Staples (not wire) (manufacture)		28750	Cabinets made of metal (not designed for placing on the floor) (manufacture)
	28730	Stranded un-insulated wire made of copper (manufacture)			
				28750	Cable drum made of metal (manufacture)
	28740	Stud link chain (manufacture)		28750	Canisters made of metal (manufacture)
	28730	Tack made of metal (not wire) (manufacture)		28750	Cash boxes made of metal (manufacture)
	28740	Torsion bar spring (manufacture)		28750	Cigarette cases made of metal (manufacture)
	28730	Uninsulated metal cable or insulated cable not useable as a conductor of electricity (manufacture)		28750	Clasps (manufacture)
				28750	Clothes hook (manufacture)
	28730	Welding rods (coated/covered with flux) made of steel (manufacture)		28750	Colanders made of metal (manufacture)
				36639	Combs of metal (manufacture)
	28730	Wire cable made of steel (manufacture)		28750	Condiment set made of metal (manufacture)
	28730	Wire fabric made of steel (manufacture)		28750	Cooking utensils made of metal (manufacture)
	28730	Wire products (manufacture)		28750	Crane hook (manufacture)
	28730	Wire products made of uninsulated copper (manufacture)		28750	Deed box (manufacture)
				28750	Desk tray made of metal (manufacture)
	28730	Wire strands made of aluminium (manufacture)		28750	Domestic hollow ware made of metal (manufacture)
25940		**Manufacture of fasteners and screw machine products**		28750	Domestic utensils made of aluminium (manufacture)
				28750	Dustbins made of metal (manufacture)
	28740	Belleville washer (manufacture)		28750	Dustpan made of metal (manufacture)
	28740	Bifurcated rivet (manufacture)		28750	Egg boxes made of metal (manufacture)
	28740	Bolt (manufacture)		28750	Eyelet (manufacture)
	28740	Bolt end (manufacture)		28750	Filing cabinet made of metal (not designed to be placed on the floor) (manufacture)
	28740	Coach bolts and screws (manufacture)			
	28740	Cotter pin (manufacture)		28750	Flat ware made of base metal (manufacture)
	28740	Fasteners made of metal (manufacture)		28750	Foil bags (manufacture)
	28740	Linchpin (manufacture)		28750	Frying pans (non-electric) (manufacture)
	28740	Lock washer (manufacture)		28750	Grapnel (manufacture)
	28740	Nut (manufacture)		36639	Hair grips and pins made of metal (manufacture)
	28740	Precision screw (manufacture)		28750	Hollow ware (domestic) made of metal (manufacture)
	28740	Rivet (manufacture)			
	28740	Roofbolts, plates and accessories (manufacture)			

UK SIC 2007	UK SIC 2003	Activity	SIC 2007	SIC 2003	Activity
	28750	Hook and eye (manufacture)		28750	Sinks made of metal (other than cast iron) (manufacture)
	28750	Household articles made of metal (manufacture)		36639	Slide fasteners made of metal (manufacture)
	28750	Household utensils made of metal (manufacture)		28750	Small metal hand-operated kitchen appliances and accessories (manufacture)
	28750	Kettles (non-electric) (manufacture)			
	28750	Ladders made of metal (manufacture)		28750	Staples for office use (manufacture)
	28750	Ladders made of metal for fire-fighting vehicles (manufacture)		28750	Statuettes and other ornaments made of base metal (manufacture)
	28750	Left luggage lockers (manufacture)		28750	Steel wool for domestic use (manufacture)
	28750	Manufacture of metal combs (manufacture)		28750	Steps made of metal (manufacture)
	28750	Marine screw propeller (manufacture)		28750	Stewpans (manufacture)
	28750	Metal badges and metal military insignia (manufacture)		28750	Stillage made of metal (manufacture)
				28750	Street furniture (manufacture)
	28750	Metal dinnerware bowls (manufacture)		28750	Striking of medals (manufacture)
	28750	Metal dinnerware platters (manufacture)		28750	Strong box (manufacture)
	28750	Metal goods for office use (manufacture)		28750	Strong room (manufacture)
	28750	Metal hair curlers (manufacture)		28750	Strong room door (manufacture)
	28750	Metal hollow ware pots (manufacture)		28750	Tableware made of base metal (manufacture)
	28750	Metal hollowware kettles (manufacture)		28750	Tea sets made of base metal (manufacture)
	28750	Metal plates (manufacture)		28750	Teapots made of base metal (manufacture)
	28750	Metal road signs (manufacture)		28750	Tensional steel strapping (manufacture)
	28750	Metal spinning		28750	Toast racks made of base metal (manufacture)
	28750	Metal umbrella handles and frames (manufacture)		28750	Trays made of base metal (manufacture)
	28750	Metal vacuum jugs and bottles (manufacture)		28750	Trellis work made of metal (manufacture)
	28750	Milk pan (manufacture)		28750	Wall mountings made of metal (manufacture)
	28750	Nameplates made of metal (manufacture)		28750	Wash basins made of metal (manufacture)
	28750	Night safe (manufacture)		36639	Zip fasteners made of metal (manufacture)
	28750	Omelette pan (manufacture)			
	28750	Pallets made of metal (manufacture)	**26110**		**Manufacture of electronic components**
	28750	Paper clips made of metal (manufacture)		32100	Amplifying valve (manufacture)
	28750	Paper fasteners made of metal (manufacture)		32100	Bare printed circuit boards (manufacture)
	28750	Pedal bins made of metal (manufacture)		32100	Capacitor for electronic apparatus (manufacture)
	28750	Percolators made of metal (non-electric) (manufacture)		32100	Cathode ray tube (manufacture)
				24660	Chemical elements in disk form for use in electronics (manufacture)
	31620	Permanent magnets (metallic) (manufacture)			
	28750	Pewter ware (manufacture)		31100	Choke and coil (electronic) (manufacture)
	28750	Plan chests made of metal (manufacture)		31200	Circuit protection device (electronic) (manufacture)
	28750	Plumbing and pipe fittings made of metal (not cast) (manufacture)		32100	Cold cathode valve or tube (manufacture)
				32100	Colour television tubes (manufacture)
	36639	Press stud (manufacture)		32100	Diode (manufacture)
	36639	Press-fasteners made of metal (manufacture)		32100	Display components (plasma, polymer, LCD) (manufacture)
	28750	Pressure cooker (manufacture)			
	28750	Railway track fixtures of assembled metal (manufacture)		32100	Electron tubes (manufacture)
				32100	Electronic active components (manufacture)
	28750	Rolls made of metal for cable, hose, etc. (manufacture)		32100	Electronic condensers (manufacture)
				31200	Electronic connectors (manufacture)
	28750	Sacrificial anodes made of zinc, magnesium or other non-ferrous metal (manufacture)		32100	Electronic crystals and crystal assemblies (manufacture)
				32100	Electronic integrated circuits (manufacture)
	28750	Safe (manufacture)		32100	Electronic micro-assemblies of moulded module, micromodule or similar types (manufacture)
	28750	Sanitary ware and fittings made of metal (manufacture)			
				32100	Electronic passive components (manufacture)
	28750	Saucepan (manufacture)		32100	Electronic tube (manufacture)
	28750	Saucepans made of aluminium (manufacture)		32100	Electronic valve (manufacture)
	28750	Scourers made of metal (manufacture)		31100	Ferrite parts for electronic apparatus (manufacture)
	28750	Scouring pads made of metal (manufacture)			
	28750	Serving dishes made of base metal (manufacture)		24660	Finished or semi-finished dice, semiconductor (manufacture)
	28750	Shackle (manufacture)			
	28750	Ship propeller blades of metal (manufacture)		24660	Finished or semi-finished wafers, semiconductor (manufacture)
	28750	Ship propellers (manufacture)			
	28750	Shower cabinets made of metal (manufacture)		32100	Hybrid integrated circuits (electronic) (manufacture)
	28750	Sign plates made of metal (manufacture)		32100	Image converters and intensifiers (manufacture)

UK SIC 2007	UK SIC 2003	Activity	SIC 2007	SIC 2003	Activity
	31100	Inductor (electronic) (manufacture)	26120		**Manufacture of loaded electronic boards**
	31100	Industrial transformer (electronic) (manufacture)		32100	Controllers interface cards (manufacture)
	32100	Insulated monolithic, hybrid and passive circuit (manufacture)		32100	Interface cards (manufacture)
	32100	Integrated circuits (analogue or digital) (manufacture)		32100	Loaded electronic boards (manufacture)
	31200	Isolating and make or break switches (electronic) (manufacture)		32100	Loaded printed circuit boards (manufacture)
				32100	Modem interface cards (manufacture)
	32100	Klystron (manufacture)		32100	Network interface cards (manufacture)
	32100	Light emitting diodes (LED) (manufacture)		32100	Phonecards and similar cards containing integrated circuits (smart cards) (manufacture)
	32100	Magnetron (manufacture)		32100	Smart cards (manufacture)
	32100	Microchip (manufacture)		32100	Sound interface cards (manufacture)
	32100	Microcircuit (manufacture)		32100	Video interface cards (manufacture)
	32100	Microprocessors (manufacture)			
	32100	Microwave components (manufacture)	26200		**Manufacture of computers and peripheral equipment**
	32100	Microwave tube (manufacture)		30020	Aluminium coating inside pc cases (manufacture)
	31300	Monitor cables (manufacture)		30020	Analogue computer (manufacture)
	32100	Monolithic integrated circuits (manufacture)		30020	Automatic teller machines (ATMs) computer terminals not mechanically operated (manufacture)
	32100	Mounted piezo-electric crystals (manufacture)		30020	Barcode readers and other optical readers (manufacture)
	32100	Photo diode (manufacture)		30020	Braille pads and other output devices (manufacture)
	32100	Photo electric cell (manufacture)		30020	Central processing units for computers (manufacture)
	32100	Photo semi-conductor device (manufacture)		30020	Computer (electronic) (manufacture)
	32100	Photo-cathode valves or tubes (manufacture)		30020	Computer (manufacture)
	32100	Photosensitive semi-conductor devices (manufacture)		30020	Computer peripheral equipment (manufacture)
				32300	Computer projectors (video beamers) (manufacture)
	32100	Piezo electric crystal (manufacture)		30020	Computer store (manufacture)
	31200	Plug (electronic) (manufacture)		30020	Computer system (manufacture)
	32100	Printed circuit (manufacture)		30020	Computer terminal unit (manufacture)
	31300	Printer and monitor connectors (manufacture)		30020	Control units for computers (manufacture)
	31300	Printer cables (manufacture)		30020	Converter for computer (manufacture)
	32100	Quartz crystal (manufacture)		30020	Data processing equipment (electronic (other than electronic calculators)) (manufacture)
	32100	Receiver or amplifier valves or tubes (manufacture)		30020	Desktop computers (manufacture)
	32300	Record cutters (manufacture)		30020	Digital computer (manufacture)
	31100	Rectifier plant (electronic) (manufacture)		30020	Digital machines (manufacture)
	31200	Relays for electronic and telecommunications use (manufacture)		30020	Floppy disk drives (manufacture)
	32100	Resistor (electronic) (manufacture)		30020	Graphics tablets and other input devices (manufacture)
	32100	Semi-conductor (not power) (manufacture)		30020	Hand-held computers PDA (manufacture)
	31100	Semi-conductor control equipment (converters) (manufacture)		30020	Hard disk drives (manufacture)
	24660	Semi-finished dice or wafers, semiconductor (manufacture)		30020	Hybrid computer (manufacture)
				30020	Hybrid machines (manufacture)
	32100	Solar panels (photovoltaic cell type) (manufacture)		30020	Information processing equipment
	32100	Solenoids for electronic apparatus (manufacture)		30020	Keyboards for computers (manufacture)
	32100	Solid state circuit (manufacture)		30020	Laptop computers (manufacture)
	32100	Stabilising valve (manufacture)		30020	Light pens (manufacture)
	31200	Switches and transducers for electronic applications (manufacture)		30020	Machines for transcribing data media in coded form (manufacture)
	32100	Television camera tubes (manufacture)		30020	Magnetic card readers (manufacture)
	32100	Television picture tube (manufacture)		30020	Magnetic card storage units (manufacture)
	32100	Television scan coil (manufacture)		30020	Magnetic disk drives (manufacture)
	31200	Terminals for electronic apparatus (manufacture)		30020	Magnetic flash drives (manufacture)
	32100	Thermionic valves or tubes (manufacture)		30020	Magnetic or optical readers (manufacture)
	32100	Thyratron (manufacture)		30020	Magnetic storage devices for computers (manufacture)
	31100	Transformer for electronic apparatus (manufacture)			
	32100	Transistors (manufacture)		30020	Mainframe computers (manufacture)
	32100	Travelling wave tube (manufacture)		30020	Memory store for computers (manufacture)
	32100	Tube (electronic) (manufacture)		30020	Mice, joysticks and trackballs (manufacture)
	31300	USB cables (manufacture)			
	32100	Valves (electronic) (manufacture)			

UK SIC 2007	UK SIC 2003	Activity
	30020	Micro-computers (manufacture)
	30020	Mini-computers (manufacture)
	30020	Monitors for computers (manufacture)
	30020	Multi-function office equipment (manufacture)
	30020	Network interface (manufacture)
	30020	Optical CD-RW, CD-ROM, DVD-ROM, DVD-RW disk drives (manufacture)
	30020	Optical disk drives (manufacture)
	30020	Peripheral equipment for computer uses including card punches and verifiers (manufacture)
	30020	Personal computers and workstations (manufacture)
	30020	Point-of-sale (pos) computer terminals not mechanically operated (manufacture)
	30020	Printer ink cartridges (manufacture)
	30020	Printer ink cartridges (refilling)
	30020	Printer servers (manufacture)
	30020	Printers and plotters (manufacture)
	30020	Printers for computers (manufacture)
	30020	Punchcard readers (manufacture)
	30020	Scanners for computer use (manufacture)
	30020	Servers and network servers (manufacture)
	30020	Smart card readers (manufacture)
	30020	Tape reader for computers (manufacture)
	30020	Tape streamers and other magnetic tape storage units (manufacture)
	30020	Virtual reality helmets (manufacture)
	30020	Visual display unit for computer (manufacture)
26301		**Manufacture of telegraph and telephone apparatus and equipment**
	32201	Bells for telephones (manufacture)
	32201	Bridges for telecommunications (manufacture)
	31620	Burglar alarm and system (manufacture)
	32201	Carrier equipment (manufacture)
	32202	Cellular phones (manufacture)
	32201	Communication devices using infrared signal (e.g. Remote controls) (manufacture)
	32201	Cordless telephones (except cellular) (manufacture)
	32201	Data transmission link line (manufacture)
	31620	Delay lines and networks (manufacture)
	32201	Dial for telephone (manufacture)
	32201	Entrance telephones (manufacture)
	32201	Facsimile transmission apparatus (manufacture)
	32201	Fax machines (manufacture)
	31620	Fire alarm and system (manufacture)
	31620	Fire alarm systems, sending signals to a control station (manufacture)
	32201	Gateways for telecommunications (manufacture)
	32201	Infrared remote controls (manufacture)
	32201	Line apparatus (carrier, duplex and repeater) (manufacture)
	32201	Line telegraphy apparatus (manufacture)
	32201	Line telephony apparatus (manufacture)
	32202	Mobile telephone (manufacture)
	32201	Modems (manufacture)
	32201	Multiplexers for telephone exchanges (manufacture)
	32201	Pagers (manufacture)
	32201	Picture transmitter (manufacture)
	32201	Private branch exchange (PBX) equipment (manufacture)
	32201	Routers for telecommunications (manufacture)

SIC 2007	SIC 2003	Activity
	31620	Security alarms and systems (manufacture)
	32201	Subscriber apparatus (telephone) (manufacture)
	32201	Switchboard for telecommunications (manufacture)
	32201	Switching equipment for telegraph and telex (manufacture)
	32201	Telegraph apparatus (manufacture)
	32201	Telephone (manufacture)
	32300	Telephone answering machines (manufacture)
	32201	Telephone apparatus (manufacture)
	32201	Telephone exchange equipment (manufacture)
	32201	Telephone exchanges (manufacture)
	32201	Telephone handset (manufacture)
	32201	Teleprinter (manufacture)
	32201	Telewriter (manufacture)
	32201	Telex machine (manufacture)
	32201	Terminal equipment for telegraphic and data communications (manufacture)
	32201	Transmission equipment for telephone and telegraph (manufacture)
26309		**Manufacture of communication equipment (other than telegraph and telephone apparatus and equipment**
	32300	Aerial (domestic) (manufacture)
	32300	Aerial (non-domestic) (manufacture)
	32300	Aerial reflectors (manufacture)
	32300	Aerial rotors (manufacture)
	32300	Aerial signal splitters (manufacture)
	32300	Amplifier for broadcasting studio (manufacture)
	32202	Cable television equipment (manufacture)
	32202	Camera for television (manufacture)
	32202	Closed circuit television equipment (CCTV) (manufacture)
	32202	Fixed transmitters (manufacture)
	32202	Ground station for relay satellite communication (manufacture)
	32202	Monitoring equipment for radio and television (manufacture)
	32202	Radio beacons (manufacture)
	32202	Radio communications equipment (manufacture)
	32202	Radio frequency booster stations (manufacture)
	32202	Radio-telephony apparatus (manufacture)
	32202	Reception apparatus for radio-telephony or radio-telegraphy (manufacture)
	32202	Relay link apparatus (manufacture)
	32202	Relay transmitters (manufacture)
	32202	Satellite relay (manufacture)
	32202	Television camera (manufacture)
	32300	Television receiver (manufacture)
	32202	Television transmitter (manufacture)
	32202	Transmission apparatus for radio-broadcasting (manufacture)
	32202	Transmitter-receivers (manufacture)
	32300	Transmitting and receiving antenna (manufacture)
	32202	Transponders (manufacture)
	32202	Video conferencing equipment (manufacture)
	32202	Video signalling equipment (manufacture)
26400		**Manufacture of consumer electronics**
	32300	Amplifier for audio separates (manufacture)
	32300	Amplifiers and sound amplifier sets (manufacture)
	32300	Audio separate (manufacture)

UK SIC 2007	UK SIC 2003	Activity	SIC 2007	SIC 2003	Activity
	32300	Cassette player (manufacture)		33201	Air navigation instruments and systems (electronic) (manufacture)
	32300	Cassette type recorders (manufacture)		33201	Airfield electronic controls and approach aids (manufacture)
	32300	Compact disc players (manufacture)		33201	Ammeters (electronic) (manufacture)
	32300	Dvd recorders and players (manufacture)		33201	Automatic pilots (electronic) (manufacture)
	32300	Earphone (manufacture)		33201	Balancing machines (electronic) (manufacture)
	36509	Electronic games (domestic) (manufacture)		33201	Barometer (electronic) (manufacture)
	36509	Electronic toys and games with replaceable software (manufacture)		33201	Biochemical analysers (electronic) (manufacture)
	32300	Gramophone (manufacture)		33201	Checking instruments and appliances (electronic) (manufacture)
	32300	Gramophone accessory (manufacture)		33201	Chromatographs (electronic) (manufacture)
	32300	Gramophone cabinet (manufacture)		33201	Colorimeters (electronic) (manufacture)
	32300	Headphones (manufacture)		33201	Comparators (electronic) (manufacture)
	32300	Headset (not telecommunication type) (manufacture)		33201	Co-oxymeters (electronic) (manufacture)
	32300	Hi-fi equipment (manufacture)		33201	Cross-talk meters (electronic) (manufacture)
	32300	Jukeboxes (manufacture)		33201	Current checking instruments (electronic) (manufacture)
	32300	Karaoke machines (manufacture)		33201	Cytometers (electronic) (manufacture)
	32300	Loudspeaker (manufacture)		33201	Density measuring optical equipment (electronic) (manufacture)
	32300	Magnetic recording head (manufacture)		33201	Diffraction apparatus (electronic) (manufacture)
	32300	Magnetic tape recorders (manufacture)		33201	Echo sounders (manufacture)
	32300	Megaphone (manufacture)		33201	Electricity meter (electronic) (manufacture)
	32300	Microphone (manufacture)		33201	Electron microscope (manufacture)
	32300	Monitors for videos (manufacture)		33201	Electronic aircraft engine instruments (manufacture)
	32300	Pick-up arm and cartridge for record player (manufacture)		33201	Electronic apparatus for testing physical and mechanical properties of materials (manufacture)
	32300	Public address system (manufacture)		33201	Electronic automotive emissions testing equipment (manufacture)
	32300	Public broadcasting equipment (manufacture)		33201	Electronic counter (manufacture)
	32300	Radio cabinets made of wood (manufacture)		33201	Electronic environmental controls and automatic controls for appliances (manufacture)
	32300	Radio receiving set (manufacture)		33201	Electronic flame and burner control (manufacture)
	32300	Receivers for radio broadcasting (manufacture)		33201	Electronic flight recorders (manufacture)
	32300	Receivers for television (manufacture)		33201	Electronic GPS devices (manufacture)
	32300	Record player accessories (manufacture)		33201	Electronic humidistats (manufacture)
	32300	Record player cabinets made of wood (manufacture)		33201	Electronic hydronic limit controls (manufacture)
	32300	Record players (manufacture)		33201	Electronic instruments and appliances for measuring, testing, and navigation (manufacture)
	32300	Record playing mechanism (manufacture)		33201	Electronic laboratory analytical instruments (manufacture)
	32300	Sound heads (manufacture)		33201	Electronic laboratory incubators and sundry laboratory apparatus for measuring, testing (manufacture)
	32300	Sound recording and reproducing equipment (manufacture)		33201	Electronic metal detectors (manufacture)
	32300	Speaker systems (manufacture)		33201	Electronic motion detectors (manufacture)
	32300	Stereo systems (manufacture)		33201	Electronic physical properties testing and inspection equipment (manufacture)
	32300	Stylus for record player (manufacture)		33201	Electronic pneumatic gauges (manufacture)
	32300	Tables for turn-tables (manufacture)		33201	Electronic polygraph machines (manufacture)
	32300	Tape player and recorder (audio and visual) (manufacture)		33201	Electronic pulse (signal) generators (manufacture)
	32300	Tape recorder cabinets made of wood (manufacture)		33201	Electronic tally counters (manufacture)
	32300	Television cabinets made of wood (manufacture)		33201	Electronic testing equipment (manufacture)
	32300	Television monitors and displays (manufacture)		33201	Expansion analysers (electronic) (manufacture)
	32300	Tone arms (manufacture)		33201	Exposure meter (electric) (manufacture)
	32300	Tuner (audio separate) (manufacture)		31620	Flight recorder (electric) (manufacture)
	32300	Tuner for radio and television (other than audio separates) (manufacture)		33201	Flow meters (electronic) (manufacture)
	32300	Turn-tables (record decks) (manufacture)		33201	Fluorimeter (electronic) (manufacture)
	36509	Video game consoles (manufacture)		33201	Frequency meter (electronic) (manufacture)
	32300	Video projector (manufacture)		33201	Gas meter (electronic) (manufacture)
	32300	Video recording or reproducing apparatus including camcorders (manufacture)		33201	Gauge (electronic) (manufacture)
26511		**Manufacture of electronic instruments and appliances for measuring, testing, and navigation, except industrial process control equipment**			
	33201	Absorptiometer (electronic) (manufacture)			

UK SIC 2007	UK SIC 2003	Activity	SIC 2007	SIC 2003	Activity
	33201	Geophysical instruments and appliances (electronic) (manufacture)		33201	Sextant (electronic) (manufacture)
	33201	Guided weapon launching gear and launch control post (electronic) (manufacture)		33201	Smoke detection equipment (electronic) (manufacture)
	33201	Gunnery control instrument (electronic) (manufacture)		33201	Sonar (manufacture)
	33201	Heat meters (electronic) (manufacture)		33201	Spectrofluorimeter (electronic) (manufacture)
	33201	Hydrographic instrument and apparatus (electronic) (manufacture)		33201	Spectrograph (electronic) (manufacture)
	33201	Hydrological instrument (electronic) (manufacture)		33201	Spectrometers (electronic) (manufacture)
	33201	Hydrometers (electronic) (manufacture)		33201	Spectrophotometer (manufacture)
	33201	Instruments for testing physical and mechanical properties of materials (electronic) (manufacture)		33201	Spectrum analysers (manufacture)
	33201	Laboratory type sensitive balances (electronic) (manufacture)		33201	Speedometers (electronic) (manufacture)
	33201	Level gauges (electronic) (manufacture)		33201	Surface tension instruments (electronic) (manufacture)
	33201	Level measuring and control instruments (electronic) (manufacture)		33201	Surveying instruments (electronic) (manufacture)
	33201	Manometers (electronic) (manufacture)		33201	Tachometer (electronic) (manufacture)
	33201	Measuring instruments and appliances (electronic) (manufacture)		33201	Taximeters (electronic) (manufacture)
	33201	Meteorological instruments (electronic) (manufacture)		33201	Telemetering instruments (manufacture)
	33201	Meters (other than for electricity and parking) (electronic) (manufacture)		33201	Telemetric equipment (manufacture)
				33201	Test benches (electronic) (manufacture)
	33201	Meters for electricity (electronic) (manufacture)		33201	Testing instruments and appliances (electronic) (manufacture)
	33201	Meters for liquid supply (electronic) (manufacture)		33201	Testing machines and equipment (electronic) (manufacture)
	33201	Meters for petrol pumps (electronic) (manufacture)		33201	Theodolite (electronic) (manufacture)
	33201	Meters for water (electronic) (manufacture)		33201	Thermometer (electronic) (manufacture)
	31620	Mine detectors (electronic) (manufacture)		33201	Thermostat (electronic) (manufacture)
	33201	Navigational instruments and appliances (electronic) (manufacture)		33201	Ultrasonic sounding instruments (electronic) (manufacture)
	33201	Nucleonic instrument (manufacture)		33201	Velocity measuring instruments (electronic) (manufacture)
	33201	Numerical control and indication equipment for machine tools (electronic) (manufacture)		33201	Viscometers (electronic) (manufacture)
	33201	Oceanographic or hydrological instruments (electronic) (manufacture)		33201	Voltage checking instruments (manufacture)
	33201	Optical density measuring equipment (electronic) (manufacture)		33201	Voltmeter (electronic) (manufacture)
	33201	Optical type measuring and checking appliances and instruments (electronic) (manufacture)		33201	Watt meter (electronic) (manufacture)
				33201	Wheel balancing machine (electronic) (manufacture)
	33201	Oscilloscope (manufacture)	**26512**		**Manufacture of electronic industrial process control equipment**
	33201	Pedometers (electronic) (manufacture)		33301	Industrial process control equipment (electronic) (manufacture)
	33201	Ph meters (electronic) (manufacture)		33301	Process control equipment (electronic) (manufacture)
	33201	Ph/gas blood analysers (electronic) (manufacture)			
	33201	Photogrammetric equipment (electronic) (manufacture)	**26513**		**Manufacture of non-electronic instruments and appliances for measuring, testing and navigation, except industrial process control equipment**
	33201	Polarimeters (electronic) (manufacture)			
	33201	Potentiometric recorder (electronic) (manufacture)		33202	Absorptiometer (non-electronic) (manufacture)
	33201	Pressure measuring and control instrument (electronic) (manufacture)		33202	Air navigation instruments and systems (non-electronic) (manufacture)
	33201	Pressure switch (electronic) (manufacture)		33202	Aircraft engine instruments (non-electronic) (manufacture)
	33201	Proton microscope (manufacture)		33202	Altimeter (non-electronic) (manufacture)
	33201	Radar equipment (manufacture)		33202	Ammeters (non-electronic) (manufacture)
	33201	Radiation measuring and detection instruments (electronic) (manufacture)		33202	Artificial horizon (non-electronic) (manufacture)
				33202	Automatic pilots (non-electronic) (manufacture)
	33201	Radio navigational aid apparatus (manufacture)		33202	Automotive emissions testing equipment (non-electronic) (manufacture)
	33201	Radio remote control apparatus (manufacture)		33202	Balancing machines (non-electronic) (manufacture)
	33201	Refractometers (electronic) (manufacture)		33202	Barometer (non-electronic) (manufacture)
	33201	Resistance checking instruments (electronic) (manufacture)		33202	Calculating instruments (non-electronic) (manufacture)
	33201	Revolution counters (electronic) (manufacture)		33202	Callipers (manufacture)
	33201	Seismometers (electronic) (manufacture)		33202	Checking instruments and appliances (non-electronic) (manufacture)

UK SIC 2007	UK SIC 2003	Activity	SIC 2007	SIC 2003	Activity
	33202	Colorimeters (non-electronic) (manufacture)		33202	Metal detectors (non-electronic) (manufacture)
	33202	Comparators (non-electronic) (manufacture)		33202	Meteorological instruments (non-electronic) (manufacture)
	33202	Compass (drawing) (manufacture)		33202	Meteorological optical instruments (non-electronic) (manufacture)
	33202	Compass (magnetic) (manufacture)			
	33202	Compressibility testing equipment (non-electronic) (manufacture)		33202	Meters (other than for electricity and parking) (non-electronic) (manufacture)
	33202	Counting instruments (non-electric) (manufacture)		33202	Meters for electricity (non-electronic) (manufacture)
	33202	Density measuring optical equipment (non-electronic) (manufacture)		33202	Meters for liquid supply (non-electronic) (manufacture)
	33202	Drafting tables and machines (manufacture)		33202	Meters for petrol pumps (non-electronic) (manufacture)
	33202	Drawing instrument (manufacture)		33202	Meters for water (non-electronic) (manufacture)
	33202	Elasticity testing equipment (non-electronic) (manufacture)		33202	Micrometer (manufacture)
	33202	Electricity meter (non-electronic) (manufacture)		33202	Microscopes (other than optical) (manufacture)
	33202	Environmental controls and automatic controls for appliances (non-electronic) (manufacture)		33202	Mine detectors (non-electronic) (manufacture)
				31620	Mine detectors, pulse (signal) generators; metal detectors (manufacture)
	33202	Equipment for testing physical and mechanical properties of materials (non-electronic) (manufacture)		33202	Mitre (manufacture)
				33202	Motion detectors (non-electronic) (manufacture)
	33202	Expansion meter (non-electronic) (manufacture)		33202	Nautical instrument (non-electronic) (manufacture)
	33202	Flame and burner control (non-electronic) (manufacture)		33202	Navigational instruments and appliances (non-electronic) (manufacture)
	33202	Flow measuring and control instrument (non-electronic) (manufacture)		33202	Non-electrical apparatus for testing physical and mechanical properties of materials (manufacture)
	33202	Flow meters (non-electronic) (manufacture)		33202	Numerical control and indication equipment for machine tools (non-electronic) (manufacture)
	33202	Fluorimeter (non-electronic) (manufacture)			
	33202	Frequency meter (non-electronic) (manufacture)		33202	Oceanographic or hydrological instruments (non-electronic) (manufacture)
	33202	Galvanometer (manufacture)		33202	Ohmmeter (manufacture)
	33202	Gas chromatograph (manufacture)		33202	Optical density measuring equipment (non-electronic) (manufacture)
	33202	Gas meter (non-electronic) (manufacture)			
	33202	Gauge (non-electronic) (manufacture)		33202	Optical type measuring and checking appliances and instruments (non-electronic) (manufacture)
	33202	Geophysical instruments and appliances (non-electronic) (manufacture)			
	33202	GPS devices (non-electronic) (manufacture)		33202	Pedometers (non-electronic) (manufacture)
	33202	Gunnery control instrument (non-electronic) (manufacture)		33202	Ph meters (non-electronic) (manufacture)
				33202	Photogrammetric equipment (non-electronic) (manufacture)
	33202	Hardness testing instrument (non-electronic) (manufacture)		33202	Photometers (non-electronic) (manufacture)
	33202	Heat meters (non-electronic) (manufacture)		33202	Physical properties testing and inspection equipment (non-electronic) (manufacture)
	33202	Humidistats (non-electronic) (manufacture)			
	33202	Hydrographic instrument and apparatus (non-electronic) (manufacture)		33202	Pneumatic gauges (non-electronic) (manufacture)
				33202	Polarimeters (non-electronic) (manufacture)
	33202	Hydrological instrument (non-electronic) (manufacture)		33202	Polygraph machines (non-electronic) (manufacture)
	33202	Hydrometers (non-electronic) (manufacture)		33202	Porosimeter (manufacture)
	33202	Hydronic limit controls (non-electronic) (manufacture)		33202	Potentiometric recorder (non-electronic) (manufacture)
	33202	Laboratory analytical instruments (non-electronic) (manufacture)		33202	Precision balance (non-electronic) (manufacture)
				33202	Precision drawing instrument (manufacture)
	33202	Laboratory incubators and sundry lab apparatus for measuring, testing (non-electronic) (manufacture)		33202	Pressure measuring and control instrument (non-electronic) (manufacture)
	33202	Laboratory type sensitive balances (non-electronic) (manufacture)		33202	Pulse (signal) generators (non-electronic) (manufacture)
	33202	Level gauges (non-electronic) (manufacture)		33202	Pyrometer (non-electronic) (manufacture)
	33202	Level measuring and control instruments (non-electronic) (manufacture)		33202	Radiation equipment and detection instruments (non-electronic) (manufacture)
	33202	Liquid supply meter (non-electronic) (manufacture)		33202	Range finders (non-electronic) (manufacture)
	33202	Magnetic compass (manufacture)		33202	Range finders (optical non-electronic) (manufacture)
	33202	Manometers (non-electronic) (manufacture)		33202	Refractometers (non-electronic) (manufacture)
	33202	Mathematical instrument (non-electronic) (manufacture)		33202	Revolution counters (non-electronic) (manufacture)
				33202	Scientific laboratory equipment (non-electrical or non-optical) (manufacture)
	33202	Measuring instruments and appliances (non-electronic) (manufacture)		33202	Seismometers (non-electronic) (manufacture)

UK SIC 2007	UK SIC 2003	Activity
	33202	Sextant (non-electronic) (manufacture)
	33202	Slide rules (manufacture)
	33202	Spectrofluorimeter (non-electronic) (manufacture)
	33202	Spectrograph (non-electronic) (manufacture)
	33202	Spectrometers (non-electronic) (manufacture)
	33202	Speedometers (non-electronic) (manufacture)
	33202	Spirit level (manufacture)
	33202	Stroboscope (manufacture)
	33202	Surface tension instruments (non-electronic) (manufacture)
	33202	Surveying instruments (non-electronic) (manufacture)
	33202	Surveying instruments (optical non-electronic) (manufacture)
	33202	Tachometer (non-electronic) (manufacture)
	33202	Tally counters (non-electronic) (manufacture)
	33202	Taximeters (non-electronic) (manufacture)
	33202	Temperature measuring and control instrument (non-electronic) (manufacture)
	33202	Test benches (non-electronic) (manufacture)
	33202	Testing instruments and appliances (non-electronic) (manufacture)
	33202	Theodolite (non-electronic) (manufacture)
	33202	Thermometer (non-electronic) (manufacture)
	33202	Thermostat (non-electronic) (manufacture)
	33202	Ultrasonic sounding instruments (non-electronic) (manufacture)
	33202	Vehicle motors testing and regulating apparatus (non-electronic) (manufacture)
	33202	Velocity measuring instruments (non-electronic) (manufacture)
	33202	Viscometers (non-electronic) (manufacture)
	33202	Voltmeter (non-electronic) (manufacture)
	33202	Water meters (non-electronic) (manufacture)
	33202	Watt meter (non-electronic) (manufacture)
	33202	Wheel balancing machine (non-electronic) (manufacture)
26514		**Manufacture of non-electronic industrial process control equipment**
	33302	Industrial process control equipment (non-electronic) (manufacture)
	33302	Process control equipment (electric) (manufacture)
	33302	Sensor for electric process control equipment (manufacture)
26520		**Manufacture of watches and clocks**
	33500	Alarm clock (manufacture)
	33500	Car clock (manufacture)
	33500	Cases for clocks and watches (manufacture)
	33500	Chronometer (manufacture)
	33500	Clock (electric) (manufacture)
	33500	Clock (manufacture)
	33500	Clock case made of wood (manufacture)
	33500	Components for clocks and watches (manufacture)
	33500	Electronic timer (not clock or watch) (manufacture)
	33500	Equipment for measuring and recording (manufacture)
	33500	Industrial timer (manufacture)
	33500	Instrument panel clock (manufacture)
	33500	Movements for clocks and watches (manufacture)
	33500	Parking meters (manufacture)

SIC 2007	SIC 2003	Activity
	33500	Pocket timer (manufacture)
	33500	Pocket watch (manufacture)
	33500	Process timers (manufacture)
	33500	Stop watch (manufacture)
	33500	Time clock (manufacture)
	33500	Time lock (manufacture)
	33500	Time recorder (manufacture)
	33500	Time switch (manufacture)
	33500	Time/date stamps (manufacture)
	33500	Timer for industrial use (manufacture)
	33500	Travelling clock (manufacture)
	33500	Watch (manufacture)
	33500	Watch case (manufacture)
	33500	Watchmakers' jewels (manufacture)
	33500	Wrist watch (manufacture)
26600		**Manufacture of irradiation, electromedical and electrotherapeutic equipment**
	33100	Baggage scanning equipment (manufacture)
	33100	CT scanners (manufacture)
	33100	Echocardiographs (manufacture)
	33100	Electro medical equipment (manufacture)
	33100	Electro medical pacemaker (manufacture)
	33100	Electro medical stimulator (manufacture)
	33100	Electrochemical apparatus for industrial use (manufacture)
	33100	Electro-diagnostic apparatus (manufacture)
	33100	Electro-encephalographs (manufacture)
	33100	Electromyographs (manufacture)
	33100	Electrotherapeutic equipment (manufacture)
	33100	Endoscopes (manufacture)
	33100	Hearing aid (electronic) (manufacture)
	33100	Irradiation equipment (manufacture)
	33100	Magnetic resonance imaging (MRI) equipment (manufacture)
	33100	Mammographs (manufacture)
	33100	Mask and respirator (not medical) (manufacture)
	33100	Medical laser equipment (manufacture)
	33100	Optometer (manufacture)
	33100	Pacemaker (electro medical) (manufacture)
	33100	Pet scanners (manufacture)
	33100	Scintillation scanners (manufacture)
	33100	Thermographs (manufacture)
	33100	Tomographs (manufacture)
	33100	Ultrasonic diagnostic equipment (manufacture)
	33100	X-ray apparatus for industrial use (manufacture)
	33100	X-ray apparatus for medical use (manufacture)
	33100	X-ray diffraction or fluorescence apparatus (manufacture)
	33100	X-ray or alpha, beta or gamma radiation apparatus (manufacture)
	33100	X-ray tubes (manufacture)
26701		**Manufacture of optical precision instruments**
	33402	Astronomical equipment (optical) (manufacture)
	33402	Auto correlator (optical) (manufacture)
	33402	Binoculars (manufacture)
	33402	Correlator (optical) (manufacture)
	33402	Fibre optic apparatus (manufacture)
	33402	Gunnery control instrument (optical) (manufacture)

UK SIC 2007	UK SIC 2003	Activity
	33402	Infrared systems for night vision (manufacture)
	33402	Laser (excluding complete equipment using laser components) (manufacture)
	33401	Lenses (except ophthalmic) (manufacture)
	33402	Magnifying glass (manufacture)
	33201	Meteorological optical instruments (electronic) (manufacture)
	33402	Monocular (manufacture)
	33402	Observation telescopes (manufacture)
	33402	Optical comparators (manufacture)
	33402	Optical fire control equipment (manufacture)
	33402	Optical gun sighting equipment (manufacture)
	33402	Optical instruments and appliances (other than photographic goods) (manufacture)
	33402	Optical machinist's precision tools (manufacture)
	33402	Optical magnifying instruments (manufacture)
	33402	Optical microscope (manufacture)
	33401	Optical mirrors (manufacture)
	33402	Optical positioning equipment (manufacture)
	33402	Optical projector (meteorological) (manufacture)
	33402	Periscopes (manufacture)
	33201	Photometers (electronic) (manufacture)
	33201	Range finder (optical) (manufacture)
	33201	Range finders (optical) (manufacture)
	33402	Sight telescopes (manufacture)
	33201	Surveying instruments (optical) (manufacture)
	33402	Telescope (manufacture)
	33402	Telescopic sights (manufacture)
	33402	Thread counters (manufacture)
	33402	Wedge (optical) (manufacture)
26702		**Manufacture of photographic and cinematographic equipment**
	33403	Accessories for photographic equipment (manufacture)
	33403	Cameras (manufacture)
	33403	Cine camera (manufacture)
	33403	Cinematographic equipment (manufacture)
	33403	Dark room equipment (manufacture)
	32300	Digital photographic cameras (manufacture)
	33403	Discharge lamp (electronic) and other flashlight apparatus (manufacture)
	33403	Episcope (manufacture)
	33403	Flashlight apparatus (manufacture)
	33403	Image projectors (manufacture)
	33403	Microfiche readers (manufacture)
	33403	Microfilm equipment (manufacture)
	33403	Microfilm readers (manufacture)
	33402	Microphotography equipment (manufacture)
	33402	Microprojection equipment (manufacture)
	33403	Negatoscopes (manufacture)
	33403	Overhead transparency projectors (manufacture)
	33201	Photo electric exposure meter (manufacture)
	33403	Photographic enlarger (manufacture)
	33403	Photographic equipment (manufacture)
	33403	Photographic film instrument (manufacture)
	33201	Photographic light meters (manufacture)
	33403	Projection screen (manufacture)
	33403	Projector (photographic or cinematographic) (manufacture)

SIC 2007	SIC 2003	Activity
	33403	Projector for cinema (manufacture)
	33403	Reducer (photographic) (manufacture)
	33403	Screen for cinema (manufacture)
	33403	Slide projector (manufacture)
	33403	Television camera lens (manufacture)
26800		**Manufacture of magnetic and optical media**
	24650	Blank diskettes (manufacture)
	24650	Blank optical discs (manufacture)
	24650	Computer discs and tapes (unrecorded) (manufacture)
	24650	Floppy disk (manufacture)
	24650	Hard drive media (manufacture)
	24650	Magnetic card (manufacture)
	24650	Magnetic disc (unrecorded) (manufacture)
	24650	Magnetic tape (unrecorded) (manufacture)
	24650	Media for sound or video recording (unrecorded) (manufacture)
	24650	Optical media (manufacture)
	24650	Rigid magnetic disk (manufacture)
27110		**Manufacture of electric motors, generators and transformers**
	31620	Actuator (electro-magnetic positioner) (manufacture)
	31100	Alternating current (ac) generators (manufacture)
	31100	Alternating current (ac) motors (manufacture)
	31100	Alternators (not for vehicles) (manufacture)
	31100	Arc-welding transformers (manufacture)
	31100	Battery charger (manufacture)
	31100	Converting machinery (electrical) (manufacture)
	31100	Direct current (dc) generator sets (manufacture)
	31100	Direct current (dc) motors and generators (manufacture)
	31100	Dynamo (not for vehicle) (manufacture)
	31100	Electric motors, generators (manufacture)
	31100	Fluorescent ballasts (i.e. Transformers) (manufacture)
	31100	Frequency converter (not power) (manufacture)
	31100	Generator set (manufacture)
	31100	Motor generator sets (manufacture)
	31100	Power generators (manufacture)
	31100	Power supply unit for electronic applications (manufacture)
	31100	Reactor shunt and limiting (manufacture)
	31100	Rewinding of armatures on a factory basis (manufacture)
	31100	Substation transformers for electric power distribution (manufacture)
	31100	Traction motors with or without associated control equipment (manufacture)
	31100	Transformer (generator, transmission system and distribution) (manufacture)
	31100	Transformer for industrial use (manufacture)
	31100	Transmission and distribution voltage regulators (manufacture)
	31100	Turbine for electricity generation (manufacture)
	31100	Turbo alternator (manufacture)
	31100	Universal ac/dc motors (manufacture)
27120		**Manufacture of electricity distribution and control apparatus**
	31200	Bus bar (switchgear type) (manufacture)
	31200	Circuit breaker (moulded case) (manufacture)

229

UK SIC 2007	UK SIC 2003	Activity
	31200	Circuit breaker for power (manufacture)
	31200	Electric control and distribution boards (manufacture)
	31200	Electric power switches (manufacture)
	31200	Electrical apparatus for switching or protecting electrical circuits (manufacture)
	31200	Electrical relays (manufacture)
	31200	Electricity distribution and control apparatus (manufacture)
	31200	Fuse and fusegear (power) (manufacture)
	31200	Fusebox for domestic use (manufacture)
	31200	Fuses for domestic use (manufacture)
	31200	Miniature circuit breaker (manufacture)
	31200	Motor starting and controlling gear (manufacture)
	31200	Moulded case circuit breaker (manufacture)
	31200	Power switching equipment (manufacture)
	31200	Prime mover generator sets (manufacture)
	31200	Switchgear (power) (manufacture)
	31200	Voltage limiters (manufacture)
27200		**Manufacture of batteries and accumulators**
	31400	Accumulator (manufacture)
	31400	Batteries for vehicles (manufacture)
	31400	Battery for car (manufacture)
	31400	Battery for flash lamp (manufacture)
	31400	Dry battery (non-rechargeable) (manufacture)
	31400	Electric accumulators including parts thereof (manufacture)
	31400	Lead acid batteries (manufacture)
	31400	Lithium batteries (manufacture)
	31400	Manganese dioxide cells (manufacture)
	31400	Mercuric dioxide cells (manufacture)
	31400	NiCad batteries (manufacture)
	31400	NiMH batteries (manufacture)
	31400	Primary battery (manufacture)
	31400	Primary cells (manufacture)
	31400	Secondary battery (manufacture)
	31400	Silver oxide cells (manufacture)
	31400	Traction battery (rechargeable) (manufacture)
	31400	Wet cell batteries (manufacture)
27310		**Manufacture of fibre optic cables**
	31300	Fibre optic cable for data transmission or live transmission of images (manufacture)
	31300	Optical fibre cables for coded data transmission (manufacture)
	33402	Optical fibres, optical fibre bundles and cables (manufacture)
27320		**Manufacture of other electronic and electric wires and cables**
	31300	Cable accessory (manufacture)
	31300	Cable jointing material (manufacture)
	31300	Electric cable (manufacture)
	31300	Insulated electrical cable (manufacture)
	31300	Insulated mains cable for power distribution (manufacture)
	31300	Insulated wire (manufacture)
	31300	Insulated wire and cable made of aluminium (manufacture)
	31300	Insulated wire and cable made of copper (manufacture)
	31300	Insulated wire and cable made of steel (manufacture)

SIC 2007	SIC 2003	Activity
	31300	Overhead line fittings (manufacture)
	31300	Ship's wiring (manufacture)
	31300	Submarine cable (manufacture)
	31300	Telecommunications wire (manufacture)
	31300	Winding wire and strip (manufacture)
27330		**Manufacture of wiring devices**
	31200	Bus bars, electrical conductors (except switchgear-type) (manufacture)
	31200	Electrical outlets or sockets (manufacture)
	31200	GFCI (ground fault circuit interrupters) (manufacture)
	31200	Junction box (manufacture)
	31200	Lamp holder (electric) (manufacture)
	31200	Lightning arresters (manufacture)
	25240	Non-current carrying plastic junction boxes (manufacture)
	25240	Non-current carrying plastic pole line fittings (manufacture)
	25240	Non-current carrying plastic switch covers (manufacture)
	31200	Outlet boxes for electrical wiring (manufacture)
	25239	Plastic electrical conduit tubing (manufacture)
	25240	Plastic non-current carrying face plates (manufacture)
	31200	Plug and socket (electric) (manufacture)
	31200	Sockets (electric) (manufacture)
	31200	Switch (electric) (manufacture)
	31200	Switch boxes for electrical wiring (manufacture)
	31200	Transmission pole and line hardware (manufacture)
	31200	Wiring accessories (manufacture)
	31200	Wiring devices (manufacture)
27400		**Manufacture of electric lighting equipment**
	31500	Advertising light (manufacture)
	31500	Arc lamp (manufacture)
	31620	Bug zappers without light (manufacture)
	31500	Bulb for flash lamp (manufacture)
	31500	Candelabra made of base metal (manufacture)
	31500	Candlestick (manufacture)
	31500	Carbide lanterns (manufacture)
	31500	Case for flash lamp (manufacture)
	31500	Ceiling rose (manufacture)
	31500	Chandeliers (manufacture)
	31500	Christmas tree lights (manufacture)
	31500	Discharge lamp (manufacture)
	31500	Electric fireplace logs (manufacture)
	31500	Electric insect lamps (manufacture)
	31500	Electric lanterns (manufacture)
	31500	Electric lighting equipment (manufacture)
	31500	Flash lamp case (manufacture)
	31500	Flashcubes (manufacture)
	31500	Flashlights (manufacture)
	31500	Fluorescent tube (manufacture)
	31500	Gas discharge lamp (manufacture)
	31500	Gas lanterns (manufacture)
	31500	Gasoline lanterns (manufacture)
	31500	Illuminated signs and nameplates (manufacture)
	31500	Illuminated traffic signs (manufacture)
	31500	Infrared lamps (manufacture)
	31500	Kerosene lanterns (manufacture)

UK SIC 2007	UK SIC 2003	Activity
	31500	Lamps (manufacture)
	31610	Lamps for cycles (manufacture)
	31500	Lampshades (not of glass or plastics) (manufacture)
	31500	Light bulb (manufacture)
	31500	Light bulbs including fluorescent and neon tubes (manufacture)
	31500	Lighting equipment (manufacture)
	31610	Lighting equipment for aircraft (manufacture)
	31610	Lighting equipment for boats (manufacture)
	31610	Lighting equipment for motor vehicles (manufacture)
	31500	Lighting fitting (other than glassware) (manufacture)
	31500	Lighting fixture of table lamps (manufacture)
	31500	Mercury vapour lamp (manufacture)
	31500	Miners' lamp (manufacture)
	31500	Neon tube (manufacture)
	31500	Non-electrical lighting equipment (manufacture)
	31500	Outdoor and road lighting (manufacture)
	31500	Photoflash bulb (manufacture)
	31500	Portable lamp (electric) (manufacture)
	31500	Projector lamp (manufacture)
	31500	Search light (manufacture)
	31500	Sodium vapour lamp (manufacture)
	31500	Spotlight (manufacture)
	31500	Stage lighting (manufacture)
	31500	Street lighting fixtures (manufacture)
	31500	Torch (manufacture)
	31500	Ultra-violet lamps (manufacture)
	31500	Vehicle lamps (bulb and sealed beam unit) (manufacture)
27510		**Manufacture of electric domestic appliances**
	29710	Aquarium heater (electric) (manufacture)
	29710	Blankets (electric) (manufacture)
	29710	Blenders for domestic use (electric) (manufacture)
	29710	Brush (electric) (manufacture)
	29710	Clothes airer (electric) (manufacture)
	29710	Coffee or tea makers (electric) (manufacture)
	29710	Coffee percolator (electric) (manufacture)
	29710	Combs (electric) (manufacture)
	29710	Cooker (electric) (manufacture)
	29710	Curlers (electric) (manufacture)
	29710	Deep freeze unit for domestic use (manufacture)
	29710	Dishwasher for domestic use (manufacture)
	29710	Dryers (electric) (manufacture)
	29710	Electric tea makers (manufacture)
	29710	Electricaire unit (manufacture)
	29710	Electrical appliances for domestic use (manufacture)
	29710	Electro-thermic appliances for domestic use (manufacture)
	29710	Fan (electric, domestic) (manufacture)
	29710	Fire (electric) (manufacture)
	29710	Floor polisher (electric) (manufacture)
	29710	Food freezer for domestic use (manufacture)
	29710	Food mixer (electric) (manufacture)
	29710	Frying pans (electric) (manufacture)
	29710	Gas lighter (electric) (manufacture)
	29710	Grills (electric) (manufacture)
	29710	Grinders (electric) (manufacture)
	29710	Hair clippers (electric) (manufacture)

SIC 2007	SIC 2003	Activity
	29710	Hair dryer (electric) (manufacture)
	29710	Heater for motor vehicle (manufacture)
	29710	Heating resistors (electric) (manufacture)
	29710	Hot plates (electric) (manufacture)
	29710	Immersion heater (electric) (manufacture)
	29710	Iron (electric) (manufacture)
	29710	Ironing machine for domestic use (electric) (manufacture)
	29710	Juice squeezers (electric) (manufacture)
	29710	Kettle (electric) (manufacture)
	29710	Knife sharpener (electric) (manufacture)
	29710	Microwave ovens (manufacture)
	29710	Ovens (electric) (manufacture)
	29710	Percolator (electric) (manufacture)
	29710	Plate warmers for domestic use (electric) (manufacture)
	29710	Portable space heaters (manufacture)
	29710	Radiator (electric) (manufacture)
	29710	Razor (electric) (manufacture)
	29710	Refrigerator for domestic use (electric) (manufacture)
	29710	Roasters (electric) (manufacture)
	29710	Shaver (electric) (manufacture)
	29710	Smoothing irons (manufacture)
	29710	Space heaters for domestic use (electric) (manufacture)
	29710	Spin dryer
	29710	Storage heaters (manufacture)
	29710	Tin openers (electric) (manufacture)
	29710	Toaster (electric) (manufacture)
	29710	Tooth brush (electric) (manufacture)
	29710	Towel rail (electric) (manufacture)
	29710	Trash compactor for domestic use (electric) (manufacture)
	29710	Tumble dryer for domestic use (manufacture)
	29710	Vacuum cleaners for domestic use (manufacture)
	29710	Ventilating or recycling hoods (manufacture)
	29710	Waffle irons (manufacture)
	29710	Washing machines for domestic use (manufacture)
	29710	Waste disposers (manufacture)
	29710	Water heaters for domestic use (electric) (manufacture)
27520		**Manufacture of non-electric domestic appliances**
	29720	Boiler for domestic use (oil) (manufacture)
	29720	Boiler for domestic use (solid fuel) (manufacture)
	29720	Calorifier (manufacture)
	29720	Cooker (gas) (manufacture)
	29720	Cooker (oil) (manufacture)
	29720	Cooker (solid fuel) (manufacture)
	29720	Cooking and heating appliances for domestic use (gas) (manufacture)
	29720	Cooking and heating appliances for domestic use (solid fuel) (manufacture)
	29720	Cooking appliances for domestic use (non-electric) (manufacture)
	29720	Domestic non-electric cooking equipment (manufacture)
	29720	Domestic non-electric cooking ranges (manufacture)
	29720	Domestic non-electric heating equipment (manufacture)

UK SIC 2007	UK SIC 2003	Activity	SIC 2007	SIC 2003	Activity
	29720	Fire (gas) (manufacture)		31300	Extension cords with insulated wire and connectors (manufacture)
	29720	Fire (oil) (manufacture)		31620	Indicator panel (manufacture)
	29720	Grates for domestic use (non-electric) (manufacture)		31620	Particle accelerator (manufacture)
	29720	Heat emitter (space heating equipment) (non-electric) (manufacture)		31620	Railway signalling equipment (electric) (manufacture)
	29720	Heating and cooking appliances for domestic use (oil fired) (manufacture)		32100	Rectifier, solid state (manufacture)
				32100	Rectifying valve and tube (manufacture)
	29720	Heating appliances for domestic use (non-electric) (manufacture)		32100	Resistors including rheostats and potentiometers (manufacture)
	29720	Plate warmers for domestic use (non-electric) (manufacture)		31620	Signal generator (manufacture)
	29720	Refrigerators (gas) (manufacture)		31620	Signalling equipment for road traffic (electric) (manufacture)
	29720	Space heaters (gas) (manufacture)		31620	Simulator (battle) (manufacture)
	29720	Space heaters (oil) (manufacture)		31620	Simulator (driving) (manufacture)
	29720	Stove (gas) (manufacture)		31620	Simulator (other training (except flying trainers)) (manufacture)
	29720	Stove (oil) (manufacture)			
	29720	Stove (solid fuel) (manufacture)		29430	Soldering irons (electrical, hand-held) (manufacture)
	29720	Stoves for domestic use (non-electric) (manufacture)		29430	Soldering machines (electric) (manufacture)
	29720	Warm air generator (non-electric) (manufacture)		31100	Solid state battery chargers (manufacture)
	29720	Water heaters (gas) (manufacture)		31100	Solid state fuel cells (manufacture)
	29720	Water heaters for domestic use (non-electric) (manufacture)		31100	Solid state inverters (manufacture)
				31200	Surge suppressors (manufacture)
				32100	Thyristor (manufacture)
27900		**Manufacture of other electrical equipment**		31620	Ultrasonic cleaning machines (except laboratory and dental) (manufacture)
	31300	Appliance cords with insulated wire and connectors (manufacture)		31620	Uninterruptible power supplies (UPS) (manufacture)
	31620	Backward wave oscillator (manufacture)		31620	Wave form generator (manufacture)
	31620	Bell apparatus (other than telegraphic or telephonic) (manufacture)		31620	Welding electrode (manufacture)
				29430	Welding machines (electric) (manufacture)
	31620	Bells (other than telephone type) (electric) (manufacture)			
	31620	Carbon brush (manufacture)	**28110**		**Manufacture of engines and turbines, except aircraft, vehicle and cycle engines**
	31620	Carbon or graphite electrodes (manufacture)		29110	Blocks for industrial engines (manufacture)
	31300	Contacts and other electrical carbon and graphite products (manufacture)		29110	Boiler-turbine sets (manufacture)
				34300	Camshaft for motor vehicle engine (manufacture)
	31620	Cyclotron (manufacture)		34300	Carburettor and parts for motor vehicle (manufacture)
	31620	Detection apparatus (manufacture)			
	31620	Electric bells (manufacture)		29110	Carburettors and such for all internal combustion engines (manufacture)
	31620	Electrical base metal conduit and fittings (manufacture)		29110	Carburettors for industrial engines (manufacture)
	32100	Electrical capacitors (manufacture)		29110	Compression ignition engines for industrial use (manufacture)
	31620	Electrical carbon (manufacture)			
	32100	Electrical condensers and similar components (manufacture)		29110	Compressor engine (manufacture)
				34300	Crankshaft for motor vehicle (manufacture)
	31620	Electrical conduit tubing, base metal (manufacture)		29110	Cylinder heads for industrial engines (manufacture)
	31620	Electrical door opening and closing devices (manufacture)		29110	Cylinder inserts for industrial engines (manufacture)
				34300	Cylinder liner for motor vehicle (manufacture)
	31620	Electrical insulators (except glass or porcelain) (manufacture)		29110	Cylinder liners for industrial engines (manufacture)
	31620	Electrical pedestrian signalling equipment (manufacture)		29110	Diesel engines for industrial use (manufacture)
				34300	Engine block for motor vehicle (finished) (manufacture)
	31620	Electrical signs (manufacture)			
	29430	Electrical soldering equipment (manufacture)		29110	Engines and parts for marine use (manufacture)
	31620	Electrical traffic lights (manufacture)		29110	Engines and parts for railways (manufacture)
	29430	Electrical welding equipment (manufacture)		29110	Engines for agricultural machinery (manufacture)
	31620	Electrodes for welding (manufacture)		29110	Engines for combine harvesters (manufacture)
	31620	Electromagnets (manufacture)		29110	Engines for construction equipment (manufacture)
	31620	Electronic filter (manufacture)		29110	Engines for forklift trucks (manufacture)
	31620	Electronic miscellaneous unspecified equipment (manufacture)		29110	Engines for industrial application (manufacture)
				29110	Engines for lawn mowers (manufacture)
	31620	Electronic scoreboards (manufacture)		29110	Engines for locomotives (manufacture)
	31300	Extension cords made from purchased insulated wire (manufacture)		29110	Engines for marine use (manufacture)

UK SIC 2007	UK SIC 2003	Activity
	29110	Engines for railway vehicles (manufacture)
	29110	Exhaust valves for internal combustion engines (manufacture)
	29110	Gas turbine (excluding turbo-jets and turbo-propellers) parts (manufacture)
	29110	Gas turbine (industrial engine) (manufacture)
	29110	Gas turbine for marine use (manufacture)
	29110	Generator engine (manufacture)
	29110	Hydraulic turbine (manufacture)
	29110	Hydraulic turbine and water wheel parts (manufacture)
	29110	Hydraulic turbines and parts thereof (manufacture)
	29110	Industrial engine parts (manufacture)
	29110	Industrial spark and compression ignition engine (manufacture)
	29110	Inlet valves for internal combustion engines (manufacture)
	29110	Internal combustion engines and parts (manufacture)
	29110	Internal combustion piston engines (manufacture)
	29110	Internal combustion piston marine engines (manufacture)
	29110	Internal combustion piston railway engines (manufacture)
	29110	Manifold for industrial engine (manufacture)
	29110	Marine non-propulsion engines (manufacture)
	29110	Petrol industrial engines (manufacture)
	29110	Piston for industrial engine (manufacture)
	34300	Piston for motor vehicle engine (manufacture)
	29110	Piston ring for industrial engine (manufacture)
	34300	Piston ring for motor vehicle engine (manufacture)
	29110	Piston rings for all internal combustion engines (manufacture)
	29110	Pistons for all internal combustion engines (manufacture)
	29110	Propulsion engine for marine use (manufacture)
	29110	Slow speed diesel engine for marine use (manufacture)
	29110	Spark ignition engines for industrial use (manufacture)
	29110	Steam and other vapour turbine parts (manufacture)
	29110	Steam engine (manufacture)
	29110	Steam turbine (not marine or for electricity generation) (manufacture)
	29110	Steam turbine for marine use (manufacture)
	29110	Steam turbines and other vapour turbines (manufacture)
	29110	Turbine-generator sets (manufacture)
	29110	Turbines and parts thereof (manufacture)
	34300	Valves for motor vehicle engines (manufacture)
	29110	Water-wheels (manufacture)
	29110	Waterwheels and regulators and parts thereof: (manufacture)
	29110	Wind turbines (manufacture)
28120		**Manufacture of fluid power equipment**
	29122	Actuator for hydraulic equipment (manufacture)
	29122	Air preparation equipment for use in pneumatic systems (manufacture)
	29130	Automatic process control valves (manufacture)
	29130	Choke manifolds (manufacture)
	29122	Cylinder for hydraulic equipment (manufacture)
	29122	Cylinder for pneumatic control equipment (manufacture)

SIC 2007	SIC 2003	Activity
	29122	Filter for pneumatic control equipment (manufacture)
	29130	Flaps, diaphragms and other parts of hydraulic and pneumatic valves (manufacture)
	29122	Flowline assembly (hydraulic equipment) (manufacture)
	29122	Fluid power equipment (manufacture)
	29122	Fluid power systems (manufacture)
	29122	Hydraulic and pneumatic components (manufacture)
	29122	Hydraulic and pneumatic cylinders (manufacture)
	29122	Hydraulic and pneumatic hoses and fittings (manufacture)
	29122	Hydraulic and pneumatic power engine and motor parts (manufacture)
	29130	Hydraulic and pneumatic valves (manufacture)
	29122	Hydraulic equipment for aircraft (manufacture)
	29122	Hydraulic exhauster (manufacture)
	29122	Hydraulic power engines and motors (manufacture)
	29140	Hydraulic transmission equipment (manufacture)
	29122	Hydro pneumatic device (manufacture)
	29122	Hydrostatic transmissions (manufacture)
	29122	Intensifier for hydraulic equipment (manufacture)
	29122	Intensifier for pneumatic control equipment (manufacture)
	29121	Jet pump (manufacture)
	29121	Liquid elevator parts (manufacture)
	29121	Montejus (compressed air chamber elevators) (manufacture)
	29122	Motor for hydraulic equipment (manufacture)
	29122	Motor for pneumatic equipment (manufacture)
	29122	Pneumatic equipment and systems for aircraft (manufacture)
	29122	Pneumatic power engines and motors (manufacture)
	29122	Positioner for pneumatic control equipment (manufacture)
	29121	Pump for hydraulic equipment (manufacture)
	29122	Reservoir (hydraulic) (manufacture)
	29122	Reservoir (pneumatic) (manufacture)
	29121	Steam pulsators (pulsometers) (manufacture)
	29122	Tube coupling and equipment for pneumatics (manufacture)
	29130	Valve actuators (hydraulic and pneumatic) (manufacture)
	29130	Valves for hydraulic equipment (manufacture)
	29130	Valves for pneumatic control equipment (manufacture)
28131		**Manufacture of pumps**
	29121	Air pump (manufacture)
	29121	Archimedean screw pump (manufacture)
	29121	Axial flow pump (manufacture)
	29121	Bicycle pumps (manufacture)
	29121	Centrifugal pump (manufacture)
	29121	Channel impeller pump (manufacture)
	29121	Circulator pump (manufacture)
	29121	Concrete pumps (manufacture)
	29121	Delivery pump (manufacture)
	29121	Diaphragm pump (manufacture)
	29121	Dosing and proportioning pump (manufacture)
	29121	Ejector pump (manufacture)
	29121	Electro-magnetic pumps (manufacture)

UK SIC 2007	UK SIC 2003	Activity	SIC 2007	SIC 2003	Activity
	29121	Emulsion (gas lift) pumps (manufacture)		29130	Heating taps (manufacture)
	29121	Fuel injection equipment for industrial engines (manufacture)		29130	Heating valves (manufacture)
				29130	Industrial intake taps (manufacture)
	29121	Fuel pump for industrial engine (manufacture)		29130	Industrial regulating valves (manufacture)
	29121	Fuel pump for internal combustion piston engine, for aircraft (manufacture)		29130	Industrial taps (manufacture)
	29121	Gas combustion pumps (manufacture)		29130	Knife valves (manufacture)
	29121	Gear pumps (manufacture)		29130	Needle valves (manufacture)
	29121	Hand or foot operated pumps (manufacture)		29130	Non-return, reflux and check valves (manufacture)
	29121	Helical rotor pump (manufacture)		29130	Parallel slide valves (manufacture)
	29121	Helicoidal pumps (manufacture)		29130	Penstock valves (manufacture)
	29121	Impeller pumps (manufacture)		29130	Plug valves (manufacture)
	29121	Inflator for cycle type tyres (manufacture)		29130	Pressure reducing valves (manufacture)
	29121	Liquid elevators (manufacture)		29130	Process control valves (manufacture)
	29121	Lobe pump (manufacture)		29130	Reducing valves (manufacture)
	29121	Lubricating pump (not for internal combustion engine) (manufacture)		29130	Relief valves (manufacture)
				29130	Safety valves (manufacture)
	29121	Medicinal pumps (manufacture)		29130	Sanitary taps (manufacture)
	29121	Metering pump (manufacture)		29130	Sanitary valves (manufacture)
	29121	Oil-cushion pumps (manufacture)		29130	Stopcocks for domestic use (manufacture)
	29121	Peristaltic pumps (manufacture)		29130	Tap parts (manufacture)
	29121	Petrol station pump (manufacture)		29130	Taps (manufacture)
	29121	Positive displacement pump (reciprocating) (manufacture)		29130	Taps for domestic use (manufacture)
				29130	Temperature regulators (manufacture)
	29121	Pump (not for hydraulic or for internal combustion engine) (manufacture)		29130	Valve actuators (electrical (other than electric motors)) (manufacture)
	29121	Pump parts (manufacture)		29130	Valve parts (manufacture)
	29121	Pumps and parts (non-electric) for oil, gas, petrol or water on motor vehicles (manufacture)		29130	Valves for industrial use (manufacture)
				29130	Valves for tyres (manufacture)
	29121	Pumps for liquids whether or not fitted with measuring devices (manufacture)	**28150**		**Manufacture of bearings, gears, gearing and driving elements**
	29121	Radial flow pump (manufacture)		29140	Articulated link chain (manufacture)
	29121	Rotary piston lobe-type pumps (manufacture)		29140	Ball bearing (manufacture)
	29121	Screw pump (manufacture)		29140	Bearing housings (manufacture)
	29121	Submersible motor pump (manufacture)		29140	Bearing, gearing and driving element parts (manufacture)
	29121	Vacuum pump (manufacture)		29140	Bearings (manufacture)
	29121	Vane pump (manufacture)		29140	Bicycle chain
28132		**Manufacture of compressors**		29140	Bush for bearing (manufacture)
	29122	Air compressor (manufacture)		29140	Camshaft (not for motor vehicle) (manufacture)
	29122	Axial compressor (manufacture)		29140	Chain (precision) (manufacture)
	29122	Centrifugal compressor (manufacture)		29140	Clutch (not for motor vehicle) (manufacture)
	29122	Compressor for air or other gas (manufacture)		29140	Cranks (manufacture)
	29122	Compressor for refrigerators (manufacture)		29140	Crankshaft (not for motor vehicle engine) (manufacture)
	29122	Compressor parts (manufacture)		29140	Cylindrical roller bearing (manufacture)
	29122	Gas compressor (manufacture)		29140	Driving elements (manufacture)
	29122	Reciprocating compressor (manufacture)		29140	Flywheel (not for motor vehicle engine) (manufacture)
	29122	Rotating compressors (manufacture)		29140	Gear box (not for motor vehicle) (manufacture)
	29122	Screw compressor (manufacture)		29140	Gear cutting (not for motor vehicle) (manufacture)
28140		**Manufacture of other taps and valves**		29140	Geared motor unit (manufacture)
	29130	Ball valves (manufacture)		29140	Gearing (manufacture)
	29130	Butterfly valves (manufacture)		29140	Gears (manufacture)
	29130	Check valves (manufacture)		29140	Mechanical power transmission bearing housings (manufacture)
	29130	Christmas trees and other assemblies of valves (manufacture)		29140	Mechanical power transmission camshafts (manufacture)
	29130	Cocks for industrial use (manufacture)		29140	Mechanical power transmission chain (manufacture)
	29130	Diaphragm valves (manufacture)		29140	Mechanical power transmission cranks (manufacture)
	29130	Gas cylinder outlet valves (manufacture)			
	29130	Gate valves (manufacture)			
	29130	Globe valves (manufacture)			

UK SIC 2007	UK SIC 2003	Activity
	29140	Mechanical power transmission crankshafts (manufacture)
	29140	Mechanical power transmission plain shaft bearings (manufacture)
	29140	Mechanical power transmission plant (manufacture)
	29140	Mechanical power transmission shafts (manufacture)
	29140	Needle roller bearings (manufacture)
	29140	Parts of bearings, gearing and driving elements (manufacture)
	29140	Plain bearing (manufacture)
	29140	Precision chain (manufacture)
	29140	Pulley (manufacture)
	29140	Pulley wheel (manufacture)
	29140	Roller bearing (manufacture)
	29140	Shaft bearings (manufacture)
	29140	Shaft couplings (manufacture)
	29140	Speed changers (manufacture)
	29140	Spherical roller bearings (manufacture)
	29140	Sprocket chain (manufacture)
	29140	Tapered roller bearings (manufacture)
	29140	Transmission chain (manufacture)
	29140	Transmission shafts (manufacture)
28210		**Manufacture of ovens, furnaces and furnace burners**
	29210	Annealing lehr (manufacture)
	29210	Blast furnace (manufacture)
	29210	Box furnace (manufacture)
	29210	Burners (manufacture)
	29210	Cement processing kiln (manufacture)
	29210	Dielectric heating equipment (manufacture)
	29210	Direct arc furnace (manufacture)
	29210	Electric and other industrial and laboratory furnaces (manufacture)
	29210	Electric and other industrial and laboratory incinerators (manufacture)
	29210	Electric and other industrial and laboratory ovens (manufacture)
	29710	Electric household heating equipment (permanently mounted) (manufacture)
	29210	Electric household type furnaces (manufacture)
	29210	Electric swimming pool heaters (permanently mounted) (manufacture)
	29210	Electro slag furnace (manufacture)
	29210	Forging furnace (manufacture)
	29210	Fuel burner (other than oil or gas) (manufacture)
	29210	Furnace (electric) (manufacture)
	29210	Furnace for strip processing line (manufacture)
	29210	Furnace, furnace burner and industrial oven parts (manufacture)
	29210	Furnaces and furnace burners (manufacture)
	29210	Gas burner (manufacture)
	29210	Grates (mechanical) (manufacture)
	29210	Heat treatment furnace (manufacture)
	29210	Heating/melting high frequency induction or dielectric equipment (manufacture)
	29210	Incinerator (manufacture)
	29210	Induction furnace (manufacture)
	29210	Laboratory furnace (manufacture)
	29210	Lime processing kiln (manufacture)
	29210	Mechanical ash dischargers (manufacture)

SIC 2007	SIC 2003	Activity
	29210	Mechanical stokers (manufacture)
	29210	Melting furnace (manufacture)
	29720	Non-electric household heating equipment (permanently mounted) (manufacture)
	29210	Non-electric household-type furnaces (manufacture)
	29210	Oil fuel burner (manufacture)
	29210	Ovens for industrial use (except bakery) (manufacture)
	29210	Re-heating furnace (manufacture)
	29210	Smelting furnace (manufacture)
	29720	Solar panels, domestic (other than photovoltaic cell type) (manufacture)
	29720	Solar panels, non-domestic (other than photovoltaic cell type) (manufacture)
	29210	Steelmaking furnace (manufacture)
	29210	Strip processing line furnace (manufacture)
28220		**Manufacture of lifting and handling equipment**
	29220	Aerial ropeway and cableway (manufacture)
	29220	Bucket wheel reclaimer (manufacture)
	29220	Builders' hoist (manufacture)
	29220	Cableway (manufacture)
	29220	Cableway excavator (manufacture)
	29220	Capstan (manufacture)
	29220	Chain pulley block (manufacture)
	29220	Container handling crane (manufacture)
	29220	Conveying plant (hydraulic and pneumatic) (manufacture)
	29220	Conveyor and feeder (not for agriculture or mining) (manufacture)
	29220	Conveyors (manufacture)
	29220	Crane (manufacture)
	29220	Derricks (manufacture)
	29220	Dock leveller (manufacture)
	29220	Dockside cranes (manufacture)
	29220	Drag scraper (manufacture)
	29220	Elevator (manufacture)
	29220	Escalator (manufacture)
	29220	Forklift truck (manufacture)
	29220	Goliath type crane (manufacture)
	35500	Hand truck made of metal (manufacture)
	29220	Handling plant (hydraulic and pneumatic) (manufacture)
	29220	Hand-operated lifting capstans (manufacture)
	29220	Hand-operated lifting hoists (manufacture)
	29220	Hand-operated lifting jacks (manufacture)
	29220	Hand-operated lifting pulley tackle (manufacture)
	29220	Hand-operated lifting winches (manufacture)
	29220	Hoists (electric wire or chain) (manufacture)
	29220	Hoists (hand operated pulley and sheave block) (manufacture)
	29220	Hoists, hydraulic, mechanical, pneumatic (not builders, hand operated and electrical) (manufacture)
	29220	Hydraulic and pneumatic conveying plant (manufacture)
	29220	Hydraulic and pneumatic handling plant (manufacture)
	29220	Industrial special purpose trucks (manufacture)
	29220	Industrial tractor (manufacture)
	29220	Jack for motor vehicle (manufacture)
	29220	Jacks (other than for motor vehicles) (manufacture)

UK SIC 2007	UK SIC 2003	Activity	SIC 2007	SIC 2003	Activity
	29220	Lift (manufacture)	28230		**Manufacture of office machinery and equipment (except computers and peripheral equipment)**
	29220	Lift and escalator maintenance (manufacture)			
	29220	Lift for motor vehicle (manufacture)		30010	Accounting machine (manufacture)
	29220	Lifting and handling equipment (manufacture)		30010	Adding machines (manufacture)
	29220	Lifting and handling equipment parts (manufacture)		30010	Address plate embossing machine (manufacture)
	29220	Live storage rack (manufacture)		30010	Addressing machine (manufacture)
	29220	Lorry loader (manufacture)		30010	Bank note counting machine (manufacture)
	29220	Mechanical industrial robots for lifting, handling (manufacture)		30010	Banknote dispensing machine (manufacture)
	29220	Mechanical lifting manipulators (manufacture)		30010	Binding machine (manufacture)
	29220	Mechanical manipulators (manufacture)		36631	Blackboards (manufacture)
	29220	Mobile crane (manufacture)		30010	Book keeping machine (manufacture)
	29220	Mobile lifting frames (manufacture)		30010	Calculating machine (manufacture)
	29220	Moving walkways (manufacture)		30010	Calculator (electronic) (manufacture)
	29220	Overhead runway (manufacture)		30010	Cash and credit card imprinting and embossing machine (manufacture)
	29220	Pallet hoist (manufacture)		30010	Cash dispenser (manufacture)
	29220	Pallet truck (manufacture)		30010	Cash register (manufacture)
	29220	Palletizer (manufacture)		30010	Cheque writing and signing machine (manufacture)
	29220	Passenger conveyor (manufacture)		30010	Coin sorting, wrapping and counting machines (manufacture)
	29220	Pneumatic and hydraulic conveying plant (manufacture)		30010	Collating machinery (manufacture)
	29220	Pneumatic and hydraulic handling plant (manufacture)		30010	Copying machine (xerographic) (manufacture)
	29220	Portal and pedestal jib cranes (manufacture)		30010	Counting and dating machines (manufacture)
	29220	Power-driven cranes (manufacture)		30010	Data processing equipment (non-electronic) (manufacture)
	29220	Power-driven lifting capstans (manufacture)		32300	Dictating machines (manufacture)
	29220	Power-driven lifting hoists (manufacture)		30010	Document copying equipment (manufacture)
	29220	Power-driven lifting jacks (manufacture)		30010	Document handling machine (manufacture)
	29220	Power-driven lifting pulley tackle (manufacture)		30010	Document shredder (manufacture)
	29220	Power-driven lifting winches (manufacture)		30010	Duplicating machines (excluding copiers) (manufacture)
	29220	Power-driven loading and unloading derricks (manufacture)		30010	Dyeline copying machine (manufacture)
	29220	Power-driven mobile lifting frames (manufacture)		30010	Electrostatic copying machine (manufacture)
	29220	Power-driven straddle carriers (manufacture)		30010	Envelope stuffing machine (manufacture)
	29220	Pulley block (manufacture)		30010	Franking machine (manufacture)
	29220	Pulley tackle (manufacture)		30010	Hectograph (manufacture)
	29220	Robots designed for lifting and handling in industry (manufacture)		30010	Hole punches (manufacture)
	29220	Rough terrain industrial trucks (manufacture)		30010	Invoicing machine (manufacture)
	29220	Scissor lift (manufacture)		30010	Labelling machine for office use (manufacture)
	29220	Side loader (manufacture)		30010	Laminating machine for office use (manufacture)
	29220	Ski lifts (manufacture)		30010	Letter opening machine (manufacture)
	29220	Special steelworks crane (manufacture)		30010	Listing machine (manufacture)
	29220	Stacking machines (manufacture)		30010	Mail handling machines (envelope stuffing, sealing, addressing; opening, sorting) (manufacture)
	29220	Stillage truck (manufacture)		36120	Marker boards (manufacture)
	29220	Straddle carrier (manufacture)		30010	Non electronic calculators (manufacture)
	29220	Suspension railway (manufacture)		30010	Office machinery (manufacture)
	29220	Tailboard lift (manufacture)		30010	Office machinery and equipment (manufacture)
	29220	Teleferics (manufacture)		30010	Pencil sharpeners (manufacture)
	29220	Tipper (manufacture)		30010	Pencil-sharpening machines (manufacture)
	29220	Transporter (manufacture)		30010	Photocopying machinery (manufacture)
	29220	Travelling crane (manufacture)		30010	Plastic office-type binding equipment or tape binding (manufacture)
	35500	Wheelbarrows made of metal (manufacture)		30010	Point of sale unit (manufacture)
	35500	Wheelbarrows made of plastic (manufacture)		30010	Postage franking machines (manufacture)
	35500	Wheelbarrows made of wood (manufacture)		30010	Postage meters (manufacture)
	29220	Winch (manufacture)		30010	Printing machines (sheet fed office type offset) (manufacture)
	29220	Winding device (manufacture)		30010	Punched card machine (other than for computer use) (manufacture)
	29220	Windlass (manufacture)			
	29220	Works trucks (manufacture)			

UK SIC 2007	UK SIC 2003	Activity	SIC 2007	SIC 2003	Activity
	30010	Shorthand writing machines (manufacture)		29230	Air conditioning equipment for aircraft (manufacture)
	30010	Staple removers (manufacture)		29230	Air conditioning machines (manufacture)
	30010	Staplers (manufacture)		29230	Air conditioning package (manufacture)
	30010	Stapling machines for office use (manufacture)		29230	Air filter for air conditioning equipment (manufacture)
	30010	Stencil duplicating machines (manufacture)		29230	Atmospheric pollution control plant (manufacture)
	30010	Stenography machines (manufacture)		29710	Attic ventilation fans (manufacture)
	30010	Tabulating machines (manufacture)		29230	Cold storage equipment (manufacture)
	30010	Tape dispensers (manufacture)		29230	Condenser for air conditioning equipment (air or water cooled) (manufacture)
	30010	Tape office-type binding equipment (manufacture)		29230	Condenser unit for refrigerator (manufacture)
	30010	Terminals for issuing of tickets and reservations (manufacture)		29230	Cooling equipment for non-domestic use (manufacture)
	30010	Ticket issuing machine (manufacture)		29230	Cooling tower for air conditioning (manufacture)
	30010	Ticket punch (manufacture)		29230	Display cabinet (refrigerated) (manufacture)
	30010	Toner cartridges (manufacture)		29230	Dust collector for air conditioning equipment (manufacture)
	30010	Typewriter (manufacture)		29230	Evaporator for refrigeration machinery (manufacture)
	30010	Visible record computer (tabulator) (manufacture)		29230	Fan coil unit (manufacture)
	30010	Voting machines (manufacture)		29710	Fans (domestic) (manufacture)
	36120	White boards (manufacture)		29230	Fans (non-domestic) (manufacture)
	30010	Word processing machines (manufacture)		29230	Food freezer over 12 cubic feet capacity (manufacture)
	30010	Xerographic copying machines (manufacture)		29230	Freezing industrial equipment, including assemblies of components (manufacture)
28240		**Manufacture of power-driven hand tools**		29230	Gas cleansing plant (manufacture)
	29410	Chain saw parts (manufacture)		29230	Heat exchange units for air conditioning (manufacture)
	29410	Chain saws (manufacture)		29230	Humidifier for air conditioning equipment (manufacture)
	29410	Circular or reciprocating saws (manufacture)		29230	Ice cream conservator (manufacture)
	29410	Disc cutting machines for stone, ceramics, asbestos-cement or similar (portable) (manufacture)		29230	Induction unit for air conditioning equipment (manufacture)
	29410	Drill (powered portable) (manufacture)		29230	Machinery for liquefying air or gas (manufacture)
	29410	Drills and hammer drills (manufacture)		29230	Non-domestic cooling and ventilation equipment parts (manufacture)
	29410	Grinding tools (powered portable) (manufacture)		29230	Refrigerated service cabinet (manufacture)
	29410	Hammer (portable, powered) (manufacture)		29230	Refrigerating and freezing equipment parts (manufacture)
	29410	Hand tools with non-electric motor parts (manufacture)		29230	Refrigerating equipment (manufacture)
	29410	Hand tools with self contained motor or pneumatic drive (manufacture)		29230	Refrigerating or freezing equipment for industrial use (manufacture)
	29410	Hedge trimmers (powered portable) (manufacture)		29230	Refrigerator cabinets made of wood (manufacture)
	29410	Mining tool (powered portable) (manufacture)		29230	Refrigerator for commercial use (manufacture)
	29410	Parts for portable hand held power tools (manufacture)		29230	Scrubber for air conditioning equipment (manufacture)
	29410	Pneumatic nailers (manufacture)		29230	Sectional coldroom (manufacture)
	29410	Pneumatic power tools (portable) (manufacture)		29230	Ventilating fans (manufacture)
	29410	Pneumatic rivet guns (manufacture)		29230	Ventilating unit (manufacture)
	29410	Portable power tool parts (manufacture)		29710	Ventilation equipment for domestic use (manufacture)
	29410	Power tool (portable) (manufacture)		29230	Ventilation equipment for non-domestic use (manufacture)
	29410	Power tools (portable electric) (manufacture)			
	29410	Power-driven buffers (manufacture)	**28290**		**Manufacture of other general-purpose machinery n.e.c.**
	29410	Power-driven grinders (manufacture)		29240	Aeration plant for effluent treatment (manufacture)
	29410	Power-driven impact wrenches (manufacture)		29240	Baling press (not agricultural) (manufacture)
	29410	Power-driven planers (manufacture)		29240	Base exchange plant for water treatment (manufacture)
	29410	Power-driven powder actuated nailers (manufacture)		29240	Bottle cleaning and or drying machinery (manufacture)
	29410	Power-driven routers (manufacture)			
	29410	Power-driven shears and nibblers (manufacture)			
	29410	Power-driven staplers (manufacture)			
	29410	Rock drill (portable) (manufacture)			
	29410	Sanding tool (powered portable) (manufacture)			
	29410	Saws (powered portable) (manufacture)			
28250		**Manufacture of non-domestic cooling and ventilation equipment**			
	29230	Air cleansing plant (not for air conditioning equipment) (manufacture)			

UK SIC 2007	UK SIC 2003	Activity	SIC 2007	SIC 2003	Activity
	29240	Bottling machinery (manufacture)		29240	Grit extraction plant for effluent treatment (manufacture)
	29240	Calender for rubber or plastics working (manufacture)		29430	Hot spraying electrical machines (metal or metal carbides) (manufacture)
	29240	Calendering machinery for textiles (manufacture)		29240	Ion exchange plant for water treatment (manufacture)
	29240	Calendering or other rolling machine parts (for plastic or rubber) (manufacture)		29430	Jigs (manufacture)
	29240	Canning machinery (manufacture)		29240	Jointing (precision component) (manufacture)
	29240	Capsuling machinery (manufacture)		29240	Labelling machinery (not for office use) (manufacture)
	29240	Cartoning machinery (manufacture)		33202	Levels (manufacture)
	29240	Case packing machinery (manufacture)		29240	Lubricator (manufacture)
	29240	Centrifuge (except laboratory type) (manufacture)		29240	Machinery (not containing electrical connectors) n.e.c. parts (manufacture)
	29240	Centrifuge parts (except laboratory type) (manufacture)		29240	Machinery for cleaning or drying bottles and for aerating beverages (manufacture)
	29240	Chlorination plant for water treatment (manufacture)		33202	Machinists precision tools (manufacture)
	29240	Clarification plant for water treatment (manufacture)		33202	Measuring rods and tapes (manufacture)
	29240	Cleaning, filling, packing or wrapping machine parts (manufacture)		33202	Measuring rule (manufacture)
	29240	Closing machinery (manufacture)		33202	Measuring tape (manufacture)
	29240	Comminution plant for effluent treatment (manufacture)		29240	Metal spraying machine (manufacture)
	29240	Cooling towers and similar for direct cooling by means of re-circulated water (manufacture)		29430	Non-electrical welding and soldering equipment (manufacture)
	29240	Crating and de-crating machinery (manufacture)		29240	Oil filter for motor vehicle (manufacture)
	29240	De-aeration plant for water treatment (manufacture)		29240	Oil refining industry machinery (other than plant) (manufacture)
	29240	Desalination plant for water treatment (manufacture)		29240	Oil seal (manufacture)
	29240	Dialysis plant for water treatment (manufacture)		29240	Packaging machinery (manufacture)
	29240	Dish washing machine parts (industrial type) (manufacture)		29240	Packing machinery (manufacture)
	29240	Dishwashing machine for commercial catering (manufacture)		29240	Paint spraying machine (manufacture)
	29240	Distilling machinery for potable spirits (manufacture)		29240	Petro-chemical industry machinery (other than plant) (manufacture)
	29240	Distilling or rectifying plant (manufacture)		29240	Purifying machinery (manufacture)
	29240	Distilling or rectifying plant for beverage industries (manufacture)		29240	Rectifying plant (manufacture)
	29240	Distilling or rectifying plant for chemical industries (manufacture)		29240	Refuse disposal plant (manufacture)
	29240	Distilling or rectifying plant for petroleum refineries (manufacture)		33202	Rule (measuring) (manufacture)
	29240	Dosing plant for water treatment (manufacture)		29240	Sand blasting machines (manufacture)
	29240	Effluent treatment plant (manufacture)		29240	Scales (platform) (manufacture)
	29240	Electrostatic precipitator (manufacture)		29240	Scales for domestic use (manufacture)
	29240	Equipment for dispersing liquids or powders (manufacture)		29240	Scales for postal use (manufacture)
	29240	Equipment for projecting liquids or powders (manufacture)		29240	Scales for shop use (manufacture)
	29240	Equipment for spraying liquids or powders (manufacture)		29240	Screening plant (not for mines) (manufacture)
	29240	Filling machinery (manufacture)		29240	Screening plant for effluent treatment (manufacture)
	29240	Filtering or purifying machinery parts (manufacture)		29240	Sealing machinery (manufacture)
	29240	Filtration equipment for hydraulic equipment (manufacture)		29240	Sedimentation plant for effluent treatment (manufacture)
	29240	Filtration equipment for the chemical industry (manufacture)		29240	Settlement plant for water treatment (manufacture)
	29240	Fire extinguisher (hand held) (manufacture)		29240	Sewage treatment plant (manufacture)
	29240	Fire extinguishing apparatus (excluding hand operated chemical extinguishers) (manufacture)		29430	Soldering machines (gas) (manufacture)
	29430	Flexible shaft drive tool (manufacture)		29240	Solvent extraction equipment for the chemical industry (manufacture)
	29240	Gas generators (manufacture)		29240	Solvent recovery equipment for the chemical industry (manufacture)
	29240	Gas or water generator parts (manufacture)		29240	Spray guns (manufacture)
	29240	Gasket (manufacture)		29240	Spraying machinery parts (manufacture)
	29240	General-purpose machinery parts (manufacture)		29240	Spring balance (manufacture)
	29240	Goods vending machines (manufacture)		29240	Sprinklers for fire extinguishing (manufacture)
				29240	Steam cleaning machines (manufacture)
				33202	Tape measure (manufacture)
				29240	Vending machine (manufacture)
				29240	Water softening plant (manufacture)

UK SIC 2007	UK SIC 2003	Activity	SIC 2007	SIC 2003	Activity
	29240	Water treatment plant (manufacture)		29320	Grain drier (manufacture)
	29240	Weighbridge (manufacture)		29320	Grinding mill for agricultural use (manufacture)
	29240	Weighing machine (manufacture)		29320	Harrow (manufacture)
	29240	Weights for weighing machine (manufacture)		29320	Harvesters (manufacture)
	29430	Welding machines (gas) (manufacture)		29320	Harvesting or threshing machinery harvesters, threshers, sorters (manufacture)
	29430	Welding torch (manufacture)		29320	Hay making equipment (manufacture)
	29240	Wrapping machinery (manufacture)		29320	Hedgecutter for agricultural use (manufacture)
28301		**Manufacture of agricultural tractors**		29320	Lawn mower (manufacture)
	29310	Safety frame for tractors (manufacture)		29320	Machines for cleaning, sorting or grading eggs, fruit (manufacture)
	29310	Tractor (half track) (manufacture)		29320	Manure spreader (manufacture)
	29310	Tractor (pedestrian controlled) (manufacture)		29320	Milking machine (manufacture)
	29310	Tractor (skidded unit) (manufacture)		29320	Mowers for agricultural use (manufacture)
	29310	Tractor (wheeled) (manufacture)		29320	Mowers for lawns, parks and sports grounds (manufacture)
	29310	Tractor for agricultural use (manufacture)		29530	Parts of milking machines and dairy machinery, n.e.c. (manufacturing)
	29310	Tractor for forestry use (manufacture)		29320	Pick-up baler (manufacture)
	29310	Tractor parts for wheeled and half-track tractors (manufacture)		29320	Planter for agricultural use (manufacture)
28302		**Manufacture of agricultural and forestry machinery (other than agricultural tractors)**		29320	Plough (manufacture)
	29320	Agricultural and forestry machinery parts (except tractor parts) (manufacture)		29320	Plough disc (manufacture)
	29320	Agricultural broadcaster (manufacture)		29320	Potato harvester and sorter (manufacture)
	29320	Agricultural machinery (manufacture)		29320	Poultry keeping machinery (manufacture)
	29320	Agricultural machinery for soil preparation (manufacture)		29320	Roller for agricultural use (manufacture)
	29320	Agricultural motor drawn trailer (manufacture)		29320	Root crop harvesting and sorting machinery (manufacture)
	29320	Agricultural planting machinery (manufacture)		29320	Seed cleaner or pre-cleaner for agricultural use (manufacture)
	29320	Agricultural self-loading semi-trailers (manufacture)		29320	Seed cleaning, sorting or grading machines (manufacture)
	29320	Agricultural self-loading trailers (manufacture)		29320	Seeders (manufacture)
	29320	Agricultural self-unloading semi-trailers (manufacture)		29320	Servicing of agricultural machinery (manufacture)
	29320	Agricultural self-unloading trailers (manufacture)		29320	Silage making machinery (manufacture)
	29320	Bale handler for agriculture (manufacture)		29320	Spraying machine for agricultural use (manufacture)
	29320	Baler for hay and straw (manufacture)		29320	Sugar beet harvester (manufacture)
	29320	Bee-keeping machinery (manufacture)		29320	Threshing machine (manufacture)
	29320	Broadcaster for agricultural use (manufacture)		29320	Tool bar for agricultural use (manufacture)
	29320	Chemical seed dresser for agricultural use (manufacture)		29320	Tractor hoe (manufacture)
	29320	Combine harvester (manufacture)		29320	Tractor plough (manufacture)
	29320	Conveyor for agricultural use (manufacture)		29320	Trailers and semi-trailers for agricultural use (manufacture)
	29320	Cultivator (manufacture)		29320	Transplanter for agricultural use (manufacture)
	29320	Cultivator tine (manufacture)		29320	Vegetable harvesting and sorting machinery (manufacture)
	29320	Digger (elevator and shaker) (manufacture)		29320	Winnower for agricultural use (manufacture)
	29320	Disc harrow (manufacture)			
	29320	Drill for agricultural use (manufacture)	**28410**		**Manufacture of metal forming machinery**
	29320	Egg cleaning, sorting and grading machines (manufacture)		29420	Bending machines (metal forming) (manufacture)
	29320	Elevator for agricultural use (manufacture)		29420	Boring machine (metal cutting) (manufacture)
	29320	Farmyard manure spreader (manufacture)		28620	Broach for metal working machine tools (manufacture)
	29320	Fertiliser distributor or broadcaster (manufacture)		29420	Broaching machine (metal cutting) (manufacture)
	29320	Fertilizing plough machinery (manufacture)		29420	Chemical process machine tool (metal working) (manufacture)
	29320	Fodder preparing equipment (manufacture)		29420	Computer numerically controlled (CNC) metal cutting machines (manufacture)
	29320	Forage harvester (manufacture)		29420	Cutting machine for metal (manufacture)
	29320	Forest machinery (manufacture)		29420	Die stamping machines (metal forming) (manufacture)
	29320	Fruit cleaning, sorting or grading machines (manufacture)		29420	Draw bench (metal forming) (manufacture)
	29320	Grain auger (manufacture)		29420	Drawing machine (metal forming) (manufacture)
	29320	Grain cleaning, sorting and grading machines (manufacture)			

UK SIC 2007	UK SIC 2003	Activity
	29420	Drilling machine (metal cutting) (manufacture)
	29420	Drop hammers (manufacture)
	29420	Electric discharge metal working tool (manufacture)
	29420	Electrochemical metal working machine tools (manufacture)
	29420	Electro-magnetic-pulse (magnetic-forming) metal forming machines (manufacture)
	28620	End mill (manufacture)
	29420	Filament wire spiralling machines (manufacture)
	29420	Forging machine (metal forming) (metal working) (manufacture)
	29420	Forming machine (high energy rate) (manufacture)
	29420	Gear making or finishing machines (metal cutting) (manufacture)
	29420	Grinding machines (metal cutting) (manufacture)
	29420	Honing machine (metal cutting) (manufacture)
	29420	Hydraulic brakes (manufacture)
	29420	Lapping machine (metal cutting) (manufacture)
	29420	Laser cutting or welding machine tools (metal working) (manufacture)
	29420	Lathe (metal cutting) (manufacture)
	29420	Lathe chuck (manufacture)
	29420	Machine tools (ultrasonic) (metal working) (manufacture)
	29420	Machine tools, for working metal, plasma arc (manufacture)
	29420	Machine tools, for working metals (manufacture)
	29420	Machine tools, for working metals, ultrasonic waves (manufacture)
	29420	Machine tools, for working metals, using a laser beam (manufacture)
	29420	Machine tools, for working metals, using a magnetic pulse (manufacture)
	29420	Machining centre (metal working) (manufacture)
	29420	Metal cutting machine (numerically controlled) (manufacture)
	29420	Metal cutting machine tool (manufacture)
	29420	Metal forming machine (numerically controlled) (manufacture)
	29420	Metal working machine tool (physical process) (manufacture)
	29420	Metal working machine tool parts (manufacture)
	29420	Milling machine (metal cutting) (manufacture)
	29420	Planing machine (metal cutting) (manufacture)
	29420	Press (hydraulic) (metal forming) (manufacture)
	29420	Press (mechanical) (metal forming) (manufacture)
	29420	Press (metal forming) (manufacture)
	29420	Press (pneumatic) (metal forming) (manufacture)
	29420	Punch presses (manufacture)
	29420	Punching machine (metal forming) (manufacture)
	29420	Sawing machine (metal cutting) (manufacture)
	29420	Screwing machines (metal cutting) (manufacture)
	29420	Shaping machine (metal cutting) (manufacture)
	29420	Shearing machine (metal forming) (manufacture)
	29420	Sheet metal forming machine (manufacture)
	29420	Slotting machine (metal working) (manufacture)
	29420	Spark erosion machines (metal working) (manufacture)
	29420	Spinning lathes (metal working) (manufacture)
	29420	Stamping or pressing machine tools (manufacture)
	29420	Surface tempering machines (manufacture)
	29420	Swaging machine (metal forming) (manufacture)

SIC 2007	SIC 2003	Activity
	29420	Thread rollers or machines for working wires (manufacture)
	29420	Thread rolling machines (manufacture)
	29420	Threading machine (metal cutting) (manufacture)
	29420	Turning machine (metal cutting) (manufacture)
	29420	Unit construction and transfer machine (metal working) (manufacture)
28490		**Manufacture of other machine tools**
	29430	Blowpipes (hand held, high or low pressure) (manufacture)
	29430	Brazing machines (electric) (manufacture)
	29430	Brazing machines (gas) (manufacture)
	29430	Cask assembly machines (manufacture)
	29430	Chipboard press (manufacture)
	29430	Cutter for wood (manufacture)
	29430	Cutting torch (manufacture)
	29430	Disc cutting machines for stone, ceramic, asbestos cement and similar (not portable) (manufacture)
	28620	Dividing heads and other special attachments for machine tools (manufacture)
	29430	Drilling or milling machines for stone, ceramics, asbestos-cement and similar articles (manufacture)
	31620	Electrolytic chemical process plant (manufacture)
	31620	Electroplating equipment (manufacture)
	29430	Finishing and polishing machine tools for optical, spectacle or clock or watch faces (manufacture)
	29430	Frame and carcass cramps (manufacture)
	29430	Glass cutting (shaping) machines for faceting or for cut-glass articles (cold glass) (manufacture)
	29430	Glass cutting machines of the wheel or diamond type (cold glass) (manufacture)
	29430	Glass engraving machines of the grinding wheel or diamond type (manufacture)
	29430	Glass polishing machines (manufacture)
	29430	Gluing machines (manufacture)
	29430	Grinding wheel cutting and dressing machines (manufacture)
	29430	Grinding, smoothing, polishing and graining machines for stone, ceramics, asbestos
	29430	Helical-wire cutting machines for stone, ceramics, asbestos-cement or similar (manufacture)
	29430	Log decorticators (manufacture)
	29430	Machine tool special attachments (excluding for metal working) (manufacture)
	29430	Machine tools for cork, bone, hard rubber, hard plastics or similar hard materials (manufacture)
	29430	Machine tools for working cold glass (manufacture)
	29430	Machine tools, for working wood (manufacture)
	29430	Machinery for making wooden clogs, soles and heels for shoes (manufacture)
	29430	Mortising machines (manufacture)
	29430	Moulding machine for wood, etc. (manufacture)
	29430	Nailing machines (manufacture)
	29430	Panel forming machines (manufacture)
	29430	Paring and slicing machines (manufacture)
	31620	Parts and accessories for electroplating machinery (manufacture)
	29430	Parts and accessories for machine tools for working hard materials e.g. wood, stone (manufacture)
	29430	Parts and accessories for stationary drills, machines, riveters, sheet metal cutters (manufacture)
	29430	Parts and accessories for stationary machine tools for nailing, stapling, glueing etc. (manufacture)

UK SIC 2007	UK SIC 2003	Activity	SIC 2007	SIC 2003	Activity
	29430	Pencil-making machinery (manufacture)	**28921**		**Manufacture of machinery for mining**
	29430	Planing machine for wood (not portable powered) (manufacture)		29521	Blow-out prevention apparatus (manufacture)
	29430	Plank glueing machines (manufacture)		29521	Borers (mining machinery) (manufacture)
	29430	Plywood press (manufacture)		29521	Bridge plugs (manufacture)
	29430	Post peeling machines (manufacture)		29521	Cage plant for mining (manufacture)
	29430	Press for chipboard (manufacture)		29521	Casing hangars (manufacture)
	29430	Presses for the manufacture of particle board (manufacture)		29521	Coal cutter (manufacture)
				29521	Coal plough (manufacture)
	29430	Sanding and polishing machines for wood (not portable) (manufacture)		29521	Coal preparation plant (manufacture)
				29521	Continuous miner (manufacture)
	28620	Saw blades for machines including wood cutting (manufacture)		29521	Conveyor for underground mining (manufacture)
				29521	Crushing machine for mining (manufacture)
	29430	Sawing machine for wood (manufacture)		29521	Cutting machinery for coal or rock (manufacture)
	29430	Sawing machines for stone, ceramics, asbestos-cement or similar materials (manufacture)		29521	Dinting machine for mining (manufacture)
				28620	Drill bits for well drilling (manufacture)
	29430	Splitting or cleaving machines for stone, ceramics, asbestos-cement or similar (manufacture)		29521	Drilling jars for mining (manufacture)
				29521	Earth boring machine (manufacture)
	29430	Splitting, stamping and fragmenting machines (manufacture)		29521	Elevators with continuous action for underground use (manufacture)
	29430	Squeeze presses (manufacture)		29521	Hauling engine for mining (stationary) (manufacture)
	29430	Stapling machines (machine tools) (manufacture)		29521	Heading machine for mining (manufacture)
	29430	Stapling machines for industrial use (manufacture)		29521	Iron roughnecks (manufacture)
	29430	Stationary machines for nailing, glueing or otherwise assembling wood, cork bone, etc. (manufacture)		29521	Liner hanger equipment (manufacture)
				29521	Loader for mining (manufacture)
	29430	Stationary rotary or rotary percussion drills, filing machines, sheet metal cutters (manufacture)		29521	Machinery for treating minerals by screening, sorting, separating, washing, crushing (manufacture)
	29430	Stave and cask croze cutting machines (manufacture)		29521	Mineral cutter (manufacture)
	29430	Stave jointing, planing, bending machinery (manufacture)		29521	Mineral dressing plant (manufacture)
				29521	Minerals treatment machinery (manufacture)
	29430	Tool holder (manufacture)		29521	Mining machinery (manufacture)
	29430	Trueing and grinding machines for cold working of glass (manufacture)		28620	Mining tool (bit) (manufacture)
				29521	Moles for mining (manufacture)
	29430	Turning, engraving, carving, etc. machines for stone, ceramics, and similar (manufacture)		29521	Motion compensation equipment for oil drilling rigs (manufacture)
	29430	Veneer press (manufacture)		29521	Mudline suspension and tie-back equipment (manufacture)
	29430	Veneer shearing machines (manufacture)		29521	Petroleum drilling equipment (manufacture)
	29430	Veneer splicing machines (manufacture)		29521	Pit bottom machinery (manufacture)
	29430	Wire weaving machine (manufacture)		29521	Powered roof support for mining (manufacture)
	29430	Wood sculpturing and engraving machines (manufacture)		29521	Production riser tensioners (manufacture)
				29521	Production riser tie-back equipment (manufacture)
	29430	Wooden button making machinery (manufacture)		29521	Ranging drum shearer for mining (manufacture)
	29430	Woodworking machinery (manufacture)		29521	Riser connector apparatus (manufacture)
	29430	Work holders (engineers' small tools) (manufacture)		29521	Road ripper for mining (manufacture)
	29430	Work holders for machine tools (manufacture)		29521	Rock cutting machinery (manufacture)
				29521	Rock drilling machinery (manufacture)
28910		**Manufacture of machinery for metallurgy**		29521	Rocker shovel for mining (manufacture)
	29510	Casting machines (manufacture)		29521	Roof support (hydraulic) (manufacture)
	29510	Casting machines for foundries (manufacture)		29521	Rotary tables (manufacture)
	29510	Converters for hot metal handling (manufacture)		29521	Sinking machine for mining (manufacture)
	29510	Die casting machines for foundries (manufacture)		29521	Skip plant for mining (manufacture)
	29510	Hot metals handling machinery and equipment (manufacture)		29521	Stowing machine for mining (manufacture)
				29521	Track laying tractors and other tractors used in mining (manufacture)
	29510	Ingot mould and bottom (manufacture)		29521	Tunnelling machine for mining (manufacture)
	29510	Investment casting equipment (manufacture)		29521	Well drilling equipment (manufacture)
	29510	Ladles for handling hot metals (manufacture)		29521	Wellhead running tools (manufacture)
	29510	Metal rolling mills and rolls for such mills (manufacture)		29521	Wellheads (manufacture)
	29510	Metallurgy machinery (manufacture)		29521	Winding machine for mining (manufacture)
	29510	Rolling mill for metals (manufacture)			
	29510	Rolls made of iron or steel (manufacture)			
	29510	Tube mill plant (manufacture)			

UK SIC 2007	UK SIC 2003	Activity	SIC 2007	SIC 2003	Activity
28922		**Manufacture of earthmoving equipment**		29523	Sorting, grinding and mixing machinery for earth, stones and other mineral substances (manufacture)
	29522	Angle-dozers (manufacture)		29523	Tar laying plant (manufacture)
	29522	Bucket for construction machinery (manufacture)		29523	Tar processing plant (manufacture)
	29522	Bulldozer and angle-dozer blades (manufacture)		29523	Tarmacadam laying plant (manufacture)
	29522	Bulldozers (manufacture)		29523	Tarmacadam processing plant (manufacture)
	29522	Crawler loader (manufacture)		29523	Track laying and other tractors used in construction (manufacture)
	29522	Crawler tractor (manufacture)			
	29522	Dragline excavator (manufacture)			
	34100	Dumpers for off road use (manufacture)	**28930**		**Manufacture of machinery for food, beverage and tobacco processing**
	29522	Earth leveller (manufacture)		29530	Animal fat or oils extraction and preparation machinery (manufacture)
	29522	Earth mover (construction equipment) (manufacture)		29530	Aspirator separators (manufacture)
	29522	Earth moving machinery (manufacture)		29530	Bakery machinery and ovens (manufacture)
	29522	Excavator (manufacture)		29530	Bakery moulders (manufacture)
	29522	Grab (manufacture)		29530	Bakery ovens (industrial) (manufacture)
	29522	Grader (manufacture)		29530	Beverage processing machinery (manufacture)
	29522	Land clearing equipment and machinery (manufacture)		29530	Biscuit making machinery and ovens (manufacture)
	29522	Levellers (manufacture)		29530	Blenders for the food industry (manufacture)
	29522	Loading shovel (manufacture)		29530	Blenders for the grain milling industry (manufacture)
	29522	Mechanical shovels (manufacture)		29530	Bran cleaners (manufacture)
	29522	Powered barrow (manufacture)		29530	Bran cleaners for the grain milling industry (manufacture)
	29522	Rear digger (manufacture)		29530	Breading rolls or mills (manufacture)
	29522	Rear digger unit (manufacture)		29530	Brewing machinery and plant (manufacture)
	29522	Ripper (manufacture)		29530	Butter churns (manufacture)
	29522	Rooter (not agricultural) (manufacture)		29530	Butter workers (manufacture)
	29522	Scraper (earth moving equipment) (manufacture)		29530	Cake depositing machines (manufacture)
	29522	Shovel loaders (manufacture)		29530	Catering equipment (electric) (manufacture)
	29522	Tractor shovel (manufacture)		29530	Cheese making machines (manufacture)
	29522	Tractor winch (manufacture)		29530	Cheese moulding machinery (manufacture)
	29522	Trencher (manufacture)		29530	Cheese press (manufacture)
	29522	Walking draglines (manufacture)		29530	Cheese pressing machinery (manufacture)
				29530	Chocolate and cocoa making machinery (manufacture)
28923		**Manufacture of equipment for concrete crushing and screening roadworks**		29530	Cider making machinery (manufacture)
	29523	Asphalt laying plant (manufacture)		29530	Cigar making machinery (manufacture)
	29523	Asphalt processing plant (manufacture)		29530	Cigarette making machinery (manufacture)
	29523	Bitumen spreaders (manufacture)		29530	Cleaning, sorting or grading machines for seeds, grain or dried leguminous vegetables (manufacture)
	29523	Chippers for road surfacing (manufacture)		29530	Coffee processing machinery (manufacture)
	29523	Concrete mixer (manufacture)		29530	Confectionery machines and processing equipment (manufacture)
	29523	Concrete placing machinery (manufacture)			
	29523	Concrete surfacing machinery (manufacture)		29530	Cooking appliance for commercial catering (non-electric) (manufacture)
	29523	Construction equipment (manufacture)		29530	Cooking equipment for commercial catering (electric) (manufacture)
	29523	Crushing plant (not for mines) (manufacture)			
	29523	Equipment for concrete crushing and screening roadworks (manufacture)		29530	Cream separator for industrial use (manufacture)
	29523	Grinding and other mineral processing machinery (manufacture)		29530	Crusher (food or drink machinery) (manufacture)
	29523	Gritting machine (manufacture)		29530	Cyclone separators (manufacture)
	29523	Mortar mixers (manufacture)		29530	Dairy industry machinery (manufacture)
	29523	Mortar spreaders (manufacture)		29530	Dairy machinery and plant (not agricultural) (manufacture)
	29523	Paving machinery (manufacture)			
	29523	Pile driving equipment (manufacture)		29530	Dairy moulding machinery (manufacture)
	29523	Pile-drivers (manufacture)		29530	Dough dividers (manufacture)
	29523	Pile-extractors (manufacture)		29530	Dough making machinery (manufacture)
	29523	Planers for road surfacing (manufacture)		29530	Drink processing including combined processing and packaging or bottling machinery (manufacture)
	29523	Pulverising machinery (not for mines) (manufacture)			
	29523	Road roller (manufacture)			
	29523	Snow blowers (manufacture)		29530	Dryers for agriculture (manufacture)
	29523	Snow ploughs (manufacture)			

UK SIC 2007	UK SIC 2003	Activity	SIC 2007	SIC 2003	Activity
	29530	Edible oil and fat processing machinery (manufacture)		29530	Sugar making and refining machinery (manufacture)
	29530	Feeders (manufacture)		29530	Tea processing machinery and plant (manufacture)
	29530	Filtering and purifying machinery for the industrial preparation of food or drink (manufacture)		29530	Tobacco processing machinery (manufacture)
	29530	Fish processing machines and equipment (manufacture)		29530	Vegetable fats or oils extraction or preparation machinery (manufacture)
	29530	Flour and meal manufacturing machinery (manufacture)		29530	Vegetable processing machines and equipment (manufacture)
	29530	Flour confectionery machinery (manufacture)		29530	Vinegar processing machinery (manufacture)
	29530	Food and drink press (manufacture)		29530	Wine making machinery (manufacture)
	29530	Food preparation machinery for hotels and restaurants (manufacture)		29530	Wine, cider, fruit juice, etc. press (manufacture)
	29530	Food processing equipment (industrial) (manufacture)		29530	Winnowers for milling (manufacture)
	29530	Food processing machinery incl. combined processing, packaging or bottling machinery (manufacture)	**28940**		**Manufacture of machinery for textile, apparel and leather production**
	29530	Fruit juice preparation machinery (manufacture)		29540	Apparel and leather production machinery (manufacture)
	29530	Fruit processing machines and equipment (manufacture)		29540	Apparel production machinery (manufacture)
	29530	Grain brushing machines (manufacture)		29540	Automatic stop motions (textile machinery) (manufacture)
	29530	Grain milling industry machinery (manufacture)		29540	Backing and curling machinery (carpet making) (manufacture)
	29530	Grain processing machinery and plant (manufacture)		29540	Bale breakers (manufacture)
	29530	Grinding mills (manufacture)		29540	Beaming machinery (textile) (manufacture)
	29530	Homogenisers (manufacture)		29540	Bleaching machinery (textile) (manufacture)
	29530	Irradiators (manufacture)		29540	Blowroom machinery (textile) (manufacture)
	29530	Macaroni, spaghetti or similar products machinery (manufacture)		29540	Bobbins for textile machinery (manufacture)
	29530	Meat processing machinery (manufacture)		29540	Card tape reader for textile machinery (manufacture)
	29530	Milk converting machinery (manufacture)		29540	Carders (manufacture)
	29530	Milk pasteurisation plant (manufacture)		29540	Carding machinery for textiles (manufacture)
	29530	Milk processing machinery (manufacture)		29540	Carpet making machinery (manufacture)
	29530	Milling machine (food processing) (manufacture)		29540	Coating machinery for textiles (manufacture)
	29530	Moulders for bakery (manufacture)		29540	Comb for textile machinery (manufacture)
	29530	Moulding machines for dairies (manufacture)		29540	Combing machinery for textiles (manufacture)
	29530	Mustard processing machine (manufacture)		29540	Cots (textile machinery accessory) (manufacture)
	29530	Nut processing machines and equipment (manufacture)		29540	Cotton gins (manufacture)
	29530	Oven (food processing) machine (manufacture)		29540	Cotton spreaders (manufacture)
	29530	Pastry roller food preparation machinery (manufacture)		29540	Crabbing machinery for textiles (manufacture)
	29530	Pea splitters (manufacture)		29540	Cropping machinery for textiles (manufacture)
	29530	Poultry processing machines and equipment (manufacture)		29540	Curing machinery for textiles (manufacture)
	29530	Press for food and drink (manufacture)		29540	Cutting machine for textile fibres (manufacture)
	29530	Press used to make wine, cider, fruit juices, etc. (manufacture)		29540	Cutting machinery for textile fabrics (manufacture)
	29530	Rice hullers (manufacture)		29540	Decatising machinery for textiles (manufacture)
	29530	Sea-food processing machines and equipment (manufacture)		29540	De-sizing machinery for textiles (manufacture)
	29530	Separators for the grain milling industry (manufacture)		29540	Dobbies (textile machinery) (manufacture)
	29530	Shell fish processing machines and equipment (manufacture)		29540	Doubling machinery for textiles (manufacture)
	29530	Sieving belts (manufacture)		29540	Drawing machinery for textiles (manufacture)
	29530	Sifters (manufacture)		29540	Dry cleaning machinery (manufacture)
	29530	Slaughterhouse machinery (manufacture)		29540	Drying machinery (commercial) for textiles (manufacture)
	29530	Slicers for bakeries (manufacture)		29540	Drying machines for laundries or dry cleaners (manufacture)
	29530	Soft drinks machinery (manufacture)		29540	Dye cycle controller (textile machinery) (manufacture)
	29530	Sterilisation equipment for food and drink (manufacture)		29540	Dyeing machinery for textiles (manufacture)
	29530	Sugar confectionery making machinery (manufacture)		29540	Dyesprings (textile machinery accessory) (manufacture)
				29540	Embossing machinery for textiles (manufacture)
				29540	Extruding machinery for textiles (manufacture)
				29540	Fabric processing machinery (manufacture)
				29540	Faller (textile machinery accessory) (manufacture)
				29540	Felt or non-woven fabric production or finishing machines (manufacture)

UK SIC 2007	UK SIC 2003	Activity	SIC 2007	SIC 2003	Activity
	29540	Felts or non-wovens production and finishing machines (manufacture)		29540	Steaming machinery for textiles (manufacture)
	29540	Film reader for textile machinery (manufacture)		29540	Stentering machinery for textiles (manufacture)
	29540	Finishing machinery for textiles (manufacture)		29540	Teasel rod (textile machinery accessory) (manufacture)
	29540	Flatwork machine for laundry (manufacture)		29540	Textile dressing or impregnating machinery (manufacture)
	29540	Footwear making or repairing machinery (manufacture)		29540	Textile fabric making machinery (manufacture)
	29540	Fusing presses (manufacture)		29540	Textile fibre preparation machinery (manufacture)
	29540	Garnetters (manufacture)		29540	Textile machinery (manufacture)
	29540	Heald (textile machinery accessory) (manufacture)		29560	Textile printing machinery (manufacture)
	29540	Heat setting machinery for textiles (manufacture)		29540	Textile unreeling, folding, or pinking machinery (manufacture)
	29540	Hides and skins preparation, tanning, working or repairing machinery (manufacture)		29540	Textile yarn preparation machinery (manufacture)
	29540	Hosiery knitting machinery (manufacture)		29540	Texturing and softening machinery for textiles (manufacture)
	29540	Ironing machine for non-domestic use (manufacture)		29540	Thread guide (textile machinery accessory) (manufacture)
	29540	Jacquard machinery for carpet making (manufacture)		29540	Tufting machinery for carpet making (manufacture)
	29540	Jacquard textile machinery (manufacture)		29540	Twisting machinery for textiles (manufacture)
	29540	Knitting machine (manufacture)		29540	Warpers for preparing textile yarns (manufacture)
	29540	Knotted net, tulle, lace, braid etc. making machines (manufacture)		29540	Warping machinery (manufacture)
	29540	Laundry machinery (manufacture)		29540	Washing machines (laundry) (manufacture)
	29540	Leather working machine (manufacture)		29540	Washing machines (textile) (non-domestic) (manufacture)
	29540	Loom (manufacture)		29540	Weaving machinery (looms) (manufacture)
	29540	Loom winder (manufacture)		29540	Weaving machines (manufacture)
	29540	Man-made textile fibre or yarn producing machinery (manufacture)		29540	Winding machine for textiles (manufacture)
	29540	Mercerising machinery for textiles (manufacture)		29540	Wool carbonisers (manufacture)
	29540	Milling machinery for textiles (manufacture)		29540	Wool scourers (manufacture)
	29540	Needles for sewing machines (manufacture)	**28950**		**Manufacture of machinery for paper and paperboard production**
	29540	Pirns (textile machinery accessory) (manufacture)		29550	Board making machinery (except chipboard) (manufacture)
	29540	Plaiting machinery for textiles (manufacture)		29550	Cardboard box making machine (manufacture)
	29540	Printing machinery for textiles (manufacture)		29550	Carton making machinery (manufacture)
	29540	Programmer for textile machinery (manufacture)		29550	Embossing machines for working paper and board (manufacture)
	29540	Raising machinery for textiles (manufacture)		29550	Envelope making machine (manufacture)
	29540	Reaching-in machinery for textiles (manufacture)		29550	Folding machinery for paper and board (not for office use) (manufacture)
	29540	Reed (textile machinery accessory) (manufacture)		29550	Fourdrinier (manufacture)
	29540	Reeling machinery for textiles (manufacture)		29550	Laminating machinery (paper working) (manufacture)
	29540	Rigging machinery for textiles (manufacture)		29550	Paper and paperboard production machinery (manufacture)
	29540	Ring traveller for textile machinery (manufacture)		29550	Paper bag making machinery (manufacture)
	29540	Roller (textile machinery accessory) (manufacture)		29550	Paper making machinery (manufacture)
	29540	Roving frames (manufacture)		29550	Pulp making machinery (manufacture)
	29540	Scouring machinery for textiles (manufacture)		29550	Slitting machine for paper (manufacture)
	29540	Sewing machine (manufacture)		29550	Stock preparation plant for paper and board (manufacture)
	29540	Sewing machine heads (manufacture)			
	29540	Shearing machinery for textiles (manufacture)	**28960**		**Manufacture of plastics and rubber machinery**
	29540	Shrinking machines for textiles (manufacture)		29560	Blow moulding machine for rubber or plastic (manufacture)
	29540	Shuttle (textile machinery accessory) (manufacture)		29560	Extruder for rubber or plastics (manufacture)
	29540	Shuttle changing machinery (manufacture)		29560	Forming machine for rubber or plastics (manufacture)
	29540	Singeing machinery for textiles (manufacture)		29560	Injection moulding equipment (for rubber or plastic) (manufacture)
	29540	Sizing machinery for textiles (manufacture)		29560	Mixing machine for working rubber or plastic (manufacture)
	29540	Slitting machinery for textiles (manufacture)			
	29540	Sliver can (textile machinery accessory) (manufacture)			
	29540	Spindles (textile machinery accessories) (manufacture)			
	29540	Spindles and spindle flyers (manufacture)			
	29540	Spinning machinery for textiles (manufacture)			
	29540	Spinning machines (manufacture)			
	29540	Spool (textile machinery accessory) (manufacture)			
	29540	Spooling machinery for carpet making (manufacture)			
	29540	Spotting table (manufacture)			

UK SIC 2007	UK SIC 2003	Activity
	29560	Moulders for rubber or plastic (manufacture)
	29560	Moulding machine for rubber or plastic (manufacture)
	29560	Plastic product making machines (manufacture)
	29560	Plastics working machinery (manufacture)
	29560	Pneumatic tyre making or retreading machines (manufacture)
	29560	Press machinery for working rubber or plastics (manufacture)
	29560	Pressure forming machines for rubber or plastic (manufacture)
	29560	Roll mill for rubber or plastic (manufacture)
	29560	Rubber working machinery (manufacture)
	29560	Transfer moulding press for rubber or plastics (manufacture)
	29560	Vulcanizing machines for working rubber and plastics (manufacture)
28990		**Manufacture of other special-purpose machinery n.e.c.**
	29560	Accumulator for hydraulic equipment (manufacture)
	35300	Aircraft carrier catapults and related equipment (manufacture)
	35300	Aircraft launching gear and related equipment (manufacture)
	33403	Apparatus and equipment for automatically developing photographic film
	29560	Battery making machine (manufacture)
	29560	Blackboard chalk making machinery (manufacture)
	29560	Blocking machine (manufacture)
	29560	Bookbinding machine (manufacture)
	29560	Bow thruster (manufacture)
	36501	Bowling alley equipment (manufacture)
	29560	Brick making machinery (manufacture)
	29560	Cable making machine (manufacture)
	29560	Carpet shampoo appliance (not domestic electric) (manufacture)
	29560	Carpet sweeper (industrial) (manufacture)
	35300	Catapult for launching aircraft (manufacture)
	29560	Cement block making machine (manufacture)
	29560	Central greasing systems (manufacture)
	29560	Centrifugal clothes dryer (manufacture)
	29560	Ceramic making machine (manufacture)
	29560	Ceramic pastes production machinery (shaped) (manufacture)
	29560	Coating machine for bookbinding or paper working (manufacture)
	29560	Collating machine for bookbinders (manufacture)
	29560	Composing room equipment (manufacture)
	29560	Concrete block making machinery (manufacture)
	29560	Creasing machine for bookbinding (manufacture)
	29560	Crystalliser for chemical industry (manufacture)
	29560	Cutting machine for bookbinding (manufacture)
	35300	Deck arresters for aircraft (manufacture)
	29560	Decompression chamber (manufacture)
	29560	Diving equipment (excluding breathing apparatus) (manufacture)
	29560	Dryers for wood, paper pulp, paper or paperboard (manufacture)
	29560	Drying machine for the chemical industry (manufacture)
	36639	Fairground rides (manufacture)
	29560	Flexographic printing machine (manufacture)

SIC 2007	SIC 2003	Activity
	29560	Forming machine for glass working (multi-head) (manufacture)
	29560	Foundry moulds forming machinery (manufacture)
	29560	Foundry moulds (for rubber or plastic) production machines (manufacture)
	29560	Gathering machine (paper working) (manufacture)
	29560	Glass, glassware and glass fibre or yarn production machinery (manufacture)
	29560	Gluing machinery for bookbinders, etc. (manufacture)
	29560	Granulator for the chemical industry (manufacture)
	29560	Graphite electrodes production machinery (manufacture)
	29560	Gravure printing machine (manufacture)
	29560	Grinding machines for glass (manufacture)
	33202	Gyroscope (manufacture)
	29560	Hatch cover (mechanically operated) (manufacture)
	29560	Hot glass working machinery (manufacture)
	29560	Industrial carpet sweeper (manufacture)
	29560	Industrial mixing equipment for the chemical industry (manufacture)
	29560	Industrial robots for multiple uses (manufacture)
	29560	Isotopic separation machinery or apparatus (manufacture)
	29560	Lamp making machine (manufacture)
	35300	Launching gear for aircraft (manufacture)
	29560	Letterpress printing machine (manufacture)
	29560	Machines for the assembly of electric or electronic lamps, tubes (valves) or bulbs (manufacture)
	29560	Offset litho printing machine (manufacture)
	29560	Photo-engraving machine (manufacture)
	29560	Photogravure machine (manufacture)
	29560	Photolitho machine (manufacture)
	33403	Photolithography equipment for the manufacture of semi-conductors (manufacturing)
	29560	Pipe making machinery (manufacture)
	29560	Polishing machine for glass (manufacture)
	29560	Pottery making machinery (manufacture)
	29560	Printing machine or press (manufacture)
	29560	Pulverising machinery for the chemical industry (manufacture)
	29560	Reduction gear for marine use (manufacture)
	29560	Robots for multiple industrial uses (manufacture)
	29560	Rope making machines (manufacture)
	36639	Roundabout for fairground (manufacture)
	29560	Ruling machinery for printing (manufacture)
	29560	Sand handling, mixing, treatment or reclamation plant for foundries (manufacture)
	29560	Scanner (printing machinery) (manufacture)
	29560	Sewing machinery for bookbinding (manufacture)
	36639	Shooting galleries (manufacture)
	29560	Size reduction equipment for the chemical industry (manufacture)
	29560	Size separation equipment for the chemical industry (manufacture)
	29560	Soap making machinery (manufacture)
	29560	Stabiliser for ship (manufacture)
	29560	Steering gear for marine use (manufacture)
	29560	Stern gear (manufacture)
	29560	Stitching machine for bookbinding (manufacture)
	29560	Surge damper for hydraulic equipment (manufacture)

UK SIC 2007	UK SIC 2003	Activity	SIC 2007	SIC 2003	Activity
	36639	Swings (fairground equipment) (manufacture)		34100	Mobile x-ray unit (not trailer) (manufacture)
	31620	Tanning beds (manufacture)		34100	Motor car (manufacture)
	29560	Thermo forming machines (manufacture)		34100	Motor coach (manufacture)
	29560	Tile making machine (not plastic working) (manufacture)		34100	Motor vehicle reconditioning by manufacturer (manufacture)
	29560	Trawl door (manufacture)		34100	Motor vehicles (manufacture)
	29560	Tubes (valves) or bulbs producing machinery (manufacture)		34100	Motor vehicles for commercial use (manufacture)
	29560	Type setting machine (manufacture)		34100	Motorised caravans (manufacture)
	29560	Tyre alignment and balancing equipment (except wheel balancing) (manufacture)		34100	Passenger cars (manufacture)
	29560	Vacuum cleaners for industrial and commercial use (manufacture)		34100	Refrigerated lorry (manufacture)
	29560	Vacuum forming machine (manufacture)		34100	Refuse disposal vehicle (manufacture)
	29560	Wire coiling machine (manufacture)		34100	Road tanker (not trailer) (manufacture)
	29560	Wire rope making machine (manufacture)		34100	Road tractor unit (manufacture)
				34100	Snow mobiles (manufacture)
29100		**Manufacture of motor vehicles**		34100	Spraying lorry (manufacture)
	34100	Ambulance (manufacture)		34100	Station wagon (manufacture)
	34100	Amphibious vehicles (manufacture)		34100	Street sweepers (manufacture)
	34100	Armoured car (except military fighting vehicles) (manufacture)		34100	Street sweeping lorry (manufacture)
	34100	ATVs, go-carts and similar including race cars (manufacture)		34100	Taxi (manufacture)
	34100	Battery powered electric commercial vehicle (manufacture)		34100	Tractors for semi-trailers (manufacture)
	34100	Breakdown lorry (manufacture)		34100	Travelling libraries, banks, etc. (not trailers) (manufacture)
	34100	Bus (manufacture)		34100	Trolley bus (manufacture)
	34100	Cars (manufacture)		34100	Truck (commercial vehicle) (manufacture)
	34100	Chassis fitted with engine (manufacture)		34100	Van (manufacture)
	34100	Chassis for motor vehicles (manufacture)			
	34100	Chassis with engine for commercial vehicle (manufacture)	**29201**		**Manufacture of bodies (coachwork) for motor vehicles (except caravans)**
	34100	Coach (manufacture)		34201	Bodies (coachworks) for motor vehicles (manufacture)
	34100	Coach engine (manufacture)		34201	Body building for motor vehicles (manufacture)
	34100	Commercial vehicles (manufacture)		34201	Body for bus (manufacture)
	34100	Concrete-mixer lorries (manufacture)		34201	Body for car (manufacture)
	34100	Crane lorry (manufacture)		34201	Body for coach (manufacture)
	34100	Dump truck (manufacture)		34201	Body for commercial vehicle (manufacture)
	34100	Electrically powered commercial vehicles (manufacture)		34201	Body shell for motor vehicle made of fibre glass (manufacture)
	34100	Engine for motor vehicle (manufacture)		34201	Body shell for motor vehicle made of plastic (manufacture)
	34100	Engines (internal combustion) for motor vehicles (manufacture)		34201	Cabs for motor vehicles (manufacture)
	34100	Estate car (manufacture)		34201	Coachwork for motor vehicles (manufacture)
	34100	Factory rebuilding of motor vehicle engines (manufacture)		34201	Outfitting of all types of motor vehicles (except caravans) (manufacture)
	34100	Fire engine (manufacture)			
	34100	Fire tender (manufacture)	**29202**		**Manufacture of trailers and semi-trailers**
	34100	Golf carts (manufacture)		34202	Bowsers (tanks on wheels) (manufacture)
	34100	Heavy goods vehicle (manufacture)		34202	Containers for carriage by one or more modes of transport (manufacture)
	34100	Internal combustion engine for motor vehicles (manufacture)		34202	Containers for freight (manufacture)
	34100	Internal combustion engines for tractors (manufacture)		34202	Flat trailer (motor drawn) (manufacture)
	34100	KD sets for cars at least 50% of value of complete vehicle (manufacture)		34202	Horse box trailers (manufacture)
				34202	Low loader trailer (manufacture)
	34100	KD sets for commercial vehicles at least 50% of value of complete vehicle (manufacture)		34202	Outfitting of tankers and removal trailers for transport of goods (manufacture)
	34100	Lorry (manufacture)		34202	Outfitting of trailers and semi-trailers for transport of goods (manufacture)
	34100	Minibus (manufacture)		34202	Platform trailer (motor drawn) (manufacture)
	34100	Mobile library (not trailer) (manufacture)		34202	Road tractor trailer (manufacture)
				34202	Semi-trailers (manufacture)
				34202	Skeletal trailer (motor drawn) (manufacture)
				34202	Tanker trailer (motor drawn) (manufacture)

UK SIC 2007	UK SIC 2003	Activity	SIC 2007	SIC 2003	Activity
	34202	Trailer (motor drawn) (manufacture)	29320		**Manufacture of other parts and accessories for motor vehicles**
	34202	Trailers and semi-trailers (manufacture)		34300	Accessories and parts for motor vehicles and their engines (manufacture)
29203		**Manufacture of caravans**		34300	Airbags for motor vehicle (manufacture)
	34203	Caravan (manufacture)		34300	Anti-roll bars for motor vehicles (manufacture)
	34203	Caravan chassis (manufacture)		36110	Arm rest for motor vehicle (manufacture)
	34203	Caravan trailers (manufacture)		34300	Auto spare parts (manufacture)
	34203	Conversion of complete vehicles to motor caravans (manufacture)		34300	Axle for motor vehicle (manufacture)
	34203	Mobile bank (not self propelled) (manufacture)		34300	Brakes and parts (excluding linings for motor vehicles) (manufacture)
	34203	Mobile canteen (not self propelled) (manufacture)		34300	Bumpers for motor vehicles (manufacture)
	34203	Outfitting of caravans (manufacture)		34300	Caps for petrol, oil or radiator for motor vehicle (manufacture)
	34203	Permanent residential caravan (manufacture)		34300	Car body parts (manufacture)
	34203	Special purpose caravans (manufacture)		34300	Car components (manufacture)
	34203	Touring caravan (manufacture)		34300	Catalyzers (manufacture)
29310		**Manufacture of electrical and electronic equipment for motor vehicles**		34300	Chassis and parts for coaches (manufacture)
	31610	Alternator for vehicle (manufacture)		34300	Clutch and parts for motor vehicles (manufacture)
	31610	Auto electrical equipment (manufacture)		34300	Coupling for articulated motor vehicle (manufacture)
	31610	Coil ignition (manufacture)		34300	Differential unit for motor vehicle (manufacture)
	31610	Dashboard instruments (electric) (manufacture)		34300	Disc brakes (manufacture)
	31610	Defrosting and demisting equipment for vehicles (manufacture)		34300	Doors for motor vehicles (manufacture)
	31610	Demisters (electrical) (manufacture)		34300	Drive shaft for motor vehicles (manufacture)
	31610	Dynamo for vehicle (manufacture)		34300	Exhaust pipes for motor vehicles (manufacture)
	31610	Dynamo lighting set for cycles (manufacture)		34300	Exhaust systems and components for motor vehicles (manufacture)
	31610	Electrical and electronic equipment for motor vehicles (manufacture)		34300	Fuel tank for motor vehicle (manufacture)
	31610	Electrical equipment for engines and vehicles (manufacture)		34300	Gear box for motor vehicle (manual or automatic) (manufacture)
	31610	Electrical equipment for vehicles and aircraft (manufacture)		34300	Half shaft (manufacture)
	31610	Generators (dynamos and alternators) (manufacture)		34300	Independent suspension units for motor vehicles (manufacture)
	31610	Glow plugs (manufacture)		34300	KD sets for vehicles if the value is less than half the value of the complete vehicle (manufacture)
	31610	Horns for motor vehicle (electric) (manufacture)		34300	Panels for motor vehicle bodywork, made of metal or fibreglass (manufacture)
	31610	Ignition coil (manufacture)		34300	Parts for motor vehicles (not electric) (manufacture)
	31610	Ignition equipment (other than coils and magnetos) (manufacture)		34300	Pipes for motor vehicles (manufacture)
	31610	Ignition magnetos (manufacture)		34300	Propeller shaft for motor vehicle (manufacture)
	31610	Indicating measuring instrument for vehicles and aircraft (electric) (manufacture)		34300	Radiator for motor vehicle (manufacture)
	31610	Insulating fittings (other than ceramic for vehicles and aircraft) (manufacture)		34300	Radiator grill for motor vehicle (manufacture)
	31610	Magneto (manufacture)		34300	Registration plate for motor vehicle (manufacture)
	31610	Magneto-dynamos (manufacture)		34300	Road wheels for motor vehicle (manufacture)
	31610	Motor vehicle electrical generators (manufacture)		34300	Running gear for motor vehicles (manufacture)
	31610	Motor vehicle electrical ignition wiring harnesses (manufacture)		34300	Safety belts for cars (manufacture)
	31610	Motor vehicle electrical power window and door systems (manufacture)		36110	Seats for motor vehicles (manufacture)
	31610	Motor vehicle purchased gauges into instrument panels (manufacture)		34300	Shock absorber for motor vehicle (manufacture)
	31610	Sirens for motor vehicles (manufacture)		34300	Silencer for motor vehicle (manufacture)
	31610	Sound or visual signalling equipment for cycles and motor vehicles (manufacture)		34300	Spring suspension for motor vehicles (manufacture)
	31610	Sparking plug (manufacture)		34300	Steering box for motor vehicle (manufacture)
	31610	Starter motor for vehicle (manufacture)		34300	Steering column for motor vehicle (manufacture)
	31610	Traffic indicators for motor vehicles (manufacture)		34300	Steering equipment components for motor vehicles (manufacture)
	31610	Voltage regulators for vehicles (manufacture)		34300	Steering wheels for motor vehicle (manufacture)
	31610	Windscreen wipers (manufacture)		34300	Suspension shock absorbers for motor vehicle (manufacture)
	31610	Wiring sets (manufacture)		34300	Suspension springs for motor vehicle (manufacture)
				34300	Tipping gear and parts thereof for motor vehicles (not hydraulic) (manufacture)
				34300	Track rods for motor vehicles (manufacture)

UK SIC 2007	UK SIC 2003	Activity	SIC 2007	SIC 2003	Activity
	34300	Universal joints for motor vehicles (manufacture)		35110	Offshore floating drilling rig (manufacture)
	34300	Wheels and hubs for motor vehicles (manufacture)		35110	Offshore support vessel (manufacture)
	34300	Window winding gear for motor vehicles (not electric) (manufacture)		35110	Oil platform fabrication of steel plate (manufacture)
	34300	Windscreen wipers for motor vehicles (non-electric) (manufacture)		35110	Oil platform structural sections (manufacture)
				35110	Passenger cargo liner (manufacture)
30110		**Building of ships and floating structures**		35110	Passenger vessel building (manufacture)
	35110	Barge (manufacture)		35110	Platform for drilling rig (manufacture)
	35110	Beacon for shipping (manufacture)		35110	Pontoons construction (manufacture)
	35110	Beacons for ships (manufacture)		35110	Refitting of ships (manufacture)
	35110	Building of commercial vessels (manufacture)		35110	Research vessel (manufacture)
	35110	Building of ferry-boats (manufacture)		35110	Salvage vessel (manufacture)
	35110	Building of fish-processing factory vessels (manufacture)		35110	Sea going luxury yachts of 100 gross tons or more (manufacture)
	35110	Building of passenger vessels (manufacture)		36110	Seats for ships and floating structures (manufacture)
	35110	Bulk carrier (cargo ship) (manufacture)		35110	Sections for ships and floating structures (manufacture)
	35110	Buoys (not plastic) (manufacture)		35110	Shipbuilding (manufacture)
	35110	Buoys made of plastic (manufacture)		35110	Sludge vessel (manufacture)
	35110	Cable ship (manufacture)		35110	Submarine (manufacture)
	35110	Caisson (manufacture)		35110	Tanker (ship) (manufacture)
	35110	Cargo ship (manufacture)		35110	Trawler (manufacture)
	35110	Cementing of ships (manufacture)		35110	Tubular modules for oil rigs (manufacture)
	35110	Coffer-dam construction (manufacture)		35110	Tug (manufacture)
	35110	Construction of drilling platforms, floating or submersible (manufacture)		35110	Warship (manufacture)
	35110	Construction of non-recreational inflatable rafts		35110	Whaler (manufacture)
	35110	Container ship (manufacture)			
	35110	Conversion of ships (manufacture)	**30120**		**Building of pleasure and sporting boats**
	35110	Deck for oil platform (manufacture)		35120	Boat kits for assembly (manufacture)
	35110	Decking for ships (manufacture)		35120	Boatbuilding (manufacture)
	35110	Dredger (manufacture)		35120	Building of motor boats (manufacture)
	35110	Drilling ship (manufacture)		35120	Building of recreation-type hovercraft (manufacture)
	35110	Ferry manufacture		35120	Building of sailboats with or without auxiliary motor (manufacture)
	35110	Fishing boats (manufacture)		35120	Canal cruiser (manufacture)
	35110	Fishing vessel (manufacture)		35120	Canoe (manufacture)
	35110	Fleet tender (manufacture)		35120	Catamaran (manufacture)
	35110	Floating crane (manufacture)		35120	Cleat for pleasure boat (manufacture)
	35110	Floating docks construction (manufacture)		35120	Collapsible boat (not inflatable dinghy) (manufacture)
	35110	Floating harbour (manufacture)		35120	Dinghy made of rubber (manufacture)
	35110	Floating landing stages construction (manufacture)		35120	Folding boat made of rubber (manufacture)
	35110	Floating tanks construction (manufacture)		35120	Houseboat (manufacture)
	35110	Flotation vessel of steel, for oil platform (manufacture)		35120	Inflatable boats (manufacture)
	35110	Hovercraft (manufacture)		35120	Inflatable dinghy made of rubber (manufacture)
	35110	Hydrofoil (manufacture)		35120	Inflatable liferaft made of rubber (manufacture)
	35110	Jacket leg of steel plate for oil platform (manufacture)		35120	Inflatable motor boats (manufacture)
	35110	Landing stage (floating) (manufacture)		35120	Inflatable rafts (manufacture)
	35110	Lifeboat (manufacture)		35120	Kayak building (manufacture)
	35110	Lifebuoy made of rubber (manufacture)		35120	Leisure craft made of rubber (manufacture)
	35110	Liferaft (not rubber inflatable) (manufacture)		35120	Masts and spars for pleasure boats (manufacture)
	35110	Lighter (ship) (manufacture)		35120	Motor boats (manufacture)
	35110	Lightship (manufacture)		35120	Personal watercraft (manufacture)
	35110	Liner (ship) (manufacture)		35120	Pleasure and sporting boats (manufacture)
	35110	Masts and spars for ships (manufacture)		35120	Power boats of all types (manufacture)
	35110	Model ship made by shipbuilder (manufacture)		35120	Punt (manufacture)
	35110	Modules for oil platform (manufacture)		35120	Refitting of pleasure craft (manufacture)
	35110	Naval dockyard (shipbuilding and repairing) (manufacture)		35120	Rowing boat (manufacture)
	35110	Naval ships of all types (manufacture)		35120	Sailboat building (manufacture)

UK SIC 2007	UK SIC 2003	Activity
	35120	Sailing boats less than 100 gross tons (manufacture)
	35120	Skiff building (manufacture)
	35120	Yacht building (manufacture)
30200		**Manufacture of railway locomotives and rolling stock**
	35200	Airfield mechanical or electro-mechanical signalling, safety, traffic control equipment(manufacture)
	35200	Assembled rail sections (manufacture)
	35200	Axles and wheels for locomotive and rolling stock (manufacture)
	35200	Axles for rail and tramway vehicles (manufacture)
	35200	Axles made of steel for railway and tramway vehicles (manufacture)
	35200	Body for locomotive (manufacture)
	35200	Bogie for locomotive (manufacture)
	35200	Bogies (manufacture)
	35200	Brakes and parts of brakes for locomotive and rolling stock (manufacture)
	35200	Braking systems for locomotives (manufacture)
	35200	Buffers and buffer parts (manufacture)
	35200	Chassis for locomotive (manufacture)
	35200	Crane vans (manufacture)
	35200	Diesel electric locomotive (manufacture)
	35200	Diesel locomotive (manufacture)
	35200	Electric rail locomotives (manufacture)
	35200	Fog signalling equipment (manufacture)
	35200	Goods van for railways (manufacture)
	35200	Goods wagon for railways (manufacture)
	35200	Hooks and coupling devices (manufacture)
	35200	Inland waterways signalling, safety or traffic control equipment (manufacture)
	35200	Level crossing control gear (manufacture)
	35200	Locomotive (manufacture)
	35200	Locomotive parts and accessories (manufacture)
	35200	Luggage van for railway (manufacture)
	35200	Mining locomotives and mining rail cars (manufacture)
	35200	Passenger carriage for railways (manufacture)
	35200	Point locks (manufacture)
	35200	Post van for railways (manufacture)
	35200	Rail locomotives (manufacture)
	35200	Railbrakes (manufacture)
	35200	Railway and tramway coaches (manufacture)
	35200	Railway and tramway locomotives and specialised parts (manufacture)
	35200	Railway and tramway rolling stock (manufacture)
	36110	Railway car seats (manufacture)
	35200	Railway coach (manufacture)
	35200	Railway locomotives and rolling stock (manufacture)
	35200	Railway or tramway not self-propelled passenger coaches, goods vans and other wagons (manufacture)
	35200	Railway self-propelled car (manufacture)
	35200	Railway signalling equipment (mechanical) (manufacture)
	35200	Railway test wagon (manufacture)
	35200	Railway track equipment (mechanical) (manufacture)
	35200	Railway wagon (manufacture)
	35200	Railway wagon axle box and axle lubricator (manufacture)
	35200	Refrigerated wagon for railway (manufacture)

SIC 2007	SIC 2003	Activity
	35200	Rolling stock (manufacture)
	35200	Self-propelled railway car (manufacture)
	35200	Self-propelled railway or tramway coaches, vans, maintenance or service vehicles (manufacture)
	35200	Shock absorbers; bodies; etc. of railway or tramway locomotives or of rolling stock (manufacture)
	35200	Signal box equipment (manufacture)
	35200	Signalling equipment for airports, inland waterways, ports, roads and tramways (manufacture)
	31620	Signalling equipment for railways (electro mechanical) (manufacture)
	35200	Signalling, safety and traffic control equipment for railways (manufacture)
	31620	Signalling, safety or traffic control equipment (electrical) (manufacture)
	35200	Special purpose railway wagon (manufacture)
	35200	Steam rail locomotives (manufacture)
	35200	Tanker wagon for railway (manufacture)
	35200	Tenders (manufacture)
	35200	Test wagons for railway (manufacture)
	31620	Traffic control equipment for roads and inland waterways (manufacture)
	35200	Wagon and locomotive frames (manufacture)
	35200	Workshop wagon for railways (manufacture)
30300		**Manufacture of air and spacecraft and related machinery**
	35300	Aero engine manufacture (all types) (manufacture)
	35300	Aero engine parts and sub assemblies (manufacture)
	35300	Aeroplanes (manufacture)
	35300	Aerospace equipment (manufacture)
	35300	Air cushion vehicle (manufacture)
	35300	Aircraft (manufacture)
	35300	Aircraft brake (not brake lining) (manufacture)
	35300	Aircraft galley (manufacture)
	35300	Aircraft parts and sub assemblies (not electric) (manufacture)
	35300	Aircraft propeller (manufacture)
	36110	Aircraft seat (manufacture)
	35300	Airframe (manufacture)
	35300	Airframe parts and sub assemblies (not electric) (manufacture)
	35300	Airscrew (manufacture)
	35300	Airship (manufacture)
	35300	Anti-icing equipment and systems for aircraft (manufacture)
	35300	Auxiliary power unit for aircraft (manufacture)
	35300	Balloon (not toy) (manufacture)
	35300	Control surfaces for aircraft (manufacture)
	35300	De-icing equipment for aircraft (manufacture)
	35300	Dirigibles (manufacture)
	35300	Doors for aircraft (manufacture)
	35300	Ejector seat for aircraft (manufacture)
	35300	Engines for aircraft (manufacture)
	35300	Flight simulator (electronic) (manufacture)
	35300	Fuel tanks for aircraft (manufacture)
	35300	Fuselage for aircraft (manufacture)
	35300	Glider (manufacture)
	35300	Ground effect vehicles (manufacture)
	35300	Ground equipment for spacecraft (excluding electronic or telemetric equipment) (manufacture)

UK SIC 2007	UK SIC 2003	Activity	SIC 2007	SIC 2003	Activity
	35300	Ground flying trainers (manufacture)		35410	Handlebar for motorcycle (manufacture)
	35300	Hang glider (manufacture)		35410	Internal combustion engines for motorcycles (manufacture)
	35300	Helicopter (manufacture)		35410	Moped (manufacture)
	35300	Helicopter rotors for aircraft (manufacture)		35410	Motor scooter (manufacture)
	35300	Hot air balloons (manufacture)		35410	Motor tricycle and parts (manufacture)
	29600	Intercontinental ballistic missiles (ICBM) (manufacture)		35410	Motorcycle (manufacture)
	35300	Jet engine (manufacture)		35410	Motorcycle parts and accessories (manufacture)
	35300	Kite (not toy) (manufacture)		35410	Pillion seats for motorcycles (manufacture)
	35300	Landing gear for aircraft (manufacture)		35410	Saddles for motorcycles (manufacture)
	35300	Launch vehicle for spacecraft (manufacture)		35410	Sidecars for motorcycles (manufacture)
	35300	Motors and engines of a kind typically found on aircraft (manufacture)		35410	Wheels for motorcycles (manufacture)
	35300	Motors for aircraft (manufacture)	**30920**		**Manufacture of bicycles and invalid carriages**
	35300	Nacelles for aircraft (manufacture)		36639	Baby carriage (manufacture)
	35300	Nozzle for gas turbine aero engine (manufacture)		35430	Bath chair (manufacture)
	35300	Orbital stations (manufacture)		35420	Bicycles (non-motorised) (manufacture)
	35300	Overhaul and conversion of aircraft or aircraft engines (manufacture)		35420	Bicycles and parts (manufacture)
	35300	Planetary probes (manufacture)		35430	Body shell for invalid carriage (manufacture)
	35300	Power control for aircraft (manufacture)		35430	Chassis for powered invalid carriage (manufacture)
	35300	Propeller for aircraft (manufacture)		35420	Children's bicycles and tricycles (non-motorised) (manufacture)
	35300	Propeller rotor blades (manufacture)		36639	Children's carriage (manufacture)
	35300	Rocket (aerospace) (manufacture)		35420	Crank wheel for pedal cycle (manufacture)
	35300	Rocket motor (manufacture)		35420	Cycles (non-motorised) (manufacture)
	35300	Rotor blades for aircraft (manufacture)		35420	Cycles and parts (manufacture)
	35300	Safety belt or harness for aircraft crew or passengers (manufacture)		35420	Cyclometer (manufacture)
	35300	Sailplane (manufacture)		35420	Delivery tricycles (non-motorised) (manufacture)
	35300	Satellites (manufacture)		36639	Folding perambulator (manufacture)
	36110	Seats for aircraft (manufacture)		35420	Frame for pedal cycle (manufacture)
	35300	Space shuttles (manufacture)		35420	Frame for pedal tricycle (manufacture)
	35300	Spacecraft (manufacture)		35420	Free wheel for pedal cycle (manufacture)
	35300	Turbo-jets and parts for aircraft (manufacture)		35420	Gear for pedal cycle (manufacture)
	35300	Turbo-propellers and parts for aircraft (manufacture)		35420	Handlebar for pedal cycle (manufacture)
	35300	Wings for aircraft (manufacture)		35430	Invalid carriage (electrically propelled) (manufacture)
30400		**Manufacture of military fighting vehicles**		35430	Invalid carriage (manually propelled) (manufacture)
	29600	Armoured amphibious military vehicles (manufacture)		35430	Invalid carriage (power operated) (manufacture)
	29600	Bridgelayer (tracked military) (manufacture)		35430	Invalid carriage parts and accessories (manufacture)
	29600	Military fighting tanks (manufacture)		35420	Pedals for bicycles (manufacture)
	29600	Military fighting vehicles (manufacture)		36639	Perambulator (manufacture)
	29600	Personnel carrier (armoured fighting vehicle) (manufacture)		36639	Push chair (manufacture)
	29600	Recovery vehicle (tracked military type) (manufacture)		35420	Saddles for pedal cycles (manufacture)
	29600	Tanks (tracked armoured fighting vehicles) (manufacture)		35430	Steering column and gear for powered invalid carriage (manufacture)
	29600	Tanks and other fighting vehicles (manufacture)		35420	Tandems (manufacture)
	29600	Troop carrier (armoured) (manufacture)		35420	Tricycles (including delivery tricycles) (manufacture)
30910		**Manufacture of motorcycles**		35420	Tricycles and parts (manufacture)
	35410	Autocycle (manufacture)		35430	Wheel chair (manufacture)
	35410	Axle for motorcycle (manufacture)		35430	Wheelchair (manufacture)
	35410	Cycles fitted with an auxiliary engine (manufacture)		35420	Wheels for pedal cycles (manufacture)
	35410	Engines (internal combustion) for motorcycles (manufacture)	**30990**		**Manufacture of other transport equipment n.e.c.**
	34100	Engines for motorcycles (manufacture)		35500	Donkey-carts (manufacture)
	35410	Frame for motor tricycle (manufacture)		35500	Handcarts (manufacture)
	35410	Frame for motorcycle (manufacture)		35500	Hearses drawn by animals (manufacture)
	35410	Gear box for motorcycle (manufacture)		35500	Horse drawn trailer (manufacture)
				35500	Horse drawn truck (manufacture)
				35500	Luggage trucks (hand propelled) (manufacture)
				35500	Manually propelled trucks (manufacture)

UK SIC 2007	UK SIC 2003	Activity	SIC 2007	SIC 2003	Activity
	35500	Shopping carts (hand-propelled) (manufacture)		36120	Show case (manufacture)
	35500	Sledges (hand-propelled) (manufacture)		36120	Tables for office or school (manufacture)
	35500	Vehicles drawn by animals (manufacture)	**31020**		**Manufacture of kitchen furniture**
31010		**Manufacture of office and shop furniture**		36130	Cabinets for kitchens (manufacture)
	36120	Bookcase (non-domestic) (manufacture)		36130	Cupboards for kitchens (manufacture)
	36120	Cabinets (non-domestic) (manufacture)		36130	Dressers for kitchens (manufacture)
	36120	Chairs and seats for offices, workrooms, hotels, restaurants and public premises (manufacture)		36130	Furniture for kitchens (manufacture)
	36120	Coat stand (non-domestic) (manufacture)		36110	Kitchen seating (fitted) (manufacture)
	36120	Counters for shops (manufacture)	**31030**		**Manufacture of mattresses**
	35500	Decorative restaurant carts, dessert cart, food wagons (manufacture)		36150	Beds (mattress and mattress support) (manufacture)
	36120	Desks (manufacture)		36150	Box spring mattress (manufacture)
	36120	Display cases for shops (manufacture)		36150	Cot mattress (manufacture)
	36120	Drawers (non-domestic) (manufacture)		36150	Divan bed (mattress and mattress support) (manufacture)
	36120	Easel (manufacture)		36150	Interior sprung mattress (manufacture)
	36120	Filing cabinet (manufacture)		36150	Mattress base (manufacture)
	36120	Fittings and furnishing for hotels (manufacture)		36150	Mattress made of plastic foam (manufacture)
	36120	Fittings and furnishings for banks (manufacture)		36150	Mattress made of sponge (manufacture)
	36120	Fittings and furnishings for bars (manufacture)		36150	Mattress support made of metal (manufacture)
	36120	Fittings and furnishings for laboratories (manufacture)		36150	Spring wire mattress (manufacture)
	36120	Fittings and furnishings for libraries (manufacture)		36150	Upholstered base for mattress (manufacture)
	36120	Fittings and furnishings for museums (manufacture)	**31090**		**Manufacture of other furniture**
	36120	Fittings and furnishings for offices (manufacture)		36110	Armchair (manufacture)
	36120	Fittings and furnishings for public houses (manufacture)		36140	Bamboo furniture (other than seating) (manufacture)
	36120	Fittings and furnishings for restaurants (manufacture)		36110	Basket chair (manufacture)
	36120	Fittings and furnishings for shops (manufacture)		36140	Basket furniture (manufacture)
	36120	Furniture for churches (manufacture)		36140	Bed heads made of metal (manufacture)
	36120	Furniture for cinemas (manufacture)		36110	Bed settee (manufacture)
	36120	Furniture for drawing offices (manufacture)		36140	Bedsteads made of metal (manufacture)
	36120	Furniture for laboratories (manufacture)		36140	Bedsteads made of wood (manufacture)
	36120	Furniture for libraries (manufacture)		36110	Bench seat (manufacture)
	36120	Furniture for museums (manufacture)		36140	Bentwood furniture (manufacture)
	36120	Furniture for offices (manufacture)		36140	Bookcase made of wood (manufacture)
	36120	Furniture for public houses (manufacture)		36140	Bunk beds made of metal (manufacture)
	36120	Furniture for restaurants (manufacture)		36140	Bunk beds made of wood (manufacture)
	36120	Furniture for schools (manufacture)		36140	Cabinet case made of wood (manufacture)
	36120	Furniture for ships (manufacture)		36140	Cabinets for sewing machines (manufacture)
	36120	Furniture for shops (manufacture)		36140	Cabinets for televisions (manufacture)
	36120	Furniture for workrooms (manufacture)		36140	Camp furniture made of wood (manufacture)
	36120	Laboratory benches, stools, and other laboratory seating (manufacture)		36110	Cane chair (manufacture)
	36120	Laboratory furniture, cabinets and tables (manufacture)		36140	Cane furniture (manufacture)
	36120	Lectern (manufacture)		36110	Chair (non-upholstered) (manufacture)
	36120	Locker (manufacture)		36110	Chair (upholstered) (manufacture)
	36110	Office seating (manufacture)		36110	Chair frames made of wood (manufacture)
	36120	Partitions (non-domestic) (manufacture)		36110	Chair seating made of cane or wicker (manufacture)
	36120	Pulpit (manufacture)		36110	Chairs made of metal (manufacture)
	36120	Racking (manufacture)		36110	Chairs made of plastic (manufacture)
	36120	Screens (non-domestic) (manufacture)		36110	Chaise longue (manufacture)
	36110	Seating for office or school (manufacture)		36140	Chest of drawers (manufacture)
	36110	Seats for cinema (manufacture)		36140	Cocktail cabinet (manufacture)
	36110	Seats for theatre (manufacture)		36140	Coffee table (manufacture)
	36120	Shelving (non-domestic) (manufacture)		36110	Convertible furniture (manufacture)
	36120	Shop fixtures for display and storage of goods (manufacture)		36140	Cot frame (manufacture)
				36110	Couch (manufacture)
				36140	Cupboard (manufacture)
				36110	Deck chairs made of wood (manufacture)

UK SIC 2007	UK SIC 2003	Activity
	36110	Dining chair (manufacture)
	36140	Dining table (manufacture)
	36140	Display cabinet for domestic use (manufacture)
	36140	Dressing table (manufacture)
	36110	Finishing of chairs and seats
	36140	Fire screen (manufacture)
	36140	Folding bed (manufacture)
	36140	Frames for mattress support made of wood (manufacture)
	36140	French polishing (manufacture)
	36140	Fume cupboards (manufacture)
	36140	Furniture components made of wood (manufacture)
	36140	Furniture finishing (except chairs and seats) (manufacture)
	36140	Furniture for bedrooms (other than mattresses and mattress supports) (manufacture)
	36140	Furniture for gardens (manufacture)
	36140	Furniture for living rooms (manufacture)
	36140	Furniture for nurseries (manufacture)
	36140	Furniture kit (manufacture)
	36140	Furniture made of bamboo (other than seating) (manufacture)
	36140	Furniture parts made of wood (manufacture)
	36110	Garden chairs (manufacture)
	36110	Garden seating (manufacture)
	36140	Hall stand (manufacture)
	36140	Ice box made of wood (manufacture)
	36140	Ice chest made of wood (manufacture)
	36140	Insulated cabinets made of wood (manufacture)
	36140	Lacquering, varnishing and gilding of furniture (manufacture)
	36110	Metal framed upholstery for seating (manufacture)
	36140	Other plastic furniture (manufacture)
	36110	Ottoman (manufacture)
	36140	Outdoor furniture (non-upholstered) (manufacture)
	36140	Outdoor furniture made of metal (manufacture)
	36110	Outdoor seating (manufacture)
	36140	Painting of furniture (manufacture)
	36110	Pew (manufacture)
	36110	Plastic shell upholstery (manufacture)
	36140	Room divide system (manufacture)
	36140	Rustic furniture (manufacture)
	36110	Seating (manufacture)
	36110	Seating made of metal (not for road vehicle or aircraft) (manufacture)
	36110	Settee (manufacture)
	36140	Sideboard (manufacture)
	36110	Sofa (manufacture)
	36110	Sofa sets (manufacture)
	36110	Sofabeds (manufacture)
	36140	Spraying of furniture (manufacture)
	36140	Storage cabinets for domestic use (manufacture)
	36110	Studio couch (manufacture)
	36140	Tables for domestic use (manufacture)
	36140	Unit furniture (non-upholstered) (manufacture)
	36110	Unit seating for domestic use (upholstered) (manufacture)
	36140	Upholstered furniture (other than chairs and seats) (manufacture)
	36110	Upholsterer (not repair and maintenance) (manufacture)

SIC 2007	SIC 2003	Activity
	36110	Upholstery of chairs and seats (manufacture)
	36140	Wall unit (manufacture)
	36140	Wardrobes (manufacture)
	36140	Wardrobes made of metal (manufacture)
	36140	Wicker furniture (manufacture)
	36140	Work benches made of wood (manufacture)
	36140	Woven fibre furniture (manufacture)
32110		**Striking of coins**
	36210	Coin striking (manufacture)
	36210	Coins (manufacture)
	36210	Royal Mint (manufacture)
32120		**Manufacture of jewellery and related articles**
	36220	Articles for religious use of base metals clad with precious metals (manufacture)
	36220	Cutlery made of precious metal (manufacture)
	36220	Diamond cutting (manufacture)
	36220	Diamond working (manufacture)
	36220	Dinnerware flatware hollowware of base metals clad with precious metals (manufacture)
	36220	Engraving (personalised) on precious metal (manufacture)
	36220	Engraving of personal non-precious metal products (manufacture)
	36220	Findings and stampings made of precious metals for jewellery (manufacture)
	36220	Gold and silver braid (manufacture)
	36220	Gold and silver embroidery (manufacture)
	36220	Gold and silver mounting (manufacture)
	36220	Gold laceman (manufacture)
	36220	Gold leaf (manufacture)
	36220	Goldsmiths' articles (manufacture)
	36220	Goldsmiths' articles of base metals clad with precious metals (manufacture)
	36220	Goldsmiths' articles of precious metals (manufacture)
	36220	Industrial quality stones (manufacture)
	36220	Jet ornaments and jewellery (manufacture)
	36220	Jewellery (gold or silver plated) (manufacture)
	36220	Jewellery engraving (not distributive trades) (manufacture)
	36220	Jewellery made of platinum (manufacture)
	36220	Jewellery made of precious metal (manufacture)
	36220	Jewellery made of semi-precious stones (manufacture)
	36220	Jewellery of base metal clad with precious metal (manufacture)
	36220	Jewellery polishing (manufacture)
	36220	Jewellery with precious stones (manufacture)
	36220	Office or desk articles of base metals clad with precious metals (manufacture)
	36220	Ornaments made of precious metal (manufacture)
	36220	Ornaments that are gold or silver plated (manufacture)
	36220	Pearl drilling (manufacture)
	36220	Pearl stringing (manufacture)
	36220	Pearls production (manufacture)
	36220	Precious and semi-precious stones in the worked state production (manufacture)
	36220	Precious stone cutting (manufacture)
	36220	Silver burnishing (manufacture)

UK SIC 2007	UK SIC 2003	Activity	SIC 2007	SIC 2003	Activity
	36220	Silversmiths' work (manufacture)		36300	Metronome (electronic or mechanical) (manufacture)
	36220	Synthetic precious and semi-precious stones (manufacture)		36300	Mouth blown signalling instruments (manufacture)
	36220	Tableware made of precious metal (manufacture)		36300	Mouth organ (manufacture)
	36220	Teapots made of precious metal (manufacture)		36300	Music box mechanisms (manufacture)
	36220	Toilet articles of base metals clad with precious metals (manufacture)		36300	Musical box (manufacture)
	33500	Watchbands, wristbands and watch straps, of precious metal (manufacture)		36300	Musical instrument parts and accessories (manufacture)
	36220	Worked pearls (manufacture)		36300	Musical instrument tuners (electronic) (manufacture)
32130		**Manufacture of imitation jewellery and related articles**		36300	Musical instruments including electronic (manufacture)
	36610	Bags made of chain (manufacture)		36300	Organ tuning (manufacture)
	36610	Britannia metalware (manufacture)		36300	Percussion instrument (manufacture)
	36610	Buhl cutting (manufacture)		36300	Piano (manufacture)
	36610	Costume jewellery (manufacture)		36300	Pipe organ (manufacture)
	36610	Costume or imitation jewellery (manufacture)		36300	Pitch pipes (manufacture)
	36610	Cuff link (not of, or clad in, precious metal, or of precious/semi precious stones) (manufacture)		36300	Recorders made of plastic or wood (manufacture)
	36610	Fashion jewellery (manufacture)		36300	Reed for musical instrument (manufacture)
	36610	Findings and stampings made of base metal for jewellery (manufacture)		36300	Rolls for automatic mechanical instruments (manufacture)
	36610	Gilt (manufacture)		36300	Stringed instruments (manufacture)
	36610	Imitation jewellery (manufacture)		36300	Strings for musical instruments (manufacture)
	36610	Imitation pearls (manufacture)		36300	Tuning fork (manufacture)
	36610	Jewellery (gilded and silvered) (manufacture)		36300	Viola (manufacture)
	36610	Jewellery containing imitation gem stones (manufacture)		36300	Violin (manufacture)
	36610	Jewellery made of ceramic (manufacture)		36300	Whistles (manufacture)
	36610	Rings, bracelets, necklaces made from base metals plated with precious metals (manufacture)		36300	Wind instrument (manufacture)
	36610	Tie pin (not of, or clad in, precious metal, or of precious/semi precious stones) (manufacture)		36300	Woodwind instruments (manufacture)
	33500	Watch straps, bands and bracelets made of non-precious metal (manufacture)	**32300**		**Manufacture of sports goods**
32200		**Manufacture of musical instruments**		36400	Archery equipment (manufacture)
	36300	Accordion (manufacture)		36400	Athletic equipment (manufacture)
	36300	Automatic pianos (manufacture)		36400	Badminton shuttlecock (manufacture)
	36300	Bagpipes and reeds (manufacture)		36400	Balls for all sports (finished) (manufacture)
	36300	Call horns (manufacture)		36400	Basins for swimming and paddling pools (manufacture)
	36300	Calliopes (manufacture)		36400	Bats (manufacture)
	36300	Cards for automatic mechanical instruments (manufacture)		36400	Bowls and bowls equipment (manufacture)
	36300	Cello (manufacture)		36400	Bows (manufacture)
	36300	Concertina (manufacture)		36400	Boxing glove (manufacture)
	36300	Discs for automatic mechanical instruments (manufacture)		36400	Climbing frame (manufacture)
	36300	Double bass (manufacture)		36400	Clubs (manufacture)
	36300	Drum (musical instrument) (manufacture)		36400	Cricket ball and equipment (manufacture)
	36300	Electronic musical instrument (manufacture)		36400	Crossbows (manufacture)
	36300	Fairground organs (manufacture)		36400	Fish hook (manufacture)
	36300	Guitar (manufacture)		36400	Fishing tackle (manufacture)
	36300	Harmoniums (manufacture)		36400	Fitness centre equipment and appliances (manufacture)
	36300	Harmoniums with free metal reeds (manufacture)		36400	Fly dressing (manufacture)
	36300	Harpsichord (manufacture)		36400	Football case made of leather (manufacture)
	36300	Horns (musical) (manufacture)		36400	Golf ball (finished) (manufacture)
	36300	Keyboard instruments (manufacture)		36400	Golf club (manufacture)
	36300	Keyboard pipe organs with free metal reeds (manufacture)		36400	Gymnasium equipment and appliances (manufacture)
	36300	Keyboard stringed instruments (manufacture)		36400	Headgear for sports (manufacture)
				36400	Hockey stick (manufacture)
				36400	Hunting requisites (manufacture)
				36400	Ice-skates (manufacture)
				36400	Landing nets (manufacture)
				36400	Mountaineering equipment (manufacture)
				36400	Nursery equipment (manufacture)

UK SIC 2007	UK SIC 2003	Activity	SIC 2007	SIC 2003	Activity
	36400	Playground equipment (manufacture)		36509	Dolls' clothes (manufacture)
	36400	Racket and racket frames (manufacture)		36509	Dolls' cots (manufacture)
	36400	Rock climbing equipment (manufacture)		36509	Dolls' houses (manufacture)
	36400	Roller skates (manufacture)		36509	Dolls made of rubber (manufacture)
	36400	Sailboards (manufacture)		36509	Dolls' prams (manufacture)
	36400	Skateboards (manufacture)		36509	Draughts set (manufacture)
	36400	Ski bindings and poles (manufacture)		36509	Games and toys (manufacture)
	36400	Ski-boots (manufacture)		36509	Indoor game (manufacture)
	36400	Skiing equipment (manufacture)		36509	Jigsaw puzzle (manufacture)
	36400	Skis (manufacture)		36509	Model kit (manufacture)
	36400	Sport fishing requisites (manufacture)		36509	Models for recreational use (manufacture)
	36400	Sports and outdoor and indoor games, articles and equipment (manufacture)		36509	Plastic bicycles and tricycles designed to be ridden (manufacture)
	36400	Sports equipment made of plastic (manufacture)		36509	Plastic game (manufacture)
	36400	Sports gloves (specialist) (manufacture)		36509	Playballs made of rubber (manufacture)
	36400	Sports goods (manufacture)		36509	Playing cards (manufacture)
	36400	Squash racket (manufacture)		36509	Puppet (manufacture)
	36400	Surfboard (manufacture)		36509	Push toy on wheels (manufacture)
	36400	Swings for playgrounds (manufacture)		36509	Puzzles (manufacture)
	36400	Table tennis ball (manufacture)		36509	Reduced-size (scale) models (manufacture)
	36400	Table tennis equipment (manufacture)		36509	Reduced-size scale model electrical trains (manufacture)
	36400	Tennis balls (finished) (manufacture)		36509	Reduced-size scale models construction sets (manufacture)
	36400	Tennis racket (manufacture)			
	36400	Water sports equipment (manufacture)		36509	Scooters for children (manufacture)
	36400	Waterwings made of rubber (manufacture)		36509	Soft toys (manufacture)
				36509	Stuffed toy (manufacture)
32401		**Manufacture of professional and arcade games and toys**		36509	Table or parlour games (manufacture)
	36501	Amusement machines (manufacture)		36509	Technical toy (manufacture)
	36501	Billiard ball (manufacture)		36509	Toy animal (manufacture)
	36501	Billiard cue (manufacture)		36509	Toy balloon (manufacture)
	36501	Billiard table (manufacture)		36509	Toy car circuit (electric) (manufacture)
	36501	Coin operated games (manufacture)		36509	Toy cars (electric) (manufacture)
	36501	Cue for billiards or snooker (manufacture)		36509	Toy cars (pedal) (manufacture)
	36501	Dart (manufacture)		36509	Toy furniture (manufacture)
	36501	Dartboard (manufacture)		36509	Toy guns (not operated by compressed air) (manufacture)
	36501	Funfair articles (manufacture)			
	36501	Funfair games (manufacture)		36509	Toy musical instruments including electronic (manufacture)
	36501	Games for professional and arcade use (manufacture)		36509	Toy perambulators and pushchairs (manufacture)
	36501	Gaming (automatic slot) machines (manufacture)		36509	Toy push cart (manufacture)
	36501	Pin-tables (manufacture)		36509	Toy trains (electric) (manufacture)
	36501	Tables for casino games (manufacture)		36509	Toy wheelbarrow (manufacture)
	36501	Toys for professional and arcade use (manufacture)		36509	Toys (mechanical) (manufacture)
				36509	Toys and games (electronic) with fixed (non replaceable software) (manufacture)
32409		**Manufacture of games and toys (other than professional and arcade games and toys) n.e.c.**		36509	Toys and games made of paper (manufacture)
	36509	Action figures (manufacture)		36509	Toys and games made of wood (manufacture)
	36509	Bagatelle board (manufacture)		36509	Toys made of cardboard (manufacture)
	36509	Bicycles for children (manufacture)		36509	Toys made of metal (manufacture)
	36509	Board game (manufacture)		36509	Toys made of plastic (manufacture)
	36509	Boxed game (manufacture)		36509	Toys made of rubber (manufacture)
	36509	Chess (electronic) (manufacture)		36509	Tricycles for children (manufacture)
	36509	Chess set (manufacture)		36509	Video games machines (manufacture)
	36509	Construction model (manufacture)		36509	Wheeled toys designed to be ridden (manufacture)
	36509	Construction sets (manufacture)			
	36509	Constructional toy (manufacture)	**32500**		**Manufacture of medical and dental instruments and supplies**
	36509	Dolls (manufacture)		24422	Absorbable haemostatics (manufacture)
	36509	Dolls and doll garments, parts and accessories (manufacture)		24422	Adhesive plaster and surgical bandage (manufacture)
				33100	Anaesthetic equipment (manufacture)

UK SIC 2007	UK SIC 2003	Activity	SIC 2007	SIC 2003	Activity
	33100	Artificial eye (manufacture)		33401	Lens (unmounted) (manufacture)
	33100	Artificial limb (manufacture)		33100	Lithotriptors (manufacture)
	33100	Artificial parts for the heart (manufacture)		33100	Massage apparatus (manufacture)
	33100	Artificial respiration equipment (manufacture)		33100	Mechano-therapy appliances (manufacture)
	33100	Artificial teeth (manufacture)		33100	Medical and dental instruments and supplies (manufacture)
	33100	Aseptic hospital furniture (manufacture)		33100	Medical appliances (manufacture)
	33100	Audiometers (manufacture)		33100	Medical instrument (non-optical) (manufacture)
	33100	Autoclaves (manufacture)		33100	Medical nucleonic apparatus (manufacture)
	33100	Bone plates and screws (manufacture)		25130	Medical rubber dressings (manufacture)
	24422	Bone reconstruction cements (manufacture)		25130	Medical rubber goods (not dressings) (manufacture)
	33100	Breathing apparatus for diving (manufacture)		33202	Medical thermometers (manufacture)
	33100	Bridges made in dental labs (manufacture)		33100	Medical, surgical, dental or veterinary examination tables (manufacture)
	33100	Cannulae (manufacture)		33100	Medical, surgical, dental or veterinary operating tables (manufacture)
	33100	Catheter (manufacture)			
	33100	Cautery and light unit (manufacture)		24422	Medicated dressings (manufacture)
	29240	Centrifuge (laboratory type) (manufacture)		33100	Mirrors for medical use (manufacture)
	33401	Colour filter (unmounted) (manufacture)		33401	Monocle (manufacture)
	33401	Contact lens (manufacture)		33100	Myograph (manufacture)
	33401	Corrective glasses (manufacture)		33100	Needles used in medicine (manufacture)
	24422	Cotton wool and tissues (manufacture)		33100	Nuclear magnetic resonance apparatus (manufacture)
	33100	Crutches (manufacture)		33100	Nucleonic medical apparatus (manufacture)
	33100	Cutter for dental use (manufacture)		33100	Operating tables (manufacture)
	33100	Dental brush (manufacture)		33401	Ophthalmic eyeglasses (manufacture)
	24422	Dental cement (manufacture)		33100	Ophthalmic instrument (manufacture)
	33100	Dental chair (manufacture)		33401	Ophthalmic lenses ground to prescription, safety goggles (manufacture)
	33100	Dental drill engines (manufacture)			
	24422	Dental filling (manufacture)		33401	Optical element (mounted, (not photographic)) (manufacture)
	33100	Dental instrument (manufacture)			
	29240	Dental laboratory furnaces (manufacture)		33401	Optical element (unmounted) (manufacture)
	33100	Dental laboratory instruments and equipment (manufacture)		33100	Orthopaedic appliances (not footwear) (manufacture)
	33100	Dental mirror (manufacture)		33100	Orthopaedic footwear (manufacture)
	33100	Dental surgical instruments and equipment (manufacture)		33100	Orthopedic and prosthetic devices (manufacture)
	24422	Dental wax and other dental plaster preparations (manufacture)		33100	Oxygen breathing equipment for medical use (manufacture)
	33100	Denture (manufacture)		33100	Oxygen therapy apparatus (manufacture)
	33100	Diathermy apparatus (manufacture)		33100	Ozone therapy apparatus (manufacture)
	33100	Dissecting instrument (manufacture)		24422	Pharmaceutical non-medicament products (manufacture)
	33100	Foot support (manufacture)			
	33100	Furniture for medical, surgical, dental or veterinary use (manufacture)		24422	Plaster bandages (manufacture)
				33401	Polarising elements (manufacture)
	24422	Gauze (surgical) (manufacture)		33401	Prisms (mounted, (not photographic)) (manufacture)
	33100	Glass eyes (manufacture)		33401	Prisms (unmounted) (manufacture)
	33401	Goggles (manufacture)		33401	Protective glasses (manufacture)
	33401	Grating (mounted, (not photographic)) (manufacture)		33100	Psychological testing apparatus (manufacture)
	33401	Grating (unmounted, optical) (manufacture)		33100	Psychology testing apparatus (manufacture)
	33100	Hospital beds with mechanical fittings (manufacture)		33401	Reading glasses (manufacture)
	33100	Hyperbaric chambers (manufacture)		33100	Reflectors used in medicine (manufacture)
	33100	Hypodermic syringe and equipment (manufacture)		33100	Respirator and mask for medical use (manufacture)
	33100	Instep support (manufacture)		33100	Resuscitation equipment (manufacture)
	33100	Instruments and apparatus used for medical, surgical, dental or veterinary purposes (manufacture)		25130	Rubber gloves (medical) (manufacture)
				33100	Scintigraphy apparatus (manufacture)
	29240	Laboratory sterilisers (manufacture)		24422	Sheep and cattle dressings (manufacture)
	29240	Laboratory type distilling apparatus (manufacture)		17403	Shroud and cerement (manufacture)
	29240	Laboratory ultrasonic cleaning machinery (manufacture)		33401	Spectacle frames (manufacture)
	24422	Laminaria (manufacture)		33401	Spectacle lens (manufacture)
	33100	Laser surgical apparatus (manufacture)		33401	Spectacle mounts (manufacture)
	33401	Lens (mounted, (not photographic)) (manufacture)			

255

UK SIC 2007	UK SIC 2003	Activity	SIC 2007	SIC 2003	Activity
	33401	Spectacles (manufacture)		36620	Mops for household use (manufacture)
	33100	Splints (manufacture)		36620	Nail brush (manufacture)
	33100	Sterilising equipment for medical use (manufacture)		36620	Paint brush (manufacture)
	33100	Sterilizers (manufacture)		36620	Paint pads (manufacture)
	24422	Sticking plaster (surgical) (manufacture)		36620	Paste brush (manufacture)
	33401	Sunglasses (manufacture)		36620	Pastry brush (manufacture)
	24422	Surgical bandage (manufacture)		36620	Plastic brush (complete) (manufacture)
	33100	Surgical belts (manufacture)		36620	Polishing mop (manufacture)
	33100	Surgical boot (manufacture)		36620	Rollers for paint (manufacture)
	33100	Surgical corset (manufacture)		36620	Scrubbing brush (manufacture)
	17403	Surgical drapes and sterile string and tissue (manufacture)		36620	Shaving brush (manufacture)
	24422	Surgical dressing (manufacture)		36620	Shoe brush (manufacture)
	33100	Surgical equipment (manufacture)		36620	Squeegees (manufacture)
	24422	Surgical gauze (manufacture)		36620	Toilet brush (manufacture)
	25130	Surgical goods made of rubber (manufacture)		36620	Tooth brush (not electric) (manufacture)
	24422	Surgical gut string (manufacture)		36620	Whitewash brush (manufacture)
	33100	Surgical hosiery (manufacture)		36620	Wire brush (manufacture)
	33100	Surgical implants (manufacture)			
	33100	Surgical instrument (manufacture)	**32990**		**Other manufacturing n.e.c.**
	24422	Surgical lint (manufacture)		36639	Amber turning (manufacture)
	24422	Surgical sutures (manufacture)		36639	Artificial flowers and fruit made of paper (manufacture)
	33100	Surgical truss (manufacture)		36639	Artificial flowers and fruit made of plastic (manufacture)
	24422	Surgical wadding (manufacture)		36639	Artificial flowers and fruit made of textiles (manufacture)
	33100	Syringes (manufacture)		36631	Ballpoint pen and refill (manufacture)
	33100	Traction or suspension devices for medical beds (manufacture)		36639	Bedfolder (manufacture)
	33100	Transfusion apparatus (manufacture)		36639	Bladder dressing (manufacture)
	33100	Transfusion pods (manufacture)		36639	Boiler covering (not asbestos or slag wool) (manufacture)
	33100	Urine bottle holders and other accessories for medical beds (manufacture)		36639	Boiler packing (not asbestos or slag wool) (manufacture)
	33100	Vascular prostheses (manufacture)		36639	Bone working (manufacture)
	33100	Veterinary equipment (manufacture)		36639	Briar pipe (manufacture)
	33100	Zimmer frames and other walking aids (manufacture)		36639	Buttons (manufacture)
				36639	Buttons made of glass (manufacture)
32910		**Manufacture of brooms and brushes**		36639	Candle (manufacture)
	36620	Artists' brush (manufacture)		36631	Carbon ribbon (manufacture)
	36620	Besom (manufacture)		36639	Carnival article (manufacture)
	36620	Birch broom (manufacture)		36639	Carry cot (manufacture)
	36620	Bristle dressing for brushes (manufacture)		36631	Cartridge refill for fountain pen (manufacture)
	36620	Broom (manufacture)		36639	Catgut (manufacture)
	36620	Brooms and brushes for household use (manufacture)		36631	Chalk for drawing or writing (manufacture)
	36620	Brush (manufacture)		36639	Cigarette lighter (manufacture)
	36620	Brush (not electrical) for machines (manufacture)		20510	Coffin board (manufacture)
	36620	Brush for cosmetics (manufacture)		20510	Coffins (manufacture)
	36620	Clothes brush (manufacture)		36639	Collar stud (manufacture)
	36620	Distemper brush (manufacture)		36639	Combs (other than of hard rubber, plastic or metal) (manufacture)
	36620	Feather duster (manufacture)		36639	Conjuring apparatus (manufacture)
	36620	Fibre dressing for brushes (manufacture)		20520	Cork life preservers (manufacture)
	36620	Floor sweepers (hand operated mechanical) (manufacture)		36631	Crayon (manufacture)
	36620	Flue brush (manufacture)		18243	Cut, make, trim of fire-resistant and protective safety clothing, fee or contract basis (manufacture)
	36620	Hair brush (manufacture)		36639	Cutlery handles made of horn, ivory, tortoise shell, etc. (manufacture)
	36620	Hair dressing for brushes (manufacture)		36631	Date sealing stamps (manufacture)
	36620	Hearth brush (manufacture)		36631	Date stamp and accessories (manufacture)
	36620	Industrial broom and mop (manufacture)		36639	Devotional article (manufacture)
	36620	Industrial brush (manufacture)			
	36620	Laundry brush (manufacture)			
	36620	Mop (manufacture)			

UK SIC 2007	UK SIC 2003	Activity	SIC 2007	SIC 2003	Activity
	36400	Ear and noise plugs (e.g. for swimming and noise protection) (manufacture)		36639	Nightlight (manufacture)
	36631	Embossing devices (hand operated) for labels (manufacture)		36631	Numbering stamps (manufacture)
	36639	False beard (manufacture)		36639	Parasol (manufacture)
	36639	False eyebrow (manufacture)		36631	Pastel (manufacture)
	36639	Feather curling (manufacture)		36631	Pen nibs (manufacture)
	36639	Feather ornament (manufacture)		36631	Pencil (manufacture)
	36639	Feather purifying (manufacture)		36631	Pencil leads (manufacture)
	36639	Feather sorting (manufacture)		36631	Penholder (manufacture)
	36631	Felt tipped pen (manufacture)		36631	Pens for writing or drawing (manufacture)
	36631	Fibre tipped pen (manufacture)		28750	Personal safety devices of metal (manufacture)
	18100	Fire resistant and protective safety clothing of leather (manufacture)		36639	Plaster cast (manufacture)
	18249	Fire-fighting protection suits (manufacture)		36631	Prepared typewriter ribbons (manufacture)
	36639	Firelighter (manufacture)		36639	Press-fasteners (manufacture)
	18249	Fire-resistant and protective safety clothing (manufacture)		36639	Press-studs (manufacture)
	36639	Flint for lighters (manufacture)		36631	Printing devices (hand operated) (manufacture)
	36631	Fountain pen (manufacture)		36631	Propelling pencil (manufacture)
	36631	Fountain pen nib (manufacture)		18249	Protective gloves for industrial use (manufacture)
	33100	Gas masks (manufacture)		18241	Protective headgear (manufacture)
	18249	Gauntlet (protective) (manufacture)		18241	Protective headgear for industrial use (manufacture)
	22110	Globes (manufacture)		36631	Ribbon (inked) (manufacture)
	36639	Gut for musical instruments and sports goods (manufacture)		18241	Riding caps (manufacture)
	36639	Gut scraping and spinning (manufacture)		36631	Roller pens and refills (manufacture)
	36639	Hair pad making (manufacture)		28750	Safety headgear made of metal (manufacture)
	36639	Hair preparation for wig making (manufacture)		25240	Safety helmets made of plastic (manufacture)
	36639	Hair slides (manufacture)		36639	Scent sprays (manufacture)
	36631	Hand printing sets (manufacture)		36639	Scientific models for educational and exhibition purposes (manufacture)
	36639	Hand riddles (manufacture)		36631	Sealing stamps (manufacture)
	36639	Hand sieves (manufacture)		36631	Seals for use with sealing wax (manufacture)
	25240	Hard hats and other personal safety equipment of plastics (manufacture)		36639	Seat-sticks (manufacture)
	36639	Horn and tortoise shell working (manufacture)		36631	Slates for writing (manufacture)
	36639	Horn pressing (manufacture)		36639	Slide fasteners (manufacture)
	18241	Industrial protective headgear (manufacture)		36639	Smokers' requisites (manufacture)
	36631	Ink pad (manufacture)		36639	Smoking pipes (manufacture)
	36639	Instruments for educational or exhibition purposes (manufacture)		36639	Snap fasteners (manufacture)
	36639	Ivory working (manufacture)		36639	Sponge bleaching (manufacture)
	36639	Jokes and novelties (manufacture)		36639	Sponge trimming (manufacture)
	25130	Life vests (manufacture)		36631	Stamps made of rubber (manufacture)
	25130	Lifebelts (manufacture)		36631	Stylographic pen (manufacture)
	20520	Lifebelts made of cork (manufacture)		36639	Sun car (manufacture)
	20520	Lifebuoy made of cork (manufacture)		36639	Sunshade (manufacture)
	25130	Lifejacket (manufacture)		36639	Sun-umbrellas (manufacture)
	20520	Lifejacket made of cork (manufacture)		36631	Tailors' chalk (manufacture)
	36639	Lighter fuel in containers not exceeding 300 cc (liquid or liquefied gas) (manufacture)		36639	Tailors' dummy (not plastic) (manufacture)
	19200	Linemen's safety belts and other belts for occupational use (manufacture)		36639	Tapers and the like (manufacture)
				36639	Taxidermy activities (manufacture)
	36631	Marker pen (manufacture)		36639	Teaching aids (electronic) (manufacture)
	36639	Models for educational or exhibition purposes (manufacture)		36639	Toothpicks made of bone (manufacture)
	36639	Models for geographical use made of wax or plaster (manufacture)		36639	Trainer (electronic training equipment) (manufacture)
				36631	Typewriter ribbons (manufacture)
	36639	Models made of plaster (manufacture)		36639	Umbrella (manufacture)
	36639	Models made of wax (manufacture)		18241	Uniform helmets (manufacture)
	36639	Natural sponge preparation (manufacture)		36639	Vacuum flask (complete) (manufacture)
				36639	Vacuum jar (manufacture)
				36639	Vacuum vessels for personal or household use (manufacture)
				36639	Walking sticks (manufacture)
				36639	Whalebone cutting and splitting (manufacture)

UK SIC 2007	UK SIC 2003	Activity	SIC 2007	SIC 2003	Activity
	36639	Wig (manufacture)		29523	Repair and maintenance of construction machinery (except earth moving type) (manufacture)
	36631	Writing instrument sets (manufacture)		29522	Repair and maintenance of earth-moving and excavating equipment (manufacture)
33110		**Repair of fabricated metal products**		29110	Repair and maintenance of engines and turbines (except aircraft, vehicle and cycle) (manufacture)
	28520	Mobile welding repair of fabricated metal products (not of machinery) (manufacture)		29523	Repair and maintenance of equipment for concrete crushing and screening and roadworks (manufacture)
	28220	Platework repair of central heating boilers and radiators (manufacture)		29122	Repair and maintenance of fluid power machinery (compressors) (manufacture)
	28520	Repair and maintenance for pipes and pipelines (manufacture)		29121	Repair and maintenance of fluid power machinery (pumps) (manufacture)
	28300	Repair and maintenance of auxiliary plant for steam collectors and accumulators (manufacture)		29320	Repair and maintenance of forestry and logging machinery (manufacture)
	28300	Repair and maintenance of auxiliary plant for use with steam condensers (manufacture)		29210	Repair and maintenance of furnaces and furnace burners (manufacture)
	28300	Repair and maintenance of auxiliary plant for use with steam economisers (manufacture)		29110	Repair and maintenance of gas turbines (manufacture)
	28300	Repair and maintenance of auxiliary plant for use with steam generators (manufacture)		29240	Repair and maintenance of general purpose machinery n.e.c. (manufacture)
	28300	Repair and maintenance of auxiliary plant for use with steam superheaters (manufacture)		29210	Repair and maintenance of industrial process furnaces (manufacture)
	34202	Repair and maintenance of containers for freight (manufacture)		29230	Repair and maintenance of industrial refrigeration equipment, air purifying equipment (manufacture)
	29600	Repair and maintenance of firearms and ordnance (manufacture)		29230	Repair and maintenance of industrial type air conditioning (manufacture)
	28220	Repair and maintenance of non-domestic central heating boilers (manufacture)		29220	Repair and maintenance of lifting and handling equipment (manufacture)
	28300	Repair and maintenance of nuclear reactors (manufacture)		29220	Repair and maintenance of lifting, handling equipment, elevators, moving walkways etc. (manufacture)
	28300	Repair and maintenance of parts for marine or power boilers (manufacture)		29220	Repair and maintenance of lifts and escalators (not in buildings or civil engineering) (manufacture)
	35500	Repair and maintenance of shopping carts (manufacture)		29560	Repair and maintenance of machinery for bookbinding (manufacture)
	29600	Repair and maintenance of sporting and recreational guns (manufacture)		29530	Repair and maintenance of machinery for food, beverage and tobacco processing (manufacture)
	28300	Repair and maintenance of steam generators (manufacture)		29510	Repair and maintenance of machinery for metallurgy (manufacture)
	28300	Repair and maintenance of steam or other vapour generators (manufacture)		29521	Repair and maintenance of machinery for mining (manufacture)
	28210	Repair and maintenance of tanks, reservoirs and containers of metal (manufacture)		29550	Repair and maintenance of machinery for paper and paperboard production (manufacture)
	29600	Repair and maintenance of weapons and weapon systems (manufacture)		29560	Repair and maintenance of machinery for printing (manufacture)
	28610	Repair of cutlery (manufacture)		29540	Repair and maintenance of machinery for textile, apparel and leather production (manufacture)
	28630	Repair of locks and hinges (manufacture)		29560	Repair and maintenance of machinery for working rubber or plastics (manufacture)
	28110	Repair of metal structures (manufacture)		29110	Repair and maintenance of marine engines (manufacture)
	28710	Repair of steel shipping drums (manufacture)		29420	Repair and maintenance of metal cutting or metal forming machine tools and accessories (manufacture)
	28620	Repair of tools (manufacture)		29420	Repair and maintenance of metal working machine tools (manufacture)
33120		**Repair of machinery**		29230	Repair and maintenance of non-domestic cooling and ventilating equipment (manufacture)
	28520	General mechanical maintenance and repair of machinery (manufacture)		29560	Repair and maintenance of non-domestic machinery for drying wood, paper pulp, etc. (manufacture)
	72500	Repair and maintenance of accounting machinery (manufacture)		72500	Repair and maintenance of office machinery (other than computers)
	29320	Repair and maintenance of agricultural machinery (manufacture)		29521	Repair and maintenance of oil and gas extraction machinery (manufacture)
	29310	Repair and maintenance of agricultural tractors (manufacture)		29430	Repair and maintenance of other machine tools (except metal working) (manufacture)
	72500	Repair and maintenance of cash registers (manufacture)			
	29240	Repair and maintenance of commercial-type general purpose machinery (manufacture)			
	29122	Repair and maintenance of compressors (manufacture)			
	29522	Repair and maintenance of construction machinery (earth moving type) (manufacture)			

UK SIC 2007	UK SIC 2003	Activity	SIC 2007	SIC 2003	Activity
	29420	Repair and maintenance of other machine tools (metal working) (manufacture)		33403	Repair and maintenance of professional photographic and cinematographic equipment (manufacture)
	29550	Repair and maintenance of papermaking machinery (manufacture)		33201	Repair and maintenance of radiation detection and monitoring instruments (manufacture)
	72500	Repair and maintenance of photocopy machines (manufacture)		33201	Repair and maintenance of surveying instruments (manufacture)
	29560	Repair and maintenance of plastic and rubber working machinery (manufacture)		33402	Repair of binoculars (manufacture)
	29121	Repair and maintenance of pumps (manufacture)		33301	Repair of electronic industrial process control equipment (manufacture)
	29110	Repair and maintenance of railway diesel engines (manufacture)		33402	Repair of electronic optical equipment (manufacture)
	72500	Repair and maintenance of reprographic machinery		31100	Repair of electronic transformers (solid state), coils, chokes, and other inductors (manufacture)
	29560	Repair and maintenance of special purpose machinery n.e.c. (manufacture)		32100	Repair of electronic valves and tubes and other electronic components (manufacture)
	29110	Repair and maintenance of steam turbines (manufacture)		32300	Repair of heads (pickup, recording, read/write, etc.), phonograph needles (manufacturing)
	29130	Repair and maintenance of taps (manufacture)		29240	Repair of laboratory distilling apparatus, centrifuges, ultrasonic cleaning machinery (manufacture)
	29130	Repair and maintenance of valves (manufacture)			
	29240	Repair and maintenance of vending machines (manufacture)		33402	Repair of microscopes (manufacture)
	29240	Repair and maintenance of weighing equipment (manufacture)		31620	Repair of mine detectors (manufacture)
	72500	Repair of calculators (manufacture)		33302	Repair of non-electronic industrial process control equipment (manufacture)
	29140	Repair of gearing and driving elements (manufacture)		33403	Repair of photographic equipment (manufacture)
	29410	Repair of other power-driven hand-tools (manufacture)		33401	Repair of prisms and lenses (manufacture)
	72500	Repair of typewriters (manufacture)		33402	Repair of telescopes (manufacture)
	29130	Repair of valves for machinery (manufacture)			
33130		**Repair of electronic and optical equipment**	**33140**		**Repair of electrical equipment**
	33201	Repair and maintenance of aircraft engine instruments (manufacture)		31500	Repair and maintenance of electric lighting equipment (manufacture)
	33201	Repair and maintenance of automotive emissions testing equipment (manufacture)		31620	Repair and maintenance of electrical signalling equipment (manufacture)
	33500	Repair and maintenance of clocks in church towers and the like		31200	Repair and maintenance of electricity distribution and control apparatus (manufacture)
	33100	Repair and maintenance of electrocardiographs (manufacture)		33100	Repair and maintenance of medical equipment, electrical (manufacture)
	33100	Repair and maintenance of electromedical endoscopic equipment (manufacture)		31100	Repair and maintenance of motor generator sets (manufacture)
	33201	Repair and maintenance of electronic equipment for measuring, checking, testing, etc. (manufacture)		33202	Repair and maintenance of non-electronic measuring, checking, testing etc. equipment (manufacture)
	33100	Repair and maintenance of hearing aids (manufacture)		31400	Repair and maintenance of primary and storage batteries (manufacture)
	33500	Repair and maintenance of industrial time measuring instruments and apparatus (manufacture)		31200	Repair and maintenance of relays and industrial controls (manufacture)
	33100	Repair and maintenance of irradiation apparatus (manufacture)		31620	Repair and maintenance of road and other non-domestic exterior lighting equipment (manufacture)
	33100	Repair and maintenance of magnetic resonance imaging equipment (manufacture)		31200	Repair and maintenance of switchgear and switchboard apparatus (manufacture)
	33201	Repair and maintenance of materials' properties testing and inspection equipment (manufacture)		31200	Repair and maintenance of wiring devices for wiring electrical circuits (manufacture)
	33100	Repair and maintenance of medical and surgical equipment and apparatus (manufacture)		31100	Repair and rewiring of armatures (manufacture)
	33100	Repair and maintenance of medical ultrasound equipment (manufacture)		31100	Repair, maintenance and rewinding of electric motors, generators and transformers (manufacture)
	33201	Repair and maintenance of meteorological instruments (manufacture)		31100	Rewind electric motor (manufacture)
	33402	Repair and maintenance of optical precision instruments (manufacture)	**33150**		**Repair and maintenance of ships and boats**
	33100	Repair and maintenance of pacemakers (manufacture)		35110	Painting of ships
				35120	Repair and maintenance of boats (manufacture)
	29710	Repair and maintenance of professional electric appliances (manufacture)		35120	Repair and maintenance of pleasure and sporting craft (manufacture)
				35110	Repair and maintenance or alteration of ships (manufacture)
				35120	Repair and routine maintenance performed on boats by floating dry docks (manufacture)

UK SIC 2007	UK SIC 2003	Activity	SIC 2007	SIC 2003	Activity
	35110	Repair and routine maintenance performed on ships by floating dry docks (manufacture)	33200		**Installation of industrial machinery and equipment**
				28520	Activities of millwrights (manufacture)
33160		**Repair and maintenance of aircraft and spacecraft**		33301	Assembling of electronic industrial process control equipment (manufacture)
	35300	Repair and maintenance of aero-engine parts and sub assemblies (manufacture)		33302	Assembling of non-electronic industrial process control equipment (manufacture)
	35300	Repair and maintenance of aero-space equipment (manufacture)		28520	Dismantling of large-scale machinery and equipment (manufacture)
	35300	Repair and maintenance of air cushion vehicles (manufacture)		31400	Installation of accumulators, primary cells and primary batteries (manufacture)
	35300	Repair and maintenance of aircraft (manufacture)		29320	Installation of agricultural and forestry machinery (manufacture)
	35300	Repair and maintenance of aircraft engines of all types (manufacture)		34100	Installation of assemblies and sub-assemblies and the like, into motor vehicles (manufacture)
	35300	Repair and maintenance of helicopters (manufacture)		29140	Installation of bearings, gears, gearing and driving elements (manufacture)
	35300	Repair and maintenance of spacecraft (manufacture)		36501	Installation of bowling alley equipment (manufacture)
33170		**Repair and maintenance of other transport equipment n.e.c.**		29530	Installation of catering equipment (manufacture)
	35500	Repair and maintenance of animal-drawn buggies and wagons (manufacture)		26400	Installation of ceramic pipes, conduit, guttering and pipe fittings (manufacture)
	35430	Repair and maintenance of invalid carriages (manufacture)		33500	Installation of clocks (manufacture)
	35200	Repair and maintenance of railway cars (manufacture)		29122	Installation of compressors (manufacture)
	35200	Repair and maintenance of railway locomotives (manufacture)		35110	Installation of drilling platforms and the like (manufacture)
	35200	Repair and maintenance of railway rolling stock (major) (manufacture)		29522	Installation of earth-moving and excavating equipment (manufacture)
	35200	Repair and maintenance of tramway rolling stock (manufacture)		31100	Installation of electric motors, generators and transformers (manufacture)
	35200	Repair and maintenance of transmissions and other parts for locomotives (manufacture)		31610	Installation of electrical equipment for engines and vehicles n.e.c. (manufacture)
	34203	Repair of caravans		31200	Installation of electricity distribution and control apparatus (manufacture)
	35430	Wheelchair repair and maintenance (manufacture)		32100	Installation of electronic valves and tubes and other electronic components (manufacture)
33190		**Repair of other equipment**		29110	Installation of engines and turbines (except aircraft, vehicle and cycle) (manufacture)
	26400	Repair and maintenance of ceramic pipes, etc. And systems thereof in industrial plants (manufacture)		29523	Installation of equipment for concrete crushing and screening and roadworks (manufacture)
	26150	Repair and maintenance of glass tubes, etc. And systems thereof in industrial plants (manufacture)		28750	Installation of fabricated metal products (manufacture)
	26810	Repair and maintenance of millstones, grindstones, polishing stones and the like (manufacture)		28520	Installation of factory assembly lines (manufacture)
	25130	Repair and maintenance of other rubber products (excluding tyres) (manufacture)		28740	Installation of fasteners, screw machine products, chain and springs (manufacture)
	25210	Repair and maintenance of plastic tubes etc. And systems thereof in industrial plants (manufacture)		29210	Installation of furnaces and furnace burners (manufacture)
	20510	Repair and maintenance of wooden products n.e.c. (manufacture)		29240	Installation of general purpose machinery n.e.c. (manufacture)
	36300	Repair and rebuilding of organs (in factory) (manufacture)		33100	Installation of industrial irradiation and electromedical equipment (manufacture)
	17402	Repair of camping goods made of canvas (manufacture)		28110	Installation of industrial machinery and equipment (manufacture)
	17520	Repair of fishing nets (manufacture)		30020	Installation of industrial mainframe and similar computers (manufacture)
	36300	Repair of pianos (in factory) (manufacture)		33201	Installation of instruments and apparatus for measuring, checking, testing, navigating (manufacture)
	36501	Repair of pinball machines and other coin-operated games (manufacture)		31300	Installation of insulated wire and cable (manufacture)
	25240	Repair of plexiglas plane windows (manufacture)		28220	Installation of large scale central heating boilers e.g. For large residential blocks (manufacture)
	17520	Repair of rigging (manufacture)		29220	Installation of lifting and handling equipment (except lifts and escalators) (manufacture)
	17520	Repair of ropes (manufacture)		28630	Installation of locks and hinges (manufacture)
	17402	Repair of sails (manufacture)		29420	Installation of machine tools (metal working) (manufacture)
	17402	Repair of tarpaulins (manufacture)			
	20400	Repair or reconditioning of wooden pallets, shipping drums or barrels (manufacture)			
	36300	Restoring of organs and other historical musical instruments (manufacture)			

UK SIC 2007	UK SIC 2003	Activity	SIC 2007	SIC 2003	Activity
	29430	Installation of machine tools (other than metal working) (manufacture)		40110	Electricity production from diesel and renewables generation facilities
	29560	Installation of machinery for bookbinding (manufacture)		40110	Electricity production from hydroelectric generation facilities
	29530	Installation of machinery for food, beverage and tobacco processing (manufacture)		40110	Electricity production from thermal generation facilities
	29510	Installation of machinery for metallurgy (manufacture)		40110	Generating station
	29521	Installation of machinery for mining (manufacture)		40110	Hydro electric power station
	29550	Installation of machinery for paper and paperboard production (manufacture)		40110	Nuclear power station
	29560	Installation of machinery for printing (manufacture)		40110	Power station
	29540	Installation of machinery for textile, apparel and leather production (manufacture)		40110	Wind farms
	29560	Installation of machinery for working rubber or plastics and making products of these (manufacture)	**35120**		**Transmission of electricity**
				40120	Electricity transmission
	30020	Installation of mainframe and similar computers and peripheral equipment (manufacture)	**35130**		**Distribution of electricity**
	33100	Installation of medical and surgical equipment and apparatus (manufacture)		40130	Electricity distribution
	28210	Installation of metal tanks (manufacture)		40130	Electricity distribution operations by lines, poles, meters, and wiring
	26810	Installation of millstones, grindstones, polishing stones and the like (manufacturing)		33202	Repair and maintenance of electricity meters
	29230	Installation of non-domestic cooling and ventilating equipment (manufacture)	**35140**		**Trade of electricity**
	29560	Installation of non-domestic machinery for drying wood, paper pulp etc. (manufacture)		40130	Electricity power agents
				40130	Electricity power brokers
	33202	Installation of non-electronic instruments and appliances for measuring, testing, etc. (manufacture)		40130	Electricity sales
	30010	Installation of office machinery (manufacture)		40130	Electricity sales agents
	33402	Installation of optical precision instruments (manufacture)		40130	Electricity sales to the user
	31620	Installation of other electrical apparatus (manufacture)		40130	Operation of electricity and transmission capacity exchanges for electric power
	33403	Installation of photographic and cinematographic equipment (manufacture)	**35210**		**Manufacture of gas**
	32300	Installation of professional radio, television, sound and video equipment (manufacture)		40210	Benzole (crude) from gas works
				40210	Coal tar (crude) from gas works
	29121	Installation of pumps (manufacture)		40210	Coke gas
	32202	Installation of radio and television transmitters (manufacture)		40210	Gas production for the purpose of gas supply
	36110	Installation of seats in aircraft, ships, trains and the like		40210	Gas works
				40210	Sulphate of ammonia from gas works
	29560	Installation of special purpose machinery n.e.c. (manufacture)		40210	Town gas production
	28300	Installation of steam generators (not central heating boilers) incl. related pipework (manufacture)	**35220**		**Distribution of gaseous fuels through mains**
				40220	Distribution and supply of gaseous fuels of all kinds through a system of mains
	28710	Installation of steel drums and similar (manufacture)		40220	Natural gas booster/compression site
	29130	Installation of taps and valves (manufacture)		40220	Natural gas distribution
	32201	Installation of telecommunications equipment		40220	Natural gas storage
	28620	Installation of tools (manufacture)		33202	Repair and maintenance of gas meters
	26150	Installation of tubes and pipes of glass, including installation of glass pipe (manufacture)		33202	Repair and maintenance of gas meters (non-electronic) (manufacture)
	25210	Installation of tubes, pipes and hoses, of plastics (manufacture)		40220	Town gas distribution
	33401	Installation of unmounted lenses (manufacture)	**35230**		**Trade of gas through mains**
	29600	Installation of weapons and weapon systems (manufacture)		40220	Commodity and transport capacity exchanges for gaseous fuels
	28730	Installation of wire products (manufacture)		40220	Gas agents (mains gas)
	28520	Machine rigging (manufacture)		40220	Gas brokers (mains gas)
				40220	Sale of gas to the user through mains
35110		**Production of electricity**	**35300**		**Steam and air conditioning supply**
	40110	Electricity generation		40300	Chilled water for cooling purposes production and distribution
	40110	Electricity generation by gas turbine		40300	Community heating plant
	40110	Electricity production		40300	Compressed air production and distribution

UK SIC 2007	UK SIC 2003	Activity	SIC 2007	SIC 2003	Activity
	40300	Hot water production and distribution		90020	Rubbish collection
	40300	Hydraulic power production and distribution		90020	Skip hire (waste transportation)
	15980	Ice (for human consumption)		90020	Trash collection
	40300	Ice (not for human consumption)		90020	Waste collection
	40300	Production and distribution of cooled air		90020	Waste collection centre
	40300	Production of ice for cooling purposes (manufacture)	38120		**Collection of hazardous waste**
	15980	Production of ice for food (manufacture)		23300	Collection and treatment of radioactive nuclear waste
	40300	Steam production, collection and distribution		90020	Collection of bio-hazardous waste
36000		**Water collection, treatment and supply**		90020	Collection of nuclear waste
	41000	Collection of rain water		90020	Collection of used oil from shipment or garages
	41000	Collection of water from rivers, lakes, wells		90020	Hazardous waste collection (e.g. Batteries, used cooking oils, etc.)
	41000	Desalting of sea or ground water to produce water		90020	Operation of waste transfer stations for hazardous waste
	41000	Distribution of water through mains, by trucks or other means		40110	Spent (irradiated) fuel elements (cartridges) of nuclear reactors
	41000	Operation of irrigation canals	38210		**Treatment and disposal of non-hazardous waste**
	41000	River management		14500	Ashes and residues of incineration of mining and quarrying waste
	41000	Sea water desalination		90020	Disposal of non-hazardous waste by combustion or incineration or other methods
	41000	Treatment of water for industrial and other purposes		90020	Landfill for the disposal of refuse and waste
	41000	Water authority (headquarters and water supply)		90020	Local authority refuse disposal
	41000	Water collection, purification and distribution		90020	Operation of landfills for the disposal of non-hazardous waste
	41000	Water company		90020	Production of compost from organic waste
	41000	Water conservation		90020	Refuse disposal plant or tip (local authority or municipally owned)
37000		**Sewerage**		90020	Refuse disposal service (not especially for agriculture)
	90010	Cesspools emptying and cleaning		90020	Refuse disposal tip operator
	90010	Chemical toilets servicing		90020	Treatment and disposal of non-hazardous waste
	90010	Drains maintenance		90020	Treatment of organic waste for disposal
	90010	Human waste water collection and transport by sewers, collectors, tanks and other means of transport		90020	Waste disposal
				90020	Waste incineration
	90010	Local authority drainage services		90020	Waste treatment by composting of plant materials
	90010	Local authority sewage services	38220		**Treatment and disposal of hazardous waste**
	90010	Rain water collection and transportation by sewers, collectors, tanks and other means of transport		23300	Disposal of nuclear waste
	90010	Septic tanks emptying and cleaning		90020	Disposal of sick or dead animals (toxic)
	90010	Sewage farm		90020	Disposal of used goods such as refrigerators to eliminate harmful waste
	90010	Sewage works		23300	Encapsulation, preparation and other treatment of nuclear waste for storage
	90010	Sewerage system maintenance and operation		90020	Incineration of hazardous waste
	90010	Treatment of human waste water by means of physical, chemical and biological processes		90020	Operation of facilities for treatment of hazardous waste
	90010	Treatment of waste water (human, industrial, from swimming pools etc.) by various processes		90020	Toxic waste treatment service
38110		**Collection of non-hazardous waste**		90020	Treatment and disposal of foul liquids (e.g. leachate)
	90020	Building debris removal		90020	Treatment and disposal of hazardous waste
	90020	Collection and removal of non hazardous debris and rubble		23300	Treatment and disposal of nuclear waste
	90020	Collection of non hazardous construction and demolition waste		90020	Treatment and disposal of radioactive waste from hospitals, etc.
	90020	Collection of non hazardous recyclable materials		90020	Treatment and disposal of toxic live or dead animals and other contaminated waste
	90020	Collection of non hazardous waste output of textile mills		23300	Treatment and disposal of transition radioactive waste
	90030	Collection of refuse in litter bins in public places		23300	Treatment, disposal and storage of radioactive nuclear waste
	90020	Construction and demolition waste collection			
	90020	Dustman			
	90020	Garbage collection			
	90020	Litter box refuse collection (public)			
	90020	Operation of waste transfer facilities for non-hazardous waste			
	90020	Refuse collection by local authority cleansing department			

UK SIC 2007	UK SIC 2003	Activity	SIC 2007	SIC 2003	Activity
38310		**Dismantling of wrecks**		90030	Oil spill pollution control services
	37100	Dismantling of automobile wrecks for materials recovery		90030	Oil spills at sea containment, dispersion and clean up services
	37100	Dismantling of computers for materials recovery		90030	Remediation activities and other waste management services
	37100	Dismantling of ship wrecks for materials recovery		90030	Specialised pollution control activities
	37100	Dismantling of televisions for materials recovery	41100		**Development of building projects**
	37100	Dismantling of wrecks for materials recovery		70110	Developing building projects for commercial buildings hotels, stores, shopping malls, restaurants
38320		**Recovery of sorted materials**		70110	Development of building projects for residential buildings
	37200	Crushing, cleaning and sorting of demolition waste to obtain secondary raw materials		70110	Housing association (building houses for later sale)
	37200	Crushing, cleaning and sorting of glass		70110	Land and building company
	37200	Crushing, cleaning and sorting of glass to produce secondary raw materials		70110	Land investment company
	37100	Mechanical crushing of metal waste (cars, washing machines, etc.) with sorting and separation		70110	Property developer
	37100	Mechanical crushing of metal waste from used bikes		70110	Property investment company
	37100	Mechanical reduction of large iron pieces such as railway wagons into secondary raw materials		70110	Real estate project development
	37200	Non-metal waste and scrap recycling into secondary raw materials	41201		**Construction of commercial buildings**
	37200	Processing cleaning, melting, grinding of plastic or rubber waste to granulates		45211	Arts, cultural or leisure facilities buildings construction
	37200	Processing of food, beverage and tobacco waste and residual substances into secondary raw materials		45211	Assembly and erection of prefabricated non-residential constructions on the site
	37200	Processing of used cooking oils and fats into secondary raw materials		28110	Assembly and installation of self-manufactured commercial buildings of metal on site
	37100	Reclaiming metals out of photographic waste, e.g. Fixer solution or photographic films and paper		25239	Assembly and installation of self-manufactured commercial buildings of plastic on site
	37200	Reclaiming of chemicals from chemical waste		20300	Assembly and installation of self-manufactured commercial buildings of wood on site
	37200	Reclaiming of rubber to produce secondary raw materials		45211	Builder and contractor for commercial buildings
	37100	Scrap metal recycling into new raw materials (except remelting of ferrous waste and scrap)		45211	Building maintenance and restoration commercial buildings
	37100	Shredding of metal waste, end-of-life vehicles		45230	Car park construction
	37200	Sorting and pelleting of plastics to produce secondary raw material for tubes, flower pots, pallets		45211	Churches and other ecclesiastical buildings construction
	37200	Sorting and pelleting of plastics to produce secondary raw materials		45211	Commercial buildings construction
39000		**Remediation activities and other waste management services**		45211	Construction factories
	45250	Asbestos removal work		45211	Construction of airport buildings
	90030	Asbestos, lead paint, and other toxic material abatement		45211	Construction of arts, cultural or leisure facilities buildings
	90030	Cleaning up oil spills and other pollutions in ocean and seas including coastal areas		45211	Construction of assembly plants
	90030	Cleaning up oil spills and other pollutions in surface water		45211	Construction of commercial buildings
	90030	Cleaning up oil spills and other pollutions on land		45211	Construction of hospitals
	90030	Decontamination and cleaning up of surface water following accidental pollution		45211	Construction of indoor sports facilities
	90030	Decontamination of industrial plants or sites		45211	Construction of office buildings
	90030	Decontamination of nuclear plants and sites		45211	Construction of parking garages
	90030	Decontamination of soil and groundwater		45211	Construction of primary, secondary and other schools
	90030	Decontamination of soils and groundwater at the place of pollution using biological methods		45211	Construction of religious buildings
	90030	Decontamination of soils and groundwater at the place of pollution using chemical methods		45211	Construction of warehouses
	90030	Decontamination of soils and groundwater at the place of pollution using mechanical methods		45211	Construction of workshops
	90030	Decontamination of surface water		45230	Flatwork for sport and recreational installations (commercial buildings)
	90030	Minefield clearance		45211	Office and shop construction
	90030	Oil spill clearance on land		45211	Prefabricated constructions (commercial) assembly and erection
				45211	Primary, secondary and other schools construction
				45230	Swimming pools construction
			41202		**Construction of domestic buildings**
				28110	Assembly and installation of self-manufactured domestic buildings of metal on site

UK SIC 2007	UK SIC 2003	Activity
	25239	Assembly and installation of self-manufactured domestic buildings of plastic on site
	20300	Assembly and installation of self-manufactured domestic buildings of wood on site
	45212	Builder and contractor for domestic buildings
	45212	Building maintenance and restoration domestic buildings
	45212	Construction of domestic buildings
	45212	Construction of housing association and local authority housing
	45212	Construction of multi-family buildings, including high-rise buildings
	45212	Construction of residential buildings:
	45212	Construction of single-family houses
	45212	House building and repairing
	45212	Housing association (building work)
	45212	Local authority house building and maintenance
	45212	Local authority or new town direct labour department (domestic dwellings)
	45212	Prefabricated constructions (domestic) assembly and erection
	45212	Remodelling or renovating existing residential structures
	45212	Scottish Special Housing Association (building work)
42110		**Construction of roads and motorways**
	45230	Airfield runway construction
	45230	Airport runway construction
	45230	Asphalt paving of roads
	45230	Asphalting contractor (civil engineering)
	45230	Construction of other vehicular and pedestrian ways
	45230	Erection of roadway barriers
	45230	Ground work contracting
	45230	Highway construction
	45230	Installation of crash barriers
	45230	Installation of non-illuminated road signs, bollards etc.
	45230	Local authority highways construction and maintenance
	45230	Local authority road construction and major repairs
	45230	Motorway and other dual carriageway construction
	45230	Parking lot markings painting
	45230	Paving contractor
	45230	Pedestrian ways construction
	45230	Road construction and repair
	45230	Road surface markings painting
	45230	Street construction
	45230	Surface work on elevated highways, bridges and in tunnels
	45230	Surface work on streets, roads, highways, bridges or tunnels
	45230	Tar spraying contractor (civil engineering)
	45230	Tarmacadam laying contracting
42120		**Construction of railways and underground railways**
	45230	Cable supported transport systems construction
	45230	Construction of underground railways
	45230	Railway construction
	45213	Railway tunnel construction
	45213	Railway tunnelling contractor
	45213	Subway construction

SIC 2007	SIC 2003	Activity
	45213	Supply line (third rail) for railway construction
	45230	Tramways construction
	45213	Transmission line construction
42130		**Construction of bridges and tunnels**
	45213	Bridge construction
	45213	Construction of tunnels
	45213	Elevated highways construction
	45213	Viaduct construction
42210		**Construction of utility projects for fluids**
	45250	Artesian well contractor
	45213	Construction of civil engineering constructions for long-distance and urban pipelines
	45240	Construction of irrigation systems (canals)
	45213	Construction of pumping stations
	45213	Construction of sewage disposal plants
	45213	Construction of sewer systems
	45213	Gas offshore pipeline laying
	45213	Installation of offshore pipelines from oil or gas wells
	45240	Irrigation system construction
	45213	Offshore oil pipeline laying
	45213	Pipeline construction
	45213	Pipeline contracting
	45213	Repair of sewer systems
	45240	Reservoir construction
	45213	Sewerage construction
	45213	Urban pipelines construction
	45213	Water main and line construction
	45240	Water treatment plant construction
	45250	Water well drilling
42220		**Construction of utility projects for electricity and telecommunications**
	45213	Cable laying
	45213	Civil engineering constructions for long-distance communication
	45213	Civil engineering constructions for power lines
	45213	Civil engineering constructions for power plants
	45213	Civil engineering constructions for urban communication
	45213	Communication lines construction
	45213	Construction of utility projects for electricity
	45213	Construction of utility projects for telecommunications
	45213	Overhead line construction
	45213	Power line construction
	45213	Repair and maintenance of above-ground telecommunication lines
	45213	Repair and maintenance of underground communication lines
	45213	Urban communication and powerlines construction
42910		**Construction of water projects**
	45240	Aqueduct construction
	45240	Coastal defence construction
	45240	Dam construction
	45240	Dredging contractor
	45240	Dredging for water projects
	45240	Dredging of waterways
	45240	Dry dock construction

UK SIC 2007	UK SIC 2003	Activity	SIC 2007	SIC 2003	Activity
	45240	Dyke construction		45110	Mining site preparation and overburden removal
	45240	Dykes and static barrages construction		45110	Overburden removal and other development of mineral properties and sites
	45240	Floodgates, movable barrages and hydro-mechanical structures construction		45110	Rock removal
	45240	Harbour construction		45110	Site preparation
	45240	Lock construction		45110	Top soil stripping work
	45240	Marina construction		45110	Trench digging
	45240	Pleasure port construction			
	45240	River work construction	**43130**		**Test drilling and boring**
	45240	Water project construction		45120	Borehole drilling
	45240	Waterway construction		45120	Core sampling for construction
				45120	Geological test drilling, test boring and core sampling
42990		**Construction of other civil engineering projects n.e.c.**		45120	Geophysical test drilling, test boring and core sampling
	45213	Civil engineering construction		45120	Test boring for construction
	45213	Civil engineering contractor		45120	Test drilling for construction
	45213	Construction of chemical plants (except buildings)			
	45213	Construction of industrial facilities (except buildings)	**43210**		**Electrical installation**
	45230	Construction of outdoor sports facilities (except buildings)		45310	Aerial erection (domestic)
	45213	Construction of refineries (except buildings)		45310	Connecting of electric appliances and household equipment, including baseboard heating
	45213	Constructional engineering		45310	Electrical contractor (construction)
	45230	Foot and cycle path construction		45310	Electrical wiring of buildings
	45230	Golf course construction		45310	Installation of aerials and residential antennas
	45213	Government department (building and civil engineering works division)		45340	Installation of airport runway lighting
	45230	Land subdivision with land improvement (e.g. Adding of roads etc.)		45310	Installation of burglar alarm systems
	45213	Local authority civil engineering department		45310	Installation of cables
	45213	Local authority engineer's department		45310	Installation of computer network cabling and other telecommunications system cables
	45213	Prefabricated constructions (civil engineering) assembly and erection		45310	Installation of electrical systems in cable television wiring, including fibre optic
	45213	Public works contractor		45310	Installation of electrical systems in computer network, including fibre optic
	45230	Sport facilities construction		45310	Installation of electrical wiring and fittings
	45230	Sports and recreation grounds, laying out		45310	Installation of fire alarms
	45230	Stadium construction		45340	Installation of illuminated road signs and street furniture
	45230	Tennis courts construction		45340	Installation of illumination and signalling systems for roads, railways, airports and harbours
				45310	Installation of intelecommunications wiring
43110		**Demolition**		45310	Installation of lighting systems
	45110	Building demolition and wrecking		45310	Installation of office switchboards and telephone lines
	45110	Demolition contracting		45340	Installation of outdoor transformer and other outdoor electrical distribution apparatus
	45110	Demolition or wrecking of buildings and other structures		45340	Installation of roadway traffic monitoring and guidance equipment
				45310	Installation of satellite dishes
43120		**Site preparation**		45310	Installation of security alarms
	45110	Agricultural land drainage		45340	Installation of street lighting and electrical signals
	45110	Blasting and associated rock removal work		45310	Installation of telecommunications wiring systems
	45110	Blasting of construction sites		45310	Installation of telephone lines
	45110	Building site drainage		45340	Local authority street lighting
	45110	Building sites clearance		45340	Sign (electric) erection and maintenance
	45110	Development and preparation of mineral properties and sites			
	45110	Drainage of agricultural or forestry land	**43220**		**Plumbing, heat and air-conditioning installation**
	45110	Earth moving excavation		45330	Air conditioning contracting
	45110	Earthmoving contractor		45330	Heat and air-conditioning installation
	45110	Excavation		45330	Heating and plumbing contracting
	45110	Forestry land drainage		45330	Heating engineering (buildings)
	45110	Land drainage contractor			
	45110	Land reclamation work			
	45110	Landfill for construction			
	45110	Levelling and grading of construction sites			

UK SIC 2007	UK SIC 2003	Activity	SIC 2007	SIC 2003	Activity
	45330	Heating service contracting		45410	Exterior stucco application in buildings or other constructions incl. related lathing materials
	45330	Hot water engineer		45410	Interior plaster application in buildings or other constructions incl. related lathing materials
	45330	Installation of air conditioning equipment and ducts		45410	Interior stucco application in buildings or other constructions incl. related lathing materials
	45330	Installation of air conditioning plant		45410	Plastering contractor
	45330	Installation of coldrooms		45410	Stucco application in buildings
	45330	Installation of cooling towers			
	45330	Installation of duct work	43320		**Joinery installation**
	45310	Installation of electric solar energy collectors		45420	Builder and joiner
	45310	Installation of electrical heating systems (except baseboard heating)		45420	Carpenter n.e.c.
	45330	Installation of fire sprinkler systems		45420	Carpentry (not structural)
	45330	Installation of furnaces		45420	Completion of ceilings
	45330	Installation of gas fittings		28120	Installation (erection) work of self-manufactured builders' ware of metal
	45330	Installation of gas heating systems		25239	Installation (erection) work of self-manufactured builders' ware of plastic
	45330	Installation of gas meters		20300	Installation (erection) work of self-manufactured builders' ware of wood
	45330	Installation of heating and ventilation apparatus		45420	Installation of built-in furniture
	45330	Installation of lawn sprinkler systems		45420	Installation of ceilings
	45330	Installation of non-electric solar energy collectors		45420	Installation of doors
	45330	Installation of oil heating systems		45420	Installation of furniture
	45330	Installation of plumbing		45420	Installation of joinery
	45330	Installation of refrigeration		45420	Installation of metal grilles and gates
	45330	Installation of sanitary equipment		45420	Installation of metal partitioning
	45330	Installation of sprinkler systems		45420	Installation of metal shutters
	45330	Installation of steam piping		45420	Installation of movable wooden partitions
	45330	Installation of ventilation		45420	Installation of suspended ceilings
	45330	Plumbing contractor		45420	Installation of windows made of any material
	45330	Repair and maintenance of domestic air conditioning		45420	Installation of wooden door-frames
	45330	Repair and maintenance of domestic boilers		45420	Installation of wooden fitted kitchens
	45330	Repair and maintenance of office, shop and computer centre air conditioning		45420	Installation of wooden shop fittings
	45330	Sanitary engineering for buildings		45420	Installation of wooden staircases
				45420	Installation of wooden wall coverings
43290		**Other construction installation**		45420	Metal window fixing
	45320	Acoustical engineering		45420	Shop fitter
	45320	Cavity wall insulation			
	45340	Fencing contractor (not agricultural)	43330		**Floor and wall covering**
	45320	Fireproofing work		45430	Carpet fitter
	45340	Installation in buildings of fittings and fixtures n.e.c.		45430	Ceramic stove fitting
	45310	Installation of automated and revolving doors		45430	Claddings (internal)
	45340	Installation of blinds and awnings		45430	Floor covering laying
	45310	Installation of elevators		45430	Flooring contractor
	45310	Installation of escalators		45430	Hanging or fitting wooden wall coverings
	45310	Installation of lifts		45430	Installation of false floors and computer floors
	45310	Installation of lightning conductors		45430	Laying or fitting carpets and linoleum floor coverings including of rubber or plastic
	45340	Installation of outdoor pumping or filtration equipment		45430	Laying or fitting other wooden floor coverings
	45320	Installation of sound insulation		45430	Laying tiling or fitting marble, granite or slate floor coverings
	45320	Installation of thermal insulation		45430	Laying tiling or fitting terrazzo, marble, granite or slate wall coverings
	45310	Installation of vacuum cleaning systems		45430	Laying, tiling, hanging or fitting ceramic wall or floor tiles
	45320	Installation of vibration insulation		45430	Laying, tiling, hanging or fitting concrete stone wall or floor tiles
	45320	Insulating contractor (buildings)		45430	Laying, tiling, hanging or fitting cut stone wall or floor tiles
	45320	Insulating work activities		45430	Laying, tiling, hanging or fitting floor and wall covering
	29220	Repair and maintenance of elevators and escalators			
	28120	Repair of automated and revolving doors in buildings and civil engineering works (manufacture)			
	45320	Roof insulation contractor			
43310		**Plastering**			
	45410	Exterior plaster application in buildings or other constructions incl. related lathing materials			

UK SIC 2007	UK SIC 2003	Activity	SIC 2007	SIC 2003	Activity
	45430	Linoleum laying		45220	Carpentry (structural)
	45430	Paperhanging		45250	Chimney construction
	45430	Parquet floor laying (not by manufacturer)		45250	Claddings (external)
	45430	Terrazzo work (building)		45250	Concrete work (building)
	45430	Tiles laying or fitting		45500	Construction machinery and equipment rental with operator
	45430	Tiling contractor (floors and walls)		45211	Construction of outdoor swimming pools
	45430	Wall covering		45220	Damp proofing of buildings
	45430	Wallpaper hanging		45250	De-humidification of buildings
43341		**Painting**		45500	Demolition equipment rental with operator
	45440	Anti-corrosive coatings application work		45250	Diamond drilling of concrete and asphalt
	45440	Builder and decorator (own account)		45250	Drying out of buildings (incl. Water damage)
	45440	Buildings painting		45500	Earth moving equipment rental with operator
	45440	Civil engineering structure painting		45250	Ferro concrete bar bending and fixing contractor
	45440	Decorating of buildings		45250	Flare stack and flareboom erection work
	45440	Exterior painting of buildings		45250	Floor screeding
	45440	Interior painting of buildings		45250	Formwork (civil engineering)
	45440	Non-specialised painting of metal structures (including ships)		45250	Foundations construction
	45440	Painting contractor		45250	Grouting contractor (building)
	45440	Protective coatings application work		45250	Grouting contractor (civil engineering)
43342		**Glazing**		45250	Helideck erection work
	45440	Glazing		45240	Hydraulic construction (subsurface work)
	45440	Glazing contractor		45250	Industrial ovens erection
	45440	Installation of glass		45250	Mason (building)
	45440	Installation of mirrors		45250	Mine sinking
43390		**Other building completion and finishing**		45250	Oil production platform (fixed concrete or composite steel/concrete) construction
	45450	Building completion work		45450	Outdoor private swimming pools
	45450	Ornamentation fitting work		45250	Pile driving
	45450	Sandblasting of buildings		45250	Piling (building)
	45450	Shotblasting of buildings		45250	Piling contractor (civil engineering)
	45450	Steam cleaning of buildings		45500	Plant hire for construction rental with operator
	45450	Stonework cleaning and renovation		45250	Pylon erection
43910		**Roofing activities**		45250	Reinforced concrete engineer (civil engineering)
	45220	Building and roofing contractor		45500	Renting of cranes with operator
	45220	Erection of roofs		45500	Renting of other building equipment with operator
	20300	Installation of builders carpentry and joinery (roofing materials)		45250	Retort setting
	45220	Roof covering		45250	Sand blasting for building exteriors
	45220	Roof covering erection		45250	Screed laying
	45220	Roofing contractor		45250	Shaft drilling (civil engineering)
	45220	Thatching		45250	Shaft sinking
43991		**Scaffold erection**		45250	Steam cleaning for building exteriors
	45250	Renting of scaffolds and work platforms with erection and dismantling		45250	Steel bending
	45250	Scaffolding hiring and erecting		45250	Steel elements (not self-manufactured) erection
	45250	Scaffolds and work platform erecting and dismantling		45250	Steelwork erection (building)
43999		**Specialised construction activities (other than scaffold erection) n.e.c.**		45250	Steelwork erection (civil engineering)
	45250	Aerial mast (self supporting) erection		45250	Steeplejacking
	45250	Boring (civil engineering)		45250	Stone carving
	45250	Brick furnace construction		45250	Stone setting
	45250	Brick kiln construction		45250	Stone walling
	45250	Bricklaying		45250	Stonemasonry (building)
	45220	Carpenter on building site		45250	Structural steelwork erection (building)
				45250	Structural steelwork erection (civil engineering)
				45240	Subsurface work
				45220	Waterproofing of buildings
				45250	Well sinking (except gas or oil)
				45250	Work with specialist access requirements necessitating climbing skills and related equipment

UK SIC 2007	UK SIC 2003	Activity	SIC 2007	SIC 2003	Activity
45111		**Sale of new cars and light motor vehicles**		50102	Lorries (used) (wholesale)
	50101	Ambulances with a weight not exceeding 3.5 tonnes (new) (retail)		50101	Lorries (wholesale)
	50101	Ambulances with a weight not exceeding 3.5 tonnes (new) (wholesale)		50102	Motor homes (used) (retail)
	50101	Four wheel drive vehicles with a weight not exceeding 3.5 tonnes (new) (retail)		50102	Motor homes (used) (wholesale)
				50102	Motor vehicle (used) exporter
	50101	Four wheel drive vehicles with a weight not exceeding 3.5 tonnes (new) (wholesale)		50101	Motorhomes (retail)
	50101	Minibuses with a weight not exceeding 3.5 tonnes (new) (retail)		50101	Motorhomes (wholesale)
	50101	Minibuses with a weight not exceeding 3.5 tonnes (new) (wholesale)		50101	Off-road motor vehicles with a weight exceeding 3.5 tonnes (new) (retail)
	50101	Motor vehicle with a weight not exceeding 3.5 tonnes (new) exporter		50101	Off-road motor vehicles with a weight exceeding 3.5 tonnes (new) (wholesale)
	50101	Motor vehicle with a weight not exceeding 3.5 tonnes (new) importer		50102	Off-road motor vehicles with a weight exceeding 3.5 tonnes (used) (retail)
	50101	Motor vehicles with a weight not exceeding 3.5 tonnes (new) (retail)		50102	Off-road motor vehicles with a weight exceeding 3.5 tonnes (used) (wholesale)
	50101	Motor vehicles with a weight not exceeding 3.5 tonnes (new) (wholesale)		50101	Semi-trailers (retail)
	50101	Off-road motor vehicles with a weight not exceeding 3.5 tonnes (new) (retail)		50102	Semi-trailers (used) (retail)
				50102	Semi-trailers (used) (wholesale)
	50101	Off-road motor vehicles with a weight not exceeding 3.5 tonnes (new) (wholesale)		50101	Semi-trailers (wholesale)
				50101	Trailers (new) (wholesale)
				50101	Trailers (retail)
				50102	Trailers (used) (retail)
				50102	Trailers (used) (wholesale)
45112		**Sale of used cars and light motor vehicles**	**45200**		**Maintenance and repair of motor vehicles**
	50102	Ambulances with a weight not exceeding 3.5 tonnes (used) (retail)		50200	Anti-rust treatment of motor vehicles
	50102	Ambulances with a weight not exceeding 3.5 tonnes (used) (wholesale)		50200	Automobile association service centres
				50200	Car valeting
	50102	Car auctions		50200	Car wash
	50102	Four wheel drive vehicles with a weight not exceeding 3.5 tonnes (used) (retail)		50200	Installation of motor vehicle parts and accessories (not as part of production process)
	50102	Four wheel drive vehicles with a weight not exceeding 3.5 tonnes (used) (wholesale)		50200	Installation of motor vehicle parts and accessories, not part of the manufacturing process
	50102	Garage selling used motor vehicles (retail)		50200	Maintenance of motor vehicles
	50102	Internet car auctions		50200	Motor repair depot
	50102	Minibuses with a weight not exceeding 3.5 tonnes (used) (retail)		50200	Motor vehicle painting and body repairing
				50200	Motor vehicle servicing
	50102	Minibuses with a weight not exceeding 3.5 tonnes (used) (wholesale)		50200	Motor vehicle spraying
	50102	Motor vehicle with a weight not exceeding 3.5 tonnes (used) importer		50200	Omnibus repair depot
				50200	Painting of motor vehicles
	50102	Motor vehicles with a weight not exceeding 3.5 tonnes (used) (retail)		50200	Panel beating services
				50200	REME workshop
	50102	Motor vehicles with a weight not exceeding 3.5 tonnes (used) (wholesale)		50200	Repair and maintenance of auto electricals
	50102	Off-road motor vehicles with a weight not exceeding 3.5 tonnes (used) (retail)		50200	Repair and maintenance of commercial vehicles
				50200	Repair and maintenance of trailers and semi-trailers
	50102	Off-road motor vehicles with a weight not exceeding 3.5 tonnes (used) (wholesale)		50200	Repair and servicing in garages, of motor vehicles
				50200	Repair of car bodies
				50200	Repair of car electrical systems
45190		**Sale of other motor vehicles**		50200	Repair of car electronics
	50101	Camping vehicles (retail)		50200	Repair of fuel injection systems for motor vehicles
	50102	Camping vehicles (used) (retail)		50200	Repair of motor cars (except roadside assistance)
	50102	Camping vehicles (used) (wholesale)		50200	Repair of motor vehicle parts
	50101	Camping vehicles (wholesale)		50200	Repair of motor vehicle seats
	50102	Caravan (used) (retail)		50200	Repair of motor vehicle windows
	50102	Caravan (used) (wholesale)		50200	Repair of motor vehicle windscreens
	50101	Caravans (retail)		50200	Repair of motor vehicles (except roadside assistance)
	50101	Caravans (wholesale)		50200	Repair of tyres and tubes (fitting or replacement)
	50101	Lorries (retail)		50200	Repair to bodywork of motor vehicles
	50102	Lorries (used) (retail)		50200	Servicing of motor vehicles
				50200	Windscreen replacement services

UK SIC 2007	UK SIC 2003	Activity	SIC 2007	SIC 2003	Activity
45310		**Wholesale trade of motor vehicle parts and accessories**		51120	Bottled gas (commission agent)
	50300	Motor accessories dealer (wholesale)		51120	Carbonates (commission agent)
	50300	Motor vehicle parts and accessories (wholesale)		51120	Catalytic preparations (commission agent)
				51120	Chemical elements in disk form and compounds doped for use in electronics (commission agent)
45320		**Retail trade of motor vehicle parts and accessories**		51120	Chemical products and residual products of the chemical or allied industries (commission agent)
	50300	Car batteries (retail)		51120	Chemically modified animal or vegetable fats and mixtures (commission agent)
	50300	Exhaust sales and fitting centre (retail)		51120	Coal (commission agent)
	50300	Mail order sales of motor vehicle parts and accessories (retail)		51120	Coal factor (commission agent)
	50300	Motor accessories dealer (retail)		51120	Coke or semi-coke of coal (commission agent)
	50300	Motor vehicle parts and accessories (retail)		51120	Colloidal precious metals (commission agent)
	50300	Tyre dealer (retail)		51120	Compound plasticisers and stabilisers for rubber or plastics (commission agent)
45400		**Sale, maintenance and repair of motorcycles and related parts and accessories**		51120	Compounds of rare earth metals, yttrium or scandium (commission agent)
	50400	Moped sales (retail)		51120	Compounds with nitrogen functions (commission agent)
	50400	Moped sales (wholesale)		51120	Dental wax and other preparations for use in dentistry with a basis of plaster (commission agent)
	50400	Motorcycle exporter (wholesale)		51120	Diols, polyalcohols, cyclical alcohols and their derivatives (commission agent)
	50400	Motorcycle importer (wholesale)		51120	Disinfectants (commission agent)
	50400	Motorcycle parts and accessories (retail)		51120	Distilled water (commission agent)
	50400	Motorcycle parts and accessories (wholesale)		51120	Enriched uranium and plutonium and their compounds (commission agent)
	50400	Motorcycle sales (retail)		51120	Enzymes and other organic compounds (commission agent)
	50400	Motorcycle sales (wholesale)		51120	Essential oils and mixtures of odiferous substances (commission agent)
	50400	Repair and maintenance of motor cycles		51120	Ethers, organic peroxides, epoxides, acetals and hemiacetals and their compounds (commission agent)
46110		**Agents involved in the sale of agricultural raw materials, live animals, textile raw materials and semi-finished goods**		51120	Fertilisers (commission agent)
	51110	Agricultural raw materials (commission agent)		51120	Finishing agents, dye carriers and similar industrial chemical products (commission agent)
	51110	Beer or distilling dregs for animal feed (commission agent)		51120	Fire extinguisher charges and preparations (commission agent)
	51110	Coffee husks and skins (commission agent)		51120	Fireworks (commission agent)
	51110	Corn exchange (commission agent)		51120	Flux (commission agent)
	51110	Corn factor (commission agent)		51120	Fuel (commission agent)
	51110	Cotton broker (commission agent)		51120	Fungicides, rodenticides and similar products (commission agent)
	51110	Grain broker (commission agent)		51120	Glues (commission agent)
	51110	Hide and skin broker (commission agent)		51120	Glycerol (commission agent)
	51110	Leather (commission agent)		51120	Halogen or sulphur compounds of non-metals (commission agent)
	51110	Live animals (commission agent)		51120	Heterocyclic compounds (commission agent)
	51110	Textile raw materials (commission agent)		51120	Hydraulic brake fluids (commission agent)
	51110	Tobacco refuse (commission agent)		51120	Hydrocarbon derivatives (commission agent)
	51110	Wool broker (commission agent)		51120	Hydrogen chloride (commission agent)
	51110	Wool exchange (commission agent)		51120	Hydrogen, argon, nitrogen, oxygen and rare gases (commission agent)
	51110	Wool grease including lanolin (commission agent)		51120	Industrial chemicals (commission agent)
46120		**Agents involved in the sale of fuels, ores, metals and industrial chemicals**		51120	Industrial fatty acids (commission agent)
	51120	Acrylic polymers in primary forms (commission agent)		51120	Industrial monocarboxylic fatty acids (commission agent)
	51120	Acyclic and cyclic hydrocarbons (commission agent)		51120	Inorganic acids (commission agent)
	51120	Aldehyde function compounds (commission agent)		51120	Insecticides, herbicides, plant growth regulators and anti-sprouting products (commission agent)
	51120	Alkili or alkaline earth metals (commission agent)		51120	Iron (commission agent)
	51120	Amine function compounds (commission agent)		51120	Isotopes and compounds thereof (commission agent)
	51120	Amino resins, phenolic resins and polyurethanes in primary forms (commission agent)			
	51120	Ammonium chloride, nitrites and carbonates (commission agent)			
	51120	Anti-freezing preparations and prepared de-icing fluids (commission agent)			
	51120	Anti-knock preparations and additives for mineral oils and similar products (commission agent)			

UK SIC 2007	UK SIC 2003	Activity	SIC 2007	SIC 2003	Activity
	51120	Ketone and quinone function compounds (commission agent)		51120	Propellant powders and prepared explosives (commission agent)
	51120	Light, medium and heavy petroleum oils (commission agent)		51120	Raw earth metals (commission agent)
	51120	Lignite or peat (commission agent)		51120	Refined sulphur (commission agent)
	51120	Liquefied gas for motor purposes (commission agent)		51120	Retort carbon (commission agent)
	51120	Liquid and compressed air (commission agent)		51120	Roasted iron pyrites (commission agent)
	51120	Lubricating oils (commission agent)		51120	Safety fuses, detonating fuses, caps, igniters and electric detonators (commission agent)
	51120	Man-made fibres and yarns (commission agent)		51120	Salts of oxometallic or perometallic acids (commission agent)
	51120	Metal broker (not scrap) (commission agent)		51120	Scandium, yttrium and mercury (commission agent)
	51120	Metal oxides, hydroxides and peroxides (commission agent)		51120	Signalling flares, rain rockets, fog signals and other pyrotechnic articles (commission agent)
	51120	Metal waste and scrap (commission agent)		51120	Silicon and sulphur dioxide (commission agent)
	51120	Metallic halogenates (commission agent)		51120	Silicones in primary forms (commission agent)
	51120	Metalloids (commission agent)		51120	Sodium nitrate (commission agent)
	51120	Metals (commission agent)		51120	Synthetic and organic colouring matter, colouring lakes and preparations (commission agent)
	51120	Modelling pastes (commission agent)		51120	Synthetic or reconstructed precious or semi-precious unworked stones (commission agent)
	51120	Monohydric alcohols (commission agent)			
	51120	Motor and aviation spirit (commission agent)		51120	Synthetic rubber (commission agent)
	51120	Natural uranium and plutonium and their compounds (commission agent)		51120	Synthetic, organic or inorganic tanning extracts and preparations (commission agent)
	51120	Nitrates of potassium (commission agent)		51120	Urea, thiourea and melamine resins in primary forms (commission agent)
	51120	Nitric acid, sulphonitric acid and ammonia (commission agent)		51120	Vegetable or resin product derivatives (commission agent)
	51120	Oil trading (commodity broking) (commission agent)		51120	Vulcanized and unvulcanized rubber and articles thereof (commission agent)
	51120	Ores (commission agent)			
	51120	Peptones/protein substances and derivatives (commission agent)		51120	Waste, parings and scrap of rubber (commission agent)
	51120	Petroleum coke, bitumen and other residues of petroleum (commission agent)		51120	Wood charcoal for fuel (commission agent)
	51120	Petroleum gases and gaseous hydrocarbons (excluding natural gas) (commission agent)	**46130**		**Agents involved in the sale of timber and building materials**
	51120	Petroleum jelly and paraffin wax (commission agent)		51130	Bituminous construction materials (commission agent)
	51120	Phenols, phenol-alcohols and phenol derivatives (commission agent)		51130	Building materials (commission agent)
	51120	Phosphates of triammonium (commission agent)		51130	Ceramic articles used in construction (commission agent)
	51120	Phosphides, carbides, hydrides, nitrides, azides, silicides and borides (commission agent)		51130	Clay (commission agent)
	51120	Phosphinates, phosphonates, phosphates and polyphosphates (commission agent)		51130	Construction materials made of glass (commission agent)
	51120	Phosphoric esters and esters of other inorganic acids, their salts and derivatives (commission agent)		51130	Containers made of wood (commission agent)
				51130	Paints and varnishes (commission agent)
	51120	Pickling preparations (commission agent)		51130	Pallets, pallet boards and other load boards made of wood (commission agent)
	51120	Piezo-electric quartz (commission agent)			
	51120	Polyamides in primary forms (commission agent)		51130	Prefabricated buildings made of wood (commission agent)
	51120	Polycarbonates, alkyd and epoxide resins (commission agent)		51130	Prepared pigments, opacifiers and colours (commission agent)
	51120	Polyethers and polyesters (commission agent)		51130	Timber (commission agent)
	51120	Polymers of ethylene in primary forms (commission agent)		51130	Timber broker (commission agent)
	51120	Polymers of propylene and other olefins in primary forms (commission agent)		51130	Vitrifiable enamels and glazes, englobes, liquid lustres and glass frit (commission agent)
	51120	Polymers of styrene in primary forms (commission agent)		51130	Wood (commission agent)
	51120	Polymers of vinyl acetate, other vinyl esters and vinyl polymers in primary forms (commission agent)	**46140**		**Agents involved in the sale of machinery, industrial equipment, ships and aircraft**
	51120	Polymers of vinyl chloride and other halogenated olefins in primary forms (commission agent)		51140	Agricultural machinery (commission agent)
	51120	Prepared binders for foundry moulds or cores (commission agent)		51140	Aircraft (commission agent)
				51140	Computer equipment (commission agent)
	51120	Prepared rubber accelerators (commission agent)		51140	Industrial equipment (commission agent)
	51120	Printing ink (commission agent)		51140	Machine broker (commission agent)

UK SIC 2007	UK SIC 2003	Activity	SIC 2007	SIC 2003	Activity
	51140	Machinery (commission agent)	46180		**Agents specialised in the sale of other particular products**
	51140	Office machinery (commission agent)		51180	Adhesive dressings, catgut and similar materials (commission agent)
	51140	Ships (commission agent)		51180	Amides and their derivatives and salts (commission agent)
46150		**Agents involved in the sale of furniture, household goods, hardware and ironmongery**		51180	Amusement goods (commission agent)
	51150	Bicycles (commission agent)		51180	Antibiotics (commission agent)
	51150	Cutlery (commission agent)		51180	Antisera and vaccines (commission agent)
	51150	Domestic electrical appliances (commission agent)		51180	Artists', students' and sign board painters' colours, modifying tints and similar (commission agent)
	51150	Furniture (commission agent)		51180	Beauty, make-up and skin-care preparations including sun tan preparations (commission agent)
	51150	Hardware (commission agent)		51180	Chemical contraceptive preparations based on hormones or spermicides (commission agent)
	51150	Household goods (commission agent)			
	51150	Ironmongery (commission agent)		51180	Chemical preparations and sensitized emulsions for photographic use (commission agent)
	51150	Statuettes and other ornaments made of wood (commission agent)		51180	Cosmetics (commission agent)
	51150	Tableware and kitchenware made of wood (commission agent)		51180	Diagnostic reagents and other pharmaceutical products (commission agent)
46160		**Agents involved in the sale of textiles, clothing, fur, footwear and leather goods**		51180	Diamond broker (commission agent)
	51160	Bed and table linen (commission agent)		51180	First aid boxes (commission agent)
	51160	Carpets (commission agent)		51180	Glands and other organs, extracts thereof and other human or animal substances (commission agent)
	51160	Clothing (commission agent)		51180	Glycosides, vegetable alkaloids, their salts, ethers, esters and other derivatives (commission agent)
	51160	Curtains, drapes and interior blinds (commission agent)		51180	Hormones and their derivatives (commission agent)
	51160	Footwear (commission agent)		51120	Jewellery (commission agent)
	51160	Fur broker (commission agent)		51180	Lip make-up and eye make-up preparations (commission agent)
	51160	Furs (commission agent)		51180	Lysine, glutamic acid and their salts (commission agent)
	51160	Kits for embroidery, etc. (commission agent)			
	51160	Leather goods (commission agent)		51180	Manicure and pedicure preparations (commission agent)
	51160	Parachutes and rotochutes (commission agent)		51180	Man-made fibre waste (commission agent)
	51160	Sacks and bags used for packing of goods (commission agent)		51180	Medical goods (commission agent)
	51160	Sport nets (commission agent)		51180	Medicaments containing hormones but (not antibiotics) (commission agent)
	51160	Tarpaulins, awnings, sunblinds and circus tents (commission agent)		51180	Medicaments containing penicillins or other antibiotics (commission agent)
	51160	Textiles (commission agent)		51180	Musical instruments (commission agent)
	51160	Workwear (commission agent)		51180	Natural cork in plates, sheets, strips, crushed, granulated or ground (commission agent)
	51160	Yarn (commission agent)			
46170		**Agents involved in the sale of food, beverages and tobacco**		51180	Oral and dental hygiene preparations including denture fixative pastes and powders (commission agent)
	51170	Alcoholic beverages (commission agent)		51180	Perfumery, cosmetic and toilet and bath preparations (commission agent)
	51170	Beverages (commission agent)		51180	Pharmaceutical goods (commission agent)
	51170	Cigars, cheroots, cigarillos and cigarettes (commission agent)		51180	Photographic equipment (commission agent)
	51170	Confectionery (commission agent)		51180	Photographic paper (commission agent)
	51170	Fish factor (commission agent)		51180	Photographic plates and film and instant print film (commission agent)
	51170	Food (commission agent)		51180	Posphoaminolipids (commission agent)
	51170	Fruit and vegetables (commission agent)		51180	Powders for cosmetic or toilet use (commission agent)
	51170	Herb infusions (commission agent)			
	51170	Non-alcoholic beverages (commission agent)		51180	Provitamins, vitamins and their derivatives (commission agent)
	51170	Peptic substances, mucilages and thickeners (commission agent)		51180	Quaternary ammonium salts and hydroxides (commission agent)
	51170	Provision exchange (commission agent)			
	51170	Spice broker (commission agent)		51180	Sails for boats, sailboards or landcraft (commission agents)
	51170	Tea exchange (commission agent)			
	51170	Tobacco (commission agent)		51180	Salicylic acids, o-acetylsalicylic acid and their salts and esters (commission agent)
	51170	Tobacco broker (commission agent)			
	51170	Vegetable saps and extracts (commission agent)			
	51170	Yeasts and prepared baking powders (commission agent)			

UK SIC 2007	UK SIC 2003	Activity	SIC 2007	SIC 2003	Activity
	51180	Shampoos, hair lacquers and permanent waving or straightening preparations (commission agent)		51250	Tobacco importer (unmanufactured) (wholesale)
	51180	Shaving preparations, personal deodorants and antiperspirants (commission agent)		51210	Tulip bulbs (wholesale)
	51180	Sleeping bags (commission agent)	**46220**		**Wholesale of flowers and plants**
	51180	Soap and organic surface-active products and preparations for use as soap (commission agent)		51220	Bulbs (wholesale)
	51180	Soft goods (commission agent)		51220	Flower and plants exporter (wholesale)
	51180	Sporting goods (commission agent)		51220	Flower and plants importer (wholesale)
	51180	Stationery (commission agent)		51220	Flower salesman (wholesale)
	51180	Steroids used primarily as hormones (commission agent)		51220	Flowers (wholesale)
	51180	Sugar ethers, sugar esters and their salts and chemically pure sugar (commission agent)		51220	Plants (wholesale)
	51180	Sulphonamides (commission agent)	**46230**		**Wholesale of live animals**
	51180	Tents and other camping goods (commission agent)		51230	Horses (wholesale)
	51180	Unrecorded media for sound recording or similar recording of other phenomena (commission agent)		51230	Live animal exporter (wholesale)
	51180	Waste and scrap (commission agent)		51230	Live animal importer (wholesale)
46190		**Agents involved in the sale of a variety of goods**		51230	Live animals (wholesale)
	51190	Export confirming house, general or undefined (commission agent)		51230	Live poultry (wholesale)
	51190	Export purchasing, general or undefined (commission agent)		51230	Livestock (wholesale)
				51230	Pig jobber (wholesale)
				51230	Pigs (wholesale)
				51230	Sheep (wholesale)
46210		**Wholesale of grain, unmanufactured tobacco, seeds and animal feeds**	**46240**		**Wholesale of hides, skins and leather**
	51210	Animal feed (wholesale)		51241	Furskins (wholesale)
	51210	Animal hair (wholesale)		51241	Furskins exporter (wholesale)
	51210	Compound feed stuff (wholesale)		51241	Furskins importer (wholesale)
	51210	Corn (wholesale)		51249	Hides (wholesale)
	51210	Corn chandler (wholesale)		51249	Hides, skins and leather exporter (wholesale)
	51210	Corn merchant (wholesale)		51249	Hides, skins and leather importer (wholesale)
	51210	Cotton (wholesale)		51249	Leather (wholesale)
	51210	Flax (wholesale)		51249	Skins (wholesale)
	51210	Fodder (wholesale)	**46310**		**Wholesale of fruit and vegetables**
	51210	Forage (wholesale)		51310	Edible nuts (wholesale)
	51210	Grain (wholesale)		51310	Fresh fruit (wholesale)
	51210	Grain, seeds and animal feeds exporter (wholesale)		51310	Fruit and vegetable exporter (wholesale)
	51210	Grain, seeds and animal feeds importer (wholesale)		51310	Fruit and vegetable importer (wholesale)
	51210	Hay (wholesale)		51310	Fruit and vegetable market porterage (wholesale)
	51210	Hemp (unprocessed) (wholesale)		51380	Fruit and vegetables (processed) (wholesale)
	51210	Hops (wholesale)		51310	Fruit and vegetables (unprocessed) (wholesale)
	51210	Horsehair (wholesale)		51310	Fruit salesman (wholesale)
	51210	Materials, residues and by-products used as animal feed (wholesale)		51310	Fruiterer (wholesale)
	51210	Oil cake (wholesale)		51310	Herbs (wholesale)
	51210	Oil seeds (wholesale)		51310	Mushrooms (wholesale)
	51210	Oleaginous fruits (wholesale)		51380	Potato products (wholesale)
	51210	Poultry spice (wholesale)		51310	Potatoes (wholesale)
	51210	Prepared feeds for farm animals (wholesale)		51380	Preserved fruit (wholesale)
	51210	Provender (wholesale)		51310	Pulses (wholesale)
	51210	Raw wool (wholesale)		51310	Vegetables (unprocessed) (wholesale)
	51210	Seed potatoes (wholesale)		51310	Vegetables (wholesale)
	51210	Seeds (wholesale)	**46320**		**Wholesale of meat and meat products**
	51210	Sponge importer (wholesale)		51320	Game (wholesale)
	51210	Straw (wholesale)		51320	Meat (wholesale)
	51250	Tobacco (unmanufactured) (wholesale)		51320	Meat and meat products exporter (wholesale)
	51250	Tobacco exporter (unmanufactured) (wholesale)		51320	Meat and meat products importer (wholesale)
				51320	Meat porter (wholesale)
				51320	Meat salesman (wholesale)
				51320	Offal salesman (wholesale)

UK SIC 2007	UK SIC 2003	Activity	SIC 2007	SIC 2003	Activity
	51320	Pork butcher (wholesale)		51350	Cigarette importer (wholesale)
	51320	Poultry (wholesale)		51350	Cigarette merchant (wholesale)
	51320	Processed meat and meat products (wholesale)		51350	Tobacco merchant (wholesale)
	51320	Sausage skins (wholesale)		51350	Tobacco products exporter (wholesale)
				51350	Tobacco products importer (wholesale)
46330		**Wholesale of dairy products, eggs and edible oils and fats**		51350	Tobacconist (wholesale)
	51331	Butter (wholesale)		51350	Tobacconists' sundriesman (wholesale)
	51331	Cheese (wholesale)			
	51331	Cream (wholesale)	**46360**		**Wholesale of sugar and chocolate and sugar confectionery**
	51331	Dairy produce exporter (wholesale)		51360	Bakery products (wholesale)
	51331	Dairy produce importer (wholesale)		51360	Bread (wholesale)
	51331	Dairy produce n.e.c (wholesale)		51360	Chocolate and sugar confectionery (wholesale)
	51333	Edible oils and fats (wholesale)		51360	Confectionery (wholesale)
	51333	Edible oils and fats exporter (wholesale)		51360	Flour confectionery (wholesale)
	51333	Edible oils and fats importer (wholesale)		51360	Ice cream (wholesale)
	51333	Edible oils and fats of animal or vegetable origin (wholesale)		51360	Sugar (wholesale)
	51332	Egg grading and packing (wholesale)		51360	Sugar and chocolate and sugar confectionery exporter (wholesale)
	51332	Egg packing station (wholesale)		51360	Sugar and chocolate and sugar confectionery importer (wholesale)
	51333	Egg products (wholesale)			
	51332	Eggs (wholesale)	**46370**		**Wholesale of coffee, tea, cocoa and spices**
	51332	Eggs exporter (wholesale)		51370	Cocoa (wholesale)
	51332	Eggs importer (wholesale)		51370	Coffee (wholesale)
	51333	Herring oil (wholesale)		51370	Coffee, tea, cocoa and spices exporter (wholesale)
	51333	Lard (wholesale)		51370	Coffee, tea, cocoa and spices importer (wholesale)
	51333	Margarine (wholesale)		51370	Spice (wholesale)
	51331	Milk (wholesale)		51370	Tea (wholesale)
	51333	Palm oil (wholesale)		51370	Tea merchant (wholesale)
	51333	Whale oil (wholesale)			
	51331	Yoghurt (wholesale)	**46380**		**Wholesale of other food, including fish, crustaceans and molluscs**
				51380	Baby food (wholesale)
46341		**Wholesale of fruit and vegetable juices, mineral waters and soft drinks**		51380	Crustaceans (wholesale)
	51341	Fruit and vegetable juices (wholesale)		51380	Dietetic foods (wholesale)
	51341	Fruit and vegetable juices exporter (wholesale)		51380	Dried fish (wholesale)
	51341	Fruit and vegetable juices importer (wholesale)		51380	Dried fruit (wholesale)
	51341	Mineral water exporter (wholesale)		51380	Eels (wholesale)
	51341	Mineral water importer (wholesale)		51380	Fish (wholesale)
	51341	Mineral waters (wholesale)		51380	Fish distribution (wholesale)
	51341	Soft drinks (wholesale)		51380	Fishmonger (wholesale)
	51341	Soft drinks exporter (wholesale)		51380	Flour (wholesale)
	51341	Soft drinks importer (wholesale)		51380	Food n.e.c. including fish, crustaceans and molluscs exporter (wholesale)
				51380	Food n.e.c. including fish, crustaceans and molluscs importer (wholesale)
46342		**Wholesale of wine, beer, spirits and other alcoholic beverages**		51380	Herrings (wholesale)
	51342	Alcoholic beverages (wholesale)		51380	Honey (wholesale)
	51342	Alcoholic beverages exporter (wholesale)		51380	Meat for domestic animals (wholesale)
	51342	Alcoholic beverages importer (wholesale)		51380	Molluscs distribution (wholesale)
	51342	Beer, wines and liqueurs (wholesale)		51380	Oysters (wholesale)
	51342	Buying of wine in bulk and bottling without transformation (wholesale)		51380	Pet animal food (wholesale)
	51342	Cider merchant (wholesale)		51380	Processed fruit (wholesale)
	51342	Liqueurs (wholesale)		51380	Processed vegetables (wholesale)
	51342	Spirits (wholesale)		51380	Shellfish (wholesale)
	51342	Wine and spirit merchant (wholesale)		51380	Shrimps (wholesale)
	51342	Wine importer (wholesale)		51380	Starch (wholesale)
				51380	Wet fish dealer (wholesale)
46350		**Wholesale of tobacco products**		51380	Yeast (wholesale)
	51350	Cigar importer (wholesale)			
	51350	Cigar merchant (wholesale)			

UK SIC 2007	UK SIC 2003	Activity	SIC 2007	SIC 2003	Activity
46390		**Non-specialised wholesale of food, beverages and tobacco**		51429	Clothing importer (wholesale)
	51390	Beverages (non-specialised) (wholesale)		51429	Clothing outfitter (wholesale)
	51390	Cash and carry predominantly food (wholesale)		51423	Footwear (wholesale)
	51390	Food (non-specialised) (wholesale)		51423	Footwear exporter (wholesale)
	51390	Food, beverages and tobacco (non-specialised) exporter (wholesale)		51423	Footwear importer (wholesale)
	51390	Food, beverages and tobacco (non-specialised) importer (wholesale)		51429	Fur articles (wholesale)
				51421	Fur clothing for adults (wholesale)
	51390	Meat and fish market porterage (wholesale)		51421	Fur merchant (wholesale)
	51390	Provisions (wholesale)		51421	Furrier (wholesale)
	51390	Quick frozen foods (wholesale)		51429	Gloves (wholesale)
	51390	Wholesale grocer (wholesale)		51421	Gloves made of fur or leather (wholesale)
				51429	Hat materials (wholesale)
46410		**Wholesale of textiles**		51429	Hosiery (wholesale)
	51410	Awnings and sun blinds (wholesale)		51422	Infants' clothing (wholesale)
	51410	Cloth (wholesale)		51421	Leather clothing for adults (wholesale)
	51410	Cloth merchant (wholesale)		51429	Men's clothing (wholesale)
	51410	Clothing textiles (wholesale)		51429	Millinery (wholesale)
	51410	Draper (wholesale)		51429	Millinery importer (wholesale)
	51410	Fabrics (wholesale)		51423	Shoes (wholesale)
	51410	Flock (wholesale)		51429	Sports clothes (wholesale)
	51410	Haberdashery (wholesale)		51429	Straw and felt hats (wholesale)
	51410	Hand knitting yarns (wholesale)		51429	Ties (wholesale)
	51410	Hand mending yarns (wholesale)		51429	Umbrellas (wholesale)
	51410	Hessian (wholesale)		51429	Walking sticks and seat sticks (wholesale)
	51410	Household linen (wholesale)		51429	Women's clothing (wholesale)
	51410	Household textiles (wholesale)			
	51410	Linen and linen goods (wholesale)	46431		**Wholesale of gramophone records, audio tapes, compact discs and video tapes and of the equipment on which these are played**
	51410	Merchant converter (textiles) (wholesale)		51431	Audio separates (wholesale)
	51410	Needles, etc. for sewing (wholesale)		51431	Compact discs (recorded) (wholesale)
	51410	Oilcloth (wholesale)		51431	Dvds (recorded) (wholesale)
	51410	Piece goods (wholesale)		51431	Gramophone records (wholesale)
	51410	Rope (new) (wholesale)		51431	Gramophone records, recorded tapes, CDs etc. and equipment for playing them, exporter (wholesale)
	51410	Sacks and bags (wholesale)		51431	Gramophone records, recorded tapes, CDs etc. and equipment for playing them, importer (wholesale)
	51410	Sewing thread, etc. (wholesale)			
	51410	Silk yarn and fabrics (wholesale)		51431	Record players (wholesale)
	51410	Tarpaulins (wholesale)		51431	Recorded audio and video tapes, CDs, DVDs and the equipment on which these are played (wholesale)
	51410	Textile converter (wholesale)			
	51410	Textiles (wholesale)		51431	Tapes (recorded) (wholesale)
	51410	Textiles exporter (wholesale)		51431	Videos (recorded) (wholesale)
	51410	Textiles importer (wholesale)			
	51410	Thread (wholesale)	46439		**Wholesale of radio and television goods and of electrical household appliances (other than of gramophone records, audio tapes, compact discs and video tapes and the equipment on which these are played) n.e.c.**
	51410	Twine (wholesale)			
	51410	Woollen flock (wholesale)			
	51410	Woollens (wholesale)			
	51410	Yarn (wholesale)		51439	Burglar and fire alarms for household use (wholesale)
46420		**Wholesale of clothing and footwear**		51439	Clothes washing and drying machines for domestic use (wholesale)
	51421	Adults' fur and leather clothing exporter (wholesale)		51439	Co-axial cable and co-axial conductors for domestic use (wholesale)
	51421	Adults' fur and leather clothing importer (wholesale)			
	51429	Braces (wholesale)		51439	Dish washing machines for domestic use (wholesale)
	51422	Children's clothing (exporter) (wholesale)		51439	Domestic machinery (wholesale)
	51422	Children's clothing (importer) (wholesale)		51439	Electric blankets (wholesale)
	51422	Children's clothing (wholesale)		51439	Electrical heating appliances (wholesale)
	51429	Clothing (wholesale)		51439	Electrical household appliances (excluding radios, televisions, etc) (wholesale)
	51429	Clothing accessories (wholesale)			
	51421	Clothing accessories made of fur or leather (wholesale)		51439	Electrical installation equipment for domestic use (wholesale)
	51429	Clothing exporter (wholesale)			

UK SIC 2007	UK SIC 2003	Activity
	51439	Electro-thermic hair-dressing or hand drying apparatus (wholesale)
	51439	Fans and ventilating or recycling hoods for domestic use (wholesale)
	51439	Immersion heaters (wholesale)
	51439	Instantaneous or storage water heaters (electric) (wholesale)
	51479	Instantaneous or storage water heaters (non-electric) (wholesale)
	51439	Microwave ovens (wholesale)
	51479	Optical goods (wholesale)
	51439	Ovens, cookers, cooking plates, boiling rings, grills and roasters (electric) (wholesale)
	51476	Photographic flashbulbs and flashcubes (wholesale)
	51476	Photographic goods (wholesale)
	51476	Photographic goods exporter (wholesale)
	51476	Photographic goods importer (wholesale)
	51439	Plugs, sockets and other apparatus for protecting electrical circuits for domestic use (wholesale)
	51439	Primary cells and primary batteries for domestic use (wholesale)
	51439	Radio, television and electrical household equipment n.e.c. exporter (wholesale)
	51439	Radio, television and electrical household goods n.e.c. importer (wholesale)
	51439	Radios and televisions (wholesale)
	51439	Refrigerators and freezers for domestic use (wholesale)
	51439	Shavers and hair clippers with self-contained motors (wholesale)
	51439	Smoothing irons (electric) (wholesale)
	51439	Space heating and soil heating apparatus (electric) (wholesale)
	51439	Television cameras for domestic use (wholesale)
46440		**Wholesale of china and glassware and cleaning materials**
	51440	China (wholesale)
	51440	China, glassware, wallpaper and cleaning materials exporter (wholesale)
	51440	China, glassware, wallpaper and cleaning materials importer (wholesale)
	51440	Cleaning materials (wholesale)
	51440	Glassware (wholesale)
	51440	Pottery (wholesale)
	51440	Washing products (e.g. washing powder) (wholesale)
46450		**Wholesale of perfume and cosmetics**
	51450	Cosmetics (wholesale)
	51450	Hairdressers' sundriesman (wholesale)
	51450	Perfume (wholesale)
	51450	Perfumes and cosmetics exporter (wholesale)
	51450	Perfumes and cosmetics importer (wholesale)
	51450	Soap (wholesale)
	51450	Toilet preparations (wholesale)
46460		**Wholesale of pharmaceutical goods**
	51460	Adhesive dressings, catgut and similar materials (wholesale)
	51460	Antibiotics (wholesale)
	51460	Antisera and vaccines (wholesale)
	51460	Artificial joints and parts for the body (wholesale)
	51460	Artificial teeth and dental fittings (wholesale)
	51460	Barbers' and similar chairs (wholesale)

SIC 2007	SIC 2003	Activity
	51460	Breathing appliances for medical use (wholesale)
	51460	Catgut and similar materials (wholesale)
	51460	Chemically pure sugars, sugar ethers and sugar esters and their salts (wholesale)
	51460	Contraceptive chemical preparations based on hormones or spermicides (wholesale)
	51460	Diagnostic reagents n.e.c. (wholesale)
	51460	Druggists' sundries (wholesale)
	51460	Druggists' sundriesman (wholesale)
	51460	Drugs (wholesale)
	51460	Electro-diagnostic apparatus for medical use (wholesale)
	51460	First aid boxes (wholesale)
	51460	Glands, other organs and their extracts and other human or animal substances n.e.c. (wholesale)
	51460	Glycoside, vegetable alkaloids and their salts, ethers, esters and other derivatives (wholesale)
	51460	Hearing aids (wholesale)
	51460	Hormones and their derivatives (wholesale)
	51460	Instruments and appliances for dental science (wholesale)
	51460	Instruments and devices for doctors and hospitals (wholesale)
	51460	Invalid carriages with or without motor (wholesale)
	51460	Lactones (wholesale)
	51460	Lysine and glutamic acid and salts thereof (wholesale)
	51460	Medical goods (wholesale)
	51460	Medical, surgical, dental and veterinary furniture (wholesale)
	51460	Medicaments containing alkaloids or their derivatives (wholesale)
	51460	Medicaments containing hormones (wholesale)
	51460	Medicaments containing penicillins or other antibiotics (wholesale)
	51460	Orthopaedic goods (wholesale)
	51460	Pace-makers (wholesale)
	51460	Patent medicines (wholesale)
	51460	Pharmaceutical chemist (wholesale)
	51460	Pharmaceutical goods exporter (wholesale)
	51460	Pharmaceutical goods importer (wholesale)
	51460	Posphoaminolipids, amides and their salts and derivatives (wholesale)
	51460	Prosthesis and orthopaedic appliances (wholesale)
	51460	Provitamins, vitamins and their derivatives (wholesale)
	51460	Quaternary ammonium salts and hydroxides (wholesale)
	51460	Sterilisers for medical, surgical or laboratory use (wholesale)
	51460	Steroids used primarily as hormones (wholesale)
	51460	Sulphonamides (wholesale)
	51460	Surgical and dental instruments and appliances (wholesale)
	51460	Syringes, needles, catheters and cannulae (wholesale)
	51460	Ultra-violet and infra-red apparatus for medical use (wholesale)
	51460	Veterinary drugs (wholesale)
46470		**Wholesale of furniture, carpets and lighting equipment**
	51479	Antiques (wholesale)
	51479	Carpets (wholesale)

UK SIC 2007	UK SIC 2003	Activity	SIC 2007	SIC 2003	Activity
	51439	Christmas tree lights (wholesale)		51479	Greeting cards (wholesale)
	51471	Furnishing contractor (wholesale)		51479	Gymnasium and athletic articles and equipment (wholesale)
	51471	Furniture (wholesale)			
	51471	Furniture exporter (wholesale)		51479	Handbags (wholesale)
	51471	Furniture importer (wholesale)		51479	Household goods n.e.c. exporter (wholesale)
	51439	Lighting equipment (wholesale)		51479	Household goods n.e.c. Importer (wholesale)
	51479	Non-electrical lamps and light fittings (wholesale)		51479	Household non-electrical appliances (wholesale)
	51479	Rugs (wholesale)		51479	Inflatable boats for pleasure or sports (wholesale)
	51439	Wires and switches for domestic lighting use (wholesale)		51479	Inflatable vessels for pleasure or sports (wholesale)
				51479	Jewellers' materials (wholesale)
46480		**Wholesale of watches and jewellery**		51479	Leather goods (wholesale)
	51479	Clocks (wholesale)		51479	Liquid or liquefied-gas fuels for lighters in containers (300 cc or more) (wholesale)
	51474	Imitation jewellery (wholesale)			
	51474	Imitation jewellery exporter (wholesale)		51479	Magazines (wholesale)
	51474	Imitation jewellery importer (wholesale)		51479	Matches (wholesale)
	51474	Imitation pearls (wholesale)		51440	Metalware for domestic use (wholesale)
	51473	Jewellery (wholesale)		51479	Newspapers (wholesale)
	51473	Jewellery exporter (wholesale)		51479	Ovens, cookers, cooking plates, boiling rings, grills and roasters (non-electric) (wholesale)
	51473	Jewellery importer (wholesale)			
	51479	Watches and clocks (wholesale)		51479	Pencils, crayons, leads, drawing charcoals, writing or drawing chalks and tailors chalk (wholesale)
46491		**Wholesale of musical instruments -**		51479	Perambulators (wholesale)
	51475	Musical instruments (wholesale)		51477	Playing cards (wholesale)
	51475	Musical instruments exporter (wholesale)		51479	Propelling or sliding pencils (wholesale)
	51475	Musical instruments importer (wholesale)		51477	Puzzles (wholesale)
46499		**Wholesale of household goods (other than musical instruments) n.e.c.**		51479	Saddlery and leather goods (wholesale)
				51479	Sailboats for pleasure or sports (wholesale)
	51479	Air heaters and hot air distributors (non-electric) (wholesale)		51479	Sails (wholesale)
				51477	Scale models (wholesale)
	51479	Artificial flowers, foliage and fruit (wholesale)		51479	Scuba diving breathing equipment (wholesale)
	51479	Baby carriages (wholesale)		51479	Sealing or numbering stamps (wholesale)
	51479	Balloons, dirigibles and other non-powered aircraft (wholesale)		51479	Smoking pipes and cigarette and cigar holders (wholesale)
	51479	Ball-point, felt-tipped and other porous-tipped pens and markers (wholesale)		51479	Snow skis, ice skates and roller skates (wholesale)
				51479	Special sports footwear such as ski boots (wholesale)
	51479	Bicycles and their parts and accessories (wholesale)			
	51479	Books (wholesale)		51479	Sports goods (wholesale)
	51479	Brooms and brushes for domestic use (wholesale)		51479	Stationers' sundries (wholesale)
	51479	Candles and tapers (wholesale)		51479	Stationery (wholesale)
	51479	Cigarette lighters (wholesale)		51479	Sun umbrellas and garden umbrellas (wholesale)
	51479	Collectors stamps and coins (wholesale)		51479	Swimming and paddling pools (wholesale)
	51479	Combs, hair-slides, hairpins, curling pins (wholesale)		51479	Tents (wholesale)
	51477	Construction and constructional toys (wholesale)		51479	Tooth brushes (wholesale)
	51479	Cork goods (wholesale)		51477	Toy trains and accessories (wholesale)
	51440	Cutlery (wholesale)		51477	Toys (wholesale)
	51479	Cycles (wholesale)		51477	Toys and games exporter (wholesale)
	51477	Dolls (wholesale)		51477	Toys and games importer (wholesale)
	51477	Dolls' carriages (wholesale)		51478	Travel accessories (wholesale)
	51479	Domestic ironmongery (wholesale)		51478	Travel and fancy goods exporter (wholesale)
	51440	Earthenware (wholesale)		51478	Travel and fancy goods importer (wholesale)
	51478	Fancy goods (wholesale)		51479	Typewriter ribbons (wholesale)
	51477	Festive, carnival or other entertainment articles, conjuring tricks and novelty jokes (wholesale)		51479	Vessels for pleasure or sports (wholesale)
				51477	Video games (wholesale)
	51479	Fishing rods, line fishing tackle and articles for hunting or fishing (wholesale)		51477	Video games of a kind used with a television receiver (wholesale)
	51479	Fountain pens, Indian ink drawing pens, stylograph pens and other pens (wholesale)		51479	Watch and clock movements (wholesale)
				51479	Water-skis, surf-boards, sail-boards and other water-sport equipment (wholesale)
	51477	Games and toys (wholesale)			
	51479	Gliders and hang-gliders (wholesale)		51477	Wheeled toys designed to be ridden by children (wholesale)

UK SIC 2007	UK SIC 2003	Activity	SIC 2007	SIC 2003	Activity
	51479	Whips and riding crops (wholesale)		51860	X-ray, alpha, beta or gamma radiation apparatus (wholesale)
	51479	Wickerwork (wholesale)			
	51479	Wooden ware (wholesale)	**46610**		**Wholesale of agricultural machinery, equipment and supplies**
	51479	Writing implement sets (wholesale)		51880	Agricultural and forestry machinery (wholesale)
	51479	Yachts (wholesale)		51880	Agricultural machinery and accessories and implements, including tractors exporter (wholesale)
46510		**Wholesale of computers, computer peripheral equipment and software**		51880	Agricultural machinery and accessories and implements, including tractors importer (wholesale)
	51840	Computers and peripheral equipment (wholesale)		51880	Bee-keeping machines (wholesale)
	51840	Laser printers (wholesale)		51880	Dairy farm machinery (wholesale)
	51840	Magnetic or optical readers (wholesale)		51880	Forestry machinery, accessories and implements (wholesale)
	51840	Software (non customised) (wholesale)		51880	Gardening tools (wholesale)
46520		**Wholesale of electronic and telecommunications equipment and parts**		51880	Harvesters (wholesale)
				51880	Horticultural machinery (wholesale)
	51431	Blank audio and video tapes and diskettes, magnetic and optical disks (CDs, DVDs) (wholesale)		51880	Lawn mowers however operated (wholesale)
				51880	Manure spreaders, seeders (wholesale)
	51860	Cathode-ray oscilloscopes and cathode-ray oscillographs (wholesale)		51880	Milking machines (wholesale)
				51880	Ploughs (wholesale)
	51860	Cathode-ray television picture tubes, television camera tubes, other cathode-ray tubes (wholesale)		51880	Poultry-keeping machines (wholesale)
				51880	Threshers (wholesale)
	51860	Diodes, transistors, thyristors, diacs and triacs (wholesale)		51880	Tractors (wholesale)
				51880	Tractors for agricultural use (wholesale)
	51860	Direction finding compasses, other navigational instruments and appliances (electronic) (wholesale)		51880	Tractors used in forestry (wholesale)
				51880	Wheeled tractor (wholesale)
	51860	Electronic and telecommunications equipment and parts (wholesale)	**46620**		**Wholesale of machine tools**
	51860	Electronic components (wholesale)		51810	Computer numerically controlled (CNC) machine tools (wholesale)
	51860	Electronic integrated circuits and micro-assemblies (wholesale)		51810	Computer-controlled machine tools (wholesale)
	51860	Electronic machinery, apparatus and materials for professional use (wholesale)		51810	Jigs and gauges (wholesale)
				51810	Machine tools (wholesale)
	51860	Electronic tubes (wholesale)		51810	Machine tools exporter (wholesale)
	51860	Electronic valves (wholesale)		51810	Machine tools importer (wholesale)
	51860	Instruments and apparatus for measuring electrical quantities (electronic) (wholesale)	**46630**		**Wholesale of mining, construction and civil engineering machinery**
	51860	Instruments for measuring flow, level, pressure etc. of liquids or gassed (electronic) (wholesale)		51820	Blades for bulldozers and angle-dozers (wholesale)
				51820	Boring and sinking machinery (wholesale)
	51860	Integrated circuits (wholesale)		51820	Bulldozers and angle-dozers (wholesale)
	51860	Light emitting diodes (wholesale)		51820	Civil engineering machinery and equipment (wholesale)
	51860	Line telegraphy or telegraphy apparatus (wholesale)		51820	Coal or rock cutters and tunnelling machinery (wholesale)
	51860	Magnetrons, klystrons and microwave tubes (wholesale)		51820	Construction machinery (wholesale)
				51820	Construction machinery exporter (wholesale)
	51860	Microchips (wholesale)		51820	Construction machinery importer (wholesale)
	51860	Mounted piezo-electric crystals (wholesale)		51820	Contractors' plant (wholesale)
	51860	Printed circuits (wholesale)		51820	Front-end shovel loaders (wholesale)
	51860	Radar apparatus (wholesale)		51820	Graders and levellers (wholesale)
	51860	Radio navigational aid apparatus (wholesale)		51820	Mechanical shovels, shovel loaders, excavators, with 360 degree revolving superstructure (wholesale)
	51860	Radio remote control apparatus (wholesale)			
	51860	Reception apparatus for radio telephony or telegraphy for professional use (wholesale)		51820	Mining machinery and equipment (wholesale)
				51820	Pit-head winding gear (wholesale)
	51860	Satellite navigation (wholesale)		51820	Pumps for concrete (wholesale)
	51860	Semi-conductor devices (wholesale)		51820	Road rollers (wholesale)
	51860	Telecommunication instruments and apparatus (wholesale)		51820	Scrapers (wholesale)
				51820	Tamping machines (wholesale)
	51860	Telecommunications equipment (wholesale)		51820	Winches specially designed for use underground (wholesale)
	51860	Telecommunications machinery, equipment and materials for professional use (wholesale)			
	51860	Telephone and communications equipment (wholesale)			
	51860	Television cameras for professional use (wholesale)			
	51860	Transmission kit for radio-telephony and telegraphy, radio or television broadcasting (wholesale)			

UK SIC 2007	UK SIC 2003	Activity	SIC 2007	SIC 2003	Activity
46640		**Wholesale of machinery for the textile industry and of sewing and knitting machines**		51870	Ballasts for discharge lamps or tubes (wholesale)
	51830	Auxiliary machinery for use with machines for working textiles (wholesale)		51870	Bearing housings and plain shaft bearings (wholesale)
	51830	Cleaning, wringing, ironing, pressing, dyeing, etc. Machines for textile yarn and fibres (wholesale)		51870	Bombs, missiles and other projectiles (wholesale)
	51830	Computer-controlled machinery for sewing and knitting machines (wholesale)		51870	Book-binding and book-sewing machinery (wholesale)
	51830	Computer-controlled machinery for the textile industry (wholesale)		51870	Buckets, grabs, shovels and grips for cranes, excavators and the like (wholesale)
	51830	Extruding, drawing, texturing or cutting machinery for man-made textile materials (wholesale)		51870	Carbon electrodes and other articles of graphite or other carbon for electrical purposes (wholesale)
	51830	Felt finishing machinery (wholesale)		51870	Cartridges and other ammunition (wholesale)
	51830	Footwear manufacturing and repairing machinery (wholesale)		51870	Centrifuges (wholesale)
	51830	Knitting machines (wholesale)		51870	Cereal and dried vegetable milling or working machinery (wholesale)
	51830	Leather, hides and skins working machinery (wholesale)		51870	Clutches and shaft couplings including universal joints (excluding motor vehicles) (wholesale)
	51830	Sewing machines (wholesale)		51870	Co-axial cable and co-axial conductors for industrial use (wholesale)
	51830	Stitch bonding machinery (wholesale)		51870	Compressors for refrigerating equipment (wholesale)
	51830	Textile fibre preparation machinery (wholesale)		51870	Compressors for use in civil aircraft (wholesale)
	51830	Textile industry machinery (wholesale)		51870	Containers designed for carriage by one or more means of transport (wholesale)
	51830	Textile spinning, doubling or twisting machinery (wholesale)		51870	Converters, ladles, ingot moulds and casting machines (wholesale)
	51830	Textile weaving machinery (wholesale)		51870	Cooking or heating equipment for non-domestic use (wholesale)
	51830	Textile winding or reeling machinery (wholesale)		51870	Crown corks and stoppers (wholesale)
	51830	Tufting machinery (wholesale)		51870	Dairy machinery (not farm) (wholesale)
46650		**Wholesale of office furniture**		51870	Derricks, cranes, mobile lifting frames (wholesale)
	51850	Furniture for offices (wholesale)		51870	Direction finding compasses and other navigational instruments and appliances (wholesale)
	51850	Furniture for schools (wholesale)		51870	Dish washing machines for commercial use (wholesale)
	51850	Office furniture (wholesale)		51870	Distilling or rectifying plant (wholesale)
46660		**Wholesale of other office machinery and equipment**		51870	Drafting tables and other drawing, marking out or mathematical calculating instruments (wholesale)
	51850	Adding machines (wholesale)		51870	Dry cleaning machines and laundry type washing machines (wholesale)
	51850	Armoured or reinforced safes, strong boxes and doors made of base metal (wholesale)		51870	Dryers for wood, paper pulp, paper or paperboard (wholesale)
	51850	Automatic typewriters and word-processing machines (wholesale)		51870	Drying machines with a capacity exceeding 10 kgs (wholesale)
	51850	Calculating and accounting machines (wholesale)		51870	Electrical apparatus for line telephony or telegraphy (wholesale)
	51850	Cash registers and similar machines incorporating a calculating device (wholesale)		51870	Electrical appliances, accessories and fittings for industry (wholesale)
	51850	Machinery and equipment for offices (wholesale)		51870	Electrical insulators and insulating fittings for electrical machines or equipment (wholesale)
	51850	Offset sheet fed printing machinery for offices (wholesale)		51870	Electrical machinery and apparatus and materials for professional use (wholesale)
	51850	Photo-copying apparatus (wholesale)		51870	Electrical motors (wholesale)
	51850	Typewriters (wholesale)		51870	Engineers' plant and stores (wholesale)
46690		**Wholesale of other machinery and equipment**		51870	Engines for aircraft (wholesale)
	51870	AC and DC electrical motors (single or multi-phase) (wholesale)		51870	Filters for oil, petrol and air for internal combustion engines (not motor vehicle) (wholesale)
	51870	Acetylene gas generators (wholesale)		51870	Filtration equipment and apparatus (wholesale)
	51870	Air conditioning machines (wholesale)		51870	Fire extinguishers (excluding motor vehicle) (wholesale)
	51870	Aircraft (wholesale)			
	51870	Aircraft launching gear and deck arresters (wholesale)		51870	Firearms, sporting, hunting or target shooting rifles (wholesale)
	51870	Articulated link chain (except for motor vehicles and bicycles) (wholesale)		51870	Flywheels and pulleys including pulley blocks (wholesale)
	51870	Automatic regulating or controlling instruments and apparatus (wholesale)		51870	Food and beverage machinery (wholesale)
	51870	Bakery ovens (wholesale)			
	51870	Balances and scales (wholesale)			
	51870	Ball and roller bearings (wholesale)			

UK SIC 2007	UK SIC 2003	Activity	SIC 2007	SIC 2003	Activity
	51870	Food and drink preparation and manufacturing machinery for industrial use (wholesale)		51870	Moulds, moulding boxes for metal foundries, mould bases and moulding patterns (wholesale)
	51870	Food, beverage and tobacco industry machinery (wholesale)		51870	Navigation machinery (wholesale)
	51870	Forklift trucks (wholesale)		51870	Ophthalmic instruments (wholesale)
	51870	Furnace burners, mechanical stokers and grates and mechanical ash dischargers (wholesale)		51870	Optical fibre cables made up of individually sheathed fibres (wholesale)
	51870	Furnaces, ovens, incinerators, for industrial or laboratory use (excluding bakery ovens) (wholesale)		51870	Paper and paperboard production machinery (wholesale)
	51870	Fuses, relays and apparatus for protecting electrical circuits (wholesale)		51870	Permanent magnets and electro-magnetic couplings, clutches and brakes (wholesale)
	51870	Garage tools (wholesale)		51870	Physical or chemical analysis instruments and apparatus (non-electronic) (wholesale)
	51870	Gaskets (excluding motor vehicle) (wholesale)		51870	Plugs, sockets and the like, for switching or protecting industrial electrical circuits (wholesale)
	51870	Gears, gearing, ball screws, gear boxes, other speed changers (excluding motor vehicles) (wholesale)		51870	Pneumatic and other continuous action elevators and conveyors for goods or materials (wholesale)
	51870	Generating sets (wholesale)		51870	Primary cells and primary batteries for industrial use (wholesale)
	51870	Ground flying trainers (wholesale)			
	51870	Hand or foot operated air pumps (wholesale)		51870	Printing block and plates preparation and production machinery, equipment and apparatus (wholesale)
	51870	Heat exchange units and machinery for liquefying air and other gases (wholesale)		51870	Printing machinery (wholesale)
	51870	Helicopters (wholesale)		51870	Process control valves, gate valves, globe valves and other valves (wholesale)
	51870	Hydraulic and pneumatic power engines and motors (wholesale)		51870	Producer gas or water gas generators (wholesale)
	51870	Hydraulic turbines and water wheels (wholesale)		51870	Production line robot (wholesale)
	51870	Hydrometers, non-medical thermometers, pyrometers, barometers, hygrometers and the like (wholesale)		51870	Public address equipment (wholesale)
				51870	Pulley tackle and hoists (wholesale)
	51870	Illuminated signs and name-plates (wholesale)		51870	Pumping plant (wholesale)
	51870	Induction or dielectric heating equipment for industrial or laboratory use (wholesale)		51870	Pumps for liquids (excluding motor vehicles) (wholesale)
	51870	Instruments for measuring or checking flow, level, pressure etc. of liquids or gases (wholesale)		51870	Railway or tramway coaches, vans and wagons (wholesale)
	51870	Insulated winding wire (wholesale)		51870	Reciprocating displacement compressors (wholesale)
	51870	Jacks and hoists of a kind used for raising vehicles (wholesale)		51870	Reciprocating positive displacement pumps for liquids (wholesale)
	51870	Laundry-type washing and dry-cleaning machines (wholesale)		51870	Refrigerating and freezing equipment and heat pumps (commercial) (wholesale)
	51870	Lifting and handling equipment (wholesale)		51870	Reservoirs, tanks and containers of metal (not for central heating) (300 litres or more) (wholesale)
	51870	Lifts, skip hoists, escalators and moving walkways (wholesale)		51870	Revolution and production counters (wholesale)
	51870	Light metal containers (wholesale)		51870	Revolvers, pistols and other firearms and similar devices (wholesale)
	51870	Linear acting (cylinders) hydraulic and pneumatic power engines and motors (wholesale)		51870	Rotary displacement compressors (single or multi shaft) (wholesale)
	51870	Liquid dielectric transformers (wholesale)		51870	Rotary positive displacement pumps for liquids (wholesale)
	51870	Locomotives (wholesale)			
	51870	Machinery and apparatus for filtering or purifying gases (wholesale)		51870	Rubber or plastics working machinery (wholesale)
	51870	Machinery n.e.c., for use in trade, navigation and other services (wholesale)		51870	Scene lighting, road lighting and other lighting (not domestic) (wholesale)
	51870	Machinery n.e.c., for industrial use (except mining, construction, textile) (wholesale)		51870	Ships and boats for the carriage of passengers or goods (wholesale)
	51870	Machinery n.e.c., for industrial use (except mining, construction, textile) exporter (wholesale)		51870	Ships and boats propellers and blades (wholesale)
	51870	Machinery n.e.c., for industrial use (except mining, construction, textile) importer (wholesale)		51870	Snow-ploughs and blowers (wholesale)
	51870	Machines and appliances for testing the mechanical properties of materials (wholesale)		51870	Spark ignition and compression ignition engines (except motor vehicle and outboard) (wholesale)
	51870	Magnetic lifting heads (wholesale)		51870	Special purpose machinery n.e.c. (wholesale)
	51870	Marine propulsion engines (wholesale)		51870	Speed indicators and tachometers (wholesale)
	51870	Measuring instruments and equipment (wholesale)		51870	Static converters (wholesale)
	51870	Metal rolling mills (wholesale)		51870	Steam and sand blasting machinery and appliances (excluding agricultural) (wholesale)
	51870	Microscopes (except optical) and diffraction equipment (wholesale)		51870	Steam generators (wholesale)
	51870	Motorised tanks and other armoured fighting vehicles (wholesale)		51870	Straddle carriers and works trucks fitted with a crane (wholesale)

UK SIC 2007	UK SIC 2003	Activity
	51870	Stranded wires, cables, plaited bands, slings and the like (not electrically insulated) (wholesale)
	51870	Stroboscopes (wholesale)
	51870	Surveying, hydrographic, oceanographic, hydrological and meteorological instruments (wholesale)
	51870	Switches (wholesale)
	51870	Swords, cutlasses, bayonets, lances and similar arms (wholesale)
	51870	Tanks, casks, drums, cans, boxes and similar containers (excluding for gas) (wholesale)
	51870	Taximeters (wholesale)
	51870	Therapeutic instruments and appliances (wholesale)
	51870	Tobacco preparation and making-up machinery (wholesale)
	51870	Tractors of a type used on railway station platforms (wholesale)
	51870	Transformers (wholesale)
	51870	Transport equipment (except motor vehicles, motorcycles and bicycles) (wholesale)
	51870	Tugs and pusher craft (wholesale)
	51870	Turbo-compressors (wholesale)
	51870	Type-setting machinery, equipment and apparatus (wholesale)
	51870	Vacuum pumps (wholesale)
	51870	Vending machines (wholesale)
	51870	Weighing machines and scales for commercial use (wholesale)
	51870	Winches and capstans (wholesale)
	51870	Wine, cider, fruit beverage production machinery (wholesale)
	51870	Wire for industrial use (wholesale)
	51870	Wire, switches and other installation equipment for industrial use (wholesale)
	51870	Works trucks (wholesale)
46711		**Wholesale of petroleum and petroleum products**
	51511	Butane (wholesale)
	51511	Crude oil (wholesale)
	51511	Crude petroleum (wholesale)
	51511	Diesel fuel (wholesale)
	51511	Fuel oil bulk distribution (wholesale)
	51511	Gaseous petroleum fuels (wholesale)
	51511	Gasoline (wholesale)
	51511	Liquefied petroleum gases (wholesale)
	51511	Liquid petroleum fuels (wholesale)
	51511	Motor spirit distribution (wholesale)
	51511	Petroleum and petroleum products exporter (wholesale)
	51511	Petroleum and petroleum products importer (wholesale)
	51511	Petroleum products distribution (wholesale)
	51511	Propane gas (wholesale)
	51511	Refined petroleum products (wholesale)
46719		**Wholesale of fuels and related products (other than petroleum and petroleum products)**
	51519	Charcoal (wholesale)
	51519	Coal (wholesale)
	51519	Coal depot (wholesale)
	51519	Coal merchant (wholesale)
	51519	Coke (wholesale)

SIC 2007	SIC 2003	Activity
	51519	Coke merchant (wholesale)
	51519	Culm (wholesale)
	51519	Enriched uranium supply to nuclear reactors (wholesale)
	51519	Fuels (other than petroleum) (wholesale)
	51519	Fuels exporter (other than petroleum) (wholesale)
	51519	Fuels importer (other than petroleum) (wholesale)
	51519	Gas bottling and distribution (wholesale)
	51519	Gas oil (wholesale)
	51519	Gaseous fuels (other than petroleum) (wholesale)
	51519	Greases (wholesale)
	51519	Heating oil (wholesale)
	51519	Kerosene (wholesale)
	51519	Liquid fuels (other than petroleum) (wholesale)
	51519	Lubricants (wholesale)
	51519	Lubricating oils and greases (wholesale)
	51519	Oil merchant (wholesale)
	51519	Patent fuel (wholesale)
	51519	Peat (wholesale)
	51519	Solid fuels (wholesale)
46720		**Wholesale of metals and metal ores**
	51520	Copper (wholesale)
	51520	Ferrous and non-ferrous metal ores (wholesale)
	51520	Ferrous and non-ferrous metals in primary forms (wholesale)
	51520	Ferrous and non-ferrous semi-finished metal products n.e.c. (wholesale)
	51520	Galvanised sheets (wholesale)
	51520	Gold and other precious metals (wholesale)
	51520	Iron (wholesale)
	51520	Iron yard (wholesale)
	51520	Lead (wholesale)
	51520	Metal stockholder (wholesale)
	51520	Metals (wholesale)
	51520	Metals and metal ores exporter (wholesale)
	51520	Metals and metal ores importer (wholesale)
	51520	Ores (wholesale)
	51520	Spelter (wholesale)
	51520	Steel (wholesale)
	51520	Steel stockholder (wholesale)
	51520	Tinplate (wholesale)
	51520	Zinc (wholesale)
46730		**Wholesale of wood, construction materials and sanitary equipment**
	51530	Baths (wholesale)
	51530	Builders' carpentry and joinery of metal (wholesale)
	51530	Cement (wholesale)
	51530	Clay (wholesale)
	51530	Construction materials (wholesale)
	51530	Doors (wholesale)
	51530	Flagstone merchant (wholesale)
	51530	Flat glass (wholesale)
	51479	Floor coverings (wholesale)
	51530	Ganister (wholesale)
	51530	Granite (wholesale)
	51530	Gravel (wholesale)
	51530	Hardwoods (wholesale)
	51530	Limestone (wholesale)

UK SIC 2007	UK SIC 2003	Activity	SIC 2007	SIC 2003	Activity
	51479	Linoleum (wholesale)		51540	Plumbing equipment and supplies (wholesale)
	51530	Logged timber (wholesale)		51540	Reservoirs, tanks and containers of metal for central heating (300 litres or more) (wholesale)
	51530	Marble (wholesale)		51540	Sanitary installation connections, rubber pipes (wholesale)
	51530	Mastics and sealants (wholesale)			
	51530	Paint and varnish (wholesale)		51540	Sanitary installation equipment (wholesale)
	51530	Paint, varnish and lacquer (wholesale)		51540	Sanitary installation taps, t-pieces (wholesale)
	51530	Pit props (wholesale)		51540	Sanitary installation tubes, pipes, fittings (wholesale)
	51530	Plaster (wholesale)		51540	Saws and sawblades (wholesale)
	51530	Plasterboards (wholesale)		51540	Saws, screwdrivers and similar hand tools (wholesale)
	51530	Plywood (wholesale)			
	51530	Precast concrete products (wholesale)		51540	Screwdrivers (wholesale)
	51530	Prefabricated buildings (wholesale)		51540	Threaded and non-threaded fasteners (wholesale)
	51530	Sand (wholesale)		51540	Tools (wholesale)
	51530	Sand and gravel merchant (wholesale)		51540	Ventilation equipment (wholesale)
	51530	Sanitary porcelain (wholesale)		51540	Wire netting (wholesale)
	51530	Sanitary ware (wholesale)			
	51530	Sheet glass merchant (wholesale)	**46750**		**Wholesale of chemical products**
	51530	Slate (wholesale)		51550	Acids (wholesale)
	51530	Slate slabs (wholesale)		51550	Acrylic polymers in primary forms (wholesale)
	51479	Slates and boards (wholesale)		51550	Agro-chemical products (wholesale)
	51530	Sleepers (wholesale)		51550	Aldehyde function compounds (wholesale)
	51530	Stones (wholesale)		51550	Amino resins, phenolic resins and polyurethanes in primary forms (wholesale)
	51530	Tiles (wholesale)			
	51530	Timber importer (wholesale)		51550	Aniline (wholesale)
	51530	Timber merchant (wholesale)		51550	Anti-sprouting products (wholesale)
	51530	Timber yard (wholesale)		51550	Carbonates (wholesale)
	51530	Toilets (wholesale)		51550	Chemical glues (wholesale)
	51530	Varnishes (wholesale)		51550	Chemical products (wholesale)
	51530	Wall boards (wholesale)		51550	Chemical products exporter (wholesale)
	51440	Wallpaper (wholesale)		51550	Chemical products importer (wholesale)
	51530	Washbasins (wholesale)		51550	Chromium, manganese, lead and copper oxides and hydroxides (wholesale)
	51530	Wood (wholesale)			
	51530	Wood in the rough (wholesale)		51550	Colouring matter (wholesale)
	51530	Wood products of primary processing (wholesale)		51550	Compounds of rare earth metals (wholesale)
	51530	Wood, construction materials and sanitary equipment exporter (wholesale)		51550	Compounds of yttrium and scandium (wholesale)
				51550	Cotton size (wholesale)
	51530	Wood, construction materials and sanitary equipment importer (wholesale)		51550	Cyanides, cyanide oxides and complex cyanides (wholesale)
				51550	Cyclic hydrocarbons (wholesale)
46740		**Wholesale of hardware, plumbing and heating equipment and supplies**		51550	Depleted uranium and thorium and their compounds (wholesale)
	51540	Barbed wire and stranded wire (wholesale)		51550	Diols, polyalcohols, cyclical alcohols and their derivatives (wholesale)
	51540	Drainpipes (wholesale)			
	51540	Fittings and fixtures (wholesale)		51550	Distilled water (wholesale)
	51540	Hammers (wholesale)		51550	Dyes (wholesale)
	51540	Hand tools (wholesale)		51550	Enzymes (wholesale)
	51540	Hardware equipment and supplies (wholesale)		51550	Essential oils (wholesale)
	51540	Hardware, plumbing and heating equipment (wholesale)		51550	Essential oils merchant (wholesale)
				51550	Ethers, organic peroxides, epoxides, acetals and hemiacetals and their derivatives (wholesale)
	51540	Hardware, plumbing and heating equipment and supplies exporter (wholesale)		51550	Explosives (wholesale)
				51550	Fertilisers (wholesale)
	51540	Hardware, plumbing and heating equipment and supplies importer (wholesale)		51550	Flavourings (wholesale)
				51550	Fulminates, cyanates and thiocyanates (wholesale)
	51540	Heat insulated tanks for central heating (wholesale)		51550	Fungicides, rodenticides and similar products (wholesale)
	51540	Heating equipment and supplies (wholesale)			
	51540	Hot water heaters (wholesale)		51550	Guano (wholesale)
	51540	Ironmonger (wholesale)		51550	Gums (wholesale)
	51540	Locks (wholesale)		51550	Halogen or sulphur compounds of non-metals (wholesale)
	51540	Nails, tacks, drawing pins and staples (wholesale)			
	51540	Plumbers' merchant (wholesale)			

UK SIC 2007	UK SIC 2003	Activity	SIC 2007	SIC 2003	Activity
	51550	Herbicides and insecticides (wholesale)		51550	Titanium oxide (wholesale)
	51550	Heterocyclic compounds (wholesale)		51550	Urea, thiourea and melamine resins in primary forms (wholesale)
	51550	Hydrogen chloride (wholesale)			
	51550	Hydrogen peroxide (wholesale)		51550	Vegetable or resin product derivatives (wholesale)
	51550	Hypochlorites, chlorates and perchlorates (wholesale)		51550	Wood charcoal (not fuel) (wholesale)
	51550	Indigo (wholesale)		51550	Zinc oxide and peroxide (wholesale)
	51550	Industrial chemicals (wholesale)	**46760**		**Wholesale of other intermediate products**
	51550	Industrial dyes (wholesale)		51560	Fluorspar (wholesale)
	51550	Industrial fatty alcohols (wholesale)		51560	Moss litter (wholesale)
	51550	Industrial gases (wholesale)		51560	Mungo (carded or combed) (wholesale)
	51550	Industrial monocarboxylic fatty acids and acid oils from refining (wholesale)		51560	Noil (wholesale)
				51560	Other intermediate products exporter (wholesale)
	51550	Industrial salt (wholesale)		51560	Other intermediate products importer (wholesale)
	51550	Insecticides (wholesale)		51560	Paper bags (wholesale)
	51550	Ketone and quinone function compounds (wholesale)		51560	Paper boards (wholesale)
	51550	Liquid and compressed air (wholesale)		51560	Paper in bulk (wholesale)
	51550	Manure (wholesale)		51560	Paper merchant (wholesale)
	51550	Metal oxides, hydroxides and peroxides (wholesale)		51550	Plastic materials in primary forms (wholesale)
	51550	Metallic halogenates (wholesale)		51560	Plastic packaging (wholesale)
	51550	Metalloids (wholesale)		51550	Polyamides in primary forms (wholesale)
	51550	Methanol (wholesale)		51550	Polyethers, polyesters, polycarbonates, alkyd and epoxide resins (wholesale)
	51550	Monohydric alcohols (wholesale)			
	51550	Nitrate of soda importer (wholesale)		51550	Polymers of ethylene and styrene in primary forms (wholesale)
	51550	Oils and other products of distilling of high temperature coal tar, pitch and pitch tar (wholesale)		51560	Precious stones (wholesale)
				51550	Rubber (wholesale)
	51550	Organo-sulphur and other organo-inorganic compounds (wholesale)		51560	Shoddy (carded or combed) (wholesale)
				51550	Silicones in primary forms (wholesale)
	51550	Paraffin (wholesale)		51550	Synthetic or reconstructed precious or semi-precious unworked stones (wholesale)
	51550	Petroleum coke, bitumen and other residues of petroleum oils (wholesale)			
				51550	Synthetic rubber (wholesale)
	51550	Phenols, phenol-alcohols and derivatives of phenols (wholesale)		51560	Tallow (wholesale)
				51560	Textile fibres (wholesale)
	51550	Phosphides, carbides, hydrides, nitrides azides, silicides and borides (wholesale)		51560	Tops (wholesale)
				51550	Waste, parings and scrap of plastic (wholesale)
	51550	Phosphinates, phosphonates, phosphates and polyphosphates (wholesale)		51560	Woodpulp and paper making materials (wholesale)
	51550	Phosphoric esters and esters of other inorganic acids and their derivatives (wholesale)		51560	Wool merchant (wholesale)
			46770		**Wholesale of waste and scrap**
	51550	Piezo-electric quartz (wholesale)		51570	Car dismantlers (wholesale)
	51550	Plant growth regulators (wholesale)		51570	Collecting, sorting, separating, stripping of used goods to obtain reusable parts (wholesale)
	51550	Polymers of vinyl chloride and other halogenated olefins in primary forms (wholesale)			
				51570	Construction materials from demolished buildings (wholesale)
	51550	Printing ink (wholesale)			
	51550	Radioactive residues (wholesale)		51570	Cotton rags (wholesale)
	51550	Refined sulphur (wholesale)		51570	Cotton waste (wholesale)
	51550	Roasted iron pyrites (wholesale)		51570	Dismantling of automobiles, computers, and other equipment for re-sale of usable parts (wholesale)
	51550	Scents (wholesale)			
	51550	Silicates, borates, perborates and other salts of inorganic acids or peroxoacids (wholesale)		51570	Engine cleaning waste (wholesale)
				51570	Glass waste (wholesale)
	51550	Silicon and sulphur dioxide (wholesale)		51570	Marine store waste (wholesale)
	51550	Soda (wholesale)		51570	Metal and non-metal waste and scrap and materials for recycling (wholesale)
	51550	Sodium nitrate (wholesale)			
	51550	Starch derivatives (wholesale)		51570	Packing, repacking, storage and delivery of waste and scrap without transformation (wholesale)
	51550	Sulphides, sulphites and sulphates (wholesale)			
	51550	Sulphur (wholesale)		51570	Rag and bone dealer (wholesale)
	51550	Synthetic organic colouring matter and colouring lakes and preparations based on them (wholesale)		51570	Rag merchant
				51570	Sawdust (wholesale)
	51550	Synthetic resin (wholesale)		51570	Scrap (wholesale)
	51550	Tanning preparations (wholesale)		51570	Scrap iron (wholesale)

UK SIC 2007	UK SIC 2003	Activity
	51570	Scrap leather (wholesale)
	51570	Scrap merchant (general dealer) (wholesale)
	51570	Scrap metal (wholesale)
	51570	Tailors' trimmings (wholesale)
	51570	Textile waste (wholesale)
	51570	Upholsterers' trimmings (wholesale)
	51570	Waste (wholesale)
	51570	Waste and scrap exporter (wholesale)
	51570	Waste and scrap importer (wholesale)
	51570	Waste paper (wholesale)
	51570	Waste rubber (wholesale)
	51570	Waste string (wholesale)
	51570	Woollen rag (wholesale)
46900		**Non-specialised wholesale trade**
	51900	Educational supplies (except furniture) (wholesale)
	51900	Feathers (wholesale)
	51900	General dealer (wholesale)
	51900	Goods n.e.c. exporter (wholesale)
	51900	Goods n.e.c. importer (wholesale)
	51900	Industrial materials (general or undefined) (wholesale)
	51900	Machinery (undefined) (wholesale)
	51900	Machinery stockist (undefined) (wholesale)
	51900	Ships chandler (wholesale)
	51900	Wholesale of a variety of goods without any particular specialisation (wholesale)
47110		**Retail sale in non-specialised stores with food, beverages or tobacco predominating**
	52113	Cinema kiosk (retail)
	52111	Confectioners, tobacconists and newsagents (CTN's) (retail)
	52113	Food (general) (retail)
	52113	Frozen food store (retail)
	52112	General store with predominant sale of food beverages or tobacco products (licensed) (retail)
	52113	General store with predominant sale of food beverages or tobacco products (unlicensed) (retail)
	52112	Grocer with alcohol licence (retail)
	52113	Grocer without alcohol licence (retail)
	52112	Hypermarket selling mainly foodstuffs with alcohol licence (retail)
	52113	Hypermarket selling mainly foodstuffs without alcohol licence (retail)
	52112	NAAFI shop with alcohol licence (retail)
	52113	NAAFI shop without alcohol licence (retail)
	52112	Supermarket (selling mainly foodstuffs) with alcohol licence (retail)
	52113	Supermarket (selling mainly foodstuffs) without alcohol licence (retail)
	52112	Superstore (selling mainly foodstuffs) with alcohol licence (retail)
	52113	Superstore (selling mainly foodstuffs) without alcohol licence (retail)
	52112	Village general store (selling mainly foodstuffs) with alcohol licence (retail)
	52113	Village general store (selling mainly foodstuffs) without alcohol licence (retail)
47190		**Other retail sale in non-specialised stores**
	52120	Department stores (retail)

SIC 2007	SIC 2003	Activity
	52120	General stores in which the sale of food beverages or tobacco products is not predominant (retail)
	52120	Household stores (retail)
	52120	Mixed business retailing both food and non food goods but non-food predominating (retail)
47210		**Retail sale of fruit and vegetables in specialised stores**
	52210	Edible nuts (retail)
	52210	Fruit shop (retail)
	52210	Fruiterer (retail)
	52210	Greengrocer (retail)
	52210	Herb seller (food) (retail)
	52210	Herbalist (food) (retail)
	52210	Mushrooms (retail)
	52210	Potatoes (retail)
	52270	Preserved fruit and vegetables (retail)
	52210	Vegetables (retail)
47220		**Retail sale of meat and meat products in specialised stores**
	52220	Butchers shop (retail)
	52220	Cooked meats (retail)
	52220	Game (retail)
	52220	Meat and meat products (retail)
	52220	Meat and meat products in specialised stores (retail)
	52220	Meat dealer (retail)
	52220	Pork butcher (retail)
	52220	Poultry and game (retail)
	52220	Tripe (retail)
47230		**Retail sale of fish, crustaceans and molluscs in specialised stores**
	52230	Crustaceans (retail)
	52230	Eels (retail)
	52230	Fish (retail)
	52230	Fishmonger (retail)
	52230	Molluscs (retail)
	52230	Seafood and seafood products (retail)
	52230	Shellfish (retail)
	52230	Wet fish (retail)
47240		**Retail sale of bread, cakes, flour confectionery and sugar confectionery in specialised stores**
	52240	Baker (retail)
	52240	Bakery (selling main activity) (retail)
	52240	Bread (retail)
	52240	Cakes (retail)
	52240	Chocolate and sweets (retail)
	52240	Flour confectionery (retail)
	52240	Pastry (retail)
	52240	Sugar confectionery (retail)
	52240	Sweets (retail)
47250		**Retail sale of beverages in specialised stores**
	52250	Alcoholic beverages (retail)
	52250	Beer (retail)
	52250	Non-alcoholic beverages (retail)
	52250	Off licence (not public house) (retail)
	52250	Soft drinks (retail)
	52250	Wine and spirit merchant (retail)

UK SIC 2007	UK SIC 2003	Activity	SIC 2007	SIC 2003	Activity
	52250	Wine and spirits (retail)		52450	Compact disc players (retail)
				52450	Dvd players (retail)
47260		**Retail sale of tobacco products in specialised stores**		52450	Hi-fi equipment (retail)
	52260	Cigarettes (retail)		52450	Installation of radios in motor vehicles (retail)
	52260	Cigars (retail)		52450	Radio sets and equipment (retail)
	52260	Smokers' requisites (retail)		52450	Record players (retail)
	52260	Tobacco (retail)		52450	Tape recorders (retail)
	52260	Tobacconist (retail)		52450	Television goods (retail)
				52450	Television sets and equipment (retail)
47290		**Other retail sale of food in specialised stores**		52450	Video equipment (retail)
	52270	Dairy grocer's shop (retail)		52450	Video recorders (retail)
	52270	Dairy products (retail)			
	52270	Dairyman (retail)	**47510**		**Retail sale of textiles in specialised stores**
	52270	Delicatessen shop (retail)		52410	Awnings and sunblinds (retail)
	52270	Eggs (retail)		52410	Bed linen (retail)
	52270	Health foods (retail)		52410	Bedding (retail)
	52270	Tea and coffee grocer (retail)		52410	Clothing fabrics (retail)
	52270	Tea merchant (retail)		52410	Draper (retail)
	52270	Vegetarian foods (retail)		52410	Dress materials (retail)
				52410	Embroidery making materials (retail)
47300		**Retail sale of automotive fuel in specialised stores**		52410	Fabrics (retail)
	50500	Cooling products for motor vehicles		52410	Haberdashery (retail)
	50500	Filling station (motor fuel and lubricants)		52410	Hand knitting yarns (retail)
	50500	Fuel for motor vehicles and motorcycles		52410	Household textiles (retail)
	50500	Lubricating products for motor vehicles		52410	Knitting yarn (retail)
	50500	Motorway services petrol filling station		52410	Needles for sewing (retail)
	50500	Petrol filling station		52410	Piece goods (retail)
				52410	Rug making materials (retail)
47410		**Retail sale of computers, peripheral units and software in specialised stores**		52410	Sewing thread (retail)
	52482	Calculating machines (retail)		52410	Sheets made of textiles (retail)
	52485	Computer games (retail)		52410	Table linen (retail)
	52482	Computer peripheral equipment (retail)		52410	Table-cloths (retail)
	52482	Computers and non-customised software (retail)		52410	Tapestry making materials (retail)
	52482	Office equipment (retail)		52410	Tarpaulins (retail)
	52482	Photocopiers (retail)		52410	Textiles (retail)
	52482	Software (non-customised) (retail)		52410	Towels (retail)
	52482	Typewriters (retail)		52410	Woollen draper (retail)
	52482	Video game consoles (retail)			
			47520		**Retail sale of hardware, paints and glass in specialised stores**
47421		**Retail sale of mobile telephones in specialised stores**		52460	Building materials such as bricks, wood, sanitary equipment (retail)
	52488	Car telephones (retail)		52460	Construction materials (retail)
	52488	Cellular telephones (retail)		52460	DIY equipment (retail)
	52488	Installation of car telephones (retail)		52460	DIY materials (retail)
	52488	Mobile telephones (retail)		52460	Doors (retail)
	52488	Mobile telephones for motor vehicles (retail)		52460	Double glazing (retail)
				52460	Flat glass (retail)
47429		**Retail sale of telecommunications equipment (other than mobile telephones) n.e.c., in specialised stores**		52460	Gardening tools (retail)
	52482	Telecommunications equipment other than mobile telephones (retail)		52460	Glass (retail)
				52460	Hardware (retail)
47430		**Retail sale of audio and video equipment in specialised stores**		52460	Ironmonger (retail)
	52450	Audio equipment (retail)		52460	Lacquers (retail)
	52450	Audio/visual cassettes (retail)		52460	Lawn mowers (retail)
	52450	Audio/visual equipment (retail)		52460	Paint and varnish (retail)
	52450	CD, DVD recorders (retail)		52460	Sanitary equipment (retail)
				52460	Saunas (retail)
				52460	Tiles for wall or floor made of ceramic (retail)
				52460	Tools (not machine tools) (retail)

UK SIC 2007	UK SIC 2003	Activity	SIC 2007	SIC 2003	Activity
	52460	Varnishes (retail)	47610		**Retail sale of books in specialised stores**
				52470	Books (retail)
47530		**Retail sale of carpets, rugs, wall and floor coverings in specialised stores**	**47620**		**Retail sale of newspapers and stationery in specialised stores**
	52481	Carpet tiles (retail)		52470	Chart seller (retail)
	52481	Carpets (retail)		52470	Greetings cards (retail)
	52440	Curtain material (retail)		52470	Map seller (retail)
	52440	Curtains (retail)		52470	Newspapers (retail)
	52481	Floor coverings (retail)		52470	Office supplies (retail)
	52481	Floor tiles (not ceramic) (retail)		52470	Paper (retail)
	52481	Lino tiles (retail)		52470	Pencils (retail)
	52440	Net curtains (retail)		52470	Pens (retail)
	52481	Rugs (retail)		52470	Stationery (retail)
	52489	Wallpaper (retail)			
47540		**Retail sale of electrical household appliances in specialised stores**	**47630**		**Retail sale of music and video recordings in specialised stores**
	52450	Domestic electrical appliances (retail)		52450	Audio and video recordings (retail)
	52450	Electrical appliances, accessories and fittings (retail)		52450	Audio tapes and cassettes (retail)
	52450	Electrical household appliances (retail)		52450	Blank tapes and discs (retail)
	52450	Knitting machines (retail)		52450	Dvds (retail)
	52450	Sewing machines (retail)		52450	Gramophone records (retail)
				52450	Records, compact discs and tapes (retail)
47591		**Retail sale of musical instruments and scores in specialised stores**		52450	Video tapes (retail)
	52450	Music shop (retail)	**47640**		**Retail sale of sporting equipment in specialised stores**
	52450	Musical instruments (retail)		52485	Bicycles (retail)
	52450	Musical scores (retail)		52485	Boats (retail)
	52450	Pianofortes (retail)		52485	Camping goods (retail)
	52450	Sheet music (retail)		52485	Cycle accessories (retail)
				52485	Cycle agent (retail)
47599		**Retail sale of furniture, lighting equipment and other household articles (other than musical instruments) n.e.c., in specialised stores**		52485	Fishing gear (retail)
				52485	Fishing tackle (retail)
				52485	Sports goods (retail)
	52440	Beds (retail)		52485	Sports outfitter (retail)
	52440	China (retail)			
	52440	Cork goods (retail)	**47650**		**Retail sale of games and toys in specialised stores**
	52440	Crockery (retail)		52485	Games and toys (retail)
	52440	Cutlery (retail)		52485	Games apparatus (retail)
	52440	Domestic furniture (retail)		52485	Toys (retail)
	52440	Earthenware (retail)			
	52482	Electrical security alarm systems e.g. Safes and vaults (not installed or maintained) (retail)	**47710**		**Retail sale of clothing in specialised stores**
				52421	Articles of fur (retail)
	52440	Gas appliances (retail)		52424	Braces (retail)
	52440	Glassware (retail)		52422	Children's wear (retail)
	52440	House furnisher (retail)		52421	Clothing accessories made of fur or leather (retail)
	52440	Household articles and equipment n.e.c. (retail)		52422	Clothing accessories, children's, other than of fur or leather (retail)
	52440	Household furnishing articles made of textile materials (retail)		52424	Clothing accessories, men's, other than of fur or leather (retail)
	52440	Household furniture (retail)		52423	Clothing accessories, women's, other than of fur or leather (retail)
	52440	Household non-electrical appliances (retail)		52423	Corsetiere (retail)
	52440	Household utensils (retail)		52423	Dress shop (retail)
	52440	Kitchen units (retail)		52423	Dressmaker (retail)
	52440	Lampshades (retail)		52421	Furrier (retail)
	52440	Lighting equipment (retail)		52421	Gloves made of fur or leather (retail)
	52482	Office furniture (retail)		52422	Gloves, children's, other than of fur or leather (retail)
	52440	Pottery (retail)		52424	Gloves, men's, other than of fur or leather (retail)
	52440	Soft furnishings (retail)			
	52440	Wickerwork goods (retail)			
	52440	Window blinds (retail)			
	52440	Wooden ware (retail)			

UK SIC 2007	UK SIC 2003	Activity	SIC 2007	SIC 2003	Activity
	52423	Gloves, women's, other than of fur or leather (retail)		52330	Toilet goods (retail)
	52422	Infants' clothing (retail)			
	52422	Juvenile outfitter (retail)	**47760**		**Retail sale of flowers, plants, seeds, fertilisers, pet animals and pet food in specialised stores**
	52422	Ladies' outfitter (retail)		52489	Fertilisers (retail)
	52421	Leather clothing for adults (retail)		52489	Florist (retail)
	52424	Men's bespoke tailor (retail)		52489	Flowers (retail)
	52424	Men's clothier and outfitter (retail)		52489	Garden centre (retail)
	52424	Men's clothing (retail)		52489	Garden seeds and plants (retail)
	52424	Men's outfitter (retail)		52489	Pet animals (retail)
	52424	Men's wear dealer (retail)		52489	Pet food (retail)
	52423	Millinery dealer (retail)		52489	Pet shop (retail)
	52424	Ties (retail)		52489	Plants (retail)
	52423	Women's bespoke tailor (retail)		52489	Seeds (retail)
	52423	Women's clothier and outfitter (retail)			
	52423	Women's clothing (retail)	**47770**		**Retail sale of watches and jewellery in specialised stores**
	52423	Women's clothing accessories (retail)		52484	Clocks (retail)
	52423	Women's hats (retail)		52484	Jewellery (retail)
	52423	Women's hosiery (retail)		52484	Watches and clocks (retail)
	52423	Women's outfitter (retail)			
	52423	Women's tailor (retail)	**47781**		**Retail sale in commercial art galleries**
	52423	Women's wear (retail)		52486	Art (retail)
				52486	Commercial art gallery (retail)
47721		**Retail sale of footwear in specialised stores**		52486	Works of art (retail)
	52431	Boots and shoes (retail)			
	52431	Footwear (retail)	**47782**		**Retail sale by opticians**
	52431	Shoes (retail)		52487	Contact lenses (retail)
				52487	Dispensing ophthalmic optician (retail)
47722		**Retail sale of leather goods in specialised stores**		52487	Dispensing optician (retail)
	52432	Handbags (retail)		52487	Dispensing optometrist (retail)
	52432	Leather goods (retail)		52487	Spectacles (retail)
	52432	Saddlery (retail)			
	52432	Travel accessories made of leather and leather substitutes (retail)	**47789**		**Other retail sale of new goods in specialised stores (other than by opticians or commercial art galleries), n.e.c**
	52432	Travel goods (retail)		52489	Ammunition (retail)
	52432	Wallets (retail)		52489	Baby carriages (retail)
				52489	Cleaning materials (retail)
47730		**Dispensing chemist in specialised stores**		52489	Coal and coke (retail)
	52310	Dispensing chemist (retail)		52485	Coins (retail)
	52310	Drug store (retail)		52489	Copper goods (retail)
	52310	Druggist (retail)		52489	Cordage, rope, twine and nets (retail)
	52310	Drugs (retail)		52489	Craftwork (retail)
	52310	Medicine dealer (retail)		52489	Curios (retail)
	52310	Pharmaceutical chemist (retail)		52489	Fancy goods (retail)
	52310	Pharmacy (retail)		52489	Firewood (retail)
				52489	Fuel oil for household use (retail)
47741		**Retail sale of hearing aids in specialised stores**		52489	Fuel wood (retail)
	52321	Hearing aids (retail)		52489	Gas (bottled) (retail)
				52489	Gift shop (retail)
47749		**Retail sale of medical and orthopaedic goods (other than hearing aids) n.e.c., in specialised stores**		52489	Handicrafts shop (retail)
	52329	Invalid carriages with or without motor (retail)		52485	Medals (new) (retail)
	52329	Medical goods (retail)		52485	Numismatist (retail)
	52329	Orthopaedic appliances (retail)		52489	Oil merchant (retail)
	52329	Surgical appliances (retail)		52482	Optical and precision goods (retail)
				52489	Oriental goods (retail)
47750		**Retail sale of cosmetic and toilet articles in specialised stores**		52489	Packaging products for food e.g. aluminium foil, plastics foil, bags, etc. (retail)
	52330	Cosmetics (retail)		52489	Paraffin (retail)
	52330	Perfume (retail)		52489	Perambulators (retail)

UK SIC 2007	UK SIC 2003	Activity
	52485	Philatelist (retail)
	52482	Photo pick-up services (retail)
	52482	Photographic goods (retail)
	52489	Picture framing (retail)
	52489	Picture postcards (retail)
	52489	Religious goods (retail)
	52489	Silverware (retail)
	52489	Souvenirs (retail)
	52485	Stamp dealer (retail)
	52485	Stamps (retail)
	52489	Umbrellas (retail)
	52489	Weapons (retail)
	52489	Yacht chandler (retail)
47791		**Retail sale of antiques including antique books, in stores**
	52501	Antique books (retail)
	52501	Antiques (retail)
	52501	Incunabula (retail)
	52630	Retail sale of antiques including antique books in retail auction houses (except internet auctions)
47799		**Retail sale of second-hand goods (other than antiques and antique books) in stores**
	52509	Pawnshops (principally dealing in second-hand goods) (retail)
	52630	Retail auction house, second-hand goods other than antiques (except internet auctions) (retail)
	52509	Second-hand books (retail)
	52509	Second-hand clothing (retail)
	52509	Second-hand furniture (retail)
	52509	Second-hand general goods (retail)
47810		**Retail sale via stalls and markets of food, beverages and tobacco products**
	52620	Fish stall (retail)
	52620	Fruit stall keeper (retail)
	52620	Grocery stall (retail)
	52620	Sale of food beverages and tobacco via stalls and markets (retail)
47820		**Retail sale via stalls and markets of textiles, clothing and footwear**
	52620	Sale via stalls and markets of clothing
	52620	Sale via stalls and markets of footwear (retail)
	52620	Sale via stalls and markets of textiles (retail)
47890		**Retail sale via stalls and markets of other goods**
	52620	Newsvendor (retail)
	52620	Sale via stalls and markets of books (retail)
	52620	Sale via stalls and markets of carpets and rugs (retail)
	52620	Sale via stalls and markets of consumer electronics (retail)
	52620	Sale via stalls and markets of games and toys (retail)
	52620	Sale via stalls and markets of household appliances (retail)
	52620	Sale via stalls and markets of music and video recordings (retail)
47910		**Retail sale via mail order houses or via Internet**
	52630	Internet auctions (retail)
	52610	Internet retail sales (retail)

SIC 2007	SIC 2003	Activity
	52610	Mail order (retail)
	52610	Radio direct sales (retail)
	52610	Telephone direct sales (retail)
	52610	Television direct sales (retail)
47990		**Other retail sale not in stores, stalls or markets**
	52630	Auction houses (except antiques and second hand goods) (retail)
	52630	Credit trader (retail)
	52630	Direct selling of firewood to the customers premises (retail)
	52630	Direct selling of fuel to the customers premises (retail)
	52630	Direct selling of heating oil to the customers premises (retail)
	52630	Door-to-door sales (retail)
	52630	Milk roundsman (not farmer) (retail)
	52630	Milkman (not farmer) (retail)
	52630	Non-store auctions (retail)
	52630	Non-store commission agents (retail)
	52630	Vending machine sales (retail)
49100		**Passenger rail transport, interurban**
	60101	Passenger transport by inter-city rail services
	60109	Passenger transport by inter-urban railways (other than inter-city services)
	60109	Rail transport (inter-urban)
	60101	Railway dining cars operated as an integrated operation of railway companies
	60101	Railway sleeping cars operated as an integrated operation of railway companies
49200		**Freight rail transport**
	60109	Freight transport by inter-urban railways
	60109	Freight transport on mainline rail networks
	60109	Shortline freight railways
49311		**Urban, suburban or metropolitan area passenger railway transportation by underground, metro and similar systems**
	60213	Elevated railways (scheduled passenger transport)
	60213	Metropolitan area passenger railway transportation by underground, metro and similar systems
	60213	Suburban area passenger railway transportation by underground, metro and similar systems
	60213	Town-to-airport transport by rail
	60213	Town-to-station transport by rail
	60213	Underground railways (scheduled passenger transport)
	60213	Urban area passenger railway transportation by underground, metro and similar systems
49319		**Urban, suburban or metropolitan area passenger land transport other than railway transportation by underground, metro and similar systems**
	60219	Aerial cable-ways operation
	60219	Bus service
	60219	Bus transport (other than inter-city services) (scheduled passenger transport)
	60219	Funicular railway
	60219	Local authority road passenger transport services
	60219	Local authority transport department

UK SIC 2007	UK SIC 2003	Activity	SIC 2007	SIC 2003	Activity
	60219	Metropolitan scheduled passenger land transport other than underground, metro and similar systems		60249	Bulk road haulage
	60219	Motor bus scheduled passenger transport		60249	Car delivery service (by independent contractors)
	60219	Municipal bus service		60249	Car delivery service (by motor manufacturers)
	60219	Omnibus service		60249	Carrier (for general hire or reward)
	60219	Operation of aerial cableways as part of urban, suburban or metropolitan transit		60249	Cartage contractor
				60249	Commercial vehicle hire with driver
	60219	Operation of funicular railways as part of urban, suburban or metropolitan transit		60249	Concrete haulage by a unit which is (not the manufacturer)
	60219	Passenger scheduled land transport (other than interurban railways or inter-city coach services)		60249	Freight transport by animal-drawn vehicles
				60249	Freight transport by man-drawn vehicles
	60219	Public service vehicle operator		60249	Freight transport operation by road
	60219	Rack railway		60249	Haulage in tanker trucks by road
	60219	Street car (scheduled passenger transport)		60249	Haulage of automobiles by road
	60219	Suburban scheduled passenger land transport other than by underground, metro and similar systems		60249	Haulage of logs by road
				60249	Heavy haulage by road
	60219	Town-to-airport transport by bus		60249	Milk collection by tanker
	60219	Town-to-station transport by bus		60249	Motor vehicle collection
	60219	Tramway (scheduled passenger transport)		60249	Refrigerated haulage by road
	60219	Trolley bus (scheduled passenger transport)		60249	Road haulage contracting for general hire or reward
	60219	Urban area scheduled passenger land transport other than by underground, metro and similar systems		60249	Road haulage contractor
				60249	Stock haulage by road
				60249	Transport of waste and waste materials
49320		**Taxi operation**		60249	Truck rental (with driver)
	60220	Cab hire			
	60220	Car rental with driver	**49420**		**Removal services**
	60220	Chauffeur driven service		60241	Furniture removal
	60220	Motor bike taxi service		60241	Removal by road transport
	60220	Private hire car with driver		60241	Removal contractor
	60220	Taxi cab service			
			49500		**Transport via pipeline**
49390		**Other passenger land transport n.e.c.**		60300	Coal pumping station
	60219	Airline coach service (scheduled)		60300	Gas transport via pipelines
	60231	Airport shuttles		60300	Liquids transport via pipelines
	60239	Animal drawn vehicle transport		60300	Offshore natural gas pipeline operation
	60231	Buses operated for transport of employees		60300	Offshore oil pipeline operation
	60231	Charters, excursions and other occasional coach services		60300	Oil pipeline terminal operating (for petroleum)
				60300	Pipeline operator
	60231	Coach hire with driver		60300	Pump stations operation
	60231	Coach services (non-scheduled)		60300	Slurry transport via pipelines
	60211	Express coach service on scheduled routes		60300	Transport of gases slurry via pipelines
	60219	Factory bus service		60300	Transport of water via pipelines
	60211	Inter-city coach services on scheduled routes		60300	Transport via pipeline
	60239	Lift operating company			
	60239	Lift operator	**50100**		**Sea and coastal passenger water transport**
	60231	Motor coach service		61101	Boat rental for passenger conveyance with crew (except for inland waterway services)
	60231	Motor coach with driver (private hire)			
	60239	Non-scheduled passenger transport n.e.c.		61101	Coastal water transport for passengers
	60239	Operation of teleferics, funiculars		61101	Excursion, cruise or sightseeing boat operation (except for inland waterway service)
	60239	Passenger transport by animal-drawn vehicles			
	60239	Passenger transport by man-drawn vehicles		61101	Hovercraft operator between UK and international ports (passenger)
	60211	Scheduled long-distance bus services			
	60219	School bus service		61101	Passenger ferry between UK and international ports
	60231	Sightseeing buses		61101	Passenger ferry on domestic or coastal routes
	60239	Ski and cable lifts operation		61101	Passenger shipping service (sea and coastal)
	60239	Trishaw (cycle rickshaw) taxi service		61101	Pleasure boat rental with crew (e.g. for fishing cruises) (except for inland waterway service)
49410		**Freight transport by road**		61101	Sea ferry (passenger)
	60249	Bulk haulage in tanker trucks		61101	Transport of passengers over seas and coastal waters
	60249	Bulk haulage milk collection at farms		61101	Transport of passengers over water (except for inland waterway service)

UK SIC 2007	UK SIC 2003	Activity	SIC 2007	SIC 2003	Activity
	61101	Water taxis operation (except for inland waterway service)	50400		**Inland freight water transport**
				61209	Barge lessee or owner (freight)
50200		**Sea and coastal freight water transport**		61209	Boat rental for transport of freight with crew (inland waterway service)
	61102	Anchor handling services		61209	Canal carrier (freight)
	61102	Anti-pollution vessel services		61209	Freight ferry (river or estuary)
	61102	Barge transport (except for inland waterway) freight service		61209	Freight ferry transport (inland waterway service)
	61102	Boat rental for transport of freight with crew (except for inland waterway service)		61209	Freight vessel rental with crew (inland waterway service)
	61102	Cable-laying vessel services		61209	Inland water transport (freight)
	61102	Coastal water transport (freight)		61209	Local authority freight ferry services on rivers, canals and lakes
	61102	Freight ferry (domestic or coastal)		61209	Transport of freight via canals
	61102	Freight ferry (sea going)		61209	Transport of freight via lakes
	61102	Freight shipping service (except for inland waterway service)		61209	Transport of freight via ports
	61102	Freight shipping service (sea and coastal)		61209	Transport of freight via rivers
	61102	Heavy lift vessel services		61209	Water transport of freight inside harbours and docks
	61102	Launch barge services			
	61102	Marine tow out services	51101		**Scheduled passenger air transport**
	61102	Merchant Navy		62101	Air-transport equipment renting with operator, for scheduled passenger transportation
	61102	Offshore supply vessel services		62101	Passenger air transport (scheduled)
	61102	Oil-rig transportation by towing or pushing		62101	Passenger air transport over regular routes
	61102	Rental of vessels with crew for coastal freight water transport			
	61102	Rental of vessels with crew for sea freight water transport	51102		**Non-scheduled passenger air transport**
				62201	Air charter service (passenger)
	61102	Royal Fleet Auxiliary		62201	Air taxi service
	61102	Standby vessel services		62201	Air transport of passengers by aero clubs for instruction or pleasure
	61102	Transport by towing or pushing of barges (except inland waterway)		62201	Charter flights (passenger)
	61102	Transport of freight over seas and coastal waters (whether scheduled or not)		62201	Helicopter passenger services (non-scheduled)
				62201	Non-scheduled air transport of passengers
	61102	Waterborne freight transport (except for inland waterway service)		62201	Passenger air transport (non-scheduled)
				62201	Passenger aircraft rental services with crew (non-scheduled)
50300		**Inland passenger water transport**		62201	Regular charter flights for passengers
	61201	Barge lessee or owner (passenger) (inland waterway service)		62201	Scenic and sightseeing flights
	61201	Boat rental for passenger conveyance with crew (inland waterway service)	51210		**Freight air transport**
				62209	Air charter service (freight)
	61201	Canal carrier (passenger)		62109	Air transport of freight over regular routes
	61201	Excursion, cruise or sightseeing boats operation (inland waterway service)		62209	Charter flights (freight)
				62209	Freight air transport (non-scheduled)
	61201	Ferry transport for passengers (inland waterway service)		62109	Freight air transport (scheduled)
				62209	Freight aircraft rental services with crew (non-scheduled)
	61201	Inland water transport (passenger)			
	61201	Lake steamer service		62209	Helicopter freight services (non-scheduled)
	61201	Local authority passenger ferry services on rivers, canals and lakes	51220		**Space transport**
				62300	Satellite launching
	61201	Passenger ferry (river or estuary)		62300	Space transport
	61201	Passenger ferry transport (inland waterway)		62300	Space transport of freight
	61201	Rental of pleasure boats with crew for inland water transport		62300	Space transport of passengers
				62300	Space vehicle launching
	61201	Transport of passengers over water (inland waterway service)			
			52101		**Operation of warehousing and storage facilities for water transport activities of division 50**
	61201	Transport of passengers via canals		63121	Blast freezing for water transport activities
	61201	Transport of passengers via inside harbours		63129	Bonded store, vault or warehouse for water transport activities
	61201	Transport of passengers via lakes			
	61201	Transport of passengers via ports		63122	Bulk liquid and gases storage services for water transport activities
	61201	Transport of passengers via rivers			
	61201	Water taxis operation (inland waterway service)			

UK SIC 2007	UK SIC 2003	Activity	SIC 2007	SIC 2003	Activity
	63129	Coal stockyard for water transport activities		63129	Film bonded warehouse for land transport activities
	63121	Cold store for water transport activities		63121	Frozen and refrigerated goods storage services, for land transport activities
	63129	Cotton warehouse for water transport activities		63129	Furniture repository for land transport activities
	63129	Film bonded warehouse for water transport activities		63123	Grain silos operation for land transport activities
	63121	Frozen and refrigerated goods storage services, for water transport activities		63123	Grain warehouse for land transport activities
	63129	Furniture repository for water transport activities		63123	Granary for land transport activities
	63123	Grain silos operation for water transport activities		63121	Refrigerated warehouses operation for land transport activities
	63123	Grain warehouse for water transport activities		63129	Safe deposit company
	63123	Granary for water transport activities		63129	Storage facilities n.e.c. for land transport activities
	63121	Refrigerated warehouses operation for water transport activities		63129	Storage of goods in foreign trade zones for land transport activities
	63129	Repository for water transport activities		63122	Storage tanks operation for land transport activities
	63129	Storage facilities n.e.c. for water transport activities		63129	Tea warehouse for land transport activities
	63129	Storage of goods in foreign trade zones for water transport activities		63129	Warehouse (general) operation for land transport activities
	63122	Storage tanks operation for water transport activities		63129	Wool warehouse for land transport activities
	63129	Tea warehouse for water transport activities	**52211**		**Operation of rail freight terminals**
	63129	Warehouse (general) operation for water transport activities		63210	Rail freight terminals operation
	63129	Wool warehouse for water transport activities	**52212**		**Operation of rail passenger facilities at railway stations**
52102		**Operation of warehousing and storage facilities for air transport activities of division 51**		63210	Operation of rail passenger facilities at railway stations
	63121	Blast freezing for air transport activities	**52213**		**Operation of bus and coach passenger facilities at bus and coach stations**
	63129	Bonded store, vault or warehouse for air transport activities		63210	Bus and coach passenger facilities at bus and coach stations
	63122	Bulk liquid and gases storage services for air transport activities		63210	Terminal facilities operation (land transport)
	63121	Cold store for air transport activities	**52219**		**Other service activities incidental to land transportation, n.e.c. (not including operation of rail freight terminals, passenger facilities at railway stations or passenger facilities at bus and coach stations)**
	63129	Cotton warehouse for air transport activities			
	63129	Film bonded warehouse for air transport activities			
	63121	Frozen and refrigerated goods storage services, for air transport activities			
	63129	Furniture repository for air transport activities		50200	Automobile association road patrols
	63129	General merchandise warehouses for air transport activities		63210	Bicycle parking operations
	63123	Grain silos operation for air transport activities		63210	Bridge operation
	63123	Grain warehouse for air transport activities		63210	Bus station operation
	63123	Granary for air transport activities		63210	Car park
	63121	Refrigerated warehouses operation for air transport activities		63210	Caravan winter storage
	63129	Storage facilities n.e.c. for air transport activities		63210	Clamping and towing away of vehicles
	63129	Storage of goods in foreign trade zones for air transport activities		63210	Commercial vehicle park
	63122	Storage tanks operation for air transport activities		63210	Garage (parking)
	63129	Tea warehouse for air transport activities		11100	Gas liquefaction for land transportation purposes
	63129	Warehouse (general) operation for air transport activities		63210	Goods handling station operation
	63129	Wool warehouse for air transport activities		63210	Lessee of tolls
52103		**Operation of warehousing and storage facilities for land transport activities of division 49**		63210	Local authority car parks
				63210	Motive power depot (railway)
	63121	Blast freezing for land transport activities		50200	Motorists' organisation (road patrol)
	63129	Bonded store, vault or warehouse for land transport activities		63210	Motorway maintenance unit
	63122	Bulk liquid and gases storage services for land transport activities		63210	Parking lot operation
				63210	Parking meter services
	63129	Coal stockyard for land transport activities		63210	Radio despatch offices for taxis, bicycle couriers etc.
	63121	Cold store for land transport activities		63210	Railway running shed
	63129	Cotton warehouse for land transport activities		63210	Railway station operation
				63210	Repair and maintenance of rolling stock (minor)
				63210	Roads operation
				50200	Roadside assistance for motor vehicles

UK SIC 2007	UK SIC 2003	Activity	SIC 2007	SIC 2003	Activity
	50200	Royal Automobile Club road patrols		63220	Tug boat service for sea barge or off-shore well
	60109	Switching and shunting		63220	Tug boat service for sea barges on domestic coastal routes
	60101	Switching and shunting on railways		63220	Tug lessee or owner for inland waterways service
	63210	Toll bridge, road or tunnel		63220	Tug owner or lessee for in port service or salvage
	50200	Towing and road side assistance		63220	Vessel laying up and storage services
	63210	Towing away of vehicles		63220	Vessel registration services
	63210	Tunnels operation		63220	Water transport (supporting activities)
	63210	Weighbridge services		63220	Waterway locks operation
				63220	Wharfinger
52220		**Service activities incidental to water transportation**		63220	Wreck raising
	63220	Berthing activities	**52230**		**Service activities incidental to air transportation**
	63220	Bunkering services		63230	Aerodrome
	63220	Canal maintenance		63230	Air terminal operated by airline
	63220	Canal operation		63230	Air traffic control activities
	63220	Cargo superintendent		63230	Air traffic control centre
	63220	Cargo terminal		63230	Air transport supporting activities
	63220	Commissioners of northern lighthouses		63230	Aircraft refuelling services
	63220	Diving contracting (non leisure)		63230	Airfield ground service activities
	63220	Dock authority		63230	Airport activities
	63220	Floating bridge company		63230	Airport fire fighting and fire-prevention services
	11100	Gas liquefaction for water transportation purposes		63230	Airway terminals operation
	63220	Harbour authority		63230	British Airports Authority
	63220	Harbour operation		63230	Communication centre (civil air)
	63220	Ice breaking services		63230	Local authority municipal airport
	63220	Landing stage		63230	Service activities incidental to air transportation
	63220	Lighter lessee or owner		62300	Service activities incidental to space transportation
	63220	Lighterage activities		63230	Terminal facilities operation (air transport)
	63220	Lighthouse activities			
	63220	Lighthouse Authority	**52241**		**Cargo handling for water transport activities of division 50**
	63220	Lightship		63110	Cargo handling for water transport activities
	63220	Local authority canal services		63110	Container handling services for water transport activities
	63220	Local authority docks and harbours		63110	Loading and unloading of goods travelling via water transport
	63220	Local authority lighthouse service		63110	Loading and unloading of passengers' luggage travelling via water transport
	63220	Marine cargo lighterage		63110	Stevedoring
	63220	Marine cargo superintendent			
	63220	Marine salvage	**52242**		**Cargo handling for air transport activities of division 51**
	63220	Navigation activities		63110	Air passenger baggage handling services
	63220	Offshore positioning services		63110	Cargo handling for air transport activities
	63220	Passenger terminal services		63110	Loading and unloading of goods travelling via air transport
	63220	Pier operation (not amusement)		63110	Loading and unloading of passengers' luggage travelling via air transport
	63220	Pier owner or authority (not amusement)			
	63220	Pilotage activities			
	63220	Port Authority	**52243**		**Cargo handling for land transport activities of division 49**
	63220	Port of London Authority		63110	Cargo handling for land transport activities
	63220	Salvage activities supporting water transport activities		63110	Loading and unloading of freight railway cars
	63220	Shore base (sea transport)		63110	Loading and unloading of goods travelling via rail transportation
	63220	Terminal facilities operation (water transport)		63110	Loading and unloading passengers' luggage travelling via rail transportation
	61209	Towing services for distressed freight vessels in inland waters			
	61102	Towing services for distressed freight vessels on sea and coastal waters	**52290**		**Other transportation support activities**
	61201	Towing services for distressed passenger vessels in inland waters		63400	Air cargo agents activities
	61101	Towing services for distressed passenger vessels on sea and coastal waters			
	63220	Trinity House			
	63220	Tug boat service for inland waterways			
	63220	Tug boat service for offshore installations			

UK SIC 2007	UK SIC 2003	Activity	SIC 2007	SIC 2003	Activity
	63400	Air freight agent or broker	55201		**Holiday centres and villages**
	63400	Customs clearance agents activities		55231	Chalets in holiday centres and holiday villages (provision of short-stay lodging in)
	63400	Export packer		55231	Cottages and cabins without housekeeping services, provided in holiday centres and holiday villages
	63400	Forwarding agents			
	63400	Freight broker		55231	Cottages in holiday centres and holiday villages (provision of short-stay lodging in)
	63400	Freight contractor			
	63400	Freight forwarding		55231	Flats in holiday centres and holiday villages (provision of short-stay lodging in)
	63400	Goods agent (not transport authority)			
	63400	Goods handling operations		55231	Holiday and other short stay accommodation, provided in holiday centres and holiday villages
	63400	Maritime agent			
	63400	Packer and shipper		55231	Holiday camp
	63400	Packing service incidental to transport		55231	Holiday centres and villages
	63400	Railway agent (not transport authority)		55231	Visitor flats and bungalows provided in holiday centres and holiday villages
	63400	Railway wagon agent			
	63400	Sea freight forwarder activities	55202		**Youth hostels**
	63400	Shipping agent or broker		55210	Mountain refuges
	63400	Textile packing incidental to transport		55210	Youth hostel
	63400	Transport documents issue and procurement			
	63400	Transport operations arranging or carrying out by air	55209		**Other holiday and other short stay accommodation (not including holiday centres and villages or youth hostels) n.e.c.**
	63400	Transport operations arranging or carrying out by road		55232	Apartment letting (self catering) other than holiday centres and holiday villages or youth hostels
	63400	Transport operations arranging or carrying out by sea		55232	Chalets, not holiday centres, holiday villages or youth hostels (self catering short-stay lodging)
53100		**Postal activities under universal service obligation**		55239	Children's and other holiday homes
	64110	Mail distribution and delivery		55232	Cottages, not in holiday centres, holiday villages, youth hostels (self catering short-stay lodging)
	64110	Parcels, distribution and delivery of, by the post office		55239	Farmhouse short stay accommodation
	64110	Post activities		55232	Flats, not in holiday centres, holiday villages or youth hostels (self catering short-stay lodging)
	64110	Post office regional headquarters			
	64110	Postal headquarters		55239	Guest house (licensed)
	64110	Postal sorting office		55239	Guest house (unlicensed)
	64110	Poste restante		55232	Holiday home (not charitable)
	64110	Sub post office (principally devoted to post office business)		55239	Hydro (accommodation)
				55239	Inns with letting rooms (short-stay lodgings)
53201		**Licensed Carriers**	55300		**Camping grounds, recreational vehicle parks and trailer parks**
	64120	Courier activities (other than national post activities) licensed		55210	Accommodation in protective shelters or plain bivouac facilities for tents and/or sleeping bags
	64120	Licensed carriers			
	64120	Messenger licensed		55220	Camping sites
	64120	Messenger service licensed		55220	Caravan holiday site operator, owner or proprietor
	64120	Parcels delivery service (not post office) licensed		55220	Caravan sites
				55220	Recreational vehicle parks and trailer parks
53202		**Unlicensed Carriers**	55900		**Other accommodation**
	64120	Courier activities (other than national post activities) unlicensed		55239	Boarding house
	64120	Messenger service unlicensed		55239	Boarding school accommodation
	64120	Messenger unlicensed		55239	Halls of residence
	64120	Parcels delivery service (not post office) unlicensed		55239	Hostels (not social work)
	64120	Unlicensed carriers		55239	Landlord (boarding house)
				55239	Lodging house (private)
55100		**Hotels and similar accommodation**		55239	Migrant worker accommodation
	55101	Hotel (licensed with restaurant)		55239	Pension (accommodation)
	55102	Hotel (unlicensed with restaurant)		55239	Private lodging house
	55103	Hotel without restaurant		55239	Railway sleeping cars when operated by separate units from the transport provider
	55101	Motel (licensed with restaurant)			
	55102	Motel (unlicensed with restaurant)		55239	Rooming and boarding houses
	55103	Motel without restaurant		55239	School dormitories
				55239	Student house accommodation

UK SIC 2007	UK SIC 2003	Activity	SIC 2007	SIC 2003	Activity
	55239	Student residences	**56210**		**Event catering activities**
	55239	University halls of residence		55520	Banquet catering
	55239	Workers' hostels		55520	Catering contractor
	55239	YMCA hostel		55520	Corporate hospitality catering
				55520	Wedding catering
56101		**Licensed restaurants**			
	55301	Buffet (licensed)	**56290**		**Other food service activities**
	55301	Cafeteria (licensed)		55520	Airline catering
	55301	Chop house (licensed)		55510	Factory or office canteens
	55301	Civic restaurant (licensed)		55520	Industrial canteen (run by catering contractor)
	55301	Fast-food restaurants, licensed		55510	Local authority school meals service
	55301	Function room (licensed)		55510	Luncheon club
	55301	Luncheon bar (licensed)		55520	Meals on wheels catering
	55301	Oyster bar (licensed)		55510	NAAFI canteen
	55301	Railway dining car or buffet (licensed)		55520	NAAFI headquarters
	55301	Restaurant (licensed)		55510	Officers' messes
	55301	Self-service restaurant (licensed)		55510	Other ranks' messes
	55301	Steak houses (licensed)		55520	Pleasure steamer caterer
				55510	Refreshment club
56102		**Unlicensed restaurants and cafes**		55520	Refreshment contracting
	55302	Burger bar restaurant (unlicensed)		55510	School canteen
	55302	Cafeteria (unlicensed)		55520	School canteen (run by catering contractor)
	55302	Coffee bar, room or saloon (unlicensed)		55510	Sergeants' messes
	55302	Dining room (unlicensed)		55520	Staff canteen (run by catering contractor)
	55302	Fast food outlet with restaurant (unlicensed)		55510	University canteen
	55302	Fast-food restaurants, unlicensed		55510	University dining halls
	55302	Local authority restaurants, cafes, snack bars, etc. (unlicensed)	**56301**		**Licensed clubs**
	55302	Motorway services cafeteria (unlicensed)		55401	Discotheques (licensed to sell alcohol)
	55302	Refreshment room (unlicensed)		55401	Night-clubs (licensed to sell alcohol)
	55302	Restaurant (unlicensed)		55401	Social clubs (licensed to sell alcohol)
	55302	Self-service restaurant (unlicensed)		55401	Working men's clubs (licensed to sell alcohol)
	55302	Supper bar or room (unlicensed)			
	55302	Tea garden (unlicensed)	**56302**		**Public houses and bars**
	55302	Tea room or shop (unlicensed)		55402	Beer gardens (independent)
	55302	Temperance buffet		55404	Beer gardens (managed)
				55403	Beer gardens (tenanted)
56103		**Take away food shops and mobile food stands**		55402	Beer halls (independent)
	55303	Burger bar (take-away)		55404	Beer halls (managed)
	55304	Burger stand		55403	Beer halls (tenanted)
	55303	Eel pie shop		55402	Licensed bars (independent)
	55303	Fish and chip shop		55404	Licensed bars (managed)
	55304	Fish and chip stand		55403	Licensed bars (tenanted)
	55304	Food preparation in market stalls		55402	Licensed victualler (independent)
	55303	Fried fish shop		55404	Licensed victualler (managed)
	55304	Hot dog vendor		55403	Licensed victualler (tenanted)
	55303	Ice cream parlour		55404	NAAFI clubs
	55303	Ice cream retailer (take away)		55402	Public houses (independent)
	55304	Ice cream vans		55404	Public houses (managed)
	55303	Jellied eel shop		55403	Public houses (tenanted)
	55303	Meat pie shop		55402	Taverns (independent)
	55303	Milk bar		55404	Taverns (managed)
	55304	Mobile food carts		55403	Taverns (tenanted)
	55303	Pea and pie vendor			
	55303	Sandwich bar	**58110**		**Book publishing**
	55303	Snack bar		22110	Architectural drawing publishing
	55303	Take away food shop		22110	Atlas publishing
	55304	Take-out eating places		22110	Audio book publishing
	55304	Tea bar		22110	Book publishing

UK SIC 2007	UK SIC 2003	Activity	SIC 2007	SIC 2003	Activity
	22110	Brochure publishing		22150	Photo and engraving publishing
	22110	Chart publishing		22150	Postcard publishing
	22110	Dictionary publishing		22150	Poster publishing
	22110	Electronic publishing of books		22150	Printed matter publishing
	22110	Encyclopaedia publishing		22150	Publishers (other than of newspapers, books and periodicals)
	22110	Engineering drawing publishing			
	22110	Geographical publishing		22220	Publishing of stamps, banknotes, advertising material, catalogues and other printed matter n.e.c.
	22110	Leaflet publishing			
	22110	Map and plan publishing		22150	Reproduction of works of art publishing
	72400	On-line book publishing		22150	Timetable publishing
	22110	Pamphlet publishing			
	22110	Publishing on CD- ROM	**58210**		**Publishing of computer games**
	22110	Religious tract publishing		72210	Computer games publishing
	22110	The Stationery Office		72400	On-line computer games publishing
58120		**Publishing of directories and mailing lists**	**58290**		**Other software publishing**
	22110	Directories and compilations (in print) publishing		72210	Licensing for the right to reproduce, distribute and use computer software
	22110	Mailing lists (in print) publishing		72400	On-line software publishing (except computer games on-line publishing)
	72400	On-line database publishing			
	72400	On-line directory publishing		72210	Publishing of software for business
	72400	On-line mailing list publishing		72210	Publishing of software for operating systems
	22110	Telephone books (in print) publishing		72210	Ready-made software publishing (except computer games publishing)
58130		**Publishing of newspapers**			
	22120	Advertising newspaper publishing		72210	Software (ready-made) publishing (except computer games publishing)
	22120	Newspaper publishing			
	72400	On-line newspaper publishing	**59111**		**Motion picture production activities**
				92111	Advertising film production
58141		**Publishing of learned journals**		92111	Animated film production
	22130	Learned journal publishing		92111	Cartoon film production
	72400	On-line learned journal publishing		92111	Film producer (own account)
				92111	Film production for projection or broadcasting
58142		**Publishing of consumer, business and professional journals and periodicals**		92111	Film studios
				92111	Motion picture production
	22130	Amusement guide periodical publishing		92111	Training film production
	22130	Journal and periodical publishing			
	22130	Magazine publishing	**59112**		**Video production activities**
	72400	On-line journal (other than learned journals) and periodical publishing		92111	Advertising video production
				92111	Animated video production
	22130	Periodical publishing		92111	Cartoon video production
	22130	Review publishing		92111	Training video production
	22130	Trade journal publishing		92111	Video producer (own account)
				92111	Video production
58190		**Other publishing activities**		92111	Video studios
	22150	Advertising material publishing			
	22150	Art publishing	**59113**		**Television programme production activities**
	22150	Catalogue publishing		92111	Production of theatrical and non-theatrical television programmes
	22150	Forms publishing			
	22150	Greeting cards publishing		92202	Television programme production activities
	22150	Law publishing		92202	Television studio
	22150	Letterpress publishing			
	72400	On-line advertising material publishing	**59120**		**Motion picture, video and television programme post-production activities**
	72400	On-line catalogue publishing			
	72400	On-line forms publishing		92119	Cinematographic film colouring, developing, printing or repairing
	72400	On-line greeting cards and postcards publishing			
	72400	On-line publishing n.e.c.		92119	Film cutting activities
	72400	On-line publishing of posters and reproductions of works of art		92119	Film editing
				92119	Film processing activities
	72400	On-line publishing of statistics and other information		92119	Film sound track dubbing and synchronisation
				92119	Film title printing
	72400	On-line publishing photos and engravings		92119	Motion picture film laboratory activities

UK SIC 2007	UK SIC 2003	Activity
	92119	Motion pictures post-production activities
	92119	Photographic film processing activities (for the motion picture and television industries)
	92119	Post production film activities
	92119	Stock footage, film library activities
	92202	Television post-production activities
	92119	Video post-production activities
59131		**Motion picture distribution activities**
	92120	Film broker
	92120	Film distribution
	92120	Film distribution rights acquisition
	92120	Film hiring agency
	92120	Film library
	92120	Film rental
	92120	Motion picture distribution activities
	92120	Motion pictures distribution to other industries
59132		**Video distribution activities**
	92120	Dvd distribution
	92120	Video tape and dvd distribution rights acquisition
	92120	Video tapes distribution to other industries
59133		**Television programme distribution activities**
	92202	Television distribution rights acquisition
	92120	Television programme distribution activities
59140		**Motion picture projection activities**
	92130	Cine club
	92130	Cinema
	92130	Cinema club
	92130	Motion picture projection
	92130	Video tape projection
59200		**Sound recording and music publishing activities**
	22140	Compact disc sound recording publishing
	74879	Copyright acquisition and registration for musical compositions
	22140	Gramophone record publishing
	22140	Matrice for record production
	22140	Music (printed) publishing
	72400	Music downloads (on-line publishing with provision of downloaded content)
	22140	Music tape publishing
	92119	Printing of sound tracks
	22140	Publishing of music and sheet books
	92201	Recording studio (radio)
	22140	Sound recording publishing
	92119	Sound recording studios
	92201	Taped radio programming production
60100		**Radio broadcasting**
	92201	British Broadcasting Corporation (radio broadcasting)
	92201	Data broadcasting integrated with radio
	92201	Independent Broadcasting Authority (radio broadcasting)
	72400	Internet radio broadcasting
	92201	Local radio station (broadcasting)
	92201	Radio broadcasting station

SIC 2007	SIC 2003	Activity
	64200	Radio programme transmission
	64200	Radio station (telecommunications)
	92201	Radio studio
	92201	Transmission of aural programming via over-the-air broadcasts, cable or satellite
60200		**Television programming and broadcasting activities**
	92202	British Broadcasting Corporation (television broadcasting)
	92202	Data broadcasting integrated with television
	72400	Image with sound internet broadcasting
	92202	Recording studio (television)
	92202	Television broadcasting station
	92202	Television channel programme creation from purchased and/or self produced programme components
	64200	Television programmes transmission
	64200	Television relay service
	92202	Television service
61100		**Wired telecommunications activities**
	64200	Cable service
	64200	Data network management and support services (wired telecommunications)
	64200	Data transmission (via cables, broadcasting, relay or satellite)
	64200	Dedicated business telephone network services (wired telecommunications)
	64200	Electronic mail services (wired telecommunications)
	64200	Hull telephone service
	64200	Image transmission via cables, broadcasting, relay or satellite
	64200	Shared business telephone network services (wired telecommunications)
	64200	Sound transmission via cables, broadcasting, relay or satellite
	64200	Telecommunication network maintenance (wired telecommunications)
	64200	Teleconferencing services (wired telecommunications)
	64200	Telegram service
	64200	Telegraph communication
	64200	Telephone communication (wired telecommunications)
	64200	Telephone exchange
	64200	Telephone service (wired telecommunications)
	64200	Telephone service operation of 0898 numbers
	64200	Teletext and other electronic message and information services (wired telecommunications)
	64200	Telex service (wired telecommunications)
61200		**Wireless telecommunications activities**
	64200	Cellular network operations
	64200	Electronic bulletin board services (wireless telecommunications)
	64200	Electronic message and information services (wireless telecommunications)
	64200	Internet access providers (wireless telecommunications)
	64200	Mobile telephone services
	64200	Paging activities and maintenance
	64200	Paging services

UK SIC 2007	UK SIC 2003	Activity	SIC 2007	SIC 2003	Activity
	64200	Purchase of wireless access and network capacity	62090		**Other information technology and computer service activities**
	64200	Radio relay service		72600	Computer related activities (other)
	64200	Wireless telecommunications activities		72220	Data archiving and backup services
61300		**Satellite telecommunications activities**		30020	Installation of personal computers and peripheral equipment
	64200	Satellite circuit rental services		72220	Software disaster recovery services
	64200	Telecommunications satellite relay station		72220	Software installation services
61900		**Other telecommunications activities**	**63110**		**Data processing, hosting and related activities**
	64200	Communications telemetry		72300	Batch processing
	64200	Dial-up internet access provision		72300	Data conversion
	64200	Radar station operation		72300	Data preparation services
	64200	Satellite terminal stations		72300	Data processing
	64200	Satellite tracking		72400	Data storage services
	64200	Telecommunications resellers		72400	Database running activities
	64200	Telephone and internet access in public facilities		72300	Tabulating service
	64200	VOIP (voice over internet protocol) provision		72300	Time sharing services (computer)
				72300	Web hosting
62011		**Ready-made interactive leisure and entertainment software developmen**	**63120**		**Web portals**
	72220	Computer games design		72400	Web search portals
	72400	Designing of structure and content of an interactive leisure and entertainment software database	**63910**		**News agency activities**
	72220	Ready-made interactive leisure and entertainment software development		92400	News agency activities
				92400	Stock market reporting service
62012		**Business and domestic software development**	**63990**		**Other information service activities n.e.c.**
	72220	Business and domestic software development		74879	Information search services on a fee or contract basis
	72220	Custom software development		92400	News clipping service
	72220	Data analysis consultancy services		92400	Press clipping service
	72400	Database structure and content design		92400	Press cutting agency
	72400	Designing of structure and content of business and domestic software databases		74879	Telephone based information services
	72400	Made-to-order software	**64110**		**Central banking**
	72220	Programming services		65110	Bank of England
	72220	Software house	**64191**		**Banks**
	72220	Software systems maintenance services		65121	Banks (deposit taking)
	72220	System maintenance and support services		65121	Discount houses (monetary intermediation)
	72220	Systems analysis (computer)		65121	Money order activities
	72220	Web page design		65121	National savings bank
				65121	Postal giro and postal savings bank activities
62020		**Computer consultancy activities**		65121	Savings bank
	72100	Computer audit consultancy services			
	72220	Computer consultancy (software)	**64192**		**Building societies**
	72100	Computer hardware acceptance testing services		65122	Building societies
	72100	Computer site planning services	**64201**		**Activities of agricultural holding companies**
	72100	Hardware consultancy		74159	Holding company in agricultural sector
	72100	Hardware disaster recovery services			
	72100	Hardware installation services	**64202**		**Activities of production holding companies**
	72220	Information systems strategic review and planning services		74158	Holding company in production sector
	72220	Information technology consultancy activities	**64203**		**Activities of construction holding companies**
	72220	Software consultancy		74153	Holding company in construction sector
	72220	System software acceptance testing consultancy services	**64204**		**Activities of distribution holding companies**
	72220	Systems and technical consultancy services		74155	Holding company in motor trades sector
62030		**Computer facilities management activities**		74157	Holding company in retail sector
	72300	Computer facilities management activities		74151	Holding company in wholesale sector
	72300	Management and operation on a continuing basis of data processing facilities belonging to others			

UK SIC 2007	UK SIC 2003	Activity	SIC 2007	SIC 2003	Activity
64205		**Activities of financial services holding companies**		65223	Ship mortgage finance company
	65234	Bank holding companies	64929		**Other credit granting (not including credit granting by non-deposit taking finance houses and other specialist consumer credit grantors and activities of mortgage finance companies) n.e.c.**
	65234	Holding company activities for banks			
	65239	Holding company in financial services sector			
64209		**Activities of other holding companies (not including agricultural, production, construction, distribution and financial services holding companies) n.e.c.**		65229	Banking institutions in Channel Islands and Isle of Man activities (not in UK banking sector)
				65229	Banks not authorised by the FSA as deposit taking
	74156	Activities of service trades holding companies not engaged in management		65229	Channel Islands and Isle of Man banking institutes
	74154	Holding company in catering sector		65229	Export finance company (other than in banks' sector)
	74159	Holding company in property sector		65229	Finance corporation for industry
	74152	Holding company in transport sector		65229	Mortgage corporation for agriculture
	74159	Holding company n.e.c.	64991		**Security dealing on own account**
64301		**Activities of investment trusts**		65233	Bill broker on own account (other than discount house)
	65231	Investment fund activities		65233	Bullion broker on own account
	65231	Investment trusts activities		65233	Bullion dealer in investment grades
64302		**Activities of unit trusts**		65233	Commodities dealing for investment purposes
	65232	In-house trust activities		65233	Dealer in stocks and shares on own account
	65232	Unit trust activities		65233	Financial futures, options and other derivatives dealing in on own account
64303		**Activities of venture and development capital companies**		65233	Inter-dealer brokers
				65233	Securities dealer on own account
	65235	Development capital company		65233	Share dealer on own account
	65235	Venture and development capital companies and funds activities		65233	Stock exchange money broker activities
64304		**Activities of open-ended investment companies**	64992		**Factoring**
				65222	Debt purchasing
	65237	Open-ended investment companies		65222	Discount company (e.g. Debt factoring)
64305		**Activities of property unit trusts**		65222	Factoring company (buying book debts)
	65238	Property unit trusts		65222	Invoice discounting
64306		**Activities of real estate investment trusts**	64999		**Other financial service activities, except insurance and pension funding, (not including security dealing on own account and factoring) n.e.c.**
	65231	Real estate investment trusts			
64910		**Financial leasing**		65239	Financial intermediation n.e.c.
	65210	Financial leasing		65239	Securitization activities
64921		**Credit granting by non-deposit taking finance houses and other specialist consumer credit grantors**		65239	Swaps, options and other hedging arrangements
				65239	Underwriter (stock and share issues)
				65239	Viatical settlement company activities
	65221	Building societies' personal finance subsidiaries activities	65110		**Life insurance**
	65221	Check trader activities		66011	Assurance company (life)
	65221	Consumer credit granting company (other than banks or building societies)		66011	Friendly society (not collecting society)
				66011	Life assurance
	65221	Credit card issuer (sole activity – requiring full payment at end of credit period)		66011	Provident fund (life)
				66011	Underwriter (life insurance)
	65221	Credit unions	65120		**Non-life insurance**
	65221	Finance house activities (non-deposit taking)		66031	Accident insurance
	65221	Hire purchase company (other than in banks' sector)		66031	Aviation insurance
	65221	Loan company (other than in banks' sector)		66031	Benevolent society (insurance)
	65221	Money lender		66031	Boiler insurance
	65221	Pawnbroker (principally lending money)		66031	Collecting society
	65221	Vehicle fuel credit card services		66031	Contingency insurance
64922		**Activities of mortgage finance companies**		66031	Export credit guarantee department
				66031	Fire insurance
	65223	Mortgage finance companies' activities (other than banks and building societies)		66031	Health insurance
				66031	Hospital contribution scheme

UK SIC 2007	UK SIC 2003	Activity	SIC 2007	SIC 2003	Activity
	66031	Hospital saving association		67122	Dealing in securities on behalf of others
	66031	House insurance		67130	Foreign exchange broker
	66031	Industrial insurance		67122	Investment broking
	66031	Injury insurance		67130	Money changer
	66031	Insurance (non-life)		67122	Securities broking
	66031	Liability insurance		67122	Securities dealing on behalf of others
	66031	Livestock insurance		67122	Security and commodity contracts dealing activities
	66031	Lloyd's underwriter (non-life)		67122	Share dealer on behalf of others
	66031	Marine insurance		67122	Trustees
	66031	Motor insurance			
	66031	Pecuniary loss insurance	**66190**		**Other activities auxiliary to financial services, except insurance and pension funding**
	66031	Personal injury insurance		67130	Bullion broking on behalf of others
	66031	Plate glass insurance		67130	Company promoting
	66031	Pluvium insurance		67130	Credit or finance broking
	66031	Property insurance		67130	Deposit broker
	66031	Provident fund (non-life)		67130	Financial advisor
	66031	Sports insurance		67130	Financial transactions centre
	66031	Transport insurance		67130	Independent financial advisors (not specialising in insurance or pensions advice)
	66031	Travel insurance		67130	Interbank (worldwide financial telecommunications society)
	66031	Underwriter (non-life insurance)		67130	Investment advisory services
	66031	Weather insurance		67130	Issuing house
				67130	Mortgage agent
65201		**Life reinsurance**		67130	Mortgage broker activities
	66012	Life re-insurance		67130	Nominee company
				67130	Organisation and development of electronic money circulation
65202		**Non-life reinsurance**		67130	Paying agent
	66032	Non-life reinsurance		67130	Public trust office
	66020	Non-life reinsurance related to pension funding		67130	Trustee, fiduciary and custody services on a fee or contract basis
	66032	Re-insurance company (non-life)			
			66210		**Risk and damage evaluation**
65300		**Pension funding**		67200	Adjuster (insurance)
	66020	Employee benefit plans		67200	Average adjuster
	66020	Pension fund (autonomous)		67200	Damage evaluators activities
	66020	Pension funding except compulsory social security		67200	Insurance risk evaluators activities
	66020	Pension funds and plans		67200	Loss adjuster
	66020	Retirement incomes provision			
	66020	Retirement plans	**66220**		**Activities of insurance agents and brokers**
	66020	Superannuation fund (autonomous)		67200	Insurance agent (not employed by insurance company)
				67200	Insurance agents activities
66110		**Administration of financial markets**		67200	Insurance broker (not employed by insurance company)
	67110	Bankers' clearing house		67200	Insurance brokers activities
	67110	Clearing house (banking)		67200	Insurance consultancy services
	67110	Commodity contracts exchanges administration		67200	Underwriting brokers
	67110	Company registration agent			
	67110	Financial markets administration	**66290**		**Other activities auxiliary to insurance and pension funding**
	67110	Futures commodity contracts exchanges administration		67200	Actuarial services
	67110	Securities exchanges administration		67200	Pension consultancy services
	67110	Stock exchange activities		67200	Pension consultants (own account)
	67110	Stock or commodity options exchanges administration			
			66300		**Fund management activities**
66120		**Security and commodity contracts brokerage**		67121	Fund management activities
	67122	Bill broking on behalf of others (other than discount house)		67121	Mutual funds management
	67130	Bureaux de change activities		67121	Pension fund management
	67122	Commodity contracts brokerage		67121	Portfolio management services
	67122	Corporate finance companies			
	67130	Currency broking			
	67122	Custodians and settlement services			
	67122	Dealing in finance markets for others (e.g. Stock broking), related activities (not fund management)			

UK SIC 2007	UK SIC 2003	Activity	SIC 2007	SIC 2003	Activity
68100		**Buying and selling of own real estate**		70310	House agent
	70120	Apartment buildings buying and selling		70310	House letting agency
	70120	Building sales and purchase		70310	Intermediation in buying selling and renting of real estate
	70120	Dwellings buying and selling		70310	Land agent
	70120	Land buying and selling		70310	Land valuer or surveyor
	70120	Real estate buying and selling		70310	Property consultant (own account)
	70120	Real estate owner		70310	Real estate agencies
	70120	Static caravan sales		70310	Real estate escrow agents activities
68201		**Renting and operating of Housing Association real estate**		70310	Surveyor and valuer (real estate)
				70310	Valuer (real estate)
	70209	Housing association (social housing for rental)	**68320**		**Management of real estate on a fee or contract basis**
	70209	Renting and operating of housing association real estate		70320	Property management (as agents for owners)
68202		**Letting and operating of conference and exhibition centres**		70320	Real estate management on a fee or contract basis
	70201	Conference centre letting (self owned)		70320	Rent collecting agencies
	70201	Exhibition centre letting (self owned)	**69101**		**Barristers at law**
68209		**Letting and operating of own or leased real estate (other than Housing Association real estate and conference and exhibition services) n.e.c.**		74112	Advocate of the Scottish bar
				74112	Barrister
				74112	Queen's counsel
	70209	Agricultural land letting	**69102**		**Solicitors**
	70209	Apartment buildings letting		74113	Attorney
	70209	Caravan residential site letting		74113	Lawyer
	70209	Development for building projects for own operation		74113	Solicitor (own account)
	70209	Dwellings letting	**69109**		**Activities of patent and copyright agents; other legal activities (other than those of barristers and solicitors) n.e.c.**
	70209	Estate company (owning and managing)			
	70209	Factory letting		74119	Arbitrators legal activities
	70209	Flat letting		74119	Bailiffs activities
	70209	Freeholder of leasehold property		74111	Copyrights preparation
	70209	Garage letting (lock up)		74119	Deeds preparation
	70209	Ground, landlord		74119	Law agent
	70209	House letting (private)		74119	Law writing
	70209	Industrial estate letting		74119	Legal activities n.e.c.
	70209	Land letting		74119	Legal documentation and certification activities
	70209	Landlord of real estate		74119	Legal examiner activities
	70209	Maisonettes letting		74119	Legal services in connection with the disposal of assets by auction
	70209	Mansions letting		74119	Notary activities
	70209	Mobile home letting (residential)		74119	Notary public
	70209	Non-residential buildings letting		74119	Parliamentary agent
	70209	Offices letting		74111	Patent agent
	70209	Property leasing (other than conference centres and exhibition halls)		74111	Patents preparation
	70209	Renting and operating of self-owned or leased real estate		74119	Process server
				74119	Referees legal activities
	70209	Residential chambers letting		74119	Scrivenery
	70209	Residential mobile home sites operation		74119	Sheriff's officer
	70209	Service flat letting		74111	Software copyright consultancy activities
	70209	Shop letting		74111	Trade mark agent
	70209	Timeshare operations (real estate)		74119	Trusts preparation
68310		**Real estate agencies**		74119	Wills preparation
	70310	Advisory services in connection with buying, selling and renting of real estate		74119	Writer to the signet
	70310	Conference centre letting (not self-owned)	**69201**		**Accounting, and auditing activities**
	70310	Estate agent		74121	Accountancy services
	70310	Exhibition centre letting (not self-owned)		74121	Accounting activities
	70310	Flat letting agency			

UK SIC 2007	UK SIC 2003	Activity
	74121	Attestations, valuations and preparation of pro forma statements
	74121	Auditing activities
	74121	Compilation of financial statements
	74121	Cost accountant
	74121	Cost draughtsman (legal)
69202		**Bookkeeping activities**
	74122	Book-keeping activities
	74122	Company secretary
	74122	Payroll bureau
69203		**Tax consultancy**
	74123	Tax consultancy
70100		**Activities of head offices**
	74154	Head office of catering company
	74153	Head office of construction company
	74155	Head office of motor trades company
	74159	Head office of other non-financial company
	74158	Head office of production company
	74157	Head office of retail company
	74156	Head office of service trades company
	74152	Head office of transport company
	74151	Head office of wholesale company
70210		**Public relations and communication activities**
	74141	Lobbying activities
	74141	Public relations and communication
	74141	Public relations consultant (not advertising agency)
70221		**Financial management**
	74142	Accounting systems design
	74142	Analysis of capital investment proposals consultancy services
	74142	Budgetary control procedures design
	74142	Business valuation services prior to mergers and/or acquisitions
	74142	Capital structure consultancy services
	74142	Chartered secretary (firm acting as)
	74142	Cost accounting programmes design
	74142	Financial management consultancy services (except corporate tax)
	74142	Insolvency management
	74142	Working capital and liquidity management consultancy services
70229		**Management consultancy activities (other than financial management)**
	74149	Advisory, conciliation and arbitration service
	74143	Advisory, guidance and operational assistance services concerning business policy and strategy
	74149	Arbitrators between management and labour
	74143	Business consultancy activities
	74143	Business consultant
	74143	Business systems consultant
	74143	Business turnaround consultancy services
	74143	Economist
	05010	Fish stock management consultancy services
	74149	Human resources management consultancy services
	74149	Industrial development consultancy services
	74143	Management audits consultancy services

SIC 2007	SIC 2003	Activity
	74143	Management consultancy activities
	74149	Marketing management consultancy activities
	74143	Overall planning, structuring and control of organisation consultancy services
	74143	Policy formulation consultancy services
	74149	Production management consultancy services (other than for construction)
	74143	Profit improvement programmes consultancy services
	74143	Strategic business plan consultancy services
	74149	Tourism development consultancy services
71111		**Architectural activities**
	74201	Advisory and pre-design architectural activities
	74201	Architectural activities and related technical consultancy
	74201	Architectural draughtsman
	74201	Building design and drafting
	74201	Construction supervision
	74201	Making of architectural maquettes
71112		**Urban planning and landscape architectural activities**
	74202	Landscape architecture
	74202	Town and city planning
	74202	Urban planning activities
71121		**Engineering design activities for industrial process and production**
	74205	Automotive production design
	74205	Consultant design engineer
	74205	Consultant engineer (other than civil or structural)
	74205	Design consultant for industrial process and production
	74205	Design office for industrial process and production
	74205	Draughtsman for industrial process and production
	74205	Electrical power systems instrumentation design activities
	74205	Electronic design consultant
	74205	Engineering design services for industrial process and production
	74205	Engineers' draughtsman
	74205	Equipment layout and other plant design services
	74205	Industrial consultants
	74205	Industrial design consultants
	74205	Industrial design service
	74205	Machinery and industrial plant design
	74205	Motor vehicle design
	74205	Naval architect
71122		**Engineering related scientific and technical consulting activities**
	74206	Aerial survey
	74206	Borehole surveying
	74206	Boundary surveying activities
	74206	Cartographic and spatial information activities
	74206	Chartered surveyor
	74206	Core preparation and analysis activities
	74206	Crude oil exploration
	74206	Digital mapping activities
	74206	Dimensional survey activities
	74206	Exploration for gas or oil

UK SIC 2007	UK SIC 2003	Activity	SIC 2007	SIC 2003	Activity
	74206	Fire and explosion protection and control consultancy activities		74204	Mechanical, industrial and systems engineering design projects
	74206	Geodetic surveying activities		74204	Mining engineering design projects
	74206	Geological and prospecting activities		74204	Pipeline design activities
	74206	Geological surveying for petroleum or natural gas (not geological consultancy)		74209	Process engineering contractor
	74206	Geologist (consultant)		74209	Project management
	74206	Geophysical consultancy activities (engineering related)		74204	Traffic engineering design projects
	74206	Geophysical, geologic and seismic surveying		74204	Water management projects design
	74206	Health and safety and other hazard protection and control consultancy activities	71200		**Technical testing and analysis**
	74206	Hydrographic surveying activities		74300	Acoustics and vibration testing
	74206	Hydrologic surveying activities		74300	Agricultural grain electrophoresis
	74206	Land surveying activities		74300	Air measuring related to cleanness
	74206	Land surveyor (not valuer)		74300	Aircraft certification
	74206	Magnetometric (subsurface) surveying activities		74300	Analytical chemist
	74206	Micropalaeontogical analysis activities		74300	Assay office
	74206	Mineral surveyor		74300	Bacteriologist (non medical)
	74206	Natural gas exploration		74300	Chemist
	74206	Petroleum exploration		74300	Composition and purity testing and analysis
	74206	Petroleum geologist		74300	Electrophoresis
	74206	Petrophysical interpretation activities		74300	Food hygiene testing activities
	74206	Seismic surveying for petroleum		74300	Integrated mechanical and electrical system testing and analysis
	74206	Sewage treatment consultancy activities		74300	Leak testing and flow monitoring activities
	74206	Sub-surface surveying activities		74300	Lloyd's Register of Shipping
	74206	Surveying activities (industrial and engineering)		74300	Marine cargo surveyor
	74206	Surveyor (other than valuer)		74300	Marine insurance survey activities
	74206	Telecommunications consultancy activities		74300	Marine surveyor
	74206	Water divining and other scientific prospecting activities		74300	Metallurgist (private practice)
				74300	MOT testing station
71129		**Other engineering activities (not including engineering design for industrial process and production or engineering related scientific and technical consulting activities)**		74300	Motor vehicles certification
				74300	Nuclear plant certification
	74204	Acoustical engineering design		74300	Performance testing of complete machinery
	74204	Architectural engineering activities		74300	Physicist
	74204	Building structure design for ancillary services		74300	Pipeline and ancillary equipment testing activities
	74204	Chemical engineering design projects		75240	Police laboratories
	74204	Consultant civil or structural engineer		74300	Pollution measuring
	74204	Consultant engineer (civil or structural)		74300	Pressurised containers certification
	74204	Corrosion engineering activities		74300	Public analyst
	74204	Design consultant for civil and structural engineering		74300	Quality control
	74204	Design office for civil and structural engineering		74300	Radioactivity measuring
	74204	Electrical and electronic engineering design projects		74300	Radiographic testing of welds and joints
	74209	Engineering contractor responsible for complete process plant		74300	Road-safety testing of motor vehicles
				74300	Ship surveyor
	74204	Engineering design activities for the construction of civil engineering works		74300	Ships certification
				74300	Strength and failure testing
	74204	Environmental consultants		74300	Sworn timber measurer
	74204	Environmental engineering consultancy activities		74300	Sworn weigher
	74204	Geotechnical engineering consultancy activities		74300	Technical and non-destructive testing services
	74204	Heating systems for buildings design activities		74300	Technical automobile inspection activities
	74204	Hydraulic engineering design projects		74300	Technical inspection services of buildings
	74209	Integrated engineering activities for turnkey projects		74300	Technical testing of bridges and other engineering structures
	74204	Jacket substructure design and other foundation design services		74300	Technical testing of lifting and handling equipment
				74300	Testing and measuring of environmental indicators: air and water pollution
	74204	Mechanical and electrical installation for buildings design activities		74300	Testing of calculations for building elements
				74300	Testing of composition and purity of minerals
				74300	Testing of physical characteristics and performance of materials

UK SIC 2007	UK SIC 2003	Activity	SIC 2007	SIC 2003	Activity
	74300	Testing or analysing laboratory		73100	Palaeontologist (consultant)
	74300	Timber measurer		73100	Physical sciences research and experimental development
	74300	Water measuring related to cleanness		73100	Physics research and experimental development
72110		**Research and experimental development on biotechnology**		73100	Research and development consultants (other than biotechnological)
	73100	Biology research and experimental development		73100	Research association (other than biotechnological)
	73100	Biotechnology research and experimental development		73100	Research chemist (private practice)
	73100	Research and experimental development on bioinformatics		73100	Research institution (other than biotechnological)
	73100	Research and experimental development on cell and tissue culture and engineering		73100	Research laboratory (other than biotechnological)
	73100	Research and experimental development on DNA/RNA		73100	Royal Observatory
	73100	Research and experimental development on gene and RNA vectors		73100	Science research council
	73100	Research and experimental development on nanobiotechnology	**72200**		**Research and experimental development on social sciences and humanities**
	73100	Research and experimental development on process biotechnology techniques		73200	Economic and Social Research Council
	73100	Research and experimental development on proteins and other molecules		73200	Economics, research and experimental development
72190		**Other research and experimental development on natural sciences and engineering**		73200	Educational research and experimental development
	73100	Agricultural research (other than biotechnological)		73200	Humanities research and experimental development
	73100	Astronomy research and experimental development		73200	Languages research and experimental development
	73100	Atomic energy research and experimental development		73200	Law research and experimental development
	73100	Building research and experimental development		73200	Legal sciences research and experimental development
	73100	Cancer research and experimental development (other than biotechnological)		73200	Linguistics research and experimental development
	73100	Chemistry research and experimental development		73100	Multi-disciplinary research and development predominantly on social sciences and humanities
	73100	Civil engineering research and experimental development		73200	National foundation for educational research
	73100	Computer hardware research and experimental development		73200	Psychology research and experimental development
	73100	Earth sciences research and experimental development (other than biotechnological)		73200	Social Science Research Council
	73100	Engineering research and experimental development		73200	Social sciences research and experimental development
	73100	Environmental pollution research and experimental development		73200	Sociology research and experimental development
	73100	Explorer	**73110**		**Advertising agencies**
	73100	Government research establishment (other than biotechnological)		74402	Advertising agencies
	73100	Hydraulic research station		74402	Advertising campaign creation and realisation
	73100	Life sciences research and experimental development (other than biotechnological)		74402	Advertising consultants
	73100	Mathematical research and experimental development		74402	Advertising contractor
	73100	Medical research establishment not attached to hospital (other than biotechnological)		74409	Advertising material or samples delivery or distribution
	73100	Medical sciences research and experimental development (other than biotechnological)		74409	Aerial and outdoor advertising services
	73100	Mining research establishment		74402	Bill posting agency
	73100	Ministry of Defence research and development (other than biotechnological)		74402	Bus carding
	73100	Multi-disciplinary research and development (not biotechnological or social sciences and humanities)		74402	Commercial artist
	73100	National physical laboratory		74402	Copywriter
	73100	Natural environment research council		74402	Creating and placing advertising
	73100	Natural sciences research and experimental development (other than biotechnological)		74402	Creation of stands and other display structures and sites
				74402	Marketing campaigns
				74402	Poster advertising
				74402	Showroom design
				74402	Signwriting
				74402	Window dressing
			73120		**Media representation**
				74401	Advertising space or time (sales or leasing thereof)
				74401	Media representation
			73200		**Market research and public opinion polling**
				74130	Export consultant

UK SIC 2007	UK SIC 2003	Activity	SIC 2007	SIC 2003	Activity
	74130	Market research agency	**74209**		**Other photographic activities (not including portrait and other specialist photography and film processing) n.e.c.**
	74130	Market research consultant		74819	Fashion photography
	74130	Market research organisation		74819	Photographing of live events such as weddings, graduations, conventions, fashion shows, etc.
	74130	Market, social and economic research services		74819	Photography for commercials, publishers or tourism purposes
	74130	Public opinion polling		92400	Photojournalists
	74130	Survey analysis and other social and economic intelligence services		92400	Picture agency
	74130	Survey design services		92400	Press photographers
74100		**Specialised design activities**		74819	Real estate photography
	74872	Boot and shoe designing		74819	Street photographer
	74872	Calico printers' designing		74819	Videoing of live events such as weddings, graduations, conventions, fashion shows, etc.
	74872	Clothes designer		74819	Wedding photography
	74872	Costume designing			
	74872	Fashion designing	**74300**		**Translation and interpretation activities**
	74872	Furniture designing		74850	Interpreter
	74872	Graphic designer		74850	Transcribing services from tapes, discs, etc.
	74872	Interior decor design		74850	Translation activities
	74872	Interior decorator activities			
	74872	Interior designers	**74901**		**Environmental consulting activities**
	74872	Jewellery designing		74206	Energy efficiency consultancy activities
	74872	Lace designing		74206	Environmental project consultancy activities
	74872	Textile or wallpaper printing designing		74206	Noise control consultancy activities
74201		**Portrait photographic activities**	**74902**		**Quantity surveying activities**
	74812	Passport photography		74203	Cost draughtsman (quantity surveyor)
	74812	Photographic studio		74203	Quantity surveying activities
	74812	Portrait photographer			
	74812	Portrait photography	**74909**		**Other professional, scientific and technical activities (not including environmental consultancy or quantity surveying) n.e.c.**
	74812	School photography		74879	Agents and agencies in entertainment
74202		**Other specialist photography (not including portrait photography)**		74879	Agents and agencies in motion picture theatrical production
	74813	Aerial photography (other than for cartographic and spatial activity purposes)		74879	Agents and agencies in placement of artworks with publishers
	74206	Aerial photography for cartographic and spatial activity purposes		74879	Agents and agencies in placement of books with publishers
	74813	Downhole photography services		74879	Agents and agencies in placement of photographs publishers
	74813	Medical and biological photography		74879	Agents and agencies in placement of plays with publishers producers
	74813	Micro-filming activities		74879	Agents and agencies in sports attractions
	74813	Photomicrography		74879	Agricultural valuer
	74813	Underwater photography services		74149	Agronomy consulting
	74813	X-ray and other speciality photography activities		74879	Appraiser and valuer (not insurance or real estate)
74203		**Film processing**		74879	Band agency
	74814	Film copying (not motion picture)		63400	Bill auditing and freight rate information
	74814	Film processing		74879	Business brokerage activities
	74814	Motion picture developing		74879	Business brokerage and appraisal activities
	74814	One hour photo shop (not part of camera shop)		74879	Business transfer agent
	74814	Photograph colouring		74879	Franchisers
	74814	Photograph copying		74879	Literary agency
	74814	Photograph developing		74206	Marine consultant (other than environmental consultancy)
	74814	Photograph enlarging		74879	Partnership agent
	74814	Photograph finishing		74879	Patent broker
	74814	Photograph mounting		74879	Publicans' broker
	74814	Photograph printing		74149	Quality assurance consultancy activities
	74814	Photographic colour printing			
	74814	Restoration, copying and retouching of photographs			
	74814	Slide and negative duplicating			
	74814	Slide mounting			

UK SIC 2007	UK SIC 2003	Activity	SIC 2007	SIC 2003	Activity
	74602	Security consultancy for industrial, household and public services		71401	Fancy dress hire
	74879	Theatrical agency		71401	Journal rental
	74879	Valuer (any trade except real estate)		71401	Juke boxes leasing
	74879	Variety agency		71401	Magazines rental
	63400	Way-bills issue and procurement		71401	Marquee hire
	74206	Weather forecasting activities		71401	Musical instrument rental
				71401	Photographic equipment hire
75000		**Veterinary activities**		71401	Piano hire
	85200	Animal ambulance activities		71401	Pleasure boats rental
	85200	Animal health care and control activities for farm animals		71401	Recreational and sports goods renting and leasing
	85200	Animal health care and control activities for pet animals		71401	Scenery rental
	85200	Animal hospital (RSPCA, PDSA, Blue Cross)		71401	Skis renting and leasing
	85200	Animal hospital run by veterinary surgeon		71401	Sound equipment rental
	85200	Animal hospital supervised or run by registered veterinarian		71401	Sports equipment rental
	85200	Clinico-pathological and other diagnostic activities pertaining to animals		71401	Theatrical costumes rental
	85200	People's dispensary for sick animals (not animal care units)	**77220**		**Renting of video tapes and disks**
				71404	CDs and disks rental
	85200	Registered veterinarian		71405	Dvd rental
	85200	Veterinary activities		71404	Records rental
	85200	Veterinary assistants or other auxiliary veterinary personnel		71405	Video tape rental
	85200	Veterinary laboratory	**77291**		**Renting and leasing of media entertainment equipment**
	85200	Veterinary surgery		71403	Media entertainment equipment renting and leasing
				71403	Radio (domestic) hire
77110		**Renting and leasing of cars and light motor vehicles**		71403	Television (domestic) hire
	71100	Automobile rental (self drive)		71403	Video recorder/player (domestic) hire
	71100	Car hire (self drive)			
	71100	Car leasing	**77299**		**Renting and leasing of other personal and household goods (other than media entertainment equipment) n.e.c.**
	71100	Car rental (self drive)		71409	Books, journals and magazine renting and leasing
	71100	Contract car hire (self drive)		71409	Clothes hire
	71100	Light motor vehicle (not exceeding 3.5 tonnes) renting or leasing		71409	Crockery hire
				71409	Cutlery hire
	71100	Van rental (self drive not exceeding 3.5 tonnes)		71401	Do-it-yourself machinery and equipment hire
				71409	Electrical domestic appliance rental and leasing
77120		**Renting and leasing of trucks**		71409	Electronic equipment for household use renting leasing
	71219	Commercial vehicle (light) hire (without driver)		71409	Flowers and plants rental and leasing
	71219	Commercial vehicle (medium and heavy type) contract hire (without driver)		71409	Footwear rental and leasing
	71219	Commercial vehicle (medium and heavy type) hire (without driver)		71409	Furniture rental and leasing
	71219	Freight container hire		71401	Garden tool hire
	71219	Haulage tractors rental (without driver)		71409	Glass rental and leasing
	71219	Recreational vehicles renting and leasing		71409	Household goods hire
	71219	Road trailer hire		71409	Housewares rental and leasing
	71219	Trailers and semi-trailers rental		71409	Jewellery rental and leasing
	71219	Truck rental (without driver)		71409	Kitchenware rental and leasing
	71219	Trucks and other heavy vehicles exceeding 3.5 tonnes renting and leasing		71409	Machinery and equipment used by amateurs or as a hobby e.g. Home repair tools rental and leasing
	71219	Van hire (exceeding 3.5 tonnes without driver)		71409	Medical and paramedical equipment (e.g. crutches) rental and leasing
				71409	Musical instruments rental and leasing
77210		**Renting and leasing of recreational and sports goods**		71409	Pottery rental and leasing
	71401	Beach chairs and umbrellas renting and leasing		71409	Tableware rental and leasing
	71401	Bicycle hire		71409	Textiles rental and leasing
	71401	Book rental		71409	Theatrical costumes leasing
	71401	Camera hire		71409	Theatrical scenery leasing
	71401	Canoes and sailboats renting and leasing		71409	Wearing apparel rental and leasing

UK SIC 2007	UK SIC 2003	Activity	SIC 2007	SIC 2003	Activity
77310		**Renting and leasing of agricultural machinery and equipment**		71229	Freight water transport equipment leasing (without operator)
	71310	Agricultural and forestry machinery and equipment operational leasing (without operator)		71229	Ship hire for freight (without crew)
	71310	Agricultural machinery and equipment rental and leasing (without operator)		71229	Water freight transport equipment rental (without operator)
	71310	Agricultural machinery hire (without operator)	**77351**		**Renting and leasing of passenger air transport equipment**
	71310	Forestry machinery and equipment rental and leasing (without operator)		71231	Air passenger transport equipment rental and leasing without operator
	71310	Horticultural machinery hire (without operator)		71231	Aircraft hire for passengers without crew
	71310	Tractor hire for agriculture (without driver)	**77352**		**Renting and leasing of freight air transport equipment**
77320		**Renting and leasing of construction and civil engineering machinery and equipment**		71239	Air transport equipment for freight rental (without operator)
	71320	Civil engineering machinery and equipment rental (without operator)		71239	Aircraft hire for freight (without crew)
	71320	Construction machinery and equipment rental (without operator)		71239	Freight air transport equipment operational leasing (without operator)
	71320	Crane hire (without operator)	**77390**		**Renting and leasing of other machinery, equipment and tangible goods n.e.c.**
	71320	Crane lorries rental (without operator)		71320	Accommodation container renting
	71320	Earthmoving equipment hire (without operator)		71340	Amusement machine hire
	71320	Hazard warning lamp hire		71340	Animal rental (herds, racehorses)
	71320	Ladder hire		71340	Audio/visual equipment for professional use hire
	71320	Plant hire for construction rental (without operator)		71219	Barrow hiring
	71320	Portable road sign hire for construction		71340	Burglar alarm hire
	71320	Scaffolding hire (without staff)		71211	Campers (transport) rental (self drive)
	71320	Tools for construction hire (without operator)		71211	Caravans (touring) rental
	71320	Work platform rental without erection and dismantling		71340	Catering equipment hire
77330		**Renting and leasing of office machinery and equipment (including computers)**		71340	Cereal milling and working machinery renting (non agricultural)
	71330	Accounting machinery and equipment rental and operating leasing		71340	Circus tent rental
	71330	Automatic data processing equipment hire		71340	Commercial machinery rental and operating leasing
	71330	Cash register hire		71340	Communication equipment rental and operating leasing
	71330	Computing machinery and equipment rental and operating leasing		71219	Container rental
	71330	Duplicating machines rental and operating leasing		71340	Dairy machinery rental (non agricultural)
	71330	Office equipment hire		71340	Engine rental and operating leasing
	71330	Office furniture hire		71340	Fire alarm hire
	71330	Office machinery and equipment leasing		71340	Floor cleaning equipment for industrial use leasing
	71330	Office machinery and equipment rental and operating leasing		71219	Freight land transport equipment rental (without driver)
	71330	Ticket machine hire		71340	Fuel bunkers leasing
	71330	Typewriter rental and operating leasing		71340	Gaming machine hire
	71330	Word processing machines rental and operating leasing		71219	Hand cart hire
77341		**Renting and leasing of passenger water transport equipment**		71340	Herd leasing
	71221	Boat hire for passengers (without crew) (not linked with recreational service)		71340	Machine tools rental and operating leasing
	71221	Passenger water transport equipment leasing (without operator)		71340	Machinery for industrial use rental and operating leasing
	71221	Ship hire for passengers (without crew)		71340	Measuring and controlling equipment rental and operating leasing
	71221	Ship rental for passengers (without operator)		71340	Mining equipment rental and operating leasing
	71221	Water passenger transport equipment rental (without operator)		71340	Motion picture production equipment rental and leasing
77342		**Renting and leasing of freight water transport equipment**		71211	Motorcycle hire
	71229	Boat hire for freight (without crew)		71211	Motorcycle rental
				71320	Office container renting
				71340	Oil field equipment rental and operating leasing
				71340	Packaging machinery leasing
				71219	Pallet rental

UK SIC 2007	UK SIC 2003	Activity
	71211	Passenger land transport equipment self drive rental (other than motor vehicles)
	71340	Plant and equipment for industrial use hire
	71340	Radio equipment for professional use, rental and operating leasing
	71211	Railroad passenger vehicles rental
	71219	Railway freight vehicle hire
	71211	Railway passenger vehicle hire
	71340	Scientific machinery rental and operating leasing
	71340	Slot machine rental
	71340	Telephone hire (other than by public telephone undertakings)
	71340	Television equipment (not domestic) rental and operating leasing
	71340	Tools for mechanics or engineers hire
	71340	Turbine rental and operating leasing
	71219	Wheelbarrow hire
77400		**Leasing of intellectual property and similar products, except copyrighted works**
	74879	Leasing of intellectual property and the like (not copyrighted works e.g. Books and software)
	74879	Leasing of non-financial intangible assets
	74879	Receiving royalties or licensing fees for the use of brand names
	74879	Receiving royalties or licensing fees for the use of franchise agreements
	74879	Receiving royalties or licensing fees for the use of mineral exploration and evaluation
	74879	Receiving royalties or licensing fees for the use of patented entities
	74879	Receiving royalties or licensing fees for the use of trademarks or service marks
78101		**Motion picture, television and other theatrical casting**
	92721	Casting activities for motion pictures, television and theatre
	74500	Casting agencies and bureaux
78109		**Activities of employment placement agencies (other than motion picture, television and other theatrical casting) n.e.c.**
	74500	Employment consultants
	74500	Employment placement agencies
	74500	Executive employment placement and search activities
	74500	Executive recruitment consultant
	74500	Labour recruitment
	74500	On-line employment placement agencies
	74500	Sales management recruitment consultant
78200		**Temporary employment agency activities**
	74500	Commercial or industrial workers (supply) (temporary employment agency)
	74500	Domestic agency (temporary employment agency)
	74500	Employment agency (temporary)
	74500	Executive personnel (supply) (temporary employment agency)
	74500	Medical personnel (supply) (temporary employment agency)
	74500	Nursing agency (supplying nurses) (temporary employment agency)
	74500	Office support personnel (supply) (temporary employment agency)

SIC 2007	SIC 2003	Activity
	74500	Personnel provision (temporary employment agency)
	74500	Supply and provision of personnel (temporary employment agency)
	74500	Teaching personnel (supply) (temporary employment agency)
	74500	Temporary employment agency activities
78300		**Other human resources provision**
	74500	Human resources provision on a long term or permanent basis
79110		**Travel agency activities**
	63301	Automobile association touring department
	63301	Chartered rail travel
	63301	Excursion agency
	63301	Motorists' organisation touring department
	63301	Packaged tour sales
	63301	Passage agent
	63301	Passenger agent (not transport authority)
	63301	Royal Automobile Club touring department
	63301	Ticket agencies for travel
	63301	Travel agency activities
79120		**Tour operator activities**
	63302	Tour operator activities
79901		**Activities of tourist guides**
	63303	Courier (travel)
	63303	Tourist guide activities
79909		**Other reservation service and related activities (not including activities of tourist guides)**
	63309	British Tourist Authority
	63309	Holiday information centre
	63309	Information bureau for tourists
	92320	Theatre ticket agency
	92349	Ticket agencies for other entertainment activities
	92320	Ticket agencies for theatre
	92629	Ticket sales activities for sports events
	92729	Ticket sales for other recreational activities
	63309	Time-share exchange services
	63309	Tourism promotion activities
	63309	Tourist assistance activities n.e.c.
	63309	Tourist board or information service
	63309	Visitor assistance services
80100		**Private security activities**
	74602	Armoured car services
	74602	Bodyguard activities
	74602	Dog training for security purposes
	74602	Fingerprinting services
	74602	Guard activities
	74602	Polygraph services
	74602	Private security activities
	74602	Security activities (not government)
	74602	Security delivery of prisoners
	74602	Security guard services
	74602	Security shredding of information on any media
	74602	Security transport of valuables and money
	74602	Store detective activities
	74602	Street patrol

UK SIC 2007	UK SIC 2003	Activity	SIC 2007	SIC 2003	Activity
	74602	Surveillance activities		74704	Telephone cleaning and sterilising service
	74602	Watchman activities		74704	Telephone sterilising
				74704	Ventilation ducts cleaning
80200		**Security systems service activities**	**81223**		**Furnace and chimney cleaning services**
	74602	Alarm monitoring activities		74705	Boiler cleaning and scaling
	74602	Burglar and fire alarm monitoring including installation and maintenance		74705	Chimney cleaning
	74602	Installation and repair of electronic locking devices with monitoring		74705	Fire-places cleaning
	74602	Installation and repair of electronic safes security vaults with monitoring		74705	Furnace cleaning
				74705	Incinerator cleaning
	74602	Installation and repair of mechanical locking devices with monitoring		74705	Stove cleaning
	74602	Installation and repair of mechanical safes and security vaults with monitoring	**81229**		**Building and industrial cleaning activities (other than window cleaning, specialised cleaning and furnace and chimney cleaning services) n.e.c.**
	45310	Installation of fire and burglar alarm systems if together with monitoring of the same systems		74709	Cleaning of industrial machinery
	74602	Monitoring activities by mechanical or electrical protective devices		74709	Exterior cleaning of buildings
	74602	Monitoring of electronic security alarm systems	**81291**		**Disinfecting and extermination services**
	74602	Security systems service activities		74703	Disinfecting of dwellings and other buildings
80300		**Investigation activities**		74703	Exterminating of insects, rodents and other pests (except agricultural)
	74879	Enquiry agency		74703	Fumigation services
	74879	Inquiry agency		74703	Local authority pest control department
	74601	Internet abuse monitoring		74703	Pest control services (except agricultural)
	74601	Investigation activities		74703	Pest destruction service (not especially for agriculture)
	74601	Private detective		74703	Rat catcher (not especially for agriculture)
	74601	Private investigator activities		74703	Rodent destroying (not agricultural)
81100		**Combined facilities support activities**		74703	Ship disinfecting and exterminating activities
	75140	Centralised supply and purchasing services (public sector)		74703	Ship fumigating and scrubbing
	70320	Combined facilities support activities		74703	Train disinfecting and exterminating activities
	70320	Facilities management		74703	Vermin destroying (not agricultural)
	75140	Operational services of government owned or occupied buildings (public sector)		74703	Wood rot preventative treatment service
	70320	Residents' property management		74703	Wood worm preventative treatment service
81210		**General cleaning of buildings**	**81299**		**Cleaning services (other than disinfecting and extermination services) n.e.c**
	74701	Building cleaning activities		74709	Aeroplane cleaning (non-specialised)
	74701	Cleaning service for factory, office or shop		90030	Airport runways clearing of snow and ice
	74701	Commercial cleaner		74709	Bottle cleaning
	74701	Contract cleaning service		74709	Bus cleaning (non- specialised)
	74701	Factory cleaning contractor		74709	Cleaning of the inside of road and sea tankers
	74701	General cleaning of houses or apartments		74709	Cleaning services n.e.c.
	74701	Industrial cleaning		90030	Highway cleaning of snow and ice
	74701	Janitorial services		90030	Hygiene contracting
	74701	Office cleaning contractor		74709	Industrial equipment cleaning (non-specialised)
81221		**Window cleaning services**		90030	Outdoor sweeping and watering of parking lots, streets, paths, public spaces etc.
	74702	Window cleaning		90030	Rental of lavatory cubicles
81222		**Specialised cleaning services**		90030	Sanding or salting of highways, etc.
	74704	Cleaning of heat and air ducts		90030	Snow and ice clearing of highways, etc.
	74704	Cleaning services for computer rooms		90030	Street cleaning and watering
	74704	Cleaning services for hospitals		74709	Swimming pool cleaning and maintenance activities
	74704	Decontamination services		74709	Train cleaning (non-specialised)
	74704	Reservoir and tank cleaning		74709	Transport equipment cleaning (non-specialised)
	74704	Specialised cleaning of tanks and reservoirs		74709	Underground train cleaning (non-specialised)
	74704	Sterilisation of objects or premises (e.g. operating theatres)	**81300**		**Landscape service activities**
				01410	Gardens and sport installations, planting and maintenance on a fee or contract basis

UK SIC 2007	UK SIC 2003	Activity	SIC 2007	SIC 2003	Activity
	01410	Hedge trimming on a fee or contract basis (except as an agricultural service activity)	82912		**Activities of credit bureaus**
	01410	Landscape contracting		74871	Credit bureau
	01410	Landscape gardening		74871	Credit rating
	01410	Landscape measures for protecting the environment		74871	Financial and credit reporting
	01410	Park laying out, planting and maintenance on a fee or contract basis	82920		**Packaging activities**
				74820	Aerosol filling on a fee or contract basis
	01410	Planting, laying out and maintenance of gardens, parks and green areas for sports installations		74820	Beverage bottling on a fee or contract basis
				74820	Blister packaging foil-covered packaging
	01410	Tree pruning, replanting, on a fee or contract basis (except as an agricultural service activity)		74820	Bottling on a fee or contract basis
				74820	Food bottling and packaging on a fee or contract basis
82110		**Combined office administrative service activities**		74820	Labelling, stamping and imprinting on a fee or contract basis
	74850	Combined office admin. Services (e.g. Reception, billing, record keeping, personnel, mail services)		74820	Liquids bottling on a fee or contract basis
				74820	Packaging activities on a fee or contract basis
	74850	Combined secretarial activities		74820	Packaging of solids
82190		**Photocopying, document preparation and other specialised office support activities**		74820	Packing of medicaments into edible capsules
				74820	Parcel packing and gift wrapping on a fee or contract basis
	74850	Blueprinting		74820	Security packaging of pharmaceutical preparations
	74850	Circular addressing	82990		**Other business support service activities n.e.c.**
	74850	Direct mailing		74879	Agricultural showground
	74850	Document copying service		74879	Bar code imprinting services
	74850	Document preparation		74879	Broadcasting transmission rights agent
	74850	Duplicating service		74879	Business activities n.e.c.
	74850	Envelope addressing service		74850	Court reporting services
	74850	Envelope stuffing, sealing and mailing service including for advertising		74879	Fashion agent
				74879	Fashion artist
	74850	Letter or resume writing		74879	Fundraising organisation services on a contract or fee basis
	64110	Mailbox rental		74879	Gas, water and electricity meter reading
	74850	Mailing pre-sorting		74879	Independent auctioneers
	74850	Multigraphing		74879	Information bureau (not tourist)
	74850	Photocopying		74879	Loyalty programme administration
	74850	Proof reading		74879	Luncheon voucher company
	74850	Reprographic activities (other than printing)		74879	Meter reading on a fee or contract basis
	74850	Shorthand writing		74879	Parking meter coin collection services
	74850	Transcription of documents, and other secretarial services		74879	Public record searching
				74850	Public stenography services
	74850	Typing, word processing and desk top publication service		74879	Real-time closed captioning
				74879	Repossession services
82200		**Activities of call centres**		74879	Stock control activities
	74860	Call centres undertaking market research or public opinion polling		74879	Trading stamp activities
				75130	Vehicle licence issuing on a fee or contract basis
	74860	Call centres working on a fee or contract basis	84110		**General public administration activities**
	74860	Inbound call centres		75110	Central government administration
	74860	Outbound call centres		75110	Customs administration
82301		**Activities of exhibition and fair organizers**		75110	Duty and tax collection
	74873	Exhibition contracting and organising		75110	Economic and social planning administration (public sector)
	74873	Exhibition stand design		75110	Executive and legislative administration (public sector)
	74873	Exhibition stand hire			
	74873	Fair organiser		75110	Financial services (public sector)
	74873	Trade centre		75110	Fiscal services (public sector)
	74873	Trades exhibition organiser		75110	Fundamental research administration (public sector)
82302		**Activities of conference organizers**		75140	General personnel administration and operational services (public sector)
	74874	Conference organisers			
82911		**Activities of collection agencies**			
	74871	Bill collecting			
	74871	Debt collector			

UK SIC 2007	UK SIC 2003	Activity	SIC 2007	SIC 2003	Activity
	75110	General public administration activities		75130	Manufacturing services administration and regulation (public sector)
	75110	Local government administration		75130	Mineral resource services administration and regulation (public sector)
	75110	Public debt services administration		75130	Mining services administration and regulation (public sector)
	75110	Public fund services administration		75130	Multipurpose development project services administration (public sector)
	75110	Statistical services (public sector)		75130	Tourism services administration and regulation (public sector)
	75110	Tax violation investigation services		75130	Transport services administration and regulation (public sector)
	75110	Taxation schemes			
	75110	Trust territory programme administration (public sector)			
84120		**Regulation of the activities of providing health care, education, cultural services and other social services, excluding social security**	**84210**		**Foreign affairs**
	75120	Community amenity services administration (public sector)		75210	Consular services abroad administration and operation (public sector)
	75120	Cultural services administration (public sector)		75210	Economic aid missions accredited to foreign governments (public sector)
	75120	Dental service administration (public sector)		75210	Foreign economic aid services administration (public sector)
	75120	Educational services administration (public sector)		75210	Information and cultural services abroad administration and operation (public sector)
	75120	Environmental services administration (public sector)		75210	International peace keeping forces contribution including assignment of manpower
	75120	Health care services administration (public sector)		75210	Military aid missions accredited to foreign governments (public sector)
	75120	Housing services administration (public sector)		75210	Non-military aid programmes to developing countries (public sector)
	75120	Pollution standards, dissemination and information services (public sector)		75210	Technical assistance and training programmes abroad (public sector)
	75120	Public administration of education programmes			
	75120	Public administration of environment programmes	**84220**		**Defence activities**
	75120	Public administration of health programmes		75220	Administration of defence-related research and development policies and related funds
	75120	Public administration of housing programmes		75220	Administration supervision and operation of engineering, and other non-combat forces and commands
	75120	Public administration of recreation programmes		75220	Administration supervision and operation of intelligence and other non-combat forces and commands
	75120	Public administration of research and development policies		75220	Administration supervision and operation of land, sea, air and space defence forces
	75120	Public administration of social services programmes		75220	Administration supervision and operation of military defence affairs
	75120	Public administration of sport programmes		75220	Administration supervision and operation of reserve/auxiliary forces of the defence establishment
	75120	Recreational services administration (public sector)		75220	Administrative, operational and supervisory services related to civil defence
	75120	Religious services administration (public sector)		75220	Administrative, operational and supervisory services related to military defence
84130		**Regulation of and contribution to more efficient operation of businesses**		75220	Army establishment (civilian personnel)
	75130	Agricultural services administration and regulation (public sector)		75220	Army establishment (service personnel)
	75130	Catering trade services administration and regulation (public sector)		75220	Civil defence administration
	75130	Commercial services administration and regulation (public sector)		75220	Defence activities
	75130	Communication services administration and regulation (public sector)		75220	Health activities for military personnel in the field
	75130	Construction services administration and regulation (public sector)		75220	Maritime search and rescue (military)
	75130	Distribution services administration and regulation (public sector)		75220	Military defence administration
	75130	Economic services administration and regulation (public sector)		75220	Military logistics
	75130	Energy services administration and regulation (public sector)		75220	Military ports
	75130	Fishing services administration and regulation (public sector)		75220	Ministry of Defence (civilian personnel)
	75130	Forestry services administration and regulation (public sector)		75220	Ministry of Defence (forces personnel)
	75130	Fuel services administration and regulation (public sector)		75220	Ministry of Defence Headquarters
	75130	Hunting services administration and regulation (public sector)		75220	Pay and personnel agency (armed forces)
	75130	Labour affairs services administration and regulation (public sector)		75220	Princess Mary's RAFNS
	75130	Land registry			

UK SIC 2007	UK SIC 2003	Activity	SIC 2007	SIC 2003	Activity
	75220	Queen Alexandra's RANC		75240	Alien registration administration and operation
	75220	Queen Alexandra's RNNS		75240	Border guard administration and operation
	75220	Royal Air Force Establishments (civilian personnel)		75240	Coast guards administration and operation
	75220	Royal Air Force Establishments (service personnel)		75240	Criminal investigation department
	75220	Royal Marines		75240	Law and order administration and operation
	75220	Royal Navy establishments (civilian personnel)		75240	Local authority school crossing patrols
	75220	Royal Navy establishments (service personnel)		75240	Local authority traffic wardens
	75220	Support for defence plans and exercises for civilians		75240	Metropolitan Police Commissioners Office
	75220	Territorial Army		75240	Police authorities
	75220	Women's Royal Air Force		75240	Police records maintenance
	75220	Women's Royal Army Corps		75240	Port guards administration and operation
	75220	Women's Royal Naval Service		75240	Provision of supplies for domestic emergency use in case of peacetime disasters
84230		**Justice and judicial activities**		75240	Public order and safety administration, regulation and operation
	75230	Administration and operation of administrative civil and criminal law courts		75240	Regional crime squad
	75230	Appeal Committee of the House of Lords		75240	School crossing patrols
	75230	Arbitration of civil actions		75240	Traffic regulation administration and operation
	75230	Companies court		75240	Traffic wardens
	75230	Coroners court			
	75230	Correctional services	**84250**		**Fire service activities**
	75230	County court		75250	Auxiliary fire brigade services
	75230	Court of appeal		75250	Fire authorities
	75230	Court of protection		75250	Fire brigades
	75230	Court of session (Scotland)		75250	Fire fighting and fire prevention
	75230	Court of the Lord Lyon		75250	Fire service activities
	75230	Crown court		75250	Local authority fire brigade services
	75230	Crown prosecution service		75250	Marine fireboat services
	75230	Detention centres			
	75230	High Court of Justice	**84300**		**Compulsory social security activities**
	75230	High court of Justice in Bankruptcy		75300	Compulsory social security administration concerning family and child benefits
	75230	High Court of Justiciary (Scotland)		75300	Compulsory social security administration concerning government employee pension schemes
	75230	Judge		75300	Compulsory social security administration concerning permanent loss of income due to disablement
	75230	Judge advocates		75300	Compulsory social security administration concerning retirement pensions
	75230	Justice of the Peace		75300	Compulsory social security administration concerning unemployment benefits
	75230	Local authority observation and assessment centres		75300	Compulsory social security administration of sickness, maternity or temporary disablement benefits
	75230	Lord Chancellor's Department		75300	Funding and administration of government provided retirement pensions,
	75230	Magistrates' court		75300	Funding and administration of government provided sickness, work-accident and unemployment insurance
	75230	Military tribunals administration and operation			
	75230	Official receiver			
	75230	Official solicitor			
	75230	Pensions appeal tribunal	**85100**		**Pre-primary education**
	75230	Prison administration and operation		80100	Church schools at nursery and primary level
	75230	Prisons (excluding naval and military)		80100	Hospital schools at nursery and primary level
	75230	Probate registry		80100	Kindergartens
	75230	Registrars Office (Courts of Justice)		80100	Nursery schools
	75230	Rehabilitation services for prisoners		80100	Pre-primary education
	75230	Remand centres			
	75230	Restrictive Practices Court	**85200**		**Primary education**
	75230	Sheriff's Court (Scotland)		80100	Infant schools
	75230	Stipendiary magistrates		80100	Junior schools
	75230	Supreme court of judicature		80100	Middle schools deemed primary
	75230	Tribunals		80100	Preparatory schools
	75230	Young offender centres			
84240		**Public order and safety activities**			
	75240	Administration and operation of regular and auxiliary police forces supported by public authorities			
	75240	Administration and operation of special police forces			

UK SIC 2007	UK SIC 2003	Activity	SIC 2007	SIC 2003	Activity
	80100	Primary and pre-primary education	**85410**		**Post-secondary non-tertiary education**
	80100	Primary education		80301	College of nursing
	80100	Primary schools		80301	Higher education (sub degree level)
	80100	Special schools at primary and pre-primary level		80301	Post-secondary non-tertiary vocational education
				80301	School of languages
85310		**General secondary education**		80301	Vocational education at post-secondary non-tertial level
	80210	Church schools at secondary level			
	80210	Comprehensive schools	**85421**		**First-degree level higher education**
	80210	Convent schools at secondary level		80302	College of higher education (degree level)
	80210	Grammar schools		80302	Correspondence college specialising in higher education courses (degree level)
	80210	Hospital schools at secondary level		80302	Council for National Academic Awards
	80210	Public schools		80302	Dental college or school
	80210	School examination board		80302	First-degree level higher education
	80210	Secondary level education		80302	Graduate school for business studies
	80210	Secondary modern schools		80302	Higher education at the first degree level
	80210	Secondary schools		80302	Law college
	80210	Sixth form colleges		80302	Medical school
	80210	Special schools at secondary level		80302	Military college
				80302	Open University
85320		**Technical and vocational secondary education**		80302	Performing arts schools providing tertiary education
	80220	Agricultural college		80302	Polytechnics
	80220	Apprentice school		80302	Study leading to a one year post graduate certificate of education (PGCE)
	80220	Arts and crafts school		80302	Theological college specialising in higher education course
	80220	Ballet school		80302	Universities' Central Council on Admissions
	80220	City and Guilds of London Institute		80302	University
	80220	Civil service college		80302	University college
	80220	College of agriculture		80302	University medical or dental school
	80220	College of art			
	80220	College of music	**85422**		**Post-graduate level higher education**
	80220	College of technology		80303	Higher education at post-graduate level
	80220	Commercial school		80303	Post-graduate college
	80220	Computer repair training			
	80220	Cosmetology and barber schools	**85510**		**Sports and recreation education**
	80220	Driving schools for occupational drivers e.g. Of trucks, buses, coaches		92629	Bridge instructor
	63230	Flying school (for airline pilots)		92629	Card game instruction
	63230	Flying schools for commercial airline pilots		92629	Chess instructor
	80220	Flying training for professional pilots		92629	Coaches of sport
	80220	Government training centre		92629	Gymnastics instruction
	80220	Instruction for chefs, hoteliers and restauranteurs		92629	Instructors of sport
	80220	Management training establishment		92629	Martial arts instruction
	80220	Military school		92629	Riding school
	80220	Music teacher (own account)		92629	Ski instructor (own account)
	80220	Nautical school		92629	Sport and game schools
	80220	Royal Academy of Dramatic Art		92629	Sports and recreation education
	80220	School of arts and crafts		92629	Swimming instruction
	80220	School of speech and drama		92629	Teachers of sport
	80220	Secretarial college		93059	Yoga instruction
	80220	Seminary			
	80429	Technical and vocational adult education (excl. cultural, sports, recreation education and the like)	**85520**		**Cultural education**
	80220	Technical and vocational education		80429	Art instruction
	80220	Technical and vocational secondary education		80429	Cultural education
	80220	Technical college		92341	Dancing academy (ballroom)
	80220	Tertiary college		92341	Dancing master
	80220	Tourist guide instruction		92341	Dancing school
	63220	Tuition for ships' licences for commercial certificates and permits		92341	Dancing schools and dance instructor activities
	80220	Works school (if separately identifiable)		80429	Fine arts schools (except academic)

UK SIC 2007	UK SIC 2003	Activity	SIC 2007	SIC 2003	Activity
	80429	Performing arts schools (except academic)		85112	Central sterile supply department (private sector)
	80429	Photography schools (except commercial)		85111	Central sterile supply department (public sector)
	80429	Piano teachers and other music instruction		85112	Children's hospital (private sector)
				85111	Children's hospital (public sector)
85530		**Driving school activities**		85112	Chronic sick hospital (private sector)
	80410	Driving instruction		85111	Chronic sick hospital (public sector)
	80410	Driving school activities		85112	Convalescent home (private sector providing medical care)
	80410	Flying school activities (not type rating)		85111	Convalescent home (public sector providing medical care)
	80410	Flying schools not issuing commercial certificates and permits		85112	Dental hospital (private sector)
	80410	Sailing schools not issuing commercial certificates and permits		85111	Dental hospital (public sector)
	80410	School of motoring		85112	Ear, nose and throat hospital (private sector)
	80410	Ship licence tuition (not commercial certificates)		85111	Ear, nose and throat hospital (public sector)
	80410	Shipping schools not issuing commercial certificates and permits		85111	Ear, nose and throat specialist (public sector)
				85112	Eye hospital (private sector)
85590		**Other education n.e.c.**		85111	Eye hospital (public sector)
	80429	Academic tutoring		85111	Eye specialist (public sector)
	80429	Adult education centre		85111	General hospital (public sector)
	80429	Adult education residential college		85111	General hospital psychiatric unit (public sector)
	80429	Computer training		85111	General medical consultant (public sector)
	80429	Continuation school		85112	General medicine consultant (private sector)
	80429	Correspondence college (not leading to degree level qualifications)		85111	Genito-urinary specialist (public sector)
	80429	Council for Accreditation of Correspondence Colleges		85112	Geriatric hospital (private sector)
	80429	Day continuation school		85111	Geriatric hospital (public sector)
	80429	Language instruction and conversational skills instruction		85111	Geriatrician (public sector)
	80429	Learning centres offering remedial courses		85111	Gynaecologist (public sector)
	80429	Lifeguard training		85111	Haematologist (public sector)
	80429	Mentally disabled adult training		85112	Hospice (private sector)
	80429	National institute for adult continuing education		85111	Hospice (public sector)
	80429	Other adult and other education n.e.c.		85112	Hospital activities (private sector)
	80421	Private training providers		85111	Hospital activities (public sector)
	80429	Professional examination review courses		85111	Infectious disease hospital (public sector)
	80429	Public speaking training		85111	Infectious disease specialist (public sector)
	80429	Religious instruction		85112	Infirmary (private sector)
	80429	Speed reading instruction		85111	Infirmary (public sector)
	80429	Survival training		85111	Isolation hospital (public sector)
	80429	Teacher n.e.c.		85111	Leprosaria (public sector)
	80429	Workers' Educational Association		85112	Maternity hospital (private sector)
				85111	Maternity hospital (public sector)
85600		**Educational support activities**		85111	Medical consultant (public sector)
	74149	Educational consulting		85111	Mental disability hospital (public sector)
	74149	Educational guidance counselling activities		85111	Mental health specialist (public sector)
	74149	Educational support activities		85112	Mental hospital (private sector)
	74149	Educational testing activities		85111	Mental hospital (public sector)
	74149	Educational testing evaluation activities		85111	Military base hospitals
	74149	Organisation of student exchange programmes		85111	Military hospital (public sector)
	93059	Scholastic agent		85111	Morbid anatomy specialist (public sector)
	93059	School agent		85112	Nuffield Hospital Trust
				85112	Ophthalmic hospital (private sector)
86101		**Hospital activities**		85111	Ophthalmic hospital (public sector)
	85112	Accident and emergency service (private sector)		85112	Orthopaedic hospital (private sector)
	85111	Accident and emergency service (public sector)		85111	Orthopaedic hospital (public sector)
	85111	Anaesthetist (public sector)		85112	Pre-convalescent hospital (private sector)
	85112	Asylums (private sector)		85111	Pre-convalescent hospital (public sector)
	85111	Asylums (public sector)		85111	Preventoria providing hospital type care (public sector)
	85111	Broadmoor Hospital		85111	Prison hospitals
				85112	Private hospital

UK SIC 2007	UK SIC 2003	Activity	SIC 2007	SIC 2003	Activity
	85112	Psychiatric unit (private sector)		85130	Dental clinic (health service)
	85111	Radiologist (public sector)		85130	Dental practice activities
	85111	Radiotherapist (public sector)		85130	Dental receptionist
	85111	Rampton Hospital		85130	Dental surgeon (not employed full time by a hospital)
	85112	Rehabilitation hospital (private sector)		85130	Dentist
	85111	Rehabilitation hospital (public sector)		85130	Endodontic dentistry
	85112	Sanatoria providing hospital type care (private sector)		85130	Oral pathology
	85111	Sanatoria providing hospital type care (public sector)		85130	Orthodontic activities
	85111	Smallpox hospital (public sector)		85130	Paediatric dentistry
	85111	Social medicine specialist (public sector)			
	85112	Special hospital (private sector)	**86900**		**Other human health activities**
	85111	Special hospital (public sector)		85140	Ambulance service
	85111	Surgeon (public sector)		85140	Artificial kidney unit
	85111	Tuberculosis hospital (public sector)		85140	Artificial limb and appliance centre
	85111	Urologist (public sector)		85140	Blood banks
				85140	Blood transfusion service
86102		**Medical nursing home activities**		85140	Chiropodist (NHS)
	85113	Nursing home with medical care (under the direct supervision of medical doctors)		85140	Chiropodist (private)
				85140	Chiropractor clinic (own account)
86210		**General medical practice activities**		85140	Clinic (health service)
	85120	District community physician		85140	Collection of female human urine for hormone extraction
	85120	Doctor (unspecified)		85140	Community health service
	85120	Doctors receptionist		85140	Community medical service clinics
	85120	Family doctor service		85140	Community psychiatric nurse (NHS)
	85120	General medical consultant (private practice)		85140	Dental hygienist
	85120	General medical practitioner		85140	Dental therapist
	85120	Medical group practice		85140	Disablement services centres
	85120	School medical officer		85140	District nurse
	85120	Surgery (doctor's)		85140	Family Planning Association clinics (not providing medical treatment)
				85140	Foot clinic (NHS)
86220		**Specialist medical practice activities**		85140	Foot clinic (private)
	85120	Ear, nose and throat specialist (private practice)		85140	Health centre
	85120	Eye specialist (private practice)		85140	Health visitor
	85120	Family planning centre providing medical treatment without accommodation		85140	Home nurse (NHS)
	85120	Genito-urinary specialist (private practice)		85140	Homeopath (not registered medical practitioner)
	85120	Gynaecologist (private practice)		85140	Limb fitting centre
	85120	Homeopath (registered medical practitioner)		85140	Mass radiography service
	85120	Infectious disease specialist (private practice)		85140	Maternity and child welfare services
	85120	Mental health specialist (private practice)		85140	Maternity clinic
	85120	Morbid anatomy specialist (private practice)		85140	Medical laboratories
	85120	Osteopath (registered medical practitioner)		85140	Midwife (NHS)
	85120	Physiologist		85140	Midwife (private)
	85120	Private consultants clinics		85140	Neuropath
	85120	Psychiatrist (private practice)		85140	Nurse (private)
	85120	Radiotherapy treatment centre		85140	Nursing co-operative
	85120	Social medicine specialist (private practice)		85140	Occupational therapist (private)
	85120	Specialist (not employed full time by a hospital)		85140	Ophthalmic clinic
	85120	Specialist medical consultant (private practice)		85140	Osteopath (not registered medical practitioner)
	85120	Specialist medical consultation and treatment		85140	Para-medical practitioner activities
	85120	Specialist physician and surgeon (private practice)		85140	Pathological laboratory
	85120	Surgeon (private practice)		85140	Physiotherapist (private)
	85120	Urologist (private practice)		85140	Physiotherapy clinic
				85140	Psychiatric clinic
86230		**Dental practice activities**		85140	Psychiatric day hospital
	85130	Community dental service clinics		85140	Psychologist
	85130	Dental activities in operating rooms		85140	Public health laboratory
	85130	Dental clinic			

UK SIC 2007	UK SIC 2003	Activity	SIC 2007	SIC 2003	Activity
	85140	Radiographer (private)		85312	Residential care activities (social) for substance abuse (non-charitable)
	85140	School dental nurse		85311	Residential care home for the mentally disabled (charitable)
	85140	School health service			
	85140	School medical clinic		85312	Residential care home for the mentally disabled (non-charitable)
	85140	Scottish Ambulance Service			
	85140	Speech therapist (NHS)		85311	Residential care home for the mentally ill (charitable)
	85140	Speech therapist (own account)		85312	Residential care home for the mentally ill (non-charitable)
	85140	Sperm banks			
	85140	St Andrew's ambulance brigade		85112	Residential care in alcoholism or drug addiction treatment facilities (private sector)
	85140	St John's ambulance brigade			
	85140	Transplant organ banks		85111	Residential care in alcoholism or drug addiction treatment facilities (public sector)
87100		**Residential nursing care activities**		85112	Residential care in rehabilitation centres (private sector)
	85140	Convalescent homes			
	85140	Homes for the elderly with nursing care		85111	Residential care in rehabilitation health centres (public sector)
	85140	Nursing care facilities			
	85140	Nursing homes	**87300**		**Residential care activities for the elderly and disabled**
	85140	Residential nursing care facilities			
	85113	Residential nursing care facilities (not directly supervised by medical doctors)		85311	Assisted-living facilities for the elderly or disabled (charitable)
	85140	Rest homes with nursing care		85312	Assisted-living facilities for the elderly or disabled (non-charitable)
87200		**Residential care activities for learning disabilities, mental health and substance abuse**		85311	Continuing care retirement communities (charitable)
				85312	Continuing care retirement communities (non-charitable)
	85140	Residential care (paramedical) in group homes for the emotionally disturbed (charitable)		85311	Home for the blind (charitable)
				85312	Home for the blind (non-charitable)
	85140	Residential care (paramedical) in mental health halfway houses		85311	Home for the disabled (charitable)
				85312	Home for the disabled (non-charitable)
	85140	Residential care (paramedical) in mental disability facilities		85311	Home for the elderly (charitable)
				85312	Home for the elderly (non-charitable)
	85140	Residential care (paramedical) in psychiatric convalescent homes		85311	Homes for the elderly with minimal nursing care (charitable)
	85311	Residential care (social) in group homes for the emotionally disturbed (charitable)		85312	Homes for the elderly with minimal nursing care (non-charitable)
	85312	Residential care (social) in group homes for the emotionally disturbed (non charitable)		85312	Local authority homes for the disabled and the elderly
	85311	Residential care (social) in mental health halfway houses (charitable)		85311	Old people's sheltered housing (charitable)
	85312	Residential care (social) in mental health halfway houses (non-charitable)		85312	Old people's sheltered housing (non-charitable)
	85311	Residential care (social) in mental disability facilities (charitable)		85312	Old persons' home (local authority)
				85311	Old persons' warden assisted dwellings (charitable)
	85312	Residential care (social) in mental disability facilities (non-charitable)		85312	Old persons' warden assisted dwellings (non-charitable)
	85311	Residential care (social) in psychiatric convalescent homes (charitable)		85140	Provision of residential care and treatment for the elderly and disabled by paramedical staff
	85312	Residential care (social) in psychiatric convalescent homes (non-charitable)		85311	Residential care activities for the elderly and disabled (charitable)
	85140	Residential care activities (paramedical) for mental health		85312	Residential care activities for the elderly and disabled (non-charitable)
	85140	Residential care activities (paramedical) for mental disability		85311	Residential care home for epileptics (charitable)
	85140	Residential care activities (paramedical) for substance abuse		85312	Residential care home for epileptics (non-charitable)
	85311	Residential care activities (social) for learning difficulties (charitable)		85311	Residential care home for disabled children (charitable)
	85312	Residential care activities (social) for learning difficulties (non-charitable)		85312	Residential care home for disabled children (non-charitable)
	85311	Residential care activities (social) for mental health (charitable)		85311	Rest homes without nursing care (charitable)
	85312	Residential care activities (social) for mental health (non-charitable)		85312	Rest homes without nursing care (non-charitable)
	85311	Residential care activities (social) for substance abuse (charitable)	**87900**		**Other residential care activities**
				85311	Children's boarding homes and hostels (charitable)
				85312	Children's boarding homes and hostels (non-charitable)

UK SIC 2007	UK SIC 2003	Activity	SIC 2007	SIC 2003	Activity
	85311	Children's home (charitable)		85322	Crèche (non-charitable)
	85312	Children's home (non-charitable)		85321	Day care for disabled children (charitable)
	85311	Community homes for children (charitable)		85322	Day care for disabled children (non-charitable)
	85312	Community homes for children (non-charitable)		85321	Day nursery (charitable)
	85311	Convalescent home without medical care (charitable)		85322	Day nursery (non-charitable)
	85312	Convalescent homes without medical care (non-charitable)		85321	Playgroup (charitable)
	85311	Discharged prisoners' hostel (charitable)		85322	Playgroup (non-charitable)
	85312	Discharged prisoners' hostel (non-charitable)	**88990**		**Other social work activities without accommodation n.e.c.**
	85311	Halfway group homes for persons with social or personal problems (charitable)		85321	Adoption activities (charitable)
	85312	Halfway group homes for persons with social or personal problems (non-charitable)		85322	Adoption activities (non-charitable)
	85311	Halfway homes for delinquents and offenders (charitable)		85321	Benevolent society (charitable services)
	85312	Halfway homes for delinquents and offenders (non-charitable)		85321	Charity administration
	85311	Hostel for the homeless (charitable)		85321	Child guidance centre (charitable)
	85312	Hostel for the homeless (non-charitable)		85322	Child guidance centre (non-charitable)
	85311	Juvenile correction homes (charitable)		85321	Citizens Advice Bureau
	85312	Juvenile correction homes (non-charitable)		85321	Community and neighbourhood activities (charitable)
	85312	Local authority community homes (children)		85322	Community and neighbourhood activities (non-charitable)
	85312	Local authority lodging houses		85321	Credit and debt counselling services (charitable)
	85312	Lodging house (local authority)		85322	Credit and debt counselling services (non-charitable)
	85311	Orphanages (charitable)		85321	Disaster relief organisations (charitable)
	85312	Orphanages (non-charitable)		85322	Disaster relief organisations (non-charitable)
	85311	Residential nurseries (charitable)		85321	Employment rehabilitation centre (charitable)
	85312	Residential nurseries (non-charitable)		85322	Employment rehabilitation centre (non-charitable)
	85311	Salvation army shelter (charitable)		85321	Family Planning Associations (not clinics)
	85311	Shelter (the charity)		85321	Family Welfare Association
	85311	Social work activities with accommodation (charitable)		85321	Jewish board of family and children's services
	85312	Social work activities with accommodation (non-charitable)		85322	Local authority citizen's advice bureau
	85311	Temporary accommodation for the homeless (charitable)		85322	Local authority probation service
	85312	Temporary accommodation for the homeless (non-charitable)		85322	Local authority social services department
	85311	Temporary homeless shelters (charitable)		85321	Marriage and family guidance (charitable)
	85312	Temporary homeless shelters (non-charitable)		85322	Marriage and family guidance (non-charitable)
88100		**Social work activities without accommodation for the elderly and disabled**		85321	National society for the prevention of cruelty to children
	85321	Day centres for the elderly, the physically or the mentally ill (charitable)		85321	Oxfam (not shops)
	85322	Day centres for the elderly, the physically or the mentally ill (non-charitable)		85322	Police court mission
	85321	Home help service (charitable)		85322	Probation and after care service
	85322	Home help service (non-charitable)		85321	Red Cross Society
	85322	Local authority home help service		75210	Refugee and hunger relief programmes abroad
	85322	Occupation and training centre for the mentally disordered (non-charitable)		85321	Refugee camp (charitable)
	85321	Occupation and training centres for the mentally disordered (charitable)		85322	Refugee camp (non-charitable)
	85321	Old age and sick visiting (charitable)		85321	Refugee services (charitable)
	85322	Old age and sick visiting (non-charitable)		85322	Refugee services (non-charitable)
	85321	Vocational rehabilitation (charitable)		85321	Royal Masonic Benevolent Institute
	85322	Vocational rehabilitation (non-charitable)		85322	Social Services Department
88910		**Child day-care activities**		85321	Social welfare society (charitable)
	85321	Child day-care activities (charitable)		85321	Social work activities for immigrants (charitable)
	85322	Child day-care activities (non-charitable)		85322	Social work activities for immigrants (non-charitable)
	85321	Crèche (charitable)		85321	Social work activities without accommodation (charitable)
				85322	Social work activities without accommodation (non-charitable)
				85321	Social worker (charitable)
				85322	Social worker (non-charitable)
				85321	Temperance association
				85321	Welfare and guidance activities for children and adolescents (charitable)

UK SIC 2007	UK SIC 2003	Activity	SIC 2007	SIC 2003	Activity
	85322	Welfare and guidance activities for children and adolescents (non-charitable)		92400	Journalists
				92319	Librettist
	85321	Welfare service (charitable)		92319	Lithographic artist (own account)
	85322	Welfare service (non-charitable)		92319	Lyric author
	85321	Women's Royal Voluntary Service		92319	Music composer
				92319	Music copyist and transcriber (own account)
90010		**Performing arts**		92319	Painters (artistic)
	92311	Actors		92319	Picture restoring
	92349	Aerobatic display		92319	Playwright
	92349	Animal training for circuses, etc.		92319	Poet
	92311	Ballet company		92319	Repair and restoration of works of art
	92311	Band (musical)		92319	Scenario writer
	92349	Circus		92319	Scenic artist
	92311	Concerts production		92319	Sculptors
	92311	Conjuror		92319	Song writer
	92311	Dance band		92319	Technical and training manual authors
	92311	Dance productions		92319	Wood engraver (artistic)
	92311	Impresario			
	92319	Lecturers	**90040**		**Operation of arts facilities**
	92311	Musicians		92320	Arts facilities operation
	92311	Opera production		92320	Concert halls operation
	92311	Orchestras		92320	Local authority concert halls and theatres
	92311	Organist (own account)		92320	Music hall
	92311	Pop group		92320	Opera house
	92319	Public speaker		92320	Theatre halls operation
	92311	Repertory company			
	92311	Revue company	**91011**		**Library activities**
	92311	Singer (own account)		92510	Art work lending and storage
	92319	Speakers (after dinner etc.)		92510	Book lending and storage
	92311	Stage productions		92510	British Library
	92311	Street musician or singer		92510	Cataloguing and preservation of collections
	92311	Theatrical presentations (live production)		92510	Film lending and storage
	92311	Theatrical touring company		92510	Lending and storage of books, periodicals
	92349	Travelling show		92510	Lending and storage of CDs, DVDs
	92311	Variety artiste (own account)		92510	Lending and storage of maps music
	92311	Ventriloquist		92510	Libraries
				92510	Library access to IT facilities including internet
90020		**Support activities to performing arts**		92510	Library training courses (IT, information literacy, basic skills)
	92349	Direction, production and support activities to circus performances		92510	Map lending and storage
	92311	Directors (theatre)		92510	National library for the blind
	92311	Performing arts support activities		92510	National library of Scotland
	92311	Scene shifters and lighting engineers		92510	National library of Wales
	92311	Stage set designers and builders		92510	Periodicals lending and storage
	92320	Support activities to performing arts e.g. stage set-up, costume and lighting design etc.		92510	Reading room (library)
				92510	Record lending and storage
90030		**Artistic creation**		92510	Tapes (music and video) lending and storage
	92319	Art expert			
	92319	Artist	**91012**		**Archive activities**
	92319	Author		92510	Archive activities
	92319	Cartoonists		75140	Government records and archives maintenance and storage (public sector)
	92319	Copper plate engraver (artistic)			
	92319	Designing (artistic)	**91020**		**Museum activities**
	92319	Engravers		92521	Art gallery (not dealer)
	92319	Etchers		92521	Art museums
	92319	Fine art expert		92521	Bethnal Green Museum
	92400	Freelance journalist		92521	British Museum
	92319	Heraldic painting		92521	Historical museums
	92319	Illuminating (illustrating)		92521	Imperial War Museum

UK SIC 2007	UK SIC 2003	Activity	SIC 2007	SIC 2003	Activity
	92521	Local authority art galleries and museums		92710	Virtual gambling web site operation
	92521	London Museum			
	92521	Military museums	**93110**		**Operation of sports facilities**
	92521	Museums of all kinds		92619	Activity centre (sports)
	92521	Museums of furniture, costumes, ceramics, silverware		92619	Bathing pool proprietor
				92629	Billiard room or saloon
	92521	Museums of jewellery		92629	Bowling alley
	92521	National galleries (Scotland)		92619	Bowling lanes operation
	92521	National gallery		92619	Boxing arena
	92521	National maritime museum		92619	Car, dog and horse racetrack operation
	92521	National portrait gallery		92619	Field and track stadium
	92521	National trust garden (property)		92619	Football ground
	92521	Natural history museum		92619	Football stadium
	92521	Open air museums		92619	Golf courses
	92521	Royal Scottish Museum (Edinburgh)		92619	Golf links
	92521	Science museums		92619	Greyhound racing stadium
	92521	Tate Gallery		92619	Greyhound track
	92521	Technological museums		92619	Groundsman
	92521	Victoria and Albert Museum		92629	Gymnasium
	92521	Wallace collection		92619	Hockey, cricket, rugby stadiums operation
				92611	Ice skating rink
91030		**Operation of historical sites and buildings and similar visitor attractions**		92619	Ice-hockey arenas operation
				92619	Leisure centres
	92522	Historical sites and buildings preservation		92619	Local authority football and other sports grounds
	92522	Preservation society for historic houses		92619	Local authority leisure centres
				92619	Local authority recreational facilities
91040		**Botanical and zoological gardens and nature reserve activities**		92619	Local authority sports facilities (incl. Football and other sports grounds, swimming baths, etc.)
	92530	Botanical gardens		92619	Local authority swimming pool
	92530	Children's zoos		92619	Managing and providing the staff for sports facilities
	92530	Kew gardens		92619	Newmarket Heath
	92530	Menagerie		92619	Racecourse operation
	92530	Nature reserves including wildlife preservation		92629	Rifle butts
	92530	Royal Botanical Gardens		92611	Roller skating rink
	92530	Wildlife preservation services		92629	Skittle alley
	92530	Zoological gardens		92629	Sports arenas
				92619	Sports facilities operation
92000		**Gambling and betting activities**		92629	Stadium operation
	92710	Amusement arcade		92629	Staging of sports events by organizations with their own facilities
	92710	Betting activities		92619	Swimming baths
	92710	Betting shop		92619	Swimming pools
	92710	Bingo hall		92619	Tennis court
	92710	Bookmaker		92619	Winter sport arenas and stadiums
	92710	Casino			
	92710	Coin-operated gambling machine establishments	**93120**		**Activities of sport clubs**
	92710	Floating casinos		92629	Athletic club
	92710	Football pools		92629	Badminton club
	92710	Gambling activities		92629	Billiards and snooker club
	92710	Gaming board for Great Britain		92629	Bowling clubs
	92710	Gaming club		92629	Bowls club
	92710	Horse race betting levy board		92629	Boxing clubs
	92710	Horserace totalisator board		92629	Card clubs
	92710	Lottery ticket sales		92629	Chess clubs
	92710	Off-track betting		92629	Cricket club
	92710	Racing pool		92629	Croquet club
	92710	Racing tipster		92629	Cycle club
	92710	Tic tac person		92629	Domino clubs
	92710	Totalisator		92629	Draughts clubs
	92710	Turf accountant			
	92710	Turf commission agency			

UK SIC 2007	UK SIC 2003	Activity
	92629	Field and track clubs
	92629	Flying club
	92629	Football clubs
	92629	Glider club
	92629	Golf club
	92629	Hockey club
	92629	Ice hockey club
	92629	Jockey club
	92629	MCC
	92629	National greyhound racing club
	92629	Pony Club
	92629	Racquet club
	92629	Rowing club
	92629	Shooting clubs
	92629	Snooker club
	92629	Sports club
	92629	Squash club
	92629	Swimming clubs
	92629	Tennis club
	92629	Trotting Club
	92629	Winter sport clubs
	92629	Wrestling clubs
	92629	Yacht club
93130		**Fitness facilities**
	92629	Body-building clubs and facilities
	92629	Fitness centre operation
	93040	Health clubs
93191		**Activities of racehorse owners**
	92621	Racehorse owner
	92621	The seeking of sponsorship, appearance money and prize money for horse racing
93199		**Other sports activities (not including activities of racehorse owners) n.e.c.**
	92629	Athletes
	92629	Boxing
	92629	Boxing promoter
	92629	Coursing
	92629	Dirt track racing
	92629	Dog breeding (for greyhound racing)
	92629	Dog racing
	92629	Drag hounds
	92629	Fishing (recreational)
	92629	Football Association
	92629	Greyhound training
	92629	Horse training (racehorse)
	92629	Hunt kennels
	92629	Hunt stables
	92629	Hunting for sport or recreation
	92629	Jockey
	92629	Judges of sport
	92629	Kennel master
	92629	Kennels and garages (racing)
	92629	Mountain guides
	92629	Physical culture expert
	92629	Promotion of sporting events
	92629	Racehorse trainer

SIC 2007	SIC 2003	Activity
	92629	Racing stables
	92629	Riding stables
	92629	Rugby League
	92629	Rugby Union
	92629	Speedway racing
	92629	Sports leagues and regulating bodies
	92629	Sports referees
	92629	Sportsmen and sportswomen
	92629	Timekeepers of sport
	92629	Trainer (racehorse or greyhound)
	92629	Training stables
93210		**Activities of amusement parks and theme parks**
	92330	Amusement park activities
	92330	Amusement park mechanical rides, water rides, games
	92330	Amusement park shows, theme exhibits and picnic grounds
	92330	Coconut shy
	92330	Fairground activities
	92330	Fortune telling (fairground)
	92330	Fun fair
	92330	Preserved railway operation
	92330	Switchback (fairground)
	92330	Theme park operation
93290		**Other amusement and recreation activities**
	92729	Adventure playground
	92729	Beach facilities rental (bathhouses, lockers, chairs, etc.)
	92729	Beach hut proprietor
	92729	Coin-operated games arcade (other than gaming machines)
	92729	Common (local authority or municipally owned)
	92341	Dance hall
	92349	Entertainment activities n.e.c.
	92330	Fairs and shows of a recreational nature
	92349	Firework displays
	92729	Hampton Court Gardens and Park
	92729	Hobby instructor (own account)
	92729	Hyde Park
	92729	Kensington Gardens
	92729	Local authority parks and gardens
	92629	Marinas
	92349	Marionette show
	92349	Model railway installations
	92729	Narrow gauge railway (recreational)
	92341	Operation of dance floors
	92629	Operation of recreational transport facilities e.g. marinas
	92729	Operation of ski hills
	92729	Park (local authority or municipally owned)
	92629	Pleasure boat hiring as an integral part of recreational facilities
	92729	Pleasure ground
	92729	Pleasure pier
	92729	Psychometry
	92729	Public park
	92349	Punch and Judy show
	92349	Puppet shows

UK SIC 2007	UK SIC 2003	Activity	SIC 2007	SIC 2003	Activity
	92729	Recreational activities n.e.c.		91120	Institute of Hygiene
	92729	Regent's Park and Primrose Hill		91120	Institute of Incorporated Photographers
	92729	Renting of leisure and pleasure equipment as an integral part of recreational facilities		91120	Institute of Mechanical Engineers
				91120	Journalists associations
	92729	Richmond Park		91120	Law Society
	92349	Rodeos		91120	Learned societies
	92729	Royal Park		91120	Legal associations
	92349	Shooting galleries		91120	Medical associations
	92729	St James's park		91120	National maritime board
	92349	Waxworks		91120	Nursing society
	92729	Windsor Great Park		91120	Painters and other artists associations
				91120	Performers associations
94110		**Activities of business and employers membership organisations**		91120	Performing Rights Society
	91110	Business and employers membership organisations		91120	Pharmaceutical society
	91110	Business organisations		91120	Professional organisations
	91110	Chamber of agriculture		91120	Royal Academy of Arts
	91110	Chambers of commerce organisations		91120	Royal Aeronautical Society
	91110	City guild (goldsmiths' company, stationers' company, etc.)		91120	Royal Agricultural Society of England
				91120	Royal College of Midwives
	91110	Confederation of British industry		91120	Royal College of Nursing
	91110	Employers organisations		91120	Royal College of Physicians
	91110	Federations of business and employers' membership organisations		91120	Royal College of Surgeons
				91120	Royal Geographical Society
	91110	Guilds and similar organisations		91120	Royal Institute of Chartered Surveyors
	91110	Joint organisation of employers and trade unions		91120	Royal Institute of Public Health
	91110	Negotiations of business and employer organisations		91120	Royal Society
				91120	Royal Society for Health
	91110	Property owners' association		91120	Royal Society of Medicine
	91110	Trade association		91120	Royal Statistical Society
				91120	Royal United Services Institution
94120		**Activities of professional membership organisations**		91120	Scientific organisation
	91120	Academic organisations		91120	Society of Apothecaries
	91120	Accounting associations		91120	Society of Arts
	91120	Architects associations		91120	Teachers' Registration Council
	91120	Association of corporate and certified accountants		91120	Technical organisations
	91120	British Association for the Advancement of Science		91120	Writers associations
	91120	British Computer Society	94200		**Activities of trade unions**
	91120	British Dental Association		91200	Labour organisations
	91120	British Medical Association		91200	Trade unions
	91120	Central midwives board		91200	Trades union congress
	91120	Chartered institute of secretaries			
	91120	Copyright protection society	94910		**Activities of religious organisations**
	91120	Cultural associations		91310	Army scripture readers association
	91120	Cultural organisation (professional)		91310	Baptist church
	91120	Educational association		91310	Bible society
	91120	Engineering associations		91310	British Humanist Association
	91120	Faculty of actuaries		91310	British Jews Society
	91120	General Council of the Bar		91310	Calvinistic Methodist Church
	91120	General Medical Council		91310	Catholic Apostolic Church
	91120	General Nursing Council		91310	Church Army
	91120	Inns of Court		91310	Church Commission
	91120	Institute of Actuaries		91310	Church in Wales
	91120	Institute of British Water Colour Painters		91310	Church Missionary Society
	91120	Institute of Chartered Accountants in England and Wales		91310	Church of Christ Scientist
				91310	Church of England
	91120	Institute of Chartered Accountants of Scotland		91310	Church of Ireland
	91120	Institute of Civil Engineers		91310	Church of Scotland
	91120	Institute of Cost and Management Accountants		91310	City mission

UK SIC 2007	UK SIC 2003	Activity	SIC 2007	SIC 2003	Activity
	91310	Convent (not school or orphanage)		91330	Civic trust
	91310	Crusaders' union		91330	Community centre
	91310	Episcopal Church in Scotland		91330	Community organisations
	91310	Evangelists Society		91330	Conservation and Preservation Society
	91310	Inter Varsity Fellowship of Evangelical Unions		91330	Consumer associations
	91310	Jewish synagogue		91330	Consumers' association
	91310	Lord's Day Observance Society		91330	Craft and collectors' clubs
	91310	Methodist Church		91330	Cultural organisations (hobby)
	91310	Missionary Society		91330	Ecological movements
	91310	Monastery		91330	Environmental movements
	91310	Mosque		91330	Ethnic and minority group organisations
	91310	Presbyterian Church		91330	Film and photo clubs
	91310	Presbyterian Church of Wales		91330	Fraternities
	91310	Religious funeral service activities		91330	Freemasons
	91310	Religious organisations		91330	Gardening clubs
	91310	Religious retreat activities		91330	Girl Guides Association
	91310	Roman Catholic Church		91330	Girls' Brigade
	91310	Salvation army		91330	Girls' Friendly Society
	91310	Society of Friends		91330	Grant giving activities by membership organisations or others
	91310	Spiritualist church		91330	Grant making foundations
	91310	Student Christian Movement		91330	Historical club
	91310	Synagogue		91330	Horse breeding society
	91310	Temple (for worship)		01500	Hunting and trapping (commercial) promotion activities
	91310	Theosophical Society		91330	Legal Aid Society
	91310	Unitarian Church		91330	Literature and book club
	91310	United Reform Church		91330	Lodge activities
	91310	United Society for Christian Literature		91330	Membership organisations n.e.c.
	91310	Wesleyan reform union		91330	Motorists' organisation (not road patrol or touring service)
94920		**Activities of political organisations**		91330	Music club
	91320	Conservative and Unionist Party		91330	National council for civil liberties
	91320	Conservative Association		91330	National trust (the)
	91320	Fabian Society		91330	National union of students (not trading activities)
	91320	Labour party		91330	Poetry club
	91320	Liberal democrat party		91330	Protest movement activities
	91320	Plaid Cymru		91330	Recreational organisations
	91320	Political organisations		91330	Rotary clubs
	91320	Scottish Nationalist Party		91330	Round Table
	91320	Social Democratic and Labour Party		91330	Royal Automobile Club headquarters
	91320	Unionist parties		91330	Royal Horticultural Society
	91320	Young people's auxiliaries associated with a political party		91330	Royal Scottish Automobile Club
				91330	Royal Society for the Prevention of Accidents
94990		**Activities of other membership organisations n.e.c.**		91330	Royal Society for the Prevention of Cruelty to Animals (not animal hospitals or homes)
	91330	Animal protection organisations		91330	Scout Association
	91330	Art clubs		91330	Sea scout association
	91330	Associations for the pursuit of a cultural or recreational activity or hobby		91330	Shire Horse Association
	91330	Automobile association		91330	Social club (not licensed to sell alcohol)
	91330	Bishopsgate Institute		91330	Student associations
	91330	Boy scouts		91330	Student Union
	91330	Boys brigade		91330	Touring clubs
	91330	Breed society		91330	United Nations associations
	91330	British Board of Film Classifications		91330	War veterans' associations
	91330	British Legion (other than social clubs)		91330	Working men's club (not licensed to sell alcohol)
	91330	British Safety Council		91330	YMCA (not hostel)
	91330	Carnival clubs		91330	Young persons associations
	91330	Church Lads' Brigade		91330	Youth centre
	91330	Citizens initiative or protest movements			

UK SIC 2007	UK SIC 2003	Activity	SIC 2007	SIC 2003	Activity
	91330	Youth club	95210		**Repair of consumer electronics**
	91330	Zionist organisation		52720	Repair of CD players
				52720	Repair of consumer electronics:
95110		**Repair of computers and peripheral equipment**		52720	Repair of domestic audio and video equipment
	30020	Repair and maintenance of automatic teller machines (ATMs)		52720	Repair of dvd players
	30020	Repair and maintenance of bar code scanners		52720	Repair of household-type video cameras
	30020	Repair and maintenance of computer monitors		52720	Repair of radios
	30020	Repair and maintenance of computer projectors		52720	Repair of televisions
	30020	Repair and maintenance of computer servers		52720	Repair of video cassette recorders (VCR)
	72500	Repair and maintenance of computing machinery	**95220**		**Repair of household appliances and home and garden equipment**
	30020	Repair and maintenance of dedicated computer terminals		52740	Repair and maintenance of domestic cookers
	30020	Repair and maintenance of desktop computers		52740	Repair and maintenance of domestic ovens
	30020	Repair and maintenance of hand-held computers (PDA's)		52740	Repair and maintenance of domestic ranges
	30020	Repair and maintenance of internal and external computer modems		52720	Repair of clothes dryers
	30020	Repair and maintenance of keyboards		52720	Repair of domestic electrical appliances
	30020	Repair and maintenance of laptop computers		52720	Repair of garden edger
	30020	Repair and maintenance of magnetic disk drives		52720	Repair of garden trimmers
	30020	Repair and maintenance of magnetic flash drives and other storage devices		29320	Repair of lawnmowers
	30020	Repair and maintenance of mice, joysticks, and trackball accessories		52720	Repair of refrigerators
	30020	Repair and maintenance of optical disk drives (CD-RW, CD-ROM, DVD-ROM, DVD-RW)		52720	Repair of room air conditioners
	30020	Repair and maintenance of point-of-sale (pos) terminals		52720	Repair of snow and leaf blowers
	30020	Repair and maintenance of printers		52720	Repair of stoves
	30020	Repair and maintenance of scanners		52720	Repair of washing machines
	30020	Repair and maintenance of smart card readers		52720	Servicing of clothes dryers
	30020	Repair and maintenance of virtual reality helmets		52720	Servicing of domestic electrical appliances
	30020	Repair of computers and peripheral equipment		52720	Servicing of garden edger
				52720	Servicing of garden trimmers
95120		**Repair of communication equipment**		29320	Servicing of lawnmowers
	32201	Repair and maintenance of carrier equipment modems		52720	Servicing of refrigerators
	52740	Repair and maintenance of cellular telephones		52720	Servicing of room air conditioners
	32202	Repair and maintenance of commercial TV cameras		52720	Servicing of snow and leaf blowers
	32202	Repair and maintenance of commercial video cameras		52720	Servicing of stoves
	32201	Repair and maintenance of communications transmission bridges		52720	Servicing of washing machines
	32201	Repair and maintenance of communications transmission modems	**95230**		**Repair of footwear and leather goods**
	32201	Repair and maintenance of communications transmission routers		52740	Heel replacement services (while you wait)
	32201	Repair and maintenance of fax machines		52710	Repair of boots
	32300	Repair and maintenance of professional radio, television, sound and video equipment		52710	Repair of footwear and leather goods
	32202	Repair and maintenance of radio and television transmitters		52710	Repair of handbags
	32201	Repair and maintenance of radio telephony apparatus		52710	Repair of luggage
	32300	Repair and maintenance of telephone answering machines		52710	Repair of shoes
	32201	Repair and maintenance of telephone sets	**95240**		**Repair of furniture and home furnishings**
	32201	Repair and maintenance of telex machines and other line telephony or telegraphy equipment		36110	Repair and restoration of chairs and seats
	32202	Repair and maintenance of two-way radios		36140	Repair and restoration of furniture n.e.c.
	32201	Repair of cordless telephones		36150	Repair and restoration of mattresses
	52740	Repair of mobile telephones		36130	Repair and restoration of other kitchen furniture
				36120	Repair and restoration of other office and shop furniture
				36110	Upholsterer (repair and restoration)
			95250		**Repair of watches, clocks and jewellery**
				52730	Repair of clock cases and housings of all materials
				52730	Repair of clock movements and chronometers
				52730	Repair of clocks
				52730	Repair of jewellery
				52730	Repair of watch cases and housings of all materials
				52730	Repair of watches and clocks

UK SIC 2007	UK SIC 2003	Activity	SIC 2007	SIC 2003	Activity
95290		**Repair of other personal and household goods**		93020	Facial massage
	52740	Invisible mending		93020	Hairdressing activities
	52740	Piano tuning		93020	Make-up and beauty treatment
	52740	Renovating of hats		93020	Manicurist
	52740	Repair and alteration of clothing		93020	Pedicure
	36300	Repair and reconditioning of musical instruments (except organs and historical musical instruments)		93020	Trichologist
	52740	Repair of bicycles	96030		**Funeral and related activities**
	52740	Repair of books		93030	Burial services
	52740	Repair of cycles		93030	Cemetery
	52740	Repair of domestic lighting articles		93030	Cremation board
	52740	Repair of household textile articles		93030	Cremation services
	52740	Repair of non-professional photographic equipment		93030	Crematorium
	52740	Repair of sporting and camping equipment (except tents)		93030	Funeral and related activities
	17402	Repair of tents		93030	Funeral direction
	52740	Repair of toys		93030	Funeral furnishing
	52740	Repair of umbrellas		93030	Human or animal corpse burial or incineration
	52740	Tuning of musical instruments		93030	Laying out the dead
96010		**Washing and (dry-)cleaning of textile and fur products**		93030	Local authority cemeteries
				93030	Local authority crematoriums
	93010	Carpet cleaning		93030	Local authority funeral services
	93010	Clean towel company		93030	Maintenance of graves
	93010	Curtain cleaning		93030	Maintenance of mausoleums
	93010	Curtain cleaning (not lace dressing)		93030	Mortuary
	93010	Diaper supply services		93030	Preparing the dead for burial or cremation
	93010	Drapery cleaning		93030	Sale of graves
	93010	Dry cleaner		93030	Undertaking
	93010	Dyer and cleaner	96040		**Physical well-being activities**
	93010	Glove cleaning		93040	Health farm
	93010	Hospital laundry		93040	Local authority baths (hot water and sauna)
	93010	Industrial clothing hire from laundries		93040	Massage salons
	93010	Job dyeing		93040	Physical well-being activities
	93010	Lace cleaning and mending (not net mending)		93040	Public baths
	93010	Launderette		93040	Reducing and slimming salon activities
	93010	Laundry		93040	Russian baths
	93010	Laundry receiving office		93040	Saunas
	93010	Linen hire (associated with laundry service)		93040	Solariums
	93010	Pressing and valeting		93040	Spas
	93010	Pressing of wearing apparel on a fee or contract basis		93040	Steam baths
				93040	Turkish baths
	93010	Rug shampooing	96090		**Other personal service activities n.e.c.**
	93010	Shirt and collar pressing		93059	Artists' model
	93010	Towel hire		93059	Astrologer
	93010	Towel supply company		93059	Blood pressure machine operation (coin operated)
	93010	Valet service		93059	Boarding of pet animals
	93010	Washing and cleaning of fur products		93059	Body piercing studios
	93010	Washing and dry cleaning of clothing		93059	Bootblack
	93010	Washing and dry cleaning of textile products		93059	Cats' home
	93010	Work uniforms rental from laundries		93059	Clairvoyant
96020		**Hairdressing and other beauty treatment**		93059	Cloakroom (not railway, etc.)
	93020	Barber		93059	Computer dating agency
	93020	Beard trimming		93059	Dating services
	93020	Beauty parlour		93059	Dogs' home
	93020	Beauty specialist		93059	Educational agency
	93020	Beauty treatment activities		93059	Emigration agency (not of foreign government, etc.)
	93020	Coiffeur		93059	Escort agency
	93020	Electrolysis specialist		93059	Fortune telling (not fairground)

UK SIC 2007	UK SIC 2003	Activity
	93059	Genealogical organisation services
	93059	Genealogist
	93059	Graphologist
	93059	Grooming of pet animals
	93059	Guide (other than tourist)
	93059	Historical research
	93059	Horse clipping
	93059	Jobbing waiter
	93059	Kennels (not racing)
	52740	Key cutting services (while you wait)
	93059	Knifegrinder (travelling)
	93059	Licensed porter
	93059	Marriage bureau
	93059	Master of Ceremonies
	93059	Naturalisation agent
	93059	Outside porter
	93059	Palmist
	93059	Pavement artist
	93059	Pet sitting services
	93051	Photographic machines (coin-operated)
	52740	Plastic coating services of identity cards, etc. (while you wait)
	93059	Poodle clipping
	93059	Porters
	93059	Salvation army emigration department
	93059	Shoe shiners
	93059	Spiritualists' activities
	93059	Tattooist
	93059	Toastmaster
	93059	Town crier
	93059	Training of pet animals
	93059	Valet car parkers
	93059	Weighing machine operation (coin operated)
97000		**Activities of households as employers of domestic personnel**
	95000	Activities of households as employers of domestic babysitters
	95000	Activities of households as employers of domestic butlers
	95000	Activities of households as employers of domestic caretakers
	95000	Activities of households as employers of domestic chauffeurs
	95000	Activities of households as employers of domestic cooks
	95000	Activities of households as employers of domestic gardeners
	95000	Activities of households as employers of domestic gatekeepers
	95000	Activities of households as employers of domestic governesses
	95000	Activities of households as employers of domestic laundresses
	95000	Activities of households as employers of domestic maids

SIC 2007	SIC 2003	Activity
	95000	Activities of households as employers of domestic secretaries
	95000	Activities of households as employers of domestic stable-lads
	95000	Activities of households as employers of domestic staff
	95000	Activities of households as employers of domestic tutors
	95000	Activities of households as employers of domestic valets
	95000	Activities of households as employers of domestic waiters
98100		**Undifferentiated goods-producing activities of private households for own use**
	96000	Undifferentiated goods producing activities of private households for own use
98200		**Undifferentiated service-producing activities of private households for own use**
	97000	Undifferentiated services producing activities of private households for own use
99000		**Activities of extraterritorial organisations and bodies**
	99000	Charge d'affaires
	99000	Commonwealth Armed Forces
	99000	Commonwealth Government Service
	99000	Commonwealth Institute
	99000	Commonwealth Secretariat
	99000	Commonwealth War Graves Commission
	99000	Consular office
	99000	Crown agents for overseas governments and administrations
	99000	Customs co-operation council
	99000	Diplomatic missions
	99000	Embassy
	99000	European Communities Representatives and Information Office
	99000	European Community
	99000	European Free Trade Association
	99000	Foreign armed forces
	99000	Foreign embassy
	99000	Foreign government service
	99000	International Labour Office
	99000	International Monetary Fund
	99000	International organisation (e.g. United Nations, International Labour Office)
	99000	Legation
	99000	Office of High Commissioner
	99000	Organisation for economic co-operation and development
	99000	Organisation of oil producing and exporting countries
	99000	United Nations and affiliated organisations (not United Nations Association)
	99000	World bank